SEVENTH EDITION

Introduction
to
Control System Technology

◆

Robert N. Bateson, P.E.

Prentice
Hall

Upper Saddle River, New Jersey
Columbus, Ohio

Editor in Chief: *Stephen Helba*
Executive Editor: *Debbie Yarnell*
Production Editor: *Louise N. Sette*
Production Supervisor: *TechBooks*
Design Coordinator: *Robin G. Chukes*
Cover Designer: *Thomas Borah*
Cover art: © *Kathy Hanley*
Production Manager: *Brian Fox*
Marketing Manager: *Jimmy Stephens*

This book was set in Times Roman by TechBooks. It was printed and bound by R. R. Donnelley & Sons Company. The cover was printed by Phoenix Color Corp.

Earlier editions, © 1993 by Macmillan Publishing Company and © 1989, 1980, 1973 by Merrill Publishing Company.

Prentice-Hall International (UK) Limited, *London*
Prentice-Hall of Australia Pty. Limited, *Sydney*
Prentice-Hall of Canada, Inc., *Toronto*
Prentice-Hall Hispanoamericana, S.A., *Mexico*
Prentice-Hall of India Private Limited, *New Delhi*
Prentice-Hall of Japan, Inc., *Tokyo*
Prentice-Hall Singapore Pte. Ltd.
Editora Prentice-Hall do Brasil, Ltda., *Rio de Janeiro*

10 9 8 7 6 5 4 3 2
ISBN: 0-13-030688-6

Preface

GOAL

The goal of *Introduction to Control System Technology* is to provide both a textbook on the subject and a reference that engineers and technicians can include in their personal libraries. This text can help students master the concepts and language of control and help engineers and technicians analyze and design control systems. The text covers the terminology, concepts, principles, procedures, and computations used by engineers and technicians to analyze, select, specify, design, and maintain control systems. Emphasis is on the application of established methods with the aid of examples and computer programs.

EVOLUTION OF TEXT

The writing of this text began 34 years ago when I faced the challenge of developing and teaching a control systems course in a 2-year engineering technology program. I had just entered the teaching profession after 10 years as a research engineer at General Mills, where I had become fascinated with control systems. I was especially intrigued by the combined electrical, mechanical, thermal, liquid, and gas elements in the mathematical models used to analyze and design control systems. This fascination led to the completion of an evening MSEE program with a major in control systems and a hands-on design course at Brown Institute in Philadelphia. In my course work, we used straight-line Bode diagrams to design control systems. This method works very well for processes with dead-time lags. The graphical approach gives the designer an "intuitive feel" of the way the controller changes the frequency response of the system. I found that this "feel" was very helpful working on plant start-ups. Indeed, the greatest benefit from learning the frequency-response design method was the understanding and judgment it imparted. It made me a much better engineer.

You can imagine the excitement with which I approached the teaching of my favorite subject. I wanted my students to feel that excitement. I wanted to impart some of the feeling and judgment that I acquired from frequency-response design of a control system. There was one major obstacle, however. There was no suitable text for my students. So I wrote 100 pages of notes on control fundamentals with emphasis on graphical design using straight-line Bode diagrams. These notes were the genesis of this text. The thrust of the notes was to bring students to the point where they could complete frequency-response design of control systems under my direction. My role was that of a control engineer, and my students were my engineering technicians. Those 100 pages of notes have grown to become the seventh edition of a 700-page textbook, but the thrust has not changed. The thrust of this text is to bring students to the point where they can complete computer-aided, frequency-response design of control systems under the direction of their instructor or control engineer.

Frequency-response graphs are constructed from the transfer functions of the system components. The chapters on common elements and Laplace transforms were written to give students the foundation required to determine these transfer functions. Analogies were used to develop common elements for modeling and analyzing electrical, thermal, mechanical, and fluid flow elements. These analogies helped students translate their knowledge of one type of component to components of other types. The parts on measurement, manipulation, and control extended the students' mastery of transfer functions and developed their ability to select, specify, and design measuring and manipulating systems. Finally, we reached the point where the students actually began the graphical design process. The frequency-response (or Bode) design method worked very well, but constructing and reconstructing Bode diagrams is very tedious and time consuming. By the time the students learned how to construct Bode diagrams, there was little time or energy left to learn how to design the controller. My goal was unfulfilled.

Then I had a dream. My dream was a computer program that would construct the Bode diagrams so that students, technicians, and engineers could concentrate on the design of the controller. The first attempt to realize my dream was a FORTRAN program that generated frequency-response data from the open-loop transfer function. The program made all the design decisions, and all the student had to do was input the transfer function. As an engineering tool, it was great, but as a teaching tool, it failed. Students could complete the design very well, but they did not understand the results. When design exceptions required an override of the program's results, students had no idea what to do—no pain but no gain.

The first program did too much, and the students did too little. The next program generated tables of frequency-response data that the students used to draw the Bode diagrams. It was not as good an engineering tool but was a much better teaching tool. Drawing the graphs took time, but the students did understand the design process.

The third version of the program uses the graphing capabilities of the QuickBASIC programming language. The program DESIGN is an interactive program that plots frequency-response graphs on the screen and accepts user inputs of the control modes. This program emulates the classical Bode design method with precise plots based on the transfer functions of the system components. The designer observes how each control mode changes the shape of the open-loop frequency response of the system. On-screen design decision data allow the designer to determine PID control mode values, which can be easily changed in a "what-if" analysis. This enables the designer to use a "design-by-trial" procedure to search for the best possible control system design. It worked; my dream was fulfilled. DESIGN is both a good engineering design tool and a good teaching tool.

This text was developed to facilitate the education of engineering technicians. Its purpose is to train technicians who understand the language and methods used by engineers, technicians who can use established methods to complete engineering design work under the direction of a control engineer.

I believe an essential part of the education of the engineering technician is to develop the ability to communicate with engineers using their language. Mathematical terms are an essential part of this language. The engineering technician must understand and be comfortable with terms such as derivative, integral, transfer function, frequency domain, and Laplace transform—not at the theoretical level, of course, but certainly at the applied level. A goal of this text is to develop an understanding of the language of control, including the mathematical terms mentioned.

COMPUTER DISK ANCILLARY

The disk packaged with this text contains QuickBASIC and executable versions of the four programs used in the text. No knowledge of QuickBasic is required to use the executable versions of the programs. They can be executed directly from DOS on any IBM-compatible computer. **You may make copies of these programs and distribute them as you wish.** I hope you enjoy using the programs as much as I enjoyed creating them. The following files are on the disk.

QuickBASIC Version	Executable Version	Text File
BODE.BAS	BODE.EXE	HOWTOUSE.EXT
DESIGN.BAS		DESIGN1.HLP
		DESIGN2.HLP
LIQRESIS.BAS		LIQRESIS.HLP
THERMRES.BAS		THERMRES.EXE

You can run the executable versions directly from DOS on any IBM-compatible PC. Just type the file name at the DOS prompt and press the <Enter> key. For example, the following command will run the program DESIGN from a disk in the A: drive:

```
A:\>DESIGN
```

The file HOWTOUSE.TXT is a DOS text file that contains directions on using the programs. You can print this file from DOS with the following command:

```
A:\TYPE HOWTOUSE.TXT > PRN
```

You can also use Word, WordPerfect, or any other word processor to examine and print HOWTOUSE.TXT.

CHANGES TO THIS EDITION

The primary emphasis for the seventh edition was a complete rework of the exercises at the end of each chapter. Every section was examined, and every question was reviewed. Numerous exercises were added to provide students with multiple opportunities to develop their grasp of the material.

The review also provided ample opportunities to improve the clarity and accuracy of the text. Example 16.3, a walk-through of a run of program Design, was revised to provide better integration of the text and figures.

Most of the material in Appendix C, "Binary Codes," was deleted because most reviewers felt the material was covered in other courses. Only seven binary tables were retained as reference material.

ORGANIZATION

The book consists of five parts. Part One is an introduction to the terminology, concepts, and methods used to describe control systems. Parts Two, Three, and Four cover the three operations of control: measurement, manipulation, and control. Part Five is concerned with the analysis and design of control systems. Each chapter begins with a set of learning objectives and ends with a glossary of terms. There is sufficient material for a two-semester course, and there are a number of possible sequences of selected chapters for a one-semester course. The following are some suggested sequences for one- and two-semester courses.

Suggested One-Semester Sequences

A. Process Control Analysis and Design:
   Chapters 1–5, 6.1–6.3, 9.4, 9.5, 13.1–13.3, 14, 5.1–15.9, 16.1–16.6
B. Servo Control Analysis and Design:
   Chapters 1–5, 6.1–6.3, 9.3, 10, 13, 14, 15, 16.1, 16.5–16.7
C. Sequential and PID Control:
   Chapters 1, 2, 5, 9, 10.4, 11, 12, 13, 16.1–16.4
D. Data Acquisition and Control:
   Chapters 1, 2, 5, 6, 7, 8, 13, 16.1–16.4

Suggested Two-Semester Sequence

Semester 1: Data Acquisition: Chapters 1–8
Semester 2: Control: Chapters 9–16

ACKNOWLEDGMENTS

I would like to acknowledge and thank the many people who supported me in the preparation of this book. My wife, Betty, has been wonderfully patient and understanding throughout the project. My children, Mark, Karen, and Paul, were always patient, understanding, and supportive.

Special recognition must go to the following for their suggestions, criticisms, and support over the past several editions: Ernest G. Carlson, Don Craighead, Keith D. Graham, Tom

Loftus, Jack Hunger, Samuel Kraemer, and Ed Lawrence. Thanks also to Debbie Yarnell, executive editor at Prentice Hall, and Louise Sette.

I would also like to thank the reviewers of this edition for their excellent comments and suggestions: A. G. Chassiakos, California State University–Long Beach; Dr. Lee Rosenthal, Fairleigh Dickinson University; and Kenneth Exworthy, Northeast Wisconsin Technical College.

Robert N. Bateson

Contents

PART ONE ◆ **INTRODUCTION**

CHAPTER 1   **Basic Concepts and Terminology**   **1**

     1.1   Introduction   2
     1.2   Block Diagrams and Transfer Functions   3
     1.3   Open-Loop Control   7
     1.4   Closed-Loop Control: Feedback   8
     1.5   Control System Drawings   15
     1.6   Nonlinearities   17
     1.7   Benefits of Automatic Control   20
     1.8   Load Changes   21
     1.9   Damping and Instability   22
     1.10   Objectives of a Control System   23
     1.11   Criteria of Good Control   24
     1.12   Block Diagram Simplification   26

CHAPTER 2   **Types of Control**   **36**

     2.1   Introduction   37
     2.2   Analog and Digital Control   38
     2.3   Regulator and Follow-Up Systems   39
     2.4   Process Control   39
     2.5   Servomechanisms   44
     2.6   Sequential Control   47
     2.7   Numerical Control   49
     2.8   Robotics   51

2.9   The Evolution of Control Systems   55
2.10   Examples of Control Systems   57

CHAPTER 3   **The Common Elements of System Components   69**

3.1   Introduction   70
3.2   Electrical Elements   72
3.3   Liquid Flow Elements   79
3.4   Gas Flow Elements   89
3.5   Thermal Elements   93
3.6   Mechanical Elements   100

CHAPTER 4   **Laplace Transforms and Transfer Functions   113**

4.1   Introduction   114
4.2   Input/Output Relationships   115
4.3   Laplace Transforms   124
4.4   Inverse Laplace Transforms   132
4.5   Transfer Functions   136
4.6   Initial and Final Value Theorems   140
4.7   Frequency Response: Bode Plots   141

P A R T   T W O   ◆   **MEASUREMENT**

CHAPTER 5   **Measuring Instrument Characteristics   153**

5.1   Introduction   154
5.2   Statistics   154
5.3   Operating Characteristics   155
5.4   Static Characteristics   157
5.5   Dynamic Characteristics   162
5.6   Selection Criteria   172

CHAPTER 6   **Signal Conditioning   181**

6.1   Introduction   182
6.2   The Operational Amplifier   183
6.3   Op-Amp Circuits   188
6.4   Analog Signal Conditioning   203
6.5   Digital Signaling Conditioning   224

CHAPTER 7   **Position, Motion, and Force Sensors   250**

7.1   Introduction   251
7.2   Position and Displacement Measurement   252

7.3   Velocity Measurement   270
7.4   Acceleration Measurement   273
7.5   Force Measurement   276

CHAPTER 8   Process Variable Sensors   285

8.1   Temperature Measurement   286
8.2   Flow Rate Measurement   303
8.3   Pressure Measurement   308
8.4   Liquid Level Measurement   312

PART THREE   ◆   MANIPULATION

CHAPTER 9   Switches, Actuators, Valves, and Heaters   321

9.1   Mechanical Switching Components   322
9.2   Solid-State Components   326
9.3   Hydraulic and Pneumatic Valves and Actuators   336
9.4   Control Valves   344
9.5   Electric Heating Elements   352

CHAPTER 10   Electric Motors   363

10.1   Introduction   364
10.2   AC Motors   370
10.3   DC Motors   377
10.4   Stepping Motors   396
10.5   AC Adjustable-Speed Drives   402
10.6   DC Motor Amplifiers and Drives   407

PART FOUR   ◆   CONTROL

CHAPTER 11   Control of Discrete Processes   417

11.1   Introduction   418
11.2   Time-Driven Sequential Processes   419
11.3   Event-Driven Sequential Processes   421
11.4   Time/Event-Driven Sequential Processes   435

CHAPTER 12   Programmable Logic Controllers   442

12.1   Introduction   443
12.2   PLC Hardwarev   446

12.3   PLC Programming and Operation   449
12.4   PLC Programming Functions   454

CHAPTER 13 **Control of Continuous Processes   472**

13.1   Introduction   473
13.2   Modes of Control   476
13.3   Electronic Analog Controllers   500
13.4   Digital Controllers   504
13.5   Advanced Control   509
13.6   Fuzzy Logic Controllers   513

PART FIVE ◆ ANALYSIS AND DESIGN

CHAPTER 14 **Process Characteristics   515**

14.1   Introduction   526
14.2   The Integral or Ramp Process   527
14.3   The First-Order Lag Process   531
14.4   The Second-Order Lag Process   542
14.5   The Dead-Time Process   555
14.6   The First-Order Lag Plus Dead-Time Process   558

CHAPTER 15 **Methods of Analysis   565**

15.1   Introduction   566
15.2   Overall Bode Diagram of Several Components   567
15.3   Open-Loop Bode Diagrams   570
15.4   Closed-Loop Bode Diagrams   571
15.5   Error Ratio and Deviation Ratio   575
15.6   Computer-Aided Bode Plots   578
15.7   Stability   587
15.8   Gain and Phase Margin   588
15.9   Nyquist Stability Criterion   592
15.10   Root Locus   595

CHAPTER 16 **Controller Design   613**

16.1   Introduction   614
16.2   The Ultimate Cycle Method   615
16.3   The Process Reaction Method   616
16.4   Self-Tuning Adaptive Controllers   619
16.5   Computer-Aided PID Controller Design   620
16.6   Example Design of a Three-Loop Control System   637
16.7   Control System Compensation   642

APPENDIX A **Properties of Materials   653**

Properties of Solids   653
Melting Point and Latent Heat of Fusion   654
Properties of Liquids   654
Properties of Gases   654
Standard Atmospheric Conditions   655

APPENDIX B **Units and Conversion   656**

Systems of Units   656
Conversion Factors   657

APPENDIX C **Binary Codes   659**

Powers of 2   659
Octal and Binary Equivalents   659
Decimal, Hexadecimal, and Binary Equivalents   660
One's and Two's Complements   660
The Gray Code   660
Binary Codes for Decimal Digits   661
Seven-Bit ASCII Code   661

APPENDIX D **Instrumentation Symbols and Identification   662**

Purpose   662
Scope   662
Definition   663
Outline of the Identification System   663

APPENDIX E **Complex Numbers   669**

Introduction   669
Rectangular and Polar Forms of Complex Numbers   670
Conversion of Complex Numbers   671
Graphical Representation of Complex Numbers   671
Addition and Subtraction of Complex Numbers   673
Multiplication and Division of Complex Numbers   673
Integer Power of a Complex Number   674
Roots of a Complex Number   674

APPENDIX F **Communications   676**

Communication Interfaces   676
Local Area Networks   682
Communication Protocols   686

References   691

Answers to Selected Exercises   693

Index   702

◆ CHAPTER 1

Basic Concepts and Terminology

◆ OBJECTIVES

Every profession, every subject, even every hobby has its own language: a set of concepts, symbols, and words that people in that field use to express ideas.

The purpose of this chapter is to introduce you to the fundamental language of control system technology. After completing this chapter, you will be able to

1. Define a control system
2. Sketch a block diagram of a closed-loop control system
3. Describe
   a. Transfer function
   b. Gain and phase difference
   c. Open-loop control
   d. Closed-loop control
   e. Components in a closed-loop control system
   f. Signals in a closed-loop control system
   g. P, I, and D control modes
   h. Benefits of automatic control
   i. Load changes in a control system
   j. Three control objectives
   k. Three criteria of good control
4. Read ISA control system drawings
5. Reduce block diagrams to a simpler form

1.1 INTRODUCTION

Control systems are everywhere around us and within us.* Many complex control systems are included among the functions of the human body. An elaborate control system centered in the hypothalamus of the brain maintains body temperature at 37 degrees Celsius (°C) despite changes in physical activity and external ambience. In one control system—the eye—the diameter of the pupil automatically adjusts to control the amount of light that reaches the retina. Another control system maintains the level of sodium ion concentration in the fluid that surrounds the individual cells.

Threading a needle and driving an automobile are two ways in which the human body functions as a complex controller. The eyes are the sensor that detects the position of the needle and thread or of the automobile and the center of the road. A complex controller, the brain, compares the two positions and determines which actions must be performed to accomplish the desired result. The body implements the control action by moving the thread or turning the steering wheel; an experienced driver will anticipate all types of disturbances to the system, such as a rough section of pavement or a slow-moving vehicle ahead. It would be very difficult to reproduce in an automatic controller the many judgments that an average person makes daily and unconsciously.

Control systems regulate temperature in homes, schools, and buildings of all types. They also affect the production of goods and services by ensuring the purity and uniformity of the food we eat and by maintaining the quality of products from paper mills, steel mills, chemical plants, refineries, and other types of manufacturing plants. Control systems help protect our environment by minimizing waste material that must be discarded, thus reducing manufacturing costs and minimizing the waste disposal problem. Sewage and waste treatment also requires the use of automatic control systems.

A *control system* is any group of components that maintains a desired result or value. From the previous examples it is clear that a great variety of components may be a part of a single control system, whether they are electrical, electronic, mechanical, hydraulic, pneumatic, human, or any combination of these. The desired result is a value of some variable in the system, for example, the direction of an automobile, the temperature of a room, the level of liquid in a tank, or the pressure in a pipe. The variable whose value is controlled is called the *controlled variable*.

To achieve control, there must be another variable in the system that can influence the controlled variable. Most systems have several such variables. The control system maintains the desired result by manipulating the value of one of these influential variables. The variable that is manipulated is called the *manipulated variable*. The steering wheel of an automobile is an example of a manipulated variable.

> *Definition of a Control System*
>
> A control system is a group of components that maintains a desired result by manipulating the value of another variable in the system.

---

*An excellent idea of the scope of control systems is given in an Instrument Society of America film, *Principles of Frequency Response*, 1958.

1.2 BLOCK DIAGRAMS AND TRANSFER FUNCTIONS

Although it is not unusual to find several kinds of components in a single control system, or two systems with completely different kinds of components, any control system can be described by a set of mathematical equations that define the characteristics of each component. A wide range of control problems—including processes, machine tools, servomechanisms, space vehicles, traffic, and even economics—can be analyzed by the same mathematical methods. The important feature of each component is the effect it has on the system. The *block diagram* is a method of representing a control system that retains only this important feature of each component. *Signal lines* indicate the input and output signals of the component, as shown in Figure 1.1.

Each component receives an input signal from some part of the system and produces an output signal for another part of the system. The signals can be electric current, voltage, air pressure, liquid flow rate, liquid pressure, temperature, speed, acceleration, position, direction, or others. The signal paths can be electric wires, pneumatic tubes, hydraulic lines, mechanical linkages, or anything that transfers a signal from one component to another. The component may use some source of energy to increase the power of the output signal. Figure 1.2 illustrates block representations of various components.

Block Diagrams

A block diagram consists of a block representing each component in a control system connected by lines that represent the signal paths. Figure 1.3 shows a very simple block diagram of a person driving an automobile. The driver's sense of sight provides the two input signals: the position of the automobile and the position of the center of the road. The driver compares the two positions and determines the position of the steering wheel that will maintain the proper position of the automobile. To implement the decision, the driver's hands and arms move the steering wheel to the new position. The automobile responds to the change in steering wheel position with a corresponding change in direction. After a short time has elapsed, the new direction moves the automobile to a new position. Thus, there is a time delay between a change in position of the steering wheel and the change in the position of the automobile. This time delay is included in the mathematical equation of the block representing the automobile.

The loop in the block diagram indicates a fundamental concept of control. The actual position of the automobile is used to determine the correction necessary to maintain the desired position. This concept is called *feedback*, and control systems with feedback are called

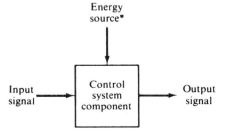

◆ **Figure 1.1**   Block representation of a component. The energy source is not shown on most block diagrams. However, many components do have an external energy source that makes amplification of the input signal possible.

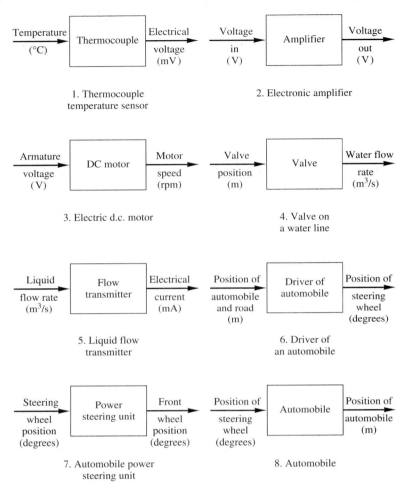

1. Thermocouple
temperature sensor

2. Electronic amplifier

3. Electric d.c. motor

4. Valve on
a water line

5. Liquid flow
transmitter

6. Driver of
an automobile

7. Automobile power
steering unit

8. Automobile

◆ **Figure 1.2** Block representations of control system components.

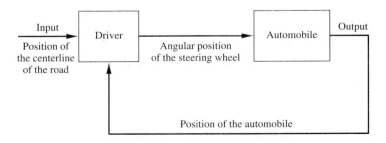

◆ **Figure 1.3** Simplified block diagram of a person driving an automobile.

closed-loop control systems. Control systems that do not have feedback are called *open-loop control systems,* because their block diagram does not have a loop and the actual condition is not used to determine a corrective action.

Transfer Functions

The most important characteristic of a component is the relationship between the input signal and the output signal. This relationship is expressed by the *transfer function* of the component, which is defined as the ratio of the output signal divided by the input signal. (Mostly, it is the Laplace transform of the output signal divided by the Laplace transform of the input signal—further details are covered in Chapter 4.) Refer to the block diagram of a thermocouple shown in Figure 1.2, item 1. If we represent the input temperature by T, the output voltage by V, and the transfer function by H, then $H = V/T$ and $V = HT$. Thus, if we know the input signal and the transfer function, then we can compute the output signal by multiplying the input by the transfer function.

The transfer function consists of two parts. One part is the *size* relationship between the input and the output. The other part is the *timing* between the input and output. For example, the size relationship may be such that the output is twice (or half) as large as the input, and the timing relationship may be such that there is a delay of 2 seconds between a change in the input and the corresponding change in the output.

If the component is linear and the input is a sinusoidal signal, then the output will also be a sinusoidal signal. Figure 1.4 illustrates a linear component with a sinusoidal input signal. Notice the mathematical functions and the graphs used to represent the input and output signals in Figure 1.4. The size relationship between the input and the output is measured by the ratio of the amplitude of the output signal, B, divided by the amplitude of the input signal, A. We call this ratio the *gain* of the component.

$$\text{Gain} = \frac{\text{amplitude of the output signal (output units)}}{\text{amplitude of the input signal (input units)}} = \frac{B}{A}$$

Observe the input and output graphs in Figure 1.4. The input signal crosses zero when $\omega t/\alpha = 0$, that is, when $t = -\alpha/\omega$ s (dimensions are radians per second for ω and radians for

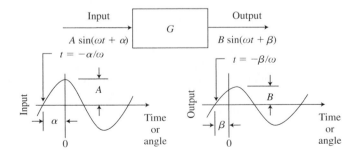

Gain = B/A
Phase difference = $\beta - \alpha$

◆ **Figure 1.4**   Gain and phase difference of a linear component.

α). The output signal crosses zero when $\omega t + \beta = 0$, that is, when $t = -\beta/\omega$ s. We could measure the timing difference between the input and the output signal by the difference between these two times.

$$\text{Timing difference} = (-\alpha/\omega) - (-\beta/\omega) = (\beta - \alpha)/\omega \text{ s}$$

It is simpler and more convenient to measure the timing difference between the input and the output by the difference between the two angles, that is, by $(\beta - \alpha)$. We call this difference the *phase difference* of the component. It is also convenient to express the phase difference in degrees rather than radians. Just imagine a translation of the time axis from seconds to degrees. One complete cycle of the sine wave becomes 360°. The distance from $t = -\alpha/\omega$ s to $t = 0$ s becomes α°; the distance from $t = -\beta/\omega$ s to $t = 0$ s becomes β°. We will use this translation of the time axis as a convenient way to represent phase angles in degrees.

The phase difference of the component is the phase angle of the output signal minus the phase angle of the input signal.

$$\text{Phase difference} = \text{output phase angle} - \text{input phase angle (degrees)}$$

Complex numbers (in polar form) are most conveniently used to represent values of the input, the output, and the transfer function. In Figure 1.4, the input is represented by the complex number $A \underline{/\alpha}$ and the output by the complex number $B \underline{/\beta}$ The transfer function, G, is represented by the complex number obtained by dividing the output, $B \underline{/\beta}$, by the input, $A \underline{/\alpha}$:

$$G = B \underline{/\beta}/A \underline{/\alpha} = (B/A) \underline{/\beta - \alpha}$$

Thus the transfer function, G, is represented by the complex number whose magnitude is the gain of the component, B/A, and whose angle is the phase of the output minus the phase of the input.

The gain of a component is often expressed as the ratio of the change in the amplitude of the output divided by the corresponding *change* in the amplitude of the input.

$$\text{Gain} = \frac{\text{change in output amplitude (output units)}}{\text{change in input amplitude (input units)}}$$

The gain of a component has the dimension of output units over input units. Thus an amplifier that produces a 10-volt (V) change in output for each 1-V change in input has a gain of 10 volts per volt (V/V). A direct-current (dc) motor that produces a change in speed of 1000 revolutions per minute (rpm) for each 1-V change in input has a gain of 1000 rpm/V. A thermocouple that produces an output change of 0.06 millivolt (mV) for each 1°C change in temperature has a gain of 0.06 mV/°C.

The gain and phase difference of a component for a given frequency are referred to as the *frequency response* of the component at that frequency. As an example, at a frequency of 1 hertz (Hz), a certain control system component has a gain of 0.995 and a phase difference of 5.71°. At a frequency of 10 Hz, the gain is 0.707 and the phase difference is 45°. At a frequency of 100 Hz, the gain is 0.0995 and the phase difference is 84.29°. These figures are given here only as an illustration. A more complete discussion is reserved for later chapters.

The transfer function of a component describes the size and timing relationship between the output signal and the input signal.

**EXAMPLE
1.1**

The input to a linear control system component is a 0.5-Hz sinusoidal signal with a peak amplitude of 5.3 V and a phase angle of 30°. The output of the component has a peak amplitude of 14 milliamperes (mA) and a phase angle of 25°. Determine the gain, the phase difference, and the transfer function for these conditions.

Solution

$$\text{Gain} = 14 \text{ mA}/5.3 \text{ V}$$
$$= 2.64 \text{ mA/V}$$
$$\text{Phase difference} = 25 - 30$$
$$= -5°$$
$$\text{Transfer function} = 2.64 \underline{/-5°} \text{ mA/V}$$

◆

1.3 OPEN-LOOP CONTROL

An open-loop control system does not compare the actual result with the desired result to determine the control action. Instead, a calibrated setting—previously determined by some sort of calibration procedure or calculation—is used to obtain the desired result.

The needle valve with a calibrated dial shown in Figure 1.5 is an example of an open-loop control system. The calibration curve is usually obtained by measuring the flow rate for several dial settings. As the calibration curve indicates, different calibration lines are obtained for different pressure drops. Assume that a flow rate of F_2 is desired and a setting of S is used. As long as the pressure drop across the valve remains equal to P_2, the flow rate will remain F_2. If the pressure drop changes to P_1, the flow rate will change to F_1. The open-loop control cannot correct for unexpected changes in the pressure drop.

The firing of a rifle bullet is another example of an open-loop control system. The desired result is to direct the bullet to the bull's-eye. The actual result is the direction of the bullet af-

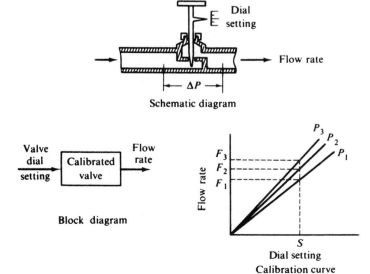

◆ **Figure 1.5** A calibrated needle valve is an example of an open-loop control system.

ter the gun has been fired. The open-loop control occurs when the rifle is aimed at the bull's-eye and the trigger is pulled. Once the bullet leaves the barrel, it is on its own: If a sudden gust of wind comes up, the direction will change and no correction will be possible.

The primary advantage of open-loop control is that it is less expensive than closed-loop control: It is not necessary to measure the actual result. In addition, the controller is much simpler because corrective action based on the error is not required. The disadvantage of open-loop control is that errors caused by unexpected disturbances are not corrected. Often a human operator must correct slowly changing disturbances by manual adjustment. In this case, the operator is actually closing the loop by providing the feedback signal.

1.4 CLOSED-LOOP CONTROL: FEEDBACK

Feedback is the action of measuring the difference between the actual result and the desired result, and using that difference to drive the actual result toward the desired result. The term *feedback* comes from the direction in which the measured value signal travels in the block diagram. The signal begins at the output of the controlled system and ends at the input to the controller. The output of the controller is the input to the controlled system. Thus the measured value signal is fed back from the output of the controlled system to the input. The term *closed loop* refers to the loop created by the *feedback* path.

Block diagrams of closed-loop control systems are shown in Figures 1.6 and 1.7. Figure 1.6 is used in the design of servomechanisms, and Figure 1.7 is used in the design of process control systems. The names of the components and variables in Figure 1.7 are used throughout this book. However, we will develop the closed-loop transfer function of both the servo control system (Figure 1.6) and the process control system (Figure 1.7). You should be thoroughly familiar with these terms and the following operations, which form the basis of a feedback control system.

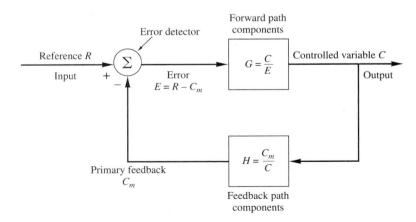

◆ **Figure 1.6**   Block diagram of a closed-loop servo control system.

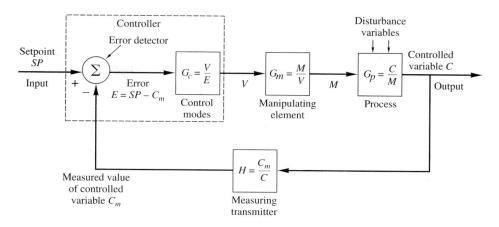

◆ **Figure 1.7**   Block diagram of a closed-loop process control system.

Operations Performed by a Feedback Control System

Measurement: measure the value of the controlled variable

Decision: compute the error (desired value minus measured value) and use the error to form a control action

Manipulation: use the control action to manipulate some variable in the process in a way that will tend to reduce the error

---

In Figure 1.6, the reference (R) is the input to the servo control system and the controlled variable (C) is the output. We now proceed to derive the transfer function, C/R, of the closed-loop servo control system. First, we write the equation for the output of the error detector in terms of reference, R, and the primary feedback, C_m.

$$E = R - C_m$$

Second, we write equations for the outputs of the other two components in terms of their input and their transfer function.

$$C = EG$$
$$C_m = CH$$

Third, we use substitution to eliminate E and C_m.

$$C = (R - C_m)G$$
$$C = (R - CH)G$$

Finally, we solve the above equation for the ratio, C/R.

$$C + CGH = RG$$
$$C(1 + GH) = RG$$

$$\frac{C}{R} = \frac{G}{1 + GH} \qquad \textbf{(1.1)}$$

Equation (1.1) is the transfer function of a closed-loop servo control system (servomechanism). The forward path transfer function (G) contains all the system components, such as motors, generators, gears, amplifiers, and so on. The feedback path transfer function (H) is usually a passive device that converts the controlled variable into a suitable signal for input to the error detector. The letter B is often used to represent the primary feedback signal in the block diagram of a servomechanism. The author prefers the notation C_m for this signal to indicate its relationship as the measured value of the controlled variable.

Occasionally, the primary feedback signal is inverted and must be added to the reference signal to form the error signal. In Figure 1.6, this positive feedback is accomplished by changing the sign of the lower input to the error detector from minus to plus. It is left as an exercise to show that the transfer function for positive feedback is given by Equation (1.2):

$$\frac{C}{R} = \frac{G}{1 - GH} \qquad \textbf{(1.2)}$$

In Figure 1.7, the setpoint (SP) is the input to the process control system, and the controlled variable (C) is the output. The feedback path consists of one component, the measuring transmitter with transfer function H. The forward path consists of three components (the control modes, the manipulating element, and the process) with transfer functions G_c, G_m, and G_p, respectively. The overall forward transfer function (G) is the product of the three component transfer functions:

$$G = G_c G_m G_p$$

The performance of a control system is usually based on a comparison between the setpoint (SP) and the measured value of the controlled variable (C_m). The reason C_m is used instead of C is that C_m is measurable and available, but C is not. We now proceed to derive the transfer function, C_m/SP, of the closed-loop process control system:

First, we write the equation for the error detector output.

$$E = SP - C_m$$

Second, we use the overall forward transfer function, G, to write the following equation for the controlled variable, C.

$$C = EG$$

Third, we write the following equation for C_m.

$$C_m = CH$$

Fourth, we use substitution to eliminate C and E.

$$C_m = EGH$$
$$C_m = (SP - C_m)GH$$

Finally, we solve the above equation for the ratio C_m/SP.

$$C_m + C_mGH = (SP)GH$$

$$C_m(1 + GH) = (SP)GH$$

$$\frac{C_m}{SP} = \frac{GH}{1 + GH} \qquad\qquad (1.3)$$

Equation (1.3) is the transfer function of a closed-loop process control system. The following is a description of each component in Figure 1.7.

Process

The *process block* in Figure 1.7 represents everything performed in and by the equipment in which a variable is controlled. The process includes everything that affects the controlled or *process variable* except the controller and the final control element. In a home heating system, for example, the process is the home and its contents. The two most important parameters of this process (the home) are the thermal resistance of the outside walls and the thermal capacitance of the air and contents inside the home.

Measuring Transmitter

The *measuring transmitter,* or *sensor,* senses the value of the controlled variable and converts it into a usable signal. Although the measuring transmitter is considered as one block, it usually consists of a primary sensing element and a signal transducer (or signal converter). The term *measuring transmitter* is a general term to cover all types of signals. In specific cases, the word *measuring* is replaced by the name of the measured signal (e.g., temperature transmitter, flow transmitter, pressure transmitter, etc.).

Figure 1.8 shows the input/output curve of a typical temperature transmitter. The primary element could be a thermocouple, a resistance element, a thermistor, or a filled thermal element. The signal transducer receives the output of the primary element and produces an electric current signal. For example, a thermocouple converts temperature into a millivolt signal,

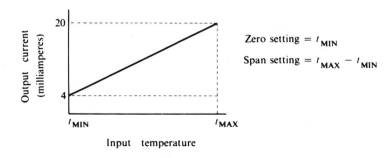

◆ **Figure 1.8** Input/output graph of a temperature-measuring transmitter.

and the thermocouple transducer converts the millivolt signal into an electric current in the range 4 to 20 mA. A resistance element converts temperature into a resistance value, and the resistance transducer converts the resistance value into an electric current signal. Other primary elements are handled in a similar manner. In our home heating system example, the term *sensor* is more appropriate than *measuring transmitter*. In many home heating systems, the temperature sensor is a bimetallic coil that converts the room temperature into the angular position of a mercury switch mounted on one end of the coil.

EXAMPLE 1.2

A temperature-measuring transmitter has an input range from -40 to $60°C$ and an output range from 4 to 20 mA. Assume the transmitter is linear and find the following:

(a) The transmitter gain, $H = C_m/C$.
(b) The equation for the output C_m in terms of input C.
(c) The output of the transmitter when input $C = 22°C$.
(d) The input to the transmitter when output $C_m = 11$ mA.

Solution

(a) $H = \dfrac{\text{output range}}{\text{input range}} = \dfrac{20 \text{ mA} - 4 \text{ mA}}{60°C - (-40°C)} = \dfrac{16 \text{ mA}}{100°C}$

$H = 0.16 \text{ mA/°C}$

(b) We will use the following slope-intercept form of the equation for C_m in terms of C:

$$C_m = mC + b$$

The gain, H, is the required slope, that is, $m = H = 0.16$.

$$C_m = 0.16C = b$$

From the data, we use $C_m = 20$ mA when $C = 60°C$ to determine the unknown, b:

$$20 = 0.16(60) + b$$

$$b = 20 - 9.6 = 10.4$$

The desired equation is

$$C_m = 0.16C + 10.4$$

(c) We use the equation from step b to find the output when $C = 22°C$:

$$C_m = 0.16(22) + 10.4 = 13.92 \text{ mA}$$

(d) We use the equation again to find the input when $C_m = 11$ mA:

$$11 = 0.16 \, (C) + 10.4$$

$$C = \frac{(11 - 10.4)}{0.16} = 3.75°C$$

◆

Controller

The controller includes the error detector and a unit that implements the control modes. The *error detector* computes the difference between the measured value of the controlled variable and the desired value (or setpoint). The difference is called the *error* and is computed according to the following equation:

$$Error = setpoint - measured\ value\ of\ controlled\ variable$$

or

$$E = SP - C_m$$

The *control modes* convert the error into a control action or *controller output* that will tend to reduce the error. The three most common control modes are the proportional mode (P), the integral mode (I), and the derivative mode (D). The three modes are defined mathematically in Chapter 13. In this chapter you need only to know the names of the three modes and have an intuitive understanding of how they work. The following discussion will show that the names of the modes suggest the types of control action that is formed.

The *proportional mode* (P) is the simplest of the three modes. It produces a control action that is proportional to the error. If the error is small, the proportional mode produces a small control action. If the error is large, the proportional mode produces a large control action. The proportional mode is accomplished by simply multiplying the error by a gain constant, K.

The *integral mode* (I) produces a control action that continues to increase its corrective effect as long as the error persists. If the error is small, the integral mode increases the correction slowly. If the error is large, the integral action increases the correction more rapidly. In fact, the rate at which the correction increases is proportional to the error signal. Mathematically, the integral control action is accomplished by forming the integral of the error signal.

The *derivative mode* (D) produces a control action that is proportional to the rate at which the error is changing. For example, if the error is increasing rapidly, it will not be long before there is a large error. The derivative mode attempts to prevent this future error by producing a corrective action proportional to how fast the error is changing. The derivative mode is an attempt to anticipate a large error and head it off with a corrective action based on how quickly the error is changing. Mathematically, the derivative control action is accomplished by forming the derivative of the error signal.

The proportional mode may be used alone or in combination with either or both of the other two modes. The integral mode can be used alone, but it almost never is. The derivative mode cannot be used alone. Thus the common control mode combinations are: P, PI, PD, and PID.

In the home heating system, the control action is built into the bimetallic coil/mercury switch described earlier. The mercury switch is attached to one end of the coil. The other end of the coil is attached to a small plate that is pivoted at the center of the coil. A pointer on the small plate indicates the temperature setting on a stationary temperature scale. We adjust the setpoint by rotating the coil assembly until the pointer indicates the desired temperature. The bimetallic coil performs an ON/OFF control action by tilting the mercury switch ON if the room temperature is slightly below the setpoint, or OFF if the temperature is slightly above the setpoint.

Manipulating Element

The *manipulating element* uses the controller output to regulate the manipulated variable and usually consists of two parts. The first part is called an *actuator,* and the second part is called the final controlling *element*. The actuator translates the controller output into an action on the final controlling element, and the final controlling element directly changes the value of the manipulated variable. Valves, dampers, fans, pumps, and heating elements are examples of manipulating elements. The valve that controls the fuel flow in a home heating system is another example of a manipulating element.

A pneumatic control valve is often used as the manipulating element in processes (see Figure 1.9). The actuator consists of an air-loaded diaphragm acting against a spring. As the air pressure on the diaphragm goes from 3 to 15 pounds per square inch (psi), the stem of the valve will move from open to closed (air to close) or from closed to open (air to open). In the home heating system, the furnace is the manipulating element. The control action opens or closes a solenoid valve that controls the flow of gas to the burner. When the solenoid valve is open, gas flows to the burner and the burner delivers heat to the home.

Variable Names

The *controlled variable (C)* is the process output variable that is to be controlled. In a process control system, the controlled variable should be a good measure of the quality of the product. The most common controlled variables are position, velocity, temperature, pressure, level, and flow rate.

The *setpoint (SP)* is the desired value of the controlled variable.

The *measured variable* (C_m) is the measured value of the controlled variable. It is the output of the measuring means and usually differs from the actual value of the controlled variable by a small amount.

The *error (E)* is the difference between the setpoint and the measured value of the controlled variable. It is computed according to the equation $E = SP - C_m$.

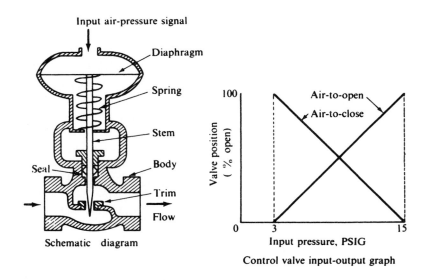

◆ **Figure 1.9** A pneumatic control valve has two possible actions: air-to-close and air-to-open. In an air-to-close valve, the valve stem moves from open to closed as the air pressure goes from 3 to 15 psi. An air-to-open valve moves from closed to open with the same change in air pressure.

The *controller output (V)* is the control action intended to drive the measured value of the controlled variable toward the setpoint value. The control action depends on the error signal (*E*) and on the control modes used in the controller.

The *manipulated variable (M)* is the variable regulated by the final controlling element to achieve the desired value of the controlled variable. Obviously, the manipulated variable must be capable of effecting a change in the controlled variable. The manipulated variable is one of the input variables of the process. Changes in the *load* on the process necessitate changes in the manipulated variable to maintain a balanced condition. For this reason, the value of the manipulated variable is used as a measure of the load on the process.

The *disturbance variables (D)* are process input variables that affect the controlled variable but are not controlled by the control system. Disturbance variables are capable of changing the load on the process and are the main reason for using a closed-loop control system.

The primary advantage of closed-loop control is the potential for more accurate control of the process. There are two disadvantages of closed-loop control: (1) closed-loop control is more expensive than open-loop control, and (2) the feedback feature of a closed-loop control system makes it possible for the system to become unstable. An unstable system produces an oscillation of the controlled variable, often with a very large amplitude. (Stability will be studied in detail later.)

> A closed-loop (or feedback) control system measures the differences between the actual value of the controlled variable and the desired value (or setpoint) and uses the difference to drive the actual value toward the desired value.

1.5 CONTROL SYSTEM DRAWINGS

The Instrument Society of America has prepared a standard, "Instrumentation Symbols and Identifications," ANSI/ISA-S5.1-1984, to "establish a uniform means of designating instruments and instrument systems used for measurement and control." This standard presents a designation system that includes symbols and an identification code that is "suitable for use whenever any reference to an instrument is required." Applications include flow diagrams, instrumentation drawings, specifications, construction drawings, technical papers, tagging of instruments, and other uses. Appendix D includes material abstracted from ANSI/ISA-S5.1-1984.

A circular symbol called a *balloon* is the general instrument symbol. The instrument is identified by the code placed inside its balloon. The identification code consists of a *functional identification* in the top half of the balloon and a *loop identification* in the bottom half. The first letter in the functional identification defines the measured or initiating variable of the control loop (e.g., flow, level, pressure, temperature). Up to three additional letters may be used to name functions of the individual instrument (e.g., indicator, recorder, controller, valve). The standard also defines symbols for instrument lines, control valve bodies, actuators, primary elements, various functions, and other devices.

Figure 1.10 illustrates the use of the standard symbols and identification code in a process control drawing. The process blends and heats a mixture of water and syrup. This system has three control loops, with loop identification numbers of 101, 102, and 103. The first two digits designate the area in the plant where this system is located. The third digit identifies a particular control loop. Loop 101 is a level control loop, as indicated by the first letter in the function code of each instrument in the loop. The meaning of each code is as follows:

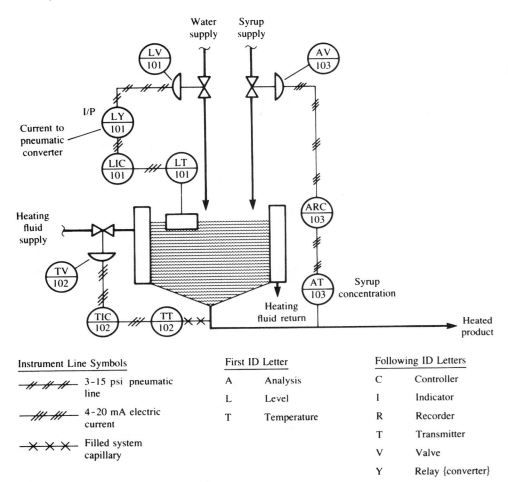

◆ **Figure 1.10** Blending and heating system instrumentation drawing.

LT-101   Level transmitter
            Uses a float to sense the level of the liquid in the tank and transduces the
            signal into an electric current in the range 4 to 20 mA.

LIC-101   Level indicating controller
            Uses the milliampere signal from the level transmitter to produce a control
            signal in the range 4 to 20 mA.

LY-101   Level current to pneumatic converter
            Converts the milliampere output from the controller into a pneumatic signal
            in the range 3 to 15 psi.

LV-101   Level control valve
            Uses the pneumatic signal from the I/P converter to position the stem of the
            level control valve.

TT-102   Temperature transmitter
              Uses a filled bulb to sense the temperature of the product leaving the blending tank and transduces the signal into an electric current in the range 4 to 20 mA.

TIC-102   Temperature-indicating controller
              Uses the milliampere signal from the temperature transmitter to produce a control signal in the range 4 to 20 mA.

TV-102   Temperature control valve
              Uses the milliampere signal from the temperature controller to position the stem of the temperature control valve.

AT-103   Analysis transmitter
              Senses the concentration of syrup in the product and transduces the signal into an electric current in the range 4 to 20 mA.

ARC-103   Analysis recording controller
              Uses the milliampere signal from the analysis transmitter to produce a control signal in the range 4 to 20 mA.

AV-103   Analysis control valve
              Uses the milliampere signal from the analysis controller to position the stem of the analysis control valve.

1.6 NONLINEARITIES

Most control system analysis and design is done with the assumption that all components in the system are linear. Actually, there are several forms of nonlinearity that occur in control system components. This section uses the input/output (or I/O) graph of a component to give you an intuitive understanding of linearity, nonlinearity, hysteresis, dead band, and saturation.*

As its name implies, the I/O graph is a plot of input values versus output values. It defines the output produced by any given input to the component. When the I/O graph is a single, perfectly straight line, the component is said to be *linear*. Figure 1.11 shows the I/O graph of a linear component, the sinusoidal input to the component, and the resulting output from the component. Observe the sinusoidal shape of the output waveform in Figure 1.11. The ability to preserve the shape of the input is an important characteristic of a linear component. In contrast, the output waveform of a nonlinear component is a distortion of the input waveform. Now observe the 45° slope of the I/O graph in Figure 1.11. With a slope of 45°, the output amplitude is equal to the input amplitude. Decreasing the slope decreases the output amplitude. Increasing the slope increases the output amplitude. The slope of the line determines the gain of the component.

A common form of *nonlinearity* is the curved I/O graph illustrated in Figure 1.12. Observe the distortion in the output waveform. We measure the nonlinearity of a component by drawing a straight line on the I/O graph positioned to minimize the deviation between it and

---

*For a more complete treatment of nonlinearity, refer to *Standards and Practices for Instrumentation,* 6th ed. (Research Triangle Park, N.C.: Instrument Society of America, 1980).

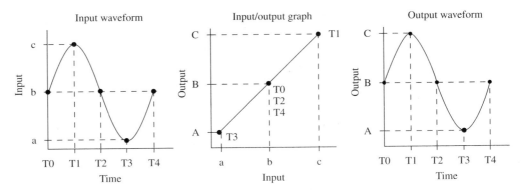

◆ **Figure 1.11**   Linear element with a sinusoidal input waveform and the resulting output waveform. Use the input/output graph to determine the value of the output for each value of the input. At time T0, input value b yields output value B; at time T1, input c yields output C; at time T2, input b yields output B; at time T3, input a yields output A; and at time T4, input b yields output B. Notice the sinusoidal shape of the output waveform. A linear element always preserves the shape of the input waveform.

the curved I/O graph. The maximum deviation between the straight line and the curved I/O graph is a measure of the nonlinearity of the component.

Dead band is the range of values through which an input can be changed without producing an observable change in the output. The dead-band effect occurs whenever the input changes direction. Backlash in gears is one example of dead band, and the term *backlash* is sometimes used in place of dead band. Figure 1.13a shows the I/O graph of a component with dead band, a sinusoidal input waveform, and the resulting output waveform. Notice the distortion in the output waveform caused by the dead band. As time goes from T0 to T2, the input increases from c to e, producing an output increase from B to D. The input begins to decrease at time T2, but dead band causes the output to remain unchanged until time T3. From T2 to T3, the input decreases from e to d, but the output remains at D. From time T3 to time

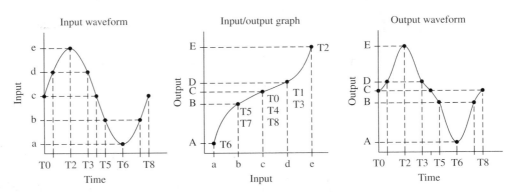

◆ **Figure 1.12**   Nonlinear element with a sinusoidal input waveform and the resulting output waveform. The input/output graph translates input values into corresponding output values. Notice how the nonlinear element distorts the shape of the input waveform, resulting in a nonsinusoidal output waveform.

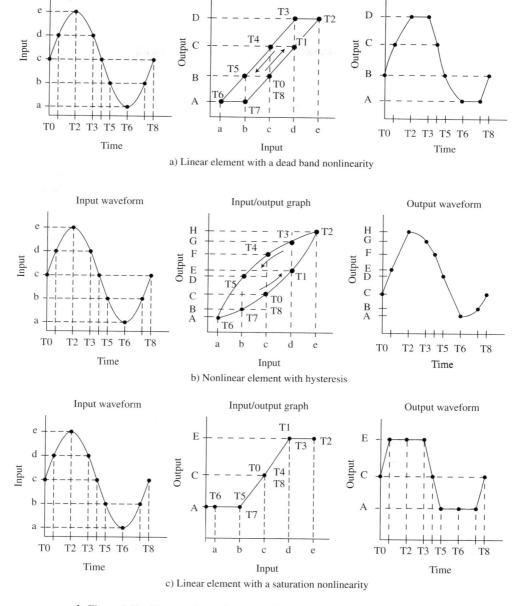

a) Linear element with a dead band nonlinearity

b) Nonlinear element with hysteresis

c) Linear element with a saturation nonlinearity

◆ **Figure 1.13** Three nonlinear elements with sinusoidal inputs and the resulting outputs. Notice how these nonlinear elements distort the output waveform, each in its own way.

T6, the input decreases from d to a, producing an output decrease from D to A. The input begins to increase at time T6, but dead band causes the output to remain unchanged until time T7. From T6 to T7, the input increases from a to b, but the output remains at A due to the dead band. From time T7 to T8, the input increases from b to c, producing an output increase from A to B. The dead-band effect occurs whenever the input changes direction. Once the dead band is crossed, it has no further effect until the input changes direction again. With dead band, each input value has a range of possible output values. For example, with an input value of b, the output could be anywhere between A and B.

Hysteresis occurs when the I/O graph follows different curved paths when the input increases and decreases. The result is an I/O graph that forms a loop, and the value of the output for any given input depends on the history of previous inputs. Figure 1.13b shows the I/O graph of a component with hysteresis, a sinusoidal input waveform, and the resulting output waveform. Notice the distortion in the output waveform caused by the hysteresis. Hysteresis is expressed as the maximum difference between the outputs for any given input value as the input traverses one complete cycle. In Figure 1.13b, the maximum output difference occurs when the input value is c (at times T0 and T4). At time T0, the output is C. At time T4, the output is F. Hysteresis is expressed as the difference between output F and output C.

Saturation refers to the limitations on the range of values for the output of a component (see Figure 1.13c). All real components reach a saturation limit when the input is increased (or decreased) beyond its limiting value. For example, a control valve may go from closed to open as the pressure in the actuator increases from 3 to 15 psi. The valve remains closed when the pressure is decreased below 3 psi. The valve remains open when the pressure is increased above 15 psi. We say that the valve reaches saturation when the pressure is below 3 psi or above 15 psi.

1.7 BENEFITS OF AUTOMATIC CONTROL

Control systems are becoming steadily more important in our society. We depend on them to such an extent that life would be unimaginable without them. Automatic control has increased the productivity of each worker by releasing skilled operators from routine tasks and by increasing the amount of work done by each worker. Control systems improve the quality and uniformity of manufactured goods and services; many of the products we enjoy would be impossible to produce without automatic controls. Servo systems place tremendous power at our disposal, enabling us to control large equipment such as jet airplanes and ocean ships.

Control systems increase efficiency by reducing waste of materials and energy, an increasing advantage as we seek ways to preserve our environment. Safety is yet another benefit of automatic control. Finally, control systems such as the household heating system and the automatic transmission provide us with increased comfort and convenience.

In summary, the benefits of automatic control fall into the following six broad categories:

1. Increased productivity
2. Improved quality and uniformity
3. Increased efficiency
4. Power assistance
5. Safety
6. Comfort and convenience

1.8 LOAD CHANGES

A control system must balance the material or energy gained by the process against the material or energy lost by the process, to maintain the desired value of the controlled variable. Usually, the material or energy loss is the load on the process and the manipulated variable must supply the balancing material or energy gain. However, sometimes the opposite condition exists and the manipulated variable must provide the material or energy loss.

To maintain the desired inside temperature, a home heating system must balance the heat supplied by the furnace against the heat lost by the house. The heat lost is the load on the control system, and the energy supplied to the furnace is regulated by the manipulated variable.

To maintain the level at the desired value, a liquid-level control system must balance the input flow rate against the output flow rate. The output flow rate is the load on the system, and the input flow rate is the manipulated variable.

To maintain the desired pump speed, the control of a variable-speed motor driving a pump must balance the input power to the motor against the power required by the pump. The power required by the pump is the load on the system, and the power input to the motor is regulated by the manipulated variable.

To maintain the desired room temperature, an air-conditioning system must balance the heat removed by the air conditioner against the heat gained by the room. The heat gained by the room is the load on the system, and the heat removed by the air conditioner is regulated by the manipulated variable.

The load on a process is always reflected in the manipulated variable. Therefore, the value of the manipulated variable is a measure of the load on the process. Every load change results in a corresponding change in the manipulated variable and, consequently, a corresponding change in the setting of the final controlling element. Consider a sudden increase in the load for the pump control system previously described. The increase in load tends to reduce the motor speed. The controller senses the reduced motor speed and produces a control action that increases the power input to the motor. In an ideal situation, the control action will cause the manipulated variable to match the increased load and the pump speed will remain at the desired value. Within the control loop, the only variable that reflects the load change is the manipulated variable. For this reason, it makes sense to define the load on the control system in terms of the manipulated variable.

> The load on a control system is measured by the value of the manipulated variable required by the process at any one time in order to maintain a balanced condition.

The load on a control system does not remain constant. Any uncontrolled variable that affects the controlled variable is capable of causing a load change. Each load change necessitates a corresponding change in the manipulated variable in order to maintain the controlled variable at the desired value. A closed-loop control system automatically makes the necessary change in the manipulated variable; an open-loop control system does not make the necessary change. Thus a closed-loop control system is necessary if automatic adjustment to load changes is desired.

There are usually several uncontrolled conditions in a process that are capable of causing a load change. Some examples of load changes are

1. A *change in demand* by the controlled medium. For example, opening the door of a house in winter necessitates more heat to keep the inside temperature at the desired value. Closing the door requires less heat. Both are load changes. In a manufacturing process, a change in production rate almost always results in a load change. In a heat exchanger, for example, a flowing liquid is continuously heated with steam. A change in the liquid flow rate is a load change because more heat is required.

2. A *change in the quality* of the manipulated variable. For example, a change in the heat content of the fuel supplied to a burner requires a change in the rate at which the fuel is supplied to the burner. In a neutralizing process, a solution of sodium bicarbonate is used to neutralize a fiber ribbon. A decrease in the concentration of sodium bicarbonate is a load change because more neutralizing solution is required.

3. A *change in ambient conditions*. For example, if the outside temperature drops, more heat is required to maintain the desired temperature in a house.

4. A *change in the amount of energy absorbed or supplied* within the process. For example, using the range to prepare supper supplies a house with a large quantity of heat. Thus less heat is required from the furnace to maintain the desired temperature. Chemical reactions often generate or absorb heat as part of the reaction; these are load changes because, as the process generates or absorbs heat, less or more heat is required from the manipulated variable.

1.9 DAMPING AND INSTABILITY

The gain of the controller determines a very important characteristic of a control system's response: the type of damping or instability that the system displays in response to a disturbance. The five general conditions are illustrated in Figure 1.14. As the gain of the controller is increased, the response changes in the following order: overdamped, critically damped, underdamped, unstable with constant amplitude, and unstable with increasing amplitude. Obviously, neither the unstable response nor the overdamped response satisfies the objective of minimizing the error. Typically, the optimum response is either critically damped or slightly underdamped. Exactly how much damping is optimum depends on the requirements of the process.

Further insight about damping can be obtained by considering a familiar oscillating system—a child bouncing a ball. The ball will continue to bounce as long as the child pushes down when the ball is moving down (i.e., the force is in the same direction as the motion of the ball). The bouncing will die down quickly if the child pushes down when the ball is moving up (i.e., the force is in opposition to the motion of the ball). The oscillations of the ball are damped out by a force in opposition to the motion. Extending this concept to control systems, *damping* is a force or signal that opposes the motion (or rate of change) of the controlled variable.

Several stabilizing techniques are used to increase the damping in a system and thereby to allow a higher gain in the controller. The general idea is to find a force or signal that will oppose changes in the controlled variable. One such signal is the rate of change of the controlled variable. In mathematics, the derivative of a variable is equal to its rate of change, and this signal is referred to as the *derivative* of the controlled variable. Damping is increased if the derivative of the controlled variable is subtracted from the error signal before it goes to the controller. This technique is sometimes called *output derivative damping*.

Another stabilizing signal is the derivative of the error signal. If the setpoint is constant, this signal is equal to the negative of the derivative of the controlled variable. Damping is in-

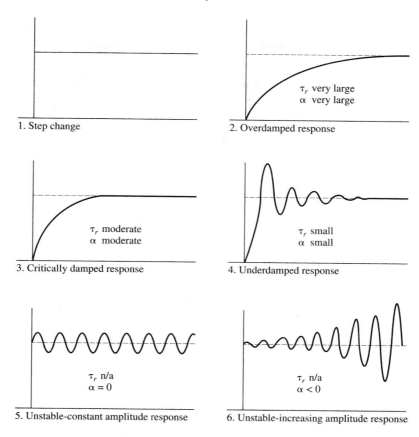

◆ **Figure 1.14**   The five types of response to a step change in load or setpoint are characterized by the rise time, τ_r, and the damping constant, α. Rise time is the time it takes a signal to go from 10 to 90% of the total change in response to the step change. Damping constant is a measure of the amount of damping in the system.

creased if the derivative of the error is added to the error signal before it goes to the controller. This technique is usually called the *derivative control mode*.

Viscous damping is a stabilizing technique sometimes used in position control systems. It operates on the fact that frictional forces always oppose motion. A simple brake, a fluid brake, or an eddy-current brake may be used to apply the damping force.

1.10 OBJECTIVES OF A CONTROL SYSTEM

At first glance, the objective of a control system seems quite simple—to maintain the controlled variable exactly equal to the setpoint at all times, regardless of load changes or setpoint changes. To do this, the control system must respond to a change before the error occurs; unfortunately, feedback is never perfect because it does not act until an error occurs. First, a load change must change the controlled variable; this produces an error. Then the con-

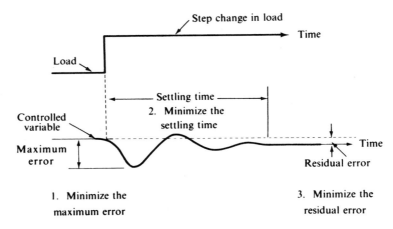

◆ **Figure 1.15** Three objectives of a closed-loop control system.

troller acts on the error to produce a change in the manipulated variable. Finally, the change in the manipulated variable drives the controlled variable back toward the setpoint.

It is more realistic for us to expect a control system to obtain as nearly perfect operation as possible. Because the errors in a control system occur after load changes and setpoint changes, it seems natural to define the objectives in terms of the response to such changes. Figure 1.15 shows a typical response of the controlled variable to a step change in load.

One obvious objective is to minimize the maximum value of the error signal. Some control systems (with an integral mode) will eventually reduce the error to zero, whereas others require a residual error to compensate for a load change. In either case, the control system should eventually return the error to a steady, nonchanging value. The time required to accomplish this is called the *settling time*. A second objective of a control system is to minimize the settling time. A third objective is to minimize the *residual error* after settling out.

Unfortunately, these three objectives tend to be incompatible. For instance, the problem of reducing the residual error can be solved by increasing the gain of the controller so that a smaller residual error is required to produce the necessary corrective control action. However, an increase in gain tends to increase the settling time and may increase the maximum value of the error as well. The optimum response is always achieved through some sort of compromise.

Control Objectives

After a load or setpoint change, the control system should

1. Minimize the maximum value of the error
2. Minimize the setting time
3. Minimize the residual error

1.11 CRITERIA OF GOOD CONTROL

To evaluate a control system effectively, two decisions must be made: (1) the test must be specified, and (2) the criteria of good control must be selected. A step change in setpoint or load is the most common test. A typical step response test is illustrated in Figure 1.16a. The

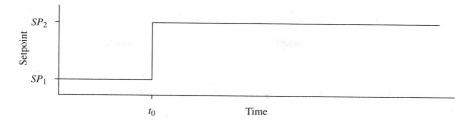

a) Step change in setpoint used to evaluate a control system

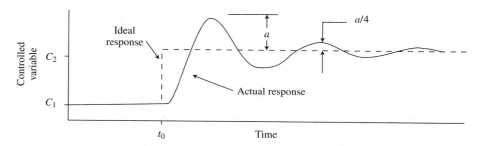

b) Quarter amplitude decay response to the step change in setpoint

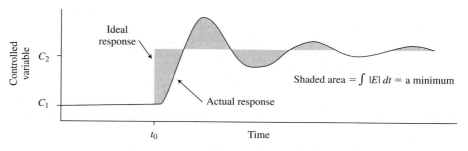

c) Minimum integral of absolute error response to the step change in setpoint

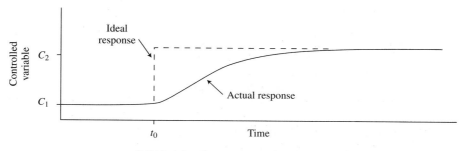

d) Critical damping response to the step change in setpoint

◆ **Figure 1.16** Step change in setpoint and the three most common criteria of good control.

three most common criteria of good control are quarter amplitude decay, integral of absolute error, and critical damping. A discussion of each criterion follows.

1. *Quarter amplitude decay.* This criterion specifies a damped oscillation in which each successive positive peak value is one fourth of the preceding positive peak value. Quarter amplitude decay is a popular criterion because it is easy to apply in the field and provides a nearly optimum compromise of the three control objectives. Figure 1.16b illustrates the quarter amplitude decay response.

    A variation of quarter amplitude decay is peak percentage overshoot (*PPO*). It is a measure of the peak overshoot of the controlled variable with respect to the size of the step change. A *PPO* of 50% is roughly equivalent to quarter amplitude decay. From Figure 1.16b, the peak percent overshoot is given by the following equation:

    $$PPO = 100\left(\frac{a}{C_2 - C_1}\right)$$

2. *Minimum integral of absolute error.* This criterion specifies that the total area under the error curve should be minimum. Figure 1.16c illustrates the minimum integral of absolute error criterion. The error is the distance between C_2 and the controlled variable curve. The integral of absolute error is the total shaded area on the curve. This criterion is easy to use when a mathematical model is used to evaluate a control system.

3. *Critical damping.* This criterion is used when overshoot above the setpoint is undesirable. Critical damping is the least amount of damping that will produce a response with no overshoot and no oscillation. Electrical instruments and some processes are critically damped. Figure 1.16d illustrates this critical damping criterion.

1.12 BLOCK DIAGRAM SIMPLIFICATION

In Section 1.4, we derived the following transfer function of the closed-loop servo control system shown in Figure 1.6.

$$\frac{C}{R} = \frac{G}{1 + GH} \qquad (1.1)$$

We can use the transfer function defined by Equation (1.1) to reduce the servo control system in Figure 1.6 to the single block shown in Figure 1.17.

We will now show how the method of block reduction can be used to simplify more elaborate block diagrams containing multiple closed loops. The method involves the reduction of portions of the block diagram until the desired simplification is obtained. The innermost loops are reduced first and are replaced with single blocks, just as the loop in Figure 1.6 was replaced by the single block in Figure 1.17. The objective of a block reduction may be to reduce the original diagram to a single block (as in Figure 1.17), to the standard form shown in Figure 1.6 or to some other simpler form.

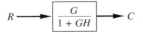

◆ **Figure 1.17** Single block representation of the closed-loop servo control system in Figure 1.6.

The reduction method is quite simple, as the following procedure illustrates:

1. Assign variable names to all signal lines in the original diagram.
2. Select the blocks you wish to reduce to a single block.
3. Use the block transfer functions to obtain the input/output equation for each block you selected in step 2.
   a. The output of a block is equal to the input to the block multiplied by the block transfer function.
   b. The output of a summing junction is equal to the algebraic sum of its inputs (i.e., add inputs marked with a plus (+) and subtract inputs marked with a minus (−)).
4. Use algebraic substitution to combine the equations into a single equation with only two signals (the input to the first selected block and the output of the last selected block).
5. Solve the equation from step 4 for the ratio of the output signal over the input signal. The right-hand side of the resulting equation is the transfer function of the single block that replaces the blocks you selected in step 2.

**EXAMPLE
1.3**

Reduce the block diagram shown in Figure 1.18a into a single block.

Solution

1. Assign the variable names as shown in Figure 1.18a.
2. Select the summing junction, block G_1, and block H_1 for reduction.
3. Write the I/O equations for each block:

$$T = S - U \qquad \textbf{(1)}$$

$$V = G_1 T \qquad \textbf{(2)}$$

$$U = H_1 V \qquad \textbf{(3)}$$

4. Use algebraic substitution to combine the three equations into a single equation in V and S.
   a. Substitute Equation (1) into Equation (2):

$$V = G_1 (S - U) \qquad \textbf{(4)}$$

   b. Substitute Equation (3) into Equation (4):

$$V = G_1 (S - H_1 V) \qquad \textbf{(5)}$$

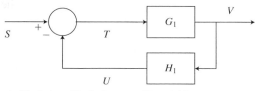

a) Single-loop block diagram with variable names assigned for use in simplification to a single block

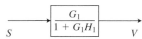

b) Single-block representation of the block diagram in part a above

◆ **Figure 1.18** Single-loop block diagram and its single-block representation (see Example 1.3).

5. Solve Equation (5) for V/S:

$$V = G_1 S - G_1 H_1 V$$

$$V + G_1 H_1 V = G_1 S$$

$$\frac{V}{S} = \frac{G_1}{1 + G_1 H_1}$$

The result of the reduction is shown in Figure 1.18b. ◆

EXAMPLE 1.4   Reduce the block diagram shown in Figure 1.19 into a single block.

Solution   The single block will be obtained by performing three simple reductions. We begin by assigning the variable names as shown in Figure 1.19.

First Reduction.   Select the inner loop for the first reduction (see Figure 1.19). This is the same loop we reduced in Example 1.3 to the following single block:

$$\frac{V}{S} = \frac{G_1}{1 + G_1 H_1}$$

The result of the first reduction is shown in Figure 1.20a.

Second Reduction.   Select the two blocks in the forward path in Figure 1.20a for the second reduction. The I/O equations are:

$$V = \left[\frac{G_1}{1 + G_1 H_1} \right] S$$

$$C = G_2 V$$

Use substitution to eliminate V in the second equation.

$$C = \left[\frac{G_1 G_2}{1 + G_1 H_1} \right] S$$

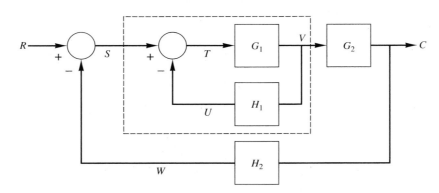

◆ **Figure 1.19**   A multiple-loop servo control system with variable names assigned to each signal line for use in block reduction of the diagram.

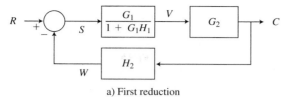

a) First reduction

b) Second reduction

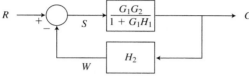

c) Final reduction

◆ **Figure 1.20**   Three steps in the reduction of the block diagram in Figure 1.19.

Solve the above equation for C/S to get the single block equivalent of the forward path.

$$\frac{C}{S} = \frac{G_1 G_2}{1 + G_1 H_1}$$

The result of the second reduction is shown in Figure 1.20b.

Third Reduction. Select all blocks in Figure 1.20b for the third reduction. The I/O equations are

$$S = R - W$$

$$C = \left[\frac{G_1 G_2}{1 + G_1 H_1}\right] S$$

$$W = H_2 C$$

Use substitution to eliminate S and W in the middle equation.

$$C = \left[\frac{G_1 G_2}{1 + G_1 H_1}\right][R - H_2 C]$$

Solve the preceding equation for C/R to get the transfer function of the single block equivalent of the original block diagram in Figure 1.19.

$$C\left[\frac{1 + G_1 G_2 H_2}{1 + G_1 H_1}\right] = \left[\frac{G_1 G_2}{1 + G_1 H_1}\right] R$$

$$C[1 + G_1 H_1 + G_1 G_2 H_2] = [G_1 G_2] R$$

$$\frac{C}{R} = \frac{G_1 G_2}{1 + G_1 H_1 + G_1 G_2 H_2}$$

This final result is shown in Figure 1.20c.

◆

◆ GLOSSARY

Actuator: An element that translates the controller output into an action on the final controlling element. (1.4*)

Closed-loop: The type of control system that uses feedback. (1.4)

Controlled variable: The process variable whose value is controlled by the control system. (1.4)

Controller: The component that computes the error signal and uses it to produce the control action. (1.4)

Controller output: The control action produced by the control modes. (1.4)

Control modes: Methods the controller uses to convert the error signal into a control action. (1.4)

Critical damping: A criterion of good control that permits no overshoot when the setpoint is changed. (1.11)

Damping: The progressive reduction or suppression of oscillation in a component. (1.9)

Dead band: The range of values through which an input can be varied without producing an observable change in the output. (1.6)

Derivative mode: A control mode that produces a control action that is proportional to the rate at which the error is changing. (1.4)

Disturbance variables: Process input variables that affect the controlled variable but are not controlled by the control system. (1.4)

Error: The signal in a controller that is obtained by subtracting the measured value of the controlled variable from the setpoint. (1.4)

Error detector: The element in a controller that computes the error signal. (1.4)

Feedback: The action of measuring the difference between the actual result and the desired result, and using that difference to drive the actual result toward the desired result. (1.4)

Gain: The ratio of the amplitude of the output signal of a component divided by the amplitude of the input signal. (1.2)

Hysteresis: The nonlinearity that causes the value of the output for a given input to depend on the history of previous inputs. (1.6)

Instability: An undesirable characteristic in which the error of a control system oscillates with constant or increasing amplitude. (1.9)

Integral mode: A control mode that produces a control action that is proportional to the accumulation of error over time. (1.4)

Load: The demand on the manipulated variable required to maintain the desired value of the controlled variable. (1.8)

Manipulated variable: The process variable that is acted on by the controller. (1.4)

Manipulating element: The component of a control system that uses the controller output to adjust the manipulated variable. (1.4)

Measured variable: A quantity or condition that is measured (e.g., temperature, flow rate). (1.4)

Measuring transmitter: The component of a control system that uses a sensing element to measure the controlled variable and converts the response into a usable signal. (1.4)

Minimum integral of absolute error: A criterion of good control that minimizes the accumulation of error over time. (1.11)

Open-loop: The type of control system that does not use feedback. (1.3)

Phase angle: An angular value that fixes the point on a sine wave where we start measuring time. It determines the value of the sinusoidal function when $t = 0$. (1.2)

Process: Everything performed in and by the equipment in which a variable is controlled. (1.4)

---

*Relevant section.

Process variable: Any variable in the process. Process controllers often refer to the controlled variable as the process variable. (1.4)

Proportional mode: A control mode that produces a control action that is proportional to the error. (1.4)

Quarter amplitude decay: A criterion of good control that progressively reduces the amplitude of oscillation by a factor of 4. (1.11)

Residual error: The error that remains after all transient responses have faded out. This is sometimes referred to as offset. (1.10)

Saturation: The characteristic that limits the range of the output of a component. (1.6)

Sensor: An element that responds to a parameter to be measured and converts the response into a more usable form. (1.4)

Setpoint: The controller signal that defines the desired value of the controlled variable. (1.4)

Settling time: The time, following a disturbance, that is required for the transient response to fade out. (1.10)

Transfer function: The mathematical expression that establishes the relationship between the input and the output of a component. (1.2)

◆ EXERCISES

Section 1.1

1.1 Write a paragraph that describes how you act as a controller during a common activity such as taking a shower.

1.2 Identify the controlled variable and the manipulated variable in your answer to Exercise 1.1.

Section 1.2

1.3 Draw a block diagram of a typical home heating system with the following components:

(a) A household thermostat where the input signal is the temperature in the living room and the output signal is either on or off.

(b) A solenoid valve where the input is the on or off signal from the thermostat and the output is the flow of gas to the furnace.

(c) A household heating furnace where the input is gas flow from the solenoid valve and the output is heat to the rooms in the house.

(d) The inside of a house where the input is heat from the furnace and the output is the temperature in the living room. *Note:* The output of the living room is also the input to the thermostat, so your diagram should form a closed loop.

1.4 Name the two parts of the relationship between the input and the output of a component, name the function that establishes this relationship, and give an example of each part from your own experience.

1.5 Determine the gain, phase difference, and transfer function for each of the following input/output conditions. The components are linear.

(a) Frequency: 0.004 Hz
Input amplitude: 10°C   phase: 15°
Output amplitude: 0.447 mV   phase: −30°

(b) Frequency: 36 KHz
Input amplitude: 0.4 V   phase: 20°
Output amplitude: 11.6 V   phase: 19°

(c) Frequency: 0.5 Hz
Input amplitude: 4.4 V   phase: 32°
Output amplitude: 600 rpm   phase: 0°
(rpm means revolutions per minute)

(d) Frequency: 0.4 Hz
Input amplitude: 5%   phase: 10°
Output amplitude: 50%   phase: 12°

(e) Frequency: 0.05 Hz
Input amplitude: 3.5 psi   phase: 27°
Output amplitude: 2.1 gpm   phase: 18°
(psi means pounds per square inch)
(gpm means gallons per minute)

(f) Frequency: 1.6×10^{-4} Hz
Input amplitude: 3.2 gpm   phase: 0°
Output amplitude: 8.4 inches   phase: −58°

Section 1.3

1.6 Write a paragraph that describes an example of an open-loop control system taken from your own experiences.

1.7 List the advantages and disadvantages of open-loop control.

Section 1.4

1.8 Write a paragraph that describes an example of a closed-loop control system taken from your own experiences.

1.9 Assume that you are explaining feedback control to a friend who knows nothing about control systems. Explain the operations performed by a feedback control system. Name each component and each signal in the system and explain how the system works.

1.10 List the advantages and disadvantages of closed-loop control.

1.11 A flow measuring transmitter has a linear I/O graph similar to Figure 1.8. The input range is 0 to 10 liters per minute (L/min); the output range is 4 to 20 mA. Find the following:

(a) Transmitter gain, $H = C_m/C$
(b) The equation for output C_m in terms of input C
(c) Output C_m when input $C = 6$ L/min
(d) Input C when output $C_m = 4$ mA

1.12 A level-measuring transmitter has a linear I/O graph similar to Figure 1.8. The input range is 5.0 to 10.0 meters (m), and the output range is 3 to 15 psi. Find the following:

(a) Output C_m when input $C = 6.3$ m
(b) Input C when output $C_m = 9.1$ psi

1.13 A pressure-measuring transmitter has a linear I/O graph similar to Figure 1.8. The input range is 0 to 200 kilopascals (kPa), and the output range is 10 to 50 mA. Find the following:

(a) Output C_m when input C = 72 kPa
(b) Input C when output $C_m = 13.7$ mA

1.14 Derive Equation 1.2.

Section 1.5

1.15 Name the component identified by each of the following component codes (see Table D.1 in Appendix D).

(a) LI-112   **(b)** AIC-113   **(c)** PDT-201
(d) AIT-113   **(e)** PAH-201   **(f)** TDAL-320
(g) LRT-112   **(h)** ST-141   **(i)** SAL-141
(j) TI-320   **(k)** FS-106   **(l)** FQI-110
(m) SAH-141   **(n)** ZZI-117   **(o)** ZXI-117
(p) TAL-320

1.16 A certain process consists of a kettle filled with liquid and heated by a gas flame. A thermocouple temperature transmitter measures the temperature of the liquid in the kettle. A control valve manipulates the flow of gas to the burner. The control system components are listed below. Name each component and sketch an instrumentation drawing for this system.

Component Code	Name	Input	Output
TT-201	_____	Temperature	4–20 mA
TRC-201	_____	4–20 mA	4–20 mA
TY-201	_____	4–20 m	3–15 psi
TV-201	_____	3–15 psi	cfm[a]

[a]Cfm stands for gas flow rate in cubic feet per minute.

Section 1.6

1.17 Test data from four components are given below. Each test consists of a complete traversal from an input value of 0 to an input value of 25. The data are listed in the order in which the traversal was made. Plot the data from each test on an input/output graph with the input on the horizontal axis and the output on the vertical axis. Use arrows to show the direction of traversal.

1.18 Identify all nonlinearities exhibited by each component in Exercise 1.17.

Component 1
Input
0   5   10   15   20   25   20   15   10   5   0
Output
0   16   38   62   84   100   84   62   38   16   0

Component 2
Input
0   5   10   15   20   25   20   15   10   5   0
Output
0   18   36   56   78   100   82   64   44   22   0

Component 3
Input
0   6   10   15   20   22   25   19   15   10   5   3   0
Output
10   10   30   55   80   90   90   90   70   45   20   10   10

Component 4
Input
0   6   10   15   20   22   25   19   15   10   5   3   0
Output
10   10   27   50   78   90   90   90   73   50   22   10   10

1.19 Determine the maximum output difference between the increasing line and the decreasing line of any component in Exercise 1.17 that has a loop in its input/output graph. To do this, find the input that has the greatest separation between the two lines. For this input, read the values of the output on the increasing and decreasing lines, and take the difference between the two output values.

Section 1.7

1.20 List six benefits of automatic control from your own experience.

Section 1.8

1.21 Select which of the following four types of load change is illustrated in each example:
1. Change in demand by controlled medium
2. Change in quality of manipulated variable
3. Change in ambient conditions
4. Change of energy supplied within the process
(a) A chemical process in plant A mixes two ingredients that combine to form a compound. Heat is generated by the reaction, and a control system is used to control the temperature of the mixture. What type of load change is a change in the amount of heat generated by the reaction?
(b) A process in plant B uses heated outside air to dry the product before packaging. A rainstorm raises the humidity of the outside air so that more heat is required to dry the product. What type of load change is this?

(c) A food process in plant C uses a dryer to toast corn flakes. What type of load change is a change in production rate from 200 pounds per hour (lb/hr) to 300 lb/hr?
(d) A food process in plant D uses a solution of sodium bicarbonate to neutralize synthetic meat fibers. The flow rate of sodium bicarbonate is the manipulated variable. What type of load change is a change in the concentration of sodium bicarbonate?
(e) In a home heating system, what type of load change is produced when the oven is used to bake a cake?
(f) In a home heating system, what type of load change is produced by a change in the heat content of the fuel used by the furnace?
(g) In a home heating system, what type of load change is produced by a sudden drop in the outside air temperature?
(h) In a home heating system, what type of load change is produced by opening a window?
(i) The block diagram of a person driving a car is shown in Figure 1.3. In this system, what type of load change is produced by a sudden crosswind?
(j) A blending and heating system is shown in Figure 1.10. In this system, what type of load change is produced by a change in the flow rate of heated product?
(k) In the blending and heating system (Figure 1.10), what type of load change occurs when the syrup supply is suddenly diluted with water?
(l) The school auditorium was empty all morning. What

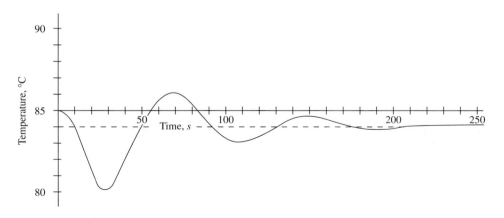

◆ **Figure 1.21** Response of a temperature control system to a sudden load change (see Exercise 1.22).

type of load change was applied to the air-conditioning system when 500 students arrive for an afternoon concert?

Section 1.9

1.22 Which of the five types of damping shown in Figure 1.14 would be most desirable for the suspension of an automobile? Explain your answer by describing the ride of the automobile with each type of damping.

Section 1.10

1.23 A temperature control system produced the response shown in Figure 1.21. Determine the maximum error, the settling time, and the residual error from the response.

Section 1.11

1.24 A step change test of a temperature control system produced an underdamped response similar to Figure

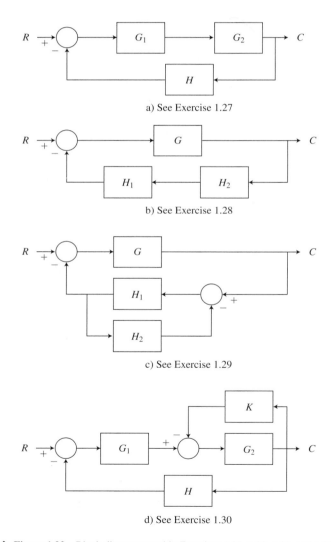

a) See Exercise 1.27

b) See Exercise 1.28

c) See Exercise 1.29

d) See Exercise 1.30

◆ **Figure 1.22** Block diagrams used in Exercises 1.27, 1.28, 1.29, and 1.30.

1.16b. The first positive peak error measured 8°C. What should the second positive peak error measure to satisfy the quarter amplitude decay criteria?

1.25 Does the response in Figure 1.21 satisfy the quarter amplitude decay criteria?

1.26 In a chemical process, two components are blended together in a large mixer. The temperature of the mixture must be maintained between 100 and 112°C. If the temperature exceeds 114°C, the finished product will not satisfy the specifications. Which of the three criteria of good control should be used for the temperature control system?

Section 1.12

1.27 Reduce the block diagram in Figure 1.22a to the standard form shown in Figure 1.20b and to a single block similar to Figure 1.20c.

1.28 Repeat Exercise 1.27 for the block diagram in Figure 1.22b.

1.29 Repeat Exercise 1.27 for the block diagram in Figure 1.22c.

1.30 Repeat Exercise 1.27 for the block diagram in Figure 1.22c.

CHAPTER 2

Types of Control

Control system technology has many facets, depending on what is controlled, how the control is accomplished, who produces the components, and who uses the control system.

The purpose of this chapter is to introduce you to various types of control systems and the language associated with each type. After completing this chapter, you will be able to describe or explain

1. Analog and digital signals
2. Regulator and follow-up control systems
3. Process control and process controller
4. Servomechanism
5. Hydraulic and dc motor position control systems
6. Sequential control
7. Event-driven operation
8. Time-driven operation
9. Timing diagram
10. Ladder diagram
11. Numerical control
12. Robot
13. Centralized control
14. Distributed control

2.1 INTRODUCTION

Control systems are classified in a number of different ways. They are classified as closed-loop or open-loop, depending on whether or not feedback is used. They are classified as analog or digital, depending on the nature of the signals—continuous or discrete. They are divided into regulator systems and follow-up systems, depending on whether the setpoint is constant or changing. They are grouped into process control systems or machine control systems, depending on the industry they are used in—processing or discrete-part manufacturing. *Processing* refers to industries that produce products such as food, petroleum, chemicals, and electric power. *Discrete-part manufacturing* refers to industries that make parts and assemble products such as automobiles, airplanes, appliances, and computers. They are classified as continuous or batch (or discrete), depending on the flow of product from the process—continuous or intermittent and periodic. Finally, they are classified as centralized or distributed, depending on where the controllers are located—in a central control room or near the sensors and actuators. Additional categories include servomechanisms, numerical control, robotics, batch control, sequential control, time-sequenced control, event-sequenced control, and programmable controllers. These general categories are summarized below.

Classifications of Control Systems

1. Feedback
   a. Not used—open-loop
   b. Used—closed-loop
2. Type of signal
   a. Continuous—analog
   b. Discrete—digital
3. Setpoint
   a. Seldom changed—regulator system
   b. Frequently changed—follow-up system
4. Industry
   a. Processing—process control
      (1) Continuous systems
      (2) Batch systems
   b. Discrete-part manufacturing—machine control
      (1) Numerical control systems
      (2) Robotic control systems
5. Location of the controllers
   a. Central control room—centralized control
   b. Near sensors and actuators—distributed control
6. Other categories
   a. Servomechanisms
   b. Sequential control
      (1) Event-sequenced control
      (2) Time-sequenced control
   c. Programmable controllers

Because we discussed open- and closed-loop control systems in Chapter 1, we do not deal with them in this chapter but cover the other classifications.

2.2 ANALOG AND DIGITAL CONTROL

The signals in a control system are divided into two general categories: analog and digital. Graphs of an analog and a digital signal are shown in Figure 2.1.

An *analog signal* varies in a continuous manner and may take on any value between its limits. An example of an analog signal is a continuous recording of the outside air temperature. The recording is a continuous line (a characteristic of all analog signals). A *digital signal* varies in a discrete manner and may take only certain discrete values between its limits. An example of a digital signal is an outdoor sign that displays the outside air temperature to the nearest degree once each minute. A graph of the signal produced by the sign does not change during an interval, but it may jump to a new value for the next interval.

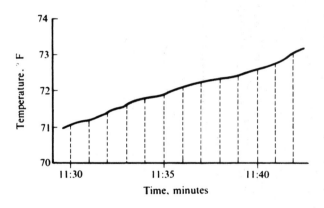

a) An analog signal of the outside air temperature

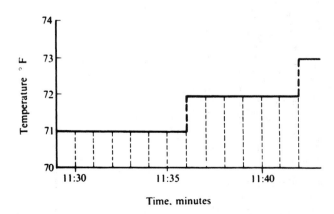

b) A digital signal of the outside air temperature

◆ **Figure 2.1** Examples of digital and analog signals of the same variable.

Analog control refers to control systems that use analog signals, and *digital control* refers to control systems that use digital signals. Examples of these control systems are shown in Section 2.4.

2.3 REGULATOR AND FOLLOW-UP SYSTEMS

Control systems are classified as regulator systems or follow-up systems, depending on how they are used. A *regulator system* is a feedback control system in which the setpoint is seldom changed; its prime function is to maintain the controlled variable constant despite unwanted load changes. A home heating system, a pressure regulator, and a voltage regulator are common examples of regulator systems. Many process control systems are used to maintain constant processing conditions and hence are regulator systems.

> A regulator control system maintains the control variable at a constant setpoint.

A *follow-up system* is a feedback control system in which the setpoint is frequently changing. Its prime function is to keep the controlled variable in close correspondence with the setpoint as the setpoint changes. In follow-up systems, the setpoint is usually called the reference variable. A ratio control system, a strip chart recorder, and the antenna position control system on a radar tracking system are examples of follow-up systems. Many servomechanisms are used to maintain a position variable in close correspondence with an input reference signal and hence are follow-up systems.

> A follow-up control system maintains the control variable at a changing setpoint.

2.4 PROCESS CONTROL

Process control involves the regulation of variables in a process. In this context, a *process* is any combination of materials and equipment that produces a desirable result through changes in energy, physical properties, or chemical properties. A continuous process produces an uninterrupted flow of product for extended periods of time. A batch process, in contrast, has an interrupted and periodic flow of product. Examples of a process include a dairy, a petroleum refinery, a fertilizer plant, a food-processing plant, a candy factory, an electric power plant, and a home heating system. The most common controlled variables in a process are temperature, pressure, flow rate, and level. Others include density, viscosity, composition, color, conductivity, pH, and hardness. Most process control systems maintain constant processing conditions and hence are regulator systems.

> A process control system maintains a variable in a process at its setpoint.

Process control systems may be either open-loop or closed-loop, but closed-loop systems are more common. The process control industry has developed standard, flexible, process controllers for closed-loop systems. Over the years these controllers have evolved from pneumatic analog controllers to electronic analog controllers to microprocessor-based digital controllers (microcontrollers). The driving force in this evolution has been increased capability and versatility, especially in microcontrollers, which tapped the power of the microprocessor.

Most process controllers share a number of common features. They show the value of the setpoint, the process variable, and the controller output in either analog or digital format. They allow the operator to adjust the setpoint and switch between automatic and manual control. When manual control is selected, they allow the operator to adjust the controller output to vary the manipulated variable in an open-loop control mode. They allow the operator to adjust the control mode settings to "tune the controller" for optimum response. Many controllers also provide for remote setting of the setpoint by an external signal, such as the output of another controller. A local/remote switch allows the operator to switch the setpoint between the local and remote settings. Figure 2.2 shows the front panel of a single-station microcontroller.

Microcontrollers provide many additional features, some unique to one vendor and others common among a number of vendors. The following is a partial list of these features:

Choice of control modes: P, I, PI, PD, and PID (see the discussion of controllers in Section 1.4)

Detects and annunciates *alarms.*

Accepts several *analog inputs* (about four).

Accepts several *digital inputs* (three or four).

Provides more than one *analog output* (can be used to manipulate process variables).

Provides several *digital outputs* (can be used for ON-OFF control of heating elements, etc.).

Direct input from a thermocouple or RTD temperature sensor.

Linearizes thermocouple inputs.

Performs *ratio, feedforward,* or *cascade* control.

Bumpless transfer between *automatic* and *manual* modes.

Bumpless transfer between *local* and *remote* modes.

Front-panel *configuration* of the controller.

Adaptive gain: Automatic adjustment of the proportional mode gain based on some combination of the process variable, the error, the controller output, and a remote input signal.

Self-tuning by process model: Determination of the control mode parameters from a model that is formed from observations of the response to step changes of the setpoint. The step change and modeling process is repeated until the model matches the actual process.

Self-tuning by pattern recognition: Automatic adjustment of the control mode parameters after a disturbance by scanning the recovery pattern and applying tuning rules that are stored in the controller's memory.

Self diagnostics: Automatic detection and annunciation of certain types of failure.

Mathematical operations, such as addition, subtraction, multiplication, division, and square root.

Digital communication with a supervisory control computer.

Figure 2.3 is a schematic diagram of an electronic analog controller in a routine process control system.

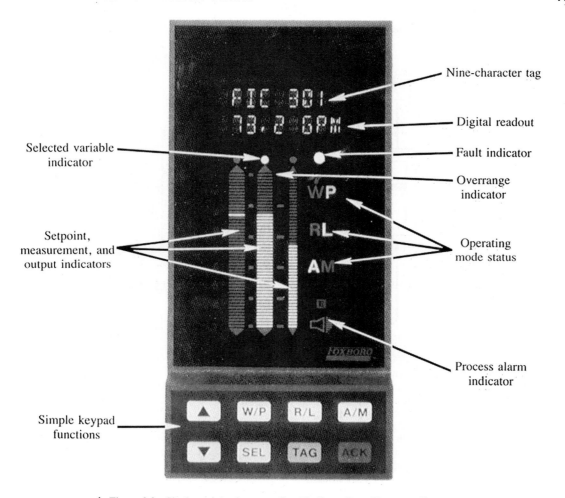

◆ **Figure 2.2**   Single-station microcontroller. The keypad provides easy adjustment of familiar controller operations and access to complete information about the process and the controller. Interactive prompting simplifies the setting of adjustable functions. The SEL key moves the selected variable indicator among the setpoint, measurement, and output indicators. The digital readout displays the value and engineering units of the setpoint, measurement, or output—whichever is chosen by the selected variable indicator. The two keys on the left side of the keypad are used to increase or decrease the setpoint or the output (depending on the position of the selected variable indicator and the operating mode status). The A/M key selects automatic or manual control. The R/L key selects remote or local setpoint. The W/P key chooses monitoring by a supervisory computer (workplace) or by an operator (panel). In the workplace setting, the controller is connected to a computer via a multidrop communication link. (Courtesy of The Foxboro Company, Foxboro, Mass.)

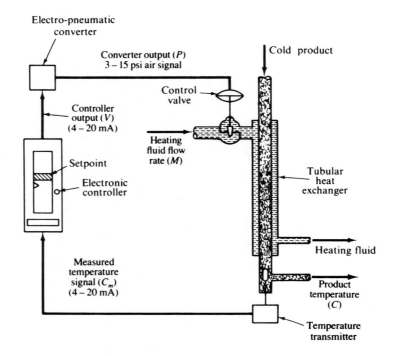

◆ **Figure 2.3** Schematic diagram of a temperature control system. An electronic analog controller regulates the temperature of a liquid product—for example, the pasteurization of milk. The heat exchanger consists of two concentric tubes; the product passes through the inner tube, which is surrounded by the heating fluid contained in the larger tube. Steam is the most common heating fluid, but hot water and hot oil are also used. The control valve manipulates the heating fluid flow rate, which determines the amount of heat transferred to the product. The temperature transmitter measures the temperature of the product as it leaves the heat exchanger. The controller compares the measured temperature with the setpoint and produces an output that manipulates the control valve to maintain the product temperature at the setpoint.

Figure 2.4 is an instrumentation diagram of a more involved process control system, a compensated mass flow control loop. The purpose of this system is to deliver a flow of gas in proportion by weight to the demand signal. The flow meter produces a signal that is proportional to the square of the *volume flow rate* of the gas. What is needed is a signal that is proportional to the *mass flow rate* of the gas. Equation (3.30) in Section 3.4 gives the mass (m) of a volume (V) of gas as a function of the absolute pressure (p), the absolute temperature (T), the universal gas constant (R), and the gram molecular weight (M) of the gas:

$$m = \left(\frac{MV}{10^3 RT} \right) p$$

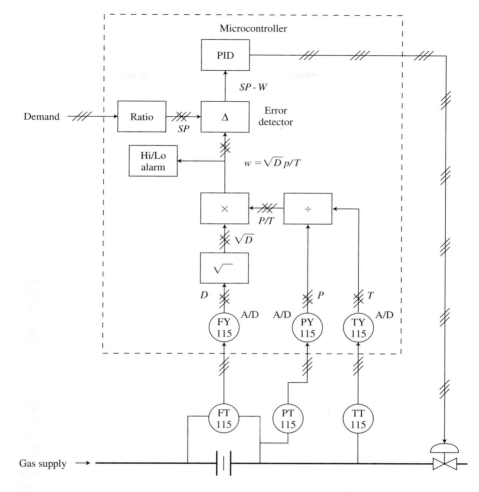

◆ **Figure 2.4** Instrumentation diagram of a compensated mass flow control loop. Before microprocessor-based digital controllers were available, this control loop required several more pieces of expensive hardware. [From *Bulletin C-404A* (Foxboro, Mass.: The Foxboro Company, November 1985), p. 4.]

The equation to convert from volume flow rate (Q) to mass flow rate (w) is obtained by

1. Dividing both sides of Equation (3.30) by time, t,

$$w = \frac{m}{t} = \frac{M}{10^3 R}\left[\frac{V}{t}\right]\left[\frac{p}{T}\right]$$

2. Substituting Q for V/t,

$$w = \frac{MQ}{10^3 R}\left[\frac{p}{T}\right] \tag{2.1}$$

where w = mass flow rate, kilogram/second
Q = volume flow rate, cubic meter/second
p = absolute pressure, pascal
T = absolute temperature, kelvin
M = gram molecular weight of the gas (see "Properties of Gases" in Appendix A), gram/mole
R = universal gas constant = 8.314 J/K · mol

The volume flow rate, Q, is equal to the square root of the output of the flow meter, D, multiplied by the flow meter proportionality constant, k_f:

$$Q = k_f\sqrt{D} \tag{2.2}$$

Substitute Equation (2.2) into (2.1) and replace $Mk_f/10^3 R$ by k to get the following conversion equation:

$$w = k\sqrt{D}\left(\frac{p}{T}\right) \tag{2.3}$$

where k = mass flow proportionality constant
D = flow meter differential pressure, pascal
p = absolute pressure of the gas, pascal
T = absolute temperature of the gas, kelvin

$$k = \frac{Mk_f}{10^3 R} \tag{2.4}$$

The controller converts the output of the flow meter (D) into a mass flow rate signal (w) by multiplying the square root of the flow meter signal by the quotient of pressure divided by temperature. The output of the multiplier is proportional to the mass flow rate and can be calibrated to the desired accuracy. This mass flow rate signal (w) is the measured variable input to the PID controller. The ratio unit multiplies the demand signal by a ratio value to form the setpoint input to the PID controller. The output of the controller is applied to the control valve to regulate the mass flow rate of the gas in a ratio to the demand signal.

2.5 SERVOMECHANISMS

Servomechanisms are feedback control systems in which the controlled variable is physical position or motion. Many servomechanisms are used to maintain an output position in close correspondence with an input reference signal and hence are follow-up systems. Servomechanisms are often part of another control system. Robotic control systems contain several servomechanisms, one for each joint in the robotic arm. Numerical control machines use servos to control the motion of the tool. Recorders use servos to position the recording pen. The driver and automobile control system in Figure 1.3 contains a power steering system, which is a servomechanism. If the car has cruise control, that is another servomechanism.

There is no theoretical difference between a servomechanism and a closed-loop process

A servomechanism controls the position or motion of some part of a system.

control system; the same mathematical elements are used to describe each system, and the same methods of analysis apply to each. However, because servo control and process control were developed independently of one another, each has evolved different design methods and a different terminology. Servomechanisms usually involve relatively fast processes—the time constants may be considerably less than 1 s. Process control involves much slower processes—the time constants are measured in seconds, minutes, and even hours. The components in a servomechanism are usually well defined mathematically, so the controller can be designed to meet the system specifications with little or no need for field adjustments. Processes are more difficult to define mathematically, so process control systems usually require field adjustments to obtain optimum response. Figures 2.5 and 2.6 provide examples of servomechanisms.

The hydraulic position control system in Figure 2.5 uses a lever to provide a mechanical feedback signal. The hydraulic valve is shown in the neutral position. If the setpoint lever is moved to the right, the valve spool also moves to the right, thus connecting the left side of the

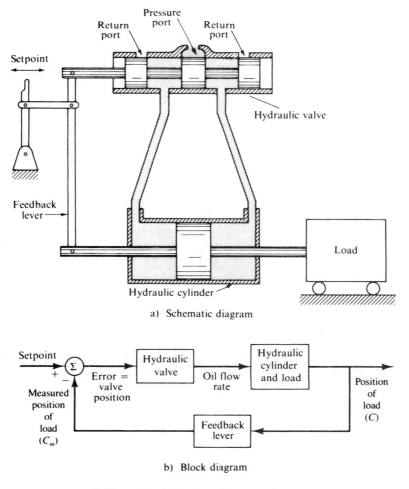

a) Schematic diagram

b) Block diagram

◆ **Figure 2.5** Hydraulic position control system.

hydraulic cylinder to the pressure port and the right side to the return port. Hydraulic fluid will flow into the left side of the cylinder, moving the piston and load to the right until the valve is back in the neutral position. If the setpoint handle were moved to the left, the load would be moved to the left. For each position of the setpoint handle, there is a corresponding position of the load that will place the valve in its neutral position. A small force will move the lever to control a large force exerted by the cylinder.

A dc motor position control system is illustrated in Figure 2.6. This system positions an antenna in response to a command voltage applied at the setpoint input. The position sensor is a 10-kilohm (kΩ) potentiometer with no stops and a 20° dead zone. The position sensor voltage output goes from $-V$ to $+V$ as the antenna rotates from its $+170°$ position to its $-170°$ position. The operational amplifier and the three resistors form a proportional (P)

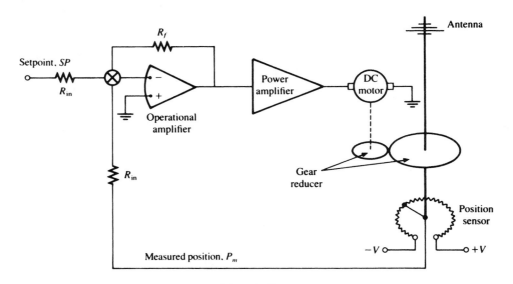

a) Schematic diagram

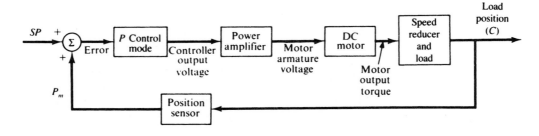

b) Block diagram

◆ **Figure 2.6** DC motor position control system.

mode controller with a proportional gain of R_f/R_{in}.

The output of the controller is $-R_f/R_{in}$ times the algebraic sum of the setpoint voltage (SP) and the measured position voltage (P_m):

$$\text{Controller output} = \frac{-R_f}{R_{in}}(SP + P_m)$$

The power amplifier inverts the controller output and increases the voltage by a factor of G_a, the gain of the power amplifier:

$$\text{Power amplifier output} = \frac{G_a R_f}{R_{in}}(SP + P_m)$$

The power amplifier output is applied to the armature of the dc motor. The motor speed is proportional to the voltage applied to the armature and the direction is such that when the armature voltage is positive, the motor drives the position sensor toward $-V$, and when the armature voltage is negative, the motor drives the antenna toward $+V$. The result is that the motor drives the position sensor in the direction that will tend to make the sum of SP and P_m equal to zero. The summing junction of the op amp is the error detector, and the term ($SP + P_m$) is the error signal ($SP - C_m$) defined in Section 1.4. Notice that $C_m = -P_m$ and the controller uses positive feedback. The negative feedback in this example is accomplished by making the sign of the position sensor voltage opposite the sign of the setpoint voltage.

2.6 SEQUENTIAL CONTROL

A *sequential control* system is one that performs a set of operations in a prescribed manner. The automatic washing machine is a familiar example of sequential control: The control system performs the operations of filling the tub, washing the clothes, draining the tub, rinsing the clothes, and spin drying the clothes. The automatic machining of castings for automobiles is another example of sequential control: A sequence of machining operations is performed on each casting to produce the finished part. Sequential control is covered in detail in Chapter 11; our objective in this chapter is to give an overview of sequential control systems.

A sequential control system performs a set of operations in a prescribed manner.

The operations in a sequential control system can be categorized according to how they are initiated and terminated. One method is to initiate or terminate an operation when some event takes place. We use the term *event-driven* for this method. The other method is to initiate or terminate an operation at a certain time or after a certain time interval. We use the term *time-driven* for this method.

An automatic washing machine is an example of a time-driven sequential control system. The washing cycle starts out with one event-driven operation—the fill operation begins when someone presses the START button and terminates when the tub is full. However, the remaining operations are all initiated and terminated by a timer. These include the wash operation, the drain operation, the rinse operation, and the spin-dry operation. Most batch process control systems are time-driven sequential systems. Time-driven systems are described by schematic diagrams and timing diagrams. Schematic diagrams show the physical configuration, and timing diagrams define the sequential operations. The timing diagram of an automatic washing machine is shown in Figure 2.7.

A traffic counter is a simple example of an event-driven system. The counter is placed at the side of the road, and the sensor, which is a long rubber tube, is stretched across the road. Each time a vehicle axle passes over the tubular sensor, the counter increases its count by 1. Thus an event (an axle passing over the sensor) drives the counter. Manufacturing industries are principal users of event-driven sequential controllers. Before 1970, large relay panels were used to control event-driven operations. In 1968, the Hydramatic Division of General Motors Corporation specified the design criteria for the first programmable logic controller (PLC). The purpose was to replace inflexible relay panels with a computer-controlled solid-state system. The project succeeded beyond anyone's dreams. Programmable logic controllers have gone beyond replacement of relay panels to include PID modules for process control and communications interfaces that make it possible to link programmable controllers into an integrated manufacturing operation.

Event-driven systems are described by ladder diagrams and Boolean equations. The symbols of components used in ladder diagrams are included in Chapter 9. The components most frequently used include switches, contacts, relays, contactors, motor starters, time-delay relays, pneumatic solenoid valves, pneumatic cylinders, hydraulic solenoid valves, and hydraulic cylinders. The pneumatic cylinder in Figure 2.8 is an example of an event-driven control system.

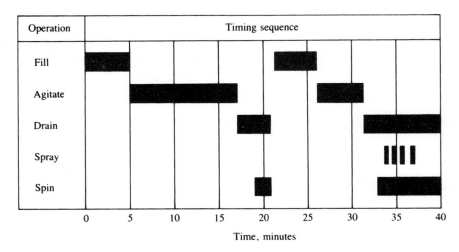

◆ **Figure 2.7**   Timing diagram of an automatic washing machine.

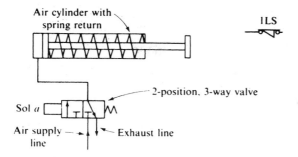

a) Schematic diagram

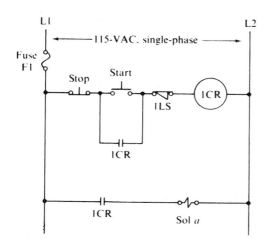

b) Electrical circuit ladder diagram

◆ **Figure 2.8**   Event-driven sequential control system for a pneumatic cylinder. The valve is shown in its deenergized position, which connects the air cylinder to the exhaust line. This allows the spring to force the piston to the retracted position shown in the diagram. The operator presses and releases the START switch to begin a cycle. This causes relay 1CR to energize, closing both contacts labeled 1CR. The 1CR contact that is connected in parallel with the START switch holds relay 1CR in the energized position. The other 1CR contact energizes solenoid a (Sol a), which moves the valve to the right, connecting the cylinder to the air supply line. The air pressure forces the cylinder to the right until it reaches and opens limit switch 1LS. When 1LS opens, relay 1CR is deenergized, opening both 1CR contacts. The valve again connects the cylinder to the exhaust line, and the spring forces the cylinder back to the retracted position, ending the cycle.

2.7 NUMERICAL CONTROL

Numerical control is a system that uses predetermined instructions to control a sequence of manufacturing operations. The instructions are coded numerical values stored on some type of input medium, such as punched paper tape, magnetic tape, or a common memory for program storage. The instructions specify such things as position, direction, velocity, and cutting speed. A *part program* contains all the instructions required to produce a desired part. A *machine program* contains all the instructions required to accomplish a desired process. Numerical control machines perform operations such as boring, drilling, grinding, milling, punching, routing, sawing, turning, winding (wire), flame cutting, knitting (garments), riveting, bending, welding, and wire processing.

A numerical control system uses a program to control a sequence of manufacturing operations.

Numerical control (NC) has been referred to as flexible automation because of the relative ease of changing the program compared with changing cams, jigs, and templates. The same machine may be used to produce any number of different parts by using different programs. The numerical control process is most justified when a number of different parts are to be produced on a particular machine; it is seldom used to produce a single part continually on the same machine. Numerical control is ideal when a part or process is defined mathematically. With the increasing use of computer-aided design (CAD), more and more processes and products are being defined mathematically. Drawings as we know them have become unnecessary—a part that is completely defined mathematically can be manufactured by computer-controlled machines. A closed-loop numerical control machine is shown in Figure 2.9.

The NC process begins with a specification (engineering drawing or mathematical definition) that completely defines the desired part or process. A programmer uses the specifica-

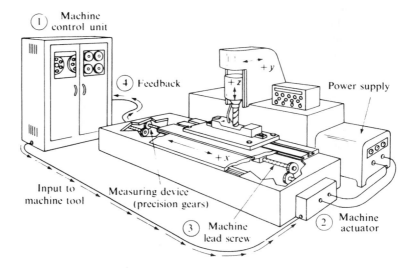

◆ **Figure 2.9** Numerical control machine that uses closed-loop systems to control *x*, *y*, and *z* positions. The *x* position controller moves the workpiece horizontally in the direction indicated by the +*x* arrow. The *y* position controller moves the milling machine head horizontally in the direction indicated by the +*y* arrow. The *z* position controller moves the cutting tool vertically as indicated by the +*z* arrow. The following actions are involved in changing the *x*-axis position. (1) The control unit reads an instruction in the program that specifies a +0.004-inch (in.) change in the *x* position. (2) The control unit sends a pulse to the machine actuator. (3) The machine actuator rotates the lead screw and advances the *x*-axis position +0.001 in. (4) The position sensor measures the +0.001-in. change in *x*-axis position and sends this information to the control unit. (5) The control unit compares the +0.004-in. required motion with the +0.001-in. measured motion and sends another pulse. Steps (1) through (5) are repeated until the measured motion equals the desired +0.004 in. [From N. O. Olesten, *Numerical Control* (New York: John Wiley & Sons, Inc., 1970), p. 12.]

tion to determine the sequence of operations necessary to produce the part or carry out the process. The programmer also specifies the tools to be used, the cutting speeds, and the feed rates. The programmer uses a special programming language to prepare a symbolic program. APT (Automatically Programmed Tools) is one language used for this purpose. A computer converts the symbolic program into the part program or the machine program. In the past, the part or machine program was stored on paper or magnetic tape. The numerical control machine operator fed the tape into the machine and monitored the operation. If a change was necessary, a new tape had to be made. Now, it is possible to store the program in a common database with provision for on-demand distribution to the numerical control machine. Graphic terminals at the matching center allow operators to review programs and make changes if necessary.

Computerized numerical control (CNC) was developed to utilize the storage and processing capabilities of a digital computer. CNC uses a dedicated computer to accept the input of instructions and to perform the control functions required to produce the part. However, CNC was not designed to provide the information exchange demanded by the recent trend toward *computer-integrated manufacturing* (CIM). The idea of CIM is to "get the right information—to the right person—at the right time—to make the right decision." "It links all aspects of the business—from quotation and order entry through engineering, process planning, financial reporting, manufacturing, and shipping—in an efficient chain of production."[*]

Direct numerical control (DNC) was developed to facilitate computer-integrated manufacturing. DNC is a system in which a number of numerical control machines are connected to a central computer for real-time access to a common database of part programs and machine programs. General Electric used a central computer connected to DNC machines through a communications network in the automation of its steam turbine-generator operations (ST-GO). "A typical turbine-generator consists of more than 100,000 parts, some of which are manufactured in thousands of different configurations to meet the specific needs of each custom-designed unit. Through the CIM system, customers can specify a needed part and receive replacement components that suit the original configuration of more than 4000 operating ST-GO installations. In some cases, the small-parts shop can now manufacture and ship some emergency parts the same day the order is received."[†]

2.8 ROBOTICS

The *industrial robot* is a programmable manipulator designed to move material, parts, tools, or other devices through a sequence of motions to accomplish a specific task. Robots are used to move parts, load NC machines, operate die-casting machines, assemble products, weld, paint, debur castings, and package products. The most common robotic manipulator is an arm with from one to six axes of motion (or degrees of freedom). The robotic arm shown in Figure 2.10 has six axes of movement:

---

[*] Searle et al., "Computer-Integrated Manufacturing System Goes Beyond CAD/CAM," *Control Engineering*, February 1985, p. 50.

[†] Searle et al., p. 51

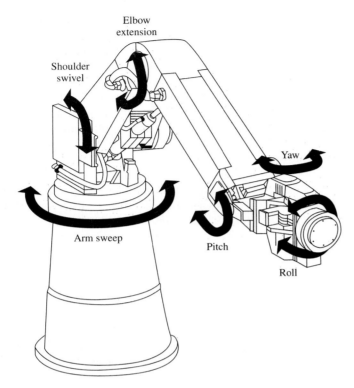

◆ **Figure 2.10**   The Cincinnati Milacron T³ robot has six axes of movement, which duplicate the movements of a human arm. (From N. M. Morris, "Where Do Robots Fit in Industrial Control?" *Control Engineering*, February 1982, p. 59.)

1. Arm sweep (left or right at the waist)
2. Shoulder swivel (up or down at the shoulder)
3. Elbow extension (in or out at the elbow)
4. Pitch (up or down at the wrist)
5. Yaw (left or right at the wrist)
6. Roll (clockwise or counterclockwise at the wrist)

> A robot is a programmable manipulator that moves various objects through a sequence of motions to accomplish a specified task.

Another type of robotic manipulator is a motorized cart that follows a programmed path to move parts from place to place in a factory.

Each axis of motion has its own actuator, connected to mechanical linkages that accomplish the motion of the joint. The actuator may be a pneumatic cylinder, a pneumatic motor, a hydraulic cylinder, a hydraulic motor, an electric servomotor, or a stepper motor. Pneumatic actuators are inexpensive, fast, and clean, but the compressibility of air limits their accuracy and ability to hold a load motionless. Hydraulic actuators can move heavy loads with precision and hold the load motionless, but they are expensive, noisy, relatively slow, and tend to leak hydraulic fluid. Electric actuators are fast, accurate, and quiet, but backlash in the gear train may limit their precision.

Industrial robots have three main parts: the controller, the manipulator, and the end effector. The *end effector* is a mechanical, vacuum, or magnetic device that is attached to the manipulator at the wrist and is used to grip parts or tools. The *manipulator* has already been described in some detail. The *controller* may be simple mechanical stops in a single-axis robot with open-loop control, or it may be a computer in a six-axis robot with closed-loop control. In any event, the controller stores the sequencing and positioning data in memory. It also initiates and stops each movement of the manipulator in the specified sequence of operations. If the controller is a computer, it may communicate with a host computer to download programs and provide management information. Each axis of motion is controlled by either an open- or closed-loop control system. The open-loop control systems may be mechanical stops on a pneumatic cylinder, cam-actuated solenoid valves on a hydraulic motor, or an electric stepper motor. The closed-loop control systems are usually follow-up position control systems (servomechanisms). However, sight, tactile sensing, and voice recognition are also being used as inputs to the controller.

The simplest type of robot is the open-loop *pick-and-place* (PNP) robot. A PNP robot picks up an object and moves it to another location. The robot's movements are usually accomplished by pneumatic actuators controlled by limit switches, cam-actuated valves, or mechanical stops. The controller initiates movement along one axis at a time in an event-driven sequence. Each movement continues until a limit is tripped, stopping the motion. The controller then initiates movement on the next axis in the sequence. Typical applications include machine loading or unloading, palletizing, stacking, and general materials-handling tasks. Open-loop PNP robots are quite accurate, but they lack coordination of the various axes.

The second level of robots uses servo control on most axes and can be programmed to move from one point to another. If the path is not critical, the robot is called *point-to-point* (PTP). If the path is critical, the robot is called continuous path (CP). A PTP robot moves from point to point and performs a function at each point. Typical PTP functions include spot welding, gluing, drilling, and deburring. A CP robot moves from point to point on a specified path and performs an operation as it moves along the path. Typical CP applications include paint spraying, seam welding, cutting, and inspection. The second-level controllers are either programmable controllers or minicomputers. A teaching pendant is used to program the robot using a simple teach-by-doing method.

The third level of robots can also be programmed to move from point to point or in a continuous path. However, in addition to on-line programming using a teaching pendant, they can also be programmed off-line using a keyboard and CRT. These robots can communicate with a host computer. They use high-level languages and artificial intelligence to process information from a CAD/computer-aided engineering (CAE) database. They are capable of integration into computer-controlled workstations.

Robotic servo control systems use position and velocity feedback signals to control movements of the manipulator. The position signal can be either absolute or incremental. The robot's controller sends a setpoint signal to each servo to move along its axis to a given position (absolute position) or through a given distance (incremental position). Position and velocity are referred to as the internal feedback signals of the servo control loop. The robotic controller has other sensory inputs (external to the servo loop) that it can use to carry out its assigned task. These external inputs include vision, tactile sensing (touch), and voice recognition. The controller uses these external sensory inputs to detect the presence of an object, the dimensions of an object, or even the identity of an object. With a sense of vision or touch, a robot can calibrate its position sensors, search a defined area to locate a part, and identify any part it finds. Sensors are covered in detail in Chapters 7 and 8.

The industrial robot comes ready to do a job but does not know how. The user must program the robot to do its assigned task. First the user must determine the sequence of operations required to do the job. Then an operator must place this sequence of operations into the robot's memory. In simple PNP applications, the operator locates the mechanical stops or limit switches and establishes the logic of the event-driven sequence. In PTP and CP applications such as welding and spray painting, the operator uses a "teach-by-doing" method to program the robot. The operator puts the controller into the TEACH mode and moves the manipulator through the desired sequence of operations. The controller *learns* what it is taught by recording in its memory the various positions of the manipulator's joints. When finished, the operator puts the controller in the RUN mode and the robot follows the position data in its memory and exactly repeats the sequence of operations it was taught.

Robot manufacturers provide a *teaching pendant* to assist the operator in programming the robot. A typical pendant looks somewhat like a pocket calculator. It has an alphanumeric display and several pushbuttons. The programmer can move the manipulator by pressing the appropriate button on the pendant or in some cases by actually moving the manipulator to the desired position. When each correct position is achieved, the programmer pushes another button to tell the computer to read and store the position of each joint in the robot's manipulator. When all positions have been recorded, the operator presses a REPEAT button to check the movements of the manipulator for any errors that might have occurred in the teaching process.

The "teach-by-doing" method works well for the simpler PTP and CP operations. However, when the operations become complex, the operator may have difficulty visualizing the program's structure. In these applications, off-line programming using a personal computer is more appropriate. One advantage of off-line programming is that valuable production time is not lost by on-line teaching of the robots. Another advantage is that the programmer can use the powerful software available for CAD and CAE. A third advantage of off-line programming is that the robot programs can be prepared before the fixtures are built, shortening the time required to put a new product into production.

In the past decade, industry has made a major effort toward factory automation. One aspect of this is the "islands of automation" created by robotic work cells. The traditional method of grouping machine tools is to put all machines of a particular type in the same department and the same location. There would be a department of milling machines, a department of drill presses, a department of grinders, and so on. The problem with this arrangement is that it proved to be very inefficient. The Comptroller General, in a 1975 report to Congress, estimated that only 5% of the time consumed in producing a part is spent on the machine itself.[*] The other 95% of the time is expended in materials handling, record keeping, and so on. The robotic work cell is an arrangement that can significantly reduce the time spent on nonproductive activities.

A *robotic work cell* is a group of machine tools and robots arranged for the efficient production of a particular type of product. In this method, products that require the same machining operations are grouped together for production in a particular workstation. A computer controls the entire workstation, so the robots and machine tools work together in the

---

[*] Morris et al., "Profitable Robotic Work Cells," *Control Engineering,* March 1985, p. 81.

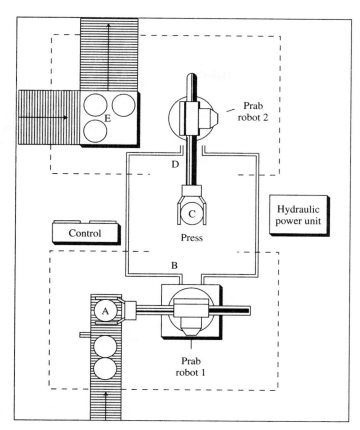

◆ **Figure 2.11** Top view of a robotic work cell for forging points. Robot 1 picks up a raw forging from (A) while robot 2 picks up a forged part at (C). Robot 1 then swings to (B) as robot 2 retracts out of the forge to (D). Robot 1 extends into the forge (C) and robot 2 swings to discharge pallet (E). Robot 1 sets the raw forging into the die (C) as robot 2 sets the forged part onto the pallet (E) in one of four positions. Robot 1 retracts from the forge and signals the forge's controller to cycle, and robot 2 swings to (D). Robot 1 then swings to the infeed conveyor (A) as robot 2 extends into the forge (C). The cycle then repeats. (From N. M. Morris, "Controlling Multiple Robot Arms," *Control Engineering,* November 1986, p. 146.)

most efficient manner. Figure 2.11 shows the top view of a robotic work cell for the production of forged parts.

2.9 THE EVOLUTION OF CONTROL SYSTEMS

Process control and machine control are going through an evolution that began with two separate and distinct systems and is approaching one integrated, distributed system of control for the entire manufacturing plant. Each step in the evolution was made possible by advancements in control system technology, and each step brought improvements to the control of manufacturing operations.

Process control began with sensing elements connected directly to recording controllers, which, in turn, were connected directly to the control valve. The control loop intelligence was distributed near the process it controlled. This distribution of loop intelligence produced good control of individual process variables, but operators could not adequately monitor all of the control loops, especially in spread-out processes such as oil refineries, paper mills, steel mills, and chemical plants. Control engineers could only dream of advanced control concepts, because there was no way to use inputs from several process variables to improve the control of critical variables in the process.

Then along came pneumatic transmitters, controllers, and valve actuators. Industry standardized on the 3- to 15-psi pneumatic signal, making it possible to obtain a measuring transmitter from one vendor, a controller from a second vendor, and a control valve from yet a third vendor. When the three components were assembled into a control loop, the system worked. Better still, the control engineer was able to collect all the controllers for a process into a single room, and the *central control room* concept was born. Operators were now able to monitor the process in a way not possible before. The central control room was the evolutionary step brought on by the 3- to 15-psi pneumatic signal. This gathering of the control loop intelligence in the control room did have a couple of serious disadvantages. One was the cost of the pneumatic lines between the process and the control room. Each control loop required two signal lines: one from the measuring transmitter to the controller, the other from the controller to the control valve. In large plants, these lines were hundreds, even thousands of feet long. Also, control was not quite as good, because of the transmission delay in the long pneumatic signal lines. Pneumatic signals travel at about 1000 ft per second (ft/s). A process loop located 500 ft from the control room would suffer a 1-s delay in the control loop—a $\frac{1}{2}$-s delay from the measuring transmitter to the controller and another $\frac{1}{2}$-s delay from the controller to the control valve.

Next came electronic analog transmitters, controllers, and electropneumatic converters. The pneumatic signal lines were replaced by 4- to 20-mA electric signal lines. Because electric signals travel at nearly the speed of light, this eliminated the second disadvantage of the control room concept (i.e., signal delay was no longer a problem). Cost of the transmission lines was still significant, and a new disadvantage was introduced. The electric signal lines in a control loop make a good antenna, and the control loops picked up noise signals from the surrounding environment. Noise reduced the accuracy of the transmitted signal and caused problems as the controller carried out the control algorithm.

The digital computer entered the process control scene in 1959 when Texaco used a TRW-300 digital computer to control a polymerization unit. At last the control engineer had a tool to implement advanced control concepts. Input of all process variables into the computer memory meant that every process variable was available for use in a control algorithm. One method of using computers for control was to replace the analog controllers with a single large digital computer. The term *direct digital control* (DDC) was used to describe this type of system. Reliability was a problem, however. If the computer failed, the entire process was out of control. This happened with enough regularity that some form of backup was in order. One backup scheme used a standby computer that was ready to take over in case the primary computer failed. In extremely critical applications such as the space program, two backup computers were used. Another method used analog controllers for backup in case the DDC computer failed. Yet another method used analog controllers for loop control with the digital computer used in a supervisory role.

Machine control began with hard-wired panels of relay logic, motor starters, fluid actuators, and solenoid valves. Servo control systems used vacuum tubes, and variable-speed drives consisted of dc generators driving dc motors (the so-called rotating amplifiers). In the 1950s, numerical control machines appeared and machine control entered the world of digital control. These NC machines were programmed off-line and used punched paper tape for storage of the program. Programmable controllers appeared about 1970, replacing hardwired relay logic with reprogrammable computer logic. Improvements in semiconductor switching elements and memory systems set the stage for the microprocessor in the 1980s. The microprocessor generated the changes that are finally bringing process control and machine control under a single umbrella of unified control system technology.

The microprocessor gave us the means to move the control loop back to the plant floor and the ability to communicate with this distributed intelligence. We can have the "short loop" advantages of distributed control and the fully informed operator of a centralized control system. Figure 2.12 shows a fully integrated, distributed control system for a hypothetical manufacturing plant that has both process control and machine control systems. The path that crosses the top and runs down the right side represents the communication network. The terms *data highway* and *local area network* (LAN) are used for different approaches to the communication network. In the 1980s, industry began a major effort to solve the problem of communicating with all the various devices in a manufacturing plant. The purpose of this effort is to develop a set of standards called the *Manufacturer's Automation Protocol* (MAP). See Appendix F for further details about communication networks.

All the control loops are closed on the plant floor. The local control units (LCUs) contain I/O modules that condition the input signals and controller modules that carry out the control algorithms. The control modules have access to all the inputs from the process, giving the control engineer complete freedom to apply advanced control techniques. The LCUs are designed to withstand the harsh environment of the plant floor, so the loop intelligence can be located close to the process it controls. The control loops are short, and the communication network is not bogged down with routine control signals. The measuring transmitters are intelligent—they contain their own microprocessor. These smart transmitters convert the analog signal into a digital signal, linearize the signal, eliminate noise, convert the signal to engineering units, store the ID tag of the transmitter, and store the date of the last calibration. They can even store values of past data for trend analysis. All modules in the system are addressed by their ID tag, and data can be obtained from any module in the system at any place in the system.

Local control units can handle continuous processes, batch processes, and robotic work cells. In addition, other units, such as programmable logic controllers (PLCs), can be interfaced with the communications network. Local operators, central control room operators, plant maintenance, plant engineering, and plant management all have access to all the information via color display and keyboards.

2.10 EXAMPLES OF CONTROL SYSTEMS

An example of the application of a microprocessor in a sequential control system is shown in Figure 2.13. This system monitors the voltage and current supplied to critical (or emergency) loads, such as a life support unit in a hospital. If a power failure occurs, the control system starts the emergency generator and switches the generator output to the critical load. This as-

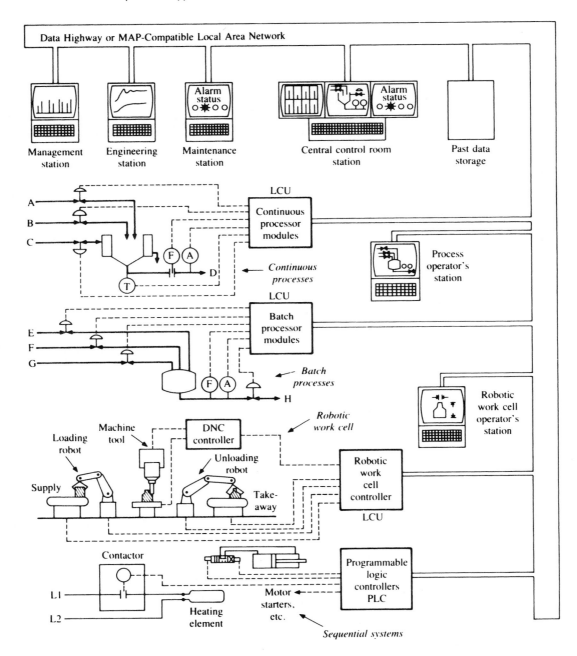

◆ **Figure 2.12** Integrated, distributed control system for a hypothetical manufacturing plant that contains both process and machine control systems.

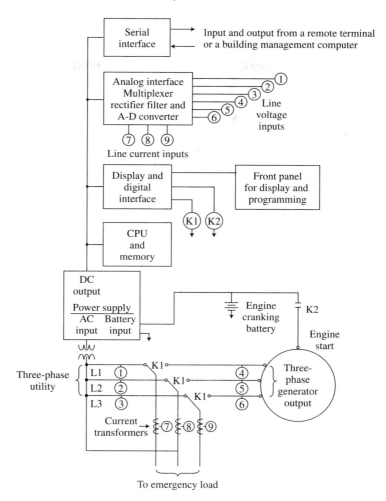

◆ **Figure 2.13**   Microprocessor control system for an emergency generator. (Courtesy of ONAN Corporation.)

sures a continuous source of power in situations where loss of power would have serious consequences.

The microprocessor consists of six units: the power supply, the central processor (CPU) and memory unit, the display and digital interface, the front panel, the analog interface, and the serial interface. The power supply provides the dc voltage necessary to operate the microprocessor. The ac line is the normal source of power, but an input from the engine cranking battery is provided to ensure uninterrupted power to the microprocessor.

The CPU controls the interpretation and execution of instructions. It consists of an arithmetic section, working storage registers, logic circuits, timing and control circuits, decoders, and parallel buses for transfer of binary data and instructions. Memory includes random access memory (RAM) and read-only memory (ROM). The RAM memory has both read and write capability and is used primarily as scratch pad memory for storage and retrieval of temporary data and instructions. The ROM is used to store the main program and permanent data.

The display and digital interface contains the electronic circuits that match the CPU to the two control relays and the front panel. Relay K1 controls the emergency load switch. The actual switching action is performed by a linear motor (not shown in Figure 2.13), and the K1 relay operates the linear motor. The K2 relay is used to start and stop the generator engine.

The human interface provided by the front panel is one of the principal advantages of digital control. The front panel contains a function switch, a digital display, and a keyboard. The function switch selects the signal that will appear on the digital display. The keyboard is used for manual input of data or instructions into memory.

The analog interface matches the CPU to the nine analog input signals. It includes a multiplexer, a rectifier, a filter, and an analog-to-digital converter. The multiplexer selects the input signals one at a time for input to the rectifier. The rectifier converts the ac analog signals into dc signals, and the filter removes the ripple in the rectifier output. The analog-to-digital converter changes the dc analog signal into a digital signal (usually an 8- to 16-bit binary number).

The serial interface matches the CPU with an external terminal or a building management computer. The ability to communicate with a remote computer or terminal is another principal advantage of digital control. This feature provides the building manager with immediate information about the status of the various systems in the building.

A variety of examples of control systems are illustrated in Figures 2.14 through 2.22. The examples include systems to control level, speed, liquid flow rate, gas pressure, engine speed, solid flow rate, sheet thickness, a hydraulic cylinder, and the composition of a liquid blend.

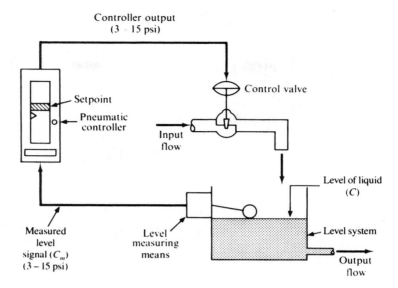

◆ **Figure 2.14**   The level in the tank remains constant when the input flow rate equals the output flow rate. The level rises when the input flow rate is greater (or drops when the input flow rate is less) than the output flow rate. The controller uses the level signal to maintain a balance between the input and output flow rates.

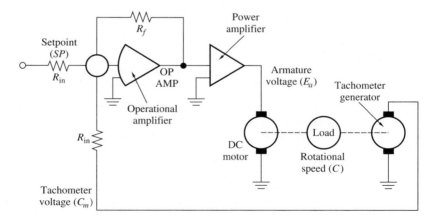

◆ **Figure 2.15**   DC motor speed control system.

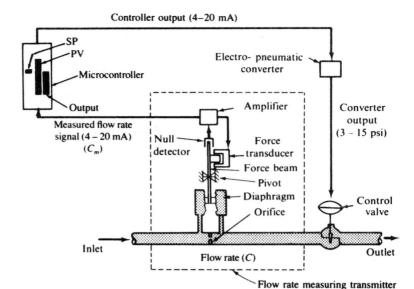

Controller output (4–20 mA)

◆ **Figure 2.16** Liquid flow rate control system. This system consists of a flow rate measuring transmitter, a microcontroller (see Figure 2.2), an electro-pneumatic converter, and a control valve.

The measuring transmitter produces a 4- to 20-mA measured flow rate signal that is proportional to the square of the flow rate. The microcontroller converts the measured flow rate signal from analog to digital, computes the square root of the digital signal, converts the square root signal to flow rate in gallons per minute, and displays the flow rate on the controller's PV indicator. The microcontroller then computes the error signal, applies the control modes to compute the controller output, and converts the controller output from digital to an analog 4- to 20-mA controller output signal.

The electro-pneumatic converter converts the 4- to 20-mA controller output signal into a 3- to 15-psi air pressure signal. The control valve positions the valve stem according to the converter output signal and, indirectly, according to the controller output signal. The control valve manipulates the flow rate as directed by the microcontroller to maintain the desired flow rate as indicated by the setpoint.

The measuring transmitter is shown in some detail to show that it has its own feedback control system. The orifice is a thin plate with a small hole positioned so that all the flowing liquid must pass through the small hole. The flow of fluid through the orifice produces a pressure differential that is proportional to the square of the flow rate.

The measuring transmitter converts the pressure drop across the orifice into a 4- to 20-mA current signal as follows: The diaphragm arrangement converts the pressure difference across the orifice into a force on the lower end of the force beam. The force transducer at the other end of the beam produces the counterbalancing force. The null detector, amplifier, and force transducer make up a closed-loop control system that maintains the force beam in the null position. The null detector senses any displacement of the force beam, and the amplifier converts this displacement into an electric current. The amplifier current passes through the force transducer, generating the counterbalancing force, which is proportional to the current supplied by the amplifier. The amplifier current, which is proportional to the pressure differential, is also the 4- to 20-mA output signal from the measuring transmitter.

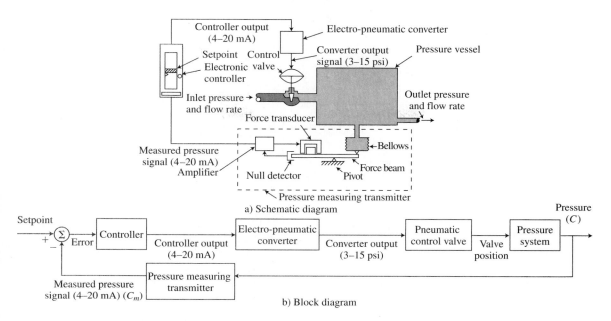

a) Schematic diagram

Setpoint

$+$ Σ Error | Controller | Controller output (4–20 mA) | Electro-pneumatic converter | Converter output (3–15 psi) | Pneumatic control valve | Valve position | Pressure system | Pressure (C)

Measured pressure signal (4–20 mA) (C_m) | Pressure measuring transmitter

b) Block diagram

◆ **Figure 2.17** A pressure control system must maintain a balance between the input and output mass flow rates. The pressure-measuring transmitter is very similar to the differential pressure transmitter in the liquid flow control system. It uses a force balance principle to convert the process pressure into a 4- to 20-mA current signal. The controller compares the measured pressure with the setpoint and manipulates the control valve to bring the two into correspondence.

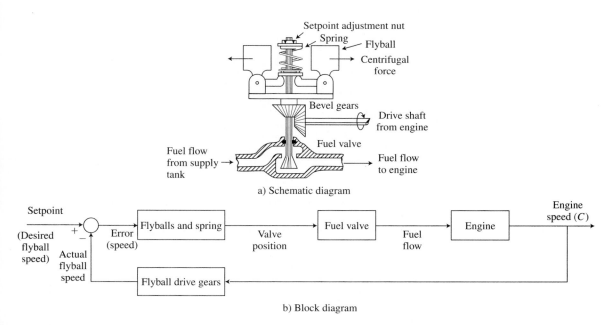

a) Schematic diagram

Setpoint

(Desired flyball speed) $+$ $-$ Error (speed) Actual flyball speed | Flyballs and spring | Valve position | Fuel valve | Fuel flow | Engine | Engine speed (C)

Flyball drive gears

b) Block diagram

◆ **Figure 2.18** A mechanical speed control system or governor is used to control the speed of gasoline engines, gas turbines, and steam engines. A drive shaft from the engine rotates the flyball and spring assembly at a speed proportional to the engine speed. The rotation of the flyballs produces a centrifugal force that compresses the spring and raises the valve stem. Thus the valve stem position is proportional to the engine speed. As the engine speed increases, the valve decreases the fuel flow. As the engine speed decreases, the valve increases the fuel flow. The engine will settle out at a speed that results in just enough fuel flow to balance the load on the engine.

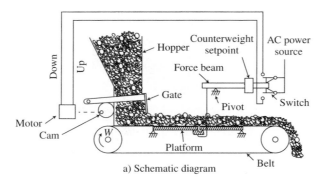

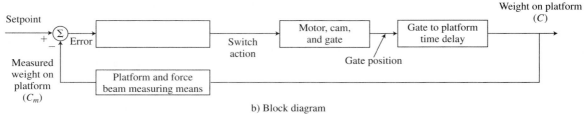

b) Block diagram

◆ **Figure 2.19** This solid flow rate control system uses a constant speed belt with a weight platform. The solid flow rate is equal to the belt speed times the weight of material per unit length of belt. The weight of material on the platform applies a downward force on the left end of the force beam; the counterweight supplies the counterbalancing force on the right side. The force beam operates the two limit switches that control the cam drive motor. If the material on the belt is too heavy, the beam will close the upper limit switch, which drives the gate down; if the material is too light, the beam will close the lower limit switch, which drives the gate up.

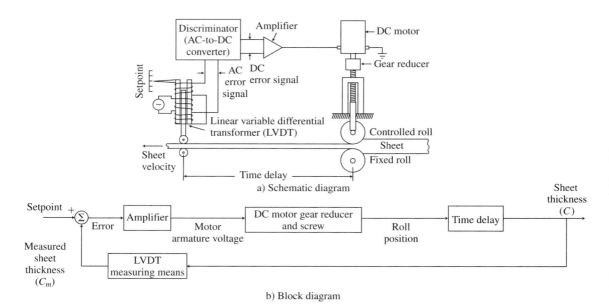

b) Block diagram

◆ **Figure 2.20** A sheet thickness control system uses a linear variable differential transformer as the thickness measuring means and error detector. The ac error signal from the LVDT is converted to a dc error signal by the discriminator (also called a phase-sensitive detector). A discriminator is an ac-to-dc converter whose output magnitude varies linearly with the input magnitude, and whose output sign depends on the relative phase of the input. The dc error signal is amplified and applied to the armature of the dc motor. The motor drives the upper roll, which determines the sheet thickness. After a change in the upper roll position, the sheet must travel to the sensor before the change in thickness is measured. This time delay is represented by a block in the block diagram.

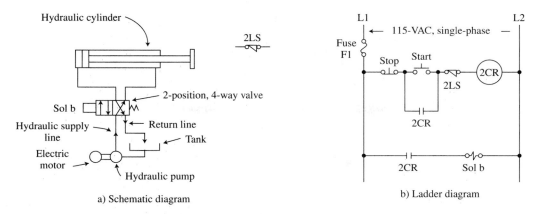

a) Schematic diagram

b) Ladder diagram

◆ **Figure 2.21** Control circuit for a hydraulic cylinder. The operator presses START to begin a cycle. Coil 2CR is energized, holding contact 2CR closed. 'Sol b' is energized, the valve moves to the right, the pump delivers fluid to the left end of the cylinder, and the piston moves to the right. When the piston reaches switch 2LS and opens it, coil 2CR is deenergized, the valve moves to the left, and the piston returns to its original position.

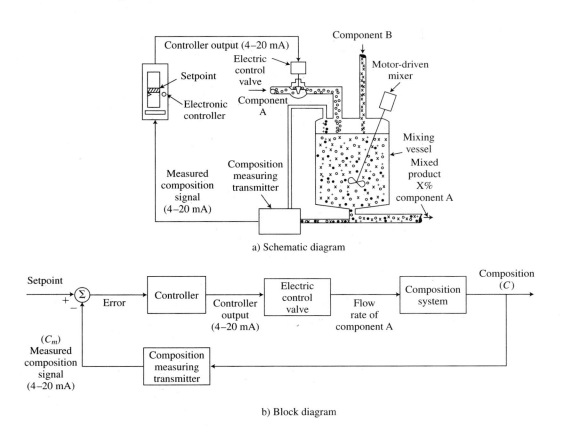

a) Schematic diagram

b) Block diagram

◆ **Figure 2.22** A composition control system maintains the desired mixture of components A and B by manipulating the input of component A. The measuring transmitter is some type of analyzer that measures the percentage of component A in the mixture. The mixing vessel blends the two components and evens out the fluctuations in the flow rates of both components.

◆ GLOSSARY

Analog signal: A signal that varies in a continuous manner and may take on any value between its limits. (2.2)

Computerized numerical control (CNC):   A numerical control system that uses a dedicated computer that accepts the input of instructions and performs the control functions required to produce a part. (2.7)

Digital signal: A signal that varies in a discrete manner and may take only certain discrete values between its limits. (2.2)

Direct numerical control (DNC):   A system in which a number of numerical control machines are connected to a central computer for real-time access to a common database of part programs and machine programs. (2.7)

Event-driven operations: Operations in a sequential control system that are initiated and terminated when some event takes place. (2.6)

Follow-up system: A feedback control system in which the setpoint is frequently changing. Its primary function is to keep the controlled variable in close correspondence with the setpoint as the setpoint changes. (2.3)

Machine program: The set of instructions required to accomplish a desired process. (2.7)

Numerical control (NC):   A system that uses predetermined instructions to control a sequence of manufacturing operations to produce a part. (2.7)

Part program: The set of instructions required to produce a desired part. (2.7)

Process control: The regulation of variables in a process. (2.4)

Regulator system: A feedback control system in which the setpoint is seldom changed. Its primary function is to maintain the controlled variable constant despite unwanted load changes. (2.3)

Robot: A programmable manipulator designed to move material, parts, tools, or other devices through a sequence of motions to accomplish a specific task. (2.8)

Sequential control: A control system that performs a set of operations in a prescribed manner. The automatic washing machine is a familiar example of a sequential control system. (2.6)

Servomechanism: A feedback control system in which the controlled variable is physical position or motion. (2.5)

Time-driven operations: Operations in a sequential control system that are initiated and terminated at a certain time or after a certain time interval. (2.6)

◆ EXERCISES

Section 2.2

2.1   The digital signal illustrated in Figure 2.1 is obtained by eliminating the decimal part of the analog temperature signal. This is referred to as truncating a signal. Thus the signals 72, 72.3, 72.56, and 72.999 are all converted to 72 by the analog-to-digital conversion. An alternative method of conversion would be to round off to the nearest integer. In this method, 72 and 72.3 would be converted to 72, and 72.56 and 72.999 would be converted to 73. Redraw Figure 2.1b using the nearest integer method of analog-to-digital conversion.

2.2   When an analog signal is converted to a digital signal, there is usually a difference between the digital and analog values. This difference is called the *quantization error*. Estimate the quantization error in Figure 2.1 at each minute from 11:30 to 11:42 (i.e., at the vertical dashed lines).

Section 2.3

2.3   Classify the following control systems as either a regulator system or a follow-up system:
(a) Home air-conditioning system
(b) Automobile power steering system

(c) Human body temperature control system

(d) Hunter aiming at a moving target

(e) Artist tracing lines to darken them

(f) Automobile cruise control

(g) Airplane autopilot

(h) Mother using a video camera to record her daughter playing in a soccer game

2.4 From your own experience, give an example of a regulator system and an example of a follow-up system. Identify the control system components in each example.

Section 2.4

2.5 Draw a block diagram of the temperature control system in Figure 2.3 and identify each block in the diagram.

2.6 Draw an instrumentation diagram of the temperature control system in Figure 2.3. Use the loop number 203 in the identification tag for each instrument.

2.7 Draw a block diagram of the compensated mass flow control loop in Figure 2.4. Show the process as a block with one manipulated input (control valve position) and three outputs (volume flow rate, pressure, and temperature).

2.8 The mass flow control system in Figure 2.4 is to be used to control the flow rate of nitrogen gas in proportion to the demand signal. A test of the flow meter revealed that a nitrogen gas flow rate of 0.002 cubic meter per second (m^3/s) produces a differential pressure of 8×10^4 pascals (Pa).

(a) Use Equation (2.2) to determine the flow meter proportionality constant, k_f.

(b) Determine the mass flow rate proportionality constant, k.

(c) Use Equation (2.3) to compute the mass flow rate given that $D = 4.0 \times 10^4$ Pa, $p = 3 \times 10^5$ Pa, and $T = 320$ kelvin (K).

Section 2.5

2.9 Classify the hydraulic position control system in Figure 2.5 as either regulator or follow-up. Explain your answer.

2.10 Classify the dc motor position control system in Figure 2.6 as either regulator or follow-up. Explain your answer.

2.11 Select an example of a servomechanism (preferably from your own experience), sketch a block diagram, and explain the operation of the system.

Section 2.6

2.12 Name and explain the two types of sequential control based on how the operations are initiated and terminated.

2.13 Give an example, from your own experience, of an event-driven control system. Describe each operation and the event that initiates it.

2.14 Give an example, from your own experience, of a time-driven control system. Draw a timing diagram for your example and describe its operation.

2.15 Give an example, from your own experience, of an event- and time-driven control system. Describe the operation of your example.

2.16 Sketch a timing diagram of a four-operation time-driven sequential control system with the following specifications:

1. Operation A is on from 0 to 6 minutes (min) and from 30 to 36 min.
2. Operation B is on from 5 to 15 min.
3. Operation C is on from 35 to 45 min.
4. Operation D is on from 30 to 32 min.

2.17 Construct a timing diagram of a series of time-driven sequential events from your own experience (e.g., a class day consists of a series of classes, which are time-driven events).

2.18 Redesign the pneumatic cylinder control system in Figure 2.8 so it will automatically cycle back and forth between the retracted position and the extended position. Draw a schematic diagram and an electric circuit diagram of your design. Write an explanation of the operation of your system similar to the caption of Figure 2.8. *Hint:* Replace the START switch with a normally open limit switch. Position the new limit switch so it will be closed when the cylinder is retracted and will open as soon as the piston moves from the retracted position.

Section 2.7

2.19 Explain the following acronyms: NC, CNC, and DNC.

Section 2.8

2.20 List the advantages and disadvantages of pneumatic actuators, hydraulic actuators, and electric actuators for robotic arms.

2.21 Describe the following types of robots: PNP, PTP, and CP.

2.22 Review the operation of the robotic work cell described in the caption of Figure 2.11. Develop a similar description for the robotic work cell shown in Figure 2.12.

Section 2.9

2.23 A pneumatic controller is located some distance from the process it controls. The pneumatic signal line from the measuring transmitter to the controller is 120 ft long. The line from the controller to the control valve is 150 ft long. Determine the transmission delay from the measuring transmitter to the control valve.

2.24 Explain distributed control and centralized control, and list advantages and disadvantages of each type.

Section 2.10

2.25 Place each control system in Figures 2.14 through 2.22 into one of the following categories: process control, servomechanism, or sequential control.

2.26 Sketch block diagrams of the control systems in Figures 2.14, 2.15, and 2.16.

2.27 Label each example of a control systems as one of the following types:

1. a regulator system
2. a follow-up system
3. a time-driven sequential system
4. an event-driven sequential system

Examples

(a) Automobile steering system
(b) Refrigerator temperature control system
(c) Oven temperature control system
(d) Automobile cruise control system
(e) Automatic door opener
(f) Automatic washing machine
(g) Human body's temperature control system
(h) Driver and automobile (Figure 1.3)
(i) Position control system (Figure 2.6)
(j) Sheet thickness control system (Figure 2.20)
(k) Hydraulic cylinder control system (Figure 2.21)

CHAPTER 3

The Common Elements of System Components

Control systems often consist of a variety of different types of components. In a temperature control loop, for example, the process and the primary sensor are thermal systems, the temperature transmitter and controller are electrical systems, and the final control element consists of fluid flow (pneumatic) and mechanical systems. This variety of components within a control system could complicate the analysis and design of the control loop. Fortunately, the behavior of these diverse systems is determined by the same four common elements: **resistance, capacitance, inertia** (or **inductance** in electrical systems), and **dead time.** The effects of these basic elements in one system are analogous to the effects of the same elements in any other system. Thus thermal resistance and capacitance have the same effect in a thermal system as electrical resistance and capacitance have in an electrical system.

The purpose of this chapter is to provide you with the means of determining the values of the four basic elements for the following types of systems: **electrical, thermal, liquid flow, gas flow**, and **mechanical.** After completing this chapter, you will be able to

1. Define the four common elements of each of the following systems
   a. Electrical
   b. Liquid flow
   c. Gas flow
   d. Thermal
   e. Mechanical
2. Use a graph to determine
   a. Electrical resistance
   b. Liquid flow resistance
   c. Gas flow resistance
   d. Mechanical resistance
3. Use a computer program to determine
   a. Liquid flow resistance
   b. Thermal resistance

4. Use a calculator to determine
   a. Liquid flow capacitance
   b. Gas flow capacitance
   c. Thermal capacitance
   d. Electrical dead-time delay
   e. Liquid flow dead-time delay
   f. Solids flow dead-time delay

3.1 INTRODUCTION

Most control system components fit into one of the following types: electrical, mechanical, liquid flow, gas flow, or thermal. A particular control system may include two, three, or even five different types of components. This mix of component types could have made the analysis and design of control systems much more difficult, but it did not. The reason for this fortunate circumstance is that the behavior of components of one type is analogous to the behavior of components of any other type. This analogous behavior is determined by four elements that are common to the five types of components. The four elements are resistance, capacitance, inertia (or inductance), and dead-time delay. Some knowledge of these "basic" elements for the five component types will be quite helpful in understanding the behavior of control system components.

This chapter provides the equations and information necessary to calculate the elements of the five types of components. The equations for each element are included in a box to make them easy to locate. All terms are defined within the box, and the units are included for each term. Examples are included to show how to complete the calculations. Computer programs are included for computing liquid flow resistance and thermal resistance.

For each type of component, the four elements are defined in terms of three variables. *The first variable defines a quantity of material, energy, or distance. The second variable defines a driving force or potential that tends to move or change the quantity variable. The third variable is time.* For example, in a liquid flow component, the quantity variable is the volume of liquid that is moved and the potential variable is the pressure drop that tends to cause the liquid to flow. Table 3.1 names the three variables for each of the five types of components. Table 3.2 lists the symbols and units for the parameters used to define the four elements for the various systems.

◆ **TABLE 3.1** The Three Variables Used to Define the Four Elements for Each Type of Component

| Type of Component | *Variable* | | |
	Quantity	Potential	Time
Electrical	Charge	Voltage	Second
Liquid flow	Volume	Pressure	Second
Gas flow	Mass	Pressure	Second
Thermal	Heat energy	Temperature	Second
Mechanical	Distance	Force	Second

◆ **TABLE 3.2**   Parameters Used to Define the Four Elements

Symbol	Name	SI Units	English Units
a	Area (of pipes)	Square meter (m^2)	Square foot (ft^2)
A	Area (of tanks)	Square meter (m^2)	Square foot (ft^2)
d	Diameter (of pipes)	Meter (m)	Foot (ft)
D	Diameter or distance	Meter (m)	Foot (ft)
e	Potential	Volt (V)	Volt (V)
f	Friction factor	Dimensionless	Dimensionless
F	Force	Newton (N)	Pound (force) (lbf)
g	Gravitational acceleration	Meter per square second (m/s^2)	Foot per square second (ft/s^2)
h	Thermal film coefficient	Watt/square meter kelvin (W/m^2 K)	Btu/square foot Fahrenheit (Btu/ft^2 F)
H	Height	Meter (m)	Foot (ft)
i	Current	Ampere (A)	Ampere (A)
l	Length	Meter (m)	Foot (ft)
m	Mass	Kilogram (kg)	Pound (mass) (lbm)
M	Molecular weight	Dimensionless	Dimensionless
P	Pressure	Pascal (Pa or N/m^2)	Pound (force) per square foot (lbf/ft^2)
q	Charge	Coulomb (C)	Coulomb (C)
Q	Heat flow rate	Watt (W)	Btu/hour (Btu/hr)
Q	Volume flow rate	Cubic meter per second (m^3/s)	Cubic foot per second (ft^3/s)
S	Specific heat	Joule/kilogram kelvin (J/kg K)	Btu/pound Fahrenheit (Btu/lb F)
t	Time	Second (s)	Second (s)
T	Temperature	Celsius or Kelvin (°C or K)	Fahrenheit or Rankin (°F or R)
υ	Velocity	Meter per second (m/s)	Foot per second (ft/s)
V	Volume	Cubic meter (m^3)	Cubic foot (ft^3)
W	Mass flow rate	Kilogram per second (kg/s)	Pound (mass) per second (lbm/s)
μ	Viscosity (absolute)[a]	Pascal second (Pa · s)	Pound (mass) per foot second (lbm/ft · s)
ρ	Density	Kilogram per cubic meter (kg/m^3)	Pound (mass) per cubic foot (lbm/ft^3)

Note: See Appendix B for alternative units and conversions.

[a] Absolute viscosity is also referred to as dynamic viscosity.

   Resistance is an opposition to the movement or flow of material or energy. It is measured by the amount of potential required to make a unit change in the quantity moved each second. Electrical resistance is the increase in the voltage across the terminals of a component required to move one more coulomb of charge through the component each second (i.e., to increase the current by 1 ampere, or 1A). Liquid flow resistance is the increase in the pressure drop between two points along a pipe required to increase the flow rate through the pipe by 1 cubic meter per second (m^3/s). Gas flow resistance is the increase in the pressure drop between two points along a pipe required to increase the flow rate through the pipe by 1 kilogram per second (kg/s). Thermal resistance is the increase in temperature difference across a wall section required to increase the heat flow through the wall section by 1 joule per second (J/s). Mechanical resistance is the change in the force applied to an object required to increase the velocity of the object by 1 m/s.

> Resistance is measured in terms of the amount of potential required to produce one unit of electric current, liquid flow rate, gas flow rate, heat flow rate, or velocity.

Capacitance is the amount of material, energy, or distance required to make a unit change in potential. It expresses the relationship between a change in quantity and the corresponding change in potential. Capacitance should not be confused with capacity, which is the total material- or energy-holding ability of a device. If we say a jug holds 1 liter (L), we are stating its capacity. If we say that 100 m³ of liquid must be added to a tank to increase the pressure at the bottom by 1 N/m² we are stating the tank's capacitance.

Electrical capacitance is the coulombs of charge that must be stored in a capacitor to increase its voltage by 1 V. Liquid capacitance is the cubic meters of liquid that must be added to a tank to increase the pressure by 1 pascal. Gas capacitance is the kilograms of gas that must be added to a tank to increase the pressure by 1 Pa. Thermal capacitance is the amount of heat energy that must be added to an object to increase its temperature by 1°C. Mechanical capacitance is the amount of compression of a spring (in meters) required to increase the spring force by 1 N.

> Capacitance is measured in terms of the amount of material, energy, or distance required to make a unit change in potential.

Inertia, inertance, or *inductance* is an opposition to a change in the state of motion. It is measured in terms of the amount of potential required to produce a unit change in the *rate* at which a quantity is moving. This element is usually important in electrical and mechanical systems, sometimes important in liquid and gas systems, and usually ignored in thermal systems. The term *inductance* is used with electrical systems, the term *inertance* is used with fluid systems, and the term *inertia* is used with mechanical systems.

Electrical inductance is the increase in voltage across an inductor required to increase the current by 1 A each second. Liquid flow inertance is the increase in the pressure drop between two points along a pipe required to accelerate the flow rate by 1 m³/s/s. Mechanical inertia is the increase in force required to produce an acceleration of 1 m/s/s.

> Inertia, inertance, or inductance is measured in terms of the amount of potential required to increase electric current, liquid flow rate, gas flow rate, or velocity by one unit per second.

Dead time is the time interval between the time a signal appears at the input of a component and the time the corresponding response appears at the output. A pure dead-time element does not change the magnitude of the signal; only the timing of the signal is changed. Dead time occurs whenever mass or energy is transported from one point to another. It is the time required for the mass or energy to travel from the input location to the output location. Dead time is also called transport time, pure delay, or distance-velocity lag. Control becomes increasingly more difficult as dead time increases.

If v is the velocity of the mass or energy and D is the distance traveled, the dead-time delay is equal to the distance divided by the velocity:

$$t_d = \frac{D}{v}, \text{s} \qquad (3.1)$$

Section 3.2 illustrates the dead-time delay on an electrical transmission line.

> The effect of dead time is to delay the input signal by the dead-time delay (t_d).

Consider the system with a dead-time delay of 5 s. Let $f_i(t)$ represent the input signal and $f_o(t)$ represent the output signal. The (t) indicates that f_i and f_o have different values at different times. If $f_i(t)$ is 5 at a time $t = 0$ s, then $f_o(t)$ will be 5 at $t = 5$ s. If $f_i(t)$ is 7 at $t = 5$ s, $f_o(t)$ will be 7 at time $t = 10$ s. Several more time intervals are indicated below.

t	$f_i(t)$	$f_o(t)$	Relationship between f_i and f_o
0	5		
5	7	5	$f_o(5) = f_i(0)$
10	8	7	$f_o(10) = f_i(5)$
15	6	8	$f_o(15) = f_i(10)$
20		6	$f_o(20) = f_i(15)$

It is interesting to note that in each case, $f_o(t) = f_i(t - 5)$ or, more generally,

$$f_o(t) = f_i(t - t_d)$$

This equation is often used to represent a dead-time process. In simple terms, it expresses the concept that the output at any time t is the same as the input was t_d before time t, that is, at time $t - t_d$.

3.2 ELECTRICAL ELEMENTS

Electrical resistance is that property of material which impedes the flow of electric current. The unit of electric resistance is the *ohm*. Good conductors have a low resistance; insulators have a very high resistance. Electrical resistance (R) is expressed by *Ohm's law*, a statement of proportionality between the applied voltage (e) and the resulting current (i) in linear resistance elements. Equation (3.2) is the usual form of Ohm's law, while Equation (3.3) is more convenient as a definition of electrical resistance.

$$e = iR, \text{ V} \tag{3.2}$$

$$R = \frac{e}{i}, \Omega \tag{3.3}$$

Expressed as $e = iR$, Ohm's law is the equation of a straight line through the origin with a slope of R (see Figure 3.1a). The resistance can be determined by finding the slope between any two points on the graph, such as point a and point b in Figure 3.1a. The resistance is equal to the increase in voltage between points a and b, Δe, divided by the increase in current between the same two points, Δi. This gives us a slightly different definition of resistance, given as

$$R = \frac{\Delta e}{\Delta i}, \Omega \tag{3.4}$$

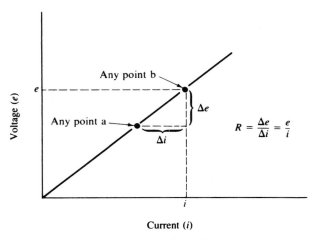

a) Resistance of a linear element

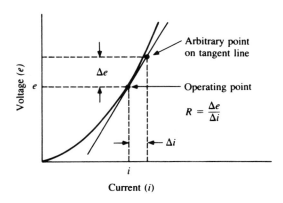

b) Resistance of a nonlinear element

◆ **Figure 3.1** Graphic determination of the resistance of linear and nonlinear electric components.

When the resistive element is nonlinear (see Section 1.6), the determination of resistance becomes more involved, and Equations (3.3) and (3.4) give us different values for the resistance. Figure 3.1b shows a nonlinear resistive element. If we use Equation (3.3) to compute the resistance, we get a value of resistance that is equal to the slope of a straight line from the origin to the operating point. We call this the *static resistance* of the element. In a control system, the operation is usually confined to small variations around the operating point. In this situation, the operation is approximately along the tangent to the curve at the operating point rather than along the line from the origin to the operating point. The resistance obtained from the slope of the tangent line is called the *dynamic resistance* of the element at the operating point. The dynamic resistance of a nonlinear element is more accurate than the static resistance for control system analysis. In the remainder of this book, the term *resistance* will mean dynamic resistance.

Dynamic resistance can be determined graphically using Equation (3.4) as shown in Figure 3.1b. Begin with a tangent line at the operating point and pick a second point *on the tangent line*. Determine the increase in voltage, Δe, and the increase in current, Δi, between the two points. The dynamic resistance is equal to Δe divided by Δi, as given by Equation (3.4).

The mathematical determination of dynamic resistance makes use of the fact that the value of the derivative at a point is equal to the slope of the tangent line at that point, as given by the equation

$$R = \frac{de}{di}, \Omega \tag{3.5}$$

Equation (3.5) would be read to mean that dynamic resistance is equal to the rate of change of the voltage with respect to current.

EXAMPLE 3.1

An electrical component is known to have a linear voltampere graph. A test with 24 V applied to the terminals of the component resulted in a measured current of 12 mA. Determine the resistance of the component.

Solution

The voltampere graph is a straight line, so Equation (3.3) applies:

$$R = \frac{e}{i} = \frac{24\,\text{V}}{0.012\,\text{A}} = 2000\,\Omega$$

◆

EXAMPLE 3.2

A light bulb is an example of an electric component with a nonlinear voltampere graph. The electrical resistance of a nonlinear component can be approximated using Equation (3.4) with a very small increment of voltage, Δe, and the resulting small increment of current, Δi. Determine the resistance of a light bulb at 6 V from the following information:

5.95 V results in 0.500 A

6.05 V results in 0.504 A

Solution

$$\Delta e = 6.05\,\text{V} - 5.95\,\text{V} = 0.10\,\text{V}$$

$$\Delta i = 0.504\,\text{A} - 0.500\,\text{A} = 0.004\,\text{A}$$

$$R = \frac{\Delta e}{\Delta i} = \frac{0.10\,V}{0.004\,A} = 25\,\Omega$$

◆

Electrical capacitance is the quantity of electric charge (q C) required to make a unit increase in the electrical potential (e V). The unit of electrical capacitance is the *farad* (F).

$$\text{Capacitance} = C = \frac{\Delta q}{\Delta e} \tag{3.6}$$

A simple manipulation of Equation (3.6) results in the following relationship between the voltage and current in a capacitance:

$$\Delta q = C\Delta e$$

$$\frac{\Delta q}{\Delta t} = I = C\frac{\Delta e}{\Delta t}$$

where Δt = time in seconds required to make the Δq change in charge

I = average current during the time interval Δt

If Δt is reduced until it approaches zero, $\Delta q/\Delta t$ becomes dq/dt, the instantaneous rate of change of charge; I becomes i, the instantaneous current; and $\Delta e/\Delta t$ becomes de/dt, the instantaneous rate of change of potential.

$$\frac{dq}{dt} = i = C\frac{de}{dt} \tag{3.7}$$

Equation (3.7) would be read to mean that the current through a capacitor is equal to the capacitance (C) times the rate of change of the voltage across the capacitor with respect to time. Equation (3.7) is used in circuit analysis as one of the models of a capacitor. The following example shows how Equation (3.7) can be used to explain the 90° phase difference between the voltage and current in a capacitor.

Let

$$e = A \sin \omega t$$

then

$$\frac{de}{dt} = A\omega \cos \omega t = A\omega \sin(\omega t + 90°)$$

and

$$i = CA\omega \sin(\omega t + 90°)$$

Thus the current i leads the voltage e by 90° in a capacitor.

EXAMPLE 3.3

A current pulse with an amplitude of 0.1 mA and a duration of 0.1 s is applied to an electrical capacitor. The voltage across the capacitor is increased from 0 V to +25 V by the current pulse. Determine the capacitance (C) of the capacitor.

Solution

$$I = C\frac{\Delta e}{\Delta t}$$

$$C = I\frac{\Delta t}{\Delta e}$$

$$C = (0.1 \times 10^{-3}\,\text{A})\left(\frac{0.1\ \text{s}}{25\ \text{V} - 0\ \text{V}}\right) = 0.4 \times 10^{-6}\,\text{A} \cdot \text{s/V} = 0.4\ \mu\text{F}$$

◆

Electrical inductance is the voltage required to produce a unit increase in electric current each second. The unit of electrical inductance is the *henry* (H). Equations (3.8) and (3.9) define the voltage–current relationship for an inductor.

$$e = L\frac{\Delta i}{\Delta t} \tag{3.8}$$

$$e = L\frac{di}{dt} \tag{3.9}$$

Equation (3.9) would be read to mean that the voltage across an inductor is equal to the inductance (L) times the rate of change of the current through the inductor with respect to time. Equation (3.9) is used in circuit analysis as one of the models of an inductor. The following example shows how Equation (3.9) can be used to explain the 90° phase difference between the voltage and current in an inductor.

Let
$$i = A \sin \omega t$$

Then
$$\frac{di}{dt} = A\omega \cos \omega t$$

and
$$e = LA\omega \sin(\omega t + 90°)$$

Thus the voltage e leads the current i by 90° in an inductor.

EXAMPLE 3.4
A voltage pulse with an amplitude of 5 V and a duration of 0.02 s is applied to an inductor. The current through the inductor is increased from 1 A to 2.1 A by the voltage pulse. Assume that the resistance of the inductor is negligible, and determine the inductance (L).

Solution
Equation (3.8) may be used to determine L.

$$e = L\frac{\Delta i}{\Delta t}$$

$$L = e\frac{\Delta t}{\Delta i}$$

$$L = (5 \text{ V})\left(\frac{0.02 \text{ s}}{2.1 \text{ A} - 1 \text{ A}}\right) = 0.0909 \text{ V} \cdot \text{s/A} = 0.0909 \text{ H} \qquad ◆$$

Electrical dead-time delay is the delay caused by the time it takes a signal to travel from the source to the destination (Figure 3.2). Although electrical signals travel at tremendous speeds (2×10^8 to 3×10^8 m/s), the transport time of an electrical signal constitutes a dead-time delay that has important consequences in some systems. In digital computer circuits, the delay on a transmission line is sometimes used to delay a signal deliberately to accomplish the desired logic function. In most control systems, however, the effect of electrical dead time is negligible because the delay is so small compared to the delay in other parts of the system.

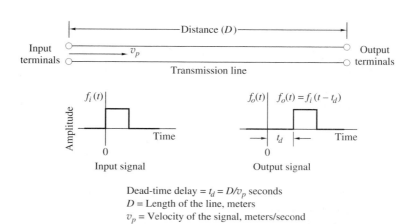

Dead-time delay = $t_d = D/v_p$ seconds
D = Length of the line, meters
v_p = Velocity of the signal, meters/second

◆ **Figure 3.2**   Electrical dead-time element: a transmission line.

The velocity of a signal on a transmission line is called the *velocity of propagation* (v_p). As mentioned before, v_p varies between 2×10^8 and 3×10^8 m/s. The dead-time delay of the line is equal to the distance the signal travels (D) divided by the velocity of propagation (v_p).

$$t_d = \frac{D}{v_p}, \text{ s}$$

EXAMPLE 3.5

(a) Determine the dead-time delay of a 600-m-long transmission line if the velocity of propagation is 2.3×10^8 m/s.

(b) Determine the dead-time delay of a signal from a space vehicle that is located 2000 km from the earth station receiving the signal. The signal travels at 3×10^8 m/s.

Solution

(a) Equation (3.1) applies.

$$t_d = \frac{600 \text{ m}}{2.3 \times 10^8 \text{ m/s}} = 2.61 \times 10^{-6} \text{ s}$$

$$= 2.61 \ \mu\text{s}$$

(b) Equation (3.1) applies.

$$t_d = \frac{2 \times 10^6 \text{ m}}{3 \times 10^8 \text{ m/s}} = 0.67 \times 10^{-2} \text{ s}$$

$$= 6.7 \text{ ms}$$

◆

ELECTRICAL EQUATIONS

Resistance

$$R = \frac{e}{i} = \frac{\Delta e}{\Delta i} = \frac{de}{di} \qquad \text{(3.3 to 3.5)}$$

Capacitance

$$C = \frac{\Delta q}{\Delta e} \qquad \text{(3.6)}$$

$$i = C \frac{de}{dt} \qquad \text{(3.7)}$$

Inductance

$$e = L \frac{\Delta i}{\Delta t} \qquad \text{(3.8)}$$

$$e = L \frac{di}{dt} \qquad \text{(3.9)}$$

Dead-Time Delay

$$t_d = \frac{D}{v_p} \qquad \text{(3.1)}$$

where C = capacitance, F
        D = distance between the input and the output, m
        e = applied voltage, V
       Δe = change in applied voltage, V
        i = instantaneous current, A
       Δi = change in current, A
        I = average current over interval Δt, A
        L = inductance, H
       Δq = change in charge, C
        R = resistance, Ω
       Δt = interval of time, s
       t_d = dead-time delay, s
       v_p = velocity of travel, m/s

3.3 LIQUID FLOW ELEMENTS

Liquid flow resistance is that property of pipes, valves, or restrictions which impedes the flow of a liquid. It is measured in terms of the increase in pressure required to make a unit increase in flow rate. The SI unit for liquid flow resistance is "pascal second/cubic meter." The English unit is "psf/cfs," where psf means "pound per square foot" and cfs means "cubic foot per second." A more practical English unit is "psi/gpm," where psi means "pound per square inch" and gpm means "gallon per minute."

Liquid resistance is determined by the relationship between the pressure drop and the flow rate as expressed by a flow equation. There are two different types of flow: *laminar* and *turbulent*. Each has a different flow equation and hence a different liquid resistance. Laminar flow occurs when the fluid velocity is relatively low and the liquid flows in layers. A colored dye injected into the center of a laminar flow will move with the liquid and remain concentrated in the center. Turbulent flow occurs when the fluid velocity is relatively high and the liquid does not flow in layers. A colored dye injected into the center of turbulent flow is soon mixed throughout the flowing fluid.

The type of flow that occurs depends on four parameters: the density of the fluid, ρ; the inside diameter of the pipe, d; the absolute viscosity of the fluid, μ; and the average velocity of the flowing fluid, v. The four parameters are arranged in a dimensionless grouping called the *Reynolds number.*[*]

$$\text{Reynolds number} = \frac{\rho v d}{\mu} \tag{3.10}$$

$$v = \frac{Q}{a} = \frac{4Q}{\pi d^2}, \text{ m/s} \tag{3.11}$$

---

[*]To show that the Reynolds number is dimensionless, we must express the viscosity unit (Pa · s) in terms of kg, m, and s as follows:

$$\text{Pa} \cdot \text{s} = [\text{N/m}^2]\text{s} = [(\text{kg} \cdot \text{m/s}^2)/\text{m}^2]\text{s} = \text{kg}/(\text{m} \cdot \text{s})$$

$$\text{Units of Reynolds number} \left(\frac{\rho v d}{\mu}\right) = \frac{(\text{kg/m}^3)(\text{m/s})(\text{m})}{\text{kg}/(\text{m} \cdot \text{s})} = \left(\frac{\text{kg}}{\text{kg}}\right)\left(\frac{\text{m}^3}{\text{m}^3}\right)\left(\frac{\text{s}}{\text{s}}\right) = \text{dimensionless}$$

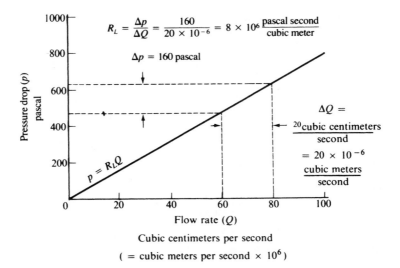

◆ **Figure 3.3** Graph of pressure versus flow rate for laminar flow is linear. Liquid flow resistance is equal to the slope of the graph and is constant for all values of flow rate.

Laminar flow occurs when the Reynolds number has a value less than 2000. Turbulent flow occurs when the Reynolds number has a value greater than 4000. A transition between laminar and turbulent flow occurs when the Reynolds number has a value between 2000 and 4000. Graphs of pressure versus flow rate for laminar and turbulent flow are shown in Figures 3.3 and 3.4.

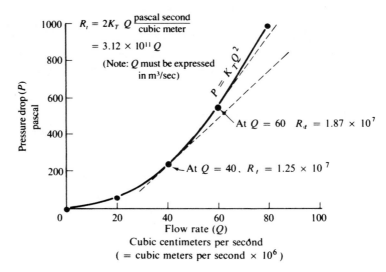

◆ **Figure 3.4** Graph of pressure versus flow rate for turbulent flow is nonlinear. Liquid flow resistance is equal to the slope of the tangent to the curve at any point. Turbulent flow resistance increases as the flow rate is increased.

Liquid flow resistance can be determined graphically in the same manner as electrical resistance. In Figures 3.3 and 3.4, the slope of the graph is the liquid flow resistance. The laminar flow graph is linear, and the resistance has a constant value for all flow rates. The turbulent flow graph is nonlinear and the resistance increases as the flow rate is increased.

The flow equation for laminar flow in a round pipe is given by Equation (3.12) (called the *Hagen–Poiseuille law*). The laminar flow resistance, given by Equation (3.13), is a constant that depends on three parameters: the absolute viscosity, μ, the inside diameter of the pipe, d, and the length of the pipe, l.

$$p = R_L Q, \text{Pa} \tag{3.12}$$

$$R_L = \frac{128\,\mu l}{\pi d^4}, \text{Pa} \cdot \text{s/m}^3 \tag{3.13}$$

The flow equation for turbulent flow in a round pipe is given by Equation (3.14) (called the *Fanning equation*). The turbulent flow resistance, given by Equations (3.15) and (3.16), is considerably more complex than laminar flow resistance. Turbulent resistance depends on the flow rate, Q, and the turbulent flow coefficient, K_t. The turbulent flow coefficient depends on four parameters: the liquid density, ρ; the length of the pipe, l; the inside diameter of the pipe, d; and a factor called the friction factor, f. The friction factor depends on the diameter of the pipe, the Reynolds number, and the smoothness of the inside of the pipe. A table of friction factors is included here as Table 3.3.

$$p = K_t Q^2, \text{Pa} \tag{3.14}$$

$$R_t = 2K_t Q, \text{Pa} \cdot \text{s/m}^3 \tag{3.15}$$

$$K_t = \frac{8\rho f l}{\pi^2 d^5}, \frac{\text{Pa} \cdot \text{s}^2}{\text{m}^6} (\text{or kg/m}^7) \tag{3.16}$$

The computation of liquid flow resistance and pressure drop begins with the determination of the Reynolds number. If the Reynolds number is less than 2000, the flow is laminar and Equations (3.12) and (3.13) are used. If the Reynolds number is greater than 4000, the flow is turbu-

◆ **TABLE 3.3**   Values of the Friction Factor (f)

Type	Diameter (cm)	Reynolds Number[a]					
		4×10^3	10^4	10^5	10^6	10^7	10^8
Smooth tubing	1–2	0.039	0.030	0.018	0.014	0.012	0.012
	2–4	0.039	0.030	0.018	0.013	0.011	0.010
	4–8	0.039	0.030	0.018	0.012	0.010	0.009
	8–16	0.039	0.030	0.018	0.012	0.009	0.008
Commercial steel pipe	1–2	0.041	0.035	0.028	0.026	0.026	0.026
	2–4	0.040	0.033	0.024	0.023	0.023	0.023
	4–8	0.039	0.030	0.022	0.020	0.019	0.019
	8–16	0.039	0.030	0.020	0.018	0.017	0.017

Source: L. F. Moody. "Friction Factors for Pipe Flow," *ASME Transactions,* Vol. 66, No. 8, 1944, p. 671.

[a] Reynolds number = $\rho v D/\mu$

lent and Equations (3.14) to (3.16) are used along with the friction factor from Table 3.3. If the Reynolds number is between 2000 and 4000, the resistance and pressure drop will be somewhere between the values from the laminar flow equations and the values from the turbulent flow equations, so both sets of equations should be used. If the flow is turbulent, exact values of the friction factor (f) must be interpolated from Table 3.3. Conversion of units may also be required.

Computations may be done using a calculator, a spreadsheet, or a computer program. Example 3.6 illustrates the use of a calculator to compute laminar flow resistance and pressure drop. Example 3.7 does the same for turbulent flow resistance and pressure drop.

LIQRESIS (see Preface) is a QuickBASIC program for computing laminar or turbulent flow resistance and pressure drop. The program accepts practical inputs in either SI or English units. It then converts the units to the correct SI units, computes the Reynolds number, selects the correct equations, computes the resistance and pressure drop, and prints the results using the system of units selected for input. Examples 3.8 through 3.10 illustrate the use of the program.

EXAMPLE 3.6

Oil at a temperature of 15°C flows through a horizontal tube that is 1 cm in diameter with a flow rate of 9.42 liters per minute (L/min). The line is 10 m long. Determine the Reynolds number, the resistance, and the pressure drop in the tube.

Solution

From "Properties of Liquids" in Appendix A,

$$\rho = 880 \text{ kg/m}^3$$

$$\mu = 0.160 \text{ Pa} \cdot \text{s}$$

Use Appendix B to convert d and Q to standard SI units:

$$d = 1 \text{ cm} = 0.01 \text{ m}$$

$$Q = 94.2 \text{ L/min} = 9.42(1.6667\text{E} - 5) = 1.57\text{E} - 4 \text{ m}^3/\text{s}$$

Note: The E notation is used in problem solutions to represent numbers that are to be entered into a computer or a pocket calculator. Large and small numbers are written with a base number equal to or greater than 1 and less than 10. The base number is followed by the letter E (for exponent), a plus or minus symbol, and the power of 10. Thus 6.72×10^{-4} is written as $6.72\text{E} - 4$.

Use Equation (3.11) to compute the average fluid velocity:

$$v = \frac{4Q}{\pi d^2} = \frac{4(1.57\text{E} - 4 \text{ m}^3/\text{s})}{\pi(0.01 \text{ m})^2} = 2.00 \text{ m/s}$$

Use Equation (3.10) to compute the Reynolds number:

$$\text{Reynolds number} = \frac{\rho v d}{\mu} = \frac{(880 \text{ kg/m}^3)(2.00 \text{ m/s})(0.01 \text{ m})}{0.160 \text{ Pa} \cdot \text{s}} = 110$$

The flow is laminar, so we use Equations (3.12) and (3.13) to compute the resistance and pressure drop:

$$R_L = \frac{128 \,\mu l}{\pi d^4} = \frac{(128)(0.160 \text{ Pa} \cdot \text{s})(10 \text{ m})}{\pi(0.01 \text{ m})^4} = 6.52 + 9 \text{ Pa} \cdot \text{s/m}^3$$

$$p = R_L Q = (6.52\text{E} + 9 \text{ Pa} \cdot \text{s/m}^3)(1.57\text{E} - 4 \text{ m}^3/\text{s}) = 1.024 + 6 \text{ Pa}$$

◆

EXAMPLE   Water at 15°C flows through a commercial steel pipe 0.4 in. in diameter with a flow rate of 6 gal/min.
3.7   The line is 50 ft long. Determine the Reynolds number, the resistance, and the pressure drop in the tube.

Solution   From "Properties of Liquids" in Appendix A,

$$\rho = 1000 \text{ kg/cm}$$

$$\mu = 0.001 \text{ Pa} \cdot \text{s}$$

When working problems with English units, our approach will be to use SI units to complete the computations and then convert the answer to the desired English units. This avoids the confusing task of converting the equations to English units. Use Appendix B to convert d, Q, and l to standard SI units:

$$d = (0.4 \text{ in.})(0.0254 \text{ m/in.}) = 0.01016 \text{ m}$$

$$Q = (6 \text{ gal/min})(6.3088\text{E} - 5 \text{ m}^3/\text{s/gpm}) = 3.7853\text{E} - 4 \text{ m}^3/\text{s}$$

$$l = (50 \text{ ft})(0.3048 \text{ m/ft}) = 15.240 \text{ m}$$

Use Equation (3.11) to compute the average fluid velocity:

$$v = \frac{4Q}{\pi d^2} = \frac{4(3.7853\text{E} - 4 \text{ m}^3/\text{s})}{\pi (0.01016 \text{ m})^2} = 4.669 \text{ m/s}$$

Use Equation (3.10) to compute the Reynolds number:

$$\text{Reynolds number} = \frac{\rho v d}{\mu} = \frac{(1000 \text{ kg/m}^3)(4.669 \text{ m/s})(0.01016 \text{ m})}{0.001 \text{ Pa} \cdot \text{s}} = 47,440$$

The flow is turbulent, so we use Equations (3.14) to (3.16) to compute the resistance and pressure drop. The complexity of the following computations can be intimidating. Fortunately, program LIQRESIS relieves us of this daunting burden in Example 3.9.

The first step is to obtain the friction factor by interpolation from Table 3.3, using the row for commercial steel pipe with a diameter of 1 to 2 in. The following variables and values will be used in the interpolation:

$$R_e = 47,440 = \text{Reynolds number from Equation (3.10)}$$

$$R_a = 10,000 = \text{largest Reynolds number column heading that is less than } R$$

$$R_b = 100,000 = \text{smallest Reynolds number column heading that is greater than } R$$

$$f = \text{friction factor for the Reynolds number } R \text{ (47,440)}$$

$$f_a = 0.035 = \text{friction factor in the } R_a \text{ column}$$

$$f_b = 0.028 = \text{friction factor in the } R_b \text{ column}$$

The interpolation formula is

$$f = f_a + (f_b - f_a)\left(\frac{R_e - R_a}{R_b - R_a}\right)$$

$$f = 0.035 + (0.028 - 0.035)\left(\frac{4.744\text{E}4 - 1\text{E}4}{10\text{E}4 - 1\text{E}4}\right)$$

$$f = 0.035 - 0.007\left(\frac{3.744}{9}\right) = 0.032088$$

Use Equation (3.16) to compute K_t:

$$K_t = \frac{8\rho f l}{\pi^2 d^5} = \frac{(8)(1000 \text{ kg/m}^3)(0.032088)(15.24 \text{ m})}{\pi^2 (0.01016 \text{ m})^5} = 3.6614\text{E}12 \text{ Pa} \cdot \text{s}^2/\text{m}^6$$

Use Equation (3.15) to compute R_t:

$$R_t = 2K_tQ = (2)(3.6614E12 \text{ Pa} \cdot \text{s}^2/\text{m}^6)(3.7853E - 4 \text{ m}^3/\text{s}) = 2.772E9 \text{ Pa} \cdot \text{s/m}^3$$

Use Appendix B to convert R_t to English units:

$$R_t = (2.7719E9 \text{ Pa} \cdot \text{s/m}^3(9.148E - 9) = 25.4 \text{ psi/gpm}$$

Use Equation (3.14) to compute p:

$$p = K_tQ^2 = (3.66148E12 \text{ Pa} \cdot \text{s}^2/\text{m}^6)(3.785E - 4\text{m}^3/\text{s})^2 = 5.25E5 \text{ Pa}$$

Use Appendix B to convert p to English units:

$$p = (5.246E5 \text{ Pa})(1.45E - 4 \text{ psi/Pa}) = 76.1 \text{ psi}$$

Answer: Rt = 25.4 psi/gpm, p = 76.1 psi ◆

EXAMPLE 3.8 Use the program LIQRESIS to solve the problem presented in Example 3.6. LIQRESIS is available on disk (see Preface).

Solution The screen produced by a run of the program follows. The first eight lines are the title and input. The last three lines are the result of the run.

```
            LIQUID FLOW RESISTANCE AND PRESSURE DROP

   Smooth tubing [T] or commercial pipe [P]──────────────────► TUBE
   SI units [S] or English units [E]────────────────────────► SI
   Flow rate in liters/minute───────────────────────────────► 9.42
   Inside diameter of TUBE in centimeters───────────────────► 1
   Length of TUBE in meters─────────────────────────────────► 10
   Absolute viscosity in Pascal seconds─────────────────────► 0.160
   Fluid density in kilograms/cubic meter───────────────────► 880

   Reynolds number = 1.099E+02, Flow is laminar
   Resistance = 6.519E+09 Pascal second/cubic meter
   Pressure drop = 1.024E+06 Pascal
```

◆

EXAMPLE 3.9 Use the program LIQRESIS to solve the problem presented in Example 3.7.

Solution

```
            LIQUID FLOW RESISTANCE AND PRESSURE DROP

   Smooth tubing [T] or commercial pipe [P]──────────────────► PIPE
   SI units [S] or English units [E]────────────────────────► ENGLISH
   Flow rate in gallons/minute──────────────────────────────► 6
   Inside diameter of PIPE in inches────────────────────────► 0.4
   Length of PIPE in feet───────────────────────────────────► 50
   Absolute viscosity in pounds/foot second─────────────────► 6.72E-04
   Fluid density in pounds/cubic foot───────────────────────► 62.43

   Reynolds number = 4.743E+04, Flow is turbulent
   Resistance = 2.536E+01 psi/gpm
   Pressure drop = 7.607E+01 psi
```

◆

EXAMPLE 3.10

A fluid with a density of 880 kg/m³ and a viscosity of 0.0053 Pa · s is flowing through a smooth tube with an inside diameter of 1 cm. The flow rate is 9.42 L/min. The line is 10 m long. Determine the Reynolds number, the resistance, and the pressure drop in the tube.

Solution

```
              LIQUID FLOW RESISTANCE AND PRESSURE DROP

Smooth tubing [T] or commercial pipe [P]─────────────────────► TUBE
SI units [S] or English units [E]────────────────────────────► SI
Flow rate in liters/minute───────────────────────────────────► 9.42
Inside diameter of TUBE in centimeters───────────────────────► 1
Length of TUBE in meters─────────────────────────────────────► 10
Absolute viscosity in Pascal seconds─────────────────────────► 0.0053
Fluid density in kilograms/cubic meter───────────────────────► 880

Reynolds number = 3.319E+03, Flow is in transition
Resistance = 2.159E+08 to 8.964E+08 Pascal second/cubic meter
Pressure drop = 3.390E+04 to 7.037E+04 Pascal
```

◆

Liquid flow capacitance is defined in terms of the increase in volume of liquid in a tank required to make a unit increase in pressure at the outlet of the tank.

$$C_L = \frac{\Delta V}{\Delta p} \tag{3.17}$$

where C_L = capacitance, m³/Pa

ΔV = increase in volume, m³

Δp = corresponding increase in pressure, Pa

The increase in pressure in the tank depends on three things: the increase in level of the liquid (ΔH), the acceleration due to gravity (g), and the density of the liquid (ρ). The relationship is expressed by the following equation:

$$\Delta p = \rho g \, \Delta H \tag{3.18}$$

The increase in level in the tank is equal to the increase in volume divided by the average surface area (A) of the liquid in the tank.

$$\Delta H = \frac{\Delta V}{A} \tag{3.19}$$

By substitution of Equations (3.18) and (3.19) into Equation (3.17),

$$\Delta p = \frac{\rho g \, \Delta V}{A}$$

$$C_L = \frac{\Delta V}{\Delta p} = \frac{\Delta V A}{\rho g \, \Delta V}$$

$$= \frac{A}{\rho g}, \text{ m}^3/\text{Pa} \tag{3.20}$$

EXAMPLE 3.11

A liquid tank has a diameter of 1.83 m and a height of 10 ft. Determine the capacitance of the tank for each of the following fluids:

(a) Water
(b) Oil
(c) Kerosene
(d) Gasoline

Solution

Equation (3.19) may be used to determine C_L.

$$C_L = \frac{A}{\rho g}$$

$$A = \frac{\pi D^2}{4} = \frac{\pi (1.83 \text{ m})^2}{4} = 2.63 \text{ m}^2$$

$$g = 9.81 \text{ m/s}^2$$

$$C_L = \frac{2.63}{9.81 \, \rho} = \frac{0.268}{\rho}$$

Density is obtained from "Properties of Liquids" in Appendix A.

(a) Water: $\rho = 1000 \text{ kg/m}^3$

$$C_L = \frac{0.268}{1000} = 2.68 \times 10^{-4} \text{ m}^3/\text{Pa}$$

(b) Oil: $\rho = 880 \text{ kg/m}^3$

$$C_L = \frac{0.268}{880} = 3.05 \times 10^{-4} \text{ m}^3/\text{Pa}$$

(c) Kerosene: $\rho = 800 \text{ kg/m}^3$

$$C_L = \frac{0.268}{800} = 3.35 \times 10^{-4} \text{ m}^3/\text{Pa}$$

(d) Gasoline: $\rho = 740 \text{ kg/m}^3$

$$C_L = \frac{0.268}{740} = 3.62 \times 10^{-4} \text{ m}^3/\text{Pa} \qquad ◆$$

Liquid flow inertance is measured in terms of the amount of pressure drop in a pipe required to increase the flow rate by 1 unit each second.

$$I_L = \frac{p}{\Delta Q/\Delta t} \qquad (3.21)$$

where I_L = inertance, $\text{Pa} \cdot \text{s}^2/\text{m}^3$
p = pressure drop in the pipe, Pa
ΔQ = change in flow rate, m^3/s
Δt = time interval, s

A more practical equation for inertance can be obtained as follows: The pressure drop acts on the cross-sectional area of the pipe (a) to produce a force (F) equal to the pressure drop (p) times the area (a):

$$F = p \times a$$

This force will accelerate the fluid in the pipe, according to Newton's law of motion:

$$F = pa = m\frac{\Delta v}{\Delta t} \tag{3.22}$$

The mass of fluid in the pipe (m) is equal to the density of the fluid (ρ) times the volume of fluid in the pipe. The volume is equal to the cross-sectional area of the pipe (a) times the length of the pipe (l).

$$m = \rho a l \tag{3.23}$$

The change in flow rate, ΔQ, is equal to the change in fluid velocity, Δv, multiplied by the area, a:

$$\Delta Q = a\,\Delta v \tag{3.24}$$

Combining Equations (3.22), (3.23), and (3.24) into Equation (3.21) results in the following equation for inertance:

$$I_L = \frac{\rho l}{a}, \text{Pa} \cdot \text{s}^2/\text{m}^3 \tag{3.25}$$

EXAMPLE 3.12   Determine the liquid flow inertance of water in a pipe that has a diameter of 2.1 cm and a length of 65 m.

Solution   Equation (3.25) applies:

$$I_L = \frac{\rho L}{a}$$

$$\rho = 1000 \text{ kg/m}^3 \text{ (Appendix A)}$$

$$a = \frac{\pi d^2}{4} = \frac{\pi (0.021)^2}{4} = 3.46\text{E} - 4 \text{ m}^2$$

$$I_L = \frac{(1000 \text{ kg/m}^3)(65 \text{ m})}{3.46\text{E} - 4 \text{ m}^2} = 1.88\text{E} + 8 \text{ Pa} \cdot \text{s}^2/\text{m}^3 \qquad\qquad ◆$$

Dead time occurs whenever liquid is transported from one point to another in a pipeline. An example of liquid flow dead time is shown in Figure 3.5. The dead-time delay is the distance traveled (D) divided by the average velocity (v) of the fluid. Equations (3.11) and (3.1) are used to compute the average velocity (v) and the dead-time delay (t_d). Since $a = \pi d^2/4$,

$$v = \frac{Q}{a} = \frac{4Q}{\pi d^2}, \text{m/s}$$

$$t_d = \frac{D}{v}, \text{s}$$

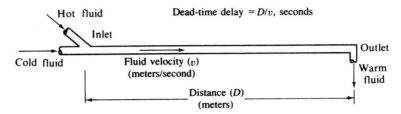

◆ **Figure 3.5**   Liquid flow dead-time element. Hot and cold fluids are combined in a
Y connection to produce a warm fluid which flows a distance *D* to the outlet of the
pipe. The input to the system is the ratio of hot fluid to cold fluid. The output is the
temperature of the fluid at the outlet end. The dead-time delay is the distance traveled
(*D*) divided by the average velocity (*v*) of the fluid.

**EXAMPLE
3.13**

Liquid flows in a pipe that is 200 m long and has a diameter of 6 cm. The flow rate is $0.0113 \text{ m}^3/\text{s}$.
Determine the dead-time delay.

Solution

Equations (3.11) and (3.1) apply.

$$v = \frac{4Q}{\pi d^2} = \frac{(4)(0.0113)}{\pi(0.06^2)} = 4 \text{ m/s}$$

$$t_d = \frac{D}{v} = \frac{200 \text{ m}}{4 \text{ m/s}} = 50 \text{ s}$$

◆

LIQUID FLOW EQUATIONS

Reynolds Number

$$\text{Reynolds number} = \frac{\rho v d}{\mu} \qquad (3.10)$$

Average Velocity

$$v = \frac{Q}{a} = \frac{4Q}{\pi d^2} \qquad (3.11)$$

Laminar Flow Resistance
(Reynolds number less than 2000)

$$p = p_1 - p_2 = R_L Q \qquad (3.12)$$

$$R_L = \frac{128 \, \mu l}{\pi d^4} \qquad (3.13)$$

Turbulent Flow Resistance
(Reynolds number more than 4000)

$$p = p_1 - p_2 = K_t Q^2 \qquad (3.14)$$

$$R_t = 2K_t Q \qquad (3.15)$$

$$K_t = \frac{8 \, \rho f l}{\pi^2 d^5} \qquad (3.16)$$

Capacitance

$$C_L = \frac{A}{\rho g}$$ (3.20)

Inertance

$$I_L = \frac{\rho l}{a}$$ (3.25)

Dead-Time Delay

$$t_d = \frac{D}{v}$$ (3.1)

where a = cross-sectional area of pipe, m^2
  A = surface area of liquid in tank, m^2
  C_L = liquid flow capacitance, m^3/Pa
  d = inside diameter of the pipe, m
  D = diameter of the tank or distance, m
  f = friction factor from Table 3.3
  g = 9.81 m/s^2 (gravitational acceleration)
  I_L = inertance, $Pa \cdot s^2/m^3$
  K_t = turbulent flow coefficient, $Pa \cdot s^2/m^6$
  l = length of the pipe, m
  p = pressure drop from inlet to outlet, Pa
  p_1 = pressure at inlet of pipe, Pa
  p_2 = pressure at outlet of pipe, Pa
  Q = flow rate, m^3/s
  R_L = laminar flow resistance, $Pa \cdot s/m^3$
  R_t = turbulent flow resistance, $Pa \cdot s/m^3$
  v = average fluid velocity in pipe, m/s
  ρ = fluid density, kg/m^3 (see "Properties of Liquids" in Appendix A)
  μ = absolute viscosity of fluid, $Pa \cdot s$ (see "Properties of Liquids" in Appendix A)

3.4 GAS FLOW ELEMENTS

Gas flow resistance is that property of pipes, valves, or restrictions that impedes the flow of a gas. It is measured in terms of the increase in pressure required to produce an increase in gas flow rate of 1 kg/s. The SI unit for gas flow resistance is pascal second per kilogram (Pa · s/kg). Gas flow in a pipe may be laminar or turbulent. In laminar flow, the pressure drop varies directly with the gas velocity. In turbulent flow, the pressure drop varies directly with the square of the gas velocity. In practice, gas flow is almost always turbulent, and the commonly used equations apply to turbulent flow.

If the pressure drop is less than 10% of the initial gas pressure, the equation for incompressible flow gives reasonable accuracy for gas flow. This is the equation we used for turbulent liquid flow with the volume flow rate (Q) replaced by mass flow rate (W):

$$p = p_1 - p_2 = K_g W^2, \text{Pa} \tag{3.26}$$

$$R_g = 2K_g W, \text{Pa} \cdot \text{s/kg} \tag{3.27}$$

$$K_g = \frac{8fl}{\pi^2 d^5 \rho}, \frac{\text{Pa} \cdot \text{s}^2}{\text{kg}^2} \left(\text{or} \frac{1}{\text{kg} \cdot \text{m}} \right) \tag{3.28}$$

Equations (3.26), (3.27), and (3.28) do not account for the expansion of the gas as the pressure decreases, and the results are not accurate when the pressure drop is greater than 10% of the initial gas pressure. Also, "Properties of Gases" Appendix A gives the density and viscosity of gases at standard atmospheric conditions. To keep things from getting unnecessarily complicated, we will limit the gas flow examples to near-atmospheric conditions with pressure drops of less than 10% of the inlet pressure. This will allow familiarization with the concept of gas flow resistance without worrying about the variations in viscosity and density caused by changes in pressure and temperature.

EXAMPLE 3.14

Figure 3.6 is a graph of mass flow rate (W) versus pressure drop (p). Determine the gas flow resistance at the point where the mass flow rate is 0.6 kg/s.

Solution

Refer to Figure 3.6 for the solution.

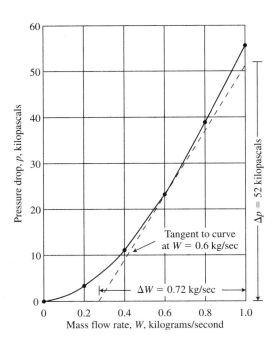

◆ **Figure 3.6** Graphical determination of gas flow resistance at the point where $W = 0.6$ kg/s. First draw a line tangent to the curve at $W = 0.6$. The desired resistance is the slope of this tangent line. The rise of the tangent line is 52,000 Pa, and the run is 0.72 kg/s. The resistance is the rise divided by the run: $R = 52,000/0.72 = 72,222$ Pa · s/kg.

EXAMPLE 3.15

A smooth tube is supplying 0.03 kg/s of air at a temperature of 15°C. The tube is 30 m long and has an inside diameter of 4 cm. Use Equations (3.26), (3.27), and (3.28) to find the pressure drop and the gas flow resistance. The outlet pressure (p_2) is 102 kPa. Find the inlet pressure (p_1).

Solution

The following alternative form of the Reynolds number equation is more convenient for this problem.

$$Re = \frac{4 W}{\pi \mu d}$$

From Appendix A,

$$\mu = 1.81E - 5 \text{ Pa} \cdot \text{s}$$

$$\rho = 1.22 \text{ kg/m}^3$$

$$Re = \frac{(4)(0.03 \text{ kg/s})}{\pi(1.81E - 5 \text{ Pa} \cdot \text{s})(0.04 \text{ m})} = 52,800$$

Interpolating from Table 3.3, we obtain friction factor f:

$$f = 0.03 + (0.018 - 0.030)\left(\frac{5.28 - 1}{10 - 1}\right) = 0.023$$

$$K_g = \frac{8 f l}{\pi^2 d^5 \rho} = \frac{(8)(0.023)(30 \text{ m})}{\pi^2(0.04 \text{ m})^5(1.22 \text{ kg/m}^3)}$$

$$= 4.469E + 6 \text{ Pa} \cdot \text{s}^2/\text{kg}^2$$

$$p = K_g W^2 = (4.469E + 6 \text{ Pa} \cdot \text{s}^2/\text{kg}^2)(0.03 \text{ kg/s})^2 = 4.02 \text{ kPa}$$

$$p_1 = p + p_2 = 4.0 + 102 = 106.0 \text{ kPa}$$

$$R_g = 2 K_g W = (2)(4.469E + 6 \text{ Pa} \cdot \text{s}^2/\text{kg}^2)(0.03 \text{ kg/s})$$

$$= 268 \text{ kPa} \cdot \text{s/kg} \qquad\qquad ◆$$

   Gas flow capacitance is defined in terms of the increase in the mass of gas in a pressure vessel required to produce a unit increase in pressure while the temperature remains constant. The SI unit of gas flow capacitance is kilogram per pascal. The ideal (or perfect) gas law may be used to determine the capacitance equation.

$$pV = nRT$$

$$pV = \left(\frac{10^3 m}{M}\right) RT \qquad\qquad \textbf{(3.29)}$$

where p = absolute pressure of the gas, Pa
   V = volume of the gas, m^3
   m = mass of the gas, kg
   n = number of moles of gas, mol
   M = gram molecular weight of the gas (see "Properties of Gases" in Appendix A), g/mol
   R = universal gas constant = 8.314 J/K · mol
   T = absolute temperature, K

Solving Equation (3.29) for m, the mass of the gas, yields

$$m = \left(\frac{MV}{10^3\,RT}\right)p \qquad (3.30)$$

Equation (3.30) applies to a pressure vessel where V is the volume of the vessel, T the absolute temperature of the gas, and M the molecular weight of the gas. For a given pressure vessel and gas, V and M are constants. The temperature (T) must also be held constant to determine the gas capacitance. When M, V, and T are all constant, the term in parentheses is constant. Under these conditions, Equation (3.30) is linear, and the relationship between a change in mass (Δm) and the corresponding change in pressure (Δp) is given by

$$\Delta m = \left(\frac{MV}{10^3\,RT}\right)\Delta p$$

$$\text{Gas flow capacitance} = C_g = \frac{\Delta m}{\Delta p}$$

$$C_g = \frac{MV}{10^3\,RT}, \text{kg/Pa} \qquad (3.31)$$

**EXAMPLE
3.16**

A pressure tank has a volume of 0.75 m³. Determine the capacitance of the tank if the gas is nitrogen at 20°C.

Solution

Equation (3.31) may be used to determine C_g.

$$C_g = \frac{MV}{10^3\,RT}$$

$M = 28.016$ g/mol for nitrogen   (Appendix A)

$V = 0.75$ m³

$T = 20°C = 293$ K

$$C_g = \frac{(28.016\text{ g/mol})(0.75\text{ m}^3)}{(10^3\text{ g/kg})(8.314\text{ J/K} \cdot \text{mol})(293\text{ K})}$$

$$= 8.6 \times 10^{-6}\text{ kg/Pa}$$

GAS FLOW EQUATIONS

Reynolds Number

$$\text{Reynolds number} = \frac{4\,W}{\pi \mu d}$$

Resistance–Low-Pressure Drop

$$p = p_1 - p_2 = K_g W^2 \qquad (3.26)$$

$$R_g = 2\,K_g W \qquad (3.27)$$

$$K_g = \frac{8fl}{\pi^2 d^5 \rho} \qquad (3.28)$$

Capacitance

$$C_g = \frac{MV}{10^3 RT}$$ (3.31)

where C_g = gas flow capacitance, kg/Pa
      d = inside diameter of the pipe, m
     K_g = turbulent flow coefficient, Pa · s²/kg²
     l = length of the pipe, m
     M = gram molecular weight of the gas (Appendix A), g/mol
     p = pressure drop from inlet to outlet, Pa
     p_1 = pressure at inlet of pipe, Pa
     p_2 = pressure at outlet of pipe, Pa
     R = universal gas constant = 8.314 J/K · mol
     R_g = gas flow resistance, Pa · s/kg
     T = temperature, K
     V = volume of gas, m³
     W = gas flow rate, kg/s
     ρ = fluid density, kg/m³
     μ = absolute viscosity of fluid, Pa · s (see "Properties of Gases" in Appendix A)

3.5 THERMAL ELEMENTS

Thermal resistance is that property of a substance that impedes the flow of heat. It is measured in terms of the difference in temperature required to produce a heat flow rate of 1 W (joule per second). Normally, heat flow occurs through a wall that separates two fluids at different temperatures. The fluids may be either liquids or gases. Heat flows through the wall from the hotter fluid to the cooler fluid. If T_o is the temperature of the outside fluid and T_i is the temperature of the inside fluid, the temperature difference is $T_o - T_i$. The heat flow rate, Q, from the outside fluid to the inside fluid is equal to the temperature difference divided by the thermal resistance, R_T.

$$Q = \frac{T_o - T_i}{R_T}, \text{W}$$ (3.32)

If the heat flow rate (Q) is negative because T_i is greater than T_o, it simply means that the heat is going from the inside fluid to the outside fluid.

    The thermal resistance, R_T, in Equation (3.32) is the resistance of the entire wall separating the two fluids. It is often convenient to express the resistance of 1 m² of the wall. We will use the term *unit resistance* and the symbol R_u for the resistance of 1 m² of the wall separating the two fluids. The term *total resistance* and the symbol R_T will be used for the resistance of the entire wall. The total resistance is equal to the unit resistance divided by the surface area of the wall between the fluids.

$$R_T = \frac{R_u}{A}, \text{K/W}$$ (3.33)

The heat flow rate, Q, is equal to the temperature difference multiplied by the surface area and divided by the unit resistance.

$$Q = \frac{(T_o - T_i)A}{R_u}, \text{W} \qquad (3.34)$$

The wall separating the two fluids is a composite, consisting of several layers, each contributing to the resistance of the wall. Figure 3.7 shows a composite wall separating two fluids that are at different temperatures. The two outside layers are thin films of stagnant fluid that form an insulating blanket around the wall. The two inner layers consist of different materials, such as wood, steel, or insulation.

The inside and outside films are always present, but their thickness varies depending on the motion of the main body of the fluid. A strong wind, for example, reduces the thickness of the air film surrounding our bodies, giving us the familiar "wind chill index." The resistance of the film depends on the thickness of the film and on the thermal resistance of the fluid. Empirical formulas have been developed to determine the resistance of fluid films under different conditions. A number of these formulas are summarized in Equations (3.40) to (3.47) at the end of this section. The formulas give the expression for the conductance of the film, represented by the letter h. Conductance is the reciprocal of the unit resistance. The film conductance, h, is referred to as the *film coefficient*. Thus h_o is the outside film coefficient and

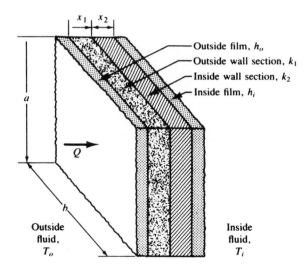

A = wall surface area = $a \times b$, square meters

R_T = thermal resistance = $\dfrac{1}{A}\left(\dfrac{1}{h_o} + \dfrac{x_1}{k_1} + \dfrac{x_2}{k_2} + \dfrac{1}{h_i}\right)\dfrac{\text{kelvin}}{\text{watt}}$

T_o = outside fluid temperature. ° Celsius

T_i = inside fluid temperature. ° Celsius

Q = heat flow rate = $(T_o - T_i)/R_T$ watts

◆ **Figure 3.7** Composite wall separating the outer fluid on the left from the inner fluid on the right. The two outside layers are thin films of stagnant fluid. The two inner layers consist of materials such as wood, steel, or insulation. The total thermal resistance of the wall is the sum of the resistance of each section.

h_i is the inside film coefficient. The unit resistance of the film is equal to the reciprocal of the film coefficient.

$$R_u(\text{film}) = \frac{1}{h}, \text{K} \cdot \text{m}^2/\text{W} \tag{3.35}$$

The unit resistance of the inner layers in the composite wall depends on the thickness of the layer (x) and the thermal conductivity (k) of the material.

$$R_u(\text{inner layer}) = \frac{x}{k}, \text{K} \cdot \text{m}^2/\text{W} \tag{3.36}$$

The value of the thermal conductivity of a number of common substances is given in Appendix A, "Properties of Solids."

The unit resistance of the wall is the sum of the unit resistances of each layer. If we use the subscripts 1, 2, ... for the inner layers, the unit resistance of a wall with n inner layers is expressed as

$$R_u(\text{wall}) = \frac{1}{h_o} + \frac{x_1}{k_1} + \frac{x_2}{k_2} + \cdots + \frac{x_n}{k_n} + \frac{1}{h_i}, \text{K} \cdot \text{m}^2/\text{W} \tag{3.37}$$

The total resistance is equal to the unit resistance divided by the surface area, A.

$$R_T(\text{wall}) = \frac{1}{A}\left(\frac{1}{h_o} + \frac{x_1}{k_1} + \frac{x_2}{k_2} + \cdots + \frac{x_n}{k_n} + \frac{1}{h_i}\right), \text{K}/\text{W} \tag{3.38}$$

The computation of thermal resistance may be done using a calculator, a spreadsheet, or a computer program. Example 3.17 illustrates the use of a calculator to compute the thermal resistance of a composite wall section. Example 3.18 shows the calculator computations of the thermal resistance of five different film conditions. THERMRES, a program for computing the thermal resistance of a composite wall section, is available on disk (see Preface).

EXAMPLE 3.17   A wall section similar to that in Figure 3.7 has two inner layers: a steel plate 1 cm thick and insulation 2 cm thick. The inside fluid is still water. The difference between the water temperature and the surface temperature (T_d) is estimated to be 10°C. The outside fluid is air, which has a velocity of 6 m/s. The water temperature is 45°C, and the air temperature is 85°C. The wall dimensions are $a = 2$ m, $b = 3$ m. Determine the unit resistance, the total resistance, the total heat flow, and the direction of the heat flow.

Solution   From "Properties of Solids" in Appendix A,

$$k_1 = 45 \text{ W/m} \cdot \text{K} \qquad \text{for steel}$$

$$k_2 = 0.036 \text{ W/m} \cdot \text{K} \qquad \text{for insulation}$$

The area of the wall section is

$$A = 2 \times 3 = 6 \text{ m}^2$$

The inside film coefficient is given by Equation (3.43):*

$$h_i = 2.26(T_w + 34.3)T_d^{0.5}$$

$$h_i = 2.26(45 + 34.3)10^{0.5}$$

$$h_i = 566.7 \text{ W/m}^2 \cdot \text{K}$$

---

*The film coefficient equations are listed at the end of this section.

The outside film coefficient is given by Equation (3.46):

$$h_o = 7.75 v_a^{0.75}$$

$$h_o = 7.75(6)^{0.75}$$

$$h_o = 29.71 \text{ W/m}^2 \cdot \text{K}$$

The unit thermal resistance is given by Equation (3.37):

$$R_u = \frac{1}{h_o} + \frac{x_1}{k_1} + \frac{x_2}{k_2} + \frac{1}{h_i}$$

$$R_u = \frac{1}{29.71 \text{ W/m}^2 \cdot \text{K}} + \frac{0.01 \text{ m}}{45 \text{ W/m} \cdot \text{K}} + \frac{0.02 \text{ m}}{0.036 \text{ W/m} \cdot \text{K}} + \frac{1}{566.7 \text{ W/m}^2 \cdot \text{K}}$$

$$R_u = 0.003366 + 0.000222 + 0.5556 + 0.001765$$

$$R_u = 0.591 \text{ K} \cdot \text{m}^2/\text{W}$$

The total thermal resistance is given by Equation (3.33):

$$R_T = R_u/A = 0.591 \text{ K} \cdot \text{m}^2/\text{W}/6 \text{ m}^2 = 0.0985 \text{ K/W}$$

The total heat flow is given by Equation (3.33):

$$Q = \frac{T_o - T_i}{R_T} = \frac{(85 - 45)\text{K}}{0.0985 \text{ K/W}} = 406 \text{ W}$$

◆

EXAMPLE 3.18

Determine the unit thermal resistance of each of the following film conditions.

(a) Natural convection in still air on a vertical surface where $T_d = 20°C$.
(b) Natural convection in still water where $T_d = 30°C$ and $T_w = 20°C$.
(c) Natural convection in oil where $T_d = 16°C$.
(d) Forced convection in air with a velocity of 4 m/s.
(e) Forced convection in water with a temperature (T_w) of 40°C, a pipe diameter of 5 cm, and a velocity of 4 m/s.

Solution

(a) Equation (3.42) applies:

$$h = 1.78 T_d^{0.25}$$

$$h = 1.78(20)^{0.25} = 3.764 \text{ W/m}^2 \cdot \text{K}$$

$$R_u = 1/h = 0.266 \text{ K} \cdot \text{m}^2/\text{W}$$

(b) Equation (3.43) applies:

$$h = 2.26(T_w + 34.3)T_d^{0.5}$$

$$h = 2.26(20 + 34.3)30^{0.5} = 672.2 \text{ W/m}^2 \cdot \text{K}$$

$$R_u = 1/h = 0.00149 \text{ K} \cdot \text{m}^2/\text{W}$$

(c) Equation (3.44) applies:

$$h = 7.0 T_d^{0.25}/\mu^{0.4}$$

$$\mu = 0.160 \text{ Pa} \cdot \text{s} \qquad \text{(from Appendix A)}$$

$$h = 7.0(16)^{0.25}/0.160^{0.4}$$
$$h = 29.14 \text{ W/m}^2 \cdot \text{K}$$
$$R_u = 1/h = 0.0343 \text{ K} \cdot \text{m}^2/\text{W}$$

(d) Equation (3.45) applies:

$$h = 4.54 + 4.1 \, v_a$$
$$h = 4.54 + 4.1(4)$$
$$h = 20.94 \text{ W/m}^2 \cdot \text{K}$$
$$R_u = 1/h = 0.0478 \text{ K} \cdot \text{m}^2/\text{W}$$

(e) Equation (3.47) applies:

$$h = \frac{20.93(68.3 + T_w)v_w^{0.8}}{d^{0.2}}$$

$$h = \frac{20.93(68.3 + 40)4^{0.8}}{0.05^{0.2}}$$

$$h = 12{,}510 \text{ W/m}^2 \cdot \text{K}$$
$$R_u = 7.99\text{E} - 5 \text{ K} \cdot \text{m}^2/\text{W}$$

EXAMPLE 3.19   Use the program THERMRES to solve the problem presented in Example 3.17. THERMRES is available on disk (see Preface).

Solution   A run of the program results in the following output.

```
                    THERMAL RESISTANCE AND HEAT FLOW
    Surface area:                       6.0 square meter
    Inside temperature:                 45 degrees Celsius
    Outside temperature:                85 degrees Celsius
    Inside film: Natural convection, water
       Td:      10 Celsius.             Tw: 45 Celsius
       hi:      5.67E+02 watt/square meter kelvin
    Outside film, Forced convection, air, smooth surface & inside pipe
       Fluid velocity:                  6.0 meter/second
       ho:      2.97E+01 watt/square meter kelvin
    INNER LAYERS:
       x (cm)    1.0                    2.0
       K (W/mk)  45.00                  0.036
    Unit resistance:                    5.91E-01 kelvin square meter/watt
    Total resistance:                   9.85E-0.2 kelvin/watt
    Total heat flow:                    4.06E+02 watts
    Heat flows from outside to inside.
```

◆

Thermal capacitance is defined in terms of the increase in heat required to make a unit increase in temperature. The SI unit of thermal capacitance is joule/kelvin (J/K). The heat capacity (or specific heat) of a substance is the amount of heat required to raise the temperature of 1 kg of the substance by 1 K. Thus the thermal capacitance (C_T) of an object is simply the product of the mass (m) of the object times the heat capacity (S_h) of its substance.

$$C_T = mS_h, \text{ J/K} \tag{3.39}$$

EXAMPLE 3.20

Determine the thermal capacitance of 8.31 m³ of water.

Solution

Equation (3.39) may be used to determine C_T.

$$C_T = mS_h$$

The mass of water (m) is equal to the density (ρ) times the volume of water (8.31 m³). From Appendix A, the density of water is 1000 kg/m³ and the specific heat is 4190 J/kg · K.

$$C_T = (1000 \text{ kg/m}^3)(8.31 \text{ m}^3)(4190 \text{ J/kg} \cdot \text{K})$$

$$= 3.48 \times 10^7 \text{ J/K}$$

◆

THERMAL EQUATIONS

Resistance

$$R_u = \frac{1}{h_o} + \frac{x_1}{k_1} + \frac{x_2}{k_2} + \cdots + \frac{x_n}{k_n} + \frac{1}{h_i} \tag{3.37}$$

$$R_T = \frac{R_u}{A} \tag{3.33}$$

$$Q = \frac{T_o - T_i}{R_T} \tag{3.32}$$

$$Q = \frac{(T_o - T_i)A}{R_u} \tag{3.34}$$

Film Coefficients (see note 1, below)

1. Natural convection in still air:
   a. Horizontal surfaces facing up:

$$h = 2.50 T_d^{0.25} \tag{3.40}$$

   b. Horizontal surfaces facing down:

$$h = 1.32 T_d^{0.25} \tag{3.41}$$

   c. Vertical surfaces:

$$h = 1.78 T_d^{0.25} \tag{3.42}$$

2. Natural convection in still water:

$$h = 2.26(T_w + 34.3)T_d^{0.5} \tag{3.43}$$

3. Natural convection in oil:

$$h = \frac{7.0T_d^{0.25}}{\mu^{0.4}}$$ (3.44)

4. Forced convection for air against smooth surfaces and inside straight pipes:

a. Air velocity, $v_a \leq 4.6$ m/s:

$$h = 4.54 + 4.1v_a$$ (3.45)

b. Air velocity, $v_a > 4.6$ m/s:

$$h = 7.75v_a^{0.75}$$ (3.46)

5. Forced convection for turbulent water flow in straight pipes:

$$h = \frac{20.93(68.3 + T_w)v_w^{0.8}}{d^{0.2}}$$ (3.47)

Capacitance

$$C_T = mS_h$$ (3.39)

where
A = area of the wall surface, m^2
C_T = thermal capacitance, J/K
d = inside diameter of pipe, m
h = film coefficient, W/m$^2 \cdot$ K (h_o = outside film, h_i = inside film)
k = thermal conductivity, W/m $\cdot$ K
m = mass, kg
Q = heat flow rate, W
R_u = unit thermal resistance, K $\cdot$ m^2/W
R_T = total thermal resistance, K/W
S_h = heat capacity, J/kg $\cdot$ K
T_d = temperature difference between the main body of the fluid and the wall surface, K or °C (see note 2 below).
T_i = inside fluid temperature, °C
T_o = outside fluid temperature, °C
T_w = water temperature, °C
v_a = velocity of air, m/s
v_w = velocity of water, m/s
x = thickness of a layer, m
μ = absolute viscosity, Pa $\cdot$ s

Notes:

1. Equations (3.40) to (3.47) are based on empirical equations in J. Kenneth Salisbury, *Kent's Mechanical Engineers' Handbook: Power,* 12th ed. (New York: John Wiley & Sons, 1950), pp. 3–17 to 3–20.
2. $T_o - T_i$ in kelvin is equal to $T_o - T_i$ in Celsius, so a temperature difference may be computed using either kelvin or Celsius units.

3.6 MECHANICAL ELEMENTS

Mechanical resistance (or friction) is that property of a mechanical system that impedes motion. It is measured in terms of the increase in force required to produce an increase in velocity of 1 m/s. The SI unit of mechanical resistance is the "newton second/meter" (N · s/m).

An automobile shock absorber and a dashpot are examples of mechanical resistance devices. Figure 3.8 illustrates the operation of a dashpot. The cylinder is stationary and the pis-

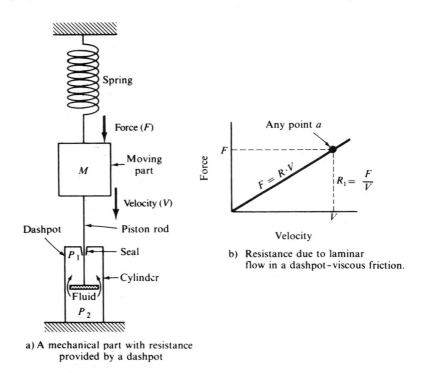

a) A mechanical part with resistance provided by a dashpot

b) Resistance due to laminar flow in a dashpot – viscous friction.

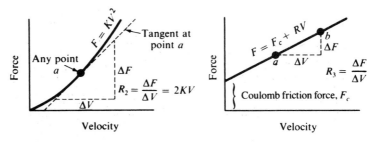

c) Resistance due to turbulent flow in a dashpot.

d) Resistance due to both viscous friction and coulomb friction.

◆ **Figure 3.8** Examples of mechanical resistance.

ton rod is attached to the moving part (M). When part M moves, the fluid in the cylinder must move through the orifice around the piston from one side to the other. The flow rate of the fluid through the orifice is proportional to the velocity of part M, and a difference of pressure $(p_2 - p_1)$ is required to force the fluid through the orifice. This difference in pressure produces a force that opposes the motion. This force is equal to the difference in pressure times the area of the piston $F = (p_2 - p_1)A$.

If the fluid flow rate through the orifice is small, the flow is laminar and the force is proportional to the velocity (Figure 3.8b). Mechanical resistance that produces a force proportional to the velocity is called *viscous friction*. If the flow rate is large, the flow is turbulent and the force is proportional to the square of the velocity (Figure 3.8c). The friction effects of the seal are neglected in the preceding examples. The seal produces a constant friction force that is independent of velocity. Mechanical resistance that produces a force independent of velocity is called *Coulomb friction*. The combination of Coulomb friction and viscous friction is illustrated in Figure 3.8d.

Pure viscous friction is the simplest to treat mathematically, and Figure 3.8b is usually used as a first approximation to mechanical resistance. The resistance is equal to the force (F) divided by the velocity (v) at any point.[*]

$$R_m = B = \frac{F}{v}, \text{N} \cdot \text{s/m} \tag{3.48}$$

In Figure 3.8c, the resistance is nonlinear, increasing as the velocity increases. At any point on the curve, the resistance can be determined by drawing a line tangent to the curve at that point. The resistance is equal to the slope of this tangent line. The slope is determined by locating two points on the tangent line that are far enough apart to give reasonable accuracy. The slope is equal to the change in force (ΔF) between the two points divided by the change in velocity (Δv) between the same two points.

$$R_m = B = \frac{\Delta F}{\Delta v}, \text{N} \cdot \text{s/m} \tag{3.49}$$

In Figure 3.8d, the resistance is linear, but Coulomb friction prevents the line from passing through the origin. Equation (3.48) cannot be used to determine the resistance. However, Equation (3.49) can be used provided the two points are located approximately as shown in the figure.

EXAMPLE 3.21

A dashpot is used to provide mechanical resistance in a packaging machine. The flow is laminar, so the viscous friction equation shown in Figure 3.8b applies. A test was conducted in which a force of 98 N produced a velocity of 24 m/s. Determine the mechanical resistance (R_m).

Solution

$$R_m = \frac{F}{v} = \frac{98 \text{ N}}{24 \text{ m/s}} = 4.08 \text{ N} \cdot \text{s/m} \qquad \blacklozenge$$

---

[*]Mechanical specs use the letter B for viscous friction. We use R_m to emphasize the analogies among the five types of systems presented in this chapter.

EXAMPLE 3.22

A mechanical system consists of a sliding load (Coulomb friction) and a shock absorber (viscous friction). The force versus velocity curve is shown in Figure 3.8d. The following data were obtained from the system.

Run	Force, F (N)	Velocity, v (m/s)
a	7.1	10.5
b	9.6	15.75

Determine the resistance (R_m) and the Coulomb friction force (F_C). Write an equation for the applied force (F) in terms of the velocity (v).

Solution

$$R_m = B = \frac{\Delta F}{\Delta v} = \frac{F_b - F_a}{v_b - v_a} = \frac{9.6 - 7.1}{15.75 - 10.5} = \frac{2.5\,\text{N}}{5.25\,\text{m/s}} = 0.476$$

$$= 0.476\,\text{N} \cdot \text{s/m}$$

$$F = F_c + 0.476v \quad \text{(from Figure 3.8d)}$$

$$7.1 = F_c + (0.476)(10.5) \quad \text{(from run a)}$$

$$F_c = 7.1 - 5.0 = 2.1\,\text{N}$$

The equation for the applied force is

$$F = 2.1 + 0.476v$$

where F = applied force, N

v = velocity, m/s ◆

Mechanical capacitance is defined as the increase in the displacement of a spring required to make a unit increase in spring force. The SI unit of mechanical capacitance is the newton/meter. The reciprocal of the capacitance is called the spring constant, K. Mechanical capacitance (C_m) is computed by dividing a change in spring displacement (Δx) by the corresponding change in spring force (ΔF).

$$C_m = \frac{\Delta x}{\Delta F} = \frac{1}{K}, \text{m/N} \tag{3.50}$$

Some useful conversion factors:

$$1\,\text{lb/in.} = 175\,\text{N/m}$$

$$1\,\text{lb/ft} = 14.6\,\text{N/m}$$

$$1\,\text{in./lb} = 0.0057\,\text{m/N}$$

$$1\,\text{ft/lb} = 0.0685\,\text{m/N}$$

EXAMPLE 3.23

A spring is used to provide mechanical capacitance in a system. A force of 100 N compresses the spring by 30 cm. Determine the mechanical capacitance and the spring constant.

Solution   Equation (3.50) may be used to determine C_m.

$$C_m = \frac{\Delta x}{\Delta F} = \frac{0.30\,\text{m}}{100\,\text{N}} = 0.003\ \text{m/N}$$

$$K = 1/C_m = 333.3\ \text{N/m}$$ ◆

   Mechanical inertia (mass) is measured in terms of the force required to produce a unit increase in acceleration. It is defined by Newton's law of motion, and the term *mass* is used for the inertia element. Equation (3.51) states Newton's law of motion in terms of the average force (F_{avg}), the mass (m), the change in velocity (Δv), and the interval of time during which the change takes place (Δt).

$$F_{\text{avg}} = m\frac{\Delta v}{\Delta t} \tag{3.51}$$

Equation (3.52) states Newton's law in terms of the instantaneous force (F), the mass (m), and the rate of change of velocity with respect to time (dv/dt).

$$F = m\frac{dv}{dt} \tag{3.52}$$

EXAMPLE   Automobile A has a mass of 1500 kg. Determine the average force required to accelerate A from 0 m/s
3.24   to 27.5 m/s in 6 s. Automobile B requires an average force of 8000 N to accelerate from 0 m/s to
                27.5 m/s in 6 s. Determine the mass of B.

Solution   Equation (3.51) may be used for both problems.

$$F_{\text{avg}} = m\frac{\Delta v}{\Delta t}$$

Automobile A:

$$F_{\text{avg}} = (1500\ \text{kg})\left(\frac{27.5\,\text{m/s}}{6\,\text{s}}\right) = 6875\ \text{N}$$

Automobile B:

$$8000 = m\left(\frac{27.5 - 0}{6}\right)$$

$$m = (8000\ \text{N})\left(\frac{6\,\text{s}}{27.5\,\text{m/s}}\right) = 1745\ \text{kg}$$ ◆

   Mechanical dead time is the time required to transport material from one place to another. A belt conveyor is frequently used to transport solid material in a process (see Figure 3.9). The dead-time delay is the time it takes for the belt to travel from the inlet end to the outlet end. Equation (3.1) may be used to compute the dead-time delay. The effect of the dead time is to delay the input flow rate by the dead-time delay (t_d).

EXAMPLE   (a) A belt conveyor is 30 m long and has a belt speed of 3 m/s. Determine the dead-time delay be-
3.25   tween the input and output ends of the belt.
                (b) Write the equation for the output mass flow rate $f_o(t)$ in terms of the input mass flow rate $f_i(t)$.

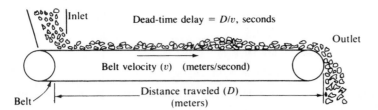

Inlet   Dead-time delay $= D/v$, seconds

Outlet

Belt velocity (v)   (meters/second)

Distance traveled (D)
(meters)

Belt

◆ **Figure 3.9**   A belt conveyor is an example of a mechanical dead-time element.

Solution

(a) Equation (3.1) may be used.

$$t_d = \frac{D}{v} = \frac{30\,\text{m}}{3\,\text{m/s}} = 10\,\text{s}$$

(b) The following equation from Section 3.1 may be used:

$$f_o(t) = f_i(t - t_d)$$
$$= f_i(t - 10)$$

◆

MECHANICAL EQUATIONS

Resistance

$$R_m = B = \frac{F}{v} \tag{3.48}$$

$$R_m = B = \frac{\Delta F}{\Delta v} \tag{3.49}$$

Capacitance

$$C_m = \frac{\Delta x}{\Delta F} = \frac{1}{K} \tag{3.50}$$

Inertia

$$F_{\text{avg}} = m\,\frac{\Delta v}{\Delta t} \tag{3.51}$$

$$F = m\,\frac{dv}{dt} \tag{3.52}$$

Dead-Time Delay

$$t_d = \frac{D}{v} \tag{3.1}$$

where C_m = mechanical capacitance, m/N
D = distance, m
F = force, N
F_{avg} = average force, N
ΔF = change in force, N

$$K = \text{spring constant, N/m}$$
$$m = \text{mass, kg}$$
$$R_m = B = \text{mechanical resistance, N} \cdot \text{s/m}$$
$$t_d = \text{dead-time delay, s}$$
$$\Delta t = \text{increment of time, s}$$
$$v = \text{velocity, m/s}$$
$$\Delta v = \text{change in velocity, m/s}$$
$$dv/dt = \text{rate of change of velocity, m/s}^2$$
$$\Delta x = \text{change in displacement, m}$$

SUMMARY OF EQUATIONS

(*Note*: Each term is followed immediately by its unit enclosed between parentheses.)

Resistance

Electrical:

$$R = \frac{\Delta e\,(\text{V})}{\Delta i\,(\text{A})}, \text{ohm}$$

Liquid flow

$$\text{Reynolds number} = \text{Re} = \frac{\rho(\text{kg/m}^3)v(\text{m/s})d(\text{m})}{\mu(\text{Pa} \cdot \text{s})}$$

Laminar flow resistance (Re < 2000):

$$R_L = \frac{128\mu(\text{Pa} \cdot \text{s})l(\text{m})}{\pi d^4(\text{m})^4}, \text{Pa} \cdot \text{s/m}^3$$

Turbulent flow resistance (Re > 4000):

$$R_T = 2K_T(\text{Pa} \cdot \text{s}^2/\text{m}^6)Q(\text{m}^3/\text{s}), \text{Pa} \cdot \text{s/m}^3$$

$$K_T = \frac{8\rho(\text{kg/m}^3)fl(\text{m})}{\pi^2 d^5(\text{m})^5}, \text{Pa} \cdot \text{s}^2/\text{m}^6$$

Gas flow:

$$R_g = 2K_g(\text{Pa} \cdot \text{s}^2/\text{kg}^2)W(\text{kg/s}), \text{Pa} \cdot \text{s/kg}$$

$$K_g = \frac{8fl(\text{m})}{\pi^2 d^5(\text{m})^5\rho(\text{kg/m}^3)}, \text{Pa} \cdot \text{s}^2/\text{kg}^2$$

Thermal:

$$R_T = \frac{1}{A(\text{m}^2)}\left(\frac{1}{h_o(\text{W/m}^2 \cdot \text{K})} + \frac{x_1(\text{m})}{k_1(\text{W/m} \cdot \text{K})} + \cdots\right.$$
$$\left. + \frac{x_n(\text{m})}{k_n(\text{W/m} \cdot \text{K})} + \frac{1}{h_i(\text{W/m}^2 \cdot \text{K})}\right), \text{K/W}$$

Mechanical:

$$R = \frac{\Delta F(\text{N})}{\Delta v(\text{m/s})}, \text{N} \cdot \text{s/m}$$

Capacitance

Electrical:

$$C = \frac{\Delta q(\text{C})}{\Delta e(\text{V})}, \text{F}$$

Liquid flow:

$$C_L = \frac{A(\text{m}^2)}{\rho(\text{kg/m}^3)g(\text{m/s}^2)}, \text{m}^3/\text{Pa}$$

Gas flow:

$$C_g = \frac{M(\text{g/mol})V(\text{m}^3)}{10^3(\text{g/kg})8.314(\text{J/K} \cdot \text{mol})T(\text{K})}, \text{kg/Pa}$$

Thermal:

$$C_T = m(\text{kg})S_h(\text{J/kg} \cdot \text{K}), \text{J/K}$$

Mechanical:

$$C_m = \frac{1}{K} = \frac{\Delta x(\text{m})}{\Delta F(\text{N})}, \text{m/N}$$

Inductance

Electrical:

$$L = \frac{e(\text{V})}{di/dt(\text{A/s})}, \text{H}$$

Inertance

Liquid flow:

$$I_L = \frac{\rho(\text{kg/m}^3)l(\text{m})}{a(\text{m}^2)}, \text{Pa} \cdot \text{s}^2/\text{m}^3$$

Inertia

Mechanical:

$$m = \frac{F(\text{N})}{dv/dt(\text{m/s}^2)}, \text{kg}$$

Dead-Time Delay

Electrical, liquid flow, and mechanical:

$$t_d = \frac{D(\text{m})}{v(\text{m/s})}, \text{s}$$

◆ GLOSSARY

Capacitance: The amount of material, energy, or distance required to make a unit increase in potential. (3.1)

Capacitance, electrical: The quantity of electric charge (q C) required to make an increase in electrical potential of 1 V. (3.2)

Capacitance, gas flow: The increase in the mass of gas in a pressure vessel required to produce an increase in pressure of 1 Pa while the temperature remains constant. (3.4)

Capacitance, liquid: The increase in volume of liquid in a tank required to make an increase in pressure of 1 Pa at the outlet of the tank. (3.3)

Capacitance, mechanical: The increase in displacement of a spring required to make an increase in spring force of 1 N. (3.6)

Capacitance, thermal: The amount of heat required to make an increase in temperature of 1 K (or °C). (3.5)

Coulomb friction: Mechanical resistance that produces a force that is independent of the velocity. (3.6)

Dead time: The time interval between the time a signal appears at the input of a component and the time the corresponding response appears at the output. (3.1)

Dead time, electrical: The delay caused by the time it takes an electrical signal to travel from the source to the destination. (3.2)

Dead time, liquid flow: The time it takes for a liquid to flow from the input end of a pipe to the output end. (3.3)

Dead time, mechanical: The time required to transport material from one place to another place. (3.6)

Fanning equation: An equation that describes turbulent flow in a round pipe. (3.3)

Hagen–Poiseuille law: An equation that describes laminar flow in a round pipe. (3.3)

Inductance, electrical: The voltage required to increase electric current at a rate of 1 A/s. (3.1, 3.2)

Inertance, liquid flow: The amount of pressure drop in a pipe required to increase the flow rate at a rate of 1 m^3/s/s. (3.3)

Inertia: The force required to increase velocity at a rate of 1 m/s/s (i.e., an acceleration of 1 m/s^2). (3.6)

Laminar flow: An orderly type of flow that occurs when the Reynolds number is less than 2000 (low flow velocity). (3.3)

Resistance: An opposition to the movement or flow of material or energy. (3.1)

Resistance, electrical: The increase in voltage required to increase current by 1 A. (3.2)

Resistance, gas flow: The increase in pressure required to produce an increase in gas flow rate of 1 kg/s. (3.4)

Resistance, liquid flow: The increase in pressure required to make an increase in liquid flow rate of 1 m^3/s. (3.3)

Resistance, mechanical: The increase in force required to produce an increase in velocity of 1 m/s. (3.6)

Resistance, thermal: The increase in temperature required to produce an increase in heat flow rate of 1 W. (3.5)

Reynolds number: A dimensionless number used to predict the type of fluid flow (laminar or turbulent). See Equation (3.10). (3.3)

Turbulent flow: A disorderly type of flow that occurs when the Reynolds number is greater than 4000 (high flow velocity). (3.3)

Viscous friction: Mechanical resistance that produces a force that is proportional to the velocity. (3.6)

◆ EXERCISES

Section 3.2

3.1 Ten components were tested to determine their electrical resistance by applying a voltage to each component and measuring the resulting current. The voltage on each component was slowly increased until the product of the voltage times the current was equal to 0.25 (the power dissipated in a component is equal to this product, so each component was dissipating $\frac{1}{4}$ watt when the measurement was made). Use Equation (3.3) to determine the static resistance of each component for the following voltage and current values.

(a)	1.04 V,	0.240 A
(b)	3.02 V,	0.0828 A
(c)	4.06 V,	0.0616 A
(d)	5.00 V,	0.0500 A
(e)	8.71 V,	0.0287 A
(f)	13.4 V,	0.0187 A
(g)	16.5 V,	0.0152 A
(h)	32.7 V,	0.00765 A
(i)	42.5 V,	0.00588 A
(j)	51.2 V,	0.00488 A

3.2 The ten components in Exercise 3.1 were tested a second time with the current reduced by 50%. Use Equation (3.3) to determine the static resistance of each component using the reduced voltage and current values listed below. Compare the reduced static resistance values with those from Exercise 3.1 and label each component as linear or nonlinear (a component is linear only if the two resistance values are equal).

(a)	0.52 V,	0.120 A
(b)	1.28 V,	0.0414 A
(c)	2.03 V,	0.0308 A
(d)	2.19 V,	0.0250 A
(e)	3.67 V,	0.0144 A
(f)	6.70 V,	0.00935 A
(g)	8.25 V,	0.0076 A
(h)	14.2 V,	0.00382 A
(i)	17.6 V,	0.00294 A
(j)	25.6 V,	0.00244 A

3.3 In Exercises 3.1 and 3.2, you computed the static resistance of 10 components using Equation (3.3). When we analyze and design control systems, we use the dynamic resistance at the operating point, not the static resistance. In this exercise, you will compute the dynamic resistances of the 10 components about the operating point values given in Exercise 3.1. Voltage and current values slightly above and below the operating point are listed below. Use Equation (3.4) to determine the dynamic resistance of each component.

	e_1	i_1	e_2	i_2
(a)	0.996 V	0.230 A	1.0825 V	0.250 A
(b)	2.831 V	0.0788 A	3.2135 V	0.0868 A
(c)	3.862 V	0.0586 A	4.257 V	0.0646 A
(d)	4.752 V	0.0480 A	5.252 V	0.0520 A
(e)	8.119 V	0.0272 A	9.322 V	0.0302 A
(f)	12.691 V	0.0177 A	14.125 V	0.0197 A
(g)	15.638 V	0.0144 A	17.376 V	0.0160 A
(h)	30.568 V	0.00725 A	34.880 V	0.00805 A
(i)	39.632 V	0.00558 A	45.443 V	0.00618 A
(j)	48.578 V	0.00463 A	53.824 V	0.00513 A

3.4 Compare the static and dynamic resistance values of the 10 components in Exercises 3.1, 3.2, and 3.3. Discuss the relationship between the static and dynamic resistances of linear and nonlinear components.

3.5 The following data were obtained in a test of a nonlinear electrical resistor.

Volts	0	5	10	15	20	25
Amperes	0	0.323	0.626	0.914	1.19	1.45

Plot a volt–ampere graph. Use the tangent line method to determine the resistance at 10 and 20 V.

3.6 An electric current, i, was applied to a capacitor for a duration of t seconds. This current pulse increased the voltage across the capacitor from e_1 to e_2 volts. Determine the capacitance, C, of the capacitor for each of the following values of i, t, e_1 and e_2.

	i	t	e_1	e_2
(a)	0.36 mA	0.25 s	0 V	16 V
(b)	0.22 mA	1.20 s	0 V	240 V
(c)	2.10 mA	0.84 s	50 V	120 V
(d)	1.50 mA	0.48 s	20 V	80 V
(e)	12.4 mA	0.09 s	0 V	90 V

3.7 A voltage pulse with an amplitude of e volts and a duration of t seconds was applied to an inductor. The current through the inductor increased from i_1 to i_2 amperes. Assume the resistance of the inductor was negligible. Determine the inductance, L, of the inductor for each of the following values of e, t, i_1 and i_2.

	e	t	i_1	i_2
(a)	12.6 V	4.0 ms	0 A	1.20 A
(b)	3.80 V	1.25 s	0 A	3.40 A
(c)	2.40 V	0.03 s	100 mA	200 mA
(d)	5.40 V	0.20 s	0.42 A	2.36 A
(e)	6.80 V	2.0 ms	0 A	0.64 A

3.8 Determine the dead-time delay in the transmission of electrical signals for the following values of distance, D, and velocity of propagation, v_p.

(a) 25 km, 2.8×10^8 m/s
(b) 2×10^5 km, 3.0×10^8 m/s
(c) 2000 ft, 2.4×10^8 m/s
(d) 12,000 miles, 2.2×10^8 m/s
(e) 200 m, 2.7×10^8 m/s

Section 3.3

3.9 Determine the Reynolds number, the liquid resistance, and the pressure drop for each of the following condition sets. The following abbreviations are used: tube for smooth tube, pipe for commercial pipe, Lpm for liters per minute, and gpm for gallons per minute.

	Fluid (15°C)	Type	Flow Rate	Inside Diameter	Length
(a)	Turpentine	tube	20 Lpm	3.8 cm	1600 m
(b)	Water	pipe	12 gpm	0.75 in.	100 ft
(c)	Ethyl alcohol	pipe	25 Lpm	1.9 cm	15 m
(d)	Gasoline	tube	36 gpm	1.25 in.	1200 ft
(e)	Glycerin	tube	2 gpm	1 in.	25 ft
(f)	Oil	pipe	10 Lpm	2.5 cm	30 m
(g)	Kerosene	tube	50 gpm	2 in.	5500 ft
(h)	Water	tube	36 Lpm	1.9 cm	26 m
(i)	Ethyl alcohol	pipe	12 gpm	0.625 in.	75 ft
(j)	Gasoline	pipe	11 Lpm	2.54 cm	16,000 m

3.10 Determine the liquid capacitance for each of the following condition sets.

	Fluid (15°C)	Tank Diameter	Tank Height
(a)	Water	0.5 m	1.5 m
(b)	Kerosene	0.8 m	1.6 m
(c)	Ethyl alcohol	0.6 m	0.8 m
(d)	Turpentine	1.0 m	1.2 m
(e)	Oil	10.0 m	5.0 m
(f)	Gasoline	1.2 m	2.0 m

3.11 Determine the liquid inertance for each of the following condition sets.

	Fluid (15°C)	Pipe Diameter	Pipe Length
(a)	Water	1.27 cm	3.66 m
(b)	Kerosene	3.175 cm	6.1 m
(c)	Mercury	0.635 cm	0.25 m
(d)	Turpentine	1.905 cm	10.0 m
(e)	Oil	15.24 cm	1600 m
(f)	Gasoline	2.54 cm	40 m

3.12 Determine the dead-time delay for each of the following liquid flow condition sets.

	Flow Rate	Pipe Diameter	Pipe Length
(a)	6.33E − 4 m³/s	3.175 cm	15 m
(b)	9.12E − 4 m³/s	2.540 cm	20 m
(c)	3.18E − 4 m³/s	1.590 cm	8 m
(d)	17.10 Lpm	1.905 cm	4 m
(e)	5.12 Lpm	0.952 cm	5 m
(f)	10.64 Lpm	1.270 cm	10 m

Section 3.4

3.13 The following data sets were obtained for gas flow systems. Use Equation (3.26) to determine the value of the turbulent flow coefficient, K_g, for each data set.

	p_1 (kPa)	p_2 (kPa)	W (kg/s)
(a)	106	101	0.054
(b)	112	102	0.0049
(c)	108	104	0.031
(d)	96	88	0.012
(e)	102	96	0.12
(f)	80	73	0.023

3.14 Determine the gas flow resistance, R_g, for each data set in Exercise 3.13. Use Equation (3.27) and the values of K_g you obtained in that exercise.

3.15 Determine new values of p_2 for the data sets given in Exercise 3.13 with the values of W replaced by the new values given below. Use your computed values of K_g and the values of p_1 from Exercise 3.13.

(a) 0.032 kg/s **(b)** 0.028 kg/s **(c)** 0.016 kg/s
(d) 0.0074 kg/s **(e)** 0.071 kg/s **(f)** 0.014 kg/s

3.16 Use Equation (3.26) to compute the turbulent flow coefficient, K_g, of a smooth tube flow meter from the following data.

$$p_1 = 102 \text{ kPa} \qquad p_2 = 92 \text{ kPa} \qquad W = 0.04 \text{ kg/s}$$

Then use your computed value of K_g to determine values of pressure drop, p, for the following gas flow rates, W_g: 0, 0.008, 0.016, 0.024, 0.032, and 0.04 kg/s.

Finally, plot your results in a calibration graph of mass flow rate (x-axis) vs pressure drop (y-axis).

3.17 The following data were obtained for a gas flow system. Plot the data and graphically determine the gas flow resistance, R_g, at flow rates of 0.002 and 0.004 kg/s. Check your results by using Equation (3.26) to compute K_g and Equation (3.27) to compute R_g.

W= gas flow rate, kg/s
p = pressure drop, kPa

W	0	0.001	0.002	0.003	0.004	0.005
p	0	0.4	1.6	3.6	6.4	10.0

3.18 In each data set below, a smooth tube is supplying the specified gas with an inlet pressure, p_1, of 120 kPa. Determine the pressure drop, p, gas flow resistance, R_g, and outlet pressure, p_2, for each data set.

	Gas (15° C)	Tube I.D. (cm)	Tube Length (m)	Flow Rate (kg/s)
(a)	Nitrogen	4.50	25.0	0.042
(b)	Carbon dioxide	2.54	8.0	0.032
(c)	Air	3.18	6.0	0.048
(d)	Helium	1.91	5.3	0.0055

3.19 Determine the gas flow capacitance for each of the following data sets.

	Gas	Temperature (°C)	Tank Volume (m³)
(a)	Nitrogen	40	1.4
(b)	Carbon dioxide	20	2.5
(c)	Air	50	3.5
(d)	Helium	30	0.8

Section 3.5

3.20 Determine the unit thermal resistance, the total thermal resistance, the total heat flow, and the direction of the heat flow for each of the following composite wall sections (see Figure 3.7).
(a) Vertical walls:
 $A = 35.7 \text{ m}^2$ $T_i = 21°C$ $T_o = -5°C$
 Three inner layers:
  (1) Wood (pine), 1.91 cm thick
  (2) Insulation, 8.89 cm thick
  (3) Wood (oak), 0.64 cm thick
 Inside film: Still air, vertical surface, estimated $T_d = 4°C$
 Outside film: Moving air, $v = 7$ m/s
(b) Ceiling:
 $A = 13.4 \text{ m}^2$ $T_i = 23°C$ $T_o = -5°C$
 Four inner layers:
  (1) Asphalt, 0.32 cm thick
  (2) Wood (pine), 1.27 cm thick
  (3) Insulation, 15.24 cm thick
  (4) Wood (pine), 1.27 cm thick
 Inside film: Still air, horizontal surface facing down, estimated $T_d = 5°C$
 Outside film: Moving air, $v = 2.75$ m/s
(c) Floor:
 $A = 13.4 \text{ m}^2$ $T_i = 20°C$ $T_o = 10°C$
 Three inner layers:
  (1) Wood (oak), 0.95 cm thick
  (2) Wood (pine), 1.27 cm thick
  (3) Insulation, 10.16 cm thick
 Inside film: Still air, horizontal surface facing up, estimated $T_d = 3°C$
 Outside film: Still air, horizontal surface facing down, estimated $T_d = 6°C$
(d) Water-cooled oil tank:
 $A = 1.8 \text{ m}^2$ $T_i = 95°C$ $T_o = 75°C$
 One inner layer: Aluminum, 0.64 cm thick
 Inside film: Still oil, estimated $T_d = 19°C$
 Outside film: Still water, estimated $T_d = 1°C$
(e) Oil-heated water tank:
 $A = 1.4 \text{ m}^2$ $T_i = 20°C$ $T_o = 90°C$

One inner layer: Brass, 0.25 cm thick
Inside film: Still water, estimated $T_d = 4.5°C$
Outside film: Still oil, estimated $T_d = 16°C$

(f) Water flowing in a straight pipe:

$A = 0.29$ m² $T_i = 10°C$ $T_o = 25°C$
One inner layer: Copper, 0.2 cm thick
Inside film: Turbulent water flow, $T_w = 10°C$, $d = 1.91$ cm, $v = 1.6$ m/s
Outside film: Still air, vertical surface, estimated $T_d = 6°C$

3.21 Cold water flows through a steel pipe that is exposed to warm air. Your job is to estimate how much heat flows through the pipe from the air to the water. The total surface area of the pipe is $\pi d_m L$, where d_m is the pipe mean diameter in meters and L is the pipe length in meters.

$d_m = 1.86$ cm $L = 20$ m $T_i = 5°C$ $T_o = 20°C$

One inner layer: Steel, 0.3 cm thick
Inside film: Turbulent water flow in straight pipes, $T_w = 5°C$, $v_w = 1.8$ m/s
Outside film: Still air, vertical surface

   Estimate the value of T_d for the outside film and try several values for T_d between 1 and 15°C to see how sensitive your estimate is.

3.22 Determine the thermal capacitance for each of the following data sets.

(a) Ethyl alcohol, $V = 1.8$ m³
(b) Gasoline, $V = 0.2$ m³
(c) Water, $V = 50$ gallons
(d) Mercury, $V = 0.01$ L
(e) Oil, $V = 0.86$ m³
(f) Aluminum, $m = 4.25$ kg
(g) Steel, $m = 5.95$ kg
(h) Copper, $m = 1.28$ kg
(i) Lead, $m = 3.64$ kg
(j) Wood (pine), $m = 10.64$ kg

Section 3.6

3.23 A technician tested six linear dashpots to determine their mechanical resistances. Coulomb friction was found to be negligible. The force and velocity test results are listed below. Determine the mechanical resistance of each dashpot.

	Force (N)	Velocity (m/s)
(a)	135	1.6
(b)	127	1.1
(c)	143	1.8
(d)	201	2.3
(e)	80	1.5
(f)	152	0.9

3.24 Assume you tested six mechanical systems that combined a sliding load (Coulomb friction) with a dashpot (viscous friction). Your test results are listed below. Determine the mechanical resistance (R_m), the Coulomb friction force (F_c), and the equation for the applied force (F) in terms of the velocity (v).

	F_1 (N)	v_1 (m/s)	F_2 (N)	v_2 (m/s)
(a)	12.0	2.0	31.5	15.0
(b)	21.2	4.2	39.7	12.6
(c)	9.1	3.8	17.7	14.3
(d)	18.6	5.4	34.4	16.7
(e)	23.3	2.9	58.2	13.8
(f)	15.6	4.6	30.8	14.1

3.25 Determine the mechanical capacitance and the spring constant for springs with the following test results.

	Force (N)	Compression (cm)
(a)	1500	8.5
(b)	2100	10.2
(c)	960	7.6
(d)	3200	14.3
(e)	1810	12.9
(f)	1670	9.6

3.26 The lift portion of a mechanical cam has two sections, A and B. Section A accelerates the load (m, kg) at a constant rate from 0 m/s to v m/s in t_A s. Section B decelerates the load at a constant rate from v m/s to 0 m/s in t_B s. Determine the inertial forces in A and B for each of the following condition sets.

	m (kg)	v (m/s)	t_A (s)	t_B (s)
(a)	2.9	0.95	0.32	0.22
(b)	1.8	1.20	0.28	0.33
(c)	3.7	0.83	0.44	0.18
(d)	4.1	1.13	0.52	0.29
(e)	2.4	1.80	0.21	0.38
(f)	1.7	0.76	0.39	0.55

3.27 Combine equations (3.22), (3.23), and (3.24) into Equation (3.21) to derive Equation (3.25).

3.28 The braking distance from 60 miles per hour (mph) is one measure of the brakes of an automobile. In this test, the brakes are applied while the car is traveling at a steady 60 mph. The braking distance is the distance from the point where the brakes were first applied to the point where the car comes to a complete stop. The average acceleration during braking is given by the following equation:

$$a = -\frac{V_o^2}{2D_b}$$

where V_o = initial velocity of the car

D_b = braking distance

Complete the following for a 2600-lb car that has a braking distance of 160 ft.

(a) Convert the initial velocity (60 mph) to feet per second and then to meters per second. Convert the mass (2600 lbm) to kilograms and the braking dis-

tance (160 ft) to meters. (The conversion factors are in Appendix B.)

(b) Compute the average acceleration in meters/second^2 and then use $f = ma$ to compute the average braking force in newtons.

(c) Convert the average braking force from newtons to pounds force (lbf).

(d) Compute the average acceleration in feet/second^2. Use the engineering fps system units (see Appendix B) to compute the average braking force in pounds. The fps system uses pounds to measure force (lbf) and pounds to measure mass (lbm). In this system, the relationship between force, mass, and acceleration is given by the following equation:

$$f = ma/g_c$$

The factor g_c is the acceleration due to gravity. The nominal value of g_c is 32.2.

(e) Compare the average braking force obtained in (d) with the converted result obtained in (c).

◆ CHAPTER 4

Laplace Transforms and Transfer Functions

◆ OBJECTIVES

Laplace transforms are used to (1) describe control systems conveniently, (2) facilitate the analysis of control systems, and (3) facilitate the design of control systems.

Control system components are described mathematically by an equation that establishes the time relationship between the input and output signals of the component. These equations are functions of time and often include derivative and/or integral terms. The Laplace transform converts these integral/differential equations into algebraic equations that are functions of frequency. These frequency-domain equations also determine a relationship between the input and output signals, namely, the frequency response of the component. When the algebraic equation is solved for the ratio of the output signal over the input signal, the result is called the transfer function. With the transfer function, we can write a computer program to compute the frequency response of the component.

The purpose of this chapter is to give you the ability to use the Laplace transform to obtain the transfer function of a component and to use the transfer function to obtain the Bode diagram (frequency response) of the component. After completing this chapter, you will be able to

1. Use Laplace transform tables to determine
   a. Laplace transform of a function of time
   b. Transfer function of a component
   c. Inverse Laplace transform
2. Use the transfer function to determine
   a. Gain and phase at a given frequency
   b. Frequency-response data (using the program BODE)
   c. Bode diagram
3. Describe how to determine frequency response by
   a. Direct measurement
   b. Hand computation
   c. Computer computation
   d. Graphical approximation

4.1 INTRODUCTION

In this chapter you will study some of the mathematical methods that control engineers use to understand, analyze, and design control systems. Your study begins with the development of the mathematical equations that describe several simple components. You will learn that the relationship between the input and the output of these simple components involves derivative and integral terms. In fact, any component that has capacitance, inertia, inertance, or inductance will require derivative and/or integral terms to describe the relationship between its input and its output. Does this mean that you must use calculus to study such components? What about the control systems formed by these components, do we need calculus to study them as well? Not necessarily. In fact, you may be amazed at how much you can learn about control systems using algebraic methods and complex numbers. You will be able to answer questions such as How does this component respond to inputs of different frequencies? or What is the frequency limit of this follow-up control system? or How can we design this control system to have the fastest possible response without being unstable?

How do we convert equations that involve derivatives and integrals into equations that use only algebraic terms? That is the job of the *Laplace transformation*. It takes *integro-differential equations* (i.e., equations that contain derivative and integral terms) and transforms them into equations that have only algebraic terms. In the process, the Laplace transformation transforms the equations from functions of time to functions of frequency, but let's worry about that later. The Laplace transformation is only one of a number of useful transformations. For example, the logarithmic transformation transforms numbers into their logarithms so that we can use the simpler methods of addition and subtraction to multiply and divide. The Laplace transformation transforms integral and differential functions of time into algebraic functions of frequency so that we can use simpler methods of algebra instead of operational calculus. Our treatment of the Laplace transformation stresses application rather than theory, use rather than development. Tables of the most often used transformations are presented and methods of using the tables are explained.

In the logarithmic transformation, we transform two numbers to their logarithms, add the logarithms, and then inverse transform the sum to obtain the product of the two numbers. In the Laplace transformation, we transform differential equations into algebraic equations, solve the algebraic equations for the output as a function of frequency, and then do an inverse Laplace transformation to obtain the output as a function of time. If it sounds complicated, don't be overly concerned. Most control system analysis and design avoids the inverse transformation and works directly with the algebraic equations. All the questions posed above can be answered without using the inverse Laplace transformation. Then why study the inverse Laplace transformation? Because it gives us a better understanding of the Laplace transformation and helps explain the Nyquist stability criterion in Chapter 15.

In Chapter 1, you learned about the transfer function as a means of describing the relationship between the input and the output of a component. Here you will learn how to obtain the transfer function of a component. It is really quite simple. Perform the Laplace transformation on the integro-differential equation that describes the input/output relationship of the component. Solve the transformed equation algebraically for the ratio of the output over the input, and you have the transfer function of the component. The fascinating thing about this subject is the many facets it has, each giving us a different perspective of the subject. One very interesting facet is coming up next.

One of the most useful things we can learn about a component is how it responds to sinusoidal inputs of different frequencies. More specifically, how do the *gain* and *phase difference* of the component vary with the frequency of the input signal? You may recall from Chapter 1 that gain is the ratio of the output amplitude over the input amplitude, and phase difference is the phase of the output minus the phase of the input. We plot gain and phase information on a pair of graphs and call them the *frequency response* graphs of the component. One graph plots the gain versus frequency, and the other plots the phase difference versus frequency. Here is the fascinating part. The transfer function of a component is a function of a frequency parameter represented by the letter s. Let's say the input to the component is sinusoidal with a radian frequency of ω radians/second (rad/s), and j is the imaginary operator (i.e., the square root of -1). If we replace the s parameter in the transfer function with $j\omega$, the result is a complex number with magnitude M and angle θ, (i.e., $M \underline{/\theta}$). The magnitude of this complex number, M, is the gain of the component, and the angle, θ, is the phase difference of the component. Given the transfer function of a component, we can use a calculator to compute frequency-response data, or we can program a computer to construct the frequency-response graphs of the component. In Section 16.5, you will learn to use DESIGN, a computer program that greatly facilitates the classical frequency-response method of control system design. DESIGN uses the transfer functions of the control system components to compute frequency-response data and constructs frequency-response graphs that provide great insight into the design process and the performance of the control system.

4.2 INPUT/OUTPUT RELATIONSHIPS

In this section we develop the mathematical equations that describe the relationship between the input and the output of five simple control system components. The five components are a self-regulating liquid tank, a nonregulating liquid tank, an electrical RC circuit, a liquid-filled thermometer, and a process control valve.

Self-Regulating Liquid Tank

Consider the self-regulating liquid tank in Figure 4.1. The liquid level in the tank remains constant when the inflow rate (q_{in}) is equal to the outflow rate (q_{out}). If the inflow rate is greater than the outflow rate, the liquid level will rise. If the inflow rate is less than the outflow rate, the level will fall. During a certain time interval (Δt), the amount of liquid in the tank will change by an amount (ΔV) equal to the average difference between the inflow rate and the outflow rate multiplied by Δt.

$$\Delta V = (q_{in} - q_{out})_{avg} \, \Delta t, \, \text{m}^3 \qquad\qquad \textbf{(4.1)}$$

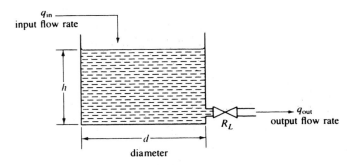

$$\tau\frac{dh}{dt} + h = Gq_{in}$$

q_{in}
input flow rate

h

R_L

q_{out}
output flow rate

d

diameter

h = height of liquid in the tank, meters

◆ **Figure 4.1** A self-regulating liquid tank has a restriction with resistance R_L at the outlet of the tank. If the liquid level in the tank increases, the flow through the restriction also increases. If the level decreases, the flow through the restriction decreases. As long as the tank is not full, the level will automatically adjust until the outflow equals the inflow.

The change in the liquid level in the tank (Δh) is equal to the change in volume (ΔV) divided by the cross-sectional area of the tank (A).

$$\Delta h = \frac{\Delta V}{A} = \frac{(q_{in} - q_{out})_{avg}\,\Delta t}{A}, \text{ m} \tag{4.2}$$

The average rate of change of the level in the tank is equal to the change in level (Δh) divided by the time interval (Δt). For example, if the level changed 0.26 m during a time interval of 100 s, the average rate of change of level would be $0.26/100 = 0.0026$ m/s.

$$\text{Average rate of change of level} = \frac{\Delta h}{\Delta t} = \frac{(q_{in} - q_{out})_{avg}}{A}, \text{ m/s} \tag{4.3}$$

Different time intervals may be used to determine the average rate of change. The average of 0.0026 m/s could have resulted from a change of 0.026 m during a time interval of 10 s or a change of 0.0026 m during a 1-s interval. When the time interval diminishes to 0 s, we call it an instant of time. As the time interval Δt diminishes to an instant of time, the average rate of change becomes the instantaneous rate of change. In mathematics, the instantaneous rate of change of liquid level is called the *derivative* of level (h) with respect to time (t) and is designated by the symbol dh/dt. If the time interval (Δt) in Equation (4.3) diminishes to an instant, the average rate of change of level becomes the instantaneous rate of change (dh/dt).

$$\text{As } \Delta t \to 0, \frac{\Delta h}{\Delta t} \to \frac{dh}{dt} = \frac{q_{in} - q_{out}}{A} \tag{4.4}$$

If the equation for flow out of the tank is linear (laminar flow), the outflow rate (q_{out}) is given by the following equation (see Chapter 3):

$$q_{\text{out}} = \frac{\rho g h}{R_L}, \text{m}^3/\text{s} \tag{4.5}$$

where q_{out} = liquid flow rate, m^3/s
ρ = liquid density, kg/m^3
g = gravitational acceleration, m/s^2
h = liquid level, m
R_L = laminar flow resistance, $\text{Pa} \cdot \text{s/m}^3$

Substituting Equation (4.5) into Equation (4.4) gives us

$$\frac{dh}{dt} = \frac{q_{\text{in}} - \rho g h / R_L}{A}$$

or

$$R_L \left(\frac{A}{\rho g}\right) \frac{dh}{dt} = \left(\frac{R_L}{\rho g}\right) q_{\text{in}} - h$$

The term $(A/\rho g)$ is the capacitance (C_L) of the liquid tank (see Chapter 3), and the entire term $R_L A / \rho g = R_L C_L$ is called the *time constant* (τ) of the liquid tank. The term $R_L/\rho g$ is the steady-state gain (G) of the system. Substituting τ and G into the preceding equation gives us the final form of the differential equation for the liquid tank.

$$\tau \frac{dh}{dt} + h = G q_{\text{in}} \tag{4.6}$$

where

$$\tau = \frac{R_L A}{\rho g}, \text{s*} \quad \text{and} \quad G = \frac{R_L}{\rho g}, \text{s/m}^2$$

Nonregulating Liquid Tank

A nonregulating liquid tank is shown in Figure 4.2. A positive-displacement pump provides a constant output flow rate, q_{out}. Equation (4.2), which was developed for the self-regulating liquid tank, also applies to the nonregulating tank.

$$\Delta h = \frac{\Delta V}{A} = \frac{(q_{\text{in}} - q_{\text{out}})_{\text{avg}} \Delta t}{A},$$

---

*The pascal unit of pressure can cause confusion in resolving the units of τ and G in liquid flow examples. The problem can be resolved by converting the Pascal unit into its equivalent kilogram, meter, and second unit as follows:

$$\text{Pa} = \text{N/m}^2 = (\text{kg} \cdot \text{m/s}^2)/\text{m}^2 = \text{kg/m} \cdot \text{s}^2$$

If we replace Pa by its equivalent $\text{kg/m} \cdot \text{s}^2$, the unit of τ reduces to s and the unit of G reduces to s/m^2.

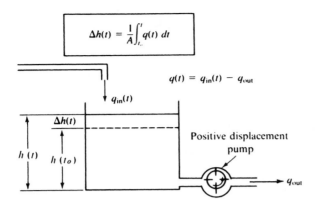

$$q(t) = q_{in}(t) - q_{out}$$

◆ **Figure 4.2** A nonregulating liquid tank has a positive-displacement pump at the outlet of the tank. The liquid level in the tank has no effect on the flow through the pump. If the pump flow rate (q_{out}) does not exactly equal the inflow rate (q_{in}), the level in the tank will change at a rate proportional to the difference, $q = q_{in} - q_{out}$. The level increases when q is positive and decreases when q is negative.

For convenience, we will define $q(t)$ as follows:

$$q(t) = q_{in} - q_{out}$$

In other words, $q(t)$ is the difference between the input flow rate and the output flow rate.

If the time interval, Δt, begins at time t_0 and ends at time t_1, then

$$\Delta t = t_1 - t_0$$

and

$$\Delta h(t) = h(t_1) - h(t_0)$$

The term $(q_{in} - q_{out})_{avg} \Delta t = q_{avg}(t) \Delta t$ represents the accumulation of liquid in the tank over the interval Δt. In calculus, this accumulation of liquid is given by the *integral* from t_0 to t_1 of $q(t)\, dt$.

$$\Delta h(t) = \frac{1}{A} \int_{t_0}^{t_1} q(t)\, dt \qquad (4.7)$$

Electrical Circuit

Consider the electrical circuit of Figure 4.3. If the output terminals are connected to a high-impedance load, then the current through the resistor and the current through the capacitor are essentially equal. Let i represent the current that passes through the resistor and the capacitor. For the resistor, the current (i) is equal to the voltage difference ($e_{in} - e_{out}$) divided by the resistance (R).

$$i = \frac{e_{in} - e_{out}}{R} \qquad (4.8)$$

For the capacitor, the current (i) is equal to the capacitance (C) times the instantaneous rate of change of the capacitor voltage.

$$i = C \frac{de_{out}}{dt} \qquad (4.9)$$

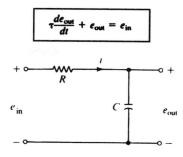

◆ **Figure 4.3**   The electrical RC circuit is sometimes used as a low-pass filter in a control system.

Equating the right-hand side of Equations (4.8) and (4.9) and setting $RC = \tau$ results in the following differential equation for the electrical circuit:

$$\tau \frac{de_{\text{out}}}{dt} + e_{\text{out}} = e_{\text{in}} \tag{4.10}$$

where

$$\tau = RC, \text{ s}$$

Liquid-Filled Thermometer

A liquid-filled thermometer is illustrated in Figure 4.4. The amount of heat (ΔQ) transferred from the fluid surrounding the bulb to the liquid inside the bulb depends on the thermal resistance (R_T) between the two fluids, the difference in temperature ($T_a - T_m$), and the time interval (Δt).

$$\Delta Q = \frac{(T_a - T_m)\, \Delta t}{R_T}$$

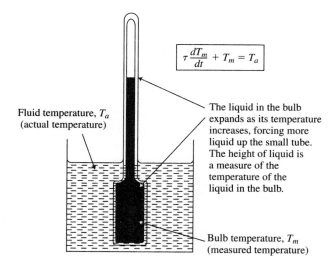

$$\tau \frac{dT_m}{dt} + T_m = T_a$$

Fluid temperature, T_a (actual temperature)

The liquid in the bulb expands as its temperature increases, forcing more liquid up the small tube. The height of liquid is a measure of the temperature of the liquid in the bulb.

Bulb temperature, T_m (measured temperature)

◆ **Figure 4.4**   A liquid-filled thermometer is similar to a liquid-filled primary element for measuring temperature in a control system.

The change in temperature of the liquid in the bulb (ΔT_m) is equal to the amount of heat added (ΔQ) divided by the thermal capacitance (C_T) of the liquid inside the bulb.

$$\Delta T_m = \frac{\Delta Q}{C_T} = \frac{(T_a - T_m)\,\Delta t}{R_T C_T}$$

Dividing both sides by Δt gives us

$$\frac{\Delta T_m}{\Delta t} = \frac{T_a - T_m}{R_T C_T}$$

or, as $\Delta t \to 0$,

$$\tau \frac{dT_m}{dt} + T_m = T_a \tag{4.11}$$

where $\tau = R_T C_T$ = time constant, s

R_T = thermal resistance between the fluid in the container and the liquid inside the bulb, K/W

C_T = thermal capacitance of the liquid in the bulb, J/K

T_a = temperature of the liquid in the container, °C or K

T_m = temperature of the liquid in the bulb, °C or K

Process Control Valve

A process control valve is illustrated in Figure 4.5. The inlet signal is air pressure that enters the diaphragm chamber through the hole at the top of the valve. The air pressure applies a downward force on the diaphragm that is equal to the air pressure (p_{in}) multiplied by the area of the diaphragm (A):

$$\text{Applied force} = F_a = p_{in} A \tag{4.12}$$

This applied force is opposed by three reaction forces: the inertial force of the moving mass (F_I), the resistive force of the seal (F_R), and the compressive force of the spring (F_C):

$$F_a = F_I + F_R + F_C \tag{4.13}$$

The inertial force is equal to the mass of the moving parts (valve stem and diaphragm plate) multiplied by the acceleration of the moving mass. If x is the position of the moving mass, the acceleration is expressed mathematically by the second derivative of x with respect to time. The acceleration (or second derivative) and the inertial force are written as follows:

$$\text{Acceleration} = \frac{d^2 x}{dt^2}$$

$$\tag{4.14}$$

$$\text{Inertial force} = F_I = m\frac{d^2 x}{dt^2}$$

$$p_{in}A = m\frac{d^2x}{dt^2} + R_m\frac{dx}{dt} + \frac{1}{C_m}x$$

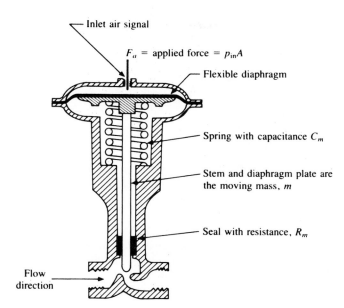

— Inlet air signal

F_a = applied force = $p_{in}A$

— Flexible diaphragm

— Spring with capacitance C_m

— Stem and diaphragm plate are the moving mass, m

— Seal with resistance, R_m

Flow direction

◆ **Figure 4.5**   Process control valve. The inlet air pressure acts on the top of the diaphragm to produce an applied force (F_a). This applied force is opposed by three reaction forces: the inertial force of the moving mass (F_I), the frictional force of the seal (F_R), and the compressive force of the spring (F_C); $F_a = F_I + F_R + F_C$.

We will use the simple viscous friction model for the resistive force, F_R. In this model the friction force is equal to the resistance multiplied by the velocity of the moving mass.

$$F_R = R_m\frac{dx}{dt} \tag{4.15}$$

The spring force is equal to the compression of the spring divided by the mechanical capacitance value of the spring (C_m). If we measure the position of the moving valve stem (x) such that the spring force is 0 when x is 0, the equation for the spring force is

$$F_C = \frac{x}{C_m} \tag{4.16}$$

Substituting Equations (4.12), (4.14), (4.15), and (4.16) into Equation (4.13) gives us the desired mathematical equation for the control valve:

$$p_{in}A = m\frac{d^2x}{dt^2} + R_m\frac{dx}{dt} + \frac{1}{C_m}x \tag{4.17}$$

Equation (4.17) is called a second-order differential equation. Equations (4.6), (4.10), and (4.11) are called first-order differential equations. First- and second-order differential equations have applications in many types of engineering systems.

As the previous examples illustrate, many components, systems, and processes are modeled by some type of differential equation. At this point, a brief discussion of differential equation terminology will be helpful.

A *differential equation* is an equation that has one or more derivatives of the dependent variable with respect to the independent variable [e.g., dh/dt in Equation (4.6)]. A *solution* of a differential equation is an expression for the dependent variable in terms of the independent variable that satisfies the differential equation. As an example, consider the following differential equation and its solution. The solution assumes that GK is a constant and $h = 0$ when $t = 0$ (the value of h at $t = 0$ is called an *initial condition* of the differential equation).

Differential equation:

$$\tau \frac{dh}{dt} + h = GK$$

Solution:

$$h = GK(1 - e^{-t/\tau})$$

The *order* of a differential equation is the order of the highest derivative that appears in the equation. Thus Equations (4.6), (4.10), and (4.11) are first-order differential equations, and Equation (4.17) is a second-order differential equation. Notice the remarkable similarity of the three first-order differential equations, Equations (4.6), (4.10), and (4.11), even though they describe three very different systems. This means these three dissimilar systems will have similar input-output relationships. By knowing the behavior of the *RC* circuit in Figure 4.3, we can predict the behavior of the thermometer in Figure 4.4 and the self-regulating liquid tank in Figure 4.1. This observation is very useful in the analysis of a control system.

A *linear* differential equation has only first-degree terms in the dependent variable and its derivatives. Thus Equations (4.6), (4.10), and (4.17) are all linear differential equations. If any term in the equation is raised to a power other than 1, the equation is nonlinear.

A linear differential equation of second order can be written in the following standard form:

$$\frac{d^2x}{dt^2} + A(t)\frac{dx}{dt} + B(t)x = F(t)$$

If $F(t) = 0$, the equation is called *homogeneous;* otherwise, it is called *nonhomogeneous.* If $A(t)$ and $B(t)$ are constants, the equation is said to have constant coefficients. Many components of a control system are modeled by linear differential equations with constant coefficients. The Laplace transform is a tool both for solving these differential equations and for converting the equations into transfer functions. Once the conversion to transfer functions is made, the operations of calculus are replaced by algebraic operations.

Often the equation that models a component includes both derivative and integral terms. Equations that have at least one integral term and one derivative term are called *integro-differential equations.* The Laplace transform method also applies to integro-differential equations.

SUMMARY OF INPUT/OUTPUT EQUATIONS

Self-Regulating Liquid Tank

$$\tau \frac{dh}{dt} + h = Gq_{\text{in}} \tag{4.6}$$

where
$$\tau = \frac{R_L A}{\rho g}, \text{ s} \quad \text{and} \quad G = \frac{R_L}{\rho g}, \text{ s/m}^2$$

Nonregulating Liquid Tank

$$\Delta h(t) = \frac{1}{A} \int_{t_0}^{t_1} q(t)\,dt \tag{4.7}$$

Electrical Circuit

$$\tau \frac{de_{\text{out}}}{dt} + e_{\text{out}} = e_{\text{in}} \tag{4.10}$$

where
$$\tau = RC, \text{ s}$$

Liquid-Filled Thermometer

$$\tau \frac{dT_m}{dt} + T_m = T_a \tag{4.11}$$

where
$$\tau = R_T C_T, \text{ s}$$

Process Control Valve

$$P_{\text{in}} A = m \frac{d^2 x}{dt^2} + R_m \frac{dx}{dt} + \frac{1}{C_m} x \tag{4.17}$$

where A = cross-sectional area of tank or control valve diaphragm, m^2
  C = electrical capacitance, F
  C_m = mechanical capacitance, m/N
  C_T = thermal capacitance, J/K
  e_{in} = input voltage, V
  e_{out} = output voltage, V
  g = gravitational acceleration, m/s^2
  h = liquid level, m
  m = moving mass, kg
  q = $q_{\text{in}} - q_{\text{out}}$, m^3/s
  q_{in} = input flow rate, m^3/s
  q_{out} = output flow rate, m^3/s
  P_{in} = inlet air pressure, Pa
  R = electrical resistance, Ω
  R_L = laminar flow resistance, Pa · s/m^3
  R_m = mechanical resistance, N · s/m
  R_T = thermal resistance, K/W
  T_a = temperature in container, °C or K
  T_m = temperature in bulb, °C or K
  x = control valve stem position, m
  ρ = liquid density, kg/m^3
  τ = time constant, s

**EXAMPLE
4.1**

The self-regulating liquid tank shown in Figure 4.1 has the following parameter values:

$$R_L = 5.62 \times 10^5 \text{ Pa} \cdot \text{s/m}^3$$
$$A = 1.85 \text{ m}^2$$
$$\rho = 880 \text{ kg/m}^3$$
$$g = 9.81 \text{ m/s}^2$$

Determine the numerical value of the coefficients of dh/dt, h, and q_{in} in Equation 4.6 and write the equation using the numerical values.

Solution

$$\tau = \frac{R_L A}{\rho g} = \frac{(5.62\text{E}5 \text{ Pa} \cdot \text{s/m}^3)(1.85 \text{ m}^2)}{(880 \text{ kg/m}^3)(9.81 \text{ m/s}^2)} = 120 \text{ s}$$

$$G = \frac{R_L}{\rho g} = \frac{(5.62\text{E}5 \text{ Pa} \cdot \text{s/m}^3)}{(880 \text{ kg/m}^3)(9.81 \text{ m/s}^2)} = 65.1 \text{ s/m}^2$$

$$120\frac{dh}{dt} + h = 65.1 \, q_{in}$$

◆

4.3 LAPLACE TRANSFORMS

In Section 4.2 we learned that even simple control system components are modeled by equations that include derivative and integral terms. The *Laplace transform* allows us to transform these integral/differential equations into simpler algebraic equations. By solving the algebraic equation for the ratio of the output over the input, we can obtain the transfer function of the component. With the transfer function, we can compute the frequency response of the component or program a computer to do the work for us. Computer programs (e.g., Mathematica, Mathlab, MathCAD) are available to assist in the performance of Laplace transforms and the related Fourier transforms. In this text, we will not derive Laplace transforms but will only apply them. Considering the widespread use of transfer functions in the control field, some knowledge of the Laplace transform is a decided asset.

The Laplace transformation takes us from a domain called the *time domain* to another domain called the *frequency domain*. Figure 4.6 shows an overview of the Laplace transformation. In the time domain, the size of the signals is given (or determined) for each instant of time. We say that the input and output signals are functions of time and append a (t) to the variable name to indicate that time is the independent variable. In Figure 4.6, $q_{in}(t)$ represents the time-domain input signal, and $h(t)$ represents the time-domain output signal. The time-domain equation of a component defines the size of the output as a function of time and the input signal. Time-domain equations often include derivative and integral terms. The equations we developed in Section 4.2 are all time-domain equations.

In the frequency domain, the size and the phase angle of the signal are given (or determined) for each value of a complex frequency parameter designated by the letter s (s has units of second^{-1}). We say that the input and output signals are functions of frequency and append an (s) to the variable name to indicate that frequency is the independent variable. In Figure 4.6, $Q_{in}(s)$ is the frequency-domain input signal and $H(s)$ is the frequency-domain output signal. The frequency-domain equation of a component defines the size and phase angle of the

Time domain   Frequency domain

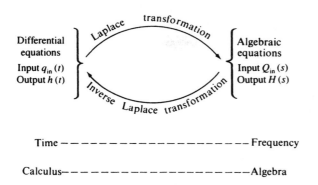

◆ **Figure 4.6** The Laplace transform takes a function from the time domain to the frequency domain. The inverse Laplace transform takes a function from the frequency domain to the time domain. Equations in the time domain determine the size of a control signal at various times, t. Equations in the frequency domain determine the amplitude and phase angle of control signals at various frequencies, s. Operations on equations in the time domain involve calculus. Operations on equations in the frequency domain involve algebra.

output signal as a function of s and the input signal. We use the Laplace transform to obtain the frequency domain equation from the time domain equation.

Lowercase letters are used to represent signals in the time domain, and uppercase letters are used for signals in the frequency domain. In Figure 4.6, for example, the input signal is represented by $q_{in}(t)$ in the time domain and by $Q_{in}(s)$ in the frequency domain. Similarly, the output signal is represented by $h(t)$ in the time domain and by $H(s)$ in the frequency domain.

Two examples will illustrate the need for the frequency domain in the analysis of systems. The first example is human hearing. If you look at an oscilloscope showing the output from a microphone, it is very difficult to distinguish similar-sounding words. Even when this time-domain waveform is digitized and a computer is used to analyze the waveform, it is still difficult to discriminate between similar-sounding words. However, if a fast Fourier transform is applied to the time-domain waveform, the computer can more easily discriminate between words. The cochlea (the spiral-shaped part of the inner ear) performs this function in human hearing. Soldiers who operated artillery in World War I had hearing damage. Autopsies of these soldiers revealed that certain cilia inside the cochlea were destroyed. The hearing damage that a noisy environment can cause is a well-documented fact, and the use of hearing protection is wise and required in most such situations.

The second example is the vibration of a motor. As a motor starts to wear out, the bearing noise increases. By the time a trained observer can hear the noise, the motor is in very bad shape. An oscilloscope view of the waveform is no better at picking up the bearing noise. However, a preventative maintenance procedure is used in which an accelerometer is connected to the motor casing and digital samples are taken with an A/D converter. A fast Fourier transform of the data clearly shows the amplitude of the higher-frequency bearing noise. When this information is available, motors can be repaired before extensive damage occurs.

Figure 4.7 illustrates the solution of a differential equation using the Laplace transformation and the inverse Laplace transformation. The Laplace transform enables us to convert a differential equation from the time domain into an algebraic equation in the frequency domain. The algebraic equation is then solved to get a frequency-domain solution. An inverse transformation converts the frequency-domain solution into a time-domain solution. Of course, we can also obtain the time-domain solution directly, using the methods of operational calculus.

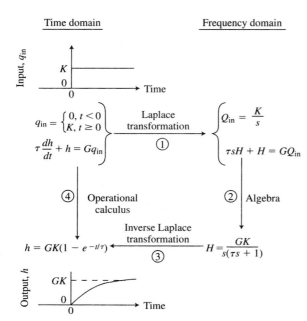

◆ **Figure 4.7** Diagram of the solution of the step response of a self-regulating liquid tank. The Laplace transform solution consists of the following three steps: (1) the Laplace transformation of $q_{in}(t)$ and ($\tau\, dh/dt + h = Gq_{in}$) to frequency domain, (2) the algebraic solution for $H(s)$, and (3) the inverse Laplace transformation of $H(s)$, to time domain $h(t)$. The calculus solution is shown as step 4.

The solution outlined in Figure 4.7 is for a step change in the inlet flow rate of the self-regulating liquid tank shown in Figure 4.1. Equation (4.6) is the time-domain equation for this component.

$$\tau \frac{dh}{dt} + h = Gq_{in}$$

Before time $t = 0$ s, the tank is empty and the input flow rate is zero (i.e., $h = q_{in} = 0$, for $t < 0$ s). At time $t = 0$, the input valve is opened and the input flow rate changes to K m³/s (i.e., $q_{in} = K$, for $t > 0$ s). This type of change is called a *step change* in the input signal (q_{in}). A graph of input (q_{in}) versus time (t) is shown in Figure 4.7. The question is: What is the level of the tank after the step change in input? [That is, $h(t) = $? for $t > 0$.]

Solution *Step 1.* Transform the input (q_{in}) and Equation (4.6) into the frequency domain. A table of Laplace transform pairs was used to make this transformation. The results of the transformation are listed below and in Figure 4.7.

Time Domain	*Frequency Domain*
$q_{in}(t) = K$	$Q_{in}(s) = \dfrac{K}{s}$
$\tau \dfrac{dh(t)}{dt} + h(t) = Gq_{in}$	$\tau sH(s) + H(s) = GQ_{in}(s)$

Step 2. Solve the two frequency-domain equations for H.

$$H(s) = \frac{GK}{s(\tau s + 1)} \tag{4.18}$$

◆ **TABLE 4.1**   Functional Laplace Transform Pairs[a]

Time Domain, f(t), t > 0	Frequency Domain, F(s)
1. K	$\dfrac{K}{s}$
2. Kt	$\dfrac{K}{s^2}$
3. Ke^{-at}	$\dfrac{K}{s + a}$
4. Kte^{-at}	$\dfrac{K}{(s + a)^2}$
5. $K \sin \omega t$	$\dfrac{K\omega}{s^2 + \omega^2}$
6. $K \cos \omega t$	$\dfrac{Ks}{s^2 + \omega^2}$
7. $Ke^{-at} \sin \omega t$	$\dfrac{K\omega}{(s + a)^2 + \omega^2}$
8. $Ke^{-at} \cos \omega t$	$\dfrac{K(s + a)}{(s + a)^2 + \omega^2}$
9. $K(1 - e^{-t/\tau})$	$\dfrac{K}{s(\tau s + 1)}$
10. $2Ke^{-at} \cos(bt + \theta)$	$\dfrac{K\underline{/\theta}}{s + a - jb} + \dfrac{K\underline{/-\theta}}{s + a + jb}$

[a]The letters a, K, and T represent any numerical constants.

Step 3. Use the table of Laplace transforms (Table 4.1) to transform Equation (4.18) into the time domain.

Time Domain	*Frequency Domain*
$h(t) = GK(1 - e^{-t/\tau})$	$H(s) = \dfrac{GK}{s(\tau s + 1)}$

The time-domain graph of $h(t)$ is shown in Figure 4.7.

Functional Laplace Transforms

Laplace transforms can be divided into two types: functional transforms and operational transforms. The first type is simply the Laplace transform of a particular function such as $\sin \omega t$ or e^{-at}. The second involves the transform of the result of some operation, such as the sum of two functions, the derivative of a function, or the integral of a function. In this section we cover *functional transforms*. The notations $\mathcal{L}\{f(t)\}$ and $F(s)$ are both used to indicate the Laplace transform of $f(t)$. The Laplace transform $F(s)$ of the function $f(t)$ is defined by the following relationship:

$$\mathcal{L}\{f(t)\} = F(s) = \int_0^\infty f(t)e^{-st}\, dt$$

As an illustration, we will use the defining equation to determine the Laplace transform of $f(t) = K$, where K is a constant.

$$F(s) = \int_0^\infty f(t)e^{-st}\, dt = \int_0^\infty Ke^{-st}\, dt$$

$$F(s) = K\frac{e^{-st}}{-s}\Big|_0^\infty = -\frac{K}{s}(e^{-s\infty} - e^0)$$

$$F(s) = -\frac{K}{s}(0 - 1) = K/s$$

Table 4.1 gives the transform pairs for the functions most often encountered in the study of control systems. Using the table to find the Laplace transform of a time-domain function is simply a matter of looking up the function and its frequency-domain mate. The table works just as well in reverse to find the inverse Laplace transform of a frequency-domain function. Several examples follow.

EXAMPLE 4.2   Use Table 4.1 to find the Laplace transform of each of the following functions:

(a) $f(t) = 12$
(b) $f(t) = 120 \cos 377t$
(c) $f(t) = 27t$
(d) $f(t) = 5e^{-2.5t}$
(e) $f(t) = 89e^{-2t}\sin 1000t$

Solution

(a) From entry 1 in Table 4.1,

$$F(s) = \frac{12}{s}$$

(b) From entry 6 in Table 4.1,

$$F(s) = \frac{120s}{s^2 + (377)^2}$$

(c) From entry 2 in Table 4.1,

$$F(s) = \frac{27}{s^2}$$

(d) From entry 3 in Table 4.1,

$$F(s) = \frac{5}{s + 2.5}$$

(e) From entry 7 in Table 4.1,

$$F(s) = \frac{89,000}{(s + 2)^2 + (1000)^2}$$

◆

EXAMPLE   Use Table 4.1 to find the inverse Laplace transform of each of the following functions:
4.3

(a) $F(s) = \dfrac{27.5}{s}$

(b) $F(s) = \dfrac{8}{s + 5}$

(c) $F(s) = \dfrac{19.6s}{s^2 + 2500}$

(d) $F(s) = \dfrac{17.4}{2s^2 + 32s + 128}$

(e) $F(s) = \dfrac{350}{(s + 2)^2 + 10{,}000}$

Solution

(a) From entry 1 in Table 4.1,

$f(t) = 27.5$

(b) From entry 3 in Table 4.1,

$f(t) = 8e^{-5t}$

(c) $F(s) = \dfrac{19.6s}{s^2 + (50)^2}$

From entry 6 in Table 4.1,

$f(t) = 19.6 \cos 50t$

(d) $F(s) = \dfrac{8.7}{s^2 + 16s + 64} = \dfrac{8.7}{(s + 8)^2}$

From entry 4 in Table 4.1,

$f(t) = 8.7te^{-8t}$

(e) $F(s) = \dfrac{(3.5)(100)}{(s + 2)^2 + (100)^2}$

From entry 7 in Table 4.1,

$f(t) = 3.5e^{-2t} \sin 100t$ ◆

Operational Laplace Transforms

Operational transforms provide the answer to questions such as: What is the Laplace transform of the sum of two or more functions when you know the transform of each function acting alone? What is the Laplace transform of the derivative or integral of a function? Table 4.2 lists the transform pairs for the most common operational transforms.

In Table 4.2, $f(t)$ can be any function of t, and $F(s)$ is the Laplace transform of $f(t)$. For example, if $f(t) = t$, then $F(s) = 1/s^2$. Entry 12 is of particular importance. It states that the

◆ **TABLE 4.2** Operational Laplace Transforms[a]

Time Domain, $f(t)$, $t > 0$	Frequency Domain, $F(s)$
11. $K(t)$	$KF(s)$
12. $f_1(t) + f_2(t) - f_3(t)$	$F_1(s) + F_2(s) - F_3(s)$
13. $\dfrac{df(t)}{dt}$	$sF(s) - f(0)$
14. $\dfrac{d^2f(t)}{dt^2}$	$s^2F(s) - sf(0) - \dfrac{df(0)}{dt}$
15. $\dfrac{d^3f(t)}{dt^3}$	$s^3F(s) - s^2f(0) - s\dfrac{df(0)}{dt} - \dfrac{d^2f(0)}{dt^2}$
16. $\int f(t)\,dt$	$\dfrac{F(s)}{s}$
17. $f(t - a)$	$e^{-as}F(s)$

[a]K represents any numerical constant.

Laplace transform of the sum of two or more functions is the sum of the transforms of each term taken alone.

Entries 13, 14, and 15 appear more complicated than they are in actual practice. The term $f(0)$, which appears in entries 13, 14, and 15, is the value of the function $f(t)$ when $t = 0$. In a mechanical system where $f(t)$ is the position of an object, $f(0)$ is the position at $t = 0$ (i.e., the initial position). The term $df(0)/dt$, which appears in entries 14 and 15, is the value of the derivative of $f(t)$ when $t = 0$. In the mechanical system, this would be the initial velocity of the object. In entry 15, the term $d^2f(0)/dt^2$ is the initial acceleration of the object. In control systems, *the usual practice is to assume that all initial conditions are zero*. The transfer function is defined as the s-domain ratio of the output over the input with all initial conditions set to zero. When using Laplace transforms to get the transfer function of a component, we *use only the first term in entries 13, 14, and 15.*

EXAMPLE 4.4

Use Tables 4.1 and 4.2 to obtain the Laplace transform of the following function:

$$f(t) = 12.3 + t + 5e^{-4t} + te^{-2t}$$

Solution

By entry 12, we know the Laplace transform of $f(t)$ will be the sum of the Laplace transforms of the four terms. Use entry 1 for the first term, entry 2 for the second term, entry 3 for the third term and entry 4 for the fourth term.

$$F(s) = \frac{12.3}{s} + \frac{1}{s^2} + \frac{5}{s + 4} + \frac{1}{(s + 2)^2}$$

EXAMPLE 4.5

Find the Laplace transform of the following function if all initial conditions are zero:

$$f(t) = 8\frac{d^2x(t)}{dt^2} + 12\frac{dx(t)}{dt} + 7x(t)$$

Solution   First we note that the Laplace transform of $x(t)$ is $X(s)$. By entry 14, the Laplace transform of the first term is

$$\mathcal{L}\left\{8\frac{d^2x(t)}{dt^2}\right\} = 8s^2X(s)$$

By entry 13, the Laplace transform of the second term is

$$\mathcal{L}\left\{12\frac{dx(t)}{dt}\right\} = 12sX(s)$$

By entry 11, the Laplace transform of the third term is

$$\mathcal{L}\{7x(t)\} = 7X(s)$$

By entry 12, the Laplace transform of $f(t)$ is

$$F(s) = 8s^2X(s) + 12sX(s) + 7X(s)$$ ◆

EXAMPLE   Repeat the problem in Example 4.5 with the following initial conditions:
4.6

$$x(0) = 6 \text{ and } \frac{dx(0)}{dt} = -4$$

Solution   By entry 14, the Laplace transform of the first term is

$$\mathcal{L}\left\{8\frac{d^2x(t)}{dt^2}\right\} = 8(s^2X(s) - s(6) - (-4))$$

$$= 8s^2X(s) - 48s + 32$$

By entry 13, the Laplace transform of the second term is

$$\mathcal{L}\left\{12\frac{dx(t)}{dt}\right\} = 12sX(s) - 72$$

The Laplace transform of the third term is unchanged.

$$\mathcal{L}\{7x(t)\} = 7X(s)$$

By entry 12, the Laplace transform of $f(t)$ is

$$F(s) = 8s^2X(s) - 48s + 32 + 12sX(s) - 72 + 7sX(s)$$
$$F(s) = 8s^2X(s) + 12sX(s) + 7X(s) - 48s - 40$$ ◆

EXAMPLE   Find the Laplace transform of the following function if all initial conditions are zero:
4.7

$$f(t) = 5.5\frac{dx(t)}{dt} + 24.2x(t) + 4.8\int x(t)\,dt$$

Solution   By entry 13, the Laplace transform of the first term is

$$\mathcal{L}\left\{5.5\frac{dx(t)}{dt^2}\right\} = 5.5sX(s)$$

By entry 11, the Laplace transform of the second term is

$$\mathcal{L}\{24.2x(t)\} = 24.2X(s)$$

By entry 16, the Laplace transform of the third term is

$$\mathcal{L}\left\{4.8 \int x(t)dt\right\} = \frac{4.8X(s)}{s}$$

By entry 12, the Laplace transform of the $f(t)$ is

$$F(s) = 5.5sX(s) + 24.2X(s) + \frac{4.8X(s)}{s} \qquad ◆$$

EXAMPLE 4.8

Entry 17 in Table 4.2 is used to obtain the Laplace transform of the dead-time delay element. If $f_i(t)$ describes the input to a component that has a dead-time delay of 5 s, then $f_o(t) = f_i(t - 5)$ will describe the output of that component. Entry 17 states that the Laplace transform of the output of a dead-time element, $F_o(s)$ is equal to the Laplace transform of the input, $F_i(s)$, multiplied by the factor e^{-as}. In entry 17, the letter a represents the amount of dead-time delay.

Find the Laplace transform of the output of dead-time delay elements with inputs and outputs as follows:

(a) $f_i(t) = 4t$ and $f_o(t) = 4(t - 6) = f_i(t - 6)$

(b) $f_i(t) = 6e^{-4t}$ and $f_o(t) = 6e^{-4(t-5)} = f_i(t - 5)$

Solution

(a) By entry 2, the Laplace transform of $f_i(t) = 4t$ is

$$F_i(s) = \frac{4}{s^2}$$

By entry 17, the Laplace transform of $f_o(t) = f_i(t - 6)$ is

$$F_o(s) = e^{-6s}\left(\frac{4}{s^2}\right)$$

(b) By entry 3, the Laplace transform of $f_i(t) = 6e^{-4}{}_t$ is

$$F_i(s) = \frac{6}{s + 4}$$

By entry 17, the Laplace transform of $f_o(t) = f_i(t - 5)$ is

$$F_o(s) = e^{-5s}\left(\frac{6}{s + 4}\right)$$

4.4 INVERSE LAPLACE TRANSFORMS

The *inverse Laplace transform* converts a frequency-domain function of (s) into a time-domain function of (t). If the frequency-domain function is in the table of Laplace transform pairs, the inverse transform is the mating time-domain function in the table. This is how the inverse transform was obtained in the example illustrated in Figure 4.7. The frequency-domain equation in Figure 4.7 is given by Equation (4.18):

$$H(s) = \frac{GK}{s(\tau s + 1)}$$

The right-hand side of Equation (4.18) is essentially the same as the frequency-domain function of entry 9. The only difference is the arbitrary constant in the numerator, which is K in entry 9 and GK in Equation (4.18). Since K and GK both represent numeric values, we can simply replace K in the time-domain function of entry 9 by GK and we have the inverse transform of the right-hand side of Equation (4.18).

$$h(t) = GK(1 - e^{-t/\tau}) \tag{4.19}$$

Of course, the inverse Laplace transform is usually more difficult than a simple table conversion. The frequency-domain functions that occur in control system analysis are usually a ratio of two polynomials that have only integer powers of s. The right-hand side of Equation (4.20) is an example of such a function, with both polynomials in factored form.

$$X(s) = \frac{8(s + 3)(s + 8)}{s(s + 2)(s + 4)} \tag{4.20}$$

If we can break the right-hand side of Equation (4.20) into a sum of terms and each term is in a table of Laplace transform pairs, we can get the inverse transform of the equation. The method for breaking a ratio of polynomials into a sum of terms is called a *partial fraction expansion*. For Equation (4.20), the partial fraction expansion takes the following form, where K_1, K_2, and K_3 are undetermined numerical constants. Notice that each term on the right-hand side of Equation (4.21) is in Table 4.1, so we can determine the inverse transform. All that remains is to determine the values of K_1, K_2, and K_3.

$$X(s) = \frac{8(s + 3)(s + 8)}{s(s + 2)(s + 4)} = \frac{K_1}{s} + \frac{K_2}{s + 2} + \frac{K_3}{s + 4} \tag{4.21}$$

In general, there will be a term on the right-hand side for each root of the polynomial in the denominator of the left-hand side. Multiple roots for factors such as $(s + 2)^n$ will have a term for each power of the factor from 1 to n. Equation (4.22) illustrates the expansion of a factor with a multiple root of multiplicity 2.

$$Y(s) = \frac{8(s + 1)}{(s + 2)^2} = \frac{K_1}{s + 2} + \frac{K_2}{(s + 2)^2} \tag{4.22}$$

Complex roots are common, and they always occur in conjugate pairs. The two constants in the numerator of the complex conjugate terms are also complex conjugates. Equation (4.23) illustrates the expansion of a polynomial with complex-conjugate roots. In Equation (4.23), j represents the square root of -1, K is a complex constant, and K^* is the complex conjugate of K.

$$Z(s) = \frac{5.2}{s^2 + 2s + 5} = \frac{K}{(s + 1 - j2)} + \frac{K^*}{(s + 1 + j2)} \tag{4.23}$$

A two-step procedure is used to find each distinct (nonmultiple) root, real or complex. Assume that you wish to evaluate the constant in the second term on the right-hand side of Equation (4.21). The first step in evaluating the constant (K_2) is to multiply both sides of the equation by the factor in the denominator of the second term, $(s + 2)$. The second step is to replace s on both sides of the equation by the root of the factor by which you multiplied in step 1 [i.e., replace s by -2, the root of the term $(s + 2)$]. All other terms on the right-hand side will be zero, leaving just the single constant, K_2. The left-hand side reduces to a numerical value, the value of K_2. This process is repeated for each distinct term on the right-hand side. Example 4.9 completes the partial expansion and inverse transformation of Equation (4.21). The two-step procedure works for distinct complex roots, but the reduction of the left-hand side involves complex numbers and the result is a complex number. Example 4.11 completes the partial expansion and inverse transformation of Equation (4.23).

The process is more involved for multiple roots, and we will limit the discussion to multiple roots with a multiplicity of 2. Example 4.10 completes the expansion and inverse transformation of Equation (4.22).

EXAMPLE 4.9   Complete the partial fraction expansion and inverse Laplace transformation of Equation (4.21).

$$X(s) = \frac{8(s + 3)(s + 8)}{s(s + 2)(s + 4)} = \frac{K_1}{s} + \frac{K_2}{s + 2} + \frac{K_3}{s + 4}$$

Solution

1. $K_1 = \left.\dfrac{8(s + 3)(s + 8)}{(s + 2)(s + 4)}\right|_{s=0} = \dfrac{8(0 + 3)(0 + 8)}{(0 + 2)(0 + 4)} = 24$

2. $K_2 = \left.\dfrac{8(s + 3)(s + 8)}{s(s + 4)}\right|_{s=-2} = \dfrac{8(-2 + 3)(-2 + 8)}{-2(-2 + 4)} = -12$

3. $K_3 = \left.\dfrac{8(s + 3)(s + 8)}{s(s + 2)}\right|_{s=-4} = \dfrac{8(-4 + 3)(-4 + 8)}{-4(-4 + 2)} = -4$

The partial fraction expansion of Equation (4.21) is

$$X(s) = \frac{24}{s} - \frac{12}{s + 2} - \frac{4}{s + 4}$$

The inverse Laplace transformation is easily seen from Table 4.1 to be

$$x(t) = 24 - 12e^{-2t} - 4e^{-4t}$$

◆

EXAMPLE 4.10   Complete the partial fraction expansion and inverse Laplace transformation of Equation (4.22).

$$X(s) = \frac{8(s + 1)}{(s + 2)^2} = \frac{K_1}{s + 2} + \frac{K_2}{(s + 2)^2}$$

Solution   Evaluate K_2 as if it were a distinct root, and then add a third step to evaluate K_1.

1. Multiply both sides of Equation (4.22) by the factor $(s + 2)^2$. Equation (4.24) is the result.

$$8(s + 1) = K_1(s + 2) + K_2 \tag{4.24}$$

2. Replace s by the root of $(s + 2)$ (i.e., -2) and solve for K_2.

$$8(-2 + 1) = K_1(-2 + 2) + K_2$$

$$K_2 = -8$$

3. Substitute K_2 into Equation (4.24) and solve for K_1. You can solve for K_1 algebraically, or you can replace s by any value except the multiple root (-2) and solve for K_1 arithmetically.

Algebraic Solution	*Arithmetic Solution*

$$8(s + 1) = K_1(s + 2) - 8 \qquad\qquad \text{let } s = 0$$

$$K_1 = \frac{8s + 16}{s + 2} \qquad\qquad 8(0 + 1) = K_1(0 + 2) - 8$$

$$= 8 \qquad\qquad 2K_1 = 16$$

$$K_1 = 8$$

The partial expansion of Equation (4.22) is

$$Y(s) = \frac{8}{s + 2} - \frac{8}{(s + 2)^2}$$

The inverse Laplace transformation is obtained from entries 3 and 4 in Table 4.1, and entry 12 in Table 4.2.

$$y(t) = 8e^{-2t} - 8te^{-2t}$$

◆

EXAMPLE 4.11   Complete the partial fraction expansion and inverse Laplace transformation of Equation (4.23).

$$Z(s) = \frac{5.2}{s^2 + 2s + 5} = \frac{K}{s + 1 - j2} + \frac{K^*}{s + 1 + j2}$$

Solution   Because K and K^* are complex conjugates, we will evaluate K and set K^* equal to the conjugate of K.

Note: $s^2 + 2s + 5 = (s + 1 - j2)(s + 1 + j2)$

1. $$\frac{5.2}{s^2 + 2s + 5}(s + 1 - j2) = \frac{5.2}{s + 1 + j2}$$

2. $$K = \frac{5.2}{s + 1 + j2}\Big|_{s = -1 + j2} = \frac{5.2}{-1 + j2 + 1 + j2}$$

$$= \frac{5.2}{j4} = -j1.3 = 1.3\underline{/-90°}$$

$$K^* = 1.3\underline{/90°}$$

The partial fraction expansion is

$$Z(s) = \frac{1.3\underline{/-90°}}{s + 1 - j2} + \frac{1.3\underline{/90°}}{s + 1 + j2}$$

The inverse Laplace transformation is

$$z(t) = 2(1.3)e^{-t}\cos(2t - 90°)$$
$$= 2.6e^{-t}\cos(2t - 90°)$$

4.5 TRANSFER FUNCTION

The transfer function was introduced in Chapter 1 as a means of describing the relationship between the input and the output of a component. It tells us how a component responds to sinusoidal inputs of different frequencies (i.e., the transfer function enables us to compute the frequency response of the component). Consider a sinusoidal input with a frequency of ω rad/s and recall that the transfer function is a function of the frequency parameter, s. If we replace s by $j\omega$, the result is a complex number, $M\underline{/\theta}$. The magnitude of this complex number, M, is the gain of the component, and the angle, θ, is the phase difference of the component. This means we can compute the frequency response of a component and program a computer to construct frequency-response graphs of any component from its transfer function. We can use a computer program to do all the "grunt work" in the classical frequency response method of control system design. With a minimum of effort, we gain all the wonderful insight afforded by this graphical design method.

A major use of the Laplace transformation in the study of control systems is to obtain the transfer function of a component. The *transfer function* is obtained by solving the frequency-domain algebraic equation for the ratio of the output signal over the input signal with all initial conditions set to zero. As an example, consider the liquid system described by Equation (4.6):

$$\tau\frac{dh}{dt} + h = Gq_{in} \tag{4.6}$$

The frequency-domain algebraic equation is obtained by applying the Laplace transformation to each term in Equation (4.6) and solving for the ratio H/Q_{in}. *Note:* It is understood that h is a function of t and H is a function of s even though the (t) and (s) are not used. We will often drop the (t) and (s) for convenience in writing long expressions.

$$\tau sH + H = GQ_{in}$$

$$H(\tau s + 1) = GQ_{in}$$

$$H = \frac{G}{\tau s + 1}Q_{in}$$

$$\frac{H}{Q_{in}} = \frac{G}{1 + \tau s} = \text{transfer function} \tag{4.25}$$

The transfer function is defined by Equation (4.25). Notice that the output (H) can be obtained by multiplying the input (Q_{in}) by the transfer function (H/Q_{in}).

> The transfer function of a component is the quotient of the Laplace transform of the output divided by the Laplace transform of the input.
>
> The Laplace transform of the output of a component is equal to the product of the Laplace transform of the input times the transfer function.

EXAMPLE 4.12   Determine the transfer function of the nonregulating liquid tank shown in Figure 4.2. The time-domain equation is Equation (4.7):

$$\Delta h(t) = \frac{1}{A} \int_{t_o}^{t_1} q(t)\, dt$$

Solution   Use entry 16 in Table 4.2 to transform the right-hand side of Equation (4.7). The term $\Delta h(t)$ will be considered to be the output variable of the component and will be transformed to $\Delta H(s)$.

$$\Delta H(s) = \frac{1}{As} Q(s)$$

$$\frac{\Delta H(s)}{Q(s)} = \frac{1}{As} = \text{transfer function} \tag{4.26}$$

◆

EXAMPLE 4.13   Determine the transfer function of the electrical circuit shown in Figure 4.3. The time-domain equation is Equation (4.10):

$$\tau \frac{de_{\text{out}}}{dt} + e_{\text{out}} = e_{\text{in}}$$

Solution   The Laplace transformation of Equation (4.10) is

$$\tau s E_{\text{out}} + E_{\text{out}} = E_{\text{in}}$$

$$\frac{E_{\text{out}}}{E_{\text{in}}} = \frac{1}{1 + \tau s} = \text{transfer function} \tag{4.27}$$

◆

EXAMPLE 4.14   Determine the transfer function of the liquid-filled thermometer shown in Figure 4.4. The time-domain equation is Equation (4.11):*

$$\tau \frac{dT_m}{dt} + T_m = T_a$$

Solution   The Laplace transformation of Equation (4.11) is

$$\tau s T_m + T_m = T_a$$

$$\frac{T_m}{T_a} = \frac{1}{1 + \tau s} = \text{transfer function} \tag{4.28}$$

◆

---

*In Equation (4.11), an exception was made to the rule of using lowercase letters for time-domain variables. Because t is used to represent time, T is used to represent temperature in both the time and frequency domains.

Notice that Equations (4.27) and (4.28) are the same, even though the first equation describes an electrical circuit and the second describes a thermometer.

EXAMPLE 4.15

Determine the transfer function of the control valve shown in Figure 4.5. The time-domain equation is Equation (4.17):

$$P_{in} A = m\frac{d^2x}{dt^2} + R_m\frac{dx}{dt} + \frac{1}{C_m}x$$

Solution

The Laplace transformation of Equation (4.17) is

$$P_{in} A = ms^2X + R_msX + \frac{1}{C_m}X$$

$$\frac{X}{P_{in}} = \frac{AC_m}{1 + R_mC_ms + mC_ms^2}$$ ◆

EXAMPLE 4.16

A control valve has the following transfer function:

$$H(s) = \frac{X(s)}{I(s)} = \frac{80.8}{s^2 + 2s + 101}$$

where $I(s)$ = Laplace transform of the input current, mA
        $X(s)$ = Laplace transform of the output position, mm

As you will soon discover, this control valve has a very underdamped response. To show the underdamped response, we will determine time-domain response of the output, $x(t)$, to the following step change in the input, $i(t)$.

$$i(t) = 0 \text{ mA}, \qquad t < 0$$

$$= 10 \text{ mA}, \qquad t \geq 0$$

Solution

1. Use entry 1 in Table 4.1 to determine the Laplace transform of the input, $I(s)$.

$$I(s) = 10/s$$

2. Multiply the control valve transfer function, $H(s)$, by the Laplace transform of the input, $I(s)$, to get the Laplace transform of the output, $X(s)$.

$$X(s) = I(s)H(s) = \frac{10}{s}\left(\frac{80.8}{s^2 + 2s + 101}\right)$$

$$X(s) = \frac{808}{s(s^2 + 2s + 101)}$$

3. Do a partial fraction expansion of $X(s)$.

$$X(s) = \frac{808}{s(s^2 + 2s + 101)} = \frac{808}{s(s + 1 - j10)(s + 1 + j10)}$$

$$= \frac{K_1}{s} + \frac{K_2}{(s + 1 - j10)} + \frac{K_2^*}{(s + 1 + j10)}$$

$$K_1 = \frac{808}{s^2 + 2s + 101}\Big|_{s=0} = \frac{808}{101} = 8$$

$$K_2 = \frac{808}{s(s + 1 + j10)}\Big|_{s=-1+j10} = \frac{808}{(-1 + j10)(-1 + j10 + 1 + j10)}$$

$$= \frac{808}{(-1 + j10)(j20)} = \frac{808}{(-200 - j20)} = \frac{808}{201\ \underline{/185.7°}}$$

$$= 4.02\ \underline{/-185.7°}$$

$$K_2^* = 4.02\ \underline{/+185.7°}$$

$$X(s) = \frac{8}{s} + \frac{4.02\ \underline{/-185.7°}}{(s + 1 - j10)} + \frac{4.02\ \underline{/+185.7°}}{(s + 1 + j10)}$$

4. Use entry 1 in Table 4.1 to do the inverse Laplace transform of the first term of $X(s)$ and entry 10 to do the last two terms.

$$x(t) = 8 + 8.04e^{-t}\cos(10t - 185.7°)\ \text{mm}, \qquad t \geq 0$$

5. Plot $i(t)$ and $x(t)$ for $0 \leq t \leq 2$. The following program will generate the necessary data.

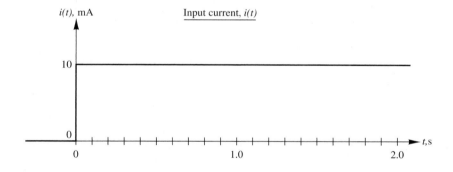

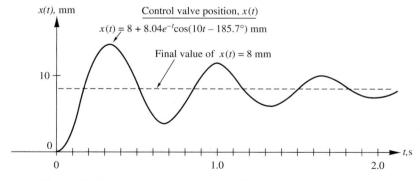

◆ **Figure 4.8**   The control valve in Example 4.16 has a very underdamped response to a step change in the input current.

```
LET Theta = 185.7 / 57.29578
For t = 0 TO 2 STEP .05
  LET x = 8 + 8.04 * EXP (−t) * COS(10 * t − Theta)
  PRINT USING "#.## ##.#"; t, x
NEXT t
```

The plots of $i(t)$ and $x(t)$ are shown in Figure 4.8. ◆

4.6 INITIAL AND FINAL VALUE THEOREMS

When we obtain a mathematical solution to a realistic problem, we should always check the solution to see whether it makes sense. Does the initial value predicted by our solution make sense? Does the final value predicted by our solution make sense? If the answer to these two questions is Yes, then we have an important confirmation of our solution. The initial and final value theorems are very helpful, because they enable us to answer these two questions without completing the time-consuming inverse Laplace transformation of our solution.

One of the criteria of good control is to minimize the residual (or steady-state) error of the control system. When we analyze the step response of a component, we determine the final (or steady state) value of the output and how this final value is reached. In both of these examples, we are interested in determining the value of the response after it has settled down to its "final" value. Mathematically, we express the final value of $x(t)$ as the limit of $x(t)$ as t approaches ∞ or simply $f(\infty)$. The final value theorem enables us to determine $x(\infty)$ from $X(s)$ without completing the inverse Laplace transform of $X(s)$. In a similar manner, the initial value theorem enables us to determine $x(0)$ from $X(s)$ without inverse transforming $X(s)$.

In simple terms, the initial value theorem states that if we multiply $X(s)$ by s and take the limit as $s \rightarrow \infty$, the result will be the initial value of $x(t)$; that is, $x(0)$.

$$\text{Initial value theorem: } x(0) = \lim_{t \to 0} x(t) = \lim_{s \to \infty} sX(s)$$

The final value theorem states that if we multiply $X(s)$ by s and take the limit as $s \rightarrow 0$, the result will be the final value of $x(t)$; that is, $x(\infty)$.

$$\text{Final value theorem: } x(\infty) = \lim_{t \to \infty} x(t) = \lim_{s \to 0} sX(s)$$

EXAMPLE 4.17

In Example 4.16 and Figure 4.8, we computed and illustrated the step response of an underdamped control valve. Use the initial and final value theorems to determine the initial and final positions of the control valve, $x(0)$ and $x(\infty)$.

Solution

$$x(0) = \lim_{s \to \infty} sX(s) = \lim_{s \to \infty} s \left[\frac{808}{s(s^2 + 2s + 101)} \right] = 0 \text{ mm}$$

$$x(\infty) = \lim_{s \to 0} sX(s) = \lim_{s \to 0} s \left[\frac{808}{s(s^2 + 2s + 101)} \right] = \frac{808}{101}$$

$$= 8 \text{ mm}$$

Both the initial and final values of $x(t)$ agree with the graph of $x(t)$ in Figure 4.8. ◆

4.7 FREQUENCY RESPONSE: BODE PLOTS

A feedback control system consists of several components connected by signal paths that form a closed loop. As the signal passes through each component, it is changed in ways that depend on the characteristics of the component. The effect that each component has on the signal is of major importance in the analysis and design of a control system. Indeed, the design of a controller consists of adding components that change the signal in a manner that provides the best possible control of the process.

Two types of input signals are used to describe the behavior of a component. The first type is a step change in the input to the component. Step changes are used to study the transient response of the component. The second type of input is the sinusoidal signal, which is used to study the steady-state response of the component. The response of a component to a sinusoidal input signal is called the frequency response of the component. A graph of the frequency response of a component is called a Bode diagram.

In Section 4.2, we developed the differential equations of five control system components. All five components are classified as linear components. The analysis of a control system is greatly simplified if all the components are linear. Fortunately, many of the systems we want to control can be modeled by linear components. A linear component has a very interesting and important characteristic: A sinusoidal input signal to a linear component will always produce a sinusoidal output signal that has the same frequency as the input. Only the amplitude and the phase angle of the signal will be changed by the component. Figure 4.9 shows typical input and output signals for a linear component.

Gain and Phase Angle

The change in amplitude of a sinusoidal signal is expressed by the ratio of the output amplitude divided by the input amplitude. This ratio is a dimensionless number called the *gain* of the component. Gain is often expressed in decibel (dB) units.

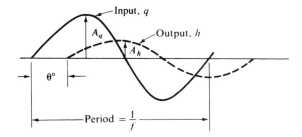

For $s = j\omega = j2\pi f$, the transfer function has a

magnitude (m) and an angle $\theta°$ such that

$m = \dfrac{A_h}{A_q}$ = amplitude ratio (change in size)

$\theta°$ = phase difference (change in timing)

◆ **Figure 4.9**   A linear component can change the amplitude and the phase of a sinusoidal signal, but the frequency of the output is the same as the input. In the diagram, the gain is A_h/A_q, and the phase angle is $-\theta$ (minus because the output lags the input).

$$\text{Gain} = \frac{\text{output amplitude}}{\text{input amplitude}}$$

$$\text{Decibel gain} = 20 \log_{10}(\text{gain}) \quad \text{(dB)}$$

The change in timing of a sinusoidal signal is expressed by the number of degrees by which the phase of the output signal differs from the phase of the input signal. The difference is called the *phase angle* or *phase difference* of the component. The phase angle is usually measured in degrees. It is positive when the output leads the input and negative when the output lags the input.

$$\text{Phase angle} = \text{output phase} - \text{input phase}$$

Bode Diagram

The *frequency response* of a component is the set of values of *gain* and *phase angle* that occur when a sinusoidal input signal is varied over a range of frequencies. For each frequency, there is a gain and a phase angle that give the characteristic response of the component *at that*

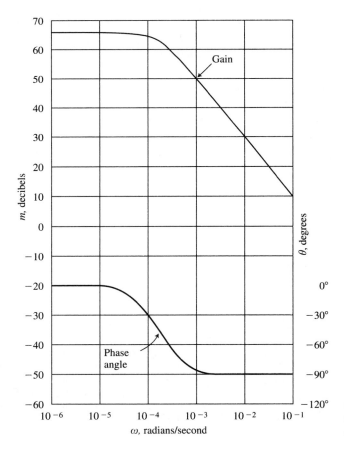

◆ **Figure 4.10** Typical Bode diagram. The frequency scale is logarithmic. The decibel gain and phase angle scales are linear.

frequency. The frequency response is plotted in a pair of graphs known as a *Bode diagram.* The two graphs share a common frequency scale on the *x*-axis. The *y*-axis of one Bode graph is the decibel gain; the other is the phase angle. Figure 4.10 shows a typical Bode diagram.

Determination of the Frequency Response

There are four ways to determine the frequency response of a component: direct measurement, hand computation, computer computation, and straight-line, graphical approximations. In the *direct measurement method,* a sinusoidal input signal is applied to the input of the component. The input and output signals are measured and the gain and phase angle are determined from the measurements. The frequency is then changed to a new value for another determination of gain and phase angle. The process is repeated until enough data are obtained to construct a Bode diagram.

In the *computation methods,* the frequency response is obtained from the transfer function by replacing *s* by *jω*. The *j* represents the square root of -1, and *ω* is the radian frequency of the sinusoidal input signal. For each value of *ω*, the transfer function reduces to a single complex number. The magnitude of this complex number is the gain of the component at that frequency. The angle of the complex number is the phase angle of the component at that frequency.

Complex numbers can be represented by a point on a two-dimensional plane called the *complex plane*. The *x*-axis is called the *real axis,* and the *y*-axis is called the *imaginary axis.* The graph of a complex number is shown in Figure 4.11. The complex number may be defined in rectangular form by its real coordinate (*a*) and its imaginary coordinate (*b*) (i.e., $N = a + jb$, where *j* identifies the imaginary coordinate). In this form, *a* is called the *real part* of the complex number, and *b* is called the *imaginary part.*

The complex number may also be defined in polar form by its distance from the origin (*m*) and the angle (*θ*). In the polar form, *m* is called the *magnitude* of the complex number and *θ* is called the *angle* of the complex number (i.e., $N = m \underline{/\theta}$). A complex number may be converted from one form to the other by using the appropriate conversion equation:

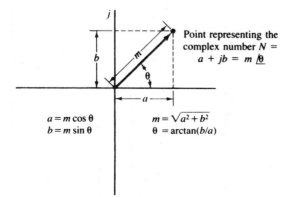

◆ **Figure 4.11**   Graph of the complex number *N*. The number *N* can be represented by its rectangular coordinates as $a + jb$, or by its polar coordinates as $m \underline{/\theta}$.

Rectangular-to-Polar Conversion

$$m = \sqrt{a^2 + b^2}$$

$$\theta = \arctan (b/a)$$

Polar-to-Rectangular Conversion

$$a = m \cos \theta$$

$$b = m \sin \theta$$

The frequency response of a component is obtained by substituting $j\omega$ for s in the transfer function. With s replaced by $j\omega$, the transfer function reduces to a complex number that can be converted to the polar form. The magnitude of the complex number is the gain of the component at the frequency ω. The angle of the complex number is the phase angle of the component at the frequency ω.

In the *straight-line, graphical approximation method* the Bode amplitude and phase plots consist of straight lines that meet at points referred to as "break points." Figure 4.12 shows the straight-line approximation of the Bode diagram in Figure 4.10. Notice that the major deviations between the straight-line plots and the exact plots occur near the break points.

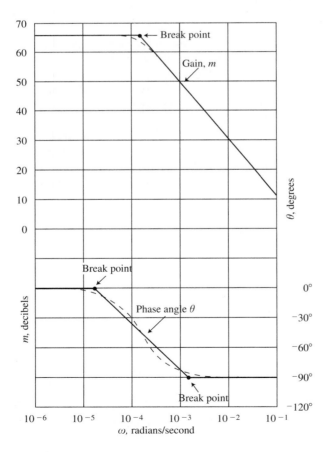

◆ **Figure 4.12** A straight-line graphical approximation of the Bode diagram in Figure 4.10. The dashed lines are the exact plot. Notice that the major deviations from the exact plot occur near the break points.

The straight-line, graphical method is a classical tool used to design control systems. A major advantage of this method is the "feel" of the system frequency response that the designer acquires during the construction of the Bode diagrams. Disadvantages of this graphical method include the time required to construct the Bode plots, inaccuracies caused by the approximations, and the excessive time required to explore various design ideas.

The *computer-aided, graphical method* eliminates all three disadvantages of the straight-line graphical method and retains the "feel" afforded by the graphical approach. This text uses the computer-aided graphical method to generate Bode diagrams and design control systems. The program BODE generates Bode diagrams for a wide variety of transfer functions. DESIGN is a program that emulates the graphical design method using Bode diagrams. A major advantage of DESIGN is the ease with which the designer can go back and change any control mode. This enables the designer to use a "what-if" analysis in a "trial-and-examine" design procedure to search for the best possible control system design. In this text, the emphasis is on computer-aided methods that emulate graphical methods, and no emphasis is placed on straight-line approximations.

EXAMPLE 4.18

A self-regulating liquid tank is described in the time domain by Equation (4.6) and in the frequency domain by its transfer function given by Equation (4.25). The input flow rate (q_{in}) varies sinusoidally about an average value with an amplitude of 0.0002 m³/s and a frequency (f) of 0.0001592 Hz. The time constant (t) is 1590 s and the gain (G) is 2000 s/m². Determine the amplitude and phase of the output (h).

Solution

1. Determine the value of $j\omega$.

$$j\omega = j2\pi f = j2\pi 0.0001592$$

$$j\omega = j0.001 \text{ rad/s}$$

2. Replace s in the transfer function by $j\omega$ and reduce it to a complex number in polar form.

$$\frac{H}{Q_{in}} = \frac{G}{1 + \tau s} = \frac{2000}{1 + 1590\,(j0.001)}$$

$$= \frac{2000}{1 + j1.59} = \frac{2000\underline{/0°}}{1.879\underline{/57.8°}} = 1065\underline{/-57.8°}$$

Gain in decibels $= 20 \log 1065 = 60.5$

The output amplitude is equal to the input amplitude times the magnitude of the transfer function.

$$\text{Output amplitude } h = 1065 \times 0.0002 = 0.213 \text{ m}$$

The phase difference is $-57.8°$, so the output (h) lags the input (q_{in}) by 57.8°.   ◆

Program BODE

The hand computation of frequency response data can become quite tedious. The large number of computations make this an ideal application for a computer. BODE (see Preface) is a program that will put a Bode diagram on the screen for any transfer function that fits the following general form:

$$\text{Transfer function} = \frac{a_0 + a_1 s + a_2 s^2 + a_3 s^3}{b_0 + b_1 s + b_2 s^2 + b_3 s^3}$$

Any of the coefficients a_0, a_1, a_2, a_3, b_0, b_1, b_2, or b_3 can have a value of 0 in a particular transfer function.

The program generates frequency response data over a range of frequencies from 1.0×10^{-6} to 5.6×10^5 rad/s. The following frequency bases are used for each power of 10: 1.0, 1.8, 3.2, and 5.6. These seem like odd numbers, but they happen to be even increments on a logarithmic scale. The Bode diagram uses a logarithmic scale for frequency, and 3.2 is just about the midpoint between 1.0 and 10.0. Also, 1.8 is just about the midpoint between 1.0 and 3.2; and 5.6 is about the midpoint between 3.2 and 10.0. This makes it easy to plot Bode diagrams on linear graph paper.

The program runs on PC compatibles with color monitors. A zoom command allows the user to expand a portion of the gain, angle, or frequency range to fill the entire graph. An unzoom command returns the original scale. The program also has an option to make a printed copy of the Bode data table.

EXAMPLE 4.19

Use BODE to generate a table of frequency-response data for the self-regulating liquid tank described in Example 4.18. BODE is available on disk (see Preface). The transfer function is as follows:

$$\frac{H}{Q_{\text{in}}} = \frac{2000}{1 + 1590s}$$

Solution

The input and a portion of the data table produced by the program are shown below.

TRANSFER FUNCTION COEFFICIENTS
$\text{TF} = \dfrac{\text{A}(0) + \text{A}(1)\text{S} + \text{A}(2)\text{S}^2 + \text{A}(3)\text{S}^3}{\text{B}(0) + \text{B}(1)\text{S} + \text{B}(2)\text{S}^2 + \text{B}(3)\text{S}^3}$
A(0) = 2000, A(1) = 0, A (2) = 0, A(3) = 0, B(0) = 1, B(1) = 1590, B (2) = 0, B(3) = 0

TRANSFER FUNCTION COEFFICIENTS			
A(0) . . A(3): 2.00E + 03	0.00E + 00	0.00E + 00	0.00E + 00
B(0) . . B(3): 1.00E + 00	1.59E + 03	0.00E + 00	0.00E + 00

BODE DATA TABLE		
Frequency, W (radian/second)	Gain (decibel)	Phase (degrees)
1.0E-06	66.0	−0.1
1.8E-06	66.0	−0.2
3.2E-06	66.0	−0.3
5.6E-06	66.0	−0.5
1.0E-05	66.0	−0.9
1.8E-05	66.0	−1.6
3.2E-05	66.0	−2.9
5.6E-05	66.0	−5.1
1.0E-04	65.9	−9.0
1.8E-04	65.7	−15.8
3.2E-04	65.0	−26.7
5.6E-04	63.5	−41.8
1.0E-03	60.5	−57.8
1.8E-03	56.5	−70.5
3.2E-03	51.8	−78.8
5.6E-03	46.9	−83.6
1.0E-02	42.0	−86.4
1.8E-02	37.0	−88.0
3.2E-02	32.0	−88.9
5.6E-02	27.0	−89.4
1.0E-01	22.0	−89.6
1.8E-01	17.0	−89.8
3.2E-01	12.0	−89.9
5.6E-01	7.0	−89.9

◆ GLOSSARY

Bode diagram: A pair of graphs that share a common log frequency scale on the *x*-axis. The *y*-axis of one graph is the decibel gain of a component; the other is its phase angle. See Frequency response. (4.7)

Derivative: A mathematical expression of the rate of change of a variable. (4.2)

Differential equation: An equation that has one or more derivative terms. (4.2)

Final value theorem: A method used to determine the final value of a function of time from its Laplace transform. (4.6)

Frequency domain: A domain in which the equation that describes a component is a function of a frequency parameter (*s*). The equation defines the amplitude and phase angle of the output of the component as a function of the frequency, gain, and phase of a sinusoidal input signal. (4.3)

Frequency response: The set of values of the input-to-output gain and phase angle of a component when a sinusoidal input signal is varied over a range of frequencies. See Bode diagram. (4.7)

Functional Laplace transform: The Laplace transform of a time function such as sin *ωt*. (4.3)

Initial value theorem: A method used to determine the initial value of a function of time from its Laplace transform. (4.6)

Integral: A mathematical expression for the accumulation of an amount (or quantity) of a variable. (4.2)

Integro-differential equation: An equation that has at least one derivative and one integral term. (4.2)

Inverse Laplace transform: A mathematical transformation that converts the solution of a differential equation in the frequency domain into a solution in the time domain. (4.4)

Laplace transform: A mathematical transformation that converts equations that involve derivatives and integrals in the time domain into equations that involve only algebraic terms in the frequency domain. (4.3)

Linear differential equation: A differential equation that has only first-degree terms (power = 1) in the dependent variable and its derivatives. (4.2)

Operational Laplace transform: The Laplace transform of a time-domain operation (e.g., addition, differentiation, integration, etc.). (4.3)

Order of a differential equation: The order of the highest derivative that appears in the equation. (4.2)

Partial fraction expansion: The result of breaking a ratio of polynomials into a sum of terms so that we can determine its inverse Laplace transform. (4.4)

Time constant (τ): A parameter that characterizes a group of components that are modeled by a linear first-order differential equation with constant coefficients. The unit of the time constant is seconds, which explains the name "time" constant. (4.2)

Time domain: A domain in which the equation that describes a component is a function of time. The equation defines the size of the output as a function of time and the input signal. (4.3)

Transfer function: The frequency-domain ratio of the output of a component over its input, with all initial conditions set to zero. (4.5)

◆ EXERCISES

Section 4.2

4.1 Equation (4.6) describes the relationship between the input flow rate (q_{in}) and the output level (h) of the self-regulating liquid tank shown in Figure 4.1. Determine the numerical values of τ and G for the following sets of parameter values, and write Equation (4.6) using the numerical values. Use a value of 9.81 m/s^2 for g.

(a) $R_L = 2.16 \times 10^6$ Pa · s/m^2, $A = 2.63$ m^2, $\rho = 1000$ kg/m^3
(b) $R_L = 5.76 \times 10^5$ Pa · s/m^2, $A = 0.35$ m^2, $\rho = 880$ kg/m^3

4.2 Equation (4.7) describes the relationship between the input flow rate (q_{in}) and the output level (h) of the nonregulating liquid tank shown in Figure 4.2. Determine the numerical value of A for the following tank diameters, and write Equation (4.7) using the numerical value of A.

(a) 1.5 m (b) 0.52 m

4.3 Equation (4.10) describes the relationship between the input voltage (e_{in}) and the output voltage (e_{out}) of the electrical circuit shown in Figure 4.3. Determine the numerical value of the time constant (τ) for the following sets of parameter values, and write Equation (4.10) using the numerical value of τ.

(a) $R_L = 22$ kΩ, $C_L = 47$ μF
(b) $R_L = 630$ kΩ, $C_L = 0.76$ μF

4.4 Equation (4.11) describes the relationship between the actual temperature (T_a) and the measured temperature (T_m) of the liquid-filled thermometer shown in Figure 4.4. Determine the numerical value of the time constant (τ) for the following sets of parameter values, and write Equation (4.11) using the numerical value of τ. Use natural convection in still water for the outside film coefficient.

	(a)	(b)
Liquid in bulb	mercury	ethyl alcohol
Bulb material	glass	glass
Bulb outside diameter, cm	1.94	0.86
Bulb outside length, cm	4.52	1.62
Bulb thickness, mm	1.4	1.1
Inside film coefficient, h_i, W/m^2 · K	4000	2210
T_w, °C	85	58
T_d, °C	20	8

4.5 Equation (4.17) describes the relationship between the inlet air pressure (p_{in}) and the position of the stem (x) of the control valve shown in Figure 4.5. Determine the numerical value of the mechanical capacitance of the valve spring for the following sets of parameters, and write Equation (4.17) using numerical values for A, m, R_m, and C_m. Hint: use Equation (3.50) to determine the mechanical capacitance: $C_m = \Delta x / \Delta F_a$ where Δx = the valve stroke (closed to open) and $\Delta F_a = \Delta p_{in} A$ change in applied force from closed to open.

	(a)	(b)
A, m^2	0.0182	0.0142
m, kg	0.561	0.387
R_m, N · s/m	9.60	9.10
ΔP_{in}, kPa	82.7	82.7
Δx, m	0.025	0.020

Section 4.3

4.6 Determine the frequency-domain function, $F(s)$, for each of the following time-domain functions, $f(t)$:

(a) $f(t) = 7.8$
(b) $f(t) = 3.2 \cos 1000t$
(c) $f(t) = 120 \sin 25t$
(d) $f(t) = 18t$
(e) $f(t) = 16e^{-8t}$
(f) $f(t) = 9e^{-3t} \sin 100t$
(g) $f(t) = 8.2te^{-2.5t}$
(h) $f(t) = 5e^{-7t} \cos 50t$
(i) $f(t) = 45e^{-5(t-6)}$
(j) $f(t) = 2 \sin(t - 6)$
(k) $f(t) = 4.8e^{-5t} \cos(400t - 36°)$

(l) $f(t) = 8\dfrac{d^2x}{dt^2} + 5\dfrac{dx}{dt}$,

where $\dfrac{dx(0)}{dt} = 8$, $x(0) = -4$

(m) $f(t) = 12 \displaystyle\int x\,dt + 17x$

4.7 Determine the time-domain function, $f(t)$, for each of the following frequency-domain functions, $F(s)$:

(a) $F(s) = \dfrac{6.7}{s^2}$

(b) $F(s) = \dfrac{25\omega}{s^2 + \omega^2}$

(c) $F(s) = \dfrac{45}{s + 72}$

(d) $F(s) = 345/s$

(e) $F(s) = \dfrac{650}{(s + 8)^2}$

(f) $F(s) = \dfrac{250\omega}{(s + 4)^2 + \omega^2}$

(g) $F(s) = \dfrac{82}{s(5s + 1)}$

(h) $F(s) = \dfrac{16(s + 5)}{(s + 5)^2 + \omega^2}$

(i) $F(s) = \dfrac{28s}{s^2 + \omega^2}$

(j) $F(s) = \dfrac{64\underline{/48°}}{s + 8 - j16} + \dfrac{64\underline{/-48°}}{s + 8 + j16}$

Section 4.4

4.8 Complete the partial fraction expansion and find the inverse Laplace transformation of each of the following functions:

(a) $\dfrac{4(s + 5)(s + 7)}{s(s + 3)(s + 6)}$ (b) $\dfrac{2(s + 5)}{(s + 1)^2}$

(c) $\dfrac{s + 2}{s^2 + 2s + 4}$

Section 4.5

4.9 The dead-time process shown in Figure 4.9 is described by the following equation:

$$f_o(t) = f_i(t - t_d)$$

where f_o = output signal, kg/s
  f_i = input signal, kg/s
  t_d = dead-time lag, s

Determine the transfer function, $F_o(s)/F_i(s)$, from the preceding equation if the dead time, t_d, is 245 s.

4.10 Determine the transfer function, $I(s)/\theta(s)$, for the temperature transmitter described by the following differential equation:

$$8.6 \frac{di}{dt} + i = 0.1\theta$$

where i = output current signal, mA

θ = input temperature signal, °C

4.11 Determine the transfer function, $X(s)/I(s)$ for a process-control valve/electropneumatic converter described by the following differential equation.

$$0.0001 \frac{d^2x}{dt^2} + 0.02 \frac{dx}{dt} + x = 0.3i$$

where x = valve stem position, in.

i = current input signal to the converter, mA

4.12 Determine the transfer function, $\theta(s)/X(s)$, for a tubular heat exchanger similar to the one shown in Figure 2.3 and described by the following differential equation.

$$25 \frac{d^2\theta}{dt^2} + 26 \frac{d\theta}{dt} + \theta = 125x$$

where x = valve position, in.

θ = temperature of the fluid leaving the heat exchanger, °C

4.13 A manufacturing plant uses a liquid surge tank to feed a positive-displacement pump. The pump supplies a constant flow rate of liquid to a continuous heat exchanger. Determine the transfer function, $H(s)/Q(s)$, if the surge tank is described by the following equation:

$$h(t) = 0.5 \int q(t) \, dt$$

where $h(t)$ = level of liquid in the surge tank, m

$q(t)$ = difference between the input flow rate and the output flow rate, m³

t = time, s

4.14 The spring-mass-damping system shown in Figure 3.8 is described by the following differential equation:

$$m \frac{d^2x}{dt^2} + R \frac{dx}{dt} + Kx = f$$

where m = mass, kg

R = dashpot resistance, N/(m/s)

K = spring constant, N/m

x = position of the mass, m

f = external force applied to the mass, N

Determine the transfer function, $X(s)/F(s)$, if

m = 3.2 kg
R = 2.0 N/(m/s)
K = 800 N/m

4.15 A PID controller is described by the following equation:

$$v = Pe + PD \frac{de}{dt} + PI \int e \, dt + v_o$$

where e = error, % of full scale (F.S.)

v = controller output, % of F.S.

v_o = controller output at $t = 0$ s, % of F.S.

P = proportional gain setting (dimensionless)

D = derivative action time constant, s

I = integral action rate, 1/s

Determine the transfer function, $V(s)/E(s)$, if P is 3.6, D is 0.008 s, I is 0.455 s, and v_o is 0.

4.16 An armature-controlled dc motor is sometimes used in speed and position control systems (see Figures 2.6 and 2.15). The dc motor operation is described by the following equations:

$$e = Ri + L \frac{di}{dt} + K_e\omega$$

$$i = \frac{q}{K_t}$$

$$q = J \frac{d\omega}{dt} + b\omega$$

where e = armature voltage, V

i = armature current, A

ω = motor speed, rad/s

q = motor torque, N · m

J = moment of inertia of the load, kg · m²

b = damping resistance of the load, N · m/ (rad/s)

R = armature resistance, Ω

L = armature inductance, H

K_e = back emf constant of the motor, V/(rad/s)

K_t = torque constant of the motor, N · m/A

A small permanent-magnet dc motor has the following parameter values:

$$J = 8 \times 10^{-4} \text{ kg} \cdot \text{m}^2$$

$b = 3 \times 10^{-4}\,\text{N} \cdot \text{m/(rad/s)}$
$R = 1.2\,\Omega$
$L = 0.020\,\text{H}$
$K_e = 5 \times 10^{-2}\,\text{V/(rad/s)}$
$K_t = 0.043\,\text{N} \cdot \text{m/A}$

Substitute these parameters into the preceding equations to obtain the exact differential equations of the dc motor. Determine the transfer function, $\Omega(s)/E(s)$, by transforming all three equations into frequency-domain algebraic equations. Use algebraic operations to obtain the ratio of Ω/E, which is the desired transfer function.

4.17 A PD controller is described by the following equation:

$$v = Pe + PD\frac{de}{dt} - \alpha D\frac{dv}{dt} + v_o$$

where e = error, % of full scale (F.S.)
   v = controller output, % of F.S.
   v_o = controller output at $t = 0$ s, % of F.S.
   P = proportional gain setting (dimensionless)
   D = derivative action time constant, s
   α = derivative limiter coefficient

Determine the transfer function, $V(s)/E(s)$ if P is 2, D is 0.025 s, α is 0.1 and v_o is 0.

4.18 The PD controller in Exercise 4.17 is one example of a control system component capable of a sudden change in output value. To show the sudden change, determine the time-domain response of the output $v(t)$ to the following step change in the input $e(t)$. Refer to Example 4.16 for further guidance.

$$e(t) = 0\,\%\,\text{F.S.}, \quad t < 0\,\text{s}$$
$$= 10\,\%\,\text{F.S.}, \quad t \geq 0\,\text{s}$$

Section 4.6

4.19 (a) Use the initial and final value theorems to determine the initial and final values of $v(t)$ for the step response of the PD controller in Exercise 4.18.
(b) Use the time-domain equations you developed in Exercise 4.18 to determine the initial and final values of $v(t)$.
(c) Compare the values from part a with the values from part b.

4.20 A temperature transmitter is calibrated so that the output is 0 mA when the input temperature is 0°C and 10 mA when the input temperature is 100°C. The temperature probe was held in ice water for a long time before time $t = 0$ s when it was suddenly plunged into water boiling at 100°C. The output current, $I(s)$, that resulted from this sudden temperature change is given by the following equation:

$$I(s) = \frac{10.0}{s(1 + 25s)},\,\text{mA}$$

Determine the initial and final values of the output current, $i(t)$. Do your initial and final values make sense? Explain your answer.

4.21 The armature-controlled dc motor in Exercise 4.16 was not moving at $t = 0$ s when the armature voltage was suddenly increased from 0 to 50 V. The motor speed, $\Omega(s)$, that resulted from this sudden voltage change is given by the following equation:

$$\Omega(s) = \frac{50}{s(0.0584 + 0.225\,s + 0.00037\,s^2)},\,\text{rad/s}$$

Determine the initial and final values of the motor speed. Do your initial and final values make sense? Explain your answer.

4.22 The liquid surge tank in Exercise 4.13 was empty at $t = 0$ s when the input flow rate, q_{in}, was suddenly changed from 0 to 0.001 m³/s. Also at $t = 0$ s, the output flow rate, q_{out}, changed from 0 to 0.00037 m³/s. The level of the liquid in the tank, $H(s)$, is given by the following equation:

$$H(s) = \left[\frac{q_{in} - q_{out}}{s}\right]\left[\frac{0.5}{s}\right],\,\text{m}$$

Determine the initial and final values of the liquid level. Does the initial value make sense? Explain the meaning of the final value. Given that the tank has a capacity of 0.5 m³, how long will it take to fill the tank at the flow rates given above?

Section 4.7

4.23 Use the program BODE to generate frequency-response data from the transfer functions obtained in Exercises 4.10, 4.11, 4.12, 4.14, 4.15, 4.16, and 4.17. Construct Bode diagrams from the data.

PART TWO   ◆   MEASUREMENT

◆ **CHAPTER 5**

Measuring Instrument Characteristics

◆ OBJECTIVES

A measuring instrument has many important characteristics. The two characteristics that have the greatest effect on the performance of a control system are accuracy and speed of response (or frequency response). These two characteristics are not completely independent. If the measured variable is changing, the accuracy of the measurement depends on how quickly the measuring instrument can follow these changes. The speed of response of a measuring instrument is related to its frequency response. In Chapter 16 you will see that the frequency response of each component in the control loop is an important factor in the design of the controller.

The purpose of this chapter is to introduce you to the characteristics of measuring instruments and the terminology associated with measurement. After completing this chapter, you will be able to

1. Explain
   a. Repeatability and reproducibility
   b. Static and dynamic characteristics
2. Determine from data or a calibration graph
   a. Sample mean and standard deviation
   b. Measured accuracy and repeatability
   c. Dead band and hysteresis
   d. Independent linearity
   e. Terminal- and zero-based linearity
3. Determine from a step response graph
   a. Time constant and 95% response time
   b. 10 to 90% rise time and 2% settling time
   c. Peak percentage of overshoot

4. Determine from a ramp response graph
   a. Dynamic lag
   b. Dynamic error
5. Construct frequency-response diagrams (using the program BODE)

5.1 INTRODUCTION

In Chapter 1, you learned that a feedback control system performs three operations: measurement, decision, and manipulation. The measuring transmitter performs the measurement operation, the controller performs the decision operation, and the manipulating element performs the manipulation operation. Measurement is an essential operation in a feedback control system. In order to control a variable, we must first measure its value. Next, we must convert the measured value into a usable signal. Only then can the controller and the manipulating element perform their tasks. The controller cannot make a decision without the measured value of the controlled variable. The manipulating element does not know how to act without a decision from the controller. Feedback control begins with the measurement operation. In this chapter you will study the characteristics and terminology of measuring instruments.

The purpose of a measuring instrument is to obtain the true value of the measured variable. The ideal measuring instrument would do this exactly, but in practice, this ideal is never achieved. There is always some uncertainty in the measurement of a variable. Indeed, there is even some uncertainty in the standards we use to calibrate a measuring instrument. For this reason, we begin the chapter with a brief review of statistics, the subject that deals with uncertainty.

The next three sections divide the characteristics of measuring instruments into three categories: operating characteristics, static characteristics, and dynamic characteristics. *Operating characteristics* include measurement details, operational details, and environmental effects. *Static characteristics* deal with accuracy when the value of the measured variable is constant or changing very slowly. In contrast, *dynamic characteristics* deal with the measurement of a variable whose value is changing rather quickly.

The selection of a measuring instrument for a particular application can be a complex and difficult procedure. In selecting a temperature transmitter, for example, a designer must consider the following possible primary elements: thermocouple, RTD, thermistor, IC sensor, liquid-filled element, vapor-filled element, gas-filled element, or radiation pyrometer. A set of selection criteria are presented in the form of questions that can be asked in the process of selecting a measuring instrument.

The final section is a glossary of terms presented in the chapter. The principal source document for the terminology used in this chapter is the Instrument Society of America standard S51.1, "Process Instrumentation Terminology."

5.2 STATISTICS

There is an uncertainty when we measure the value of a variable. This uncertainty occurs when repeated measurements under identical conditions give different results. For example, let's assume that five measurements of the temperature of a fluid result in the following measured values: 207, 204, 205, 205, and 206°C. Statistics cannot tell us what the true temperature is, but it can help us understand the uncertainty we are confronted with.

The individual measurements of a variable are called *observations,* and the entire collection of observations is called a *sample*. The simplest statistical measure of the sample is the *arithmetic average* or *mean*. The sample mean is an estimate of the expected value of the next observation. The mean is computed by summing the observations and dividing by the number of observations. The mean of a sample of n observations is given by the following equation:

$$\text{Sample mean} = \bar{x} = \frac{x_1 + x_2 + x_3 + \cdots + x_n}{n} \tag{5.1}$$

The mean gives us an estimate of the expected value of an observation, but it gives no idea of the dispersion or variability of the observations. For a measure of variability, we begin by computing the deviation between each observation and the mean.

$$\text{Deviation of observation } x_i = d_i = x_i - \bar{x}$$

The standard deviation, S_x, is a measure of variability, which is defined by the following equation:

$$S_x = \sqrt{\frac{d_1^2 + d_2^2 + d_3^2 + \cdots + d_n^2}{n - 1}} \tag{5.2}$$

The standard deviation gives us an idea of the variability of the observations in the sample. If the errors in measurement are truly random and we take a large number of observations, 68% of all observations will be within 1 standard deviation of the mean. Over 95% of all samples will be within 2 standard deviations of the mean, and almost all samples will be within 3 standard deviations of the mean.

EXAMPLE 5.1

Compute the mean and standard deviation of the following temperature measurements: 207°C, 204°C, 205°C, 205°C, and 206°C.

Solution

$$\bar{x} = \frac{207 + 204 + 205 + 205 + 206}{5} = 205.4°C$$

$$d_1^2 = (207 - 205.4)^2 = (1.6)^2 = 2.56$$

$$d_2^2 = (204 - 205.4)^2 = (-1.4)^2 = 1.96$$

$$d_3^2 = (205 - 205.4)^2 = (-0.4)^2 = 0.16$$

$$d_4^2 = (205 - 205.4)^2 = (-0.4)^2 = 0.16$$

$$d_5^2 = (206 - 205.4)^2 = (0.6)^2 = 0.36$$

$$S_x = \sqrt{\frac{2.56 + 1.96 + 0.16 + 0.16 + 0.36}{5 - 1}}$$

$$= 1.14°C \qquad \qquad ◆$$

5.3 OPERATING CHARACTERISTICS

Operating characteristics include details about the measurement by, operation of, and environmental effects on the measuring instrument.

Measurement

A measuring instrument can measure any value of a variable within its *range* of measurement. The range is defined by the *lower range limit* and the *upper range limit*. As the names imply, the range consists of all values between the lower range limit and the upper range limit. The *span* is the difference between the upper range limit and the lower range limit.

$$\text{Span} = \text{upper range limit} - \text{lower range limit}$$

Resolution, dead band, and sensitivity are different characteristics that relate in different ways to an increment of measurement. When the measured variable is continuously varied over the range, some measuring instruments change their output in discrete steps rather than in a continuous manner. The *resolution* of this type of measuring instrument is a single step of the output. Resolution is usually expressed as a percentage of the output span of the instrument. Sometimes the size of the steps varies through the range of the instrument. In this case, the largest step is the *maximum resolution*. The *average resolution,* expressed as a percentage of output span, is 100 divided by the total number of steps over the range of the instrument.

$$\text{Average resolution } (\%) = \frac{100}{N} \tag{5.3}$$

where N represents the total number of steps.

The *dead band* of a measuring instrument is the smallest change in the measured variable that will result in a measurable change in the output. Obviously, a measuring instrument cannot measure changes in the measured variable that are smaller than its dead band. *Threshold* is another name for *dead band*.

The *sensitivity* of a measuring instrument is the ratio of the change in output divided by the change in the input that caused the change in output. Sensitivity and gain are both defined as a change in output divided by the corresponding change in input. However, *sensitivity* refers to static values, whereas *gain* usually refers to the amplitude of sinusoidal signals.

EXAMPLE 5.2

A potentiometer with 1200 turns of wire is used to measure the rotational position of a shaft. The input range is from $-175°$ to $+175°$. The output range is from 0 to 10 V. Determine the span, the sensitivity in volts per degree, and the average resolution in volts and as a percentage of span.

Solution

$$\text{Span} = 175° - (-175°) = 350°$$

$$\text{Sensitivity} = \frac{(10 - 0)}{350°} = 0.0286 \text{ V/degrees}$$

$$\text{Average resolution} = \frac{100}{1200} = 0.0833\% \text{ of span}$$

$$\text{Average resolution} = \frac{10}{1200} = 0.00833 \text{ V} \qquad ◆$$

Operation

The *reliability* of a measuring instrument is the probability that it will do its job for a specified period of time under a specified set of conditions. The conditions include limits on the operating environment, the amount of overrange, and the amount of drift of the output.

Overrange is any excess in the value of the measured variable above the upper range limit or below the lower range limit. When an instrument is subject to an overrange, it does not immediately return to operation within specifications when the overload is removed. A period of time called the *recovery time* is required to overcome the saturation effect of the overload. The *overrange limit* is the maximum overrange that can be applied to a measuring instrument without causing damage or permanent change in the performance of the device. Thus one reliability condition is that the measured variable not exceed the overrange limit.

Drift is an undesirable change over a specified period of time. *Zero drift* is a change in the output of the measuring instrument while the measured variable is held constant at its lower limit. *Sensitivity drift* is a change in the sensitivity of the instrument over the specified period. Zero drift raises or lowers the entire calibration curve of the instrument. Sensitivity drift changes the slope of the calibration curve. The reliability conditions specify an allowable amount of zero drift and sensitivity drift.

Environmental Effects

The environment of a measuring instrument includes ambient temperature, ambient pressure, fluid temperature, fluid pressure, electromagnetic fields, acceleration, vibration, and mounting position. The *operating conditions* define the environment to which a measuring instrument is subjected. The *operative limits* are the range of operating conditions that will not cause permanent impairment of an instrument.

Temperature effects may be stated in terms of the zero shift and the sensitivity shift. The *thermal zero shift* is the change in the zero output of a measuring instrument for a specified change in ambient temperature. The *thermal sensitivity shift* is the change in sensitivity of a measuring instrument for a specified change in ambient temperature.

5.4 STATIC CHARACTERISTICS

Static characteristics describe the accuracy of a measuring instrument at room conditions with the measured variable either constant or changing very slowly. *Accuracy* is the degree of conformity of the output of a measuring instrument to the ideal value of the measured variable as determined by some type of standard. Accuracy is measured by testing the measuring instrument with a specified procedure under specified conditions. The test is repeated a number of times, and the accuracy is given as the maximum positive and negative error (deviation from the ideal value). The *error* is defined as the difference between the measured value and the ideal value:

$$\text{Error} = \text{measured value} - \text{ideal value}$$

Accuracy is expressed in terms of the error in one of the following ways:

1. In terms of the measured variable (e.g., $+1°C/-2°C$)
2. As a percentage of span (e.g., $\pm 0.5\%$ of span)
3. As a percentage of actual output (e.g., $\pm 1\%$ of output)

EXAMPLE 5.3

A tachometer-generator is a device used to measure the speed of rotation of gasoline engines, electric motors, speed control systems, and so on. The "tach" produces a voltage proportional to its rotational speed. Consider a tachometer-generator that has an ideal rating of 5.0 V per 1000 revolutions per minute (rpm) a range of 0 to 5000 rpm, and an accuracy of ±0.5%. If the output of the tach is 21 V, what is the ideal value of the speed? What are the minimum and maximum possible values of the speed?

Solution

The range of the tachometer is 0 to 25 V, corresponding to a speed range of 0 to 5000 rpm. The ideal value of the speed is equal to 200 times the output voltage.

$$\text{Ideal speed} = 21 \times 200 = 4200 \text{ rpm}$$

The accuracy is ±0.5% of full scale, or ±0.005 × 5000 = ±25 rpm. Thus the speed could be anywhere between 4200 − 25 and 4200 + 25 rpm.

$$\text{Ideal speed} = 4200 \text{ rpm}$$

$$4175 \le \text{speed} \le 4225 \text{ rpm} \qquad \qquad ◆$$

The *repeatability* of a measuring instrument is a measure of the dispersion of the measurements (the standard deviation is another measure of dispersion). Accuracy and repeatability are not the same. Figure 5.1 uses the pattern of bullet holes in a target to illustrate the difference between repeatability and accuracy. Notice that a rifle, which is repeatable but not accurate, produces a tight pattern—but that the pattern is not centered on the bull's-eye. The distance from the center of the bull's-eye to the center of the pattern is called the *bias* or *systemic error*. A shooter who is aware of the bias may adjust accordingly and produce an accurate and repeatable pattern on the next try. An experienced operator will make a similar adjustment to compensate for bias in a controller. Thus an automatic controller that is repeatable but not accurate may still be very useful.

Repeatability and reproducibility deal in slightly different ways with the degree of closeness among repeated measurements of the same value of the measured variable. *Repeatability* is the maximum difference between several consecutive outputs for the same input when approached from the same direction in full-range traversals. *Reproducibility* is the maximum difference between a number of outputs for the same input, taken over an extended period of time, approaching from both directions. Reproducibility includes hysteresis, dead band, drift, and repeatability. The measurement of reproducibility must specify the time period used in the measurement. Reproducibility is obviously more difficult to determine because of the extended time period that is required.

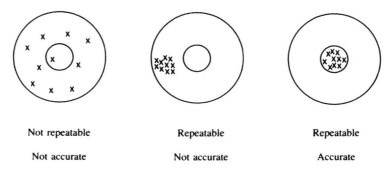

| Not repeatable | Repeatable | Repeatable |
| Not accurate | Not accurate | Accurate |

◆ **Figure 5.1** Rifle target patterns illustrate the difference between repeatability and accuracy.

EXAMPLE 5.4 The temperature measurements in Example 5.1 were obtained in a test for repeatability. The temperature transmitter used in the test has an upper range limit of 200°C and a lower range limit of 100°C. Each measurement was obtained by moving the probe from a container of oil at 100°C to a container of oil at 200°C. The equilibrium temperature was recorded, and the probe was moved back to the cooler oil for the next test. Determine the bias and repeatability in degrees Celsius and as a percentage of span.

Solution

$$Span = 200°C - 100°C = 100°C$$

$$Bias = 205.4°C - 200°C = 5.4°C$$

$$Bias = 100\frac{5.4}{100} = 5.4\% \text{ of span}$$

$$Repeatability = 207°C - 204°C = 3°C$$

$$Repeatability = 100\frac{3}{100} = 3\% \text{ of span} \qquad \blacklozenge$$

The procedure of determining the accuracy of a measuring instrument is called *calibration*. It consists of three full-range traversals of the measuring instrument. The data from the calibration of a measuring instrument is presented in tabular form in a *calibration report*. Table 5.1 is an example of a calibration report. The input is usually expressed as a percentage of the input span. The output and the error are usually expressed as a percentage of the ideal output span.

◆ **TABLE 5.1** Calibration Report (Percent)

Input or Ideal Output	Cycle 1	Cycle 2	Cycle 3	Average	Error
0	−0.06	−0.05	−0.04	−0.05	−0.05
10	9.80	9.82	9.84	9.82	−0.18
20	19.69	19.70	19.72	19.70	−0.30
30	29.65	29.67	29.69	29.67	−0.33
40	39.70	39.71	39.72	39.71	−0.29
50	49.85	49.86	49.87	49.86	−0.14
60	60.02	60.03	60.06	60.04	0.04
70	70.14	70.16	70.17	70.16	0.16
80	80.21	80.23	80.23	80.22	0.22
90	90.19	90.22	90.23	90.21	0.21
100	100.08	100.09	100.11	100.09	0.09
90	90.24	90.25	90.27	90.25	0.25
80	80.26	80.28	80.29	80.28	0.28
70	70.22	70.24	70.27	70.24	0.24
60	60.12	60.13	60.14	60.13	0.13
50	49.96	49.97	49.98	49.97	−0.03
40	39.78	39.80	39.82	39.80	−0.20
30	29.73	29.75	29.77	29.75	−0.25
20	19.76	19.78	19.79	19.78	−0.22
10	9.84	9.85	9.86	9.85	−0.15
0	−0.05	−0.04	−0.04	−0.04	−0.04

The measured accuracy and repeatability of the measuring instrument are taken directly from the calibration report. *Measured accuracy* is the maximum negative and positive errors in any of the readings in the calibration report. Repeatability is the greatest difference between the readings for the three cycles on any given line of data in the calibration report.

The calibration data may also be presented in graphical form in a *calibration curve*. Figures 5.2 and 5.3 are two forms of the calibration curve. In Figure 5.2, the error is plotted versus the input to the measuring instrument. This gives a detailed view of the error over the input range of the instrument. In Figure 5.3, the output is plotted versus the input. This gives a good overall view of the performance of the instrument.

The calibration curves in Figures 5.2 and 5.3 both form a loop as the upscale and downscale curves follow different paths. The loop formed by an upscale–downscale cycle is a combination of *hysteresis* and *dead band*. Hysteresis is defined as the dependence of the value of a variable on the history of past values and the current direction of traversal. Hysteresis is determined by subtracting the dead band from the maximum difference between corresponding upscale and downscale readings for a full-cycle traversal (see Figure 5.3).

EXAMPLE 5.5

The measuring instrument that was calibrated in Table 5.1 has a dead band of 0.02% of the ideal output span. Determine the measured accuracy, repeatability, and hysteresis of the measuring instrument from the calibration report in Table 5.1. Use the average traversal column to determine the combined hysteresis and dead band.

Solution

The maximum negative error occurs in cycle 1 at 30% on the upscale part of the cycle (i.e., top half of the table).

$$\text{Maximum negative error} = 29.65 - 30 = -0.35\%$$

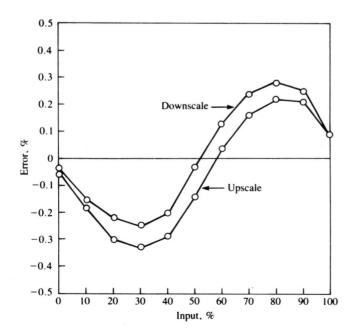

◆ **Figure 5.2** Calibration curve of the error from the calibration report in Table 5.1. Plotting the error instead of the output gives a more detailed view of the nature of the error over the range of the instrument.

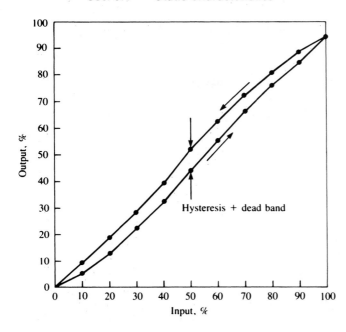

Input, %

◆ **Figure 4.3** Calibration curve of output versus input. Plotting the output gives a good overview of the performance of the measuring instrument. The data in this graph were altered to emphasize the hysteresis to make a better illustration. Hysteresis is usually much less than that shown here. Table 5.1 is a more realistic example of hysteresis in a measuring instrument.

The maximum positive error occurs in cycle 3 at 80% on the downscale side of the cycle (i.e., bottom half of the table).

$$\text{Maximum positive error} = 80.29 - 80 = 0.29\%$$

Measured accuracy is the combination of the maximum negative and positive errors.

$$\text{Measured accuracy} = -0.35\% \text{ to } +0.29\%$$

The maximum difference between readings on a given line in Table 5.1 occurs at an input of 70% in the downscale traversal. The outputs are 70.22% for cycle 1, 70.24% for cycle 2, and 70.27% for cycle 3.

$$\text{Repeatability} = 70.27\% - 70.22\% = 0.05\%$$

The maximum difference between the average upscale and downscale readings occurs at an input of 50%. The upscale reading is 49.86%, and the downscale reading is 49.97%. The combined hysteresis and dead band is the difference between these two readings.

$$\text{Combined hysteresis and dead band} = 49.97\% - 49.86\% = 0.11\%$$

Subtracting the dead band gives us the hysteresis.

$$\text{Hysteresis} = 0.11\% - 0.02\% = 0.09\%$$ ◆

The ideal measuring instrument would produce a perfectly straight calibration curve. On a graph of input versus output, the ideal straight line would pass through the points (0,0) and (100,100). The ideal never actually occurs, and the closeness of the calibration data to a straight line is called the *linearity* of the measuring instrument. When linearity is stated, the method of determining the straight line must also be specified. The most common types of linearity are *independent linearity, terminal-based linearity, zero-based linearity*, and *least-squares linearity*.

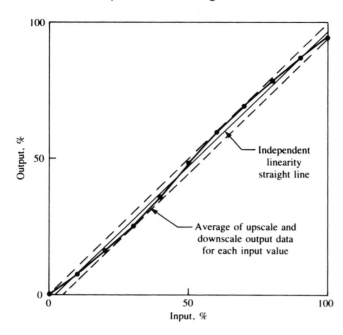

◆ **Figure 5.4** The calibration curve shown here is the average of the upscale and downscale outputs for each input value. The independent linearity straight line can be determined on the calibration curve. First determine the location of the two parallel lines that are the closest together while enclosing the entire calibration curve. These are the two dashed lines in the graph. The independent linearity straight line is the line midway between the two dashed lines.

When calibration data are compared to a straight line, attention focuses on the deviation or error between each output reading and the output determined by the straight line for the same input value. All methods of picking the straight line attempt to minimize the amount of the error. Independent linearity is illustrated in Figure 5.4. The straight line is placed such that it minimizes the absolute value of the maximum error. A pair of parallel lines helps to locate the independent linearity line on the calibration curve. The two lines must enclose the entire calibration curve, and they must be as close together as possible. The independent linearity line will be midway between the two parallel lines. The line does not necessarily pass through the origin, and it does not necessarily pass through either endpoint on the calibration curve.

The zero-based and terminal-based linearity lines are shown in Figure 5.5. The zero-based line passes through the minimum point of the calibration curve and has a slope such that it minimizes the maximum error. The terminal-based line is the easiest to construct. It simply connects the two endpoints on the calibration curve.

The least-squares fit (or linear regression curve) is the most common method used with spread sheet generated graphs. The least-squares line minimizes the sum of the squares of the deviations. The result is usually quite close to the independent linearity line.

5.5 DYNAMIC CHARACTERISTICS

Dynamic characteristics describe the performance of a measuring instrument when the measured variable is changing rapidly. Most sensors do not give an immediate, complete response to a sudden change in the measured variable. A measuring instrument requires a certain

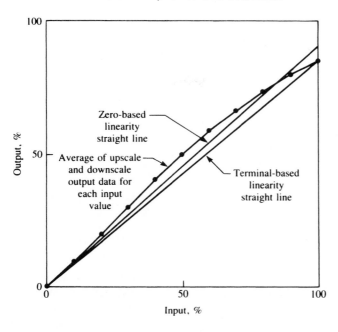

◆ **Figure 5.5** The zero-based and terminal-based straight lines can also be determined on the calibration curve. The terminal-based straight line simply connects the two endpoints of the calibration curve. The zero-based straight line passes through the zero point on the calibration curve and minimizes the absolute value of the maximum error.

amount of time before the complete response is indicated by the output. The amount of time required depends on the resistance, capacitance, mass or inertance, and dead time of the instrument. Dynamic characteristics are stated in terms of the step response, ramp response, and frequency response of the measuring instrument.

Step Response

A step change in the measured variable is easy to apply, and it produces a worst-case response. For these reasons, the response of a measuring instrument to a step change in the measured variable is often used to define its dynamic characteristics. A step change occurs when the measured variable is suddenly changed from one steady-state value to a second steady-state value. For example, a step change in temperature can be achieved by quickly moving the temperature probe from ice water to boiling water. This constitutes a step change from 0 to 100°C. The step response curve is a plot of the output of the measuring instrument from the time the step change occurred until the time the output reaches its new steady-state value. The response curve is normalized by expressing the initial steady-state value as 0%, the final steady-state value as 100%, and the output of the measuring instrument as a percentage of change. Terminology defined in terms of the percentage of output change can easily be applied to step changes of any given size.

The step response of an instrument can be classified as overdamped, critically damped, or underdamped (see Figure 1.14). The step response of overdamped or critically damped instruments is stated in terms of response time and rise time. The step response of underdamped instruments is stated in terms of rise time, peak percentage of overshoot, and settling time.

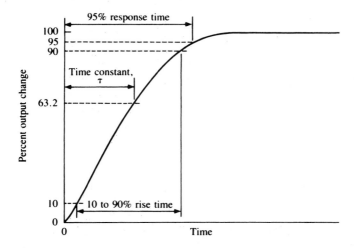

◆ **Figure 5.6**   Typical step response curve for an overdamped or critically damped measuring instrument. The response is stated in terms of the response time and the rise time.

A typical overdamped or critically damped step response curve is shown in Figure 5.6. *Response time* is the time required for the output to reach a designated percentage of the total change. The percentage is stated as a prefix. The 95% response time shown in Figure 5.6 is the time required for the output to reach 95% of the total change. A 98% response time (not shown) is the time required for the output to reach 98% of the total change. The 63.2% response time is given the special name *time constant* and symbol τ. This is the same time constant that we encountered several times in Chapter 4 [see Equations (4.6), (4.10), and (4.11)]. The following are examples of time constants of temperature-measuring instruments:

Typical Time Constants

Bare thermocouple (in air)	35 s
Thermocouple in glass well (in air)	66 s
Thermocouple in porcelain well (in air)	100 s
Thermocouple in iron well (in air)	120 s
Bare thermocouple (in still liquid)	10.0 s
Thermometer bulb (water flowing at 2 ft/min)	6.0 s
Thermometer bulb (water flowing at 60 ft/min)	2.4 s

Rise time is the time required for the output to go from a small percentage of change to a large percentage of change. The two percentages are stated as a prefix. Thus a 10 to 90% rise time is the time required for the output to go from 10% to 90% of the total change. A 5 to 95% rise time is the time required for the output to go from 5% to 95% of the total change. If the percentages are omitted, they are assumed to be 10% and 90%.

A typical underdamped step response curve is shown in Figure 5.7. The rise time is defined the same as it was for the overdamped or critically damped response. The peak percentage of overshoot (PPO) is defined as follows:

$$\text{PPO} = \frac{(\text{peak level}) - (\text{final ss level})}{(\text{final ss level}) - (\text{initial ss level})} \times 100$$

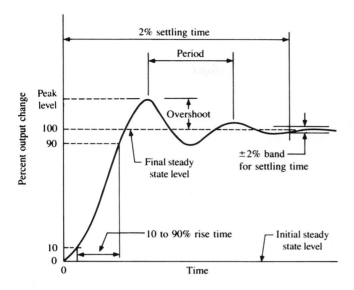

◆ **Figure 5.7**   Typical step response curve for an underdamped measuring instrument. The response is stated in terms of the rise time, peak percentage of overshoot, and settling time.

For the normalized curve, the final steady-state (ss) level is 100, the initial steady-state level is zero, and the peak percentage of overshoot reduces to the following:

$$\text{PPO} = (\text{peak level}) - (\text{final ss level})$$

The *settling time* is the time it takes for the response to remain within a small band above and below the 100% change line. The 2% settling time is shown in Figure 5.7.

Both first- and second-order differential equations are used to model the behavior of a measuring instrument. If a measuring instrument is dominated by a single capacitance, it is modeled by a first-order differential equation. The first-order model produces a step response similar to that in Figure 5.6. The transfer function and step response of the first-order model were developed in Chapter 4 [see Figure 4.7 and Equations (4.6) and (4.25)].

If the behavior of a measuring instrument is dominated by two capacitances (or one capacitance and one mass or inductance element), it is modeled by a second-order differential equation. The second-order model is more complex than the first-order model, and the step response may be overdamped, critically damped, or underdamped. The overdamped and critically damped second-order models produce step responses similar to those in Figure 5.6. The underdamped model produces a step response similar to that in Figure 5.7. The transfer function and step response of an underdamped model were developed in Chapter 4 (see Example 4.15, Example 4.16, and Figure 4.8).

FIRST-ORDER MODEL OF A MEASURING INSTRUMENT

Transfer Function

$$\frac{C_m(s)}{C(s)} = \frac{G}{1 + \tau s} \qquad\qquad \textbf{(5.4)}$$

Step Input

$$C(s) = \frac{K}{s} \tag{5.5}$$

Step Response

$$C_m(s) = \frac{KG}{s(1 + \tau s)} \tag{5.6}$$

where C = measuring instrument input, (input units)

C_m = measuring instrument output, (output units)

G = steady-state gain, $\dfrac{\text{(output units)}}{\text{(input units)}}$

K = step size, (input units)

τ = time constant, s

SECOND-ORDER MODEL OF A MEASURING INSTRUMENT

Transfer Function

$$\frac{C_m(s)}{C(s)} = \frac{G\omega_0^2}{\omega_0^2 + 2\alpha s + s^2} \tag{5.7}$$

Step Input

$$C(s) = \frac{K}{s} \tag{5.5}$$

Step Response

$$C_m(s) = \frac{KG\omega_0^2}{s(\omega_0^2 + 2\alpha s + s^2)} \tag{5.8}$$

Condition	*Type of Response*
$\alpha > \omega_0$	overdamped
$\alpha = \omega_0$	critically damped
$\alpha < \omega_0$	underdamped

where C = measuring instrument input, (input units)

C_m = measuring instrument output, (output units)

G = steady-state gain, $\dfrac{\text{(output units)}}{\text{(input units)}}$

K = step size, (input units)

α = damping coefficient, 1/s

ω_0 = resonant frequency, 1/s

The following is a summary of the important parameters of an underdamped second-order system.

$$\text{Resonant frequency} = \omega_0, \text{rad/s}$$

$$\text{Damping coefficient} = \alpha, 1/s$$

$$\text{Damping ratio} = \zeta = \alpha/\omega_0$$

$$\text{Natural frequency} = \omega_d = \sqrt{\omega_0^2 - \alpha^2}, \text{rad/s}$$

$$= \omega_0\sqrt{1 - \zeta^2}, \text{rad/s}$$

The natural frequency is also called the damped natural frequency.

EXAMPLE 5.6   The following data were obtained from a thermometer with a protective well, which was plunged into a moving liquid at time $t = 0$ s. The thermometer and well were maintained at 50°C before the test. The liquid temperature was 150°C. Plot the step response curve and determine the rise time, the time constant, and the 95% response time.

Time (s)	Temperature (°C)
0	50
20	64
40	78
60	92
80	104
100	113
120	120
140	126
160	130
180	133
200	135
300	143
400	147
500	150

Solution   The normalized response curve is plotted in Figure 5.8. The desired values can be read directly from the graph.

$$\text{Time constant, } \tau = 100 \text{ s}$$

$$95\% \text{ reponse time} = 330 \text{ s}$$

$$10 \text{ to } 90\% \text{ rise time} = 235 \text{ s} \qquad ◆$$

EXAMPLE 5.7   Determine the time constant of a mercury thermometer similar to Figure 4.4 and defined by Equation (4.11). The bulb has a diameter of 0.5 cm and a length of 1.5 cm. The outside film coefficient is estimated to be 30 W/m² · K. The inside film coefficient and the resistance of the thermometer wall are both negligible compared to the outside film coefficient. The thickness of the wall may also be neglected.

Solution   From Equation (4.11), the time constant is

$$\tau = R_T C_T$$

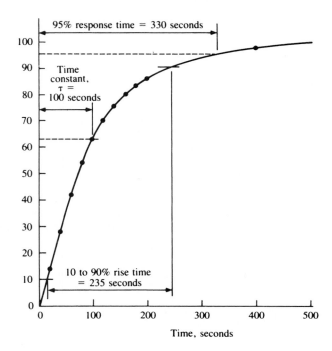

◆ **Figure 5.8** Step response curve for Example 5.6. The time constant, response time, and rise time are read directly from the graph.

1. Determine the thermal resistance of the bulb.

$$\text{Bulb surface} = A = \frac{\pi d^2}{4} + \pi dL$$

$$A = \frac{\pi (0.005)^2}{4} + \pi (0.005)(0.015)$$

$$= 2.55\text{E} - 4 \text{ m}^2$$

$$R_T = \frac{1}{Ah} = \frac{1}{(2.55\text{E} - 4)(30)}$$

$$= 131 \text{ K/W}$$

2. Determine the thermal capacitance of the bulb.

$$\text{Bulb volume} = \frac{\pi d^2 L}{4}$$

$$= \frac{\pi (0.005)^2 (0.015)}{4}$$

$$= 2.95\text{E} - 7 \text{ m}^3$$

From Appendix A, the density and specific heat of mercury are

$$\rho = 13{,}600 \text{ kg/m}^3$$

$$S_h = 140 \text{ J/kg} \cdot \text{K}$$

The thermal capacitance of the bulb is

$$C_T = \rho V S_h$$

$$= (1.36E + 4)(2.95E - 7)(140)$$

$$= 0.56 \text{ J/K}$$

3. Determine the time constant of the bulb.

$$\text{Time constant} = \tau = R_T C_T$$

$$= (131)(0.56) = 73.2 \text{ s}$$

Ramp Response

The ramp response of a measuring instrument, although not widely used, does give additional insight into the dynamic characteristics of a measuring instrument. Figure 5.9 shows a graph of the results of a ramp response test. The input temperature was increased at a steady rate until it was allowed to "settle in" at a higher value. The measured temperature lagged behind the input temperature and caught up some time after the input had settled in at the new temperature. There are two ways to view the ramp response curve. One way leads us to the concept of dynamic error, the other to the concept of dynamic lag.

If we draw a horizontal line at a temperature of T_1 on the dynamic response graph, the line intersects the input curve at time t_1, and the output curve at time t_2. The difference between these two times, $t_2 - t_1$, is the *dynamic lag* at temperature T_1. Dynamic lag is the amount of time that elapses between the time the input reaches a certain temperature and the time the output reaches that same temperature.

If we draw a vertical line at a time of t_2 on the ramp response graph, the line intersects the output curve at temperature T_1, and the input curve at temperature T_2. The difference between these two temperatures is the *dynamic error* at time t_2. Dynamic error is the difference between the input temperature and the output temperature at a given time.

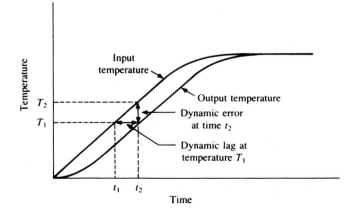

◆ **Figure 5.9** Typical ramp response curve showing the dynamic error and the dynamic lag.

**EXAMPLE
5.8**

A temperature sensor is used to measure the temperature of an oil bath. The input temperature is increasing at a constant rate of 1.2°C/min, and the dynamic lag is 2.3 min.

(a) What is the dynamic error in degrees Celsius?

(b) If the range of the temperature sensor is 75 to 125°C, what is the dynamic error as a percentage?

Solution

(a) Let T_1 represent the input temperature at time t_1. At time $t_2 = t_1 + 2.3$ min, the input temperature will be $T_2 = T_1 + (1.2)(2.3) = T_1 + 2.76$°C. Thus, at time t_2, the input temperature will be $T_1 + 2.76$°C and the output temperature will be T_1. The dynamic error is the difference between the input temperature and the output temperature at time t_2.

$$\text{Dynamic error} = T_1 + 2.76 - T_1 = 2.76°C$$

(b) The range of the sensor is $125 - 75 = 50$°C. The dynamic error in percent is $100 \times 2.76/50 = 5.52\%$.

$$\text{Dynamic error} = 5.52\% \qquad ◆$$

Frequency Response

When it comes to control system analysis and design, frequency response and the transfer function are the most useful methods for defining the dynamic response of a measuring instrument. In Chapter 4 we introduced you to frequency response and presented a computer program that produces frequency-response data from the transfer function. In this section we apply to measuring instruments the methods presented in Chapter 4.

The liquid-filled thermometer in Figure 4.4 has a transfer function that is typical of many measuring instruments. A component with this transfer function is called a first-order lag because the highest derivative in the time-domain equation is first order. The transfer function for the liquid-filled thermometer is given by Equation (4.28), which is reproduced here for convenience:

$$\frac{T_m}{T_a} = \frac{1}{1 + \tau s}$$

In Example 5.4, we calculated a time constant of 73.2 s for a mercury-filled thermometer. The frequency response of this component is plotted in Figure 5.10. Notice how the gain is near 0 decibels (dB) at frequencies below 0.002 rad/s and how it drops at a rate of 20 dB per decade for frequencies above 0.1 rad/s. The low-frequency gain curve is a straight line at 0 dB. The high-frequency gain curve is also a straight line with a slope of -20 dB per decade increase in frequency. Extend the two lines and you will see that they intersect at the point on the 0-dB line marked with a diamond and labeled ω_b. This point is called the *break point,* and the radian frequency of the break point is equal to the reciprocal of the time constant, τ.

$$\text{Break-point frequency} = \omega_b = \frac{1}{\tau} \qquad \textbf{(5.9)}$$

All first-order lag components have frequency-response curves just like the one in Figure 5.10. The only difference is the break-point frequency, ω_b, which is given by Equation (5.9).

Some measuring instruments have a more complex transfer function than the first-order lag we just discussed. Figure 5.11 shows a liquid-filled thermometer enclosed in a protective sheath (or well). The sheath presents a second thermal resistance and the fluid in the well is

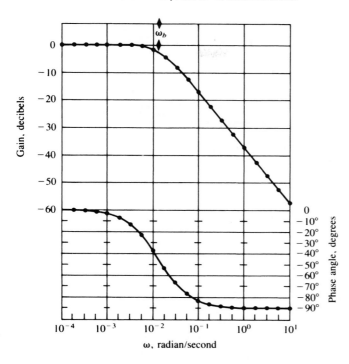

◆ **Figure 5.10**   Bode diagram of a mercury-filled thermometer with a time constant of 73.2 s (Example 5.7).

a second thermal capacitance. The two resistances and two capacitances form what is known as a two-capacity, interacting system. The differential equation for this system has both a second-order and a first-order derivative term, so this system is also called a *second-order lag*. The transfer function is given by Equation (5.10).

$$\frac{T_m}{T_a} = \frac{1}{1 + (\tau_1 + \tau_2 + \tau_2 R_1/R_2)s + (\tau_1\tau_2)s^2} \qquad (5.10)$$

where $\tau_1 = R_1C_1$ = time constant of the well

$\tau_2 = R_2C_2$ = time constant of the thermometer

R_1 = thermal resistance between the measured fluid and the inside of the protective sheath, K/W

R_2 = thermal resistance between the inside of the protective sheath and the liquid inside the bulb, K/W

C_1 = thermal capacitance of the sheath, J/K

C_2 = thermal capacitance of the bulb, J/K

The *frequency response* of a second-order system is more complex, and more interesting, than that of a first-order system, especially when the damping ratio, ζ, is less than 1. Figure 5.12 shows the Bode diagram of a second-order lag for values of ζ from 0.05 to 20. Notice that at low frequencies, the gain is near 0 dB, just like the first-order lag. At high frequencies, the second-order lag gain drops 40 dB per decade increase in frequency, twice the rate of the first-order lag. However, it is near the resonant frequency, ω_0, where the most interesting

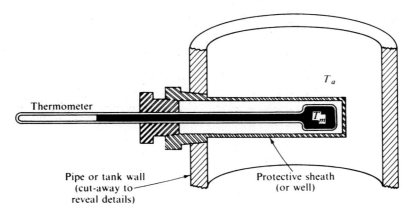

a) A cut-away view of a thermometer and protective sheath

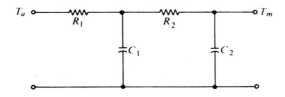

b) The equivalent electrical circuit for the thermometer and protective sheath.

◆ **Figure 5.11** Temperature sensor in a protective sheath and the electrical equivalent.

things happen. As the damping ratio drops below 1, the gain at the break point increases until it reaches 20 dB with a damping ratio of 0.05. A gain of 20 dB is a tenfold increase. In other words, the output is 10 times as large as the input at the resonant frequency when the damping ratio is 0.05. This phenomenon has many useful applications in electronic systems.

5.6 SELECTION CRITERIA

Selecting the right measuring instrument for a job is complicated by the fact that usually there is no single correct choice. The designer must choose from among a number of good choices, each with its own set of advantages and disadvantages. We have already mentioned some of the choices for temperature measurement: thermocouple, RTD, thermistor, IC sensor, liquid-filled element, vapor-filled element, gas-filled element, and radiation pyrometer. Flow measurement presents the following set of choices for the primary element: magnetic, orifice, venturi, vortex, pitot tube, turbine, thermal, ultrasonic, and variable area. Similar choices exist for other measurements, such as force, velocity, level, pressure, position, and displacement. There is no set procedure for selecting a measuring instrument, and judgment is an im-

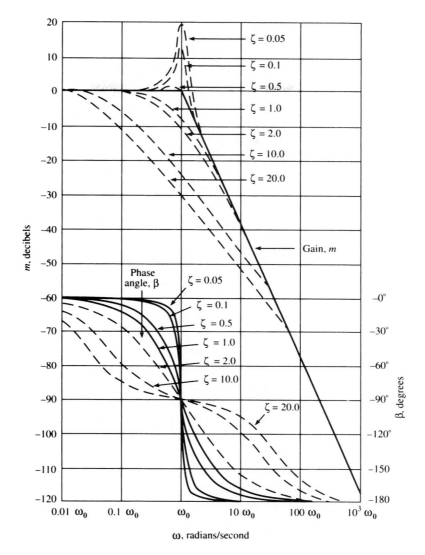

◆ **Figure 5.12** Bode diagram of second-order components.

portant part of the process. Some questions the designer might consider in the process of selecting a measuring instrument are listed here:

A. Questions about the measured variable
   1. What is the measured variable?
   2. What accuracy is required?
      a. At 100% of the input range?
      b. At 75% of the input range?
      c. At 50% of the input range?
      d. At 25% of the input range?

3. What fluid or solid is being measured?

4. What are the upper and lower range limits?

5. What overrange might occur?

6. What type of dynamic changes might occur in the measured variable?

    a. What is the maximum rate of change?

    b. What is the maximum frequency at which changes occur?

B. Questions about the measuring instrument

1. What is the primary sensor?

2. What effect will the primary sensor have on the measured variable?

3. What is the purpose of the output: indicate, transmit, record, totalize, other?

4. Is the output signal analog or digital?

5. What is the output signal: 3–15 psi, 4–20 mA, 0–5 V, 8-bit TTL digital, other?

6. What type of signal conditioning is provided?

    a. Amplification or attenuation?

    b. Signal conversion: resistance to current, millivolt to current, analog to digital, digital to analog, etc.?

    c. Filters: low pass, high pass, band pass?

    d. Linearization of the signal?

    e. Square-root function?

    f. Other special functions?

    g. Conversion to engineering units?

7. What is the operating principle of the transmitter?

8. What environmental conditions will the instrument encounter?

9. What are the thermal zero drift and thermal sensitivity drift ratings of the instrument?

10. What are the long-term zero drift and sensitivity drift ratings of the instrument?

11. What are the dead-band, hysteresis, and linearity ratings of the instrument?

C. Questions about economics

1. What is the initial cost of the instrument?

2. What is the installation cost?

3. What is the maintenance cost?

4. What is the expected life of the instrument?

5. What is the anticipated replacement cost?

◆ GLOSSARY

Accuracy: The degree of conformity of the output of a measuring instrument to the ideal value of the measured variable. (5.4)

Accuracy, measured: The maximum positive and negative errors in a calibration report. (5.4)

Bias: The difference obtained by subtracting the ideal value from the average of repeated measurements of a measured variable. (5.4)

Calibration: The procedure of determining the accuracy of a measuring instrument. (5.4)

Calibration curve: A graphical presentation of calibration data. (5.4)

Calibration report: A tabular presentation of calibration data. (5.4)

Damping coefficient (designated by α):   A number that determines the type of damping of a second-order system. The response is underdamped if $\alpha < \omega_0$, critically damped if $\alpha = \omega_0$, and over-damped if $\alpha > \omega_0$. (5.5)

Damping ratio (designated by ζ):   The ratio of damping coefficient (α) over resonant frequency (ω_0). The response is underdamped if $\zeta < 1$, critically damped if $\zeta = 1$, and overdamped if $\zeta > 1$. (5.5)

Dead band: The smallest change in the measured variable that will result in a measurable change in the output. (5.3)

Drift: An undesirable change over a specified period of time. (5.3)

Drift, sensitivity: A change over time in the sensitivity of a measuring instrument. (5.3)

Drift, zero: A change over time in the output of a measuring instrument when the measured variable is at the lower range limit. (5.3)

Dynamic error: The difference between the input temperature and the output temperature at a given time. (5.5)

Dynamic lag: The amount of time that elapses between the time the input reaches a certain temperature and the time the output reaches that same temperature. (5.5)

Error: The difference determined by subtracting the ideal value from the measured value. (5.4)

Hysteresis: The dependence of the value of a variable on the history of past values and the current direction of traversal. (5.4)

Linearity: The closeness of the calibration curve to a straight line. (5.4)

Linearity, independent: Comparison of the calibration data with a straight line that minimizes the maximum value of the deviation between the data and the line. (5.4)

Linearity, least-squares: Comparison of calibration data with the straight line that minimizes the sum of the squares of the deviations between the data and the straight line. (5.4)

Linearity, terminal-based: Comparison of the calibration data with a straight line that connects the two ends of the calibration curve. (5.4)

Linearity, zero-based: Comparison of the calibration data with a straight line that passes through the zero point of the calibration curve and minimizes the maximum value of the deviation between the data and the line. (5.4)

Natural frequency (designated by ωd):   The frequency of oscillation of an underdamped second-order system. (5.5)

$$\omega_d = \sqrt{\omega_0^2 - \alpha^2} = \omega_0 \sqrt{1 - \zeta^2}$$

Operating conditions: The environment in which a measuring instrument operates. (5.3)

Operative limits: The end values of the range of operating conditions that will not cause permanent impairment of a measuring instrument. (5.3)

Overrange: Any excess in the value of the measured variable above the upper range limit or below the lower range limit. (5.3)

Overrange limit: The maximum overrange that can be applied to a measuring instrument without causing damage or permanent change in performance. (5.3)

Overshoot: The maximum height of the step response curve measured above the 100% change line. (5.5)

Range: The values of the measured variable that can be measured by the measuring instrument. The range includes all values between the *lower range limit* and the *upper range limit*. (5.3)

Reliability: The probability that a component will do its job for a specified time period under a specified set of conditions. (5.3)

Repeatability: The maximum deviation from the average of repeated measurements of the same static variable. (5.4)

Reproducibility: The maximum difference between a number of outputs for the same input, taken over an extended period of time, approaching from both directions. Reproducibility includes hysteresis, dead band, drift, and repeatability. (5.4)

Resolution: A single step of output in a measuring instrument whose output changes in discrete steps. (5.3)

Resonant frequency (designated by ω_0): The break-point frequency of an underdamped second-order component. (5.5)

Response time: The time required for the output to reach a designated percentage of the total change, after a step change in input. (5.5)

Rise time: The time required for the output to go from a small percentage of change to a large percentage of change, after a step change in input. Unless otherwise specified, the change is from 10% to 90%. (5.5)

Sensitivity: The ratio of the change in output divided by the change in input that caused the change in output. (5.3)

Settling time: The time it takes for the response of an underdamped component to remain within a small band above and below the 100% change line. (5.5)

Span: The size of the range, equal to the upper range limit minus the lower range limit. (5.3)

Systemic error: Another name for bias. (5.4)

Thermal sensitivity shift: A change in the sensitivity of a measuring instrument caused by a specified change in the ambient temperature. (5.3)

Thermal zero shift: A change in the zero output of a measuring instrument caused by a specified change in the ambient temperature. (5.3)

Time constant (designated by r): The time required for the output of a component to reach 63.2% of the total change after a step change in input. (5.5)

◆ EXERCISES

Section 5.1

1.1 Compute the mean and standard deviation of the following sets of measurements:

(a) 76°C, 77°C, 75°C, 76°C, 75°C
(b) 2.3 gpm, 2.5 gpm, 2.5 gpm, 2.3 gpm, 2.4 gpm
(c) 7.4 lpm, 7.3 lpm, 7.5 lpm, 7.6 lpm, 7.4 lpm
(d) 410 kPa, 411 kPa, 413 kPa, 412 kPa, 413 kPa, 413 kPa
(e) 32.6 psi, 32.5 psi, 32.8 psi, 32.6 psi, 32.9 psi, 32.8 psi
(f) 4.26 m, 4.25 m, 4.26 m, 4.25 m, 4.23 m, 4.24 m

Section 5.2

5.2 A pressure sensor is tested for repeatability by increasing the input pressure from 0 to 15 psi 10 times and recording the output reading each time the input reaches 15 psi. The output readings are as follows:

15.45, 15.53, 15.61, 15.42, 15.55
15.47, 15.51, 15.59, 15.46, 15.60

(a) Calculate the mean and standard deviation of the 10 readings.
(b) Use three standard deviations to estimate the range of readings you would expect in a very large number of test runs.

5.3 Repeat Exercise 5.2 for the following output readings:

14.64, 14.72, 14.69, 14.62, 14.61
14.67, 14.71, 14.78, 14.75, 14.80

Section 5.3

5.4 Determine the span of each of the following measuring transmitters:

Type	Upper Range Limit	Lower Range Limit
(a) Temperature	150°F	−50°F
(b) Pressure	100 psi	0 psi
(c) Level	4 ft	2 ft
(d) Flow rate	15 gpm	5 gpm
(e) Force	210 lbf	200 lbf
(f) Temperature	250°C	50°C
(g) Pressure	200 kPa	0 kPa
(h) Level	1.5 m	1.0 m
(i) Flow rate	12 lpm	8 lpm
(j) Force	10 N	8 N

5.5 A potentiometer has 800 turns of wire, a lower range limit of 0 V, and an upper range limit of 20 V. Determine the span in volts, the average resolution as a percentage of the span, and the average resolution in volts.

5.6 A pressure gage with a range from 3 to 15 psi was tested for dead band. When the input pressure was increased from 9.00 to 9.05 psi, the output remained stationary at 9.10 psi. When the input was increased from 9.00 to 9.06 psi, the output changed from 9.10 to 9.11 psi. Express the dead band in pounds per square inch, as a percentage of the lower input value, and as a percentage of the span.

5.7 A potentiometer is used to measure the position of a shaft. The input range of shaft position is from $-160°$ to $+160°$. The corresponding output range is from -20 to $+20$ V. What is the span of this position sensor? What is its sensitivity in volts per degree?

5.8 The following results were obtained from the calibration of a force transducer:

Input Force (N)	Output Voltage (V)
0	0.06
2	0.63
4	1.20
6	1.77
8	2.35
10	2.94
12	3.55
14	4.17
16	4.80
18	5.43
20	6.06

Determine the sensitivity of the force transducer in volts per newton at 20 and 80% of full scale. Hint: Divide the change in output voltage between 2 and 6 N by the change in force (i.e., $6 - 2 = 4$ N) to determine the sensitivity at 20%.

5.9 At 70°F, an input of 0 psi to a pressure transmitter produces an output of 4 mA, and an input of 100 psi produces an output of 20 mA. The transmitter has a thermal zero shift of 0.02 ma/°F and a thermal sensitivity shift of 0.0004 (mA/psi)/°F. Determine the following:

(a) The sensitivity of the transmitter at 70°F.

(b) The sensitivity of the transmitter at 20°F and 120°F.

(c) The 0-psi output of the transmitter at 20°F and 120°F.

(d) The 100-psi output of the transmitter at 20°F and 120°F.

5.10 A temperature transmitter has an input range of 0 to 100°C, an output range of 4 to 20 mA, and an accuracy of $\pm0.5\%$. The ideal transmitter has a linear output from 4 mA at 0°C to 20 mA at 100°C. If the output of the transmitter is 13.6 mA, what are the ideal, minimum, and maximum possible temperatures?

5.11 A pressure transmitter has an input range of 100 kPa to 150 kPa, an output range of 4 to 20 mA, and an accuracy of $\pm1\%$. The ideal transmitter has a linear output from 4 mA at 100 kPa to 20 mA at 150 kPa. If the output is 15.2 mA, what are the ideal, minimum, and maximum possible pressures?

Section 5.4

5.12 The accuracy of the position sensor in Exercise 5.7 is $\pm1\%$ of span. If the output is $+8$ V, what is the ideal position? What are the minimum and maximum possible positions?

5.13 The measurements in Exercise 5.1 were obtained in tests for repeatability. Each measurement was obtained by traversing the sensor from the lower range limit to the ideal measurement and recording the equilibrium measurement of the sensor. Determine the bias and repeatability of each sensor in terms of the measured variable and as a percentage of span.

	Range Limit		Ideal
	Upper	Lower	Measurement
(a)	100°C	0°C	75°C
(b)	10 gpm	0 gpm	2.7 gpm
(c)	20 lpm	0 lpm	7.8 lpm
(d)	500 kPa	400 kPa	416 kPa
(e)	40 psi	30 psi	32 psi
(f)	5.0 m	4.0 m	4.2 m

5.14 Determine the following from the test of the pressure sensor in Exercise 5.2:

(a) The bias or systemic error in pounds per square inch and in percentage of span.

(b) The repeatability in pounds per square inch and in percentage of span.

5.15 Repeat Exercise 5.14 for the pressure sensor test in Exercise 5.3.

5.16 Explain the difference between repeatability and reproducibility.

5.17 Determine the measured accuracy of the pressure sensor in Exercise 5.2. Express the maximum positive and negative errors in pounds per square inch and in percentage of span.

5.18 Determine the measured accuracy of the pressure sensor in Exercise 5.3. Express the maximum positive and negative errors in pounds per square inch and in percentage of span.

5.19 The following data are the average upscale and downscale values from a calibration report. Both the input and the output are expressed in terms of percentage of span. Plot the calibration curve and determine the combined hysteresis and dead band.

	Upscale		*Downscale*	
Input	Output	Input	Output
0	2.5	100	90.2
10	7.2	90	84.8
20	13.4	80	78.0
30	20.6	70	70.0
40	29.1	60	61.3
50	38.3	50	52.2
60	47.8	40	43.1
70	57.1	30	33.9
80	67.0	20	24.0
90	77.5	10	14.0
100	90.2	0	2.5

5.20 Draw the calibration curve for the force transducer in Exercise 5.8. Then draw the straight lines for independent linearity, terminal-based linearity, and zero-based linearity. Determine the maximum error for each straight line.

5.21 The following data are the averages of the upscale and downscale readings from four calibration reports. For each report, draw the calibration curve and then draw the straight lines for independent linearity, terminal-based linearity, and zero-based linearity. Determine the maximum error for each straight line.

	Average Output			
Input	(a)	(b)	(c)	(d)
0	2.8	4.0	3.0	0.0
10	9.6	15.7	10.5	9.9
20	17.4	26.7	18.2	18.8
30	26.5	37.1	27.7	26.7
40	38.7	46.9	38.2	33.6
50	50.5	56.0	50.2	40.0
60	61.1	64.5	61.9	47.4
70	70.0	72.3	72.4	54.8
80	76.8	79.5	81.7	64.2
90	83.2	86.0	89.5	75.1
100	88.8	92.0	97.0	88.0

Section 5.5

5.22 Explain the difference between static characteristics and dynamic characteristics.

5.23 The following data were obtained from four temperature probes that were plunged from a liquid bath maintained at 50°C into a second bath maintained at 100°C. For each probe, plot the response curve and determine the 95% response time, the time constant, and the 10 to 90% rise time.

	Temperature (°C)			
Time (s)	(a)	(b)	(c)	(d)
0	50.0	50.0	50.0	50.0
10	66.0	59.1	62.7	60.6
20	76.8	66.5	72.2	68.9
30	84.2	72.6	79.3	75.5
40	89.3	77.5	84.6	80.7
50	92.7	81.6	88.5	84.8
60	95.0	84.9	91.4	88.0
70	96.6	87.7	93.6	90.6
80	97.7	89.9	95.2	92.6
90	98.4	91.7	96.5	94.1
100	98.9	93.2	97.4	95.4
110	99.3	94.5	98.0	96.4
120	99.5	95.5	98.5	97.1
130	99.7	96.3	98.9	97.7
140	99.8	97.0	99.2	98.2
150	99.8	97.5	99.4	98.6
160	99.9	98.0	99.6	98.9
170	99.9	98.3	99.7	99.1
180	100.0	98.6	99.8	99.3
190	100.0	98.9	99.8	99.5
200	100.0	99.1	99.9	99.6

5.24 The following data were obtained from a step response test of an underdamped component. Plot the response curve and determine the 10 to 90% rise time, the overshoot, and the 2% settling time.

Time (s)	Output (%)
0	0.0
5	36.5
10	74.5
13.5	100.0
15	110.0
18.4	120.0
21.8	110.0
24.0	100.0
25.0	96.0

28.4	91.0
30.0	92.0
32.6	96.0
34.0	100.0
36.0	103.0
39.5	105.0
42.5	103.0
44.4	100.0
50.0	97.8
55.0	100.0
60.0	101.0

Measured Variable	Steady State Gain, G^*	Damping Constant, α (1/s)	Resonant Frequency ω_0
(a) Position	0.001	0.5	10
(b) Voltage	1	2100	2100
(c) Liquid level	0.41	0.0007	0.0014
(d) Motor speed	22	1.26	12.6
(e) Temperature	1	0.2	0.02
(f) Pressure	0.2	22	11

$^*G$ is dimensinless (% full scale out/% full scale in)

5.25 Determine the time constant of an alcohol-filled thermometer. The bulb has a diameter of 0.25 cm and a length of 0.85 cm. The film coefficient is estimated to be 12 W/m^2 · K. Ignore the resistance and thickness of the bulb wall.

5.26 For each of the following ramp response tests, determine the dynamic error in terms of the measured variable and as a percentage of span.

Measured Variable	Rate of Increase	Dynamic Lag	Span
(a) Temperature	2°C/min	3.6 min	100°C
(b) Pressure	0.2 psi/s	0.8 s	10 psi
(c) Flow rate	0.25 lpm/s	1.4 s	20 lpm
(d) Speed	10 rpm/ms	4 ms	5000 rpm

5.27 Use the program BODE from Chapter 4 to generate frequency-response data for each of the following first-order measuring instruments. Use the data to construct Bode diagrams for each instrument. See Example 4.19.

Measured Variable	Steady State Gain, G^*	Time Constant, τ (s)
(a) Temperature	1	800
(b) Level	0.82	7020
(c) Flow rate	2	10
(d) Gas pressure	2	2.8
(e) Blending	1	1000
(f) Voltage	1	0.5

$^*G$ is dimensionless (% full scale out/% full scale in)

5.28 Use the program BODE from Chapter 4 to generate frequency-response data for each of the following second-order measuring instruments. Use the data to construct Bode diagrams for each instrument. See Equation (5.7). Compute the value of the damping ratio for each instrument and compare your Bode diagrams with Figure 5.12.

5.29 A liquid-filled thermal element is enclosed in a protective well. The following thermal resistances and capacitances were computed for the element and well.

$$R_1 = 125 \text{ K/W}$$

$$C_1 = 0.45 \text{ J/K}$$

$$R_2 = 65 \text{ K/W}$$

$$C_2 = 0.72 \text{ J/K}$$

The transfer function is given by Equation (5.10). Determine the coefficients of s^2 and s, and write the transfer function. Use the program BODE to generate frequency response data and plot the Bode diagram.

5.30 A measuring instrument has a resonant frequency (ω_0) of 5 rad/s, a damping coefficient (α) of 3 1/s, and a steady-state gain (G) of 0.2 mA/°C. Determine the following:
(a) The transfer function, C_m/C (see Equation 5.7).
(b) The frequency-domain response, C_m, to a step change in input, C, from 0°C to 100°C (i.e., multiply the transfer function by $C = 100/s$). Put the result in the form of Equation (5.8).
(c) Determine the roots of the quadratic term in the denominator of the equation you obtained in step b. Verify that the response is underdamped.
(d) Complete the partial fraction expansion of C_m.
(e) Use transform pairs 1 and 10 in Table 4.1 to get the time-domain expression, c_m for the step response.
(f) Use a spreadsheet or a program similar to the following to generate a table of $c_m(t)$ versus t values from the time-domain expression you obtained in part e.

```
10  FOR T = 0 to 3 STEP .05
20  CM(T) = 10 + 12*EXP (−4*T)
    *COS (6*T) + 110.03/57.3)
30  LPRINT USING ´´ ##.## ##.##´´;
    T, CM(T)
```

```
40  NEXT  T
50  End
```

(g) From the table you produced in part f, determine the 10 to 90% rise time, the peak percentage of overshoot, and the 2% settling time. You can do this by plotting a graph and reading the values from the graph, or you can get the values by interpolation directly from the data table.

◆ CHAPTER 6   **Signal Conditioning**

◆ OBJECTIVES

A signal conditioner prepares a signal for use by another component. The input to a signal conditioner is usually the output from a sensor (or primary element). The operations performed by a signal conditioner include isolation, impedance conversion, noise reduction, amplification, linearization, and conversion.

The purpose of this chapter is to give you the ability to analyze, specify, or design signal conditioning systems. After completing this chapter, you will be able to

1. Explain and/or specify
   a. Ideal and practical models of an op amp
   b. Common-mode voltage
   c. Common-mode gain
   d. Common-mode rejection ratio
   e. Acquisition time of a sample-and-hold circuit
   f. Isolation and impedance transformation
   g. Quantization error of an A/D converter
2. Design or construct the following
   a. Inverting amplifier
   b. Noninverting amplifier
   c. Summing amplifier
   d. Differential amplifier
   e. Instrumentation amplifier
   f. Low-pass filter
   g. Notch filter
   h. I/O table or graph of a linearizer
   i. Piecewise-linear approximation of a nonlinear function
   j. Calibration graph of an RTD/Wheatstone bridge circuit

6.1 INTRODUCTION

A measuring transmitter consists of two parts: a primary element (or sensor) and a signal conditioner. The primary element uses some characteristic of its material and construction to convert the value of a measured variable into an electrical, mechanical, or fluidic signal. The output of the primary element may be a small voltage or an electrical resistance. It may be a force, a displacement, a pressure, or some other phenomenon. No matter what form it takes, the primary element output depends on the value of the measured variable.

The output of the primary element is a measure of the variable it is sensing, but it is seldom in a form suitable for transmission and use by other components in a system. Usually, a signal conditioner is required to convert the primary element output into an electrical (or pneumatic) signal suitable for use by a controller or display device. The standard transmission signals are a 4- to 20-mA electric current or a 3- to 15-psi pneumatic pressure. For example, a thermocouple (T/C) is a primary element that converts temperature into a small voltage. A signal conditioner converts the millivolt output of the thermocouple into a 4- to 20-mA electric current. A T/C temperature transmitter includes both the thermocouple and the signal conditioner. The following are some of the tasks performed by a signal conditioner:

1. Isolation and impedance conversion
2. Amplification and analog-to-analog conversion
3. Noise reduction (filtering)
4. Linearization
5. Data sampling
6. Digital-to-analog conversion
7. Analog-to-digital conversion

The operational amplifier is the heart of many modern signal conditioners. For this reason, our study of signal conditioning begins with the operational amplifier and its application in various signal conditioning circuits. You should complete the section on op-amp circuits before moving on to the section on analog signal conditioning.

One of the objectives of a measuring transmitter is to minimize the effect of the measurement on the measured variable. In some situations, this means electrical isolation of the primary element with an isolation amplifier. In other situations, it means providing a high impedance at the input to the signal conditioner. Our study of signal conditioning begins with *isolation and impedance conversion.*

Another objective of a measuring transmitter is to convert the output from the primary element into a signal suitable for use by a controller or display device. Some primary elements convert changes in the measured variable into small changes in resistance. Strain gages and resistance temperature detectors (RTDs) are two examples. A bridge circuit is often used to convert the change in resistance into a small change in voltage. This small voltage is then amplified and converted to an electric current signal suitable for use by the controller. Our study of signal conditioning continues with *amplification and analog-to-analog conversion.*

Industry presents a harsh environment for making precise measurements. There are many sources of electrical interference, and noise is always a concern when electrical signals are transmitted. *Noise* reduction with various types of filters is another topic for our study.

Linearization is the next topic. The ideal measuring transmitter output is a linear signal that goes from 0 to 100% as the measured variable goes from the lower range limit to the upper range limit. Sometimes the signal is close to the ideal. At other times, the signal is very nonlinear. An orifice flow element, for example, produces a pressure drop that is proportional to the square of the flow rate. Linearization of this type of signal is a very desirable goal. Thermocouples also produce a nonlinear signal, although not nearly as nonlinear as the flow orifice. Linearization of thermocouple signals is also desirable.

Digital signal conditioning is our final topic. *Data sampling* is a process in which a switch momentarily closes to connect an analog signal to a sample-and-hold circuit. The sample-and-hold circuit maintains the analog signal until the next momentary switch closure. While the analog signal is held, an *analog-to-digital converter* converts the analog signal into an equivalent digital signal. Once the signal is converted to digital, a microprocessor can use the digital samples to perform digital signal conditioning such as filtering, linearization, calibration, and conversion to engineering units. A digital controller receives the digital samples and produces a digital output signal. A *digital-to-analog converter* converts the digital output of the controller into an analog signal suitable for use by the final control element.

6.2 THE OPERATIONAL AMPLIFIER

An *operational amplifier* is a high-gain dc amplifier with two inputs and one output. The output is equal to the difference between the voltages on the two inputs multiplied by the gain of the amplifier. The gain of different op amps ranges from 10^4 to 10^7, with 10^5 being a typical value. An op amp requires both a positive and a negative voltage supply, labeled $+V_{cc}$ and $-V_{cc}$, respectively. Two more inputs are provided for offset null adjustments, bringing the total number of terminals to seven. Op amps come in several different IC packages, including the popular eight-lead MINIDIP package.

The operating characteristics of an operational amplifier circuit depend almost entirely on components external to the amplifier. The gain, input impedance, output impedance, and frequency response depend on the external resistors and capacitors in the circuit, not on the gain, input impedance, or output impedance of the amplifier. This means that the behavior of an op-amp circuit can be made to fit a particular application by proper selection and placement of a few resistors and capacitors. Also, the stability of these components can be selected to meet the requirements of the application. With this versatility and ease of design, it is not hard to understand why the op amp has become such a popular component in control systems.

Op-Amp Equivalent Circuit

Figure 6.1 shows the circuit symbol and an equivalent circuit of an operational amplifier. The two inputs are called the *inverting input* and the *noninverting input*. We will use the symbol v_1 for the voltage at the inverting input, v_2 for the voltage at the noninverting input, v_{out} for the voltage at the output terminal, V_{cc} for the positive supply voltage, and $-V_{cc}$ for the negative supply voltage. The currents of interest are the current entering the inverting input, i_1, the current entering the noninverting input, i_2, and the current leaving the output terminal, i_{out}. Notice that $i_1 = -i_2$, and we needed to define only one of the two currents. However, it is useful in analyzing op-amp circuits to be able to refer to both i_1 and i_2.

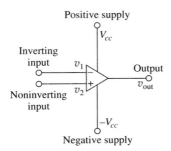

a) Operational amplifier circuit symbol

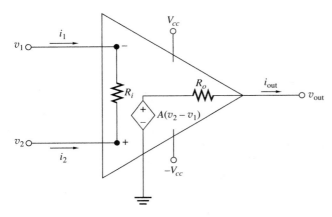

b) Operational amplifier equivalent circuit

◆ **Figure 6.1** Circuit symbol and equivalent circuit for an operational amplifier. Typical values for a 741 op amp are $R_i = 2\ M\Omega$, $R_o = 75\ \Omega$, and $A = 10^5$. In many applications, the op amp can be considered to be ideal, with R_i equivalent to an open circuit, R_o equal to zero, and A approaching infinity.

The gain of the op amp, A, determines the output voltage as follows:

$$v_{out} = A(v_2 - v_1) \tag{6.1}$$

The output voltage of an op amp is limited by the supply voltages. When the output reaches its positive or negative limit, we say that the op amp has saturated. For most op amps, the saturation voltage level is about 80% of the supply voltage.

$$-V_{sat} \leq v_{out} \leq +V_{sat} \tag{6.2}$$

$$-V_{sat} = -0.8V_{cc} \text{ (approximately)}$$

$$+V_{sat} = +0.8V_{cc} \text{ (approximately)}$$

When the op amp is not saturated, its output voltage is between the two saturation voltages, and we say that it is operating in its linear region. The high gain of the op amp and the relatively low supply voltage (usually ±20 V) combine to limit the difference between v_1 and v_2 to very small values. Let us illustrate with Example 6.1.

EXAMPLE   Determine the value of $v_2 - v_1$ that will saturate an op amp if the gain, A, is 10^5, and the supply volt-
6.1   ages are -20 V and $+20$ V.

Solution   The amplifier will saturate at 80% of the supply voltages, about -16 V and $+16$ V. At $+16$ V, the dif-
                ference is

$$v_2 - v_1 = \frac{16}{10^5} = 0.16 \text{ mV}$$

At -16 V, the difference is

$$v_2 - v_1 = \frac{-16}{10^5} = -0.16 \text{ mV}$$

◆

Ideal Op-Amp Model

Figure 6.2 shows a graph of the open-loop response of an op amp. The sloping straight line
in the center of the graph is called the *linear operating region*. The horizontal lines on either
side make up the *saturation region*. Some op-amp applications utilize the saturation region,
but most op-amp circuits are designed to operate in the linear region.

Five simplifying assumptions are used to define the model of an ideal op amp operating
in the linear region. These assumptions greatly simplify the analysis of linear op-amp circuits,
often with negligible effect on the accuracy of the results. The five assumptions are as follows:

Assumption	*Result*
1. Infinite gain, $A = \infty$	$v_1 = v_2$
2. Infinite input resistance, $R_i = \infty$	$i_1 = i_2 = 0$ A
3. Zero output resistance	$R_o = 0 \ \Omega$
4. Infinite slew rate	No frequency limit
5. No offsets	Ignore offset error

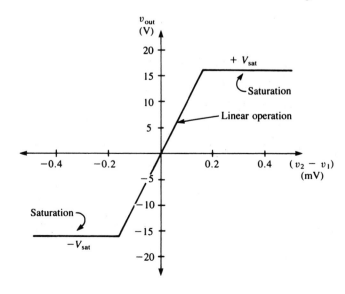

◆ **Figure 6.2**   Open-loop response of an op
amp with a gain of 10^5 and supply voltages of
± 20 V. The straight line in the middle of the
graph is called the linear operating region. The
straight lines at the top and bottom of the graph
are called the saturation region.

Example 6.1 showed us that the difference between v_1 and v_2 is only a fraction of a millivolt. Assumption 1 states that we can neglect this difference in most applications. The practical result of assumption 1 is stated in Equation (6.3):

$$v_1 = v_2 \tag{6.3}$$

The internal resistance between the two input terminals, R_i, is large, ranging from 50 kΩ to over 100 MΩ. A typical value of R_i is 2 MΩ. The small value of $v_2 - v_1$ and the large value of R_i combine to limit i_2 to extremely small values. For the conditions in Example 6.1 with $R_i = 2$ MΩ, the value of i_2 at positive saturation is

$$i_2 = \frac{v_2 - v_1}{R_i}$$

$$= \frac{1.6 \times 10^{-4}}{2 \times 10^6} = 8 \times 10^{-11} \text{ A}$$

The value of current i_2 is so small that we say it is *virtually* zero.

Assumption 2 states that we can neglect currents i_1 and i_2 in most applications. The practical result of assumption 2 is stated in Equation (6.4):

$$i_1 = i_2 = 0 \text{ A} \tag{6.4}$$

Assumption 3 is stated in Equation (6.5):

$$R_o = 0 \ \Omega \tag{6.5}$$

The *slew rate* is the maximum rate at which the output of an op amp can change. Assumption 4 states that there is no such limit and the output can change instantaneously from one value to another. Assumption 5 states that there are no offset errors in the op-amp circuit.

The five assumptions define what is called the *ideal* operational amplifier. We will use this model frequently in analyzing op-amp circuits.

Common-Mode Rejection Ratio

In an ideal op amp, the output depends only on the difference between v_2 and v_1, not on the level of the two voltages. Consider the following two sets of conditions:

Condition 1:
$$v_1 = -50 \ \mu V$$
$$v_2 = +50 \ \mu V$$
$$A = 10^5$$

Condition 2:
$$v_1 = 99.95 \text{ mV}$$
$$v_2 = 100.05 \text{ mV}$$
$$A = 10^5$$

In both conditions, $v_2 - v_1 = 100 \ \mu V$ and $A(v_2 - v_1) = 10.0$ V. Thus the output of an ideal op amp is 10.0 V for both conditions 1 and 2.

In real life, operational amplifiers are not ideal, and the output will be different for the two conditions. The reason for the difference is the slight differences in the way the op amp handles the two inputs. The inputs are amplified by slightly different gains, with the result that the common level of the two signals is amplified and added to the output. We call this common level the *common-mode voltage*, v_c. The common-mode voltage is simply the average of v_1 and v_2:

$$v_c = \frac{v_1 + v_2}{2} \tag{6.6}$$

In a practical op amp, the common-mode voltage is multiplied by the common-mode gain, A_c, and then added to the output of the op amp. The output consists of the following two components:

$$v_{\text{out}} = A_c v_c + A(v_2 - v_1) \tag{6.7}$$

Let's continue our example of the two conditions by assuming a common-mode gain, A_c, of 10. The outputs including the common-mode component are

Condition 1: $v_c = \dfrac{50 + -50}{2} = 0 \, \mu\text{V}$

$v_{\text{out}} = A_c v_c + A(v_2 - v_1)$

$= 10(0) + 10^5 [50 - (-50)] \times 10^{-6} = 10.0 \text{ V}$

Condition 2: $v_c = \dfrac{99.5 + 100.5}{2} = 100 \text{ mV}$

$v_{\text{out}} = A_c v_c + A(v_2 - v_1)$

$= 10(0.1) + 10^5 (100.05 - 99.95) \times 10^{-3} = 11.0 \text{ V}$

Thus a common mode of only 0.1 V results in a 10% increase in the output voltage. If the common mode is raised to 1 V, the op amp will saturate. Clearly, common-mode voltage can be a problem in applications where it is present. The ability of an op amp to minimize the influence of the common-mode voltage is measured by the ratio of the differential gain of the op amp, A, divided by the common-mode gain, A_c. This ratio is called the *common-mode rejection ratio (CMRR)*:

$$\text{CMRR} = \frac{A}{A_c} \tag{6.8}$$

The *common mode rejection* (CMR) is the logarithm of CMRR expressed in decibel units:

$$\text{CMR} = 20 \log_{10}\left(\frac{A}{A_c}\right) \tag{6.9}$$

The CMRR and CMR of our example are

$$\text{CMRR} = \frac{10^5}{10} = 10^4$$

$$\text{CMR} = 20 \log_{10}(10^4) = 80 \text{ dB}$$

EXAMPLE 6.2

In a certain application of an op amp, a differential voltage ($v_2 - v_1$) of 120 μV has a common-mode voltage that varies from 0 to 2 V. The amplifier differential gain, A, is 10^5. The common-mode error must not be greater than 1.5% of the output when the common-mode voltage is 0. Specify the common-mode rejection ratio for this application.

Solution

First, determine the output with no common-mode voltage; second, determine the maximum allowable output voltage with a common mode of 2 V; third, determine the common-mode gain; and fourth, determine the value of CMRR.

$$v_{out}|_{v_c=0} = A(v_2 - v_1) = (1E + 5)(120E - 6) = 12V$$

$$v_{out}|_{v_c=2} = 12\left(\frac{101.5}{100}\right) = 12.18V$$

$$A_c = \frac{12.18 - 12}{v_c} = \frac{0.18}{2} = 0.09$$

$$CMRR = \frac{1E + 5}{0.09} = 1.11 \times 10^6$$

$$CMR = 121 \text{ dB} \qquad \blacklozenge$$

6.3 OP-AMP CIRCUITS

This section covers a number of op-amp circuits that are used in control systems. The op-amp model used in the circuits assumes that $i_1 = i_2 = 0$, $v_1 = v_2$, and $R_o = 0$.

Comparator

A *comparator* is a circuit that accepts two input voltages and indicates which voltage is greater. It is the simplest operational-amplifier circuit. All we need to make a comparator is an op amp in the open-loop configuration with no feedback or input resistors.

The comparator circuit in Figure 6.3 uses a 12-V indicator lamp to indicate the output of the comparator. An analysis of this circuit will show that the lamp is OFF when v_2 is less than v_1, and it is ON when v_2 is greater than v_1 by more than a fraction of a millivolt. The op amp in Figure 6.3 has a gain of 105, a positive supply voltage (V_{cc}) of 15 V, and a negative supply voltage ($-V_{cc}$) of 0 V. The supply voltages limit the op-amp output to the following voltage range:

$$0 \text{ V} \leq v_{out} \leq (0.8)(15) \text{ V}$$

$$0 \text{ V} \leq v_{out} \leq 12V$$

The graph in Figure 6.3 shows the output voltage versus the difference between the input voltages ($v_2 - v_1$). The graph can be divided into the following three regions:

1. Negative saturation region: ($v_2 - v_1$) < 0 V
   a. v_2 is less than v_1
   b. $v_{out} = 0$ V
   c. lamp is OFF

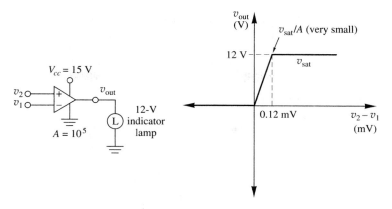

◆ **Figure 6.3**   An op-amp comparator signals which of two input voltages is greater. The 12-V indicator lamp is OFF when $v_2 < v_1$. The lamp goes ON as v_2 increases from v_1 to $v_1 + 0.12$ mV and stays ON when $v_2 > v_1 + 0.12$ mV.

2. Linear operating region: $0\ V \le (v_2 - v_1) \le 0.12$ mV
   a. v_2 is slightly greater than v_1
   b. v_{out} goes from 0 V to 12 V as $(v_2 - v_1)$ increases
   c. lamp goes from OFF to ON as $(v_2 - v_1)$ increases
3. Positive saturation region: $(v_2 - v_1) > 0.12$ mV
   a. v_2 is more than 0.12 mV greater than v_1
   b. $v_{out} = 12$ V
   c. lamp is ON

The effect of the linear operating region is small enough that we can make the following practical statements about the comparator circuit in Figure 6.3.

PRACTICAL OPERATION OF THE COMPARATOR CIRCUIT
IN FIGURE 6.3

The lamp is OFF when v_2 is less than v_1.
The lamp is ON when v_2 is greater than v_1.

One practical application of a comparator circuit is the voltage-level-indicators illustrated in Figure 6.4. The high-level indicator in Figure 6.4a uses a voltage divider and a 10-V source to produce an 8-V reference voltage. The op amp compares the input voltage with the 8-V reference and operates the HI lamp as follows.

PRACTICAL OPERATION OF THE HIGH-LEVEL INDICATOR
IN FIGURE 6.4A

When $v_{in} \le 8$ V, the HI lamp is OFF.
When $v_{in} > 8$ V, the HI lamp is ON.

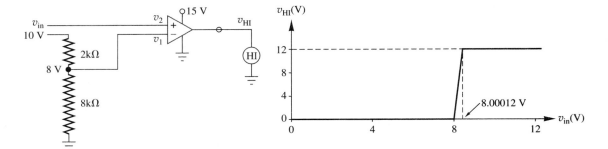

a) A high-level indicator. When $v_{in} < 8$ V, the HI lamp is OFF. As v_{in} increases from 8 V to 8.00012 V, the HI lamp goes from OFF to ON. When $v_{in} > 8.00012$ V, the HI lamp is ON.

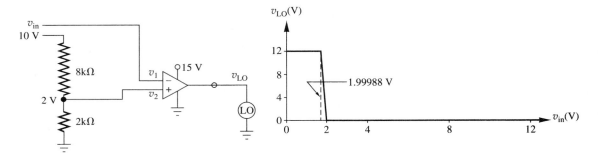

b) A low-level indicator. When $v_{in} < 1.99988$ V, the LO lamp is ON. As v_{in} increases from 1.99988 V to 2 V, the LO lamp goes from ON to OFF. When $v_{in} > 2$ V, the LO lamp is OFF.

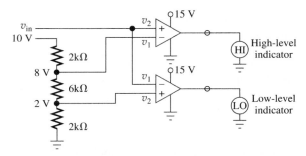

c) A high-low-level indicator combines the high-level and low-level indicators into a single circuit. In this circuit, the HI and LO lamps operate the same as they do in (a) and (b) above.

◆ **Figure 6.4** These voltage-level-indicator circuits use a voltage divider to produce reference voltages of 2 V and 8 V.

A low-level indicator is shown in Figure 6.4b. Notice the reversed position of the two inputs of the op amp. In this circuit, the input voltage is connected to the negative input of the op amp and the 2-V reference voltage is connected to the positive input. The op amp compares the input voltage with the 2-V reference and operates the LO lamp as follows.

PRACTICAL OPERATION OF THE LOW-LEVEL INDICATOR
IN FIGURE 6.4B

When $v_{in} < 2$ V, the LO lamp is ON.
When $v_{in} \geq 2$ V, the LO lamp is OFF.

The high-low-level indicator in Figure 6.4c is simply a combination of the two circuits in Figures 6.4a and 6.4b. The HI and LO lamps operate exactly the same as described in the preceding boxes.

Voltage Follower

A *voltage follower* is another simple application of an op amp. The output voltage is equal to the input voltage. At this point, the obvious question is: If the voltage is not changed, why use it? The answer is that although the voltage is the same, the impedance is not.

We can look at impedance changes from two perspectives. Consider placing or not placing a voltage follower between a primary sensing element and a signal conditioning amplifier. If the voltage follower is not used, the primary sensor "sees" the input impedance of the amplifier. This might be 50 kΩ. With the voltage follower in place, the primary sensor sees the input impedance of the op amp, which could be 100 MΩ. So the effect of the voltage follower is to increase the impedance as seen by the primary sensor. This higher load impedance is a decided advantage because it reduces the current output of the primary sensor, thus reducing self-heating errors and nonlinearities caused by high current levels.

Now let's examine the second perspective. The amplifier also sees an impedance when it looks toward the primary element. If the voltage follower is not used, the amplifier sees the Thévenin equivalent of the primary element. This could be 1, 10, or 100 kΩ. With the voltage follower in place, the amplifier sees the output impedance of the op amp, which could easily be less than 100 Ω. Just as a high load impedance is good for the primary element, a low source impedance is good for the amplifier. The advantage of the voltage follower is that it transforms impedances in both directions.

The operation of a voltage follower is limited to the linear operating region of the op amp. This is also true of the remaining op-amp circuits in this section. The following analysis applies only to an op amp in the linear operating region.

A voltage follower is shown in Figure 6.5a. Notice that the output terminal is connected to the inverting input terminal, making $v_1 = v_{out}$. Also, the input voltage is connected to the noninverting input, making $v_2 = v_{in}$. In an ideal op amp, $v_1 = v_2$ which, in turn, makes $v_{out} = v_{in}$.

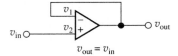

a) Voltage follower

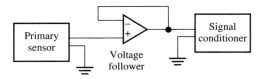

b) Practical application of a voltage follower

◆ **Figure 6.5** An op-amp voltage follower has unity voltage gain and a very high input impedance (typically 100 MΩ). A voltage follower may be inserted between the primary sensing element and the signal conditioner of a measuring instrument, thereby reducing the current produced by the primary element. The smaller primary element current reduces errors by self-heating and loading effects.

VOLTAGE FOLLOWER
IN THE LINEAR OPERATING REGION

$$v_{out} = v_{in} \qquad\qquad (6.10)$$

Figure 6.5b shows a voltage follower used as a bidirectional impedance transformer in a measuring instrument.

Inverting Amplifier

The *inverting amplifier* changes the sign and the level of the input signal. It can either increase, decrease, or not change the size of the signal, based on the values of two resistors, one between the input signal and the inverting input, the other between the inverting input and the output terminal. The gain of the inverting amplifier is the ratio of the second resistor over the first. The diagram of an inverting amplifier is shown in Figure 6.6.

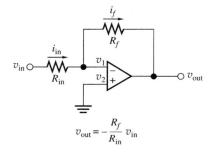

◆ **Figure 6.6** An inverting amplifier has a gain that can be less than 1, equal to 1, or greater than 1.

The analysis of the equivalent circuit of the inverting amplifier begins with the observation that voltage v_2 is equal to zero and the current through resistor R_{in} is equal to the current through resistor R_f. Then we apply Ohm's law to replace the currents by voltage drops divided by resistance values.

$$v_1 = v_2 = 0$$

$$i_{in} = i_f$$

$$\frac{v_{in} - v_1}{R_{in}} = \frac{v_1 - v_{out}}{R_f} \tag{6.11}$$

$$\frac{v_{in}}{R_{in}} = -\frac{v_{out}}{R_f}$$

**INVERTING AMPLIFIER
IN THE LINEAR OPERATING REGION**

$$v_{out} = -\frac{R_f}{R_{in}} v_{in} \tag{6.12}$$

**EXAMPLE
6.3**

A primary element produces an output voltage that ranges from 0 to 100 mV as the measured variable goes from the lower range limit to the upper range limit. Design an inverting amplifier that will take the output of the primary element and produce an output range of 0 to -5 V.

Solution

The required gain is 5/0.1 = 50. If we choose the value of R_{in} to be 1 kΩ, the required value of R_f is

$$R_f = 50R_{in} = 50(1000) = 50 \text{ k}\Omega \qquad ◆$$

Noninverting Amplifier

The *noninverting amplifier* circuit can increase the size of the signal, but it cannot decrease the size. In the extreme case, it can leave the size of the signal unchanged, but that reduces the circuit to the simple voltage-follower circuit.

Figure 6.7 shows the circuit diagram of a noninverting amplifier. Notice that the positions of the input voltage and the ground connection are interchanged from their positions in the inverting amplifier. The analysis of the equivalent circuit is similar to the analysis of the inverting amplifier.

$$v_1 = v_2 = v_{in}$$

$$i_{in} = i_f$$

$$\frac{v_1}{R_{in}} = \frac{v_{out} - v_1}{R_f} \tag{6.13}$$

$$\frac{v_{in}}{R_{in}} + \frac{v_{in}}{R_f} = \frac{v_{out}}{R_f}$$

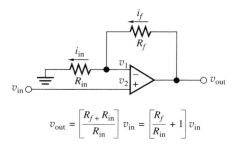

$$v_{\text{out}} = \left[\frac{R_f + R_{\text{in}}}{R_{\text{in}}}\right] v_{\text{in}} = \left[\frac{R_f}{R_{\text{in}}} + 1\right] v_{\text{in}}$$

◆ **Figure 6.7** A noninverting amplifier has a gain that cannot be less than 1.

NONINVERTING AMPLIFIER
IN THE LINEAR OPERATING REGION

$$v_{\text{out}} = \left(\frac{R_f + R_{\text{in}}}{R_{\text{in}}}\right) v_{\text{in}} = \left(\frac{R_f}{R_{\text{in}}} + 1\right) v_{\text{in}} \qquad \textbf{(6.14)}$$

EXAMPLE 6.4

Repeat the amplifier design from Example 6.3 using the noninverting amplifier.

Solution

The required gain is 50. We will use a 1-kΩ resistor for R_{in}. The gain is given by the first term enclosed in parentheses in Equation (6.14).

$$50 = \frac{R_f + 1000}{1000} = 50$$

$$R_f = 49 \text{ k}\Omega \qquad\qquad ◆$$

Summing Amplifier

A *summing amplifier* adds two or more input signals forming an output that is the inverse of the sum. Each input may be multiplied by a weighting factor that is formed by the ratio of two resistors. We will use the ideal op-amp model to develop the equation that defines the output of the summing amplifier circuit.

Figure 6.8a shows the circuit diagram for a two-input summing amplifier. We will develop the equation for the output of the three-input summing amplifier shown in Figure 6.8b. The development begins by applying Kirchhoff's current law at the inverting input terminal.

$$i_a + i_b + i_c - i_f - i_1 = 0$$

From Equation (6.4), $i_1 = 0$ and

$$i_f = i_a + i_b + i_c$$

From Equation (6.3), $v_1 = v_2 = 0$ (v_2 is grounded). Now apply Ohm's law at each resistor:

$$\frac{-v_{\text{out}}}{R_f} = \frac{v_a}{R_a} + \frac{v_b}{R_b} + \frac{v_c}{R_c}$$

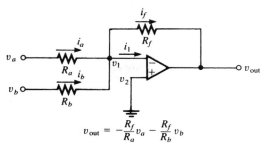

$$v_{\text{out}} = -\frac{R_f}{R_a}v_a - \frac{R_f}{R_b}v_b$$

a) Two-input summing amplifier

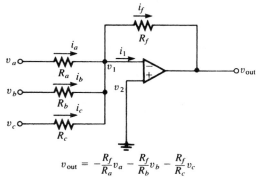

$$v_{\text{out}} = -\frac{R_f}{R_a}v_a - \frac{R_f}{R_b}v_b - \frac{R_f}{R_c}v_c$$

b) Three-input summing amplifier

◆ **Figure 6.8**   A summing amplifier forms the inverted and weighted sum of its inputs. The weighting factor is the feedback resistance, R_f, divided by the input resistor, R_a, R_b, or R_c.

SUMMING AMPLIFIER
IN THE LINEAR OPERATING REGION

$$v_{\text{out}} = -\left(\frac{R_f}{R_a}\right)v_a - \left(\frac{R_f}{R_b}\right)v_b - \left(\frac{R_f}{R_c}\right)v_c \qquad \textbf{(6.15)}$$

The resistor ratio R_f/R_a is the weighting factor for input voltage v_a, ratio R_f/R_b is the weighting factor for input voltage v_b, and so on.

EXAMPLE 6.5   Design a circuit that produces an output voltage that is the average of three input voltages. An inversion of the output signal is permissible.

Solution   The design can be implemented with a three-input summing amplifier with weighting factors of $\frac{1}{3}$ on each input. We will select a 3.33-kΩ resistor for R_f and 10-kΩ resistors for R_a, R_b, and R_c.   ◆

Integrator

An *integrator* circuit produces an output that is proportional to the integral of the input voltage. An inverting amplifier can be converted to an integrator by replacing resistor R_f by a capacitor. Figure 6.9 shows an integrator circuit and a graphical interpretation of the integral.

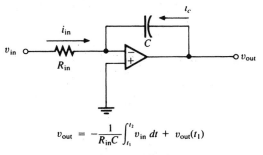

$$v_{out} = -\frac{1}{R_{in}C}\int_{t_1}^{t_2} v_{in}\, dt + v_{out}(t_1)$$

a) Integrator circuit

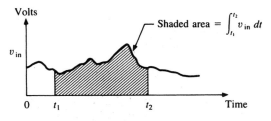

b) Graphical interpretation of an integral

◆ **Figure 6.9** An integrator produces an output that is proportional to the integral of the input voltage. On a graph of input voltage versus time, the integral between time t_1 and time t_2 is equal to the area under the graph between time t_1 and time t_2. The speedometer of a car displays the speed of the car, while the odometer displays the integral of the speed (i.e., the distance traveled).

The graph in Figure 6.9 is a plot of the input voltage, v_{in}, plotted versus time. The area under the graph between time t_1 and time t_2 is shaded. This shaded area is equal to the integral of the input voltage from time t_1 to time t_2.

$$\text{Shaded area} = \int_{t_1}^{t_2} v_{in}\, dt$$

If the input voltage represented the speed of travel of an object, the integral from t_1 to t_2 would represent the distance traveled during the time interval from t_1 to t_2. If the input voltage represented the flow rate of liquid into a tank, the integral would represent the amount of liquid that flowed into the tank from time t_1 to time t_2. In general, the integral of a rate of change of quantity is the amount of that quantity that changes between the time limits of the integration.

Refer to the integrator circuit in Figure 6.9. Applying Kirchhoff's current law at the inverting input terminal, and applying Equation (6.4) (i.e., $i_1 = 0$), we get the following equation:

$$i_{in} = -i_c$$

The current through a capacitor is equal to the capacitance value times the rate of change of the voltage across the capacitor terminals. The voltage across the capacitor is $v_{out} - v_1 = v_{out}$ because $v_1 = v_2 = 0$. Thus

$$i_c = C\frac{dv_{out}}{dt}$$

$$\frac{v_{in}}{R_{in}} = -C\frac{dv_{out}}{dt}$$

$$dv_{out} = -\left(\frac{1}{R_{in}C}\right)v_{in}\,dt$$

Integrating the equation above gives the equation for the output of the integrator as a function of the input, v_{in}.

**INTEGRATOR
IN THE LINEAR OPERATING REGION**

$$v_{out}(t_2) = \frac{-1}{R_{in}C}\int_{t_1}^{t_2} v_{in}(t)\,dt + v_{out}(t_1) \qquad\qquad \textbf{(6.16)}$$

EXAMPLE 6.6

The input to an integrator is a constant 100 mV. The input resistance is 10 kΩ and the capacitance is 1 μF.

(a) Find the expression for the output voltage at time t_2 as a function of t_1, t_2, and $v_{out}(t_1)$.

(b) If $t_1 = 5$ s and $v_{out}(t_1) = +10$ V, find the time t_2 when the op amp reaches saturation at -16 V (i.e., when $v_{out}(t_2) = -16$ V).

Solution

(a) $v_{out}(t_2) = -\dfrac{1}{R_{in}C}\displaystyle\int_{t_1}^{t_2} v_{in}(t)\,dt + v_{out}(t_1)$

$v_{out}(t_2) = -\dfrac{1}{10^4 \cdot 10^{-6}}\displaystyle\int_{t_1}^{t_2} 0.1\,dt + v_{out}(t_1)$

$v_{out}(t_2) = -100(0.1t)\Big|_{t_1}^{t_2} + v_{out}(t_1)$

$v_{out}(t_2) = -10(t_2 - t_1) + v_{out}(t_1)$

(b) $-16 = -10(t_2 - 5) + 10$

$t_2 = \dfrac{16 + 10}{10} + 5 = 7.6$ s   ◆

Differentiator

A *differentiator* circuit produces an output that is proportional to the rate of change (derivative) of the input voltage. An inverting amplifier can be converted to a differentiator by replacing resistor R_{in} by a capacitor. Figure 6.10 shows a differentiator and a graphical interpretation of the derivative.

Unfortunately, the most rapidly changing part of the input signal is often unwanted noise spikes or other forms of high-frequency noise. In this case, the differentiator magnifies the noise even more than the useful signal, making the output very jittery. The solution to the noise problem is to add a resistor in series with the capacitor to limit the amplification of high-frequency signals. Sometimes this series resistance is provided by the source impedance.

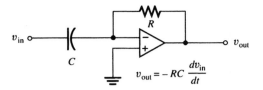

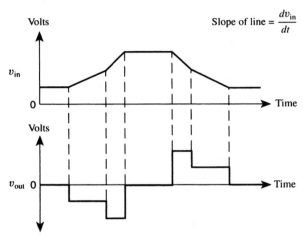

a) Differentiator circuit

b) Graphical interpretation of a derivative

◆ **Figure 6.10** A differentiator produces an output voltage that is inverted and proportional to the rate of change of the input voltage. On a graph of input voltage versus time, the derivative is equal to the slope of the graph.

Differential Amplifier

A *differential amplifier* is a circuit that amplifies the difference between two input voltages, neither of which is equal to zero. The common-mode rejection ratio is a major concern in a differential amplifier. The circuit diagram of a differential amplifier circuit is shown in Figure 6.11.

We begin the development of the differential amplifier equation by recalling that $v_1 = v_2$ and by applying the voltage-divider rule to resistors R_b and R_g.

$$v_1 = v_2 = v_b \left(\frac{R_g}{R_b + R_g} \right)$$

For resistors R_a and R_f, we have the following equation:

$$\frac{v_a - v_1}{R_a} = \frac{v_1 - v_{out}}{R_f}$$

$$v_{out} = v_1 \left(\frac{R_f}{R_a} + 1 \right) - \left(\frac{R_f}{R_a} \right) v_a \qquad (6.17)$$

$$v_{out} = \left(\frac{R_g}{R_a} \right) \left(\frac{R_f + R_a}{R_b + R_g} \right) v_b - \left(\frac{R_f}{R_a} \right) v_a$$

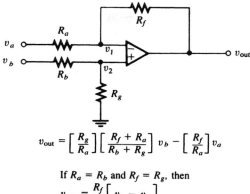

$$v_{out} = \left[\frac{R_g}{R_a} \right] \left[\frac{R_f + R_a}{R_b + R_g} \right] v_b - \left[\frac{R_f}{R_a} \right] v_a$$

If $R_a = R_b$ and $R_f = R_g$, then

$$v_{out} = \frac{R_f}{R_a} \left[v_b - v_a \right]$$

◆ **Figure 6.11** A differential amplifier is used to measure the difference between two voltages. The circuit equation simplifies considerably when $R_a = R_b$ and $R_f = R_g$.

If $R_a = R_b$, and $R_f = R_g$, Equation 6.17 reduces to

DIFFERENTIAL AMPLIFIER
IN THE LINEAR OPERATING REGION

$$v_{out} = \frac{R_f}{R_a} (v_b - v_a) \qquad\qquad (6.18)$$

Instrumentation Amplifier

An *instrumentation amplifier* (IA) is essentially a differential amplifier with high input impedance, high common-mode rejection, balanced differential inputs, and gain determined by a user-selected external resistor. In order to provide a guaranteed level of performance, all components except the external gain resistor are inside the IA package. The high input impedance of the IA minimizes the current draw from the input circuit, thus reducing self-heating and loading effects on the input circuit. Figure 6.12 shows a three-op-amp instrumentation amplifier. The input stage consists of two op amps in a modified voltage follower configuration. The second stage is the differential amplifier shown in Figure 6.11 with balanced resistors (i.e., $R_b = R_a$ and $R_g = R_f$).

Assuming ideal op amps in Figure 6.12, the same current (i) flows through resistors R_1', R_e, and R_1. By Ohm's law,

$$i = \frac{v_b - v_a}{R_1' + R_e + R_1} = \frac{v_b - v_a}{R_1 + R_e + R_1} = \frac{v_b - v_a}{2R_1 + R_e}$$

and

$$i = \frac{v_{ib} - v_{ia}}{R_e}$$

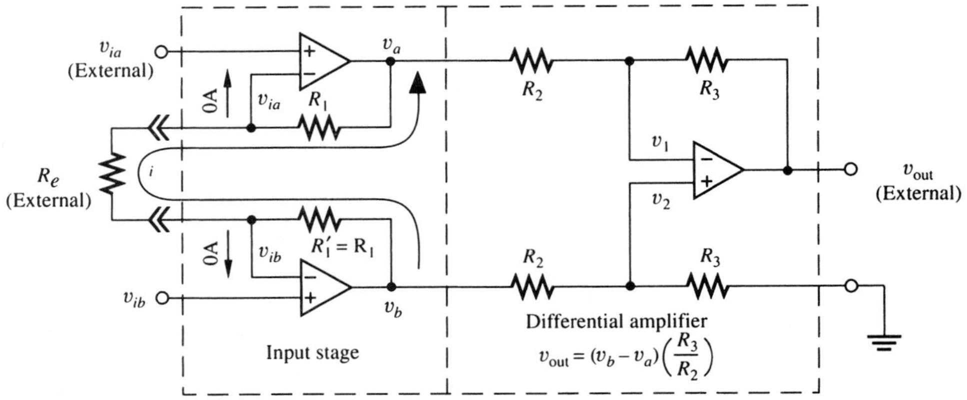

◆ **Figure 6.12** A three-op-amp instrumentation amplifier consists of an input stage and a differential amplifier. The user can adjust the size of the external resistor to obtain a wide range of gains.

Equating the right side of the two equations for i, we have

$$\frac{v_b - v_a}{2R_1 + R_e} = \frac{v_{ib} - v_{ia}}{R_e}$$

$$v_b - v_a = \left(\frac{2R_1 + R_e}{R_e}\right)(v_{ib} - v_{ia})$$

Finally, replacing $v_b - v_a$ in Equation (6.18) by the right side of the above equation gives us the following equation for the instrumentation amplifier:

INSTRUMENTATION AMPLIFIER
IN THE LINEAR OPERATING REGION

$$v_{out} = \left(\frac{2R_1 + R_e}{R_e}\right)\left(\frac{R_3}{R_2}\right)(v_{ib} - v_{ia}) \tag{6.19}$$

The user can adjust the size of external resistor R_e to obtain a wide range of gains without increasing the common-mode error signal.

EXAMPLE 6.7

An instrumentation amplifier similar to Figure 6.12 has the following precision resistor values:

$$R_1 = 1000\ \Omega,\ R_2 = 1000\ \Omega,\ \text{and}\ R_3 = 1000\ \Omega$$

Determine the value of the external resistor, R_e, that will result in a gain of 1000.

Solution

$$\text{IA gain} = \left(\frac{2R_1 + R_e}{R_e}\right)\left(\frac{R_3}{R_2}\right)$$

$$1000 = \left(\frac{2(1000) + R_e}{R_e}\right)\left(\frac{1000}{1000}\right)$$

$$1000R_e = 2000 + R_e$$

$$999R_e = 2000$$

$$R_e = \frac{2000}{999} = 2.002 \ \Omega$$

◆

Function Generator

Some signal conditioning applications require a component with an output that is a nonlinear function of its input. Components of this type are called *function generators*. The inverting amplifier can be converted into a function generator by replacing one of its two resistors with a component that has a nonlinear voltampere relationship.

A logarithmic amplifier is one example of a function generator. A transistor with a grounded base serves as the nonlinear element. The transistor replaces the feedback resistor with the collector connected to the summing junction and the emitter connected to the output terminal of the op amp. The logarithmic amplifier utilizes the logarithmic voltampere relationship of the transistor to generate an output voltage that is proportional to the natural logarithm of the input voltage.

Linearization of the signal from a differential pressure flow transmitter is another example of an application of a function generator. The flow transmitter output voltage (v_F) is equal to a constant (K_F) multiplied by flow rate (Q) squared, as shown in Equation (6.20):

$$v_F = K_F Q^2 \tag{6.20}$$

In other words, the flow transmitter output voltage is proportional to the flow rate squared. We would prefer a voltage that is proportional to the flow rate, and for that we need a function generator whose output is equal to the square root of its input. With an input voltage equal to $K_L Q^2$, a square root function generator will produce an output voltage equal to $\sqrt{K_L}(Q)$, which is the desired linear signal. Figure 6.13a shows a square-root function generator used to linearize the output signal from a flow transmitter.

An op-amp circuit for the square-root function generator is shown in Figure 6.13b. The circuit is an inverting amplifier with a nonlinear element in the feedback path. The nonlinear element has the following voltampere relationship:

$$v_{NL} = \sqrt{iR_i} \tag{6.21}$$

Recall that the summing junction of the op amp is a virtual 0 V. Thus the voltage across R_1 is v_F, and current i is easily determined by Ohm's law:

$$i = \frac{v_F}{R_1} \tag{6.22}$$

The output voltage v_{out} is also easily determined:

$$v_{out} = -v_{NL} \tag{6.23}$$

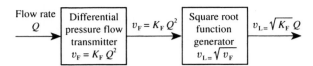

a) A square root function generator is used to linearize the output voltage of a differential pressure flow transmitter.

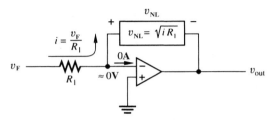

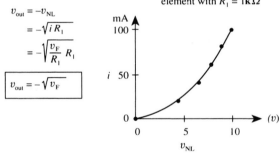

v–i graph of nonlinear element with $R_1 = 1k\Omega$

b) An inverting amplifier is used to make a square root function generator.

◆ **Figure 6.13** A nonlinear function generator implements a square-root function to linearize the output of a flow transmitter.

Combining Equations (6.21), (6.22), and (6.23) results in the following equation for v_{out} in terms of v_F:

$$v_{out} = -\sqrt{v_F} \tag{6.24}$$

EXAMPLE 6.8

A square-root function generator similar to Figure 6.13b uses a 1-kΩ resistor for R_1. Draw a voltampere graph of the nonlinear element for values of current i from 0 to 100 mA.

Solution

$$v_{NL} = -\sqrt{iR_1} = \sqrt{1000i}$$

i(mA)	0	20	40	60	80	100
v_{NL} (V)	0	4.47	6.32	7.75	8.94	10.00

The voltampere graph is shown in Figure 6.13b. ◆

6.4 ANALOG SIGNAL CONDITIONING

The type of signal conditioning required for a particular measurement depends on the electrical characteristics of the primary element and the component that will receive the signal. Typical conditioning includes galvanic isolation, common-mode isolation, impedance transformation, amplification, noise reduction, signal conversion, linearization, compensation, and calibration.

In the trend toward distributed control, the signal conditioning circuitry has moved close to the sensor on the process floor. Signal conditioning is done in the individual measuring transmitters, and it is done in signal conditioning systems using plug-in modules for various functions. The plug-in modules are rack mounted in card cages that can accept 4, 8, or 16 plug-in modules. The following is a brief description of some typical plug-in modules.[*]

A *millivolt/thermocouple* module accepts low-level dc voltage signals with a span as low as 2 mV or as high as 55 mV. The zero can be adjusted from −5 to 25 mV. The module converts the input to a high-level output such as a 4- to 20-mA current signal. The circuit provides isolation from dc and ac common-mode voltages of several hundred volts, and a common-mode rejection ratio of 140 dB at 60 Hz. The module provides noise rejection, linearization, cold junction compensation, and break detection for thermocouple inputs. The module can be factory calibrated for a specified zero and span.

A *resistance temperature detector (RTD)* module accepts input from 100-Ω platinum or 10-Ω copper RTD sensors. The input signal is converted to a high-level input such as a 4- to 20-mA current signal. The module provides common-mode isolation from ac and dc voltages of several hundred volts, and a CMR of 120 dB at 60 Hz. It also provides excitation for the RTD sensor, lead wire compensation, noise reduction, linearization, and break detection. A typical 98% response time is 0.25 s.

A *frequency input* module accepts pulse signals from digital tachometers, turbine flow meters, and other sensors that produce a series of pulses. The frequency of the input pulses is converted to a high-level voltage or current signal. The frequency range is selectable from as low as 25 Hz to as high as 12 kHz. The 98% response time varies from 10 s for the low-frequency range to 0.2 s for the high-frequency range.

The remainder of this section covers details of various signal conditioning operations.

Isolation and Impedance Conversion

To quote Lord Kelvin: "The act of measuring something destroys that which is measured." Consider the problem of measuring the temperature of a small container of water with a thermometer. Let's assume that the water temperature is 40°C and the temperature of the thermometer is 22°C before the measurement takes place. The measurement consists of placing the thermometer in the water and reading the thermometer after a temperature equilibrium has been reached between the water and the thermometer. The equilibrium temperature depends on the ratio of the heat capacities of the thermometer and the water. If the heat capacities are equal, the equilibrium temperature will be 31°C, the average. Whatever the equilibrium temperature is, it will be less than 40°C and greater than 22°C. Thus the act of measuring the temperature of the water has changed the temperature.

---

[*]*Series 1800 Analog Signal Conditioning System* (Wixom, Mich.: Acromag).

One purpose of isolation and impedance conversion is to avoid, or at least minimize, the destruction of "that which is measured." An equally important purpose is to protect the measuring instrument from "that which is measured." For example, a high common-mode voltage can easily destroy an op-amp circuit unless the circuit is isolated from the high voltage.

Impedance transformation is one method of protecting the measured variable and the measuring instrument from each other. The voltage follower, introduced in Section 6.3, does this job quite well (refer to Section 6.3 for further details). When galvanic isolation is required, special amplifiers called *isolation amplifiers* are used. There are two methods that can be used to obtain galvanic isolation: transformer coupling and optical coupling. A typical isolation amplifier with transformer coupling has a capacitance of about 10 picoFarad (pF) between the input and the output circuits, and a CMR of about 120 dB. Optical coupling uses a light beam to transmit the signal from the source circuit to the receiver circuit. This makes it possible to remove all electrical connections between the two circuits.

Amplification and Analog-to-Analog Conversion

Changing the level of an analog signal is accomplished by either an inverting amplifier, a noninverting amplifier, or a differential amplifier. Sometimes it is necessary to convert the signal from a voltage to a current or from a current to a voltage. Figure 6.14 shows simple op-amp converter circuits.

The current-to-voltage converter in Figure 6.14a is a voltage follower with a resistor connected from the noninverting input to ground. The input current, i_{in}, passes to ground through

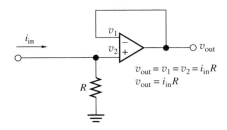

$$v_{out} = v_1 = v_2 = i_{in}R$$
$$v_{out} = i_{in}R$$

a) Current-to-voltage converter

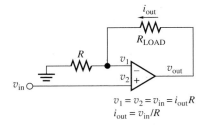

$$v_1 = v_2 = v_{in} = i_{out}R$$
$$i_{out} = v_{in}/R$$

b) Voltage-to-current converter

◆ **Figure 6.14**   Current-to-voltage and voltage-to-current converters using the ideal op-amp model.

the resistor, making voltage v_2 equal to $i_{in}R$. The output terminal is connected to the input, making $v_{out} = v_1$. Assuming that the op amp is ideal makes $v_1 = v_2$, and by proper substitution, we get the following equation for the current-to-voltage converter:

$$v_{out} = i_{in}R \tag{6.25}$$

EXAMPLE 6.9

Design a current-to-voltage converter that converts a 20-mA input signal into a 5-V output signal.

Solution

Equation (6.25) applies to this problem.

$$v_{out} = i_{in}R$$

$$R = \frac{v_{out}}{i_{in}} = \frac{5}{0.02} = 250 \ \Omega$$

The converter is the circuit in Figure 6.14a with $R = 250 \ \Omega$.   ◆

The voltage-to-current converter in Figure 6.14b is a noninverting amplifier with the load resistor placed in the feedback path where R_f is normally placed. The output current, i_{out}, passes to ground through load resistor R_{LOAD} and scaling resistor R, making voltage v_1 equal to $i_{out}R$. The input voltage is connected to the noninverting terminal, making $v_2 = v_{in}$. For an ideal op amp, $v_1 = v_2$, and we get the following equation for the voltage-to-current converter:

$$i_{out} = \frac{v_{in}}{R} \tag{6.26}$$

EXAMPLE 6.10

Design a voltage-to-current converter that converts a 12-V dc input signal into a 20-mA current signal.

Solution

Equation (6.26) applies to this problem.

$$i_{out} = \frac{v_{in}}{R}$$

$$R = \frac{v_{in}}{i_{out}} = \frac{12}{0.02} = 600 \ \Omega$$

The converter is the circuit shown in Figure 6.14b with $R = 600 \ \Omega$.   ◆

Bridge Circuits

A number of primary elements convert changes in the measured variable into small changes in the resistance of the element. Strain gage force transducers, strain gage pressure transducers, and resistance temperature detectors are three examples. A bridge circuit is the traditional method of measuring small changes in the resistance of an element. The operation of a bridge falls into two categories, balanced and unbalanced operation. In the balanced operation, the resistance of the sensor is determined from the values of three other resistors whose values are known with precision. In the unbalanced operation, the change in the sensor resistance from a base value produces a small difference between two voltages. A differential amplifier is used to amplify the difference between the two voltages.

A *balanced Wheatstone bridge* is shown in Figure 6.15. The unknown resistance of the sensor is labeled R_s. Resistors R_2 and R_4 have fixed values for a given measurement. Resistor R_3 can be continuously adjusted over a calibrated range of values. The value of R_3 is adjusted until the meter in the center reaches a null position, indicating that $v_a = v_b$ and $i_{ab} = 0$. When this occurs, the bridge is said to be *balanced*. When the bridge is balanced, current i_1 passes through resistors R_s and R_2. Also, current i_3 passes through resistors R_3 and R_4. Because $v_a = v_b$, it follows that the voltages across R_s and R_3 are equal, as well as the voltages across R_2 and R_4. Thus,

$$i_1 R_s = i_3 R_3$$

$$i_1 R_2 = i_3 R_4$$

Dividing the first equation by the second yields

$$\frac{R_s}{R_2} = \frac{R_3}{R_4} \tag{6.27}$$

$$R_s = \frac{R_2}{R_4} R_3 \tag{6.28}$$

If the value of R_3 and the ratio R_2/R_4 are known with precision, the value of R_s can be determined accurately.

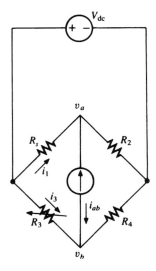

◆ **Figure 6.15** Wheatstone bridge circuit. When the circuit is in the balanced condition, $R_s = R_2 R_3/R_4$.

EXAMPLE A Wheatstone bridge is used to measure an unknown resistor, R_s, as shown in Figure 6.15. Resistor R_3
6.11 is adjusted until the bridge is balanced and the meter is in the null position. The known resistor values
are $R_2 = 500\ \Omega$, $R_3 = 226\ \Omega$, and $R_4 = 1000\ \Omega$. Find the value of the unknown resistor, R_s.

Solution Equation (6.28) applies to this problem.

$$R_s = \left(\frac{500}{1000}\right)226 = 113\ \Omega \qquad\qquad ◆$$

The Wheatstone bridge is useful for measuring the value of unknown resistors, but the
need to adjust resistors to obtain a null makes it unsuitable for use in modern control systems.
What we need is a method for automatically balancing the bridge in a manner that provides
a useful signal. The *current balance bridge* is one such method.

A self-nulling current balance bridge is shown in Figure 6.16. The adjustable resistor, R_3,
is replaced by two fixed resistors, R_{3a} and R_{3b}. The smaller resistor, R_{3a}, is also connected to
the output of a null controller. The two null voltages, v_a and v_b, are connected to the input of
an IA. The current on the IA input lines is negligible, due to the IA's very high input impedance. The output of the IA, v_{out}, is equal to its input, $v_a - v_b$, multiplied by the gain of the IA
[see Equation (6.19)]. The null controller manipulates its output current, i_N, such that it maintains the null condition (i.e., $v_a = v_b$). The null controller output current, i_N, is a measure of
the change in the sensor resistance, R_s. We will now derive an equation for the value of i_N that
will balance the bridge. The expression for v_a is easily obtained by voltage division, and for
a balanced bridge, $v_b = v_a$.

$$v_a = V_{dc}\left(\frac{R_2}{R_2 + R_s}\right) = v_b$$

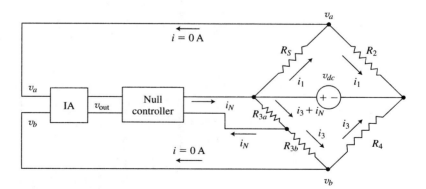

◆ **Figure 6.16** A self-nulling current balance bridge. The instrumentation amplifier (IA) has a very high input impedance, so the current from the bridge to the IA
may be neglected. The null controller manipulates current i_N to maintain the bridge
balance condition (i.e., $v_b = v_a$). See Example 6.12 for a detailed analysis.

Next we use Ohm's law and Kirchhoff's voltage law to determine expressions for i_3 and i_N:

$$i_3 = \frac{v_b}{R_4} = \left[\frac{V_{dc}\,R_2}{R_4(R_2 + R_s)}\right]$$

$$(i_3 + i_N)R_{3a} + i_3 R_{3b} + i_3 R_4 = V_{dc}$$

$$i_N R_{3a} = V_{dc} - i_3(R_{3a} + R_{3b} + R_4)$$

$$i_N = \frac{V_{dc}}{R_{3a}}\left[1 - \frac{R_2(R_{3a} + R_{3b} + R_4)}{R_4(R_2 + R_s)}\right] \tag{6.29}$$

Example 6.12 illustrates the use of a self-nulling current balance bridge to measure the resistance of an RTD.

EXAMPLE 6.12

The self-nulling current balance bridge shown in Figure 6.16 is used to condition the output of an RTD. The RTD has an input range of 0 to 100°C and an output range of 100 to 140 Ω. The RTD resistance element is labeled R_s in the bridge circuit. The other bridge resistance values and the supply voltage are

$$R_{3a} = 10\ \Omega,\ R_{3b} = 110\ \Omega,\ R_2 = R_4 = 1000\ \Omega,\ V_{dc} = 10\ \text{V}$$

The bridge is balanced when $R_s = 120\ \Omega$ and $i_N = 0$. We choose the mid-range resistance for the null current balance to minimize nonlinearity in the output current, i_N.

Determine the values of the null controller current, i_N, when the RTD temperature is 0, 25, 50, 75, and 100°C.

Determine the equation of the zero-based linearity line and express the nonlinearity as a percentage of the output span.

Solution

1. Determine v_a and v_b when $R_s = 120\ \Omega$ and $i_N = 0$ A.

$$v_a = V_{dc}\left(\frac{R_2}{R_2 + R_s}\right) = 10\left(\frac{1000}{1120}\right) = 8.9286\ \text{V}$$

$$v_b = V_{dc}\left(\frac{R_4}{R_{3a} + R_{3b} + R_4}\right) = 10\left(\frac{1000}{1120}\right) = 8.9286\ \text{V}$$

The bridge is balanced as it should be.

2. Use Equation (6.29) to determine i_N for the following values of R_s: 100, 110, 120, 130, and 140 Ω.

$$i_N = \frac{V_{dc}}{R_{3a}}\left[1 - \frac{R_2(R_{3a} + R_{3b} + R_4)}{R_4(R_2 + R_s)}\right]$$

$$i_N = \frac{10}{10}\left[1 - \frac{1000(10 + 110 + 1000)}{1000(1000 + R_s)}\right]$$

$$i_N = 1 - \frac{1120}{(1000 + R_s)}$$

The results are presented in the first three columns of the table at the end of this example.

3. The zero-based linearity line extends in both directions from the point where $i_N = 0$ mA and $R_s = 120 \ \Omega$. The line is positioned such that it is an equal distance from i_N at each extremity. We will use L_N for the y-coordinate of the zero-based line. The magnitude of L_N at each extremity is determined by averaging the magnitude of i_N at each extremity.

When $R_s = 140$, $L_N = (17.544 + 18.182)/2 = 17.863$ mA
When $R_s = 100$, $L_N = -17.863$ mA

The equation of the zero-based linearity line is determined by the following two points:

$$R_s = 120 \ \Omega \qquad L_N = 0.0 \text{ mA}$$
$$R_s = 140 \ \Omega \qquad L_N = 17.863 \text{ mA}$$

Zero-based linearity line $L_N = 0.89315R_s - 107.178$
The values of L_N, $L_N - i_N$, and the percentage of difference are in the last three columns of the table below.

T (°C)	R_s (Ω)	i_N (mA)	L_N (mA)	$L_N - i_N$	% Diff.
0	100	−18.182	−17.863	0.319	0.89
25	110	−9.009	−8.931	0.078	0.22
50	120	0.000	0.000	0.000	0.00
75	130	8.850	8.931	0.081	0.23
100	140	17.544	17.863	0.319	0.89

◆

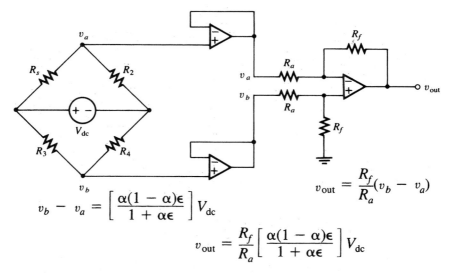

$$v_b - v_a = \left[\frac{\alpha(1 - \alpha)\epsilon}{1 + \alpha\epsilon} \right] V_{dc}$$

$$v_{out} = \frac{R_f}{R_a}(v_b - v_a)$$

$$v_{out} = \frac{R_f}{R_a} \left[\frac{\alpha(1 - \alpha)\epsilon}{1 + \alpha\epsilon} \right] V_{dc}$$

◆ **Figure 6.17** Unbalanced Wheatstone bridge and instrumentation amplifier circuit. The purpose of the circuit is to produce an output voltage that is proportional to the difference between R_3 and R_{bal}.

The *unbalanced Wheatstone bridge* is another method for obtaining a signal suitable for use in a control system. The output of the unbalanced bridge is usually connected to the input of an instrumentation amplifier as shown in Figure 6.17. The values of resistors R_2, R_3, and R_4 are set such that the bridge is balanced when R_s is at a predetermined base value. We will call this base resistance value R_{bal} for balanced resistor value. Equation (6.30) expresses R_{bal} in terms of resistors R_2, R_3, and R_4.

$$R_{bal} = \frac{R_2 R_3}{R_4} \qquad (6.30)$$

The purpose of an unbalanced bridge circuit is to produce an output voltage that is proportional to the difference between R_s and R_{bal}. To simplify and normalize the unbalanced bridge equation, we introduce two additional parameters: ϵ and α. The first parameter, ϵ, is the fractional difference between R_s and R_{bal} as defined by Equation (6.31):

$$\epsilon = \left(\frac{R_s - R_{bal}}{R_{bal}} \right) \qquad (6.31)$$

Notice that 100ϵ is the percentage difference between R_s and R_{bal}.

The second parameter, α, is the voltage divider ratio for the voltage across resistor R_3 as defined by Equation (6.32).

$$\alpha = \frac{R_3}{R_3 + R_4} \qquad (6.32)$$

The bridge equation is developed by applying the voltage divider rule to write equations for v_a, v_b, and $v_b - v_a$, reducing the equations, and using the parameters defined in Equations (6.30) to (6.32).

$$v_a = \left(\frac{R_2}{R_s + R_2} \right) V_{dc}$$

$$v_b = \left(\frac{R_4}{R_3 + R_4} \right) V_{dc}$$

$$\frac{v_b - v_a}{V_{dc}} = \left(\frac{R_4}{R_3 + R_4} - \frac{R_2}{R_s + R_2} \right) \qquad (6.33)$$

We now proceed to convert the right-hand side of Equation (6.33) to a function of α and ϵ. The first step is to combine the two terms into a single fraction:

$$\frac{v_b - v_a}{V_{dc}} = \frac{R_s R_4 + R_2 R_4 - R_2 R_3 - R_2 R_4}{(R_3 + R_4)(R_s + R_2)}$$

In the numerator, use $-R_2 R_4$ to cancel $R_2 R_4$, replace $R_2 R_3$ by $R_{bal} R_4$, and factor out R_4. In the denominator, replace R_2 by $R_{bal} R_4 / R_3$. Equation (6.30) was used for both replacements.

$$\frac{v_b - v_a}{V_{dc}} = \frac{R_4 (R_s - R_{bal})}{(R_3 + R_4)(R_s + R_{bal} R_4 / R_3)}$$

Now multiply the numerator and denominator by $R_3(R_3 + R_4)/R_{bal}$ and rearrange the terms as follows:

$$\frac{v_b - v_a}{V_{out}} = \left(\frac{R_3}{R_3 + R_4}\right)\left(\frac{R_4}{R_3 + R_4}\right)\left[\frac{(R_s - R_{bal})/R_{bal}}{\left(\dfrac{R_sR_3 + R_{bal}R_4}{(R_3 + R_4)R_{bal}}\right)}\right]$$

By Equation (6.32):

$$\frac{R_3}{R_3 + R_4} = \alpha$$

and

$$\frac{R_4}{R_3 + R_4} = (1 - \alpha)$$

By Equation (6.31):

$$\frac{R_s - R_{bal}}{R_{bal}} = \epsilon$$

Adding and subtracting $R_{bal}R_3$ in the last term in the denominator changes nothing but does facilitate its reduction.

$$\frac{R_sR_3 + R_{bal}R_4}{(R_3 + R_4)R_{bal}} = \frac{R_sR_3 + R_{bal}R_4 + R_{bal}R_3 - R_{bal}R_3}{(R_3 + R_4)R_{bal}}$$

$$= \frac{(R_3 + R_4)R_{bal} + R_3(R_s - R_{bal})}{(R_3 + R_4)R_{bal}}$$

$$= 1 + \alpha\epsilon$$

Finally,

$$\frac{v_b - v_a}{V_{dc}} = \alpha(1 - \alpha)\left(\frac{\epsilon}{1 + \alpha\epsilon}\right) \tag{6.34}$$

For very small values of ϵ, Equation (6.34) reduces to

$$\frac{v_b - v_a}{V_{dc}} = \alpha(1 - \alpha)\epsilon \tag{6.35}$$

Table 6.1 gives the percentage difference between the result using Equation (6.34) and the result using Equation (6.35). For values of ϵ less than or equal to 0.01, the difference is less than 1%.

The instrumentation amplifier in Figure 6.17 isolates the bridge with its high input impedance and multiplies the bridge output, $v_b - v_a$, by the amplifier gain, R_f/R_a. Equation (6.36) defines the amplifier output, v_{out}.

$$v_{out} = \frac{R_f}{R_a}\left[\frac{\alpha(1 - \alpha)\epsilon}{1 + \alpha\epsilon}\right]V_{dc} \tag{6.36}$$

◆ **TABLE 6.1** Values of the Percentage Difference Between Results from Equation (6.34) and Results from Equation (6.35)

ϵ	α					
	0.0	0.2	0.4	0.6	0.8	1.0
0.0001	0.000	0.002	0.004	0.006	0.008	0.010
0.0010	0.000	0.020	0.040	0.060	0.080	0.100
0.0100	0.000	0.200	0.398	0.596	0.794	0.990
0.1000	0.000	1.961	3.846	5.660	7.407	9.091

Example 6.13 illustrates the use of an unbalanced bridge to measure the output of a resistance temperature detector.

EXAMPLE 6.13

The unbalanced bridge in Figure 6.17 is used to condition the resistance of the RTD described in Example 6.12. The parameters in Figure 6.17 have the following values:

$$R_s = \text{RTD resistance element (range} = 100 \text{ to } 140 \ \Omega)$$

$$R_3 = 120 \ \Omega \qquad R_2 = R_4 = 1000 \ \Omega \qquad R_{\text{bal}} = 120 \ \Omega$$

$$V_{\text{dc}} = 1 \text{ V} \qquad R_f/R_a = 100$$

Determine the values of the v_{out} for the following RTD temperatures: 0, 25, 50, 75, and 100°C.
Determine the equation of the zero-based linearity line and express the nonlinearity as a percentage of the output span.

Solution

Equation 6.36 applies to this problem. The value of α is given by Equation (6.32) and ϵ is given by Equation (6.31).

$$\alpha = \frac{R_3}{R_3 + R_4} = \frac{120}{120 + 1000} = 0.107143$$

$$1 - \alpha = 0.892857$$

$$\alpha(1 - \alpha) = 0.09566327$$

$$V_{\text{dc}}\frac{R_f}{R_a} = 1(100) = 100$$

$$v_{\text{out}} = V_{\text{dc}}\left(\frac{R_f}{R_a}\right)\left[\frac{\alpha(1 - \alpha)\epsilon}{(1 + \alpha\epsilon)}\right]$$

$$v_{\text{out}} = 100\left[\frac{(0.107143)(0.892857)\epsilon}{(1 + \alpha\epsilon)}\right]$$

$$v_{\text{out}} = 9.5663265\left[\frac{\epsilon}{(1 + \alpha\epsilon)}\right]$$

The results are presented in the following table:

T (°C)	R_s (Ω)	ϵ	$1 + \alpha\epsilon$	v_{out} (V)
0	100	−0.16666	0.98214	−1.6234
25	110	−0.08333	0.99107	−0.8044
50	120	0.0000	1.00000	0.0000
75	130	+0.08333	1.00893	0.7901
100	140	+0.16666	1.01786	1.5664

The equation of the zero-based linearity line is

$$L_N = 0.079745R_S - 9.5694, \text{ V}$$

The values of L_N, $L_N - v_{out}$, and the percentage difference are in the table below:

T (°C)	R_s (Ω)	v_{out} (V)	L_N (V)	Diff.	% Diff.
0	100	−1.6234	−1.5949	0.0285	0.89
25	110	−0.8044	−0.7975	0.0069	0.22
50	120	0.0000	0.0000	0.0000	0.00
75	130	0.7901	0.7974	0.0073	0.23
100	140	1.5664	1.5949	0.0285	0.89

◆

Noise Reduction

Control systems exist in an environment filled with high levels of electromagnetic energy just waiting to produce noise in electric signal lines. The best noise reduction system is one that prevents the noise from ever getting to the signal. Noise prevention consists of careful grounding of signal lines and shielding of cables to keep the signals as free of noise as possible. Despite the best noise prevention effort, some noise will appear in the control signals. Special circuits called *filters* are designed to reduce the level of the noise in the signals. Actually, filters can be used to reduce everything in a specific range (or band) of frequencies. The terms *low-pass filter, band-pass filter, high-pass filter,* and *notch filter* name various filters according to the band of frequencies they allow to pass through unaffected.

Before going on, we should pause a moment and consider the frequency components of a signal. Everyone is familiar with the fact that we can tune a radio to receive different stations. Each station is assigned a carrier frequency on which it superimposes its voice and music signals. A radio antenna receives the carrier signals from all stations in its vicinity, but the radio receiver is sensitive to only one carrier at a time, depending on the position of the tuner. You select a different station by moving the tuner to a different position.

In the discussion of filters, we view control signals as a collection of signal components with different frequencies. The term *component,* as used in the following discussion, refers to a frequency component of a control signal. A filter chooses to accept certain frequencies and reject other frequencies. The Bode diagram tells us which frequency components the filter reduces and how much each component is reduced. In this section we construct Bode diagrams of a low-pass filter and a notch filter.

A low-pass filter does not affect the components of a signal that are below a certain frequency called the break-point frequency, f_b. All components of the signal that are above the break point are reduced in amplitude, and the higher the frequency, the greater the reduction. Low-pass filters are very common in signal conditioning because most of the useful information in the signal is in the low-frequency components. Because noise tends to occur at the higher frequencies, a low-pass filter can often be designed to reduce the noise without affecting the information content of the signal.

A high-pass filter is just the opposite of the low-pass filter. It does not affect the components of the signal that are above the break-point frequency. All components of the signal that are below the break point are reduced in amplitude, and the lower the frequency, the greater the reduction. High-pass filters do not make much sense in control, for the same reasons that low-pass filters are useful.

A band-pass filter has two frequency values, called the *half-power frequencies,* that are separated by a frequency range called the *bandwidth* of the filter. The band-pass filter has little or no effect on components of the signal that are between the two half-power frequencies. All components of the signal that are outside the half-power frequencies are reduced in amplitude, and the farther they are from the half-power frequency, the greater the reduction. The radio tuner is an adjustable band-pass filter that accepts one station and rejects all other stations.

A notch filter is just the opposite of the band-pass filter. The notch filter also has two half-power frequencies separated by the bandwidth. However, the notch filter reduces all components of the signal between the two half-power frequencies and does not affect the components on either side. The maximum reduction of the signal occurs at the midpoint between the two half-power frequencies. Much of the noise in a process signal occurs at particular frequencies, such as the 60-Hz noise generated by electric power lines. Notch filters are good for removing a particular frequency component, such as the 60-Hz noise component of a signal. Low-pass filters and notch filters are the most common noise reduction circuits.

Three low-pass filter circuits are shown in Figure 6.18. The first circuit is a passive RC low-pass filter. This is the same circuit we analyzed in Chapter 4 (see Figure 4.3). The time-domain equation and transfer function were also developed in Chapter 4. The two equations are given below with the letters e and E replaced by the letters v and V:

$$\tau \frac{dv_{\text{out}}}{dt} + v_{\text{out}} = v_{\text{in}} \qquad (4.10)$$

$$\frac{V_{\text{out}}}{V_{\text{in}}} = \frac{1}{1 + \tau s} \qquad (4.27)$$

where $\tau = RC$.

One problem with the passive circuit in Figure 6.18a is that the equation of the circuit changes when a load resistor is connected to the output terminal. The break-point frequency is thus dependent on the load resistor—a very undesirable situation because it means we cannot design the filter until we know the exact size of the load resistance. Any time the load resistance changes, the filter must also be changed.

The active filter in Figure 6.18b uses a voltage follower to isolate the filter circuit from the load resistance. The active filter break-point frequency is not affected by the load resistance value. If the input impedance of the source, v_{in}, is a problem, a voltage follower can be used to isolate the input as well. Equations (4.10) and (4.27) also apply to the active circuit in Figure 6.18b.

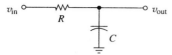

a) Passive low-pass filter

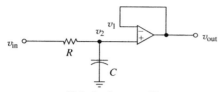

b) Active low-pass filter

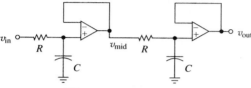

c) Two-stage active low-pass filter

◆ **Figure 6.18**   The passive *RC* circuit (a) is the simplest low-pass filter. Adding the voltage follower (b) isolates the filter from the load resistor. Adding a second isolated stage results in a two-stage active low-pass filter (c) with twice the attenuation of the single-stage filter.

A two-stage active low-pass filter is shown in Figure 6.18c. This circuit has twice the reduction power of the single-stage circuit. We will use the frequency domain to develop the transfer function of the two-stage filter from the transfer functions of each stage.

$$\frac{V_{out}}{V_{in}} = \left(\frac{V_{mid}}{V_{in}}\right)\left(\frac{V_{out}}{V_{mid}}\right)$$

$$\frac{V_{out}}{V_{in}} = \left(\frac{1}{1 + \tau s}\right)\left(\frac{1}{1 + \tau s}\right)$$

$$\frac{V_{out}}{V_{in}} = \frac{1}{(1 + \tau s)(1 + \tau s)} \tag{6.37}$$

$$\frac{V_{out}}{V_{in}} = \frac{1}{1 + 2\tau + \tau^2 s^2} \tag{6.38}$$

EXAMPLE 6.14

Use the program BODE to generate frequency response data for a two-stage active low-pass filter with a time constant, $\tau = 0.01$. Plot the Bode diagram of the filter and determine the attenuation at a frequency of 60 Hz (i.e., $\omega = 377$ rad/s).

Solution

The Bode diagram is shown in Figure 6.19. The gain at 377 rad/s is -23.6 dB, so the attenuation is 23.6 dB. The gain can also be computed from the transfer function by substituting $j377$ for ω.

$$\text{Gain (at } \omega = 377) = \frac{1}{[1 + j377(0.01)][1 + j377(0.01)]}$$

$$\text{Gain} = \frac{1}{(3.90\underline{/75°})(3.90\underline{/75°})} = 0.0657\underline{/-150°}$$

$$\text{Decibel gain} = 20 \log_{10}(0.0657) = -23.6$$

◆

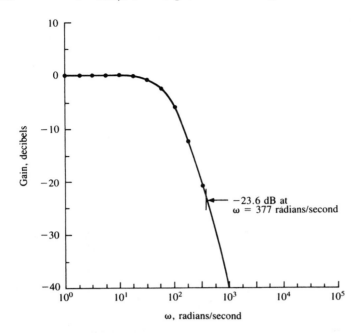

◆ **Figure 6.19** Bode diagram for the two-stage active filter circuit with a time constant $\tau = 0.01$ s (Example 6.14). The attenuation at 60 Hz (377 rad/s) is 23.6 dB.

Filter Design

The design of a low-pass filter consists of determining the filter time constant, τ [i.e., the co-efficient of s in the denominator of Equation (4.27) or Equation (6.37)]. The following rules are intended for attenuation factors of 10 or more. For a single-stage low-pass filter, the time constant (τ) is equal to the attenuation factor (A) divided by the radian frequency (ω_s) of the component you wish to attenuate. For a two-stage low-pass filter, the time constant is equal to the square root of the attenuation factor divided by ω_s. These two rules are summarized in the following equations:

$$\text{Single-stage filter: } \tau = \frac{A}{\omega_s} \tag{6.39}$$

$$\text{Two-stage filter: } \tau = \frac{\sqrt{A}}{\omega_s} \tag{6.40}$$

where τ = filter time constant, s

$A = \dfrac{\text{amplitude before filtering}}{\text{amplitude after filtering}}$

ω_s = frequency to be filtered, rad/s

EXAMPLE 6.15

Design a single-stage, active, low-pass filter that will provide an attenuation factor of 25 for a 60-Hz signal. Calculate the resistor size required if a 10-μF capacitor is used in the RC circuit.

Solution

$$\omega_s = 377 \text{ rad/s (60 Hz)}$$

$$\tau = A/\omega_s = 25/377 = 0.0663 \text{ s}$$

$$R = \tau/C = 0.0663/10\text{E} - 6 = 6.63 \text{ k}\Omega$$

◆

EXAMPLE 6.16

Design a two-stage, active, low-pass filter that will provide an attenuation factor of 200 for a 60-Hz signal. Calculate the resistor size required if 10-μF capacitors are used in the two RC circuits.

Solution

$$\omega_s = 377 \text{ rad/s (60 Hz)}$$

$$\tau = \sqrt{A}/\omega_s = \sqrt{200}/377 = 0.0375 \text{ s}$$

$$R = \tau/C = 0.0375/10\text{E} - 6 = 3.75 \text{ k}\Omega \qquad ◆$$

A notch filter produces a V-shaped curve on the Bode amplitude diagram. The notch of the V is located at the frequency of the component to be attenuated. Equation (6.41) gives the transfer function of a notch filter.

$$\frac{V_{out}}{V_{in}} = \frac{(1 + \tau_s s)(1 + \tau_s s)}{(1 + k\tau_s s)\left(1 + \frac{\tau_s}{k}s\right)} = \frac{1 + 2\tau_s s + \tau_s^2 s^2}{1 + \tau_s(k + 1/k) + \tau_s^2 s^2} \qquad \textbf{(6.41)}$$

where $\tau_s = 1/\omega_s$
  ω_s = frequency to be filtered, rad/s
  k = number determined by the attenuation factor

A value of $k = 20$ produces an attenuation factor, A, of slightly more than 10. Values of k for other attenuation factors can be determined experimentally, using the program BODE to verify the attenuation factor obtained.

EXAMPLE 6.17

Design a notch filter that will attenuate a 60-Hz signal component by a factor of 10. Use the program BODE to generate frequency response data and plot the Bode diagram amplitude curve.

Solution

The first step is to determine the coefficients in Equation (6.41) and multiply the numerator and denominator binomials to prepare the equation for use in the program BODE.

$$\omega_s = 2\pi(60) = 377 \text{ rad/s}$$

$$\tau_s = \frac{1}{377} = 0.002653 \text{ s}$$

$$k\tau_s = 20(0.002653) = 0.05305 \text{ s}$$

$$\frac{\tau_s}{20} = \frac{0.002653}{20} = 0.0001326 \text{ s}$$

$$\frac{V_{out}}{V_{in}} = \frac{(1 + 0.002653s)(1 + 0.002653s)}{(1 + 0.05305s)(1 + 0.0001326s)}$$

$$\frac{V_{out}}{V_{in}} = \frac{1 + 5.31 \times 10^{-3}s + 7.04 \times 10^{-6}s^2}{1 + 5.32 \times 10^{-2} + 7.04 \times 10^{-6}s^2}$$

The Bode diagram is shown in Figure 6.20. ◆

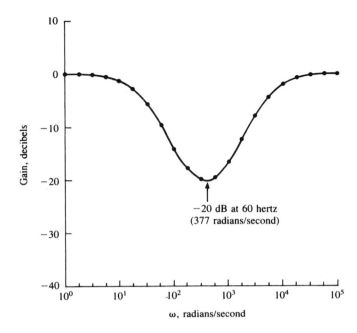

◆ **Figure 6.20** Bode diagram of a notch filter designed to produce at least 20 dB of attenuation at 60 Hz (377 rad/s). See Example 6.17 for details.

Linearization

The ideal measuring instrument produces a linear calibration curve in which the output goes from 0 to 100% of the output range as the input goes from 0 to 100% of the input range. Some primary elements, such as a platinum resistance detector, produce a very linear signal, and the right amount of amplification will produce a nearly ideal calibration curve. Other primary elements produce nonlinear outputs, and the signal must be linearized to produce a nearly ideal calibration curve. We will examine the *linearization* of two nonlinear signals. The first nonlinear signal involves the measurement of liquid flow rate with an orifice and differential pressure (Δp) transmitter. The second involves the measurement of the volume of a round liquid tank that is resting on its side.

Figure 6.21 illustrates the measurement and linearization of liquid flow rate when the flow is turbulent. The flow meter consists of the orifice and differential pressure transmitter, which produces an output, Δp, that is proportional to the square of the flow rate, Q. If we express both signals as a percentage of their full-scale range, the equation for the output of the flow meter is given by Equation (6.42).

$$\Delta p = p_1 - p_2 = 0.01Q^2 \tag{6.42}$$

The table on the left in Figure 6.21 shows corresponding values of Δp and Q over the range of the measurement. Notice the obvious nonlinearity of the output, Δp. At a flow rate of 20%, Δp is only 4%; at a flow rate of 50%, Δp is only 25%. One way to handle a nonlinear signal is to plot the calibration curve on nonlinear graph paper. A square-root scale will linearize the signal from the Δp transmitter. However, the preferred method of handling the signal is to linearize the signal with a signal conditioning component that we will call a linearizer.

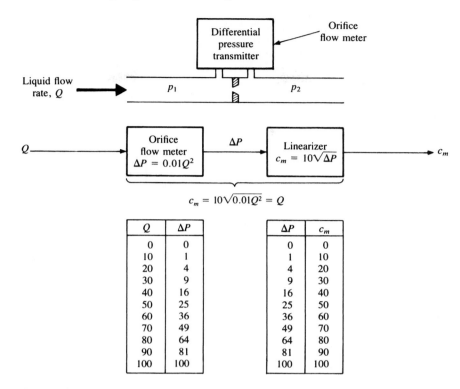

◆ **Figure 6.21**   An orifice flow meter produces a Δp signal that is proportional to the square of the flow rate, Q. By forming the square root of its input signal, the linearizer produces an output that is proportional to the flow rate, Q.

The output of the Δp transmitter, Δp, is the input to the linearizer. The output of the linearizer c_m is also expressed as a percentage of the full-scale range. The design goal of the linearizer is to produce an output that is exactly equal to the flow rate signal, Q, as given by the equation

$$c_m = Q \tag{6.43}$$

The mathematical description of the linearizer is the inverse of the function that describes the flow meter. We can obtain the inverse function by solving Equation (6.42) for Q and then substituting c_m for Q in the resulting inverse function.

$$0.01Q^2 = \Delta p$$

$$Q^2 = 100 \, \Delta p$$

$$Q = \sqrt{100 \, \Delta p}$$

$$c_m = 10 \sqrt{\Delta p} \tag{6.44}$$

The table on the right in Figure 6.21 shows corresponding values of Δp and c_m over the range of measurement. Tracing a few signals through the two tables will verify Equation (6.43). For example, a flow rate of 20% produces a Δp of 4%, which in turn produces a c_m of 20%. The input/output curves of the flow meter and the linearizer are plotted in Figure 6.22. A function and its inverse have an interesting and useful graphical feature. The two curves produce a pattern that is symmetrical about the line $y = x$. We can use this symmetry to plot the graph of the inverse of any nonlinear calibration curve. When we do this, we are producing the curve that defines the linearizer for that nonlinear signal. Another way of defining the linearizer is to produce a table of values similar to the table on the right in Figure 6.21. This table was con-

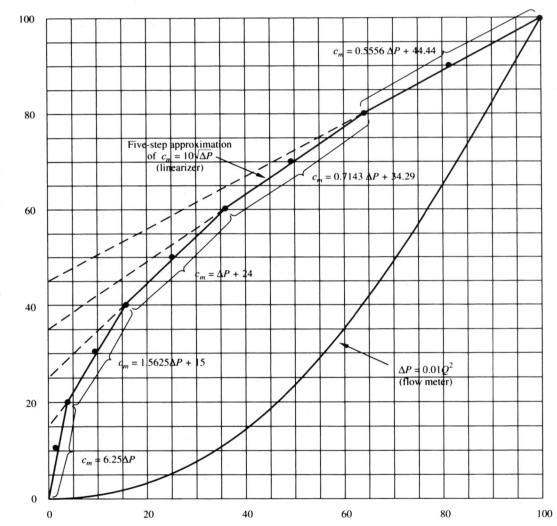

◆ **Figure 6.22** A piecewise-linear approximation is one way to implement a non-linear function. Here the inverse of the square function, $\Delta p = 0.01Q^2$, is formed by a five-segment piecewise-linear approximation. With careful selection of the lines, the error in the top 65% of the curve can be limited to about 0.5% of full-scale range.

◆ **TABLE 6.2** Comparison of the Midpoints of the Line Segments with the Corresponding Ideal Values[a]

Δp	Ideal c_m	Approximate c_m	Error (%)
1	10	6.25	3.75
9	30	29.06	0.94
25	50	49.00	1.00
49	70	69.29	0.71
81	90	89.44	0.56

[a] All values are a percentage of full-scale range.

structed by switching the two columns from the table on the left and changing the heading on the flow rate column from Q to c_m.

> The linearizer can be defined mathematically by the inverse of the function that defines the primary element. It can be defined with a table by switching the columns of the primary element table and renaming the left column. It can be defined graphically by constructing the curve that completes the symmetry about the line $y = x$.

Once the linearizer is defined by one of the preceding three methods, we are faced with the problem of implementing the definition in an electric circuit. An operational amplifier makes an excellent function generator. However, it requires an electrical component that has a nonlinear voltampere characteristic that matches either the primary element function or its inverse. Refer to Section 6.3 for details on the op-amp circuits for generating functions. Another approach to implementing the linearizer is to construct an approximate inverse function by a set of straight lines, each forming a portion of the nonlinear curve. This type of function is called a *piecewise-linear function*. Figure 6.22 uses a five-step piecewise-linear function to approximate the function $c_m = 10\sqrt{\Delta p}$. Each of the five lines covers 20% of the c_m scale. For simplicity, only one of the break points is not on the ideal curve. The point at $c_m = 20\%$ was moved from $\Delta p = 4$ to $\Delta p = 3.2$. All other endpoints are on the ideal curve. Table 6.2 of corresponding values at the midpoints illustrates the accuracy of the approximation. The errors above $c_m = 35\%$ can be reduced to about 0.5% by moving the endpoints until the maximum error is minimized. This will be left as an exercise for the student.

EXAMPLE 6.18

Construct a five-step, piecewise-linear approximation of the function defined by the input/output table below. Locate the endpoints on the curve at the following input values: 0, 20, 40, 60, 80, and 100. Compute the slope and intercept of each line segment, write the equation for each segment, and construct an error table similar to Table 6.2. Move the endpoints to reduce the largest errors and write the equations and the error table for the second set of line segments.

Input	Output	Input	Output	Input	Output
0	3.0	35	33.2	70	72.8
5	7.0	40	38.7	75	77.1
10	11.0	45	44.6	80	81.2
15	15.2	50	50.4	85	85.2
20	19.5	55	56.4	90	89.3
25	23.8	60	62.1	95	93.2
30	28.3	65	67.8	100	97.0

Solution   We will proceed as follows:

1. Use $m = (y_2 - y_1)/(x_2 - x_1)$ to compute the slope of each line segment.
2. Use $b = y_1 - mx_1$ to compute the intercept of each segment.
3. Use $y = mx + b$ to write the equation of each line.

The results are summarized in the following two tables.

Endpoints and Equations of the Line Segments

Line No.	(x_1, y_1)	(x_2, y_2)	Equation
1a	0, 3.0	20, 19.5	$y = 0.825x + 3.0$
2a	20, 19.5	40, 38.7	$y = 0.960x + 0.30$
3a	40, 38.7	60, 62.1	$y = 1.170x - 8.10$
4a	60, 62.1	80, 81.2	$y = 0.955x + 4.80$
5a	80, 81.2	100, 97.0	$y = 0.790x + 18.00$

Midpoint Error for Each Line Segment

Input	Ideal Output	Approximate Output	Error
10	11.0	11.25	−0.25
30	28.3	29.10	−0.80
50	50.4	50.40	0.00
70	72.8	71.65	1.15
90	89.3	89.10	0.20

The two large errors are in line segments 2a and 4a. One way to reduce the error is to make these two segments smaller. We chose to move the second endpoint from 20 to 30 and the fourth endpoint from 60 to 70. The results are as follows:

Endpoints and Equations of the Line Segments

Line No.	(x_1, y_1)	(x_2, y_2)	Equation
1b	0, 3.0	30, 28.3	$y = 0.843x + 3.0$
2b	30, 28.3	40, 38.7	$y = 1.040x - 2.90$
3b	40, 38.7	70, 72.8	$y = 1.137x - 6.77$
4b	70, 72.8	80, 81.2	$y = 0.840x + 14.00$
5b	80, 81.2	100, 97.0	$y = 0.790x + 18.00$

Midpoint Error for Each Line Segment

Input	Ideal Output	Approximate Output	Error
15	15.2	15.65	−0.45
35	33.2	33.50	−0.30
55	56.4	55.77	0.63
75	77.1	77.05	0.10
90	89.3	89.10	0.20

Further reduction is possible with a careful selection of the endpoints. Sometimes moving the endpoints off the line helps to reduce the error. A large, accurate graph is very helpful in determining the endpoints.

◆

In the flow example, the equation of the inverse function was easy to determine. In the second example, the inverse function is more difficult to determine, so we resort to graphical and tabular methods to define the linearizer. Figure 6.23a illustrates a liquid storage tank resting on its side with the axis in a horizontal position. The input to the sensor is the level, h, of the liquid in the tank. The output of the sensor is a nonlinear volume signal designated by V_{NL} (the NL subscript indicates that we consider the output of the sensor to be a nonlinear measurement of the volume in the tank). The graph of h versus V_{NL} forms an S-shaped curve as shown below the sensor in Figure 6.23c. The output from the sensor, V_{NL}, is the input to the linearizer. The output of the linearizer, V_{LIN}, is such that a graph of h versus V_{LIN} is the

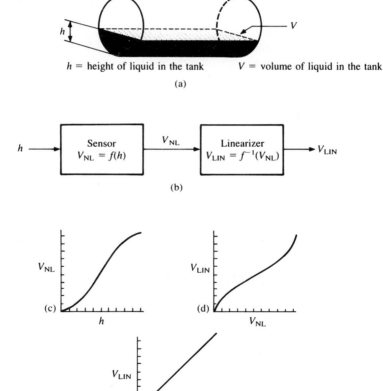

◆ **Figure 6.23**   A nonlinear signal results when liquid level is used to measure the volume of liquid in a round tank resting on its side. The graph of the liquid level, h, versus the volume in the tank forms an S-shaped curve. The linearizer must form a reverse S-shaped curve to linearize the signal.

◆ **TABLE 6.3** Values of Height, h, versus Volume, V_{NL}, for the Sensor in Figure 6.23[a]

h	V_{NL}	h	V_{NL}	h	V_{NL}	h	V_{NL}
0	0.0	25	19.6	55	56.4	80	85.8
5	1.9	30	25.2	60	62.7	85	90.6
10	5.2	35	31.2	65	68.8	90	94.8
15	9.4	40	37.3	70	74.8	95	98.1
20	14.2	45	43.6	75	80.4	100	100.0
		50	50.0				

[a] Values are a percentage of full-scale range.

straight line at the bottom in Figure 6.23e. The required graph of the linearizer is the reverse S-shaped curve shown under the linearizer in Figure 6.23d.

The equation that defines the volume in the tank as a function of the height, h, is quite complex, so complex, in fact, that we will use an equation that defines the volume of the tank from empty to half full. Then we will use the obvious symmetry of the tank to complete the curve for the top half. In the following equations, h is the height of the tank, r is the radius of the tank, and V_{NL} is the output of the level sensor. The following equations define the volume of liquid in the bottom half of the tank:

$$k = \frac{h}{r}$$

$$\alpha = \tan^{-1}\left[\frac{\sqrt{k(2 - k)}}{k - 1}\right] \qquad (6.45)$$

$$V_{NL} = \frac{50(\alpha - \sin \alpha)}{\pi} \qquad (6.46)$$

More accurate graphs of V_{NL} versus h and V_{LIN} versus V_{NL} are shown in Figure 6.24. Table 6.3 was used to construct the two curves in Figure 6.24.

6.5 DIGITAL SIGNAL CONDITIONING

Digital control systems are used in a wide variety of industries to increase productivity and efficiency. Industries that use digital control systems include food processing, oil refining, chemical processing, steel production, automobile manufacturing, and many others. Digital control systems are also used in many of the products we buy. In automobiles, for example, a microprocessor-based system controls fuel flow and spark timing to achieve optimum engine performance.

Most of the physical parameters in a process are analog signals. Digital systems must be able to convert these analog signals into digital form for processing by the computer. This is the job of the *data acquisition system* shown in Figure 6.25. In a sense, the data acquisition system is the "eyes and ears" of the digital computer. The computer must also be able to convert its digital control actions into analog form for input to the analog actuators. This is the job of the *data distribution system,* also shown in Figure 6.25. In a sense, the data distribu-

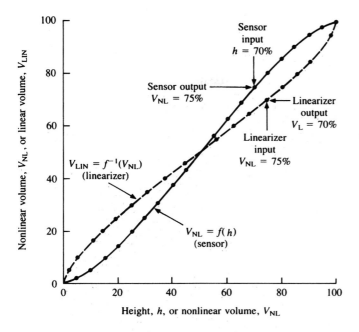

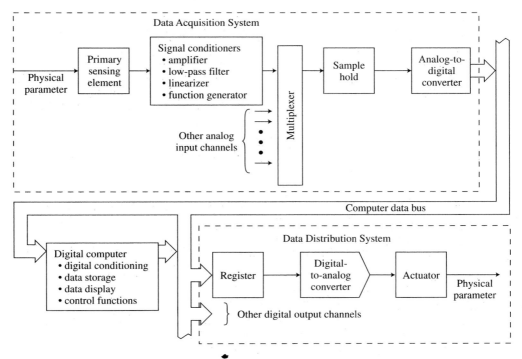

◆ **Figure 6.24** Input versus output curves for the sensor and the linearizer for measurement of the volume of liquid in a round tank resting on its side.

◆ **Figure 6.25** The data acquisition system is the "eyes and ears" of a digital control system; the data distribution system is its "hands and arms."

tion system is the "hands and arms" of the digital computer. Data acquisition and distribution involve two very important concepts: *data sampling and data conversion.*

Data Sampling

Data sampling is a process in which a switch connects momentarily to an analog signal in a sequence of pulses separated by evenly spaced increments of time called the *sampling interval.* You can also visualize data sampling as the result obtained by multiplying the analog signal by a series of pulses with an amplitude of 1, a very narrow pulse width, and a period equal to the sampling interval. Figure 6.26a shows an analog signal, Figure 6.26b shows a series of

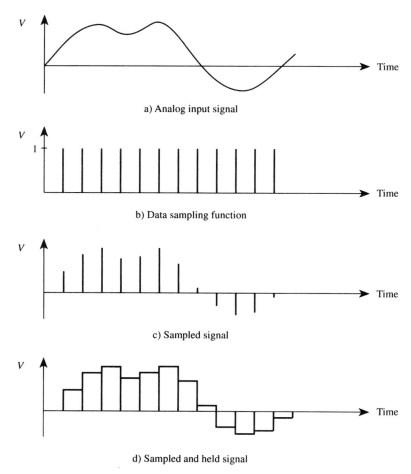

a) Analog input signal

b) Data sampling function

c) Sampled signal

d) Sampled and held signal

◆ **Figure 6.26**   A sampled signal (c) may be viewed as the product of the analog input signal (a) multiplied by the data sampling function (b). If the data samples are held until the next sample (d), the resulting signal is a stairstep approximation of the analog signal.

sampling pulses, and Figure 6.26c shows the sampled signal. If the sampling switch is replaced by a *sample-and-hold* circuit, each sampled value is held until the beginning of the next sampling interval, as shown in Figure 6.26d. The sampled and held signal gives a stairstep approximation of the original signal.

The sample-and-hold circuit is used to hold a sample of an analog signal as it is converted to a digital signal by an analog-to-digital converter. Refer to Figure 6.27 for the sample-and-hold circuit diagram. The sample-and-hold device has a signal input, a control input, and an output. It operates in two modes: the sampling or tracking mode when the switch is closed and the holding mode when the switch is open. The ideal sample and hold takes a sample in zero time and holds the signal value indefinitely with no degradation of the signal. This ideal circuit is called a zero-order sample and hold.

A practical sample-and-hold circuit varies from the ideal in a number of parameters. The following are some parameters of a practical sample-and-hold circuit:

1. The *acquisition time* is the time from the instant the sample command is given until the output is within a specified band of the input. This is determined by the size of the input resistor and the holding capacitor. These two elements form a first-order lag with a time constant equal to the product of the resistance times the capacitance value.

$$\text{Time constant} = \tau = R_{in}C$$

   An acquisition time equal to 5 times the time constant is enough for the output to reach 99.3% of the total change to the new input value. This assumes that the input is constant during the sampling period.
2. The *aperture time* is the time it takes for the switch to open. It is the time between the hold command and the time the switch is completely open.
3. The *decay rate* is the rate of change of the output in the holding mode. It is caused by leakage through the capacitor and the small current that enters the op amp through the inverting input.

Data sampling raises two obvious questions:

1. How often must we sample a signal to get all the information in the original signal?
2. What happens if we do not sample often enough?

The Nyquist criterion answers the first question, and the explanation of the Nyquist criterion will reveal the answer to the second question. The *Nyquist criterion* may be stated as follows:

> *All the information in the original signal can be recovered if it is sampled at least twice during each cycle of the highest-frequency component.*

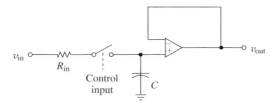

◆ **Figure 6.27**   A sample-and-hold circuit has a signal input, a control input, and an output. It operates in two modes: the sampling or tracking mode when the switch is closed and the holding mode when the switch is open.

If f_h is the highest frequency that occurs in the original signal, then the minimum *sampling rate* is given by the following equation:

$$\text{Minimum sampling rate} = f_s(\text{min}) = 2f_h \tag{6.47}$$

We can gain some insight into the Nyquist criterion by considering the frequency spectra of the original signal and the sampled signal. Figure 6.28a shows a typical *frequency spectrum* of an analog signal. A frequency spectrum is simply a plot of the maximum voltage of each possible component of a signal versus the frequency of that component. The main point in Figure 6.28a is that there are no components of the original signal with a frequency greater than f_h.

The data sampling function is represented in Figure 6.26b as a series of pulses spaced f_s apart. This sampling function can be represented by an infinite series of sinusoidal components called a Fourier series. This series has components with frequencies of f_s, $2f_s$, $3f_s$, . . . , nf_s, . . . , etc. If we visualize the sampling process as the multiplication of the original signal by the sampling function, then, in effect, we have amplitude modulation of the original signal with carrier frequencies of f_s, $2f_s$, $3f_s$, . . . , etc. The original signal is shifted and folded

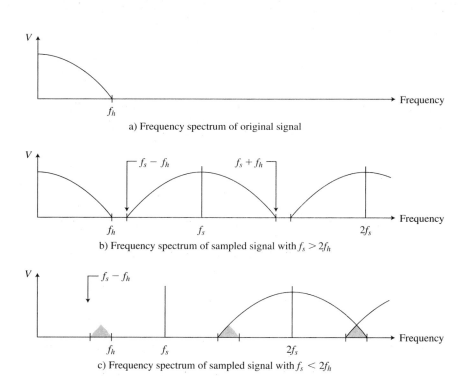

a) Frequency spectrum of original signal

b) Frequency spectrum of sampled signal with $f_s > 2f_h$

c) Frequency spectrum of sampled signal with $f_s < 2f_h$

◆ **Figure 6.28**   In a sampled signal, the frequency spectrum of the original signal is both subtracted from and added to the sample frequency, f_s, and all multiples of f_s. If $f_s > 2f_h$. the original signal can be recovered by filtering out all frequencies above f_s. When $f_s < 2f_h$, the difference frequency spectrum overlaps the original signal, and filtering will not recover the original signal.

around each carrier as shown in Figure 6.28b. The original signal can be recovered from the sampled signal shown in Figure 6.28b with a low-pass filter that removes all frequency components greater than f_h.

Figure 6.28c shows what happens when the sampling rate is less than f_h. Part of the folded signal around f_s overlaps part of the original signal. The overlapped portion of the folded signal becomes part of the original signal. It cannot be removed by a low-pass filter. This overlap of the folded signal is called *aliasing*. The effect of aliasing is that one or more frequency components are added to the original signal. Figure 6.29 illustrates how an *alias frequency* is produced by an insufficient sampling rate. The obvious solution to the aliasing problem is either to increase the sampling rate or filter the original signal to remove all components with a frequency greater than $f_s/2$.

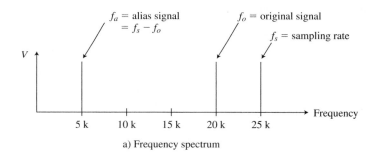

a) Frequency spectrum

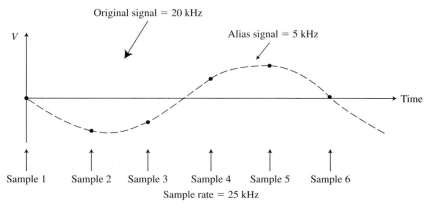

b) Voltage vs. time diagram

◆ **Figure 6.29**   Sampling a 20-kHz original signal at a sampling rate of 25 kHz results in a 5-kHz alias component in the sampled signal. In the frequency spectrum, the alias appears as the difference frequency obtained by subtracting the original frequency from the sampling rate.

Data Conversion

Data converters are the interface between the analog and digital domains. The *analog-to-digital converter* (*ADC* or *A/D*) converts analog signals into a digital code. The *digital-to-analog converter* (*DAC* or *D/A*) converts digital codes into analog voltage levels.

The input to an *A/D converter* is an analog voltage that can have any value from 0 to its full-scale (*FS*) range. The output, however, can have only a finite number of output codes, each defining a state of the ADC. Figure 6.30 shows the transfer function of an ideal 3-bit A/D converter with eight output states. The output states are assigned 3-bit binary codes from 000 to 111. We will use Figure 6.30 to help explain several important points concerning A/D converters.

The resolution of an A/D converter is defined as the number of output states and is often expressed as the number of bits in the output code. If n is the number of bits, then 2^n is the number of output states. Thus the 3-bit ADC in Figure 6.30 has $2^3 = 8$ states. An 8-bit ADC has $2^8 = 256$ states, and a 10-bit ADC has $2^{10} = 1024$ states. Between each pair of steps is a transition that makes the change from one output state to another. There is always one less transition than the number of states. The 3-bit ADC has $8 - 1 = 7$ transitions. An n-bit ADC has $2^n - 1$ transitions.

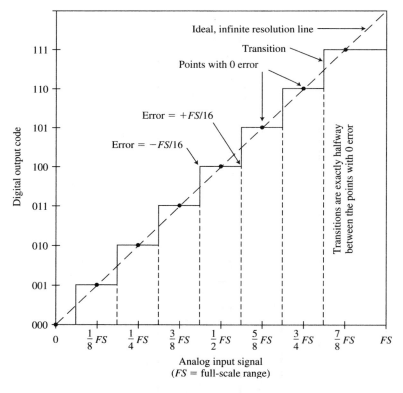

◆ **Figure 6.30**　The transfer function of an ideal 3-bit analog-to-digital converter has $2^3 = 8$ output states and $2^3 - 1 = 7$ transitions. Each step represents a range of analog input values of *FS*/8. The quantization error is 0 in the center of each step and ranges from $-FS/16$ at the front edge of the step to $+FS/16$ at the rear of the step.

At any state, there is a small range of input voltages that forms a step that extends from one transition to the next. This range of values is called the quantization size or "width" of the code. The ideal width of each interior step is designated as V_{LSB}, where LSB signifies least significant bit. Resolution is sometimes expressed in terms of V_{LSB}. The value of V_{LSB} is equal to the full-scale range of the input divided by the number of output states:

$$\text{Resolution} = V_{LSB} = FS/2^n \tag{6.48}$$

Notice that the center of each step intersects the ideal line that extends from the lower left corner to the upper right corner of the graph. These intersections are marked with dots, and they are the only points on the graph that have zero error. On each step, there is an error called the *quantization error* that ranges from $-V_{LSB}/2$ on the left end of the step to $+V_{LSB}/2$ on the right end. In Figure 6.30 the quantization error is $\pm FS/16$.

The input to a *D/A converter* is a finite number of digital input codes. The output is an equal number of discrete output voltage levels. The resolution of a DAC is sometimes specified as the number of bits in the input code (the same as for an ADC). Figure 6.31 shows the ideal transfer function of a 3-bit D/A converter.

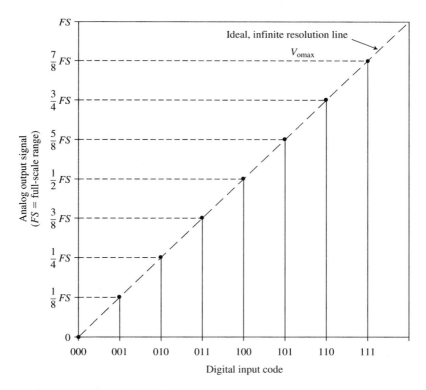

◆ **Figure 6.31**   The transfer function of an ideal 3-bit digital-to-analog converter has only eight possible input states and therefore, only eight possible output voltage levels. The eight dots on the graph define the input/output relationship for the eight states.

Unlike the A/D converter, the analog signal from a DAC does not range over an infinite number of values between 0 and *FS*. Rather, there is a discrete analog output value for each digital input code. As the input code is traversed from 000 to 111, the analog output steps from point to point on the ideal, infinite resolution line. Each step of the analog output is a change in voltage of V_{LSB}, which is also used to express the resolution of a DAC [see Equation (6.48)].

The top step in Figure 6.31 represents the maximum output voltage of the DAC. We will designate this voltage as V_{omax}. The step size (and resolution) of a DAC can also be expressed in terms of V_{omax} as defined in Equation (6.49):

$$\text{Resolution} = V_{LSB} = \frac{V_{omax}}{2^{n-1}} \qquad \textbf{(6.49)}$$

As Equations (6.48) and (6.49) indicate, resolution is often looked upon as the "step size" when the input binary word goes from one binary state to another. It is important to note that the maximum output, V_{omax}, will never reach the full-scale value, *FS*. However, as the resolution of the DAC increases, V_{omax} does approach the value of *FS*. In fact, *FS* is the limiting value of V_{omax} as the resolution approaches zero and *n* approaches infinity.

Digital-to-Analog Converters

The function of a digital-to-analog converter is to produce an output voltage that is a weighted sum of the nonzero bits in the digital input code. The least significant bit (bit 0) is given a weighting of 1. The weighting of each successive bit is 2 times the weighting of the previous bit. Thus a 4-bit DAC will have weighting factors of 1, 2, 4, and 8; an 8-bit DAC will have weighting factors of 1, 2, 4, 8, 16, 32, 64, and 128. The bit with the highest weighting factor is called the most significant bit (MSB). The weighting of the MSB of an *n*-bit DAC is $2^n - 1$.

The sum of all the weighting factors is the weighting given to the highest input code (the code with all bits equal to 1). An interesting fact about these weighting factors is that they form a series in which the sum of any number of terms is always 1 less than 2 times the largest term. For example,

$$1 + 2 + 4 + 8 = 15 = (2 \times 8) - 1$$

$$1 + 2 + 4 + 8 + 16 + 32 + 64 + 128 = 255 = (2 \times 128) - 1$$

The binary-weighted digital-to-analog converter uses a summing amplifier to form the weighted sum of all the nonzero input bits. The input resistances of the summing amplifier provide the required weighting factors. Figure 6.32 shows a 4-bit *binary-weighted DAC*. The feedback resistance (R_f), the MSB input resistance (R), and the input voltage level used for a binary 1 determine the output voltage level of the MSB. If we let V_1 represent the input voltage of binary 1, then Equation (6.50) defines the voltage level of the MSB. In Figure 6.32, the value of 5 V is used for V_1. Equation (6.51) gives the full-scale voltage, and Equation (6.52) gives the LSB voltage.

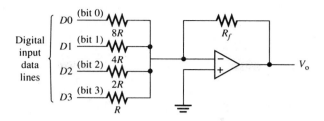

Digital inputs = 5 volts for binary 1
= 0 volts for binary 2

Digital input	V_o	Digital input	V_o
0000	0	1000	FS/2
0001	FS/16	1001	9FS/16
0010	FS/8	1010	5FS/8
0011	3FS/16	1011	11FS/16
0100	FS/4	1100	3FS/4
0101	5FS/16	1101	13FS/16
0110	3FS/8	1110	7FS/8
0111	7FS/16	1111	15FS/16

$$V_o = -\frac{R_f}{R}\left[D3 + \frac{1}{2}D2 + \frac{1}{4}D1 + \frac{1}{8}D0\right]$$

$$V_{LSB} = -\frac{R_f}{R}\left[\frac{1}{8}D0\right]$$

$$V_{MSB} = -\frac{R_f}{R}[D3]$$

$$V_{FS} = 2V_{MSB}$$

◆ **Figure 6.32** A 4-bit binary-weighted D/A converter. Weighting is provided by input resistors R, $2R$, $4R$, and $8R$. Each digital input ($D0$–$D3$) has a value of 0 or 5 V, depending on the corresponding bit in the input code. The output is the weighted sum of only those inputs that have a value of 5 V.

$$V_{MSB} = \frac{-R_f V_1}{R} \tag{6.50}$$

$$V_{FS} = 2 \times V_{MSB} \tag{6.51}$$

$$V_{LSB} = \frac{V_{MSB}}{2^n} \tag{6.52}$$

Although simple to understand, there are many disadvantages to the binary-weighted DAC. Each input resistor has a different resistance based upon the ratio of R–$2R$–$4R$–$8R$, and so on. It is difficult to build IC resistors that can be accurately matched at ratios greater than 20:1. This limits binary-weighted DACs to 5 bits or less. Precision resistors or 10-turn potentiometers could be used, but that would be expensive. In addition, each binary bit position is represented by a different Thévenin equivalent resistance, and the devices that provide the inputs to the DAC will see different loads.

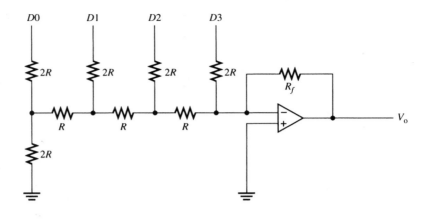

$$V_o = -\frac{R_f}{2R}\left[D3 + \frac{1}{2}D2 + \frac{1}{4}D1 + \frac{1}{8}D0\right]$$

$$V_{LSB} = -\frac{R_f}{2R}\left[\frac{1}{8}D0\right]$$

$$V_{MSB} = -\frac{R_f}{2R}[D3]$$

$$V_{FS} = 2V_{MSB}$$

◆ **Figure 6.33** A 4-bit *R–2R* D/A converter uses only two resistor sizes to provide the four weighting factors. The digital inputs are 0 V for a binary 0 and 5 V for a binary 1. The output is the weighted sum of only those inputs that have a value of 5 V.

The *R–2R ladder-type DAC* eliminates both of the disadvantages of the binary-weighted DAC. It is a resistive network consisting of only two different values of resistors, and the resultant analog output is the weighted sum of the applied digital input word. Figure 6.33 shows an *R–2R* DAC, and Figure 6.34 shows a development of the output equation of the *R–2R* DAC. This development uses the superposition principle and Thévenin's theorem to develop four equivalent circuits, one for each of the 4 bits in the digital input word.

Analog-to-Digital Converters

The function of an analog-to-digital converter is to produce a digital code word that accurately represents the level of the analog input voltage. In this section, we begin with examples of five analog-to-digital conversion techniques and end with a discussion of the major considerations in the selection of an analog-to-digital converter. Here are the five A/D conversion techniques:

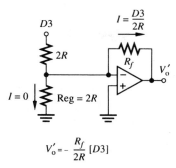

$$V_0' = -\frac{R_f}{2R}[D3]$$

a) Equivalent circuit with $D2 = 0$,
$D1 = 0$, and $D0 = 0$

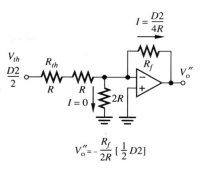

$$V_0'' = -\frac{R_f}{2R}[\frac{1}{2}D2]$$

b) Equivalent circuit with $D3 = 0$,
$D1 = 0$, and $D0 = 0$

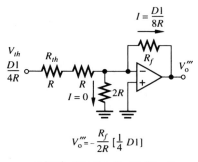

$$V_0''' = -\frac{R_f}{2R}[\frac{1}{4}D1]$$

c) Equivalent circuit with $D3 = 0$,
$D2 = 0$, and $D0 = 0$

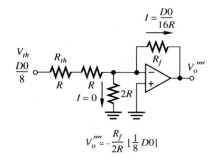

$$V_0'''' = -\frac{R_f}{2R}[\frac{1}{8}D0]$$

d) Equivalent circuit with $D3 = 0$,
$D2 = 0$, and $D1 = 0$

$$V_0 = V_0' + V_0'' + V_0''' + V_0'''' = -\frac{R_f}{2R}\left[D3 = \frac{1}{2}D2 + \frac{1}{4}D1 + \frac{1}{8}D0\right]$$

◆ **Figure 6.34** Superposition and Thévenin's theorem are used to construct the four equivalent circuits used to obtain the output equation of the $R-2R$ D/A converter.

1. The *binary counter technique* uses a binary counter connected to a D/A converter and a comparator to generate the digital word that represents an analog input. A series of clock pulses advance the counter through its binary states, producing a stairstep voltage output from the D/A converter. Both the analog input and the DAC output are input to a comparator. At the instant that the DAC output passes the analog input, the output of the comparator switches to a high level. The logic circuit latches the count in the counter. Because the latched count is the one that made the DAC output equal to the analog input, it is an accurate measure of the analog input.

   Although simple, the counter-type ADC is relatively slow. The conversion time depends on the level of the analog signal and the number of bits in the digital word. In the worst case, an 8-bit counter type ADC would require 256 clock pulses to complete the conversion. For a 12-bit ADC, the worst case would require 4096 clock pulses.

A counter-type analog-to-digital converter is illustrated in Figure 6.35. This is a relatively easy circuit to construct using the components specified in the schematic. The timing diagram is quite useful to help explain the conversion operation. The 555-clock waveform is shown at the top of the timing diagram. The output of the clock is fed into the binary counter so that each clock pulse will cause the counter to increase its count by 1 bit. The output of the counter will sequence from 0000 to 1111 in the same sequence as the binary inputs in Figure 6.32. When the counter reaches 1111, the next clock pulse restores the 0000 count, and the sequence starts over. The counter output is shown in binary form on the second section of the timing diagram.

The binary output of the counter is fed into a binary ladder DAC, producing the stairstep output shown in the third section of the timing diagram. The analog input signal (the signal that is to be converted to digital) is superimposed on the stairstep waveform to show the relative size of the two inputs to the comparator. The output of the comparator is positive when the analog input signal is greater than the analog signal from the DAC. When the relative magnitude of the two analog signals reverses, the output of the comparator reverses also.

Notice that the comparator output switches from $+V_c$ to 0 at the point that the analog input signal crosses the stairstep output from the DAC. The one-shot circuit converts the $+V_c$ to 0 change in the comparator output into a pulse that is used to trigger the quad latch. Each time the quad latch is triggered, the binary signal from the counter is transferred to the output of the latch and held there until the next triggering pulse. The result is that the output of the quad latch reflects the count for which the analog input signal was approximately equal to the stairstep voltage. The binary output signal from the analog-to-digital converter is shown on the bottom line of the timing diagram.

2. The *successive approximation technique* sequentially compares a series of binary weighted values with the analog input to generate a digital word that represents the analog input voltage. Successive approximation is a relatively fast technique, completing the conversion in just *n* steps, where *n* is the resolution in bits. This technique is best compared to the process of weighing a person on a balance scale. The person adjusts the scale, adds weights until the scale is balanced, and then announces the weight.

The successive approximation technique selects the most significant bit that is less than the analog value being converted and then adds all successive lesser bits for which the total of accepted bits does not exceed the analog value. When all bits have been considered, the analog value is approximated by the sum of the bits accepted. This is one of the more commonly used techniques when medium- to high-speed conversion is required. Figure 6.36 shows a 4-bit successive approximation A/D converter.

3. The *single slope technique* uses a fixed-rate, variable-time ramp voltage to generate a digital word that represents the analog input voltage. The ramp voltage is generated by an integrator with a reference voltage input. A comparator compares the ramp voltage to the analog input voltage and signals the instant when the ramp voltage equals the analog input voltage. The time required for the ramp voltage to equal the input voltage is measured by a clock counter. When the two values are nearly equal (depending on the resolution), the counter is disabled and the digital word in the counter's

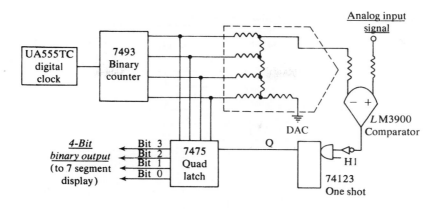

a) Schematic diagram

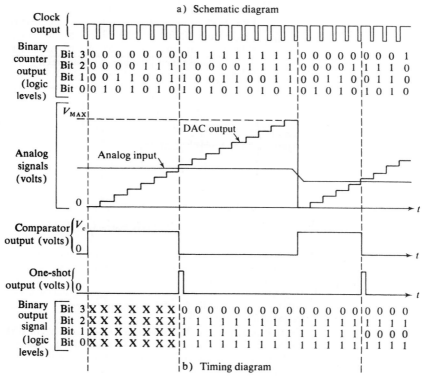

b) Timing diagram

◆ **Figure 6.35** Schematic diagram and waveforms of a counter type of analog-to-digital converter.

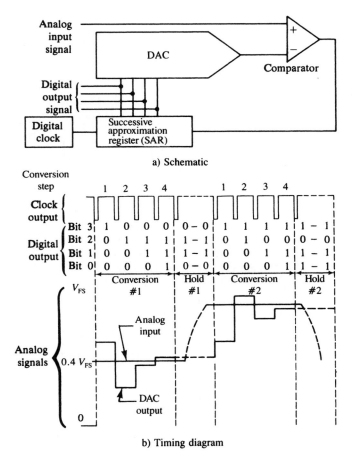

a) Schematic

b) Timing diagram

◆ **Figure 6.36** A 4-bit successive approximation A/D converter. A conversion takes place in four steps, each initiated by a negative clock pulse. The four steps for conversion 1 are as follows:

1. The SAR outputs digital word 1000, producing a DAC output of $V_{FS}/2$ V. This is higher than the analog input, so bit 3 is reset to 0 and latched for the duration of the conversion.
2. The SAR outputs word 0100, producing a DAC output of $V_{FS}/4$ V. This is lower than the analog input, so bit 2 is set to 1 and latched for the duration.
3. The SAR outputs word 0110, producing a DAC output of $3V_{FS}/8$ V. This is lower than the analog input, so bit 1 is set to 1 and latched for the duration.
4. The SAR outputs digital word 0111, producing a DAC output of $7V_{FS}/16$ V. This is higher than the analog input, so bit 0 is reset to 0 and latched for the duration.

The final digital output is 0110.

output register represents the analog value. The counter is read and then reset to prepare for the next measurement. This technique is one of many integrating analog-to-digital converters. Figure 6.37 shows a single-slope ADC.

4. The *dual slope technique* uses a variable-slope, fixed-time ramp followed by a fixed-slope, variable-time ramp to generate the digital word that represents the analog input voltage. The variable-slope ramp is generated by an integrator with the analog input voltage as input. The fixed-slope ramp is generated by the same integrator with a reference voltage as input. A timer counter is used to time the first ramp and measure the second ramp.

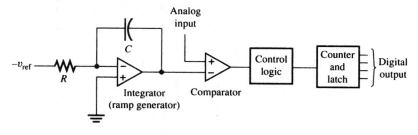

a) Schematic diagram

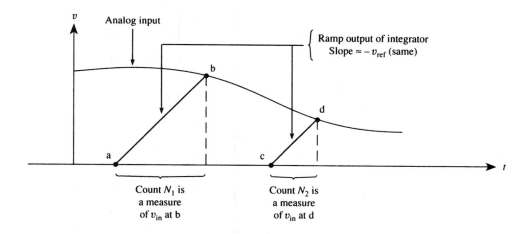

b) Waveform of two conversions
• conversion 1 begins at a and ends at b
• conversion 2 begins at c and ends at d

◆ **Figure 6.37**   A single-slope A/D converter. Conversion begins (at point a or c) with the counter reset and the integrator output at 0 V. The ramp has the same slope every time, so the count in the counter when the ramp reaches v_{in} is proportional to v_{in}. Thus count N_1 is proportional to v_{in} at point b and count N_2 is proportional to v_{in} at point d.

The dual-slope ADC has a high degree of accuracy with relatively little effect from time or temperature drifts. This is due to the inherent self-calibrating characteristics of the dual-slope method. It also has excellent noise rejection. However, the dual-slope converter is slow, so it is used primarily where accuracy is important and speed is not a primary consideration. Figure 6.38 shows a dual-slope A/D converter.

5. The *flash or parallel technique* compares the analog input voltage with a set of reference voltages to generate the digital word that represents the analog input. The analog voltage to be converted is input to the array of comparators. All of the compara-

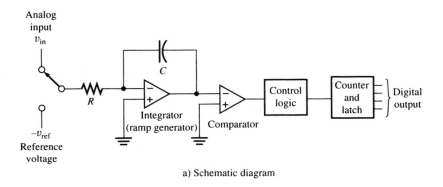

a) Schematic diagram

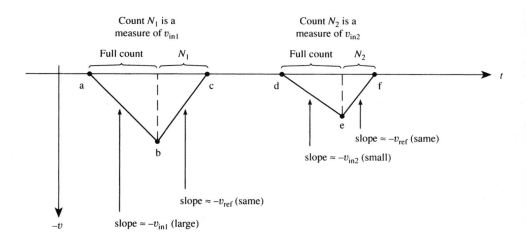

b) Waveform of two conversions with $v_{in1} > v_{in2}$

◆ **Figure 6.38**   A dual-slope A/D converter. The first slope (ab or de) is proportional to the input voltage and continues for the same time interval every time (while the counter steps through a full count). Thus the voltage levels at b and c are proportional to v_{in1} and v_{in2} respectively. The second slope is the same every time, so the count required to reach 0 V at c and f is proportional to v_{in1} and v_{in2}, respectively. The count at c is a measure of v_{in1} and the count at f is a measure of v_{in2}.

tors whose voltage reference is less than the analog value being converted will have a high output; those whose reference is greater will have a low output. The outputs of the comparators are the input to a priority encoder that generates the desired digital code word. The main advantage of the flash converter is its very short conversion time: It is very fast.

The flash converter requires a reference voltage and a comparator for every state of the digital output except zero. Thus a 3-bit flash converter will have seven reference voltages and seven comparators. A 4-bit flash converter will need 15 reference voltages and comparators. In general, an n-bit flash converter will require $2^n - 1$ reference voltages and comparators. The large number of comparators required for a reasonable resolution makes the flash technique fairly expensive. Figure 6.39 shows a 3-bit flash converter.

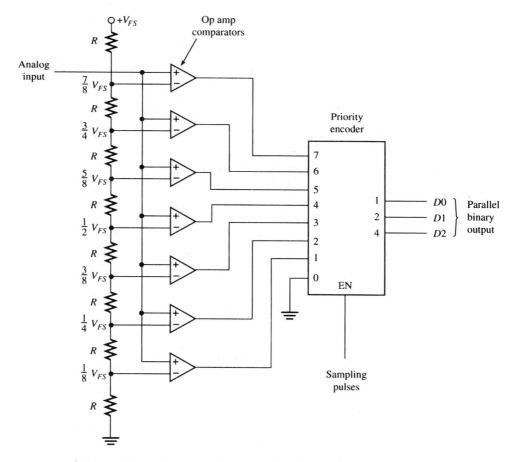

◆ **Figure 6.39**   A 3-bit flash A/D converter. The reference voltage for each comparator is set by the resistive voltage-divider network. The output of each comparator is connected to an input of the priority encoder. The encoder is sampled by a pulse on Enable input, and a 3-bit binary code representing the value of the analog input appears on the encoder's outputs. The binary code is determined by the highest-order input having a HIGH level.

Selection of an Analog-to-Digital Converter

In addition to the usual factors of cost, size, and availability, the control system designer has the following three factors to consider in the selection of an ADC:

1. *Resolution* is specified as the number of bits in the digital code of the output states. It determines the number of output states, the size of the LSB, and the quantization error ($\pm 1/2$LSB). The quantization error ranges from 0.19% for 8-bit resolution to 0.00076% for 16-bit resolution. The designer must specify the minimum resolution that reduces the quantization error to an acceptable level.

2. *Accuracy* includes other factors besides resolution, such as gain error, offset error, linearity, and missing codes. *Gain error* is a change in the slope of the infinite resolution line from the ideal shown in Figure 6.30. The line still passes through the 0-V, 000 digital code point. *Offset error* is a displacement left or right of the infinite resolution line with no change in its slope. *Linearity error* is a deviation of the infinite resolution line from a straight line. *Missing codes* is the absence of one or more expected codes in the output as the input is traversed over its full range.

   When considering the accuracy of an ADC, the designer's major concern is missing codes. The gain and offset errors can be hardware adjusted or software compensated. Linearity can also be compensated but not as easily as gain and offset. Missing codes, however, cannot be restored.

3. *Conversion speed* is determined by how fast the analog signal changes, and it dictates the type of ADC selected. If the analog signal varies at a very slow speed, there is little need for a fast converter that requires fast, expensive components. If the analog signal varies at a moderate speed, the converter will have to operate faster, requiring faster, more expensive components and conversion techniques. If the analog signal varies at a high speed, both conversion techniques and component speed are of paramount importance. In general, the conversion speed requirement will dictate the type of converter selected.

EXAMPLE 6.19

An analog-to-digital converter has a resolution of 4 bits and a quantization error of $\pm \frac{1}{2}$ bit The input voltage is 5 V at the center of the top step. Determine the quantization error in volts and the accuracy as a percentage of the full-scale range.

Solution

The binary number 0000 represents 0 V and the number 1111 represents 5 V. There will be $2^4 - 1$, or 15, steps between 0000 and 1111. A 1-bit change in the binary number is equal to a $5/15 = 0.333$-V change in the voltage it represents. The quantization error is $\pm \frac{1}{2}$ bit, which is $\pm \frac{1}{2}(0.333) = \pm 0.166$ V. The "percentage of" accuracy is given in the following equation:

$$\text{Accuracy} = \pm \left(\frac{0.166}{5} \right)(100) = \pm 3.33\% \qquad ◆$$

Digital Conditioning

Signal conditioning is not always completed when the analog signal is converted to digital. Filtering, linearization, calibration, and conversion to engineering units can all be accomplished by a microprocessor working with the digitized samples. Digital signal processing makes it possible to extract additional information from the signal. Vision systems, for example, involve high-speed processing of an array of data samples.

Digital filtering consists of computing some type of weighted average of the current sample and previous samples. Digital linearization and calibration are accomplished by storing factory calibration data in some type of read-only memory for use by the microprocessor. Additional information, such as ambient temperature or static pressure, can be included in the calibration data. The result is a precise correction for all the nonlinearities of the sensor throughout the range of temperatures and pressures for which the measuring instrument is rated. Digital calibration produces accuracies that cannot be achieved by analog calibration, with the added benefit of using the actual engineering units for the calibrated signal.

◆ GLOSSARY

Acquisition time: The time from the instant the sample command is given until the output is within a specified band of the input. (6.5)

ADC: Abbreviation of analog-to-digital converter. (6.5)

A/D converter: Abbreviation of analog-to-digital converter. (6.5)

Alias frequency: One or more frequency components that are added to the original signal when the signal is sampled at an insufficient sampling rate. (6.5)

Analog-to-digital converter: A device that produces a digital code word that accurately represents the level of an analog input voltage. (6.5)

Aperture time: The time it takes a sample-hold switch to open, measured from the time the hold command is given until the switch is completely open. (6.5)

Band-pass filter: A circuit that does not change components within a specified band of frequencies and attenuates components outside of the specified band. (6.4)

Binary counter ADC: An analog-to-digital converter that uses a binary counter and a D/A converter to generate a digital word that represents an analog input voltage. (6.5)

Binary-weighted DAC: A digital-to-analog convertor that uses a summing amplifier to form the weighted sum of all the nonzero bits in the digital input word. (6.5)

CMR: Abbreviation of common-mode rejection. (6.2)

CMRR: Abbreviation of common-mode rejection ratio. (6.2)

Common-mode rejection: The logarithm of the common-mode rejection ratio expressed in decibel units. (6.2)

Common-mode rejection ratio: The ratio of the differential gain (A) of an operational amplifier divided by its common mode gain (A_c). (6.2)

Comparator: A circuit that accepts two input voltages and signals which one is greater. (6.3)

DAC: Abbreviation of digital-to-analog converter. (6.5)

D/A converter: Abbreviation of digital-to-analog converter. (6.5)

Data acquisition system: A system that conditions a number of analog signals and converts them into digital form for processing by a computer. (6.5)

Data distribution system: A system that converts the digital representation of a number of signals into analog form for output to analog actuators. (6.5)

Data sampling: A process in which a switch connects momentarily to an analog signal in a sequence of pulses separated by evenly spaced increments of time called the sampling interval. (6.5)

Decay rate: The rate of change in the output of a sample-hold circuit when it is in the hold mode. (6.5)

Differential amplifier: A circuit that amplifies the difference between two input voltages. (6.3)

Differentiator: A circuit that produces an output voltage that is equal to the rate of change of the input voltage. (6.3)

Digital-to-analog converter: A device that produces an output voltage that is an accurate analog representation of the input digital code word. (6.5)

Dual-slope ADC: An analog-to-digital converter that uses a variable-slope, fixed-time ramp followed by a fixed-slope, variable-time ramp to generate a digital code word that represents an analog input voltage. (6.5)

Flash or parallel ADC: An analog-to-digital converter that uses a series of reference voltage/comparator pairs to generate a digital code word that represents an analog input voltage. (6.5)

Frequency spectrum: A plot of the maximum voltage of each possible component of a signal versus the frequency of that component. (6.5)

Function generator: An inverting amplifier that produces a nonlinear relationship between its input and output voltages. (6.3)

High-pass filter: A circuit that does not change components above its break-point frequency and attenuates components below its break-point. (6.4)

Instrumentation amplifier: A differential amplifier with voltage followers on each input. (6.3)

Integrator: A circuit in which the change in the output voltage over some interval of time is equal to the integral of the input voltage over that same time interval. (6.3)

Inverting amplifier: A circuit that produces an output that is equal to $-G$ times the input, where G can be any value greater than 0. (6.3)

Isolation amplifier: An amplifier that uses transformer or optical coupling to electrically separate the input and output circuits. (6.4)

Linearization: The process of conditioning a nonlinear calibration signal to produce a linear calibration signal. (6.4)

Linear operating region: The middle operating region of an operational amplifier, where the output is equal to the gain times the difference between its two input voltages ($v_2 - v_1$). (6.2)

Low-pass filter: A circuit that does not change components below its break-point frequency and attenuates components above its breakpoint. (6.4)

Noninverting amplifier: A circuit that produces an output voltage that is equal to $+G$ times its input voltage, where G can be any value greater than 1. (6.3)

Notch filter: A circuit that rejects components within a specified band of frequencies and does not change components outside of that band. (6.4)

Nyquist criterion: A sampling criterion that states that all the information in the original signal can be recovered if it is sampled at least twice during each cycle of the highest-frequency component. (6.5)

Operational amplifier: A very high-gain amplifier that has two input lines and one output line. The output voltage is equal to the product of the gain times the difference between the two input voltages. (6.2)

R–$2R$ ladder DAC: A digital-to-analog converter that uses two resistor sizes in a resistive ladder network to form the weighted sum of all the nonzero bits in the digital input word. (6.5)

Sample-and-hold: A circuit that uses a sampling switch to sample an analog voltage and a capacitor to hold the sample until the next closure of the sampling switch. (6.5)

Sampling interval: The time between the momentary closures of a sampling switch. Numerically equal to the reciprocal of the sampling rate. (6.5)

Sampling rate: The rate of momentary closures of a sampling switch. Numerically equal to the reciprocal of the sampling interval. (6.5)

Saturation region: Two regions where the output of an operational amplifier has a constant value ($-V_{sat}$ on the left or $+V_{sat}$ on the right). (6.2)

Single-slope ADC: An analog-to-digital converter that uses a fixed-rate, variable-time ramp voltage to generate a digital word that represents the analog input voltage. (6.5)

Slew rate: The maximum rate at which the output voltage of an operational amplifier can change. (6.2)

Successive approximation ADC: An analog-to-digital converter that sequentially compares a series of binary-weighted values with the analog input to generate a digital code word that represents the analog input voltage. (6.5)

Summing amplifier: A circuit with several inputs that produces an output voltage that is the inverted, weighted sum of the input voltages. (6.3)

Voltage follower: A circuit with a very high input impedance, a very low output impedance, and an output voltage that is equal to the input voltage. (6.3)

Wheatstone bridge: A circuit used to measure the value of an unknown resistor in which the un-known resistor and three other resistors are connected in a diamond configuration. (6.4)

◆ EXERCISES

Section 6.2

6.1   Determine the values of $v_2 - v_1$ that will saturate an op amp for each of the following condition sets:

(a) $A = 10^4$, $+V_{cc} = 20$ V, $-V_{cc} = -20$ V

(b) $A = 10^7$, $+V_{cc} = 20$ V, $-V_{cc} = -20$ V

(c) $A = 4 \times 10^5$, $+V_{cc} = 16$ V, $-V_{cc} = -18$ V

(d) $A = 2 \times 10^6$, $+V_{cc} = 18$ V, $-V_{cc} = -16$ V

6.2   Determine the value of $v_2 - v_1$ that will produce positive saturation of an op amp for each of the following condition sets:

(a) $A = 10^4$, $+V_{cc} = 20$ V, $A_c = 0.1$, $v_c = 8$ V

(b) $A = 10^7$, $+V_{cc} = 20$ V, $A_c = 0.1$, $v_c = 8$ V

(c) $A = 4 \times 10^5$, $+V_{cc} = 16$ V, $A_c = 0.08$, $v_c = 6$ V

(d) $A = 2 \times 10^6$, $+V_{cc} = 18$ V, $A_c = 0.12$, $v_c = 5$ V

6.3   For each of the following condition sets, determine the CMMR and CMR required to limit the common-mode error to a maximum of 2% of the output when $v_c = 0$ V

(a) $A = 10^4$, $v_2 - v_1 = 1.0$ mV, maximum $v_c = 8$ V

(b) $A = 10^7$, $v_2 - v_1 = 0.8$ μV, maximum $v_c = 8$ V

(c) $A = 4 \times 10^5$, $v_2 - v_1 = 20$ μV, maximum $v_c = 6$ V

(d) $A = 2 \times 10^6$, $v_2 - v_1 = 5$ μV, maximum $v_c = 5$ V

(e) $A = 2 \times 10^5$, $v_2 - v_1 = 80$ μV, maximum $v_c = 5$ V

Section 6.3

6.4   **(a)** Revise the high-level indicator in Figure 6.4a so that the HI lamp will be ON only when $v_{in} > 9$ V.

(b) Revise the low-level indicator in Figure 6.4b so that the LO lamp will be ON only when $v_{in} < 1$ V.

(c) Revise the high-low-level indicator in Figure 6.4c so that the LO lamp will be ON only when $v_{in} < 1$ V and the HI lamp will be ON only when $v_{in} > 9$ V.

6.5   Design a three-level indicator with three comparators and three lamps, marked (1/4), (1/2), and (3/4). The (1/4) indicator lamp will be ON only when $v_{in} > 2.5$ V. The (1/2)

indicator lamp will be ON only when $v_{in} > 5$ V. The (3/4) indicator lamp will be ON only when $v_{in} > 7.5$ V. When v_{in} is less than 2.5 V, all lamps are OFF. When v_{in} is greater than 7.5 V, all lamps are ON.

6.6   The following conditions give the input and output voltages of an inverting amplifier, and one of its two resistors, R_f or R_{in}. Determine the value of the unknown resistor.

(a) $v_{in} = 6$ V, $v_{out} = -15$ V, $R_f = 10$ kΩ

(b) $v_{in} = 1.2$ V, $v_{out} = -12$ V, $R_f = 50$ kΩ

(c) $v_{in} = 0.2$ V, $v_{out} = -8$ V, $R_{in} = 2$ kΩ

(d) $v_{in} = 0.75$ V, $v_{out} = -12$ V, $R_{in} = 4$ kΩ

(e) $v_{in} = 10$ V, $v_{out} = -2$ V, $R_f = 5$ kΩ

6.7   A primary element produces an output voltage that ranges from 0 to 22 mV. Design an inverting amplifier that produces an output voltage with a range of 0 to -4 V. Use a 1-kΩ resistor for R_{in}.

6.8   The following conditions give the input and output voltages of a noninverting amplifier, and one of its two resistors, R_f or R_{in}. Determine the value of the unknown resistor.

(a) $v_{in} = 6$ V, $v_{out} = 15$ V, $R_f = 10$ kΩ

(b) $v_{in} = 1.2$ V, $v_{out} = 12$ V, $R_f = 50$ kΩ

(c) $v_{in} = 0.2$ V, $v_{out} = 8$ V, $R_{in} = 2$ kΩ

(d) $v_{in} = 0.75$ V, $v_{out} = 2$ V, $R_{in} = 4$ kΩ

6.9   Repeat Exercise 6.7, but change the output voltage range to 0 to 4 V and use a noninverting amplifier.

6.10   Determine the output of a two-input summing amplifier for each of the following condition sets.

(a) $v_a = 2$ V, $v_b = 4$ V, $R_f = 5$ kΩ, $R_a = R_b = 10$ kΩ

(b) $v_a = 2$ V, $v_b = 4$ V, $R_f = R_a = R_b = 10$ kΩ

(c) $v_a = -2$ V, $v_b = 6$ V, $R_f = 10$ kΩ, $R_a = 8$ kΩ, $R_b = 6$ kΩ

(d) $v_a = -6$ V, $v_b = 4$ V, $R_f = 16$ kΩ, $R_a = 12$ kΩ, $R_b = 8$ kΩ

6.11 A primary element produces an output voltage that ranges from 0 to 22 mV. Design a summing amplifier that will produce an output voltage with a range of -1 to -5 V. Use a 1-kΩ resistor for R_{in}.

6.12 The summing amplifier in Figure 6.8a can be used to implement the equation $y = mx + b$ (the slope-intercept form of the equation of a straight line). Use a 10-kΩ resistor for R_f and determine the values of R_a and R_b to implement the following slope-intercept equation.

$$y = -v_{out}, x = v_a, m = R_f/R_a = 3.42, b = v_b$$

$$y = mx + b$$

$$-v_{out} = 3.42v_a + v_b$$

6.13 Given a 10-μF capacitor, compute the value of R_{in} for an integrator with the following defining equation:

$$v_{out} = -\int_{t_1}^{t_2} v_{in} \, dt + v_{out}(t_1)$$

6.14 Determine $v_{out}(t_2)$ of the integrator in Example 6.6 for each of the following condition sets.
(a) $t_1 = 5$ s, $t_2 = 7$ s, $v_{out}(t_1) = 10$ V
(b) $t_1 = 5$ s, $t_2 = 6$ s, $v_{out}(t_1) = 10$ V
(c) $t_1 = 0$ s, $t_2 = 1.2$ s, $v_{out}(t_1) = 0$ V
(d) $t_1 = 0$ s, $t_2 = 1.6$ s, $v_{out}(t_1) = 0$ V
(e) $t_1 = 0$ s, $t_2 = 2.0$ s, $v_{out}(t_1) = 0$ V

6.15 For each of the following condition sets, determine the value of the external resistor, R_e, for an instrumentation amplifier similar to Figure 6.12.
(a) gain = 100, $R_1 = 1$ kΩ, $R_2 = 10$ kΩ, $R_3 = 1$ kΩ
(b) gain = 100, $R_1 = R_2 = R_3 = 1$ kΩ
(c) gain = 500, $R_1 = R_2 = R_3 = 1$ kΩ
(d) gain = 200, $R_1 = R_2 = R_3 = 1$ kΩ

Section 6.4

6.16 Design a current-to-voltage converter that converts a 10-mA input into a 15-V output (refer to Figure 6.14a and Example 6.9).
(a) Determine the value of resistor R (Figure 6.14a).
(b) Determine the value of V_{cc} necessary for the op amp to be in the linear operating region with 1 V to spare.

6.17 Design a voltage-to-current converter that converts a 6-V input into a 20-mA output (refer to Figure 6.14b and Example 6.10).
(a) Determine the value of resistor R (Figure 6.14b).
(b) Determine the value of R_{LOAD} that will result in an output voltage (v_{out}) of 11 V when $v_{in} = 6$ V.

(c) Determine the value of v_{cc} necessary for the op amp to be in the linear operating region with 1 V to spare if R_{LOAD} has the value determined in step (b).

6.18 A constant current source is shown in Figure 6.40.
(a) Find the value of v_{out}.
(b) Find the value of v_{out}.
(c) Assume that $v_{cc} - 20$ V and $V_{sat} - 0.8\,V_{cc}$. Determine the maximum value of R_{LOAD} for which i_{out} will have the value of step (a).

6.19 Design a current-to-current converter with an input range of 0 to 0.2 mA and an output range of 0 to 1 mA (i.e., a gain of 5). Plan to use the following three resistors in your design:

R_1 (see resistor R in Figure 6.14b)
R_2 (see resistor R in Figure 6.14a)
R_{LOAD} (see resistor R_{LOAD} in Figure 6.14b)

You may use the following assumptions in your design:

$v_{out} = 10$ V when i_{out} 1 mA
$R_{LOAD} = 2$ kΩ
i_{in} can produce 12 V at the v_2 input terminal
$v_{cc} = 15$ V

Use $i_{in} = 0.2$ mA and $i_{out} = 1$ mA as your design conditions when you determine the values of R_1, v_1, v_2, and R_2.

6.20 Find the value of the unknown resistor, R_s, for each of the following balanced Wheatstone bridge conditions (see Figure 6.15).
(a) $R_2 = 5$ kΩ, $R_3 = 1.71$ kΩ, $R_4 = 10$ kΩ
(b) $R_2 = 10$ kΩ, $R_3 = 2.56$ kΩ, $R_4 = 5$ kΩ
(c) $R_2 = 100$ kΩ, $R_3 = 6.27$ kΩ, $R_4 = 1$ kΩ
(d) $R_2 = 20$ kΩ, $R_3 = 3.47$ kΩ, $R_4 = 1$ kΩ

6.21 The self-nulling current balance bridge in Figure 6.16 is used to condition the output of a strain-gage load cell. The load cell has an input force range of ± 100 N and an output resistance range of $\pm 1\ \Omega$. An input of -100 N produces an output of 149 Ω, an input of 0 N produces an output of 150 Ω, and an input of $+100$ N produces an output of 151 Ω.

The strain gage element is designated R_s in Figure 6.16. The bridge is balanced when $R_s = 150\ \Omega$ and $I_N = 0$ A. The other bridge parameters are

$$R_{3a} = 5\ \Omega \quad R_{3b} = 145\ \Omega \quad R_2 = R_4 = 2\ k\Omega \quad V_{dc} = 10\ V$$

Determine the values of I_N for the following input force conditions, F_{in}: -100, -50, 0, 50, and 100 N.

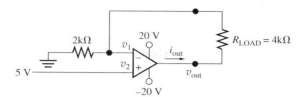

◆ **Figure 6.40** A constant current source (Exercise 6.18).

Determine the equation of the zero-based linearity line and express the nonlinearity as a percentage of the output span.

6.22 The unbalanced bridge in Figure 6.17 is used to condition the output of the strain-gage load cell described in Exercise 6.21. The parameters in Figure 6.17 have the following values:

R_s = load cell resistance (range = 149 to 150 Ω)
$R_3 = 150\ \Omega$ $R_2 = R_4 = 2\ \mathrm{k}\Omega$ $R_{bal} = 150\ \Omega$
$V_{dc} = 1\ \mathrm{V}$ $R_f/R_a = 1000$

Determine the values of V_{out} for the following input force conditions, F_{in}: -100, -50, 0, 50, and 100 N.

Determine the equation of the zero-based linearity line and express the nonlinearity of as a percentage of the output span.

6.23 Design a single-stage, active, low-pass filter that will provide an attenuation factor of 50 for a 200-Hz signal. Calculate the resistor size required if a 50-μF capacitor is used in the RC circuit.

6.24 Design a two-stage, active, low-pass filter that will provide an attenuation factor of 500 for a 400-Hz signal. Calculate the resistor size required if 0.1-μF capacitors are used in the two RC circuits.

6.25 Design (a) a single-stage low-pass filter and (b) a two-stage low-pass filter that will attenuate a 159.2-Hz signal by an attenuation factor of 100. Use the program BODE to generate frequency-response data and plot a Bode magnitude diagram for each filter. Compare the two Bode diagrams and comment.

6.26 Design a notch filter that will attenuate a 159.2-Hz signal by an attenuation factor of 20. Use the program BODE and plot a Bode magnitude diagram for the notch filter. Compare the notch filter Bode diagram with the two low-pass filter diagrams from Exercise 6.25a.

6.27 Change the endpoints of the line segments in Figure 6.22 to minimize the maximum error. Construct a table similar to Table 6.2 showing the error at the midpoints and the endpoints of the line segments.

6.28 Construct a five-step piecewise-linear approximation of the function defined by the input/output table that follows. Construct a graph similar to Figure 6.22 and a table similar to Table 6.2.

Measuring Instrument Input/Output Table

Input	Output	Input	Output	Input	Output
0	2.9	35	31.9	70	69.8
5	6.8	40	36.7	75	75.2
10	10.8	45	41.8	80	80.3
15	14.8	50	47.2	85	85.1
20	18.9	55	52.8	90	89.5
25	23.1	60	58.5	95	93.5
30	27.4	65	64.2	100	97.1

Note: All values are a percentage of full-scale range.

6.29 A plantinum RTD has the following resistance versus temperature characteristic:

Temperature, T (°C)	Resistance, R_s (Ω)
0	100.0
25	109.8
50	119.8
75	129.6
100	139.3

An unbalanced Wheatstone bridge is to be used to convert the RTD resistance into a voltage signal. The balance resistance value $R_{bal} = 119.8\ \Omega$, so the bridge output will be 0 V at 50°C.

Your assignment is to design the unbalanced bridge circuit, an instrumentation amplifier, and a summing amplifier that will give a full-scale output from 1 to 5 V as the temperature goes from 0 to 100°C. As part of your design assignment, you are to complete the following table of results from various steps of the design process:

$T(°C)$	$R_s(\Omega)$	ϵ	$v_b - v_a$	v_1	v_{out}
0	100.0				
25	109.8				
50	119.8				
75	129.6				
100	139.3				

Your design must take into account the self-heating effect of the current through the resistance element. In the steady state, the heat dissipated by the current through the resistance element causes the probe temperature (T_s) to be slightly above the temperature of the surrounding fluid (T_f). The resistance element actually measures T_s, resulting in a self-heating-effect error equal to $T_s - T_f$. The self-heating effect is expressed in terms of the *dissipation constant* of the probe, in mW/°C. The heat dissipated by the probe is equal to the product of the dissipation constant times the heating-effect error ($T_s - T_f$). The heat generated by the probe current is equal to v_s^2/R_s (where v_s is the voltage across the probe and R_s is the resistance of the probe). In the steady-state condition, the heat dissipated is equal to the heat generated, thus

$$v_s^2/R_s = \frac{(\text{dissipation constant})(\text{self-heating-effect error})}{1000}$$

The division by 1000 is necessary to convert the units of the right side from milliwatts to watts to match the units of the left side.

The design specifications provide the following information relative to the self-heating effect and the selection of the resistance values of the bridge circuit.

1. Self-heating effect of the probe: 8 m W/°C
2. Maximum self-heating-effect error
   for balanced bridge at 50°C: 0.1°C
3. Bridge voltage, V_{dc}: 1 V
4. Balance resistance value, R_{bal}: 119.8 Ω
5. Resistance ratio, R_3/R_{bal}: 1

Complete the following design steps.

(a) Determine the resistances of R_2, R_3, and R_4, and the value of α for the bridge.

1. Use $v_s^2/R_{bal} = (8 \text{ mW/°C})(0.1°C)/1000$ to compute the voltage across the resistance element, v_s, when the bridge is balanced.
2. Use v_s from the step above, $V_{dc} = 1$ V, and $R_{bal} = 119.9$ Ω to compute the resistance of R_2. As a safety factor, multiply your result by 1.1 and round up to the nearest whole ohm. Use the rounded value for R_2.

3. Use design specification 5 to determine the resistance of R_3.
4. Use Equation (6.30) to determine the resistance of R_4.
5. Use Equation (6.32) to determine α.

(b) Determine the values of ϵ.
1. Use Equation (6.31) to determine the value of ϵ for the following temperature values: 0, 25, 50, 75, 100°C.
2. Complete the ϵ column in the results table.

(c) Determine the values of $v_b - v_a$.
1. Use Equation (6.34) to determine the value of $v_b - v_a$ for the following temperature values: 0, 25, 50, 75, 100°C.
2. Complete the $v_b - v_a$ column in the results table.

(d) Design the instrumentation amplifier.
1. Figure 6.12 is a diagram of the instrumentation amplifier. The first step in the design is to switch the two input voltages to give an inverted input. This is necessary to cancel the inversion of the summing amplifier that follows the instrumentation amplifier. With the inputs reversed and with the output labeled v_1 instead of v_{out}, the output of the instrumentation amplifier is given by the following equation:

$$v_1 = \frac{-R_f}{R_a}(v_b - v_a)$$

2. Compute the gain, R_f/R_a, that will result in a span of 4 V for v_1 as the temperature goes from 0 to 100°C. This is the gain of the instrumentation amplifier. Use a 1-kΩ resistor for R_a and determine the value of R_f to give the required gain.
3. Complete the v_1 column in the results table. The values are defined by the equation for v_1 in step d1. Use the gain you computed in step d2.

(e) Design the summing amplifier.
1. Figure 6.8a is a diagram of the two-input summing amplifier. Replace the upper input in Figure 6.8a by v_1, the output of the instrumentation amplifier. Replace the lower input by v_c, an offset voltage to be provided. Rename resistors R_f and R_a in Figure 6.8a as R_1, and rename resistor R_b as R_c. With these name changes, the output of the summing amplifier is given by the following equation:

$$V_{out} = -v_1 - \frac{R_1}{R_c}v_c$$

2. Use a -1 V supply for v_c, a 1-kΩ resistor for R_c, and compute the resistance of R_1 that will result in an output voltage of 1 V when $T = 0°C$.

3. Complete the v_{out} column in the results table.

(f) Determine the terminal-based linearity of the output voltage, v_{out}.

Section 6.5

6.30   Determine the time constant and the 99.3% acquisition time of a sample-and-hold circuit if $C = 100\ \mu F$ and $R = 1$ kΩ.

6.31   Determine the minimum sampling rate to get all the information from a signal that has a maximum frequency of 1200 Hz.

6.32   Determine the alias frequency that is produced when a 16-kHz signal is sampled at a rate of 20 kHz.

6.33   The logic levels of the input to the 4-bit binary-weighted DAC in Figure 6.32 are 5 V for a logic 1 and 0 V for a logic 0. The resistance values are $R_f = 2$ kΩ and $R = 1$ kΩ.

(a) Determine V_{LSB}, V_{MSB}, and V_{FS}.

(b) Determine v_0 for each of the following binary inputs:

$$0101 \quad 1011 \quad 1111 \quad 0111$$

6.34   Use the successive approximation technique outlined in Figure 6.36 to convert each of the following analog voltages to 4-bit digital words. Use $V_{FS} = 20$ V.

(a) 4.2 V   **(b)** 19.6 V
(c) 9.25 V   **(d)** 14.71 V

6.35   Determine the quantization error in volts and percentage of the full-scale range for each of the following ADCs. The input voltage is 10 V at the center of the top step.

(a) Resolution = 4 bits; error = $\pm\frac{1}{2}$ bit
(b) Resolution = 8 bits; error = ±1 bit
(c) Resolution = 10 bits; error = $\pm\frac{1}{2}$ bit
(d) Resolution = 12 bits; error = $\pm\frac{1}{2}$ bit

CHAPTER 7

Position, Motion, and Force Sensors

◆ **OBJECTIVES**

In the late eighteenth century, a Scottish engineer named James Watt greatly improved the steam engine with several inventions. One of Watt's inventions was a governor that regulated the speed of the steam engine. Since that early beginning, the control of position or speed has become a major branch of control technology, sometimes referred to as servo control. Numerical control of machine tools requires precise positioning of a workpiece and exact control of the speed and feed of the cutting tool. Robotic arms move in specified paths that require sensing and control of position and speed. Some sequential controllers must sense the presence of a part and may even identify the part and determine its exact location. The power-steering unit in a car uses the position of the steering wheel as a setpoint and positions the wheels of the car accordingly.

The purpose of this chapter is to give you an entry-level ability to discuss, select, and specify position, motion, and force sensors. The examples selected represent a reasonable cross section of the variety of sensors used to measure these quantities. After completing this chapter, you will be able to

1. List major considerations in the selection of
   a. Position sensors
   b. Displacement sensors
   c. Speed sensors
   d. Acceleration sensors
   e. Force sensors
2. Describe the following *position* sensors
   a. Potentiometer
   b. LVDT
   c. Synchro
   d. Resolver
   e. Optical encoder
   f. Proximity
   g. Photoelectric
3. Describe the following *speed* sensors
   a. DC tachometer
   b. AC tachometer
   c. Optical tachometer

4. Describe an *accelerometer*
5. Describe the following *force* sensors
   a. Strain gage
   b. Pneumatic
6. Select a proximity sensor (using a spec sheet)   ◆◆◆◆

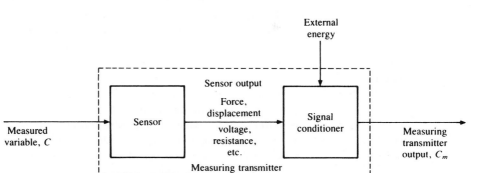

a) Block diagram

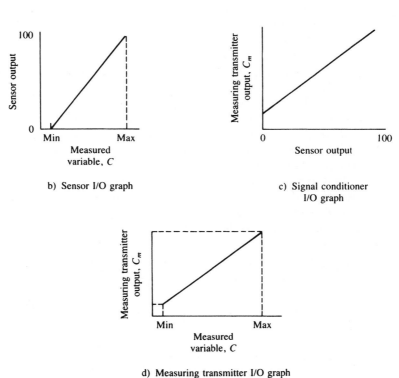

d) Measuring transmitter I/O graph

◆ **Figure 7.1**   A measuring transmitter consists of two parts: a sensor (or primary element) and a signal conditioner.

7.1 INTRODUCTION

A *sensor* is a device that converts a *measurand* (signal to be measured) into a signal in a different form. The measurand is the input to the sensor, and the signal produced by the sensor is the output. Sensors are also called *transducers* or *primary elements*. The output of the sensor may be a force, displacement, voltage, electrical resistance, or some other physical quantity. Usually, a signal conditioner is required to convert the sensor output into an electrical (or pneumatic) signal suitable for use by a controller or display device. The sensor and its signal conditioner comprise the two parts of a measuring transmitter (Figure 7.1). Typical input/output graphs of the sensor, signal conditioner, and measuring transmitter are included in Figure 7.1.

The types of sensors are too numerous to include all of them in a couple of chapters. Examples of several types of sensors are included for each of the more common controlled variables. Our purpose is to illustrate the basic principles of sensors with a series of specific examples. In a control system, the function of the transducer is to provide a signal that is a measure of the controlled variable. For this reason, the examples are classified by the variable that is measured rather than by the manner in which the output is developed.

7.2 POSITION AND DISPLACEMENT MEASUREMENT

Sensing Methods

Position and displacement measurement is divided into two types: linear and angular. *Linear* position or displacement is measured in units of length. *Angular* position or displacement is measured in radians or degrees. The primary standard of length is the international meter, which is defined in terms of the wavelength of the red-orange line of krypton 86. All other units of length are defined in terms of the primary standard. The meter is the basic unit of length in the International System of Units (SI), and the radian is the unit of angle. A radian is equal to $180/\pi$ degrees (approximately 57.3°). Linear and angular displacement are illustrated in Figure 7.2.

In continuous processes, displacement sensors are used to measure the thickness of a sheet, the diameter of a rod, the separation of rollers, or some other dimension of the product. In discrete-parts manufacturing, position sensors are used to measure the presence of a part, to identify a part, to determine the position of a part, or to measure the size of a part. The

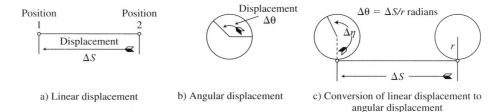

a) Linear displacement   b) Angular displacement   c) Conversion of linear displacement to angular displacement

◆ **Figure 7.2** Linear displacement can be converted into angular displacement of a wheel (c). The angular displacement in radians ($\Delta\theta$) is equal to the linear displacement (ΔS) divided by the radius of the wheel $\Delta\theta = \Delta S/r$.

acronym PIP refers to *presence, identification, and position* measurement of parts in a manufacturing operation.

Sensors used to measure the size of a product or part usually have a sensing shaft that is in mechanical contact with the object to be measured. The mechanical connection between the sensing shaft and the object is of paramount importance—the sensor actually measures the position of the sensing shaft. A number of methods are used to translate the position of the sensing shaft into a measurable quantity. For example, movement of the shaft may cause a change in capacitance, self-inductance, mutual inductance, or electrical resistance. The shaft might also be coupled to an encoder, a device that converts the movement into a digital signal.

The simplest PIP sensor, a lever-actuated switch, also makes contact with the object to be measured. However, many PIP sensors do not touch the object to be measured. Some noncontacting sensors use changes in self-inductance, reluctance, or capacitance to detect the presence of an object. Other noncontacting sensors use the interruption or blocking of a light beam to sense the presence of an object, to identify the object, or to measure the position of the object.

Potentiometers

A *potentiometer* consists of a resistance element with a sliding contact that can be moved from one end to the other. Potentiometers are used to measure both linear and angular displacement, as illustrated in Figure 7.3. The resistance element produces a uniform drop in the applied voltage, E_s, along its length. As a result, the voltage of the sliding contact is directly proportional to its distance from the reference end.

$$\text{Linear potentiometer:} \quad E_{\text{out}} = \left(\frac{x}{L}\right) E_s$$

$$\text{Angular potentiometer:} \quad E_{\text{out}} = \left(\frac{\theta}{\theta_L}\right) E_s$$

When the resistive element is wirewound, the resolution of the potentiometer is determined by the voltage step between adjacent loops in the element. If there are N turns in the element, the voltage step between successive turns is $E_T = E_s/N$, where E_s is the full-scale voltage. Expressed as a percentage of the full-scale output, the percentage resolution is given by the following relationship:

$$\text{Resolution } (\%) = \frac{100 \, E_T}{E_s} = \frac{100(E_s/N)}{E_s}$$

or

$$\text{Resolution } (\%) = \frac{100}{N} \tag{7.1}$$

Potentiometers are subject to an error whenever a current passes through the lead wire connected to the sliding contact. This error is called a *loading error* because it is caused by the load resistor connected between the sliding contact and the reference point. A potentiometer with a load resistor is illustrated in Figure 7.4. If R_p is the resistance of the poten-

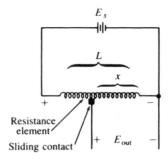

a) A linear displacement potentiometer

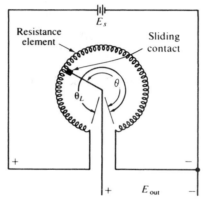

b) An angular displacement potentiometer

◆ **Figure 7.3** Two types of potentiometric displacement sensors: (a) linear; (b) angular. In both types, E_{out} is a measure of the position of the sliding contact.

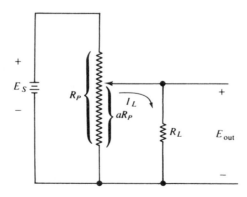

◆ **Figure 7.4** A loading error is produced in a potentiometer when a load resistor is connected between the sliding contact and the reference terminal.

tiometer and a is the proportionate position of the sliding contact, then aR_p is the resistance of the portion of the potentiometer between the sliding contact and the reference point. The load resistor, R_L, is connected in parallel with resistance aR_p. The equivalent resistance of this parallel combination is $(R_L)(aR_p)/(R_L + aR_p)$.

The resistance of the remaining portion of the potentiometer is equal to $(1 - a)R_p$, and the equivalent total resistance is the sum of the last two values.

$$R_{EQ} = (1 - a)R_p + \frac{aR_LR_p}{R_L + aR_p}$$

The output voltage, E_{out}, may be obtained by voltage division as follows:

$$E_{out} = \left[\frac{aR_LR_p/(R_L + aR_p)}{(1 - a)R_p + aR_LR_p/(R_L + aR_p)} \right] E_s$$

$$E_{out} = \left(\frac{a}{1 + ar - a^2r} \right) E_s$$

where

$$r = \frac{R_p}{R_L}$$

The loading error is the difference between the loaded output voltage (E_{out}) and the unloaded output voltage (aE_s).

$$\text{Loading error} = aE_s - E_{out}$$

$$= aE_s - \left(\frac{a}{1 + ar - a^2r} \right) E_s$$

$$= \left[\frac{a^2r(1 - a)}{1 + ar - a^2r} \right] E_s \quad V$$

The loading is usually expressed as a percentage of the full-scale range, E_s.

$$\text{Loading error} = (\%) = 100 \left[\frac{a^2r(1 - a)}{1 + ar(1 - a)} \right] \tag{7.2}$$

EXAMPLE 7.1   The potentiometer in Figure 7.4 has a resistance of 10,000 Ω and a total of 1000 turns. Determine the resolution of the potentiometer and the loading error caused by a 10,000-Ω load resistor when $a = 0.5$.

Solution   The resolution is given by Equation (7.1)

$$\text{Resolution} = \frac{100}{N} = \frac{100}{1000} = 0.1\%$$

The loading error is given by Equation (7.2)

$$\text{Loading error} = 100 \left[\frac{a^2r(1 - a)}{1 + ar(1 - a)} \right]$$

$$= 100 \left[\frac{(0.5)^2(1)(0.5)}{1 + (0.5)(1)(0.5)} \right]$$

$$= 10\% \qquad\qquad ◆$$

Linear Variable Differential Transformers

The *linear variable differential transformer (LVDT)* is a rugged electromagnetic transducer used to measure linear displacement. A diagram of an LVDT is shown in Figure 7.5. It consists of a single primary winding located between two secondary windings on a hollow cylindrical form. A movable magnetic core provides a variable coupling between the windings.

The sensing rod is attached to the magnetic core and moves the core in response to the displacement that is to be measured. An ac voltage is applied to the primary winding, and the transformer coupling results in an ac voltage across each secondary winding. When the magnetic core is in the center position, the two secondary voltages cancel each other at terminals *a–b* (i.e., at the input to the phase-sensitive detector). When the core is moved to one side, the voltage in the secondary coil with more coupling becomes larger and the other secondary voltage becomes smaller. An ac voltage appears at terminals *a–b*. We will call this ac voltage v_{a-b}. The magnitude of v_{a-b} is proportional to the displacement of the core from the null position. When the core is moved to the same distance on the other side of the null position, the ac voltage at terminals *a–b* is equal to $-v_{a-b}$. In other words, the amplitude of the ac voltage is proportional to the amount of displacement and the phase depends on the direction of the displacement. A positive displacement produces a 0° phase angle. A negative displacement produces a 180° phase angle.

The phase-sensitive detector converts the ac secondary voltage into a dc voltage, E_{out}. The magnitude of the dc voltage is proportional to the amplitude of the ac voltage. The sign of the dc voltage is positive if the ac phase angle is 0°, and negative if the ac phase angle is 180°. The result is the overall input/output graph illustrated in Figure 7.5b. In some ac systems, the ac secondary voltage is used as the error signal. The phase-sensitive detector is not required in these systems.

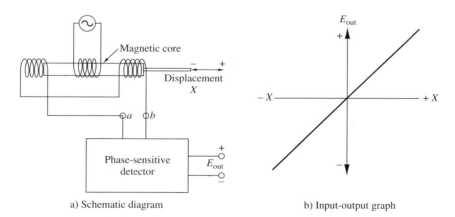

a) Schematic diagram   b) Input-output graph

◆ **Figure 7.5** The linear variable differential transformer (LVDT) and phase-sensitive detector produce a dc voltage proportional to the displacement of the movable magnetic core.

Synchro Systems

A *synchro* is a rotary transducer that converts angular displacement into an ac voltage, or an ac voltage into an angular displacement. Three different types of synchros are used in angular displacement transducers: control transmitter, control transformer, and control differential.

Synchros are used in groups of two or three to provide a means of measuring angular displacement. For example, a control transmitter and a control transformer form a two-element system that measures the angular displacement between two rotating shafts. The displacement measurement is then used as an error signal to synchronize the two shafts. The term *electronic gears* is sometimes used to describe this type of system because the two shafts are synchronized as if they were connected by a gear drive. The addition of a control differential forms a three-element system that provides adjustment of the angular relationship of the two shafts during operation.

A two-element synchro system is shown in Figure 7.6. The control transmitter is designated CX, and the control transformer is designated CT. Both the transmitter and the transformer have an H-shaped rotor with a single winding. Connections to the rotor winding are made through slip rings on the shaft. The stators each have three coils spread 120° apart and connected in a Y configuration.

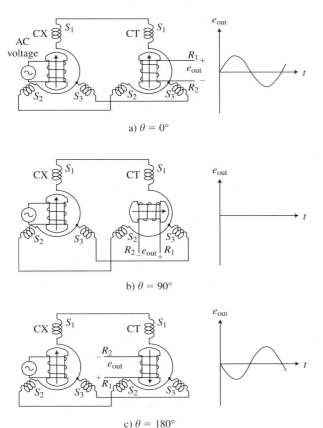

a) $\theta = 0°$

b) $\theta = 90°$

c) $\theta = 180°$

◆ **Figure 7.6**   A two-element synchro system measures the phase difference between two rotating shafts.

An ac voltage is applied to the rotor winding of the control transmitter. This voltage induces an ac voltage in the three stator windings, which are uniquely determined by the angular position of the rotor. The voltages induced in the transmitter are applied to the transformer stator windings, which, in turn, induce a voltage in the transformer rotor winding. The transformer rotor voltage is uniquely determined by the relative position of the two rotors, as shown in Figure 7.6. A graph of the transformer rotor voltage (e_{out}) is included for each of the three relative rotor positions shown. The amplitude of e_{out} is a maximum when the angular displacement of the two rotors is 0° or ±180°, and zero when the angular displacement is ±90°. Notice the change in sign of the ac voltage between the 0 and 180° positions.

A graph of the amplitude of the output voltage (e_{out}) versus the angular displacement (θ) is shown in Figure 7.7. The output voltage is described by the following mathematical relationship:

$$e_{out} = (E_m \cos \theta) \sin \omega t \tag{7.3}$$

where E_m = maximum amplitude

θ = angular displacement

ω = radian frequency of the ac voltage applied to the transmitter rotor

The sign and magnitude of the amplitude term ($E_m \cos \theta$) are uniquely determined by the angular displacement θ, as shown in Figure 7.7. The graph is reasonably linear for values of θ from 20 to 160°. The operating point is located at the center of this linear region (i.e., θ = 90°). The magnitude of e_{out} is proportional to the amount of angular displacement. The sign or phase of e_{out} is determined by the direction of the angular displacement.

A control differential may be added to the system in Figure 7.8 to provide remote adjustment of the operating point. The control differential has a three-pole rotor, with three windings connected in a Y configuration. Connections to the other end of the rotor windings are made through three slip rings on the shaft. The stator also has three windings connected in a Y. The control differential is connected between the transmitter and the transformer, as shown in Figure 7.7. The output voltage is now given by the following relationship:

$$e_{out} = [E_m \cos \theta + \theta_d] \sin \omega t \tag{7.4}$$

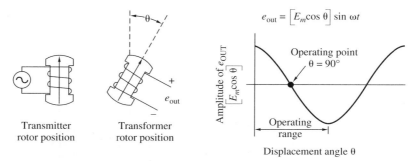

◆ **Figure 7.7** The amplitude of e_{out} varies as the cosine of the angular displacement between the two shafts.

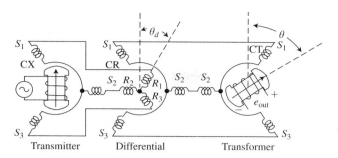

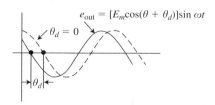

$$e_{out} = [E_m\cos(\theta + \theta_d)]\sin \omega t$$

◆ **Figure 7.8** A three-element synchro system provides remote adjustment of the operating point.

The angle θ_d is the relative displacement of the differential rotor. When θ_d is zero, the three-element system is not different from the two-element system (the dashed output curve). However, when θ_d is not zero, the entire curve is displaced by an angular amount equal to θ_d. In other words, the control differential provides a means of adjusting the operating point simply by moving the differential rotor. This is especially useful when the transmitter and transformer rotors are connected to rotating shafts.

An application of a three-element synchro system is illustrated in Figure 7.9. The process consists of two rotating rolls that must be synchronized with the capability to adjust their an-

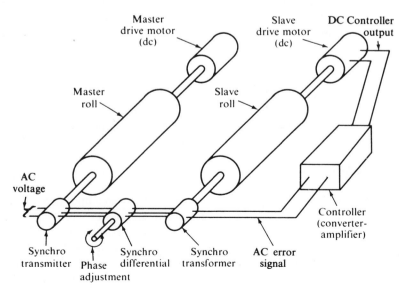

◆ **Figure 7.9** A three-element synchro system is used to synchronize two shafts with adjustment of the angular displacement between the two shafts while they are rotating.

gular relationship during operation. The control is accomplished by regulating the speed of the slave roll to maintain the desired angular relationship. The three-element synchro system acts as the measuring transmitter, setpoint, and error detector. The setpoint is the angular position of the differential rotor. It represents the desired angular relationship between the two driven rolls. The transmitter and transformer combination compares the angular relationship of the two rolls with the desired relationship, and produces an ac output signal proportional to the difference (i.e., the error signal). The converter–amplifier acts as a high-gain proportional controller. The gain is sufficiently high that the proportional offset is maintained within acceptable limits.

EXAMPLE 7.2

The synchro system in Figure 7.9 operates at a frequency of 400 Hz. The maximum amplitude of the transformer rotor voltage is 22.5 V. Determine the ac error signal produced by each of the following pairs of angular displacement:

(a) $\theta = 90°$, $\theta_d = 0°$
(b) $\theta = 60°$, $\theta_d = 0°$
(c) $\theta = 135°$, $\theta_d = -15°$
(d) $\theta = 100°$, $\theta_d = -45°$

Solution

Equation (7.4) gives the relationship between θ, θ_d, and e_{out}.

$$e_{out} = E_m \cos(\theta + \theta_d) \sin \omega t$$

At 400 Hz, $\omega = 2\pi(400) = 2570$ rad/s. The maximum amplitude $E_m = 22.5$ V.

(a) $\cos(\theta + \theta_d) = \cos(90° + 0°) = \cos 90° = 0$

$\quad e_{out} = 0$ V

(b) $\cos(\theta + \theta_d) = \cos(60° + 0°) = \cos 60° = 0.5$

$\quad e_{out} = (22.5)(0.5)\sin 2570t$

$\qquad = 11.25 \sin 2570t$ V

(c) $\cos(\theta + \theta_d) = \cos(135° - 15°) = \cos(120°) = -0.5$

$\quad e_{out} = (22.5)(-0.5)\sin 2750t$

$\qquad = -11.25 \sin 2570t$ V

or

$\quad e_{out} = (11.25)\sin(2570t + 180)$ V

(d) $\cos(\theta + \theta_d) = \cos(100 - 45) = \cos(55) = 0.574$

$\quad e_{out} = (22.5)(0.574)\sin 2570t = 12.9 \sin 2750t$ ◆

Resolvers

A *resolver* is a rotary transformer that produces an output signal that is a function of the rotor position. Figure 7.10 shows the position of the coils in a resolver. The two rotor coils are placed 90° apart. The two stator coils are also placed 90° apart. Either pair of coils can be used as the primary with the other pair forming the secondary. The following equations define the secondary voltages in terms of the primary voltages when the rotor coils are used as the primary:

$$E_1 = K(E_3 \cos \theta - E_4 \sin \theta) \tag{7.5}$$

$$E_2 = K(E_4 \cos \theta + E_3 \sin \theta) \tag{7.6}$$

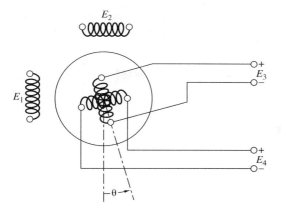

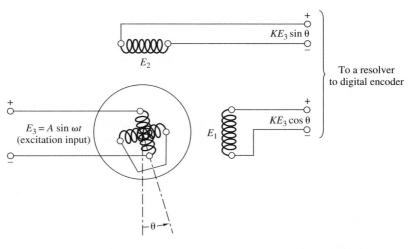

◆ **Figure 7.10**   A resolver is a transformer-type sensor that produces a signal that is a trigonometric function of shaft position, θ. The two rotor coils are placed 90° apart. The two stator coils are also 90° apart.

When a resolver is used as a sensor, one of the rotor windings is shorted as shown in Figure 7.11. If E_4 is the shorted coil, Equations (7.6) and (7.5) simplify to the following:

$$E_1 = KE_3 \cos \theta \tag{7.7}$$

$$E_2 = KE_3 \sin \theta \tag{7.8}$$

Equations (7.7) and (7.8) define the output of the resolver shown in Figure 7.11. The $\sin \theta$ term in output E_2 provides a reasonably good measurement of θ over the range $-35° \leq \theta \leq +35°$. The relationship between θ and $\sin \theta$ is slightly nonlinear. This nonlinear relationship can be represented by the following independent linearity straight line:

$$\theta_i = 60 \sin \theta$$

where θ = actual angle, °
$\qquad \theta_i$ = independent linearity angle, °

◆ **Figure 7.11**   A resolver used as a sensor has one of its rotor coils shorted. An ac voltage is applied to the other rotor coil. The two output voltages, E_1 and E_2, are trigonometric functions of the displacement angle θ.

The following table shows just how close the independent linearity angle, θ_i, is to the real angle, θ. Notice that the difference between θ and θ_i is less than 1% of the 70° span from $-35°$ to 35°.

			Comparison of Angle θ with Independent Linearity Angle θ_i				
θ	θ_i	$\theta_i - \theta$	Percentage of Span	θ	θ_i	$\theta_i - \theta$	Percentage of Span
35	34.41	−0.59	−0.84	0	0.00	0.00	0.00
30	30.00	0.00	0.00	−5	−5.23	−0.23	−0.33
25	25.36	0.36	0.51	−10	−10.42	−0.42	−0.60
20	20.52	0.52	0.74	−15	−15.53	−0.53	−0.76
15	15.53	0.53	0.76	−20	−20.52	−0.52	−0.74
10	10.42	0.42	0.60	−25	−25.36	−0.36	−0.51
5	5.23	0.23	0.33	−30	−30.00	0.00	0.00
0	0.00	0.00	0.00	−35	−34.41	0.59	0.84

The excitation voltage, E_3, is a sinusoidal voltage that can be represented as follows:

$$E_3 = A \sin \omega t$$

Substituting the equation for E_3 into the equation for E_2 and applying the independent linearity equation, we obtain the following relationship for E_2:

$$E_2 = KE_3 \sin \theta = K(A \sin \omega t)\sin \theta = K(A \sin \omega t)\left(\frac{\theta_i}{60}\right)$$

$$E_2 = \left(\frac{KA}{60}\right)(\theta_i)\sin \omega t$$

Note: $\left(\dfrac{KA}{60}\right)$ is a constant.

The last equation shows us that E_2 is a sinusoidal voltage whose amplitude varies according to θ_i and, very closely, according to θ. Now all we need is a signal conditioning circuit that will convert the voltage E_2 into a usable signal representing the position of the rotor, θ. If a digital signal is required, the signal conditioner may also convert the signal from an analog form to a digital form. We will refer to a signal conditioner that performs both functions as a *resolver-to-digital converter*.

One problem with a resolver is the necessity of brushes and slip rings to bring the excitation voltage to coil E_3 on the rotor. The brushes are subject to wear and must be protected from the dirty environment encountered in industry. The brushless resolver has been developed to solve this problem. A brushless resolver uses a transformer to couple the excitation voltage to the rotor coil, eliminating the need for brushes and slip rings.

Optical Encoders

An *encoder* is a device that provides a digital output in response to a linear or angular displacement. The resolver and digital converter discussed in the preceding section is an angular encoder. In this section we describe another type of encoder, the optical encoder.

An optical encoder has four main parts: a light source, a code disk, a light detector, and a signal conditioner. This section deals with the first three parts. Position encoders can be classified into two types: incremental encoders and absolute encoders.

An *incremental encoder* produces equally spaced pulses from one or more concentric tracks on the code disk. The pulses are produced when a beam of light passes through accurately placed holes in the code disk. Each track has its own light beam; thus an encoder with three tracks will have three light sources and three light sensors. Figure 7.12 illustrates an optical encoder with three tracks. Each track has a series of equally spaced holes in an otherwise opaque disk.* The inside track has only one hole, which is used to locate the "home" position on the code disk. The other two tracks have a series of equally spaced holes that go completely around the code disk. The holes in the middle track are offset from the holes in

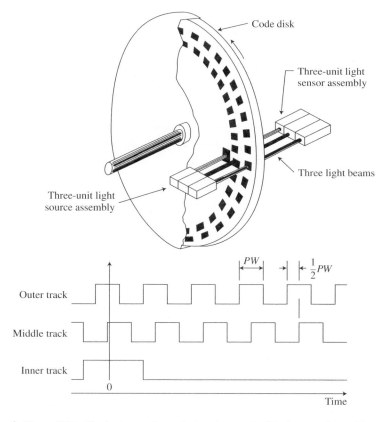

◆ **Figure 7.12**   The incremental encoder has three tracks. The inner track provides a reference signal to locate the home position. The middle track provides information about the direction of rotation. In one direction, the middle track lags the outside track; in the other direction, the middle track leads the outside track.

---

*Many optical encoders use reflective spots instead of holes in the disk. The source and sensor units are mounted such that the light beam bounces off the reflective spots onto the sensor. The source and sensor are mounted on the same side of the disk, a decided advantage in some applications.

the outside track by one-half the width of a hole. The purpose of the offset is to provide directional information. The diagram of the track pulses in Figure 7.12 was made with the disk rotating in the counterclockwise direction. Notice that the pulses from the outer track lead the pulses from the inner track by half the pulse width. If the direction is reversed to a clockwise direction, the pulses from the middle track will lead the pulses from the outer track by the same half of the pulse width.

The primary functions of the signal conditioner for an incremental encoder are to determine the direction of rotation and count pulses to determine the angular displacement of the code disk. The pulse count is a digital signal, so an analog-to-digital converter is not required for an encoder.

An angular, incremental encoder can be used to measure a linear distance by coupling the encoder shaft to a tracking wheel as shown in Figure 7.13. The wheel rolls along the surface to be measured, and the signal conditioner counts the pulses. The total displacement that can be measured in this manner is limited only by the capacity of the counter in the signal conditioner. The incremental encoder simply rotates as many times as the application requires. The measured displacement is obtained from the total pulse count as given by the following equation:

$$x = \frac{\pi d N_T}{N_R} \tag{7.9}$$

where
x = measured displacement, m
d = diameter of the tracking wheel, m
N_T = total pulse count
N_R = number of pulses in one revolution

An *absolute encoder* produces a binary number that uniquely identifies each position on the code disk. Absolute encoders may have from 6 to 20 tracks. Each track produces 1 bit of the binary number according to the code that is established by the hole pattern in the code disk. Figure 7.14 shows an absolute encoder with seven tracks that form the natural binary representation of 128 unique positions on the code disk. The number of unique positions on the code disk is related to the number of bits in the binary number (which is equal to the num-

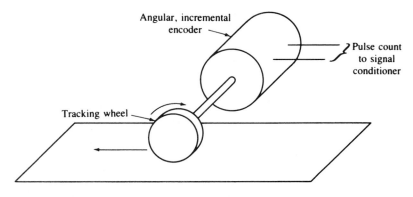

◆ **Figure 7.13** An incremental encoder coupled to a tracking wheel is used to measure linear displacement.

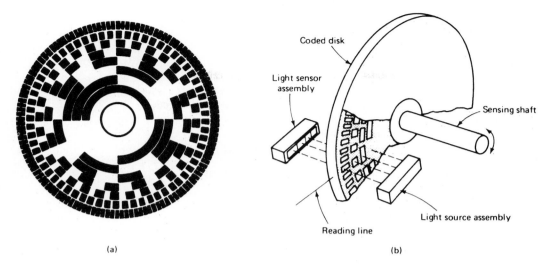

Coded disk

Light sensor
assembly

Sensing shaft

Light source assembly

Reading line

(a)   (b)

◆ **Figure 7.14**   Absolute optical encoder: (a) typical code disk; (b) encoder elements. [From H. Norton, *Sensor and Analyzer Handbook* (Englewood Cliffs, N.J.: Prentice-Hall, Inc., 1982), Fig. 1–44, p. 107.]

ber of tracks on the code disk). This also establishes the resolution of the encoder according to the following equations:

$$\text{Number of positions} = 2^N \tag{7.10}$$

$$\text{Resolution} = 1 \text{ part in } 2^N \tag{7.11}$$

where N = number of tracks = number of bits in the number

There are a number of binary codes that could be used in an encoder. The three most popular codes are the natural binary code, the Gray code, and the BCD code. Figure 7.15 shows the pattern for these three codes for the numbers from 0 to 10. The Gray code is a popular code for counters because only 1 bit changes each time the count increases by 1. Refer to Appendix C for further discussion of binary codes.

EXAMPLE 7.3

An incremental encoder is used with a tracking wheel to measure linear displacement as shown in Figure 7.13. The tracking wheel diameter is 5.91 cm, and the code disk has 180 holes in the outside track and 180 holes in the middle track. Determine the linear displacement per pulse and the displacement measured by each of the following total pulse counts:

(a) $N_T = 700$
(b) $N_T = 2220$

Solution

The linear displacement per pulse can be determined by dividing Equation (7.9) by N_T.

$$\text{Displacement per pulse} = \frac{\pi d}{N_R} = \frac{\pi(0.0591)}{180}$$

$$= 0.00103 \text{ m} = 0.103 \text{ cm}$$

$$\text{Total displacement} = 0.00103 \, N_T \text{m}$$

(a) $N_T = 700$
$$x = (0.103)(700) = 72.2 \text{ cm} = 0.722 \text{ m}$$
(b) $N_T = 2220$
$$x = (0.103)(2220) = 229 \text{ cm} = 2.29 \text{ m}$$
◆

EXAMPLE 7.4

An absolute encoder is to be used for measurements that require a resolution of at least 1 minute of arc. Determine the number of bits required to meet the specified resolution.

Solution

First determine the number of minutes in a full circle.

$$N = (60)(360) = 21,600 \text{ min/cycle}$$

Next find the smallest power of 2 that is larger than 21,600.

$$2^{14} = 16,384$$
$$2^{15} = 32,768$$

The encoder must have 15 bits to have a resolution of at least 1 minute of arc. ◆

Proximity Sensors

Proximity sensors are switches that sense the presence of an object without actually touching the object. In this section we deal with inductive and capacitive proximity sensors. The next section covers photoelectric proximity sensors.

Inductive proximity sensors are used to sense the presence of metal parts. They are also used in automated equipment to sense position and motion. Two or more proximity sensors

Arabic number	(Natural) Binary		Gray (Binary)		Binary Coded Decimal (BCD)			
	Digital number	Code pattern	Digital number	Code pattern	Digital number		Code pattern	
					Tens	Units	Tens	Units
	8 4 2 1	$2^3\ 2^2\ 2^1\ 2^0$		$G_3\ G_2\ G_1\ G_0$	8 4 2 1	8 4 2 1	2^0	$2^3\ 2^2\ 2^1\ 2^0$
0	0 0 0 0		0 0 0 0		0 0 0 0	0 0 0 0		
1	0 0 0 1		0 0 0 1			0 0 0 1		
2	0 0 1 0		0 0 1 1			0 0 1 0		
3	0 0 1 1		0 0 1 0			0 0 1 1		
4	0 1 0 0		0 1 1 0			0 1 0 0		
5	0 1 0 1		0 1 1 1			0 1 0 1		
6	0 1 1 0		0 1 0 1			0 1 1 0		
7	0 1 1 1		0 1 0 0			0 1 1 1		
8	1 0 0 0		1 1 0 0			1 0 0 0		
9	1 0 0 1		1 1 0 1		0 0 0 0	1 0 0 1		
10	1 0 1 0		1 1 1 1		0 0 0 1	0 0 0 0		

◆ **Figure 7.15** Digital code structure for absolute encoders. [From H. Norton, *Sensor and Analyzer Handbook* (Englewood Cliffs, N.J.: Prentice-Hall, Inc., 1982), Fig. 1–45, p. 108.]

can be used for simple part identification (e.g., differentiating between short, medium, and long parts).

An inductive proximity sensor consists of a coil with a ferrite core, an oscillator/detector circuit, and a solid-state switch. The oscillator creates a magnetic field in front of the sensor, centered on the axis of the coil. When a metal object enters the magnetic field, the amplitude of the oscillation diminishes due to a loss of energy to the target. The detector senses the change in amplitude and actuates a solid-state switch. When the metal object leaves the magnetic field, the oscillation returns to full amplitude and the solid-state switch returns to its deactivated state.

The *sensing range* of a proximity sensor is the distance from the sensing face within which a standard target will be detected. The *operate point* is the point at which the switch is activated as a standard target enters the magnetic field. The *release point* is the point at which the switch is deactivated as the target leaves the magnetic field. The release point is farther from the sensing face than the operate point. The difference between the operate point and the release point is called the *hysteresis* of the sensor. A typical hysteresis value is 15% of the sensing range. Table 7.1 lists representative specifications for inductive proximity sensors.

Both the sensing range and the switching speed depend on the size and material of the target. The ratings are based on a square, mild steel plate 1 mm thick, with sides equal to the diameter of the sensing face or 3 times the sensing range, whichever is larger. The sensing range must be reduced for nonferrous, metal targets such as aluminum, brass, copper, and stainless steel. The reduction factors range from 0.25 to 1.00, depending on the material and the type of sensor.

Capacitive proximity sensors can detect the presence of metal objects or nonmetallic materials such as glass, wood, paper, rubber, plastic, water, and milk. Any material that has a dielectric constant of 1.2 or greater can be detected with a capacitive sensor. The larger the dielectric constant, the easier the material is to detect. Materials with a high dielectric constant can be detected at greater distances than those with a low dielectric constant. Materials with a high dielectric constant can even be detected through a container made of material with a lower dielectric constant. Table 7.2 gives the dielectric constants of a few selected materials.

The following are a few applications of capacitive proximity sensors:

1. Detecting the flow of liquid into a container
2. Sensing the level of liquid through a glass or plastic container
3. Detecting the level of liquids or powdered materials in a bin or container
4. Detecting the presence of a sheet of material on a belt conveyor
5. Detecting a tear in a sheet of paper traveling on a belt conveyor

◆ **TABLE 7.1**   Representative Specifications for Inductive Proximity Sensors

Sensing Range (mm)	Diameter (mm)	Length (mm)	Switching Speed (Hz)	Typical Part Number
1	8	40	5000	IP-1
2	12	40	1000	IP-2
5	18	40	400	IP-3
10	30	50	200	IP-4
20	47	60	40	IP-5

◆ **TABLE 7.2** Dielectric Constants

Material	Dielectric Constant	Material	Dielectric Constant
Cereal	3–5	Salt	6
Ethanol	24	Sand	3–5
Flour	2.5–3	Sugar	3
Gasoline	2.2	Water	80
Nylon	4–5	Wood (dry)	2–6
Paper	1.6–2.6	Wood (wet)	10–30

Typical capacitive proximity sensors range in size from a diameter of 16 mm to a diameter of 95 mm. The sensing ranges for water are 5 mm for the smaller unit and 70 mm for the larger unit. The sensing distances for metals are the same as those for water. The sensing distance is reduced for other materials. For glass the reduction factor is 0.3; for wood it varies from 0.5 to 0.8, depending on the moisture content. Manufacturers of capacitive proximity sensors recommend physical tests to ensure detection of a particular material.

Photoelectric Sensors

Photoelectric sensors use a beam of light to detect the presence of objects that block or reflect the light beam. A light source provides the beam of light, and a phototransistor detects the presence or absence of light from the source. Both incandescent lamps and infrared LEDs are used as the light source. Incandescent lamps provide a wide spectrum of visible light with relatively high-power output, and the visible beam is easy to align. LEDs have a much longer life, generate less heat, and are less susceptible to shock and vibration damage. The LED has the added advantage that it can be modulated for increased noise immunity. Modulation means the LED is pulsed on and off at a very high frequency. The receiver is tuned to receive the modulated infrared light while rejecting unmodulated light. Modulation greatly reduces the influence of background lighting on the receiver. A modulated LED is called an *emitter,* and its phototransistor is called a *receiver.*

Figure 7.16 illustrates five different ways of using a light beam to detect the presence of an object. The five methods are direct scan, retroreflective scan, diffuse scan, convergent beam scan, and specular scan. In the *direct scan* method, the object to be detected passes between the light source on one side and the receiver on the other side. The object is detected when it breaks the light beam. Direct scan has the greatest range (up to 100 ft) and can handle the dirtiest environment. In the *retroreflective scan* method, the light source and the receiver are mounted in the same sensing unit. A special retroreflective target is used to reflect the light beam back to the receiver. The double distance the beam must travel and reflector losses give the retroreflective scan method a range of 10 to 30% of the range of the direct scan method.

The *diffuse scan* method is similar to the retroreflective method, but without the reflective target. The light beam strikes the surface of the object and is diffused in all directions. A portion of the diffused light reaches the receiver and actuates the switching action. In the direct and retroreflective methods, the target breaks the beam. In the diffuse method, the target

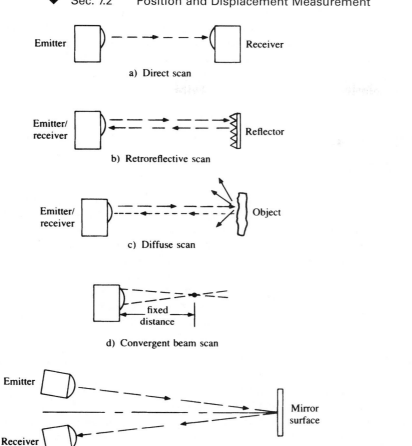

◆ **Figure 7.16** Five methods used with photoelectric sensors.

makes the beam. Photoelectric proximity sensors use the diffuse scan method. The *convergent beam* method is a variation of the diffuse scan method in which special lenses are used to focus the light beam on a point located a fixed distance from the light source. The convergent beam sensor will only detect objects that are near the point of focus. It is used to detect parts at a certain range while ignoring the background. Applications include broken wire detection and edge detection.

The *specular scan* method is used only when the object has a mirrorlike finish. The emitter and the receiver are mounted at equal angles from a line perpendicular to the mirror surface of the object to be detected. When the object is present, it reflects the light beam back to the receiver (the angle of incidence equals the angle of reflectance). The specular scan method is used to detect the difference between a dull and shiny surface. An example is detecting the presence or absence of a foil wrapping on a package.

Ultraviolet (UV) sensors are used in some interesting applications. For example, ultraviolet inks are used to print lot numbers and other information on consumer packages (e.g., soap products). This information is invisible to the customer, so it can be written on top of the visible information. The "hidden" information conserves space on the label, and it does not detract from the message on the label. In another example, American Express uses an ultra-violet coating on their charge cards. A retail clerk can detect any tampering with the card with the use of a simple UV light.

7.3 VELOCITY MEASUREMENT

Sensing Methods

Velocity is the rate of change of displacement or distance. It is measured in units of length per unit time. Velocity is a vector quantity that has both magnitude (speed) and direction. A change in velocity may constitute a change in speed, a change in direction, or both.

Angular velocity is the rate of change of angular displacement. It is measured in terms of radians per unit time or revolutions per unit time. Angular velocity measurement is more common in control systems than linear velocity measurement. When linear velocity is measured, it is often converted into an angular velocity and measured with an angular velocity transducer. Three methods of measuring angular velocity are considered in this section: dc tachometers, ac tachometers, and optical tachometers.

DC Tachometers

A *tachometer* is an electric generator used to measure angular velocity. A *brush-type dc tachometer* is illustrated in Figure 7.17. The coil is mounted on a metal cylinder called the *armature*. The armature is free to rotate in the magnetic field produced by the two permanent-magnet field poles. The two ends of the coil are connected to opposite halves of a segmented connection ring called the *commutator*. There are two segments on the commutator for each coil on the armature (only one is shown in Figure 7.17). For example, an armature with 11 coils would have a commutator with 22 segments.

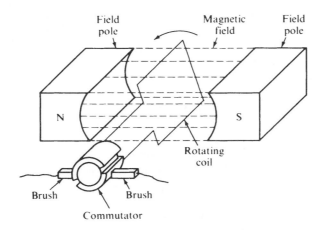

◆ **Figure 7.17** Tachometer generator.

The two carbon brushes connect the lead wires to the commutator segments. The brushes and commutator act as a reversing switch that reverses the coil connection once for each 180° rotation of the armature. This switching action converts the ac voltage induced in the rotating coil into a dc voltage. In other words, the commutator and brush constitute an ac-to-dc converter.

The tachometer produces a dc voltage that is directly proportional to the angular velocity of the armature. Equation (7.12) defines the output voltage of the dc tachometer. The constant of proportionality is called the EMF constant, K_E, and it has units of volts per revolution per minute (V/rpm). The EMF constant depends on the physical parameters of the coil and the magnetic field, as defined by Equation (7.13).

$$E = K_E S = \frac{30 K_E \omega}{\pi} \tag{7.12}$$

$$K_E = \frac{2\pi RBNL}{60} \tag{7.13}$$

where E = tachometer output, V
K_E = EMF constant, V/rpm
S = angular velocity, rpm
ω = angular velocity, rad/s
R = average radius, m
B = flux density of the magnetic field, weber/m^2 (Wb/m^2)
N = effective number of conductors
L = length of each conductor, m

A harsh industrial environment can be very hard on brush-type tachometers. Particulate contaminants can cause excessive wear in the brushes. Gaseous contaminants build up films on the commutator that cause inaccuracies. A sealed enclosure results in excessive heat buildup and thermal drift problems. A *brushless dc tachometer* solves these problems by reversing the positions of the permanent magnet and the coil. The armature is the permanent magnet and the coil is stationary. The brushes and commutator are not required because there are no electrical connections necessary to the armature. However, additional circuitry is required to sense the position of the armature and provide appropriate solid-state switching to produce a dc output. The solid-state switching circuit serves the same function as the brushes and commutator.

EXAMPLE 7.5

A dc tachometer has the following specifications:

$$R = 0.03 \text{ m}$$
$$B = 0.2 \text{ Wb/m}^2$$
$$N = 220$$
$$L = 0.15 \text{ m}$$

Determine K_E and the output voltage at each of the following speeds:

$$S = 1000, 2500, \text{ and } 3250 \text{ rpm}$$

Solution
$$K_E = \frac{2\pi RBNL}{60}$$

$$= \frac{2\pi(0.03)(0.2)(220)(0.15)}{60}$$

$$= 0.0207 \text{ V/rpm}$$

For $S = 1000$ rpm,

$$E = (0.0207)(1000) = 20.7 \text{ V}$$

For $S = 2500$ rpm,

$$E = (0.0207)(2500) = 51.8 \text{ V}$$

For $S = 3250$ rpm,

$$E = (0.0207)(3250) = 67.3 \text{ V} \qquad ◆$$

AC Tachometers

An ac tachometer is a three-phase electric generator with a three-phase rectifier on its output. The ac tachometer works well at high speeds, but the output becomes nonlinear at low speed due to the voltage drop across the rectifiers (about 0.7 V). For this reason, ac tachometers are usually limited to speed ranges of 100 to 1, compared with 1000 to 1 for dc tachometers. The ac tachometer has no brushes and has the same ability to withstand a contaminated environment as the brushless dc generator.

Comparison of DC and AC Tachometers		
Brush-Type DC Tachometer	Brushless DC Tachometer	AC Tachometer
Speed Range	1000/1	1000/1	100/1
Harsh Environment	No	Yes	Yes
Circuitry	None	Position sensor and solid-state switching	Rectifier

Optical Tachometers

An incremental encoder connected to a rotating shaft produces a sequence of pulses from which a digital velocity signal can be easily obtained. The major signal conditioning requirement is a timed counter. For example, assume that an incremental encoder has 1000 holes in the outside track and produces a new total every 10 ms. A shaft speed of 600 rpm (10 revolutions per second) will produce $10 \times 1000 = 10{,}000$ pulses per second. The counter will count $0.01 \times 10{,}000 = 100$ pulses during a 10-ms interval. Thus a count of 100 corresponds to an angular velocity of 600 rpm. Equations (7.14) and (7.15) define the relationship between the shaft speed and the timed count for an optical tachometer.

$$S = \frac{60C}{NT_c} \tag{7.14}$$

$$C = \frac{SNT_c}{60} \tag{7.15}$$

where S = shaft speed, rpm
        N = number of pulses per shaft revolution
        C = total count during time interval T_c
        T_c = counter time interval, s

When a speed measurement is obtained from an absolute encoder, the track with the greatest number of holes (least significant digit) is used in the same manner as an incremental encoder. Optical encoders can handle very large dynamic ranges with extremely high accuracy and excellent long-term stability.

EXAMPLE 7.6

An incremental encoder has 2000 pulses per shaft revolution.

(a) Determine the count produced by a shaft speed of 1200 rpm if the timer count interval is 5 ms.
(b) Determine the speed that produced a count of 224 for a timer count interval of 5 ms.

Solution

(a) Equation (7.15) applies:

$$C = \frac{(1200)(2000)(0.005)}{60} = 200$$

(b) Equation (7.14) applies:

$$S = \frac{(60)(224)}{(2000)(0.005)} = 1344 \text{ rpm}$$

◆

7.4 ACCELERATION MEASUREMENT

Sensing Methods

Acceleration is the rate of change of velocity. The measurement of linear acceleration is based on Newton's law of motion: $f = Ma$ [i.e., the force (f) acting on a body of mass (M) is equal to the product of the mass times the acceleration (a)]. Acceleration is measured indirectly by measuring the force required to accelerate a known mass (M). The units of linear acceleration are meter/second^2.

Angular acceleration is the rate of change of angular velocity. The measurement of angular acceleration is usually obtained by differentiating the output of an angular velocity transducer. Angular acceleration is expressed in terms of radian/second^2 or revolution/second^2.

Accelerometers

A schematic diagram of a linear accelerometer is shown in Figure 7.18. The *accelerometer* is attached to the object whose acceleration is to be measured and undergoes the same acceleration as the measured object. The mass M is supported by cantilever springs attached to the accelerometer frame. Motion of the mass M is damped by a viscous oil surrounding the mass.

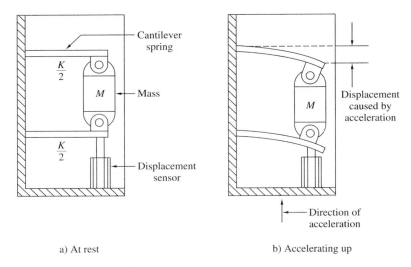

a) At rest   b) Accelerating up

◆ **Figure 7.18** Linear accelerometer.

The accelerometer is a spring-mass-damping system similar to the control valve shown in Figure 4.5 and the second-order process shown in Figure 14.9a. A second-order system is characterized by its resonant frequency (f_0) and its damping ratio (ζ), as determined by the following equations:

$$f_0 = \frac{1}{2\pi}\sqrt{\frac{K}{M}} \tag{7.16}$$

$$\zeta = \sqrt{\frac{b^2}{4KM}} \tag{7.17}$$

where f_0 = resonant frequency, Hz
     ζ = damping ratio
     K = spring constant, N/m ($K = 1/C_m$)
     M = mass, kilogram
     b = damping constant, N · s/m

Consider the situation in which the accelerometer frame in Figure 7.18b is accelerated upward at a constant rate. The mass M will deflect the cantilever springs down until the springs exert a force large enough to accelerate the mass at the same rate as the frame. When this occurs, the spring force (Kx) is equal to the accelerating force ($f = Ma$).

$$Kx = Ma$$

or

$$x = \frac{M}{K}a \tag{7.18}$$

where x = displacement of the mass, m
     M = mass, kg

K = spring constant, N/m

a = acceleration, m/s^2

Equation (7.18) indicates that the displacement x is proportional to the acceleration and may be used as a measure of the acceleration.

However, most accelerometers are used to measure accelerations that change with time. The response of the accelerometer depends on the frequency with which the measured acceleration changes. If the frequency is well below the resonant frequency (f_0), Equation (7.18) is accurate and the displacement may be used as a measure of the acceleration. At frequencies well above the resonant frequency, the mass remains stationary. The displacement is equal to the displacement of the accelerometer frame. It is not a measure of the acceleration. At frequencies near the resonant frequency, the displacement of the mass is greatly exaggerated. Again, the displacement is not a measure of the acceleration. In conclusion,

---

An accelerometer must have a resonant frequency considerably larger than the frequency of the acceleration it is measuring.

Thompson* has shown that the error in Equation (7.18) is less than 0.5% if the resonant frequency (f_0) is at least 2.5 times as large as the measured frequency when the damping ratio is 0.6.

EXAMPLE 7.7

The accelerometer in Figure 7.18 has the following specifications:

$$M = 0.0156 \text{ kg}$$

$$K = 260 \text{ N/m}$$

$$b = 2.4 \text{ N} \cdot \text{s/m}$$

$$x_{max} = \pm 0.3 \text{ cm}$$

Determine the following.

(a) The maximum acceleration that can be measured
(b) The resonant frequency, f_0
(c) The damping ratio, ζ
(d) The maximum frequency for which Equation (7.18) can be used with less than 0.5% error

Solution

(a) The maximum acceleration may be determined by using Equation (7.18):

$$x = \frac{Ma}{K}$$

$$a_{max} = \frac{x_{max}K}{M}$$

---

*W. T. Thompson, *Mechanical Vibrations* (Englewood Cliffs, N.J.: Prentice-Hall, Inc., 1948).

$$= \frac{(0.003)(260)}{0.0156}$$

$$= 50 \text{ m/s}^2$$

(b) The resonant frequency is given by Equation (7.16):

$$f_0 = \frac{1}{2\pi}\sqrt{\frac{260}{0.0156}}$$

$$= 20.6 \text{ Hz}$$

(c) The damping ratio is given by Equation (7.17):

$$\zeta = \sqrt{\frac{2.4^2}{(4)(260)(0.0156)}}$$

$$= 0.595$$

(d) The maximum frequency for an error less than 0.5% is $f_0/2.5$:

$$f_{max} = \frac{20.6}{2.5} = 8.25 \text{ Hz} \qquad \qquad ◆$$

7.5 FORCE MEASUREMENT

Sensing Methods

Force is a physical quantity that produces or tends to produce a change in the velocity or shape of an object. It has a magnitude and a direction, which are both defined by Newton's second law of motion.

$$f = Ma \tag{7.19}$$

where f = force applied to mass M, N
   M = mass, kg
   a = acceleration of mass M, m/s^2

The magnitude of force f is equal to the product of the magnitudes of M and a. The direction of force f is the same as the direction of acceleration a.

Equation (7.19) does not mean that there are no forces on a body if it is not accelerating, only that there is no net unbalanced force. Two equal and opposite forces applied to a body will balance, and no acceleration will result. All methods of force measurement use some means of producing a measurable balancing force. Two general methods are used to produce the balancing force: the null balance method and the displacement method.

A beam balance is an example of a null balance force sensor. The unknown mass is placed on one pan. Accurately calibrated masses of different sizes are placed on the other pan until the beam is balanced. The unknown mass is equal to the sum of the calibrated masses in the second pan.

A spring scale is an example of a displacement type of force sensor. The unknown mass is placed on the scale platform, which is supported by a calibrated spring. The spring is displaced until the additional spring force balances the force of gravity acting on the unknown mass. The displacement of the spring is used as the measure of the unknown force.

Two force sensors are covered in this section. The *strain gage load cell* is a *displacement* type of force sensor. The unknown force is applied to an elastic member. The displacement of the elastic member is converted to an electric signal proportional to the unknown force. The *pneumatic force transmitter* is a *null balance* type of force sensor. The unknown force is balanced by the force produced by air pressure acting on a diaphragm of known area. The air pressure is proportional to the unknown force and is used as the measured value signal.

Strain Gage Force Sensors

Strain is the displacement per unit length of an elastic member. For example, if a bar of length L is stretched to length $L + \Delta L$, the strain, ϵ, is equal to $\Delta L/L$. A *strain gage* is a means of converting a small strain into a corresponding change in electrical resistance. It is based on the fact that the resistance of a fine wire varies as the wire is stretched (strained).

There are two general types of strain gages: bonded and unbonded. *Bonded strain gages* are used to measure strain at a specific location on the surface of an elastic member. The bonded strain gage is cemented directly onto the elastic member at the point where the strain is to be measured. The strain of the elastic member is transferred directly to the strain gage, where it is converted into a corresponding change in resistance. *Unbonded strain gages* are used to measure small displacements. A mechanical linkage causes the measured displacement to stretch a strain wire. The change in resistance of the strain wire is a measure of the displacement. The displacement is usually caused by a force acting on an elastic member.

> The unbounded strain gage measures the total displacement of an elastic member. The bonded strain gage measures the strain at a specific point on the surface of an elastic member.

The *gage factor* of a strain gage is the ratio of the unit change in resistance to the strain.

$$G = \frac{\Delta R/R}{\Delta L/L} \tag{7.20}$$

where G = gage factor
ΔR = change in resistance, Ω
R = resistance of the strain gage, Ω
ΔL = change in length, m
L = length of the strain gage, m

The gage factor is usually between 2 and 4. The effective length (L) ranges from about 0.5 cm to about 4 cm. The resistance (R) ranges from 50 to 5000 Ω.

The stress on an elastic member is defined as the applied force divided by the unit area. If f is the applied force and A is the cross-sectional area, the stress is equal to f/A. In elastic materials, the ratio of the stress over the strain is a constant called the *modulus of elasticity* (E).

$$E = \frac{S}{\epsilon} \tag{7.21}$$

where E = modulus of elasticity, N/m^2

S = stress, $N/s \cdot m^2$

ϵ = strain, m/m

An example of a strain gage force sensor is illustrated in Figure 7.19. The cantilever beam is the elastic member (a diving board is a familiar example of a cantilever beam). The unknown force is applied to the end of the beam. The strain produced by the unknown force is measured by a bonded strain gage cemented onto the top of the beam. The center of the strain gage is located a distance L units from the end of the beam.

The cantilever beam assumes a curved shape that approximates a semicircle. The top surface is elongated, while the bottom surface is compressed. Halfway between these two surfaces is a neutral surface, in which there is no displacement. The stress at any point on the top surface of the cantilever beam is given by the following equation:

$$S = \frac{6fL}{bh^2} \qquad (7.22)$$

where S = stress, N/m

f = applied force, N

L = distance from the point to the end of the beam, m

b = width of the cantilever beam, m

h = height of the cantilever beam, m

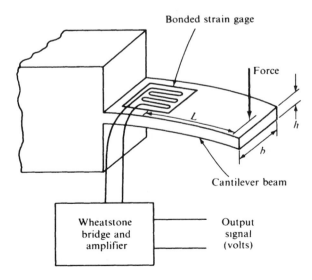

◆ **Figure 7.19** The strain gage load cell is an example of a displacement-type force sensor. The switch back path of the strain wire increases its effective length (by a factor of 6 as shown). This amplifies both R and ΔR by the same factor of 6.

Equations (7.20), (7.21), and (7.22) can be combined to obtain the following expression for the unit change in resistance produced by the unknown force f:

$$\frac{\Delta R}{R} = \left(\frac{6GL}{bh^2E}\right)f \tag{7.23}$$

where ΔR = change in resistance of the strain gage, Ω
R = unstrained resistance of the strain gage, Ω
G = gage factor of the strain gage
L = distance from the center of the strain gage to the end of the beam, m
b = width of the cantilever beam, m
h = height of the cantilever beam, m
E = modulus of elasticity of the cantilever beam, N/m^2 (see Table 7.3)

Pneumatic Force Transmitter

A pneumatic force transmitter is illustrated in Figure 7.20. The unknown force f is balanced by the force of the air pressure against the effective area of the diaphragm. The ball and nozzle is arranged such that the balance of the two forces is automatic. For example, suppose that the force f increases. The force rod moves upward, reducing the opening between the ball and nozzle. The pressure in the diaphragm chamber increases and restores the balanced condition. The air pressure (p) in the diaphragm chamber is determined by the following equation:

$$f = (p - 3)A \tag{7.24}$$

where f = unknown force, lb
p = air pressure, lb/in^2
A = effective area of the diaphragm, in^2

◆ **TABLE 7.3**   Modulus of Elasticity
of Common Metals

Material	E (N/m^2)
Aluminum	6.9×10^{10}
Beryllium	2.9×10^{11}
Copper	1.1×10^{11}
Gold	7.8×10^{10}
Steel	2.1×10^{11}
Lead	1.8×10^{10}
Molybdenum	3.5×10^{11}
Nickel	2.1×10^{11}
Silicon	7.1×10^{10}
Silver	7.7×10^{10}
Tungsten	4.1×10^{11}
Zinc	7.9×10^{10}

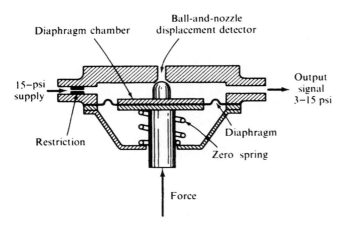

◆ **Figure 7.20** The pneumatic force transmitter is an example of a null balance force sensor.

The ($p - 3$) term simply indicates that 3 psi corresponds to a force of zero. The signal range is from 3 to 15 psi.

EXAMPLE 7.8

The strain gage force sensor in Figure 7.19 has the following specifications:

Cantilever Beam

Material: steel

$E = 2 \times 10^{11}$ N/m^2

Maximum allowable stress = 3.5×10^8 N/m^2

$b = 1$ cm

$h = 0.2$ cm

$L = 4$ cm

Strain Gage

Gage factor = 2

Nominal resistance = 120 Ω

Determine the maximum force that can be measured and the change in resistance produced by the maximum force.

Solution

The maximum force is obtained by substituting the maximum allowable stress into Equation (7.22) and solving for f_{max}.

$$f_{max} = \frac{S_{max}bh^2}{6L}$$

$$= \frac{(3.5E + 8)(0.01)(0.002)^2}{(6)(0.04)}$$

$$= 58.3 \text{ N}$$

The change in resistance is obtained from Equation (7.23).

$$\Delta R = R\left(\frac{6GL}{bh^2E}\right)f$$

$$= 120 \left[\frac{(6)(2)(0.04)}{(0.01)(0.002)^2(2E + 11)} \right] (58.3)$$

$$= 0.42 \ \Omega \qquad \qquad ◆$$

EXAMPLE   The pneumatic force transmitter in Figure 7.20 has an effective area of 2.1 in^2. Determine the force
7.9   range of the transmitter.

Solution   Equation (7.24) may be used.

$$f = (p - 3)A = (15 - 3)(2.1) = 25.2 \ \text{lb} \qquad \qquad ◆$$

◆ GLOSSARY

Absolute encoder: A device that produces a binary code that uniquely identifies the angular position of a sensing disk. (7.2)

Accelerometer: A device that measures the acceleration of an object to which it is attached. (7.4)

Bonded strain gage: A sensor that is bonded to an elastic member to measure the strain at that specific location on the surface of the elastic member. (7.5)

Convergent beam method: A variation of the diffuse scan method of photoelectric detection in which special lenses are used to focus the light beam on a fixed location. (7.2)

Diffuse scan method: A method of photoelectric detection in which the light source and the receiver are mounted in the same sensing unit. The light beam strikes the object and is diffused in all directions, but enough light reaches the receiver to actuate the switch. (7.2)

Direct scan method: A method of photoelectric detection in which the object to be detected passes between the emitter and the receiver. (7.2)

Encoder: A device that provides a digital output in response to a linear or angular displacement. (7.2)

Gage factor: The ratio of the unit change in resistance of a strain gage to the strain it is measuring. (7.5)

Incremental encoder: A device that, when rotated, produces equally spaced pulses from one or more concentric tracks on a code disk. (7.2)

Load cell: A sensor that measures force. (7.5)

Loading error: An error in a potentiometer caused by the current through the load resistor and equal to the difference between the loaded output voltage and the no-load output voltage. (7.2)

LVDT: Abbreviation for linear variable differential transformer, an electromagnetic transducer used to measure linear displacement. (7.2)

Operate point: The point at which a proximity sensor actuates when a standard target approaches the sensing face. (7.2)

Photoelectric sensor: A sensor that uses a beam of light to detect the presence of an object. The two parts of a photoelectric sensor are the emitter and the receiver. (7.2)

PIP: Abbreviation for presence, identification, and position measurement of parts in a manufacturing operation. (7.2)

Potentiometer: A resistance element with a sliding contact that can be moved from one end to the other. (7.2)

Proximity sensor: A device that senses the presence of an object without actually touching the object. (7.2)

Release point: The point at which a proximity sensor deactivates when a standard target moves away from the sensing face. (7.2)

Resolver: A rotary transformer that produces an output signal that is a function of the rotor position. (7.2)

Retroreflective scan method: A method of photoelectric detection in which the light source and the receiver are mounted in the same sensing unit. A special retroreflective target, mounted on the object to be detected, reflects the light beam from the source back to the receiver. (7.2)

Sensing range: The distance from the sensing face of a proximity sensor within which a standard target will be detected. (7.2)

Specular scan method: A method of photoelectric detection that is used only when the object has a mirror-like finish. The light source and the receiver are mounted such that the light beam reflects from the object into the receiver. (7.2)

Strain gage: A sensor that measures the displacement per unit length of an elastic member that is under stress. (7.5)

Synchro: A rotary transducer that converts angular displacement into an ac voltage or vice versa. The three types of synchro are the transmitter, the transformer, and the differential. (7.2)

Tachometer: An electric generator used to measure angular velocity. (7.3)

Unbonded strain gage: A sensor that is attached to an elastic member at two points to measure the total displacement between the two attachment points. (7.5)

◆ EXERCISES

Section 7.2

7.1   Determine the number of turns required to produce a potentiometer with each of the following resolutions.
(a) 1%   **(b)** 0.5%   **(c)** 0.2%
(d) 0.1%   **(e)** 0.01%

7.2   The potentiometer in Figure 7.4 has a resistance of 100,000 Ω. Determine the loading error caused by the following values of R_L and a.
(a) R_L = 1000 Ω; a = 0.25, 0.5, 0.75
(b) R_L = 10,000 Ω; a = 0.25, 0.5, 0.75
(c) R_L = 100,000 Ω; a = 0.25, 0.5, 0.75

7.3   The synchro system in Figure 7.6 operates at a frequency of 60 Hz. The maximum amplitude of the transformer rotor voltage is 6.2 V. Determine the ac error signal produced by each of the following angular displacements.
(a) θ = 75°
(b) θ = 45°
(c) θ = 150°
(d) θ = 110°

7.4   For the synchro system in Exercise 7.3, determine the angular displacement that will produce each of the following ac error signals.
(a) 3.1 sin 377t
(b) −4.8 sin 377t
(c) 5.5 sin 377t
(d) 2.7 sin(377t + 180°)

7.5   The synchro system in Figure 7.9 operates at a frequency of 400 Hz. The maximum amplitude of the transformer rotor voltage is 22.5 V. Determine the ac error signal produced by each of the following pairs of angular displacements.
(a) θ = 60°, θ_d = −60°
(b) θ = −30°, θ_d = −20°
(c) θ = 45°, θ_d = 20°
(d) θ = −18°, θ_d = −17°

7.6   Equations (7.5) and (7.6) define the stator voltages (E_1 and E_2) of a resolver in terms of the rotor voltages (E_3 and E_4).

$$E_1 = K(E_3 \cos \theta - E_4 \sin \theta) \qquad \textbf{(7.5)}$$

$$E_2 = K(E_4 \cos \theta + E_3 \sin \theta) \qquad \textbf{(7.6)}$$

Assume that $K = 1$ and show that Equations (7.5) and (7.6) can be rearranged to define the rotor voltages in terms of the stator voltages as follows:

$$E_3 = E_1 \cos \theta + E_2 \sin \theta$$

$$E_4 = E_2 \cos \theta - E_1 \sin \theta$$

Hint: Use the following equivalent forms of cos θ and sin θ during the manipulation of the equations.

$$\cos \theta = \frac{x}{\sqrt{x^2 + y^2}}$$

$$\sin \theta = \frac{y}{\sqrt{x^2 + y^2}}$$

7.7 Which of the following would you consider for angular measurement over a range of $\pm 160°$?
(a) potentiometer
(b) synchro
(c) resolver
(d) absolute encoder

7.8 Determine the number of positions in an absolute encoder with each of the following number of tracks:
(a) 6   (b) 8   (c) 10   (d) 12
(e) 16   (f) 20

7.9 An incremental encoder is used with a tracking wheel that has a diameter of d cm. The encoder has N_R holes in the outside track and also in the middle track. For each of the following condition sets, determine the linear displacement per pulse and the total pulse count required to measure a distance of x meters.
(a) $d = 10$ cm, $N_R = 2000$, $x = 6.3$ m
(b) $d = 5$ cm, $N_R = 500$, $x = 0.8$ m
(c) $d = 18$ cm, $N_R = 1000$, $x = 2.2$ m
(d) $d = 12$ cm, $N_R = 1600$, $x = 8.6$ m

7.10 An absolute encoder is used to measure angular position over a range of 0 to 360°. Determine the number of bits required in the counter for each of the following resolution specifications.
(a) 1° of arc   (b) 0.1° of arc
(c) 0.5 minute of arc   (d) 0.1 minute of arc

7.11 A proximity sensor will be used to sense the teeth on a rotating gear. The output of the sensor will be used to measure the rotational speed of the gear. Select a sensor from Table 7.1 for each of the following condition sets.
(a) A 20-tooth gear is operating at 1000 rpm with a minimum sensing range of 4 mm.
(b) A 10-tooth gear is operating at 300 rpm with a minimum sensing range of 8 mm.
(c) A 50-tooth gear is operating at 4200 rpm with a minimum sensing range of 0.8 mm.
(d) A 30-tooth gear is operating at 1800 rpm with a minimum sensing range of 1.6 mm.
(e) A 10-tooth gear is operating at 180 rpm with a minimum sensing range of 16 mm.

7.12 The North Star Engineering Company specializes in photoelectric systems for inspection and control. The company just received a request to bid on a system to check the label placement on 2-L plastic bottles. The bottles are 30 cm high and have a 3-cm-diameter white cap. The label wraps around the middle of the bottle and should extend from 6 cm above the bottom to 18 cm above the bottom. Tests show that the model PCS-1 convergent beam sensor can detect the presence of a bottle when a cap is at the point of focus (2 cm from the lens). The model PDS-2 direct scan sensor can see through the bottle and its contents but not through the label. Write a proposal with appropriate sketches to show how one PCS-1 and two PDS-2 sensors could be used to check the placement of labels.

7.13 The resolver shown in Figure 7.11 is used to measure the angular displacement, θ. Notice that E_4 is shorted, E_3 is a sinusoidal input voltage, and voltages E_1 and E_2 depend on the angular position θ as given in Equations (7.7) and (7.8). Notice that E_1 and E_2 are sinusoidal voltages whose amplitudes vary as $\cos \theta$ and $\sin \theta$, respectively.
(a) Assume that $E_3 = 10 \cos 2000\pi t$ V and $K = 1$. By Equations (7.7) and (7.8), the amplitude of E_1 is 10 $\cos \theta$ and the amplitude of E_2 is 10 $\sin \theta$. Sketch graphs of the amplitudes of E_1 and E_2 versus θ for values of θ from $-90°$ to $+90°$.
(b) Assume that voltage E_2 has been converted to a dc voltage equal to 10 $\sin \theta$ V. Design a five-step, piecewise-linear function that will input the voltage 10 $\sin \theta$ V and will output the voltage $\theta/9$ V.

7.14 An optical encoder is used with a 10-cm-diameter tracking wheel to measure linear displacement. The encoder generates 256 pulses per revolution (N_R).
(a) Determine the total pulse count (N_T) produced by the measurement of a linear displacement of 2 m.
(b) Determine the minimum number of bits required to store the count produced by the measurement of a distance of 20 m.

Section 7.3

7.15 Determine the output voltage (E) of a dc tachometer for each of the following condition sets:
(a) $K_E = 0.018$ V/rpm, $S = 3600$ rpm
(b) $K_E = 0.012$ V/rpm, $S = 5200$ rpm
(c) $K_E = 0.022$ V/rpm, $S = 1150$ rpm
(d) $K_E = 0.016$ V/rpm, $S = 3200$ rpm

7.16 Determine the value of K_E from each of the following tests of different dc tachometers.
(a) $E = 50$ V @ $S = 4000$ rpm
(b) $E = 60$ V @ $S = 3000$ rpm
(c) $E = 45$ V @ $S = 5000$ rpm
(d) $E = 36$ V @ $S = 4200$ rpm

7.17 A dc tachometer has the following specifications:

$$R = 0.025 \text{ m}$$
$$B = 0.22 \text{ Wb/m}^2$$
$$N = 120$$
$$L = 0.25 \text{ m}$$

Determine K_E and construct a calibration curve for a velocity range of 0 to 5000 rpm.

7.18 Determine the speed measured by each of the following incremental optical encoders:
(a) Count = 102, timer interval = 4 ms, $N = 500$ pulses/revolution
(b) Count = 2800, timer interval = 10 ms, $N = 1600$ pulses/revolution
(c) Count = 1800, timer interval = 8 ms, $N = 2000$ pulses/revolution
(d) Count = 1200, timer interval = 10 ms, $N = 1000$ pulses/revolution

7.19 Determine the count produced by each of the following incremental optical encoders:
(a) Shaft speed = 1800 rpm, timer interval = 40 ms, $N = 500$ pulses/revolution
(b) Shaft speed = 3600 rpm, timer interval = 10 ms, $N = 1800$ pulses/revolution
(c) Shaft speed = 5000 rpm, timer interval = 16 ms, $N = 2000$ pulses/revolution
(d) Shaft speed = 4400 rpm, timer interval = 8 ms, $N = 1200$ pulses/revolution

Section 7.4

7.20 The accelerometer in Figure 7.18 has the following specifications:

$$M = 0.012 \text{ kg}$$
$$K = 320 \text{ N/m}$$
$$X_{max} = \pm 0.25 \text{ cm}$$

Determine the following:
(a) The maximum acceleration that can be measured.
(b) The resonant frequency, f_0.

(c) The damping constant, b, required to produce a damping ratio of 0.6.
(d) The maximum frequency for which Equation (7.18) can be used with less than 0.5% error.

Section 7.5

7.21 The strain gage force transducer in Figure 7.19 has the following specifications:

Cantilever Beam

Material: steel
$E = 2 \times 10^{11} \text{ N/m}^2$
Maximum allowable stress = $5.0 \times 10^8 \text{ N/m}^2$
$b = 1.25 \text{ cm}$
$h = 0.25 \text{ cm}$
$L = 6 \text{ cm}$

Strain Gage

Gage factor = 2
Nominal resistance = 200 Ω

Determine the maximum force that can be measured and the change in resistance produced by the maximum force.

7.22 The pneumatic force transmitter in Figure 7.20 is to have an input of 0 to 50 lb force and an output signal range of 3 to 15 psi. Determine the required effective area.

7.23 The strain gage force transducer in Figure 7.19 has the following specifications:

Cantilever Beam

Material: beryllium
$E = 2.9 \times 10^{11} \text{ N/m}^2$
Maximum allowable stress = $10.0 \times 10^8 \text{ N/m}^2$
$b = 2.1 \text{ cm}$
$h = 0.4 \text{ cm}$
$L = 12 \text{ cm}$

Strain Gage 2

Gage factor = 2
Nominal resistance = 300 Ω

Determine the maximum force that can be measured and the change in resistance produced by the maximum force.

◆ CHAPTER 8   **Process Variable Sensors**

◆ OBJECTIVES

In Chapter 7 we examined the sensors used to measure position, motion, and force. In this chapter we examine the sensors used to measure temperature, pressure, flow rate, and level. Some of these sensors are already familiar to you. For example, the home thermostat contains a temperature sensor that measures the room air temperature. Water pumps contain a pressure sensor that actuates a switch to turn the pump on or off. Water meters measure the amount of water you use, and every home has one or two level sensors.

The purpose of this chapter is to give you an entry-level ability to discuss, select, and specify temperature, flow rate, pressure, and level sensors. After completing this chapter, you will be able to

1. List major considerations in the selection of
   a. Temperature sensors
   b. Flow rate sensors
   c. Pressure sensors
   d. Level sensors
2. Describe the following *temperature* sensors
   a. Bimetallic
   b. Resistance
   c. Filled thermal
   d. Thermister
   e. Thermocouple
   f. Radiation pyrometer
3. Describe the following *flow rate* sensors
   a. Differential pressure
   b. Turbine
   c. Vortex shedding

      d. Magnetic

      e. Positive displacement sensors

   4. Describe the following *pressure* sensors

      a. Strain gage diaphragm

      b. Bourdon element

      c. Capacitance diaphragm

      d. Bellows

   5. Describe the following *level* sensors

      a. Displacement float

      b. Static pressure

      c. Capacitance

   6. Compute the coefficients of a quadratic calibration equation

8.1 TEMPERATURE MEASUREMENT

Sensing Methods

Temperature is a measure of the degree of thermal activity attained by the particles in a body of matter. When two adjacent bodies of matter are at different temperatures, heat is transmitted from the warmer body to the cooler body until the two bodies are at the same temperature (two bodies at the same temperature are said to be in thermal equilibrium). The standard temperature scales are illustrated in Figure 8.1. The Celsius and Kelvin scales are the common and absolute temperature scales of the SI system of units.

    In the act of measuring the temperature of a body, heat is transmitted between the thermometer and the body until the two are in equilibrium. The thermometer actually measures the equilibrium temperature—not the initial temperature of the body. Thus the measuring process imposes a change in the original temperature of the body. Except when the thermometer and the measured body are at the same temperature before the measurement, 100%

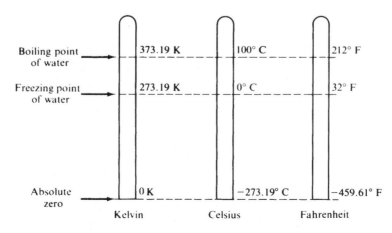

◆ **Figure 8.1**   Standard temperature scales.

accuracy in temperature measurement is impossible to attain. Additional errors are introduced by heat loss from the portion of the thermometer that is not immersed in the measured body. Careful consideration is required to minimize these measurement errors.

Differential Expansion (Bimetallic) Thermostats

A bimetallic element consists of two strips of different metals bonded together to form a leaf, coil, or helix. The two metals must have different coefficients of thermal expansion so that a change in temperature will deform the original shape. A bimetallic thermometer is formed by attaching a scale and indicator to the bimetallic element such that the indicator displacement is proportional to the temperature. A *bimetallic thermostat* is formed by replacing the dial and indicator with a set of contacts. The bimetallic thermostat is frequently used in on/off temperature control systems. Figure 8.2 is a schematic diagram of a bimetallic thermostat.

Filled Thermal Systems

A *filled thermal system* (FTS) uses a bulb filled with a liquid, gas, or vapor as the temperature sensor. A small-bore tube called a capillary connects the bulb to a spiral, helix, or bellows pressure element that converts the pressure into a usable signal, a temperature indication, or a temperature recording. The Scientific Apparatus Makers Association (SAMA) has classified filled thermal systems into four major categories according to the type of filling substance.

Class I. Liquid-filled thermal systems use the thermal expansion of a liquid to measure temperature. The filling fluid is usually an organic liquid, such as xylene, which has a very large coefficient of thermal expansion. Class I systems have the range −87 to 371°C. Desirable features include small bulb sizes, narrow spans, and relatively low cost. Less desirable features include a short capillary and compensation difficulties.

Class II. Vapor-filled thermal systems use the vapor pressure of a volatile liquid to measure temperature. The filling medium is in both the liquid and the gaseous form, and the interface must occur in the sensing bulb. A *class IIA* system can only measure temperatures above ambient—the capillary and pressure element contain the vapor while the bulb contains some vapor and all the liquid. A *class IIB* system can measure temperatures only below ambient—the capillary and pressure element contain liquid whereas the bulb contains some liquid and all the vapor. A *class IIC* system can measure temperatures above and below ambient, but the vapor and liquid must change places during a transition from one side of ambient to the other side. This transition causes a delay that makes class IIC unsuitable for measuring temperatures that pass through ambient. A *class IID* system uses a second nonvolatile fluid to overcome the transition problem of the class IIC

thermostat.

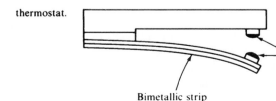

Contacts

Bimetallic strip

◆ **Figure 8.2**   Bimetallic thermostat.

system. The nonvolatile fluid fills the pressure element, the capillary, and part of the sensing bulb. The bulb contains all the volatile fluid (both the vapor and the liquid portions). The nonvolatile fluid acts as a seal and transmitter of pressure between the sensing bulb and the pressure element. Desirable features of class II systems include a long capillary, a short response time (4 to 5 s), and no need to compensate for the temperature of the capillary and pressure element. Less desirable features include a nonlinear output and no overrange capacity.

Class III. Gas-filled thermal systems use the fact that the pressure of a confined gas is proportional to its absolute temperature. Nitrogen is usually used as the fill, except for extremely low temperatures, when helium is preferred. Class III systems have the temperature range −268 to 760°C. Different filling pressures are used to obtain different temperature ranges. Desirable features of class III systems include no elevation effect due to the weight of the fluid, a large temperature range, and a large overrange (150 to 300%). Less desirable features include large bulb sizes, large required span, and low power in the pressure element.

Class V. Mercury-filled thermal systems use the thermal expansion of mercury to measure temperature. (*Note*: There is no SAMA class IV.) Mercury is separated from the other liquid-filled systems (class I) because of the unique characteristics of mercury for temperature measurement. Desirable features of class V include a very linear scale, easy compensation, rapid response, plenty of power in the pressure element, and excellent accuracy. The major objection to mercury is the possibility of mercury contamination on accidental breakage of the filled system.

A class III temperature transmitter is illustrated in Figure 8.3. The primary element consists of an inert gas sealed in a bulb that is connected by a capillary to a bellows pressure element. The bulb is immersed in the liquid to be measured until a thermal equilibrium is reached. The inert gas responds to the temperature change with a corresponding change in internal pressure. The primary-element bellows converts the gas pressure into an upward force on the right-hand side of the force beam. The air pressure in the feedback bellows produces a balancing upward force on the left-hand side of the force beam.

The feedback bellows pressure is regulated by the combination of the restriction in the supply line and the relative position of the nozzle and force beam. Air leaks from the nozzle at a rate that depends on the clearance between the force beam and the end of the nozzle. The restriction in the supply line is sized such that the feedback bellows pressure is 3 psi when the nozzle clearance is at maximum. The bellows pressure will rise to 15 psi when the nozzle is completely closed by the force beam. The left end of the force beam and the nozzle form what is called a flapper and nozzle displacement detector. This is a very sensitive detector that produces a 3- to 15-psi signal, depending on a very small displacement of the flapper.

The arrangement of the force beam is such that it automatically assumes the position that results in a balance-of-forces condition. For example, assume a rise in the measured temperature. The gas pressure increases the primary element bellows, thereby increasing the force on the right-hand side of the force beam. The increased force tends to rotate the force beam about the flexure, which acts as a fulcrum. This moves the left end of the force beam closer to the nozzle, increasing the feedback bellows pressure. The increase in the feedback bellows pressure increases the balancing force on the left-hand side of the force beam. The force beam

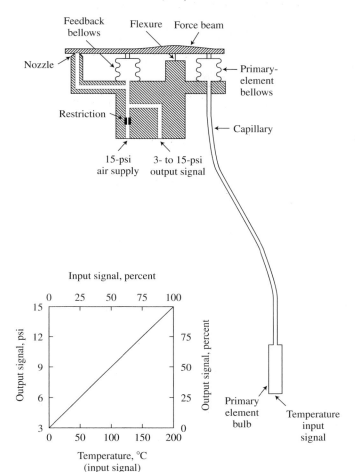

◆ **Figure 8.3** Class III temperature transmitter and its input/output graph.

is balanced by a feedback bellows pressure that is always proportional to the primary element bellows pressure. The feedback bellows pressure is an accurate measure of the primary element gas pressure, and it is used as the output signal of the pressure transducer. A typical input/output graph is illustrated in Figure 8.3.

Resistance Temperature Detectors

Resistance temperature sensors use a temperature-induced change in the resistance of a material to measure temperature. The electrical resistance of most metals increases as the temperature increases. Resistance temperature detectors (RTDs) use this property to measure temperature. A typical RTD temperature transmitter is illustrated in Figure 8.4. The sensing element is a wirewound resistor located in the end of a protecting tube. Platinum and nickel are the metals most often used to construct the sensing element. Platinum is noted for its accuracy and lin-

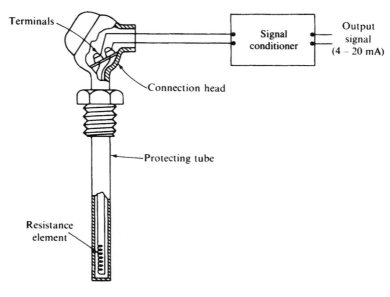

◆ **Figure 8.4** Typical RTD temperature transmitter.

earity; nickel is noted for its modest cost and relatively large change in resistance for a given change in temperature. Table 8.1 lists temperature, resistance, and output signals for a typical RTD temperature transmitter. Desirable features of RTDs include a wide temperature range (-240 to $649°C$), high accuracy, excellent repeatability, and good linearity.

◆ **TABLE 8.1** Typical Values for a Platinum RTD
Temperature Transmitter

Temperature (°C)	Resistance (Ω)	Output Signal (mA)
0	100.0	4
25	109.9	8
50	119.8	12
75	129.6	16
100	139.3	20

EXAMPLE 8.1

The resistance of a platinum RTD is approximated by the following equation:

$$R = R_0(1 + a_1 T + a_2 T^2)$$

where
R = resistance at $T°C$, Ω
R_0 = resistance at $0°C$, Ω
T = temperature, $°C$
a_1, a_2 = constants

Determine the values of R_0, a_1, and a_2 for the platinum RTD described in Table 8.1. Use the resistance values at 0, 50, and 100°C to find R_0, a_1, and a_2. Check the accuracy of the equation at 25°C.

Solution   There are three unknowns: R_0, a_1, and a_2. Therefore, three equations are required to determine the three values. These equations are obtained by substituting the following three sets of temperature and resistance values from Table 8.1 (0°, 100 Ω), (50°, 119.8 Ω), and (100°, 139.3 Ω).

$$100.0 = R_0(1 + 0a_1 + 0^2 a_2) \qquad \textbf{(a)}$$

$$119.8 = R_0(1 + 50a_1 + 50^2 a_2) \qquad \textbf{(b)}$$

$$139.3 = R_0(1 + 100a_1 + 100^2 a_2) \qquad \textbf{(c)}$$

Equation (a) is easily solved for R_0.

$$R = 100 \ \Omega \qquad \textbf{(a}'\textbf{)}$$

Substitute 100 for R_0 into Equations (b) and (c), and simplify.

$$50a_1 + 2500a_2 = 0.198 \qquad \textbf{(b}'\textbf{)}$$

$$100a_1 + 10{,}000a_2 = 0.393 \qquad \textbf{(c}'\textbf{)}$$

Equations (b′) and (c′) are easily solved for a_1 and a_2.

$$a_1 = 0.0039$$

$$a_2 = -6 \times 10^{-7}$$

The equation for R is

$$R = 100(1 + 0.00399T - 6 \times 10^{-7}T^2)$$

Check the equation at 25°C:

$$R = 100[1 + (0.00399)(25) - (6E - 7)(25)^2]$$
$$= 109.9 \ \Omega$$

The value predicted by the equation is the same as the value in Table 8.1.   ◆

  Resistance temperature detectors require a signal conditioner with a source of power and a means of measuring electrical resistance. Four methods of measuring resistance are shown in Figure 8.5. In the *direct methods,* a constant current source, i_s, passes a known current through the resistance element, R_s, producing a voltage drop, $v_s - i_s R_s$. The voltage-to-current converter converts voltage v_s into a usable 4- to 20-mA signal (see Table 8.1). In the *bridge methods,* the resistance element is compared with a known resistor in an unbalanced bridge circuit (a self-balancing bridge could also be used). The output of the unbalanced bridge is a voltage, v_s, that can also be converted to a usable 4- to 20-mA signal.

  The *two-wire direct method* in Figure 8.5a is the simplest of the four methods. The known current, i_s, passes through the resistance element and its two lead wires, producing voltage v_s. The voltage-to-current converter inputs voltage v_s and outputs a 4- to 20-mA signal similar to the one in Table 8.1. In this method, the resistance of the lead wires is included in the measurement. This results in a small error that can be corrected if the lead-wire resistance is known.

  The *four-wire direct method* in Figure 8.5b completely eliminates the lead-wire error, making it the most accurate of the four methods presented here. One pair of leads is used to carry the known current to the sensor and back; the other pair of leads is used to measure the voltage drop across the sensing element. Due to the high input impedance of the converter, the current in the voltage-sensing leads is negligible. Consequently, the error introduced by the lead-wire resistance is also negligible.

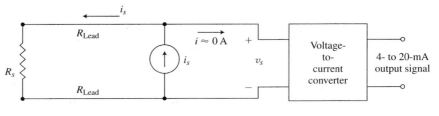

a) Two-wire direct method

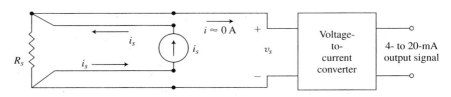

b) Four-wire direct method

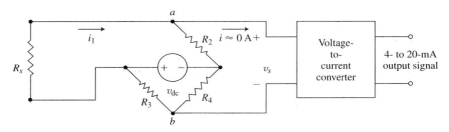

c) Two-wire bridge method

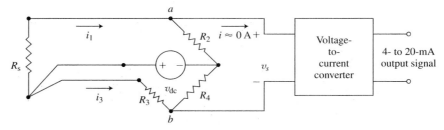

d) Three-wire bridge method

◆ **Figure 8.5** Four signal conditioning methods used to convert resistance changes into usable control signals.

The *two-wire bridge circuit* has the same lead-wire error as the two-wire direct method. However, the bridge circuit has a decided advantage. The RTD resistance is measured in terms of accurately calibrated bridge resistors, independent of the bridge voltage value. Refer to Section 6.4 for a detailed analysis of self-balancing and unbalanced bridge circuits.

The *three-wire bridge circuit* corrects most of the effect of the lead-wire resistance. The lead wires are matched so they have the same resistance. One lead wire is placed in the R_s side of the bridge, and the other is placed in the R_3 side. When R_s and R_3 are equal, the lead-wire effect is virtually eliminated. When R_s and R_3 are not equal, there will be a small lead-wire error. Many RTDs are supplied with three-wire leads.

EXAMPLE   Use the values in Table 8.1 to determine the average sensitivity of the platinum RTD over the range from
8.2   0 to 100°C. Then assume the two-wire direct method is used to measure the RTD resistance and determine the lead-wire error caused by 10 ft of each of the following sizes of copper wire:

Wire Gage	Resistance (Ω/ft)
12	0.00162
14	0.00258
16	0.00409
18	0.00651
20	0.0104
24	0.0262
28	0.0662
32	0.167

Solution

$$\text{Sensitivity} = \frac{139.9 - 100.0}{100 - 0} = 0.393 \ \Omega/°C$$

$$\text{Error} = \frac{\text{lead-wire resistance, } \Omega}{\text{sensitivity, } \Omega/°C}, °C$$

$$\text{Error (12 gage)} = \frac{10(0.00162)}{0.393} = 0.041°C$$

The results are summarized in the following table:

Wire Gage	Ten-Foot Lead-Wire Error (°C)
12	0.041
14	0.066
16	0.104
18	0.166
20	0.265
24	0.667
28	1.684
32	4.249

◆

EXAMPLE 8.3

Given the following data for the circuit in Figure 8.5b:

$$i_s = 1 \text{ mA}$$

The sensor is a platinum RTD as defined in Table 8.1

$$R_s = 100 \text{ to } 139.3 \ \Omega \text{ when } T = 0 \text{ to } 100°C$$

(a) Determine v_s when $T = 0°C$ and when $T = 100°C$.

(b) Design an op-amp circuit that will convert V_s into a 4- to 20-mA current signal as follows:

$$i_{out} = 4 \text{ mA when } T = 0°C$$

$$i_{out} = 20 \text{ mA when } T = 100°C$$

Solution

(a) Current i_s passes through the sensor resistance, R_s, and produces the following range for v_s:

$$v_s (0°C) = (0.001 \text{ A}) (100 \ \Omega) = 0.1 \text{ V}$$

$$v_s (100°C) = (0.001 \text{ A}) (139.3 \ \Omega) = 0.1393 \text{ V}$$

(b) The design, shown in Figure 8.6, uses four stages and a regulated voltage source that produces ± 1.5 V. The first stage provides the current source used by the four-wire direct method. It uses the 1.5-V source and a precision 1.5-kΩ resistor to produce the constant 1-mA source, i_s (see Exercise 6.15 and Figure 6.40.)

$$i_s = \frac{1.5 \text{ V}}{1.5 \text{ k}\Omega} = 1 \text{ mA}$$

The second stage is an instrumentation amplifier that isolates the sensor with its high input impedance and amplifies voltage v_s by a factor of 100 (see Figure 6.12 and Equation 6.19). The output of the instrumentation amplifier, v_1, has the following range:

$$v_1 (0°C) = 100 v_s (0°C) = 100 (0.1) = 10.0 \text{ V}$$

$$v_1 (100°C) = 100 v_s (100°C) = 100 (0.1393) = 13.93 \text{ V}$$

The third stage is a summing amplifier whose output, v_2, is equal to v_1 minus a constant voltage value, v_0. The value of v_0 is adjusted so that the ratio between the maximum and minimum values of v_2 is that same as the ratio between the maximum and minimum values of the output current.

$$\frac{v_2 (100°C)}{v_2 (0°C)} = \frac{13.93 - v_0}{10.0 - v_0} = \frac{20 \text{ mA}}{4 \text{ mA}} = 5$$

$$13.93 - v_0 = 5(10.0 - v_0)$$

Solving the last equation gives the following value for v_0 and range for v_2.

$$v_0 = 9.0175 \text{ V}$$

$$v_2 (100°C) = 13.93 - 9.0175 = 4.9125 \text{ V}$$

$$v_2 (0°C) = 10 - 9.0175 = 0.9825 \text{ V}$$

You may wish to verify that $v_2 (100°C)/v_2 (0°C) = 5.000$.

The fourth stage is a voltage-to-current converter (see Figure 6.14b). Resistor R is sized to produce the required 4- to 20-mA output current. Using the maximum values, R has the following value:

$$R = \frac{v_2}{i_{out}} = \frac{4.9125}{0.020} = 245.6$$

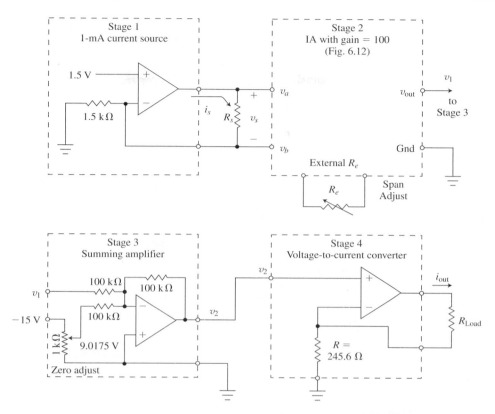

◆ **Figure 8.6** This four-stage signal conditioner converts the resistance of an RTD sensing element into a 4- to 20-mA output signal. The circuit uses the four-wire direct method for measuring resistance (see Example 8.3).

You may wish to verify that the minimum values result in the same value for resistor R.

One final observation. The adjustable resistor in stage 2 provides a means of span adjustment, and the adjustable value of v_0 provides the zero adjustment of the transmitter.   ◆

The current used to measure the value of a resistance element also dissipates power in the form of heat. The power dissipated in the element raises the temperature of the element, resulting in an error called *the self-heating error*. The size of the self-heating error depends on three factors: (1) the power dissipated in the element, (2) the internal thermal resistance of the element, and (3) the thermal resistance of the film between the probe and the surrounding fluid.

The self-heating error of an RTD is given by the dissipation constant, a number that specifies the amount of power required to raise the temperature of the sensing element by 1°C (or 1°F). The specification of the dissipation constant includes one of the following environmental conditions: still air, air moving at 1 m/s, still water, water moving at 1 m/s, still oil, or oil moving at 1 m/s. Commercial RTDs with a diameter of 1.5 to 5 mm have dissipation constants on the order of 1 to 10 mW/°C in still air, on the order of 3 to 30 mW/°C in air moving at 1 m/s, and on the order of 14 to 140 mW/°C in water moving at 1 m/s. The self-heating error is given by the following equation:

$$\text{Self-heating error} = \frac{i_s^2 R_s}{P_{\text{diss}}}, °C$$

where
i_s = sensor current, A
R_s = sensor resistance, Ω
P_{diss} = dissipation constant, W/°C

EXAMPLE 8.4

An RTD is used to measure the temperature of air flowing in a duct at 1 m/s. The dissipation constant for this environment is 20 mW/°C. The air temperature is 50°C, and the RTD has a resistance of 200 Ω at 50°C. Determine the self-heating error for each of the following sensor currents: 1 mA, 10 mA, and 20 mA.

Solution

For i_s = 1 mA:

$$\text{Self-heating error} = (0.001)^2 \left(\frac{200}{0.02}\right) = 0.01°C$$

For i_s = 10 mA:

$$\text{Self-heating error} = (0.01)^2 \left(\frac{200}{0.02}\right) = 1.0°C$$

For i_s = 20 mA:

$$\text{Self-heating error} = (0.02)^2 \left(\frac{200}{0.02}\right) = 4.0°C$$

◆

Thermistors

A *thermistor* is a semiconductor temperature sensor whose resistance changes inversely with temperature. The resistance versus temperature response of a thermistor is nonlinear and very sensitive. Small changes in temperature produce large decreases in resistance. Consequently, thermistors are often used where high sensitivity is required and accurate temperature indication is not, for example, sensing for high or low temperature limits.

Desirable features of thermistors include small size, fast response, narrow span, and high sensitivity. Less desirable features include a very nonlinear calibration graph, poor high-temperature stability, unsuitability for large spans, and high impedance. Most thermistor applications are in the range from −80 to +150°C. Table 8.2 lists resistance versus temperature values for a typical thermistor.

Signal conditioning of thermistors involves the measurement of relatively large changes in resistance. The direct methods and bridge methods in Figure 8.5 can be used, but close attention must be given to the self-heating error. Typical thermistors have dissipation constants on the order of 1 to 10 mW/°C in still air and 25 to 250 mW/°C in a well-stirred oil bath. The relatively high resistance of thermistors, especially in the low end of the temperature range, exacerbates the potential for self-heating errors. The voltage divider circuit shown in Figure 8.7 can be used as a signal conditioner for thermistors. The nonlinearity of the voltage divider compensates for the nonlinearity of the thermistor, resulting in an output voltage that is rea-

◆ **TABLE 8.2** Typical Values for a Thermistor Temperature Sensor

Temperature (°C)	Resistance (Ω)	Temperature (°C)	Resistance (Ω)
−80	1,936,000	40	1,399
−70	819,000	50	946.2
−60	369,000	60	653.3
−50	176,100	70	460.0
−40	88,400	80	329.8
−30	46,480	90	240.5
−20	25,500	100	178.3
−10	14,530	110	134.1
0	8,576	120	102.2
10	5,227	130	79.0
20	3,281	140	61.7
30	2,116	150	48.8

sonably linear and has a positive slope. You will have the opportunity to explore this simple circuit in the end-of-chapter exercises.

Thermocouples

A thermocouple consists of two dissimilar wires that are joined at one end to form a *measuring junction* and at the other end to form a *reference junction*. When the two junctions are at different temperatures, a small voltage (called an electromotive force, or emf) is generated that tends to cause a current to flow in the loop formed by the two wires. This voltage, called the Seebeck effect, is proportional to the temperature difference between the two junctions and can be used as a measure of that difference. Knowing the temperature difference, we must also know the reference junction temperature to determine the measuring junction temperature. Traditionally, the reference junction was immersed in an ice bath that maintained it at 0°C. Because the measuring junction temperature is usually greater than 0°C, the measuring junction is referred to as the *hot junction* and the reference junction as the *cold junction*.

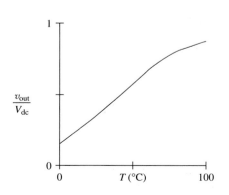

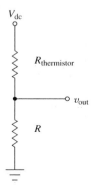

◆ **Figure 8.7** Placing a thermistor in the top branch of a voltage divider results in a reasonably linear output with a positive slope.

◆ **TABLE 8.3** American National Standard Thermocouple Type Designations

ANSI Type	Thermocouple Materials	Range (°C)
B	Platinum–6% Rhodium/Platinum–30% Rhodium	0 to 1800
E	Chromel[a]/Constantan[a]	−190 to 1000
J	Iron/Constantan[a]	−190 to 800
K	Chromel[a]/Alumel[a]	−190 to 1370
R	Platinum/Platinum–13% Rhodium	0 to 1700
S	Platinum/Platinum–10% Rhodium	0 to 1765
T	Copper/Constantan[a]	−190 to 400

Note: ANSI Standard MC 96.1, "Temperature Measurement Thermocouples," (Instrument Society of America, 1975).

[a]Alumel, Chromel, and Constantan are trade names for alloys used in thermocouples.

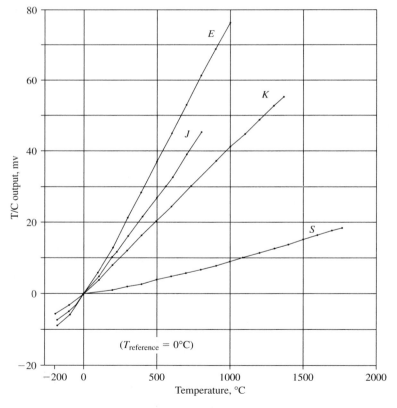

◆ **Figure 8.8** Output voltage vs. temperature graphs of four thermocouples display the different ranges and sensitivities of the four types.

The magnitude of the thermocouple emf (voltage) depends on the temperature difference between the two junctions and the material used in the two dissimilar wires. There are many possible materials for use in thermocouples, but only a limited number are used for thermocouples, and most of those are alloys that were developed specifically for such use. The most commonly used and available thermocouples have been standardized. The American National Standard (ANSI) MC 96.1 established type designations for thermocouples (see Table 8.3), thermocouple wire, and thermocouple extension wire. The platinum/platinum–10% rhodium (ANSI Type S) thermocouple is especially important because it is used to define the International Temperature Scale between 630.5 and 1063°C.

The National Bureau of Standards has published reference tables that show the emf (in millivolts) generated by the standard thermocouples in small increments of temperature over their useful range. Tables in degrees Fahrenheit and in degrees Celsius are contained in the National Bureau of Standards (NBS) Monograph 125 (U.S. Government Printing Office, Washington, DC). Figure 8.8 shows the graphs of four standard thermocouples over their useful range. Observe the different sensitivities, useful ranges, and nonlinearities displayed in the four graphs. Table 8.4 shows the millivolts generated by six standard thermocouple types in 100°C increments over their useful ranges. The table clearly shows the different sensitivities and useful ranges of the six types, but their nonlinearities are more apparent on a graph.

◆ **TABLE 8.4**   Output Voltages of Six Thermocouples Over Their Useful Range of Temperatures in 100°C Increments (The Thermocouple Reference Temperature is 0°C)

Temperature (°C)	Output Voltage (mV)					
	E	J	K	R	S	T
−190	−8.45	−7.66	−5.60			−5.379
−100	−5.18	−4.63	−3.49			−3.348
0	0.00	0.00	0.00	0.000	0.000	0.000
100	6.32	5.27	4.10	0.645	0.643	4.277
200	13.42	10.78	8.13	1.465	1.436	9.288
300	21.04	16.32	12.21	2.395	2.316	14.864
400	28.95	21.85	16.40	3.399	3.251	20.873
500	37.01	27.39	20.65	4.455	4.221	
600	45.10	33.11	24.91	5.563	5.224	
700	53.14	39.15	29.14	6.720	6.260	
800	61.08	45.53	33.30	7.925	7.329	
900	68.85		37.36	9.175	8.432	
1000	76.54		41.31	10.471	9.570	
1100			45.16	11.817	10.741	
1200			48.89	13.193	11.935	
1300			52.46	14.583	13.138	
1400			54.88[a]	15.969	14.337	
1500				17.355	15.530	
1600				18.727	16.715	
1700				20.090	17.891	
1765					18.648	

[a] The value 54.88 mV is at a temperature of 1370°C.

Each thermocouple wire must be homogeneous over the entire branch from the measuring junction to the reference junction; that is, the material must be the same for the entire branch, or it must have the same thermal emf characteristic (within an allowable tolerance). The ANSI Standard MC 96.1 includes extension wires that match the thermal emf characteristic of each thermocouple type. Extension wires make it possible to move the reference junction to a remote location without introducing errors from the additional junctions in the loop. However, a third metal can be introduced into the thermocouple loop with no effect on the emf generated provided the two junctions of the third metal are at the same temperature. This

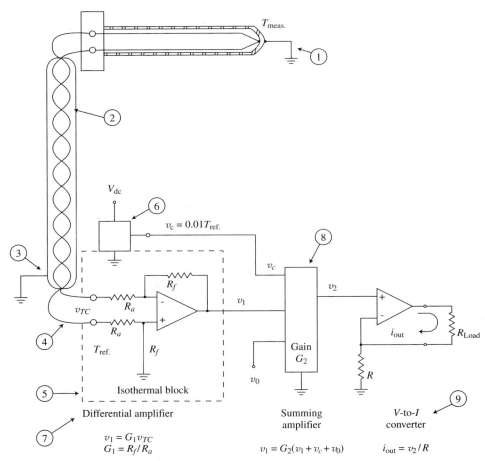

$$v_1 = G_1 v_{TC}$$
$$G_1 = R_f / R_a$$

$$v_1 = G_2(v_1 + v_c + v_0)$$

$$i_{out} = v_2 / R$$

◆ **Figure 8.9** A thermocouple signal conditioner provides noise rejection, reference junction compensation, amplification, summation, conversion, and burnout protection (not shown). Noise rejection is accomplished by (1) a grounded measuring junction, (2) twisted extension wires, (3) a grounded shield around the extension wires, and (4) high common-mode rejection. Reference junction compensation is done by (5) an isothermal block and (6) solid-state reference junction compensation. Amplification and summation are accomplished by (7) a high-gain differential amplifier and (8) a summing amplifier. Conversion is done by (9) a voltage-to-current converter.

means that electrical terminals and solder connections can be used as long as both terminals are located on an isothermal block.

A thermocouple signal conditioner is illustrated in Figure 8.9. Signal conditioning of the emf output of a thermocouple has four major concerns.

1. Reference junction temperature
2. Low-level output voltage
3. Electrical noise
4. Protection from thermocouple burnout

Three techniques are used to establish the *reference junction temperature* of a thermocouple circuit without using the traditional ice water bath.

1. A constant temperature reference block is heated or cooled to remain at some specified temperature. All thermocouple outputs are terminated in the reference block. This method is practical when many thermocouples are used and there is a suitable location for the reference box. The measuring instruments can be adjusted to use any convenient reference temperature as long as its value is known and constant.
2. A compensation circuit senses the actual reference junction temperature and makes the thermocouple output behave as if the reference junction is at 0°C. Most thermocouple transmitters provide reference junction (or cold junction) compensation.
3. A digital computer makes the reference correction. In digital systems, both the thermocouple output voltage and the reference junction temperature are converted to digital form for input to a computer. The computer is programmed to compensate for the reference junction temperature when it converts the thermocouple voltage into the measured temperature.

The *low-level output voltage* of a thermocouple (see Table 8.4) is not suitable for use by a control system. A high-gain instrumentation amplifier is used to amplify the thermocouple output voltage, and a voltage-to-current converter is used to convert the amplified voltage to a 4- to 20-mA current output signal. Current signals are relatively easy to transmit, and most commercial controllers are designed to receive a current input from the measuring transmitter. Table 8.5 shows a typical output from a thermocouple transmitter.

◆ **TABLE 8.5**   Typical Values for a Type J Thermocouple
Temperature Transmitter (see Exercise 8.17)

Temperature (°C)	EMF (mV)	Output Signal (mA)
0	0.00	4.00
20	1.02	5.51
40	2.06	7.06
60	3.11	8.62
80	4.19	10.22
100	5.27	11.82
120	6.36	13.44
140	7.45	15.06
160	8.56	16.71
180	9.67	18.35
200	10.78	20.00

The industrial environment is full of *electrical noise*, and a thermocouple circuit makes a good antenna. Noise levels picked up by a thermocouple circuit can be equal to or greater than the low-level voltage produced by the thermocouple. Noise control begins with rejection and ends with reduction. It is easier to reject noise than it is to reduce noise once it has contaminated the signal. Noise rejection and reduction steps include the following:

1. Ground the measuring junction.
2. Twist the thermocouple extension wires.
3. Wrap the extension wires in a grounded foil sheath.
4. Use an amplifier with high common-mode rejection.
5. Use a low-pass filter to reduce high-frequency noise.

Thermocouple burnout protection is a safety feature required of most thermocouple signal conditioners. This feature provides maximum output from the transmitter whenever the thermocouple circuit is open. In a control system burnout protection turns OFF the heat source if the thermocouple is open circuited for any reason.

Desirable features of thermocouples include their small size, low cost, ease of installation, ruggedness, wide range (from near absolute zero to 2700°C), reasonable accuracy, and fast response time. Less desirable features include noise pickup, low signal level, high minimum span (about 40°C), and undetectable errors caused by nonhomogeneous wires.

Integrated Circuit Temperature Sensors

Integrated circuit (IC) temperature sensors are precision solid-state devices whose output is linearly proportional to one of four temperature scales: Celsius, Kelvin, Fahrenheit, or Rankine. The LM35 series, for example, is calibrated directly in degrees Celsius. It produces an output voltage that is equal to 10 mV times the temperature in degrees Celsius over a range from $-55°C$ to $+155°C$. If T is the temperature in degrees Celsius, then the output of the LM35 series is given by the following equation:

$$v_{out} = 0.01T, \text{ V}$$

Desirable features of the IC temperature sensors include direct calibration in degrees, a linear scale factor, accuracy of 0.25% of span, low voltage requirements, low current draw, low self-heating errors (less than 0.10°C in still air), and typical nonlinearity of $\pm 0.25°$.

Radiation Pyrometers

A *radiation pyrometer* measures the temperature of an object by sensing the thermal radiation emanating from the object, without contacting the object. An optical system collects the visible and infrared energy coming from the object and focuses it on a detector. The detector converts the energy into an electrical signal, which is a complex function of the absolute temperature of the object.

The pyrometric sensing method is based on radiation laws that define the total heat radiated from an ideal *blackbody* and the relationship between the surface temperature and the wavelength spectrum of the radiated energy. These laws show that the total heat radiated from a blackbody varies as the fourth power of the absolute temperature of the surface. They also show that the wavelength of the radiated energy decreases as the surface temperature increases. This is evidenced by the changing glow of an object as its temperature increases.

Most objects emit less energy than a blackbody. The *emittance* of a body is the ratio of the radiant energy emitted from the body to the radiation emitted by a blackbody at the same temperature. The emittance of most objects is considerably less than 1. For example, the emittance of unoxidized aluminum is 0.06 and that of rough steel plate is 0.97. Most other materials fall between these two values. Emittance is a very uncertain characteristic of an object, and a major objective of a radiation pyrometer is to overcome this variable effect.

A *wide-band pyrometer* is designed to measure as much radiation from the object as possible. It has no filters and depends on an unobstructed view of the object. The presence of smoke or carbon dioxide will result in a low reading. The output must be calibrated for the emittance of the object. The calibration varies considerably with the emittance and the temperature of the object. For example, assume a wide-band pyrometer is calibrated to measure an object at 3000°F with an emittance of 0.9. If the emittance of the object changes to 0.7, the reading will be in error about 200°F.

A *narrow-band pyrometer* is designed to measure only energy in a narrow band of frequencies. The emittance usually does not vary as much over a narrow band as it does over the wide spectrum. The actual wavelength used is selected according to the application.

A *ratio pyrometer* measures the energy in two narrow bands that are very close to each other. The energy from one band is divided by the energy from the other band. The expectation is that the emittance will change by the same amount in each band. Taking the ratio should cancel the effect of equal changes in emittance. The two wavelengths are selected according to the application.

Desirable features of a radiation pyrometer include no physical contact with the object whose temperature is being measured, fast response, the ability to measure the temperature of small objects, and the ability to measure very high temperatures. Less desirable features include high cost, a nonlinear response approaching the fourth power of the temperature, variations in the emittance of the object that cause erroneous readings, and the relatively wide temperature span required.

8.2 FLOW RATE MEASUREMENT

Sensing Methods

The flow rate of liquids and gases is an important variable in industrial processes. The measurement of the flow rate indicates how much fluid is used or distributed in a process. Flow rate is frequently used as a controlled variable to help maintain the economy and efficiency of a given process.

The average flow rate is usually expressed in terms of the volume of liquid transferred in 1 s or 1 min.

$$\text{Average flow rate} = q_{avg} = \frac{\text{change in volume}}{\text{change in time}} = \frac{\Delta V}{\Delta t}$$

The instantaneous flow rate is determined by the limit of the average flow rate as Δt is reduced to zero. In mathematics, this limit is called the *derivative of V* with respect to *t* and is represented by the symbol dV/dt.

$$\text{Instantaneous flow rate} = q = \lim_{\Delta t \to 0} \frac{\Delta V}{\Delta t} = \frac{dV}{dt}$$

The flow rate in a pipe can also be expressed in terms of the average fluid velocity, v_{avg}, and the cross-sectional area of the pipe, A.

$$q_{\text{avg}} = A v_{\text{avg}}$$

The SI unit of flow rate is cubic meter/second.

Sometimes it is preferable to express the flow rate in terms of the mass of fluid transferred per unit time. This is usually referred to as the *mass flow rate*. The mass flow rate, W, is obtained by multiplying the flow rate, q, by the fluid density, ρ.

$$\text{Mass flow rate} = W = \rho q$$

The SI unit of mass flow rate is kilogram/second.

Differential Pressure Flow Meters

Differential pressure flow meters operate on the principle that a restriction placed in a flow line produces a pressure drop proportional to the flow rate squared. A differential pressure transmitter is used to measure the pressure drop, h, produced by the restriction. The flow rate, q, is proportional to the square root of the measured pressure drop.

$$q = K\sqrt{h} \tag{8.1}$$

The restriction most often used for flow measurement is the orifice plate—a plate with a small hole, which is illustrated in Figure 8.10a. The orifice is installed in the flow line in such a way that all the flowing fluid must pass through the small hole (see Figure 8.10b).

Special passages transfer the fluid pressure on each side of the orifice to opposite sides of the diaphragm unit in a differential pressure transmitter. The diaphragm arrangement converts the pressure difference across the orifice into a force on one end of a force beam. A force transducer on the other end of the beam produces an exact counterbalancing force. A displacement detector senses any motion resulting from an imbalance of the forces on the force arm. The amplifier converts this displacement signal into an adjustment of the current input to the force transducer that restores the balanced condition. The counterbalancing force produced by the force transducer is proportional to both the pressure drop and the input current, I. Thus the current, I, is directly proportional to the pressure drop across the orifice, h. This same electric current is used as the output signal of the differential pressure transducer.

In Figure 8.10, the orifice is the primary element and the differential pressure transmitter is the secondary element. The orifice converts the flow rate into a differential pressure signal, and the transmitter converts the differential pressure signal into a proportional electric current signal. A typical calibration curve is illustrated in Figure 8.10c.

Desirable features of the orifice flow meter include the fact that it is simple and easy to fabricate, it has no moving parts, a single differential pressure transmitter can be used with-

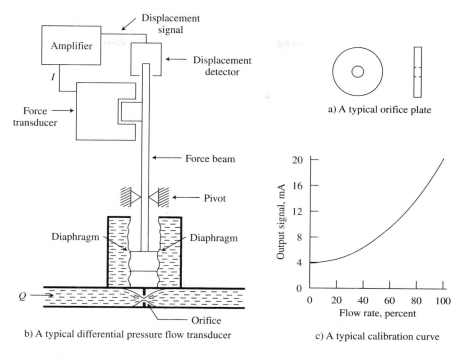

a) A typical orifice plate

b) A typical differential pressure flow transducer

c) A typical calibration curve

◆ **Figure 8.10**   A differential pressure flow transmitter is the most widely used method of flow measurement.

out regard to pipe size or flow rate, and it is a widely accepted standard. A less desirable feature is that an orifice does not work well with slurries.

Turbine Flow Meters

A *turbine flow meter* is illustrated in Figure 8.11. A small permanent magnet is embedded in one of the turbine blades. The magnetic sensing coil generates a pulse each time the magnet passes by. The number of pulses is related to the volume of liquid passing through the meter by the following equation: $V = KN$, where V is the total volume of liquid, K the volume of liquid per pulse, and N the number of pulses. The average flow rate, q_{avg}, is equal to the total volume V divided by the time interval Δt.

$$q_{avg} = \frac{V}{\Delta t} = K\frac{N}{\Delta t}$$

But $N/\Delta t$ is the number of pulses per unit time (i.e., the pulse frequency f). Thus

$$q = Kf \tag{8.2}$$

The pulse output of the turbine flow meter is ideally suited for digital counting and control techniques. Digital blending control systems make use of turbine flow meters to provide ac-

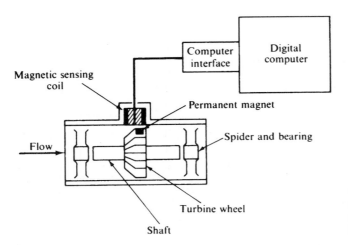

◆ **Figure 8.11** A turbine flow meter produces an accurate linear, digital flow signal. Turbine meters are used in the petrochemical and other industries for a broad range of applications.

curate control of the blending of two or more liquids. Turbine flow meters are also used to provide flow rate measurements for input to a digital computer, as shown in Figure 8.11.

EXAMPLE 8.5

A turbine flow meter has a K value of 12.2 cm^3 per pulse. Determine the volume of liquid transferred for each of these pulse counts:

(a) 220
(b) 1200
(c) 470

Also determine the flow rate, if each of the pulse counts occurs during a period of 140 s.

Solution

$$V = KN$$

(a) For 220 pulses in 140 s,

$$V = KN = (12.2)(220) = 2684 \text{ cm}^3$$

$$Q = \frac{V}{\Delta t} = \frac{2684}{140} = 19.2 \text{ cm}^3/\text{s}$$

(b) For 1200 pulses in 140 s,

$$V = (12.2)(1200) = 14{,}640 \text{ cm}^3$$

$$Q = \frac{14{,}640}{140} = 104.6 \text{ cm}^3/\text{s}$$

(c) For 470 pulses in 140 s,

$$V = (12.2)(470) = 5734 \text{ cm}^3$$

$$Q = \frac{5734}{140} = 41 \text{ cm}^3/\text{s}$$

◆

Vortex Shedding Flow Meters

A *vortex shedding flow meter* uses an unstreamlined obstruction in the flow stream to cause pulsations in the flow. The pulsations are produced when vortices (or eddies) are alternately formed and then shed on one side of the obstruction and then on the other side of the obstruction. The resulting pulsations are sensed by a piezoelectric crystal. The frequency of the pulses is directly proportional to the fluid velocity, thus forming the basis of a volumetric flow meter. Figure 8.12 illustrates a vortex shedding flow meter.

The frequency, f, of the vortex shedding is proportional to the average fluid velocity, v_{avg}, and inversely proportional to the width of the obstruction, w. The expression fw/v_{avg} is called the *Strouhal number*. The Strouhal number is constant over many ranges of Reynolds number. The relationship between the mass flow rate, W, and the vortex frequency, f, is given by the following equation:

$$W = \frac{\rho w A f}{\text{St}}$$ **(8.3)**

where A = cross-sectional area of the pipe, m^2
   f = frequency of the vortex shedding, Hz
   W = mass flow rate, kg/s
   w = width of the obstruction, m
   ρ = density of the fluid, kg/m^3
   St = Strouhal number

Desirable features of the vortex shedding meter include a linear digital output signal, good accuracy over a wide range of flow, no moving or wearing parts, and low installed cost. Less desirable features include decreasing rangeability with increasing viscosity, and practical considerations limit the size to a diameter range of 1 to 8 in.

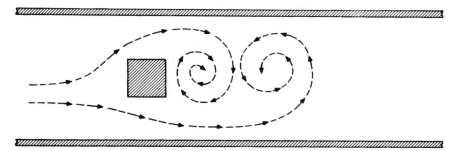

◆ **Figure 8.12**   In a vortex flow meter, vortices are alternately formed and shed on one side of an obstruction and then on the other side. The resulting pulsations are sensed by a piezoelectrical crystal, and the frequency of the pulses is directly proportional to the volumetric flow rate.

Magnetic Flow Meters

The *magnetic flow meter* has no moving parts and offers no obstructions to the flowing liquid. It operates on the principle that a voltage is induced in a conductor moving in a magnetic field. A magnetic flow meter is illustrated in Figure 8.13. The saddle-shaped coils placed around the flow tube produce a magnetic field at right angles to the direction of flow. The flowing fluid is the conductor, and the flow of the fluid provides the movement of the conductor. The induced voltage is perpendicular to both the magnetic field and the direction of motion of the conductor. Two electrodes are used to detect the induced voltage, which is directly proportional to the liquid flow rate. The magnetic flow transmitter converts the induced ac voltage into a dc electric current signal suitable for use by an electronic controller.

Desirable features of magnetic flow meters include no obstruction to the fluid flow, no moving parts, low electric power requirements, ability to handle slurries, and very low flow capabilities. Less desirable features include the fact that the fluid must have a minimum electrical conductivity, the large size and high cost of a magnetic flow meter, and the fact that periodic zero flow checks are required.

8.3 PRESSURE MEASUREMENT

Sensing Methods

Pressure is defined as the force per unit area exerted by a liquid or gas on a surface. Liquid pressure is the source of the buoyant force that supports a floating object, such as a boat or a swimmer. Pressure is also the motivating force that causes liquids and gases to flow through a pipe.

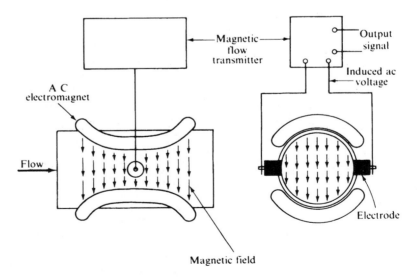

◆ **Figure 8.13** A magnetic flow meter has a completely unobstructed flow path, a decided advantage for slurries and food products.

An extremely wide range of pressures is measured and controlled in industrial processes—all the way from 0.1 Pa (about 0.001 mm of mercury) to above 100 MPa (about 10,000 psi). A great variety of primary elements have been developed to measure pressure over various parts of this extreme range. Most of these primary elements convert the measured pressure into a displacement or force. A signal conditioner then converts the force or displacement into a voltage, current, or air pressure signal suitable for use by a controller.

Pressure measurements are always made with respect to some reference pressure. Atmospheric pressure is the most common reference. The difference between the measured pressure and atmospheric pressure is called the *gage pressure*. Weather conditions and altitude both cause variations in the atmospheric pressure, and the gage pressure will vary accordingly. The standard atmospheric pressure is 101.3 kPa, or 14.7 psi. A pressure less than atmospheric is called a *vacuum*. A pressure of zero is called a *perfect vacuum*. A perfect vacuum is sometimes used as the pressure reference. The difference between a measured pressure and a perfect vacuum is called the *absolute pressure*. An arbitrary pressure is also used as the reference pressure. The difference between a measured pressure and an arbitrary reference pressure is called the *differential pressure*. A differential pressure sensor was discussed in Section 8.2, "Flow Rate Measurement."

Strain Gage Pressure Sensors

A strain gage is based on the fact that stretching a metal wire changes its resistance. The change in resistance bears an almost linear relationship to the change in length. The strain gage uses the change in resistance to measure extremely small changes in displacement. Figure 8.14 is a schematic diagram of a strain gage pressure sensor. The body acts as a mechanical transducer that converts the pressure into a displacement that increases the length of upper strain wires R_1 and R_3 and decreases the length of lower strain wires R_2 and R_4. A Wheatstone bridge converts the change in resistance of the upper and lower strain wires into an electric voltage proportional to the pressure. The voltage output of the Wheatstone bridge is then amplified to produce a usable signal. The primary element is the body and strain wire assembly. The signal conditioner consists of the Wheatstone bridge and amplifier. Strain gage pressure sensors are used in the high-pressure range (from 100 to 10,000 psi).

Strain gages are divided into two types: bonded and unbonded. Figure 8.14 is an example of an unbonded type of strain gage. The displacement is transferred to the strain wire by a mechanical linkage. Bonded strain gages are cemented directly onto the body of the transducer, as shown in Figure 7.19.

Deflection-Type Pressure Sensors

Deflection-type pressure sensors consist of a primary element, a secondary element, and a signal conditioner. The primary element converts the measured pressure signal into a proportional displacement. The secondary element converts the displacement into a change of an electrical element. The signal conditioner converts the change of the electrical element into a signal suitable for use by a controller, computer, or indicating device.

Figure 8.15 illustrates the common primary elements used in deflection-type pressure transducers. The *Bourdon element* is a flattened tube that is shaped into an incomplete circle,

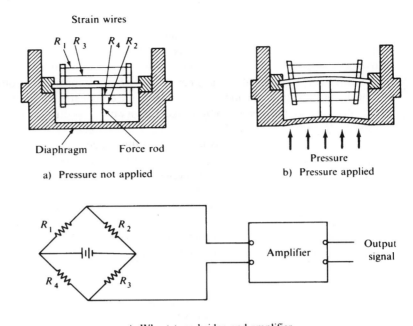

a) Pressure not applied

b) Pressure applied

c) Wheatstone bridge and amplifier

◆ **Figure 8.14** Strain gage pressure sensor.

spiral, or helix. The tube tends to straighten out as the internal pressure increases, providing a displacement proportional to the pressure. Bourdon elements of bronze, steel, or stainless steel are available to cover the range from 0 to 10,000 psi or greater.

The *bellows* element is a thin-walled metal cylinder with corrugated sides. The shape allows the element to extend as the internal pressure is balanced against a calibrated spring. The bellows displacement is proportional to the measured pressure. Bellows elements are widely used to measure pressures up to 100 psi. They are also used to make vacuum and absolute pressure measurements.

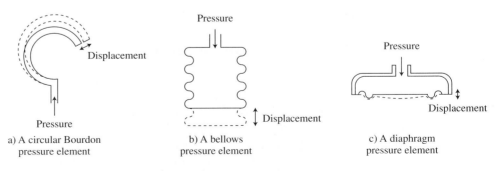

a) A circular Bourdon pressure element

b) A bellows pressure element

c) A diaphragm pressure element

◆ **Figure 8.15** Deflection-type pressure elements.

The *diaphragm* element may be flat or corrugated. The diaphragm allows sufficient movement to balance the pressure against a calibrated spring or force transducer.

The secondary element converts the primary element displacement into an electrical change that the signal conditioner can use to produce a usable signal. The conversion is accomplished by using the displacement to adjust one of the three electric circuit elements: resistance, capacitance, or inductance. The signal conditioner then produces an electric signal based on the value of the variable element. Examples of the three types of secondary elements are illustrated in Figure 8.16.

A variable resistance pressure sensor is illustrated in Figure 8.16a. The calibrated spring is displaced by an amount proportional to the pressure in the bellows. The sliding contact causes a change in the resistance between the two leads connected to the transmitter. The transmitter, in turn, produces an electrical signal based on the resistance value.

A variable inductance pressure sensor is illustrated in Figure 8.16b. The LVDT displacement transducer and demodulator produces a linear dc voltage signal proportional to the displacement of the core from a central null position. As the core moves in one direction from null, a positive voltage is produced. Movement in the other direction produces a negative volt-

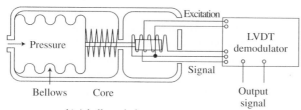

a) A bellows-resistance pressure sensor

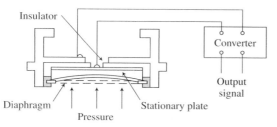

b) A bellows-inductance pressure sensor

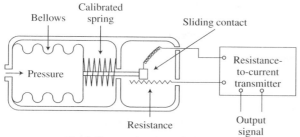

c) A diaphragm-capacitance pressure sensor

◆ **Figure 8.16**   Examples of deflection-type pressure sensors.

age. A major advantage of the LVDT is the fact that it does not touch the internal bore of the transformer. This eliminates problems of mechanical wear, errors due to friction, and electrical noise due to a rubbing action.

A variable capacitance pressure transducer is illustrated in Figure 8.16c. The diaphragm and the stationary plate form the two plates of the capacitor. The displacement of the diaphragm reduces the distance between the two plates, thereby increasing the capacitance. The signal conditioner produces an electrical signal based on the capacitance value of the primary element.

EXAMPLE 8.6

A bellows pressure element similar to Figure 8.16b has the following values:

$$\text{Effective area of bellows} = 12.9 \text{ cm}^2$$

$$\text{Spring rate of the spring} = 80 \text{ N/cm}$$

$$\text{Spring rate of the bellows} = 6 \text{ N/cm}$$

What is the pressure range of the sensor if the motion of the bellows is limited to 1.5 cm?

Solution

The total spring rate is $80 + 6 = 86$ N/cm. The force required to deflect the spring a distance of 1.5 cm is $(1.5 \text{ cm}) \times (86 \text{ N/cm}) = 129$ N. The pressure required to produce this force is $(129 \text{ N})/(12.9 \text{ cm}^2) = 10 \text{ N/cm}^2$. The range is 0 to 10 N/cm^2, or 0 to 100 kN/m^2.   ◆

8.4 LIQUID LEVEL MEASUREMENT

Sensing Methods

The measurement of the level or weight of material stored in a vessel is frequently encountered in industrial processes. Liquid level measurement may be accomplished directly by following the liquid surface, or indirectly by measuring some variable related to the liquid level. The direct methods include sight glasses and various floats with external indicators. Although simple and reliable, direct methods are not easily modified to provide a control signal. Consequently, indirect methods provide most level control signals.

Many indirect methods employ some means of measuring the static pressure at some point in the liquid. These methods are based on the fact that the static pressure is proportional to the liquid density times the height of liquid above the point of measurement.

$$p = \rho g h \tag{8.4}$$

where p = static pressure, Pa
ρ = liquid density, kg/m^3
h = height of liquid above the measurement point, m
g = 9.81 m/s^2 (acceleration due to gravity)

Thus any static pressure measurement can be calibrated as a liquid level measurement. If the vessel is closed at the top, the differential pressure between the bottom and the top of the vessel must be used as the level measurement.

The following are examples of some of the other indirect methods used to measure liquid level:

1. The displacement float method is based on the fact that the buoyant force on a stationary float is proportional to the liquid level around the float.
2. The capacitance probe method is based on the fact that the capacitance between a stationary probe and the vessel wall depends on the liquid level around the probe.
3. The gamma-ray system is based on the fact that the number of gamma rays that penetrate a layer of liquid depends on the thickness of the layer.

Displacement Float Level Sensors

A displacement float level sensor is illustrated in Figure 8.17. The float applies a downward force on the force beam equal to the weight of the float minus the buoyant force of the liquid around the float. The force on the beam is given by the following equation:

$$f = Mg - \rho g A h \tag{8.5}$$

where f = net force, N
M = mass of the float, kg
g = 9.81 m/s^2 (gravity)
ρ = liquid density, kg/m^3
A = horizontal cross-sectional area of the float, m^2
h = length of the float below the liquid surface, m

Equation (8.5) shows that the force, f, bears a linear relationship to the liquid level.

The load cell applies a balancing force on the force beam that is proportional to f and, consequently, bears a linear relationship with the liquid level. The load cell is a strain gage

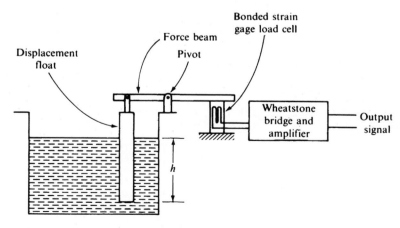

◆ **Figure 8.17**   A float and a force or displacement sensor use the buoyant force on the float as a measure of the liquid level.

force transducer that varies its resistance in proportion to the applied force. A signal conditioner converts the load cell resistance into a usable electrical signal.

Static Pressure Level Sensors

Static pressure level sensors use the static pressure at some point in the liquid as a measure of the level. They are based on the fact that the static pressure is proportional to the height of the liquid above the point of measurement. The relationship is given by Equation (8.4):

$$p = \rho g h$$

where p = static pressure, Pa
    ρ = liquid density, kg/m^3
    h = height of liquid above the measurement point, m
    g = 9.81 m/s^2

If the top of the tank is open to atmospheric pressure, an ordinary pressure gage may be used to measure the pressure at some point in the liquid. A variety of methods is used to measure the static pressure. One method is illustrated in Figure 8.18, where a bellows resistance pressure sensor and transmitter is used to measure the level. The output of the transmitter is a 4- to 20-mA current signal corresponding to a level range from 0 to 100%.

If the top of the tank is not vented to the atmosphere, the static pressure is increased by the pressure in the tank at the liquid surface. The height of the liquid above the point of measurement is proportional to the difference between the static pressure and the pressure at the top of the tank. A differential pressure measurement is required. Figure 8.19 illustrates the use of a differential pressure transducer to measure the level in a closed tank.

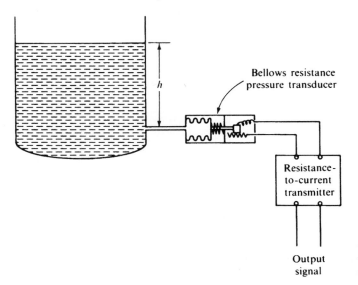

Bellows resistance
pressure transducer

Resistance-
to-current
transmitter

Output
signal

◆ **Figure 8.18**   A static pressure sensor uses the pressure near the bottom of the tank as a measure of the liquid level.

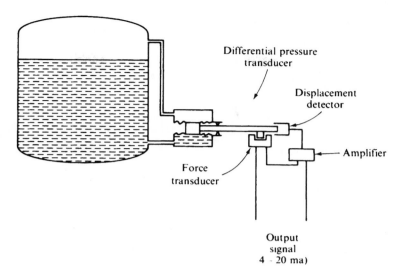

◆ **Figure 8.19**   A differential pressure sensor can be used to measure liquid level in a closed tank that is under high pressure.

Capacitance Probe Level Sensors

A capacitance probe level sensor is illustrated in Figure 8.20. The insulated metal probe is one side of the capacitor and the tank wall is the other side. The capacitance varies as the level around the probe varies. The transmitter uses a capacitance bridge to measure the change in capacitance. Solid and liquid levels up to 70 m can be measured with a capacitance probe.

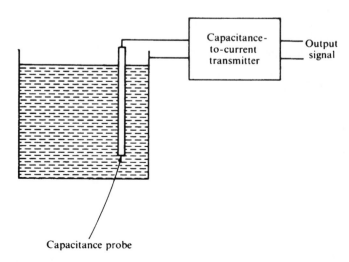

◆ **Figure 8.20**   A capacitance probe level sensor can measure solid (granular) or liquid levels up to 70 m.

EXAMPLE 8.7

The displacement float level sensor in Figure 8.17 has the following data:

Mass of the float, $M = 2.0$ kg

Cross-sectional area of float, $A = 20$ cm^2

Length of the float, $L = 2.5$ m

Liquid in the vessel, kerosene

Determine the minimum and maximum values of the force, f, applied to the force beam by the float.

Solution

From the "Properties of Liquids" table in Appendix A, the density (ρ) of kerosene is 800 kg/m^3. The force, f, is given by Equation (8.5).

$$f = Mg - \rho g A h$$

$$= (2)(9.81) - (800)(9.81)\left(\frac{20}{10^4}\right)h$$

$$= 19.62 - 15.7h \text{ N}$$

The minimum force occurs when $h = L$:

$$f_{min} = 19.62 - (15.7)(2.5)$$

$$= -19.63 \text{ N}$$

The maximum force occurs when $h = 0$:

$$f_{max} = 19.62 - (15.7)(0)$$

$$= 19.62 \text{ N}$$

The force applied by the float on the force beam ranges from 19.62 N when the vessel is empty to -19.63 N when the vessel is full. ◆

EXAMPLE 8.8

Water is to be stored in an open vessel. The static pressure measurement point is located 2 m below the top of the tank. A pressure transducer is selected to measure the liquid level. Determine the range of the pressure transducer for 100% output when the tank is full.

Solution

Equation (8.4) gives the desired relationship. The density of water is obtained from the "Properties of Liquids" table in Appendix A.

$$\rho = 1000 \text{ kg/m}^3$$

$$p = \rho g h$$

$$p_{max} = (1000)(9.81)(2)$$

$$= 1.962 \times 10^4 \text{ Pa}$$

or

$$p_{max} = 19.62 \text{ kPa}$$ ◆

◆ GLOSSARY

Absolute pressure: The difference between a measured pressure and an absolute vacuum (0 pressure). (8.3)

Bellows: A thin-walled cylinder with corrugated sides used to measure pressure. (8.3)

Bimetallic thermostat: Two strips of different metals bonded together to form a temperature-activated switch. (8.1)

Blackbody: An ideal body used to model the heat radiated from objects. A blackbody has an emittance of 1. (8.1)

Bourdon element: A flattened tube bent into an incomplete circle, spiral, or helix that is used to measure pressure. (8.3)

Differential pressure: The difference between a measured pressure and a reference pressure. (8.3)

Differential pressure flow meter: A flow meter that operates on the principle that a restriction in a flowing fluid produces a pressure drop that is proportional to the flow rate squared. (8.2)

Emittance: The ratio of the radiant energy emitted from a body to the radiation emitted from an ideal blackbody. (8.1)

Filled thermal system: A temperature sensor that uses a bulb filled with a liquid, gas, vapor, or mercury. Thermal expansion of the fluid in the bulb produces a motion that is a measure of the temperature of the fluid (8.1)

Gage pressure: The difference between a measured pressure and atmospheric pressure. (8.3)

Magnetic flow meter: A flow meter that uses the voltage induced when a conductor moves in a magnetic field (the flowing fluid is the conductor). (8.2)

Perfect vacuum: A pressure of zero. (8.3)

Radiation pyrometer: A temperature sensor that measures the temperature of an object by sensing the thermal radiation emanating from the object. (8.1)

RTD: Abbreviation of resistance temperature detector; a temperature sensor that uses the change in resistance of a conductor due to a change in temperature of the conductor. (8.1)

Strouhal number: A constant used in the flow rate equation of a vortex shedding flow meter. (8.2)

Thermistor: A temperature sensor using a semiconductor that has a large change in resistance with changes in temperature. (8.1)

Thermocouple: A temperature sensor that uses the fact that two dissimilar wires connected at each end generate a voltage that is a measure of the difference in temperature between the two ends. (8.1)

Turbine flow meter: A flow meter that uses the rotation of a turbine blade to measure fluid flow rate. (8.2)

Vacuum: A pressure that is less than atmospheric pressure. (8.3)

Vortex shedding flow meter: A flow meter that uses pulsations caused by an unstreamlined obstruction in the flow stream to measure flow rate. (8.2)

◆ EXERCISES

Section 8.1

8.1 Name the four types of fluids used in filled thermal system temperature sensors.

8.2 The following data were obtained in a calibration test of a class III FTS temperature transmitter similar to Figure 8.3:

Temperature (°C)	0	38	81	120	162	199
Output Signal (psi)	3.01	5.34	7.86	10.14	12.78	15.01

Construct a calibration graph by plotting the data points. Draw a line through the endpoints and estimate the terminal-based nonlinearity of the transmitter in degrees celsius and percentage of full scale.

8.3 Check the equation developed in Example 8.1 at 75°C.

8.4 The resistance of nickel wire at 20°C is given by the following equation:

$$R = \frac{\rho L}{A}$$

where R = resistance at 20°C, Ω

ρ = resistivity of nickel = 47.0

A = area of the wire, circular mil [circular mil = (diameter in mils)^2]

L = length of the wire, ft

A nickel resistance thermometer element is to have a resistance of 100 Ω at 20°C. Determine the length of wire required if the diameter of the wire is 0.004 in. (4 mils).

8.5 A two-wire direct method is used to measure the resistance of a nickel RTD that has a sensitivity of 0.817 Ω/°C. Determine the lead-wire error caused by 15 ft of each of the wire gages listed in Example 8.2.

8.6 Given the following data for the circuit in Figure 8.5b:

i_s = 1 mA.

The sensor is a nickel RTD.

R_s = 120.0 Ω when T = 0°C

R_s = 201.7 Ω when T = 100°C

(a) Determine v_s when T = 0°C and when T = 100°C.

(b) Design an op-amp circuit similar to Figure 8.6 that will convert v_s into a 4- to 20-mA current signal as follows:

i_{out} = 4 mA when T = 0°C

i_{out} = 20 mA when T = 100°C

8.7 An RTD is used to measure the temperature of air flowing in a duct at 1.5 m/s. The dissipation constant for this environment is 15 mW/°C. The air temperature is 60°C, and the RTD has a resistance of 120 Ω at 60°C. Determine the self-heating error for each of the following sensor currents: 1 mA, 10 mA, and 20 mA.

8.8 Plot the thermistor resistance values in Table 8.2 with R on the y-axis and T on the x-axis.

8.9 Determine the sensitivity (change in resistance per degree Celsius) of the thermistor in Table 8.2 at −70, 0, 70, and 140°C. Express the resistance change as a percentage of the resistance at the designated temperature. Use the average resistance change over the 20°C band from −80 to −60°C to compute the sensitivity at −70°C, the average change from −10 to +10°C to compute the sensitivity at 0°C, etc. The average resistance change and sensitivity at T = −70°C are computed as follows:

$$\Delta R_{avg}\,(-70°C) = \frac{R(-80°C) - R(-60°C)}{20°C}, \Omega/°C$$

$$\text{Sensitivity}\,(-70°C) = 100 \times \frac{\Delta R_{avg}\,(-70°C)}{R(-70°C)}$$

8.10 Determine the self-heating error for a thermistor in the voltage divider circuit shown in Figure 8.7. The thermistor has a dissipation constant of 6 mW/°C and a resistance of 3281 Ω @ 20°C. The circuit conditions are

$$T = 20°C$$

$$R = 3500\ \Omega$$

$$V_{dc} = 10\ V$$

$$i_{out} = 0\ (\text{zero loading effect})$$

8.11 A thermistor with the resistance values given in Table 8.2 is used in the voltage divider circuit in Figure 8.7. The output voltage, v_{out}, is easily obtained from the voltage divider rule.

$$v_{out} = \left(\frac{R}{R + R_{thermistor}}\right) V_{dc},\ V$$

(a) Compute v_{out} for T = 0, 10, 20, 30, . . . , 90, 100°C. Use the following circuit conditions in your calculations:

R = 1400 Ω V_{dc} = 1 V

Plot v_{out} versus T.

(b) Repeat part a with R = 650 Ω, V_{dc} = 1 V.

8.12 A type J (iron–constantan) thermocouple is used to check the temperature in an oven. The reference junction was placed in an ice bath at 0°C, and the measuring junction was placed in the oven. The thermocouple output voltage was measured to be 28.3 mV.

Use the data in Table 8.4 and the following interpolation formula to determine the oven temperature. Proceed as follows. In Table 8.4, locate the two consecutive lines for which the smaller voltage is less than 28.3 mV and the larger voltage is greater than 28.3. We will use L to indicate the lesser line, H to indicate the greater line, and M to indicate the measurement. Thus $V\,(L)$ = 27.39 mV, $V\,(M)$ = 28.3 mV, and $V(H)$ = 33.11 mV. Also, $T(L)$ = 500°C, $T(M)$ = measured temperature, and $T(H)$ = 600°C. Use the following interpolation formula to determine the measured temperature, $T(M)$:

$$T(M) = T(L) + [T(H) - T(L)]\left[\frac{V(M) - V(L)}{V(H) - V(L)}\right], °C$$

8.13 Repeat Exercise 8.12 for a type K thermocouple with an output voltage of 34.2 mV.

8.14 The EMF produced by a thermocouple may be approximated by the following equation:

$$E = E_0 + a_1 T + a_2 T^2$$

where E = thermocouple EMF at $T°$C, V

E_0 = thermocouple EMF at 0°C, V

T = temperature, °C

a_1, a_2 = constants

Determine the values of E_0, a_1, and a_2 for the iron–constantan thermocouple described in Table 8.5. Use the EMF values at 0, 100, and 200°C to find E_0, a_1, and a_2. Check the accuracy of the equation at 60°C. (Hint: See Example 8.1.)

8.15 Construct a graph of the temperature versus the output signal for the iron–constantan thermocouple in Table 8.5. Determine the terminal-based nonlinearity of the graph (i.e., the maximum difference between the actual curve and a straight line connecting the two endpoints).

8.16 An LM35 integrated circuit temperature sensor has the following output voltage versus temperature values:

$$v_{out} = -500 \text{ mV at } T = -50°C$$

$$v_{out} = 0 \text{ mV at } T = 0°C$$

$$v_{out} = 500 \text{ mV at } T = +50°C$$

Design an op-amp circuit that will convert the output of the LM35 into the following linear current signal:

$$i_{out} = 4 \text{ mA at } T = -50°C$$

$$i_{out} = 20 \text{ mA at } T = +50°C$$

8.17 The signal conditioner in Figure 8.9 will be used to condition the output of a type J thermocouple. The desired input/output conditions are an input temperature range of 0 to 200°C and an output current range of 4 to 20 mA. Table 8.5 lists the thermocouple voltages and the output current values for temperatures from 0 to 200°C in increments of 20°C. Determine the required values of the differential amplifier gain, G_1, the offset voltage, v_0, the summing amplifier gain, G_2, and the voltage-to-current converter resistor, R. Proceed as follows:

(a) Determine the value of G_1 that matches the solid-state compensator to the type J thermocouple over the temperature range of 0 to 40°C (the expected range of reference junction temperatures). The solid-state compensator voltage, v_c, has a slope of 10 mV/°C. Over the range of 0 to 40°C, a type J thermocouple has a slope of 2.06 mV/40°C = 0.0515 mV/°C. Determine the value of G_1 such that G_1 times the thermocouple slope is equal to the compensator slope (i.e., $0.0515G_1 = 10$).

(b) Determine the value of v_0 that matches the ratio v_2 (max)/v_2 (min) to the ratio i_{out} (max)/i_{out} (min) (i.e., to $20/4 = 5$). In this computation, assume $G_2 = 1$ and

$v_c = 0$ (i.e., $T_{ref} = 0°C$). Under these conditions, v_2 (min) = v_0 and v_2 (max) = $G_1 v_{TC}$ (200°C) + v_0. The required condition is

$$\frac{v_2 \text{ (max)}}{v_2 \text{ (min)}} = \frac{[G_1 v_{TC} (200°C) + v_0]}{v_0} = 5$$

(c) Assume $G_2 = 1$ and determine the value of R that produces the desired 4- to 20-mA current output.

(d) Compute the output current at temperatures of 0, 100, and 200°C with $T_{ref} = 0°C$.

(e) Repeat step d with the reference junction at 20°C and again at 40°C. Compare both sets of results with the results in step d by computing the percentage of difference.

8.18 Repeat Exercise 8.17 for a temperature range of 0 to 400°C and an output current range of 10 to 50 mA. Use Table 8.4 for the mV output at 400°C.

8.19 Complete the following for the iron–constantan thermocouple transmitter described in Table 8.5:

(a) Determine the terminal-based linearity of the transmitter output.

(b) Design a five-step, piecewise linear function that will linearize the transmitter output.

Section 8.2

8.20 Differential pressure flow motors are used as the sensors in six liquid flow control systems. Pretend you were assigned the task of constructing calibration graphs for the six flow controllers. Your first step was to conduct four calibration tests for each control system. In the first test, you set the controller at 25%, collected the output flow in a container of known volume, and measured the time required to fill the container. You then repeated this test at controller settings of 50%, 75%, and 100%. Your test results are summarized in the following table.

Controller Tag Number	Container Volume V (gal.)	Time in Minutes to Fill the Container (volume = V gal) at Each Controller Setting			
		25%	50%	75%	100%
(a) FIC 101	0.25	3.32	2.39	1.98	1.72
(b) FIC 112	1.00	1.74	1.22	1.03	0.88
(c) FIC 116	0.25	1.84	1.29	1.04	0.91
(d) FIC 119	5.00	1.49	1.06	0.84	0.74
(e) FIC 134	1.00	2.95	2.05	1.68	1.47
(f) FIC 148	5.00	4.56	3.14	2.66	2.26

Determine the average flow rate in gpm for each controller setting by dividing the container volume, V, by the time in minutes. Finally, use your calculated results to construct a calibration graph for each controller.

8.21 The flow control systems in Exercise 8.20 are described by the equation

$$q = K\sqrt{sp}$$

where q = flow rate, gal/min

   K = constant

   sp = controller setting, %

Calculate the value of K for each calibration test result in Exercise 8.20. Find the average K value for each control system, and use this value to express the relationship between the flow rate, q, and the square root of the controller setting, sp.

8.22 A turbine flow meter has a K value of 34.1 cm^3 per pulse. Determine the volume of liquid transferred for each of the following pulse counts:
(a) 8200   **(b)** 32,060   **(c)** 680

8.23 Determine the flow rate for each pulse count in Exercise 8.22 if the pulse counts occur during a 210-s time interval.

Section 8.3

8.24 A bellows pressure element similar to Figure 8.16b has the following values:

   Effective area of the bellows = 21 cm^2

   Spring rate of the spring = 200 N/cm

   Spring rate of the bellows = 10 N/cm

What is the pressure range of the transducer if the motion of the bellows is limited to 2 cm?

8.25 A bellows pressure element similar to Figure 8.16b has the following values:

   Effective area of the bellows = 4 cm^2

   Spring rate of the spring = 400 N/cm

   Spring rate of the bellows = 25 N/cm

What stroke of the bellows is required for a range of 0 to 50 N/cm^2?

Section 8.4

8.26 A displacement float level transducer has the following data:

   Mass of the float, M = 6.5 kg

   Area of the float, A = 50 cm^2

   Length of the float, L = 4.0 m

   Liquid in the vessel, water

Determine the minimum and maximum values of the force, F, applied to the force beam by the float.

8.27 Gasoline is to be stored in a vented vessel. The static-pressure measurement point is 8 m below the top of the vessel. A pressure sensor is to be used as the level sensor. Determine the range of the pressure sensor for 100% output when the tank is filled.

8.28 The following liquid storage tanks use a pressure sensor at the bottom of the tank as the level sensor. Determine the static pressure in kPa at the bottom of each tank when the tank is full.
(a) A 4-m-high tank full of gasoline.
(b) A 2-m-high tank full of water.
(c) A 6-m-high tank full of oil.
(d) A 1-m-high tank full of kerosene.

8.29 Convert each of the pressure readings in Exercise 8.28 from kPa to psi.

◆ CHAPTER 9

Switches, Actuators, Valves, and Heaters

◆ **OBJECTIVES**

A controller has two interfaces with the process it controls; one is the input to the controller, the other is the output from the controller. Sensors and signal conditioners handle the input side. Various types of switching elements, actuators, control valves, heaters, and electric motors handle the output side.

The purpose of this chapter is to give you an entry-level ability to discuss, select, and specify switching elements, pneumatic actuators, hydraulic actuators, process control valves, and heaters. After completing this chapter, you will be able to

1. Interpret standard symbols for
   a. Mechanical switches and relays
   b. Solid-state components
   c. Solenoid valves
2. Describe
   a. Pushbutton switch, limit switch, level switch, pressure switch, and temperature switch
   b. Relay, contactor, starter, and time-delay relay
   c. Transistor, SCR, triac, UJT, and diac
   d. Pneumatic cylinder and pneumatic motor
   e. Hydraulic cylinder and hydraulic motor
   f. Process control valve
3. Select a pneumatic or hydraulic cylinder
4. Determine the proper size of a process control valve
5. Compute the heat flow rate required to melt and heat a substance

9.1 MECHANICAL SWITCHING COMPONENTS

Switches are devices that make or break the connection in an electric circuit. The switching action may be accomplished mechanically by an actuator, electromechanically by a solenoid, or electronically by a solid-state device. All three methods of accomplishing the switching action are used in control systems. Switches turn on electric motors and heating elements, sense the presence of an object, regulate the speed of an electric motor, actuate solenoid valves that control pneumatic or hydraulic cylinders, and initiate actions in sequential control systems.

Mechanical switching components are used in the manipulating element of some process control systems. As you may recall from the control system block diagram (Figure 1.7), the manipulating element is located between the controller and the process. One application of a mechanical switching element is the electric heater control in the lower left corner of Figure 2.12. The contactor is a mechanical switching component that turns the electric heating element ON and OFF. A second application is the solids feeder control system in Figure 2.19. Here, two limit switches operate a motor to lift and lower the gate on the belt feeder.

Mechanical switching components are often used in sequential control systems to perform a set of operations in a prescribed manner. They are used to start and stop the motion of pneumatic and hydraulic cylinders in Figures 2.8 and 2.21. They are used to start and stop a large electric motor in Figure 9.5, and to control a hydraulic hoist in Figure 11.4.

The circuit diagrams in the figures mentioned in the previous paragraph are called "ladder diagrams" because they look like a ladder. A ladder diagram consists of two vertical lines and any number of horizontal circuits that connect the two vertical lines. The vertical lines represent the electrical supply lines—usually the 115-V ac hot and common lines or 12-V dc positive and negative (or ground) lines.

The horizontal lines in a ladder diagram are called "rungs" because they look like the rungs on a ladder. A rung in a ladder diagram has two parts: an *input* section and an *output* element. The input section consists of one or more contacts. The output element may be an electric motor, a heating element, a solenoid valve, or a relay coil. Only one rule is required to understand a ladder diagram: the output element is turned ON when there is a *closed path* connecting the element to the two vertical lines, and the element is OFF when the path is not closed (due to an open contact). You will be given the opportunity to construct simple ladder diagrams in the exercises at the end of this chapter.

Mechanical Switches

Mechanical *switches* use one or more pairs of contacts to make or break an electric circuit. The contacts may be normally open (NO) or normally closed (NC). A normally open switch will close the circuit path between the two terminals when the switch is actuated and will open the circuit path when the switch is deactuated. A normally closed switch will open the circuit path when the switch is actuated and will close the circuit path when the switch is deactuated. The more common actuating mechanisms include pushbuttons, toggles, levers, plungers, and rotary knobs.

The switching action may be momentary-action or maintained-action. In a *momentary-action* switch, the operator pushes the button, moves the toggle, or rotates the knob to change the position of the contacts. When the operator releases the switch, the contacts return to the normal condition. When an operator actuates a *maintained-action* switch, the contacts remain in the new position after the operator releases the actuator. In most momentary-action switches, the actuator returns to its original position when released, and in most maintained-action switches, the actuator remains in the new position when released. However, in some maintained-action switches the actuator returns to the original position, even though the contacts remain in the new position.

Mechanically actuated switches may be operated manually by an operator or automatically by fluid pressure, liquid level, temperature, flow, thermal overload, a cam, or the presence of an object. Figure 9.1 shows standard wiring diagram symbols for various types of switches. The limit switches in Figure 9.1 are actuated by a cam or some other object that engages the switch actuator. Pressure-actuated switches open or close their contacts at a given pressure. Liquid level–actuated switches open or close their contacts at a given liquid level. Temperature-actuated switches open or close their contacts at a given temperature. Flow-actuated switches open or close their contacts when a given flow rate is sensed. The overload switches are circuit breakers that open a normally closed contact when an overload condition occurs. Overload switches are intended to protect motors and other equipment from damage caused by an overload condition.

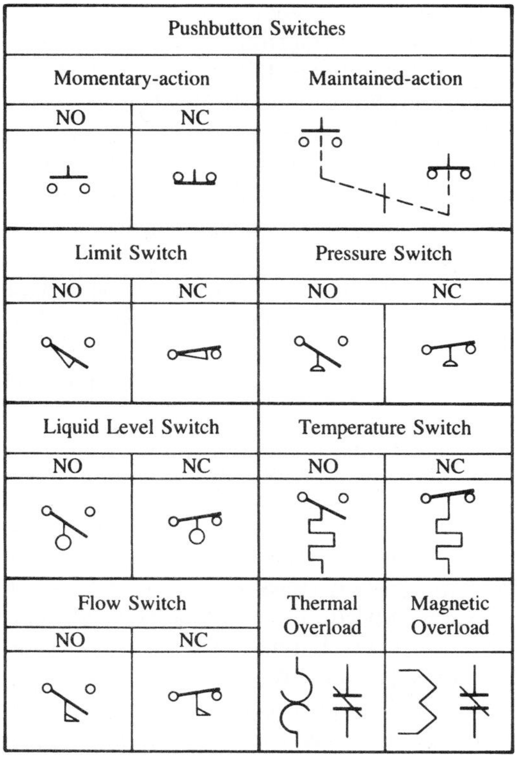

◆ **Figure 9.1**   Wiring diagram symbols for mechanical switches.

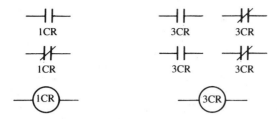

◆ **Figure 9.2** Control relays use the coil designation (e.g., 1CR, 2CR) to identify the contacts that are actuated by the coil.

Relays

A *relay* is a set of switches that are actuated when electric current passes through a coil of wire. The electric current passing through the coil of wire generates a magnetic field about the core of the coil. This magnetic field pulls a movable arm that forces the contact to open or close. The *pull-in current* is the minimum coil current that causes the arm to move from its OFF position to its ON position. The *drop-out current* is the maximum coil current that will allow the arm to move from its ON position to its OFF position. Figure 9.2 illustrates two *control relays,* one with two switches, the other with four switches. The relay coils are represented schematically by the circles with the designation 1CR and 3CR. The CR signifies a control relay, and the numbers 1 and 3 are used to distinguish between the two control relays. Each relay in a drawing must have a unique designation.

The relay is actuated (or energized) by completing the circuit branch that contains the relay coil. The relay coil designation is used to identify the contacts that are actuated by a particular relay coil. Normally open contacts are designated by a pair of parallel lines. Normally closed contacts are designated by parallel lines with a diagonal line connecting the two parallel lines. A normally open contact will close the circuit path when the relay coil is energized and will open the circuit path when the relay is deenergized. A normally closed contact will open the circuit path when the relay coil is energized and will close the circuit path when the relay coil is deenergized.

Notice that relay 1CR has one normally open contact and one normally closed contact. Relay 3CR has two normally open and two normally closed contacts. This method of using the relay coil designation to identify the relay contacts is necessary because the contacts may occur anywhere in an electric circuit diagram. The designation identifies which relay coil actuates each set of contacts.

Time-Delay Relays

Time-delay relays are control relays that have provisions for a delayed switching action (see Figure 9.3). The delay in switching is usually adjustable, and it may take place when the coil is energized or when the coil is deenergized. An arrow is used to identify the switching direction in which the time delay takes place. In relay 1TR, the delay occurs when the coil is energized. Contact 1TR delays before it closes. When coil 1TR is deenergized, contact 1TR opens immediately. In relay 2TR, the delay also occurs when the coil is energized. The 2TR contact delays before it opens and closes immediately when the coil is deenergized. In relays

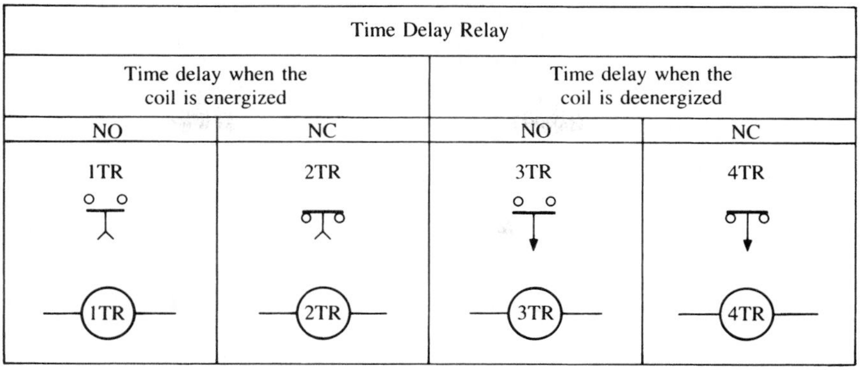

Time Delay Relay			
Time delay when the coil is energized		Time delay when the coil is deenergized	
NO	NC	NO	NC
1TR	2TR	3TR	4TR
1TR	2TR	3TR	4TR

◆ **Figure 9.3**   In time-delay relays, the arrow points in the direction in which the delayed action occurs.

3TR and 4TR, the delay occurs when the coil is deenergized. Contact 3TR closes immediately when coil 3TR is energized and delays before it opens when the coil is deenergized. Contact 4TR opens immediately when coil 4TR is energized, and delays before it closes when coil 4TR is deenergized.

> In a time-delay relay, the delay occurs in the direction in which the arrow is pointing.

Contactors and Motor Starters

Relays with heavy-duty contacts are used to switch circuits that use large amounts of electric power. When the circuit load is an electric motor, the relay is called a *motor starter;* otherwise, it is called a *contactor*. An example of a relay for switching a three-phase system is shown in Figure 9.4. The three heavy contacts are used to switch the three lines supplying electric power to the load. The two light contacts are used in the control circuit.

Figure 9.5 illustrates a circuit for starting and stopping a 480-V ac three-phase electric motor. The two momentary-action pushbutton switches in the 115-V ac circuit are the start–stop station for the motor. When the START button is pressed, coil 1M is energized, closing all four 1M contacts. The three large contacts connect the three 480-V ac lines to the motor. The small contact in parallel with the START switch is used to hold the circuit closed after the START button is released. The small contact is called a "holding contact" because it "holds" coil 1M in the energized condition after the operator releases the START button. When the operator presses the STOP button, the circuit breaks and all four 1M contacts open. The circuit remains deenergized after the STOP button is released because the 1M holding contact is open.

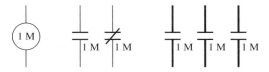

◆ **Figure 9.4**   Motor starters and contactors have three large contacts that are used to switch large amounts of electric power. The small auxiliary contacts are used in the control circuit.

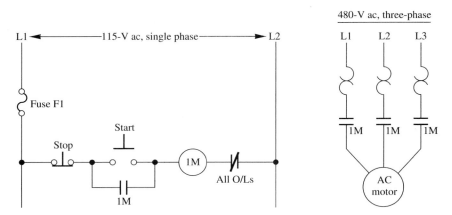

◆ **Figure 9.5**   Control circuit for starting a large ac motor.

9.2 SOLID-STATE COMPONENTS

Solid-state components are used in a variety of ways in control systems. For example, they are used to convert a dc voltage into a controlled ac voltage in Figure 9.7. They are used to vary the ac power delivered to a load in Figures 9.8 and 9.10. They are used as ON/OFF switches to control motors, solenoid valves, and a light in Figure 9.20. They are used as electronic switches in a brushless dc motor in Figure 10.11. They are used in a full-wave bridge rectifier to control large dc motors in Figure 10.23. They are used in dc power amplifiers to control dc servo motors in Figures 2.6, 2.15, and 10.22. They are used in a phase discriminator to control the thickness of a sheet in Figure 2.20. Finally, they are used in a three-phase converter and inverter to control the speed of a large ac motor in Figure 10.18.

The purpose of this section is to describe the names, symbols, and operation of the solid-state components most frequently used in control. Table 9.1 gives the symbols and voltampere characteristics of these components.

Silicon-Controlled Rectifiers

One of the major uses of solid-state components is in switching circuits of all sizes. Although transistors are used in some switching circuits, power switching is the domain of the *silicon-controlled rectifier (SCR)* and the triac. The SCR has a number of useful characteristics. First, it is a rectifier—an SCR will conduct current in only one direction. Second, the SCR is a latching switch—the SCR can be turned ON by a short pulse of control current into the gate and it remains ON as long as current is flowing from the anode to the cathode. Third, the SCR has a very high gain—the anode current is about 3000 times as large as the control current. This means that a small amount of power in the turn-on circuit is used to control a large amount of power in the main circuit. Figure 9.6 shows the voltampere graph of an SCR.

When the anode-to-cathode voltage, V_{AC}, is negative, the SCR is said to be *reverse biased*. With a reverse bias, the anode current is negligible until the breakdown occurs—typically 50 to 2000 V. When the anode-to-cathode voltage is made positive, the anode current is

Name of Solid-State Component	Graphical Symbol	Voltampere Characteristic	Typical Application
Diode	$+$ V $-$; i	i ; V	Rectifier block bypass
N P N transistor	Collector i_C $+$; Base V_{CB} ; i_B $-$; Emitter	$i_C(+)$; i_{B_5} ; $i_B > 0$; i_{B_1} ; V_{CE} $(+)$	Amplifiers, switches, oscillators
P N P transistor	Collector i_C $+$; Base V_{CB} ; i_B $-$; Emitter	$(-)$ V_{CE} ; i_{B_1} ; $i_B < 0$; i_{B_5} i_C ; $(-)$	Amplifiers, switches, oscillators
Unijunction transitor (UJT)	Base 2 ; Emitter ; i_E $+$; V_{E-B_1} $-$ Base 1	V_P ; V_{E-B_1} ; i_E	Timers, oscillators, SCR trigger
Programmable unijunction transistor (PUT)	Anode ; $+$ i_A Gate ; V_{AC} ; $-$ Cathode	i_A ; V_{AC}	Timers, oscillators, SCR trigger
Silicon-controlled rectifier (SCR)	Anode ; i_A $+$; Gate V_{AC} ; $-$; Cathode	i_A ; V_{AC}	Power switches, inverters, frequency converters
Triac	Anode 2 ; i_2 $+$; Gate V_{21} ; $-$; Anode 1	i_2 ; V_{21}	Switches, relays
Diac	i $+$; V ; $-$		Triac and SCR trigger

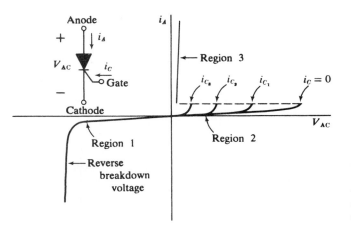

◆ **Figure 9.6** The voltampere characteristics of an SCR can be divided into three operating regions.

still negligible. However, when V_{AC} is positive and a small voltage is applied between the gate and cathode, the SCR switches ON. When the SCR is ON, it remains ON even if the triggering voltage is removed. The SCR has three operating regions.

Region 1. V_{AC} is negative and i_A is negligible.

Region 2. V_{AC} is positive and i_A is negligible.

Region 3. V_{AC} is positive and i_A is large; the SCR is ON.

There are four ways that an SCR can make the transition from region 2 to region 3. The first way is to increase the anode voltage until the forward breakdown voltage is reached and the device turns ON. The second way is by applying a positive voltage pulse to the gate input to "trigger" the SCR ON. A third method is by applying light to the gate/cathode junction. The light energy turns the SCR ON. This is called a LASCR (Light-Actuated Silicon Controlled Rectifier). The fourth method is to rapidly increase the anode-to-cathode voltage, V_{AC}. The rapid increase in voltage will also turn the SCR ON. The second method, a triggering voltage pulse at the gate, is usually used in control applications.

Once the SCR is turned ON, it will remain ON. There are three ways that an SCR can be turned OFF.

1. The anode current can be reduced below a minimum value called the *holding current*. The holding current is typically about 1% of the rated current. This causes a transition from region 3 to region 2. The SCR is still forward biased, but it is OFF instead of ON.
2. If the anode voltage is reversed (i.e., $V_{AC} < 0$), the SCR will go from region 3 to region 1. This method is used in ac circuits where the voltage reverses each half-cycle. It is also used in inverters and frequency converters where a charged capacitor is switched into the SCR circuit to force a reversal of the anode-to-cathode voltage.
3. In small current applications, the SCR can also be turned OFF by supplying a negative gate current to increase the holding current. When the increased holding current exceeds the load current, the SCR switches into region 2.

The second method, reversing V_{AC}, is the method most often used in control circuits.

One application of the SCR is the inverter circuit illustrated in Figure 9.7. An inverter is a device that converts the voltage from a dc source into an ac voltage. The dc source might be a

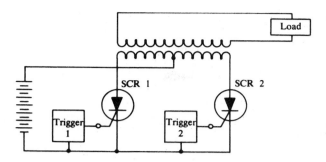

◆ **Figure 9.7** This center-tapped primary, SCR-controlled inverter converts a dc voltage into an ac voltage.

bank of storage batteries used to store electrical energy collected by solar cells or a wind generator. The ac voltage is needed to operate ac equipment such as fluorescent lights, appliances, and electric power tools. The circuit operation consists of the SCRs alternately switching the dc current from one half of the primary winding to the other half. The effect is equivalent to an alternating current in a single primary winding so that the secondary winding delivers an alternating current to the load. The term *commutation* is used to name the transfer of current from one SCR to the other SCR. Reliable commutation is one of the major problems in inverters. In Figure 9.7, additional commutation circuitry is required to turn OFF the SCRs by reverse biasing (method 2) each time the other SCR is turned ON by a triggering impulse.

Triacs

The *triac* was developed as a means of providing improved controls for ac power. The major difference between the triac and the SCR is that the triac can conduct in both directions, whereas the SCR can conduct in only one direction. The voltampere characteristic of the triac is shown in Table 9.1. A positive or negative gate current of sufficient amplitude will trigger the triac ON when V_{21} is either positive or negative. A typical triac circuit is shown in Figure 9.8. The trigger circuit produces pulses that turn the triac ON. The first trigger pulse turns the triac on during the positive half-cycle. The triac remains on until the ac voltage reverses and turns the triac OFF. The next triggering pulse turns the triac on during the negative half-cycle.

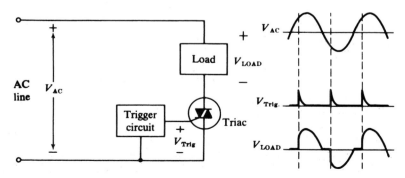

◆ **Figure 9.8** Basic triac circuit for control of an ac load.

The triac again remains ON until the ac voltage reverses and turns the triac OFF. The trigger circuit determines when the trigger pulse will turn the triac ON. This in turn determines how much current is delivered to the load.

Unijunction Transistors

A number of devices are used to produce the trigger pulse for SCRs and triacs. The *unijunction transistor* (UJT) is one device used in the triggering circuit. A basic unijunction trigger circuit and its waveforms are shown in Figure 9.9. In this circuit, the capacitor is charged by the current through R_3 until the emitter voltage reaches V_p (see Table 9.1) and the UJT turns on. This discharges the capacitor through R_2, producing the trigger spike in V_G. When the emitter voltage reaches 2 V, the UJT turns OFF and the cycle is repeated. The values of R_3 and C determine the time between the triggering pulses.

Figure 9.10 shows an example of a UJT used as an SCR trigger circuit. The circuit operates as follows:

1. The ac input is applied to the load in series with the SCR. The load voltage remains at zero until the SCR is switched on by a gate pulse.
2. Voltage e_{in} is also applied to the zener diode clipper circuit (R_S and V_Z). The zener diode breaks down at V_Z V and thus limits or clips the positive peaks of e_{in} (see the v_{xy} waveform in the figure). During the negative half-cycle of e_{in}, the zener is forward biased and maintains a near 0 voltage between x and y.
3. The clipped sine wave, v_{xy}, is the supply voltage for the UJT circuit. During the positive half-cycle of e_{in} while v_{xy} is at $+V_Z$ V, the capacitor charges through R until it reaches V_P of the UJT. When it does, the UJT turns ON and discharges C, producing a positive pulse across R_1 (see v_{B1} waveform). This pulse is fed to the gate of the SCR and turns ON the SCR.
4. Once the SCR is ON, the load voltage becomes approximately equal to e_{in} for the duration of the positive half-cycle. (See the v_{load} waveform.)
5. During the negative half-cycle, the SCR stays OFF and v_{load} stays at zero.
6. The amount of power delivered to the load is controlled by varying the RC time constant, which causes C to charge slower or faster, thereby triggering the UJT and SCR later or earlier in the positive half-cycle of e_{in}. In other words, the RC time constant

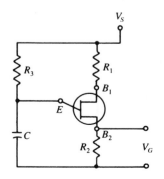
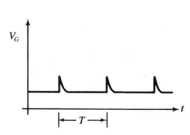

◆ **Figure 9.9** UJT trigger circuit.

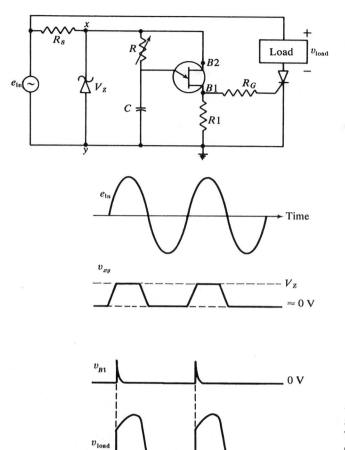

◆ **Figure 9.10** A UJT is used to trigger an SCR in this ac power control circuit. [From R. Tocci, *Fundamentals of Electronic Devices,* 4th ed. (Columbus, Ohio: Merrill/Macmillan, 1991), p. 592.]

controls the SCR trigger time and therefore the load power. As *RC* is increased, the trigger time increases and the load decreases.[*]

Diodes

Diodes are used as rectifiers in dc power supplies and as one-way "valves" to block or bypass undesired electric currents. The voltampere characteristic of the diode is shown in Figure 9.11. The diode has three operating regions depending on the anode-to-cathode voltage (*V*). If the anode-to-cathode voltage is positive, the diode is said to be forward biased; if the voltage is negative, the diode is reverse biased. In operating region 1, the diode is forward biased and conducts electric current with very low resistance. In region 2, the diode is reversed biased and blocks electric current with very high resistance. In region 3, the reverse bias has increased to the point of breakdown and the diode conducts electric current. For most applica-

---

[*]R. Tocci, *Fundamentals of Electronic Devices,* 4th ed. (Columbus, Ohio: Merrill/Macmillan, 1991), p. 592.

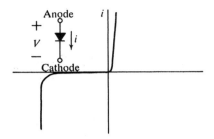

◆ **Figure 9.11** Voltampere characteristic of a diode.

tions, operation in region 3 is avoided. An exception to this rule is the zener diode, which is operated in region 3 as a voltage regulator.

A full-wave dc power supply is illustrated in Figure 9.12. The diodes are used as one-way switches that pass current only when they are forward biased. The center-tapped transformer configuration is such that the two diodes are forward biased by opposite polarities of the ac input voltage (V_{in}). When V_{in} is positive, diode D_1 is forward biased, and when V_{in} is negative, diode D_2 is forward biased. Therefore, diode D_1 conducts current during each positive half-cycle of the input voltage, and diode D_2 conducts current during each negative half-cycle. The output of each diode is connected to output terminal a, producing the full-wave rectified output voltage shown in Figure 9.12. A capacitor is usually placed across terminals a and b to produce a more constant output voltage.

Transistors

Transistors are used in control systems as switches, amplifiers, and oscillators. There are two types of transistors, the NPN type and the PNP type. The graphic symbols and voltampere characteristics of both types are included in Table 9.1. The operating characteristics of NPN and PNP transistors are the same except that the directions of the voltages and currents are opposite. The following discussion is confined to the NPN transistor.

An NPN transistor has three terminals called the collector, base, and emitter, as shown in

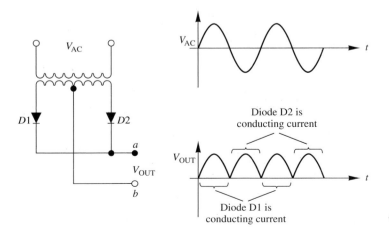

◆ **Figure 9.12** Full-wave dc power supply.

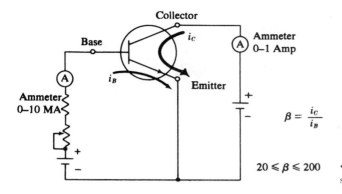

$$\beta = \frac{i_C}{i_B}$$

$$20 \leqslant \beta \leqslant 200$$

◆ **Figure 9.13**   Test circuit for an NPN transistor.

Figure 9.13. Transistors are essentially controlled-current amplifiers. A relatively small current entering the base is used to control a much larger current entering through the collector. Both currents leave through the emitter. The ratio of the collector current (i_C) to the base current (i_B) is called the β (beta) of the transistor ($\beta = i_C/i_B$). The β of a transistor ranges from 20 to 200, and it varies from one transistor to another, even if they are of the same type. Aging and changes in temperature may also cause the β to change.

A transistor may be used in one of three different amplifier configurations, depending on which terminal is common to the input and the output. The three configurations and a summary of their characteristics are shown in Figure 9.14.

The input/output graphs of a typical common-emitter amplifier are shown in Figure 9.15. Notice the signal inversion in both graphs (i.e., as the input voltage or current increases, the output voltage or current decreases). The voltage gain is determined as follows. Select two points on the straight portion of the graph that are spaced as far apart as possible. Measure the

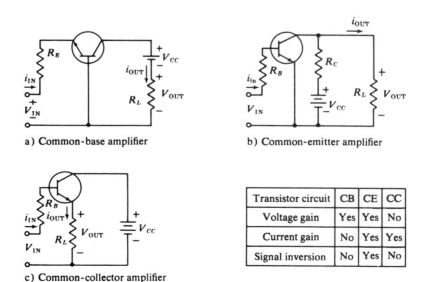

a) Common-base amplifier

b) Common-emitter amplifier

c) Common-collector amplifier

Transistor circuit	CB	CE	CC
Voltage gain	Yes	Yes	No
Current gain	No	Yes	Yes
Signal inversion	No	Yes	No

◆ **Figure 9.14**   Configuration and characteristics of transistor amplifiers.

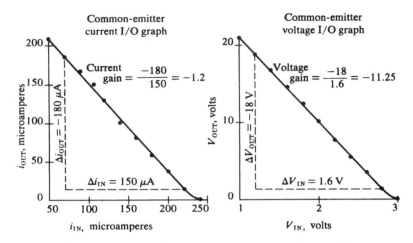

◆ **Figure 9.15** Input/output graphs of a common-emitter amplifier.

change in the input voltage (ΔV_{in}) and the change in the output voltage (ΔV_{out}) between the two points. The voltage gain is the change in output voltage divided by the change in input voltage, $\Delta V_{out}/\Delta V_{in}$. The current gain is determined in a similar manner.

In the example illustrated in Figure 9.15, the voltage gain is -11.25 and the current gain is -1.2. The negative sign is due to the signal inversion. This signal inversion is a characteristic of the common-emitter amplifier. The inversion is eliminated when two common-emitter amplifiers are connected in series, because the inverted output of the first stage is inverted again by the second stage. The overall voltage gain of a two-stage amplifier is the product of the voltage gains of the two stages. The same is true for the current gain. For example, if stage 1 has a voltage gain of -12 and stage 2 has a voltage gain of -10, the overall voltage gain is $(-12)(-10) = 120$.

The input/output graphs of a typical common-collector amplifier are shown in Fig-

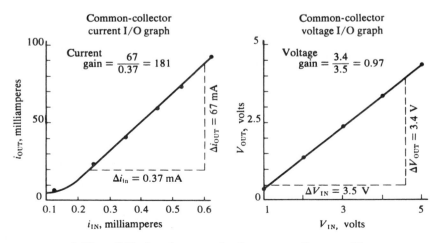

◆ **Figure 9.16** Input/output graphs of a common-collector amplifier.

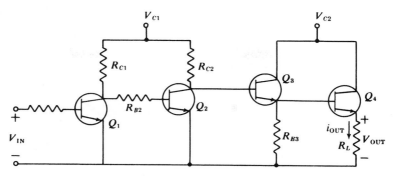

◆ **Figure 9.17** Four-stage transistor amplifier.

ure 9.16. The voltage gain of this amplifier is 0.97, and the current gain is 181. Notice the absence of the signal inversion that characterized the common-emitter amplifier. This direct relationship between input and output signals and the absence of voltage gain are characteristics of the common-collector amplifier. The common-collector amplifier is a current amplifier.

The common-emitter and common-collector amplifiers may be combined to form multistage amplifiers with increased voltage and current gains. A four-stage amplifier is shown in Figure 9.17. Transistors Q_1 and Q_2 form common-emitter amplifiers for stages 1 and 2. Transistors Q_3 and Q_4 form common-collector amplifiers for stages 3 and 4.

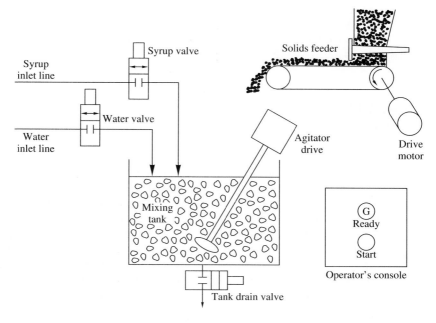

◆ **Figure 9.18** Industrial blending process.

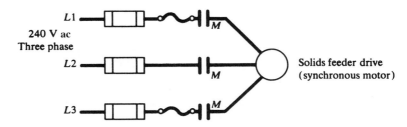

◆ **Figure 9.19** Power circuit for the solids feeder drive motor in the blending process in Figure 9.18.

EXAMPLE 9.1

A schematic diagram of an industrial process is shown in Figure 9.18. Two-position solenoid valves are used to control the water and syrup flows into the tank. A third solenoid valve is used to control the flow out of the tank. The solids feeder is driven by a synchronous motor, and a universal motor is used for the agitator drive. The operator's console completes the main components of the process.

The main contacts for starting the solids feeder drive motor are shown in Figure 9.19. The control circuit for energizing the solids feeder starter coil is shown in Figure 9.20. Triac 4 in Figure 9.20 is triggered by a dc signal controlled by bit 2 of the output port of the ICST-1 microcomputer. Triac 4 is ON when bit 2 is a logic 1 and it is OFF when bit 2 is a logic 0. Transistor T3 acts as a solid-state switch to control the triggering of the triac. Similar triac circuits are used to control the solenoid valves and the ready light.

The agitator drive motor is controlled by triac 1 in Figure 9.20. Variable resistor R_2, capacitor C_2, and the diac make up the triggering circuit. Resistor R_2 and capacitor C_2 form a variable time constant circuit that triggers triac 1 when the voltage across C_2 reaches the breakover voltage of the diac. Resistor R_1 and capacitor C_1 reduce the transient voltages developed during switching due to the inductive energy stored in the motor windings. The diac triggering circuit provides a smooth control of motor speed from an intermediate value up to full speed. Other triggering techniques are used when the full control range is required.

The control circuit shown in Figure 9.20 is simpler than the actual circuits used in an industrial process, but it illustrates some of the variety of applications of solid-state components in control systems. ◆

9.3 HYDRAULIC AND PNEUMATIC VALVES AND ACTUATORS

Actuators are an important part of every control loop. They provide the power needed to position the final control element. Hydraulic and pneumatic actuators often provide the most economical and trouble-free method for positioning the final control element. *Hydraulic actuators* are used where slow, precise positioning is required, or where heavy loads must be moved. Hydraulic actuators operate at higher pressures than pneumatic actuators, so they can provide larger forces from a given-size cylinder or motor. *Pneumatic actuators* are used where lighter loads are encountered, where an air supply is available, and where fast response is required. Pneumatic actuators are less expensive than hydraulic actuators, and they are capable of much faster movement. The speed of a hydraulic actuator is limited by the maximum flow rate provided by the hydraulic pump. Pneumatic diaphragm operators are used in most control valves, a topic covered in Section 9.4. In this section we cover hydraulic and pneumatic valves, cylinders, and motors.

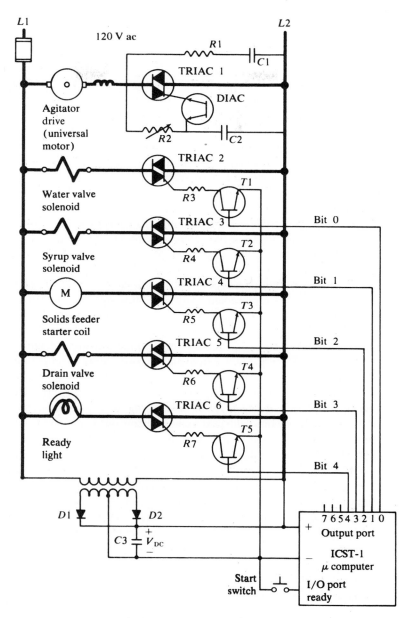

◆ **Figure 9.20** Control circuit for the blending process in Figure 9.18.

Solenoid Valves

Solenoid valves are used to control fluid flow in hydraulic or pneumatic systems. A *solenoid valve* consists of a movable spool fitted inside a housing with one or two solenoids capable of moving the spool to different positions relative to the housing. Both the spool and the housing have fluid passages. The passages in the housing are connected together or blocked in different ways, depending on the position of the spool in the housing. A valve with one solenoid has two spool positions, one position when the solenoid is deenergized, the other position when the solenoid is energized. A valve with two solenoids has three spool positions, one position when both solenoids are deenergized, a second position when solenoid *a* is energized, and a third when solenoid *b* is energized.

Each passage in the housing may be blocked, or it may be connected to the supply line, to the return line, or to a line from a cylinder. Each such connection is called a *way*. A solenoid valve is characterized by the number of positions the valve has and by the number of connecting lines (ways) it has. Thus a valve with one solenoid and two connecting lines is called a two-position, two-way valve. A valve with two solenoids and four connecting lines is called a three-position, four-way valve.

Symbols of common types of solenoid valves are illustrated in Figure 9.21. In the diagram, the spool is represented by two or three squares, one square for each position of the valve. Each connecting line or way is indicated by a line outside of the squares that represent the spool. The passages in the spool are indicated by lines within the squares. Blocked lines indicate blocked passages. Arrows indicate the direction of fluid flow. The spool is always shown in the deenergized position, and the two, three, or four connecting lines (ways) are attached to the square that is in effect when the valve is deenergized.

Figure 9.21a illustrates the operation of a two-position valve. The solid-line diagram on top shows the valve in the deenergized position, the way it would be shown in a schematic diagram. Notice that the spool square next to the spring is lined up with the three connecting lines. In this position, the cylinder is connected to the return or exhaust line, and the supply line is blocked. The dashed-line diagram in the bottom of Figure 9.21a shows the valve in the energized position. In this position, the supply line is connected to the cylinder and the return line is blocked.

Figure 9.21b illustrates the operation of a three-position valve. The solid-line diagram on top shows the valve in the deenergized position. Notice that the two cylinder lines are blocked and the supply line is connected to the return line. The middle dashed-line diagram shows the valve in the position it will be in when solenoid *a* is energized. Here the supply line is connected to the left cylinder line, and the right cylinder line is connected to the return line. The lower dashed-line diagram shows the valve in the position it will be in when solenoid *b* is energized. In this position, the supply line is connected to the right cylinder line, and the left cylinder line is connected to the return line.

Cylinders

Hydraulic and *pneumatic cylinders* are used to produce a linear motion with a definite maximum distance of travel. Cylinders may be single acting or double acting, as illustrated in Figure 9.22. Single-acting cylinders allow fluid on only one side of the cylinder. The piston is returned to the starting position by gravity or by a return spring. Double-acting cylinders

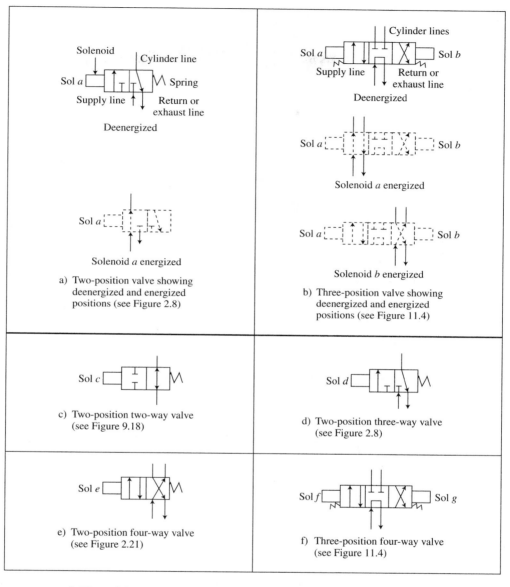

◆ **Figure 9.21** Drawing symbols for solenoid valves used for hydraulic and pneumatic systems.

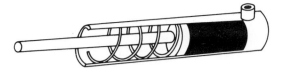

a) Single-acting spring-return cylinder

b) Double-acting cylinder

◆ **Figure 9.22**   Hydraulic and pneumatic cylinders produce large forces for linear movement of a load or final control element.

allow fluid on both sides of the cylinder, providing a power return of the piston. Table 9.2 lists the theoretical forces and speeds for typical hydraulic and pneumatic cylinders.

Motors

Hydraulic and *pneumatic motors* develop large torques to produce rotational motion in a load or final control element. The four most common types of motor are the piston motor, the gear motor, the rotary vane motor, and the turbine motor. Figure 9.23 illustrates these four types of motor.

The gear motors are used only for hydraulic fluids. The turbine motor is used only for pneumatic systems. The piston and rotary motors are used for both hydraulic and pneumatic systems.

◆ **TABLE 9.2**   Theoretical Force and Speed for Hydraulic and Pneumatic Cylinders

| Cylinder Bore (in.) | Force (lb) | | Speed (in./min) |
	Hydraulic (3000 psi)	Pneumatic (100 psi)	Hydraulic (1 gal/min)
1.5	5,301	177	131
2.0	9,425	314	74
3.0	21,206	707	33
4.0	37,699	1257	18
5.0	58,905	1963	12
6.0	84,823	2827	8.2
8.0	150,796	5027	4.6
10.0	235,619	7854	2.9

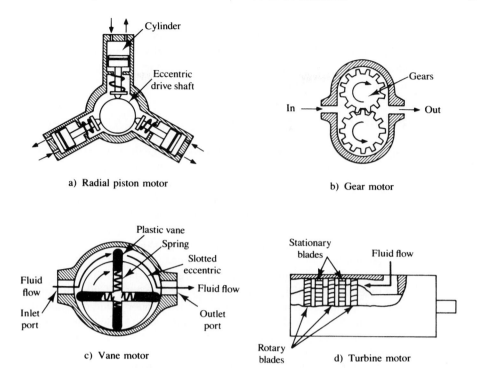

a) Radial piston motor

b) Gear motor

c) Vane motor

d) Turbine motor

◆ **Figure 9.23** The piston- and vane-type motors are used in both hydraulic and pneumatic systems. The gear motor is used only in hydraulic systems, and the turbine motor is used only in pneumatic systems.

The *radial piston motor* has pistons that radiate out from an eccentric drive shaft. Fluid enters the cylinder and forces the piston against the eccentric thrust ring, causing the shaft to rotate. Radial piston motors are usually limited to speeds below 2000 rpm.

Gear motors are the most popular hydraulic motor. The entering fluid causes the two gears to rotate in the direction indicated. The gears carry the fluid from the inlet port to the outlet port. The output of the gear motor is delivered by a shaft attached to one of the gears. The hydraulic fluid moves the gears, the output gear moves the shaft, and the shaft moves the load. Gear motors can operate at speeds up to 3000 rpm.

The *vane motor* is basically the same for hydraulic and pneumatic systems, although the motors are not interchangeable. The vane motor consists of a slotted eccentric, with spring-loaded plastic vanes. As fluid enters the motor, it applies force against the vanes, with more force against the vane that is extended. The unbalanced force on the extended vane causes the eccentric shaft to rotate. The fluid is trapped between two vanes and is carried from the inlet port to the outlet port. Hydraulic vane motors can operate at speeds up to 4000 rpm, and pneumatic vane motors can operate up to about 6500 rpm.

The *turbine motor* consists of several sets of rotating blades separated by rows of stationary blades. The rotating blades are attached to the shaft, which is also connected to the load. The fluid enters the motor and flows through the turbine blades, producing a force on the rotary blades that tends to rotate the shaft. Notice that the rotary blades are canted and the stationary blades are straight. The canted blades deflect the fluid, producing a torque that tends to rotate the shaft. The shaft transfers this torque to the load. The stationary blades straighten the fluid flow for the next set of rotary blades. Turbine motors have low starting torque but are capable of very high speeds.

Selection of a Cylinder

The first step in selecting a cylinder is to determine the force required to move the load. The second step is to determine the speed required by the application. Other considerations include the choice between hydraulic and pneumatic systems, the working fluid pressure, the flow rate (hydraulic systems), the valve size, and the oversize factor. The *oversize factor* is a number greater than 1 that is used to increase the size of the cylinder to provide the additional force necessary to accelerate the load. One supplier of hydraulic and pneumatic cylinders recommends selecting a cylinder with 25% extra force for slow-moving loads and 100% extra force for fast-moving loads.

Hydraulic cylinders are capable of much larger forces than pneumatic cylinders. However, pneumatic cylinders are capable of greater speeds than hydraulic cylinders. The speed of a hydraulic cylinder is determined by the flow rate of the fluid (see Table 9.2). The speed of an air cylinder cannot be calculated. Air cylinder speed depends on the force available to accelerate the load and the rate at which air can be vented ahead of the advancing piston. In fact, an adjustable valve on the cylinder vent port is one method of regulating the speed of a pneumatic cylinder.

The force produced by a cylinder is equal to the area of the cylinder times the working pressure in the cylinder.

$$\text{Force} = (\pi d^2 / 4)\,(\text{working pressure}) \tag{9.1}$$

Table 9.2 gives the force produced by working pressures of 100 and 3000 psi in cylinders with diameters ranging from 1.5 to 10 in. A very common design calculation is the computation of the *working pressure* required to produce a given force in a cylinder with a given diameter. The designer may use Equation (9.1) for this computation, but there is a simpler way based on Table 9.2 and the law of proportions. We now use Equation (9.1) to develop the equation for this simpler method. Here we let f_1 and WP_1 represent the force and working pressure from Table 9.2. We want the working pressure (WP_2) required to produce the given force (f_2). Using Equation (9.1), we get the following two equations:

$$f_1 = \frac{\pi d^2}{4}\,WP_1$$

and

$$f_2 = \frac{\pi d^2}{4}\,WP_2$$

Now we divide the second equation by the first:

$$\frac{f_2}{f_1} = \frac{(\pi d^2/4)\,WP_2}{(\pi d^2/4)\,WP_1} = \frac{WP_2}{WP_1}$$

Finally, multiply both sides of the equation by WP_1.

$$WP_2 = WP_1 \frac{f_2}{f_1} \tag{9.2}$$

The speed of a hydraulic cylinder is equal to the flow rate in cubic inches per minute divided by the area of the cylinder in square inches. Equation (9.3) gives the speed in inches per minute for a flow rate of Q gallons per minute and a cylinder diameter of d inches (1 gal = 231 in^3).

$$\text{Speed} = \frac{231Q}{\pi d^2/4} \tag{9.3}$$

Table 9.2 also gives the speed of hydraulic cylinders for a flow rate of 1 gal/min. Another common design calculation is the computation of the flow rate required to produce a given speed in a cylinder with a given diameter. The designer may use Equation (9.3) for this computation, but again there is a simpler method, which we develop as follows. Let s_1 and Q_1 represent the speed in Table 9.2 ($Q_1 = 1$). We want the flow rate (Q_2) required to produce the given speed (s_2). Using a development similar to the one above, we obtain the following design equation:

$$Q_2 = Q_1 \frac{s_2}{s_1} \tag{9.4}$$

The selection of the working pressure is interrelated with the size and cost of the cylinder and operating considerations such as leakage costs and pump costs. Trade-offs are involved. For example, a higher working pressure will permit a smaller cylinder and a lower flow rate to achieve a specified force and speed. The downside is that higher working pressures result in higher leakage rates and higher noise levels. Example 9.2 illustrates the process of selecting a cylinder for a given application.

EXAMPLE 9.2   A manufacturing operation requires the movement of a workpiece a distance of 10 in. in 15 s. A force of 10,000 lb is required to move the workpiece. The available space limits the cylinder diameter to a maximum of 5 in. Leakage and noise considerations make it desirable to limit the working pressure to a maximum of 1000 psi. A hydraulic pump with capacity of 2.5 gal/min is available in the surplus equipment yard. Select a cylinder for this application.

Solution   The required force is 10,000 lb. The required speed is 10/0.25 = 40 in./min. The speed is relatively slow, so a 25% oversize factor will be used.

$$\text{Required cylinder force} = 1.25 \times 10,000 = 12,500 \text{ lb}$$

The force and diameter requirements dictate a hydraulic cylinder. Three cylinder sizes will be considered: 3, 4, and 5 in.

For a 3-in.-diameter cylinder:

From Table 9.2, WP_1 = 3000-psi pressure and f_1 = 21,206 lb for a 3-in.-diameter cylinder. From Equation (9.2), the pressure required to produce a force of 12,500 lb is

$$WP = (3000)\left(\frac{12,500}{21,206}\right) = 1768 \text{ psi}$$

From Table 9.2, a flow rate of 1 gal/min produces a speed of 33 in./min in a 3-in.-diameter cylinder. From Equation (9.4), the flow rate required to produce a speed of 40 in./min is

$$Q = (1)\left(\frac{40}{33}\right) = 1.2 \text{ gal/min}$$

For a 4-in.-diameter cylinder:

From Table 9.2, 3000 psi produces a force of 37,699 lb in a 4-in. cylinder. The required working pressure is again given by Equation (9.2):

$$WP = (3000)\left(\frac{12,500}{37,699}\right) = 995 \text{ psi}$$

From Table 9.2, a flow rate of 1 gal/min produces a speed of 18 in./min in a 4-in.-cylinder. The required flow rate is given by Equation (9.4):

$$Q = (1)\left(\frac{40}{18}\right) = 2.2 \text{ gal/min}$$

For a 5-in.-diameter cylinder the required working pressure is

$$WP = (3000)\left(\frac{12,500}{58,905}\right) = 637 \text{ psi}$$

The required flow rate is

$$Q = (1)\left(\frac{40}{12}\right) = 3.3 \text{ gal/min}$$

Conclusion: The 4-in. cylinder is selected for this application. The required flow rate of 2.2 gal/min can be provided by the surplus pump. The working pressure of 995 psi is within the 1000-psi limit, and the 4-in.-diameter cylinder fits in the available space.   ◆

9.4 CONTROL VALVES

In many process control systems, the manipulated variable is the flow rate of a fluid, and the pneumatic *control valve* is the most common final control element. A typical pneumatic control valve is illustrated in Figure 9.24. The input to the valve is a 3- to 15-psi air pressure signal, which is applied to the top of the diaphragm. The diaphragm actuator converts the air pressure into a displacement of the valve stem. The valve body and trim vary the area through which the flowing fluid must pass.

The input air pressure signal may come directly from a pneumatic controller; it may come from a pneumatic valve positioner; or it may come from an electropneumatic transducer. A valve positioner is a pneumatic amplifier that provides additional power to operate the valve. The electropneumatic transducer converts the milliampere signal from an electronic controller into the 3- to 15-psi air pressure signal required by the control valve.

The diaphragm actuator consists of a synthetic rubber diaphragm and a calibrated spring.

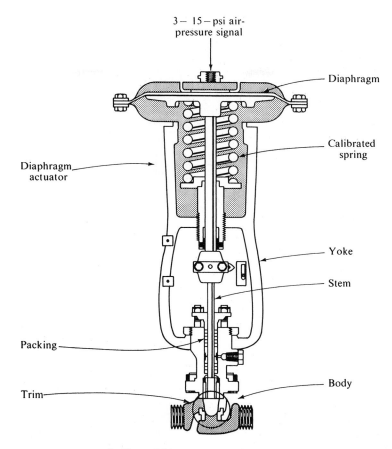

3 − 15 − psi air-
pressure signal

Diaphragm

Diaphragm
actuator

Calibrated
spring

Yoke

Stem

Packing

Body

Trim

◆ **Figure 9.24**   Pneumatic control valve.

The air pressure on the top of the diaphragm is opposed by the spring. Full travel of the valve stem is obtained with a change in the air pressure signal from 3 to 15 psi. The graph of the valve stem displacement versus the air pressure signal is nearly linear. The actuator may be described as a *direct actuator* or a *reverse actuator,* depending on the location of the calibrated spring. The two types of actuator are illustrated in Figure 9.25.

A yoke attaches the diaphragm actuator to the valve body. The valve stem passes through a packing, which allows motion to the valve stem and prevents leakage of the fluid flowing around the stem. The trim consists of the valve plug and the valve seat or port. These two parts are often produced in matched pairs to provide tight shutoff.

The valve position (P) is the displacement of the plug from the fully seated position. The valve action is the direction in which the valve plug moves as the air pressure signal increases. If increasing air pressure closes the valve, the valve action is described as air-to-close. If increasing air pressure opens the valve, the action is described as air-to-open. The variations of valve action and actuator type are illustrated in Figure 9.26.

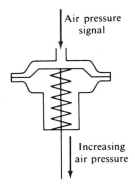

Air pressure
signal

Increasing
air pressure

a) Direct actuator

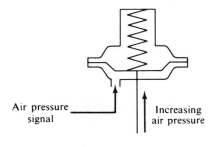

Air pressure
signal

Increasing
air pressure

b) Reverse actuator

◆ **Figure 9.25** Direct- and reverse-action valve actuators.

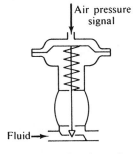

Air pressure
signal

Fluid →

a) Direct acting air-to-close

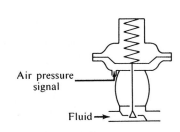

Air pressure
signal

Fluid →

b) Reverse acting air-to-close

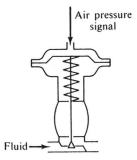

Air pressure
signal

Fluid →

c) Direct acting air-to-open

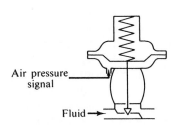

Air pressure
signal

Fluid →

d) Reverse acting air-to-open

◆ **Figure 9.26** Four combinations of valve action (air-to-open or air-to-close) and actuator type (direct or reverse).

Valve Characteristics

The characteristic of a control valve is the relationship between the valve position and the flow rate through the valve. The *inherent characteristic* of a valve is obtained when there is a constant pressure drop across the valve for all valve positions. The *installed characteristic* is obtained when the valve is installed in a system and the pressure drop across the valve depends on the pressure drop in the remainder of the system. Thus the installed characteristic depends on both the inherent characteristic of the valve and the flow characteristic of the system in which the valve is installed.

The three most common valve characteristics—quick opening, linear, and equal percentage—are illustrated in Figure 9.27.

The *quick-opening characteristic* provides a large change in flow rate for a small change in valve position. This characteristic is used for ON/OFF or two-position control systems in which the valve must move quickly from open to closed, or vice versa. A typical quick-opening characteristic is shown in Figure 9.27.

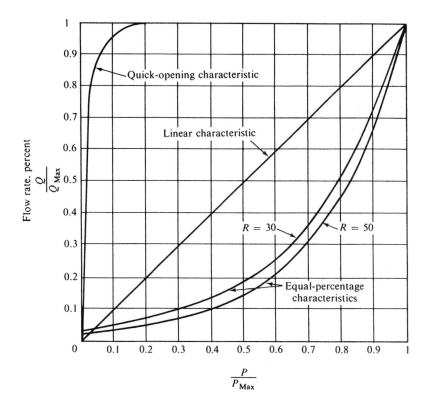

Valve position, percent

◆ **Figure 9.27**   Three types of inherent characteristics for control valves.

A *linear characteristic* provides a linear relationship between the valve position (P) and the flow rate (Q). The linear characteristic is described by the following mathematical relationship.

$$\frac{Q}{Q_{max}} = \frac{P}{P_{max}} \tag{9.5}$$

where $\quad Q$ = valve flow rate

$\quad\quad P$ = valve position (displacement of the plug from the closed position)

$\quad\quad P_{max}$ = maximum valve position

$\quad\quad Q_{max}$ = maximum flow rate

The flow rate and valve position are usually given in the English units of gallons per minute and inches. However, any convenient units may be used. The values of Q_{max} and P_{max} are usually included in the valve manufacturer's literature. The linear characteristic is usually used in installed systems in which most of the system pressure drop is across the valve.

An *equal-percentage characteristic* provides equal percentage changes in flow rate for equal changes in valve position. An equal-percentage valve is designed to operate between a minimum flow rate (Q_{min}) and a maximum flow rate (Q_{max}). The ratio Q_{max}/Q_{min} is called the *rangeability* of the valve, R.

$$R = \frac{Q_{max}}{Q_{min}} \tag{9.6}$$

Most commercial control valves have a rangeability between 30 and 50. The simplest way to express the equal-percentage characteristic is in terms of the change in flow rate, corresponding to a small change in valve position.

$$\frac{Q_2 - Q_1}{Q_1} = K\frac{P_2 - P_1}{P_{max}} \tag{9.7}$$

where $\quad Q_1$ = initial flow rate

$\quad\quad P_1$ = initial valve position

$\quad\quad Q_2$ = final flow rate

$\quad\quad P_2$ = final valve position

$\quad\quad P_{max}$ = maximum valve position

The value of the constant K depends on the rangeability and the relative size of the change in valve position, $(P_2 - P_1)/P_{max}$. For example, if $R = 50$ and $(P_2 - P_1)/P_{max} = 0.01 = 1\%$, then $K = 4$. In other words, a 1% change in valve position will produce a 4% change in flow rate. If the flow rate is 3 gal/min when the valve is 10% open, a change in valve position to 11% will increase the flow rate by 4% of 3, or 0.12 gal/min. When the valve is 50% open, the flow rate will be 14.2 gal/min. A change in valve position to 51% open will increase the flow rate by 4% of 14.2, or 0.57 gal/min.

The exact mathematical relationship for the equal percentage valve is given by the equation

$$\frac{Q}{Q_{max}} = \left(\frac{Q_{max}}{Q_{min}}\right)^{(P - P_{max})/P_{max}} = R^{(P - P_{max})/P_{max}} \tag{9.8}$$

where Q = valve flow rate

Q_{max} = maximum flow rate

Q_{min} = minimum flow rate

R = rangeability of the valve

P = valve position

P_{max} = maximum valve position

Any convenient units of flow rate and position may be used in Equation (9.8).

The *installed characteristics* of a valve depend on how much of the total system pressure drop is across the valve. If 100% of the system pressure drop is across the valve, the installed characteristic is the same as the inherent characteristic. The installed characteristics of linear and equal-percentage valves are illustrated in Figure 9.28. As the percentage pressure drop across the valve decreases, the installed characteristic of the linear valve approaches the

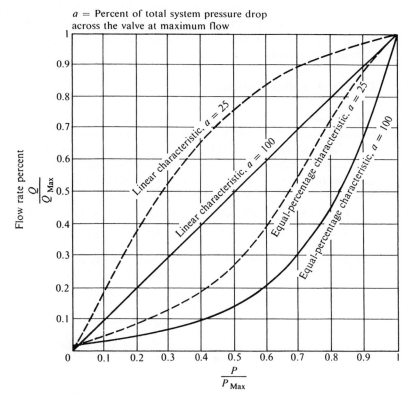

◆ **Figure 9.28**   As the percentage of total system pressure drop across the valve at maximum flow (a) decreases, the linear valve approaches a quick-opening characteristic, and the equal-percentage valve approaches a linear characteristic.

◆ **TABLE 9.3** Approximate C_v Values for Common Valve Sizes

Valve Size	C_v	Valve Size	C_v
0.25	0.3	3	108
0.5	3	4	174
1	14	6	400
1.5	35	8	725
2	55	10	1100

quick-opening characteristic, and the installed characteristic of the equal-percentage valve approaches the linear characteristic.

The proper selection of a control valve involves matching the valve characteristics to the characteristics of the process. When this is done, the control valve will contribute to the stability of the control system. Matching the valve characteristics to a particular system requires a complete dynamic analysis of the system. When a complete dynamic analysis is not justified, an equal-percentage characteristic is usually specified if less than half of the system pressure drop is across the control valve. If most of the system pressure drop is across the valve, a linear characteristic may be preferred.

Control Valve Sizing

Control valve sizing refers to the engineering procedure of determining the correct size of a control valve for a specific installation. A capacity index called the *valve flow coefficient* (C_v) has greatly reduced the difficulty in "sizing" a control valve. The valve flow coefficient is defined as the number of U.S. gallons of water per minute that will flow through a wide-open valve with a pressure drop of 1 psi. For example, a control valve with a C_v of 5 will pass 5 gal/min of water when the valve is wide open and the pressure drop is 1 psi.

The valve-sizing formulas (C_v formulas) are based on the basic equation of liquid flow.

$$Q = K\sqrt{\text{pressure drop}}$$

The following formulas may be used to determine the maximum flow rate for different flowing fluids. The approximate C_v values for standard valve sizes are given in Table 9.3.

C_v FORMULAS

$$\text{Liquids: } Q_L = C_v\sqrt{\frac{P_1 - P_2}{G_L}} \tag{9.9}$$

$$\text{Gases: } Q_g = 960C_v\sqrt{\frac{(P_1 - P_2)(P_1 + P_2)}{G_g(T + 460)}} \tag{9.10}$$

$$\text{Steam: } W = 90C_v\sqrt{\frac{P_1 - P_2}{V_1 + V_2}} \tag{9.11}$$

where C_v = valve flow coefficient

G_g = gas specific gravity (density of the gas divided by the density of air with both gases at standard conditions)

G_L = liquid specific gravity (density of the liquid divided by the density of water)

W = steam flow, lb/hr

P_1 = valve inlet pressure, psia

P_2 = valve outlet pressure, psia

Q_g = gas flow rate, ft^3/hr at 14.7 psia and 60°F

Q_L = liquid flow rate, gal/min

T = gas temperature, °F

V_1 = specific volume of the steam at the valve inlet, ft^3/lb

V_2 = specific volume of the steam at the valve outlet, ft^3/lb

EXAMPLE 9.3

Determine the size of the control valve required to control the flow rate of a liquid with a specific gravity of 0.92. The maximum flow rate is 320 gal/min. The available pressure drop across the valve at maximum flow is 60 psi. Allow a safety factor of 25% (i.e., $Q_L = 1.25Q_{max}$).

Solution

Equation (9.9) may be used:

$$Q_L = C_v \sqrt{\frac{P_1 - P_2}{G_L}}$$

$$(1.25)(320) = C_v \sqrt{\frac{60}{0.92}}$$

$$400 = 8.08 C_v$$

$$C_v = 49.5 \text{ minimum}$$

Conclusion: A 2-in. valve is required (Table 9.3).   ◆

EXAMPLE 9.4

Determine the size of the control valve required to control the flow rate of a gas with a specific gravity of 0.85. The maximum flow rate is 20,000 standard cubic feet per hour (14.7 psia and 60°F). The gas temperature, inlet pressure, and outlet pressure are 500°F, 90 psia, and 70 psia, respectively. Allow a safety factor of 25%.

Solution

Equation (9.10) may be used.

$$Q_g = 960 C_v \sqrt{\frac{(P_1 - P_2)(P_1 + P_2)}{G_g(T + 460)}}$$

$$(1.25)(20,000) = 960 C_v \sqrt{\frac{(20)(160)}{(0.85)(960)}}$$

$$25,000 = 1901 C_v$$

$$C_v = 13.15$$

Conclusion: A 1-in. valve is required (Table 9.3).   ◆

EXAMPLE 9.5

Determine the size of the control valve required to control the flow rate of steam to a process. The maximum flow rate is 600 lb/hr. The inlet and outlet steam conditions are given below.

$$P_1 = 40 \text{ psia}$$
$$V_1 = 10.5 \text{ ft}^3/\text{lb}$$
$$P_2 = 30 \text{ psia}$$
$$V_2 = 14.0 \text{ ft}^3/\text{lb}$$

Use a safety factor of 25%.

Solution

Equation (9.11) may be used.

$$W = 90C_v\sqrt{\frac{P_1 - P_2}{V_1 + V_2}}$$

$$(1.25)(600) = 90C_v\sqrt{\frac{10}{24.5}}$$

$$C_v = 13.0$$

Conclusion: A 1-in. valve is required (Table 9.3). ◆

Control Valve Transfer Function

A pneumatic control valve is essentially a second-order system, usually underdamped or critically damped. The resonant frequency (ω_0) is usually between 1 and 10 rad/s. The transfer function is given by Equation (9.12).

CONTROL VALVE TRANSFER FUNCTION

$$\frac{P}{I} = \frac{G}{1 + (2\zeta/\omega_0)s + (1/\omega_0^2)s^2} \qquad \textbf{(9.12)}$$

where G = valve gain, in./psi
   I = input air pressure signal, psi
   P = valve position, in.
   s = frequency parameter, 1/s
   ω_0 = resonant frequency, rad/s
   ζ = damping ratio

The second-order model presented here is adequate for small-signal changes. For large-signal changes, the model must include a limitation on the valve speed similar to the slew-rate limitation on an op amp.

9.5 ELECTRIC HEATING ELEMENTS

Electric heating elements convert electrical energy into heat energy. As energy converters, electric heaters are 100% efficient. All the electrical energy is converted into heat. However, some of the heat energy is lost to the surrounding environment, resulting in less than 100% efficiency of heat energy delivered to the product.

Electric heaters come in a variety of shapes and sizes to satisfy different heating requirements. The burner on an electric stove is an example of an electric heating element. Industrial heaters come in the following forms: cartridge, band, strip, cable, radiant, flexible, ceramic fiber, and various process heater assemblies. We will describe the first two forms in some detail. Further information can be obtained from manufacturers' catalogs.

A *cartridge heater* consists of a nickel–chromium wire wound on a supporting core and enclosed in a metal tube sheath (see Figure 9.29). The outside diameter of the tube ranges from 0.125 to 1 in. The length of the tube ranges from 1.25 to 12 in. for the 0.125-in.-diameter element. For the 1-in.-diameter element, the length ranges from 1.25 to 72 in. Typical wattage ratings are 25 to 50 W for the 0.125-in.-diameter by 1.25-in.-long element. A 1-in.-diameter by 36-in.-long element has a rating of 2500 W.

Cartridge heaters are immersed in the object to be heated. When used to heat solid objects, the cartridge is inserted into a hole in the object. The fit should be as tight as possible while allowing easy insertion and removal of the heating element. When used to heat liquids, the cartridge is immersed in the liquid to be heated. Immersion elements may be fitted with a threaded plug or a flange for mounting through the wall of the containing vessel (see Figure 9.30). Stock immersion heaters have a diameter of 0.625 in. Sizes vary from a 500-W unit (6.25-in. overall length) to an 18,000-W unit (35-in. overall length).

Band heaters are used to heat round objects or liquid product in round containers such as pipes, extruder barrels, drums, cylinders, and so on. The band heater wraps around the object to be heated much like a belt (see Figure 9.31). Stock band heaters range from a 100-W unit [1 in. inside diameter (ID) by 1 in. wide] to a 2300-W unit (12 in. ID by 2 in. wide). Units are available with diameters ranging from 0.875 to 44 in. and widths ranging from 0.625 to 15 in.

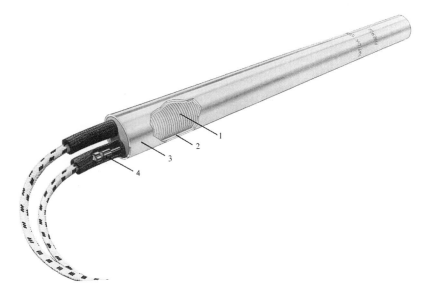

◆ **Figure 9.29**   A cartridge heater has four main parts: (1) a coil of resistance wire wound on a supporting core; (2) magnesium oxide insulation that combines high dielectric strength with efficient, fast heat transfer; (3) a corrosion-resistant sheath; and (4) electrical leads. [From *Everything You Ever Wanted to Know About Electric Heaters and Control Systems* (St. Louis, Mo.: Watlow Electric Mgf. Co.), p. 1.]

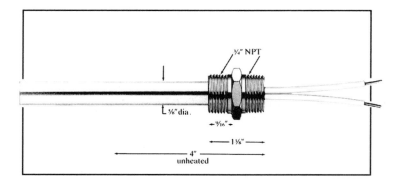

◆ **Figure 9.30** Electric immersion heaters are inserted into the heated fluid through a threaded hole in the wall of the container. Elements with a flange mounting or "over the side" mounting are also available. [From *Everything You Ever Wanted to Know About Electric Heaters and Control Systems* (St. Louis, Mo.: Watlow Electric Mfg. Co.), p. 13.]

The first step in selecting an electrical heater is to determine the heat flow rate required to do the job. The heat energy requirements fall into one of the following three categories:

1. Heat required to raise the temperature of the product and any surrounding objects
2. Heat required to change the state of the product (i.e., change the product from a solid to a liquid, or from a liquid to a gas)
3. Heat losses by conduction, convection, and radiation

The second step is to examine other considerations that will influence the selection. Examples include the following:

1. The operating environment (e.g., the size and shape of the container, temperature, humidity, etc.)
2. Life requirements

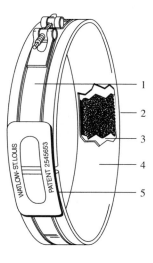

◆ **Figure 9.31** Band heaters consist of five parts: (1) a clamping strap, (2) a high-temperature heating ribbon, (3) mica insulation, (4) a rust-resistant steel sheath, and (5) post terminals for electrical connections. [From *Everything You Ever Wanted to Know About Electric Heaters and Control Systems* (St. Louis, Mo.: Watlow Electric Mfg. Co.), p. 35.]

3. Operating costs
4. Safety factor

The third step is to select the type, size, and number of heating elements.

The heat flow rate required to raise the temperature of an object is determined by the thermal capacitance of the object and the elapsed time for the temperature rise. In this discussion, an amount of liquid product will be treated as an "object" to be heated. Thermal capacitance was defined in Chapter 3 as the amount of heat required to increase the temperature by 1°C. Thus the heat flow rate required to raise the temperature of an object is given by the following equation:

$$Q_1 = \frac{C_T(T_{final} - T_{initial})}{t_e} \tag{9.13}$$

$$C_T = mS_h = \rho V S_h \tag{3.39}$$

where Q_1 = heat flow rate required to raise the temperature, W
C_T = thermal capacitance, J/K
T_{final} = final temperature, °C or K
$T_{initial}$ = initial temperature, °C or K
t_e = elapsed time, s
m = mass of the object, kg
ρ = density of the object, kg/m³
S_h = heat capacity, J/kg · K
V = volume of the object, m³

The amount of heat required to convert 1 kg of a solid into a liquid is called the latent heat of fusion (H_f). Appendix A lists the melting point and the latent heat of fusion for several substances. The amount of heat required to convert 1 kg of a liquid into a gas is called the latent heat of vaporization (H_v). Water boils at 100°C, and the latent heat of vaporization is 2.26 MJ/kg. The heat flow rate required to change the state of a given amount of product is given by the following equations:

$$Q_2 = \frac{mH_f}{t_e} \tag{9.14}$$

$$Q_3 = \frac{mH_v}{t_e} \tag{9.15}$$

where Q_2 = heat flow rate required to melt the product, W
Q_3 = heat flow rate required to vaporize the product, W
m = mass of the product, kg
H_f = heat of fusion of the product, J/kg
H_v = heat of vaporization of the product, J/kg
t_e = elapsed time, s

The heat loss by conduction is determined by the program THERMRES of Chapter 3. Exact calculations of convection and radiation losses are quite difficult. The simplest approach is to use a combined surface loss factor that gives the heat loss for a unit area of surface at a given temperature. The heat loss rate is equal to the product of the surface area and the combined surface loss factor.[*]

EXAMPLE 9.6

Oil is to be heated from 10° to 55°C and pumped at a rate of 2 L/min. Determine the theoretical heat flow rate required to heat the oil.

Solution

Equations (9.13) and (3.39) apply. An elapsed time of 1 min or 60 s will be used for the calculation.

$$t_e = 60 \text{ s}$$

The following properties of oil were obtained from Appendix A.

$$\rho = 880 \text{ kg/m}^3$$

$$S_h = 2180 \text{ J/kg} \cdot \text{K}$$

The mass of oil that must be heated in 1 min is

$$m = \left(\frac{2}{1000}\right)(880) = 1.76 \text{ kg}$$

The thermal capacitance of this amount of oil is

$$C_T = mS_h = (1.76)(2180) = 3837 \text{ J/K}$$

The heat flow rate required to heat the oil is

$$Q_1 = \frac{C_T(T_{\text{final}} - T_{\text{initial}})}{t_e}$$

$$= \frac{(3837)(55 - 10)}{60}$$

$$= 2878 \text{ W} \qquad \qquad \blacklozenge$$

EXAMPLE 9.7

A tank of paraffin is to be used for coating parts. The tank contains 85 kg of paraffin, which is at room temperature before the start of a production run. Determine the heat flow rate required to melt the paraffin and raise the temperature from room temperature (20°C) to the dipping temperature of 65°C. The desired heating time is 90 min. Assume the same value of heat capacity for solid and liquid paraffin.

Solution

Equations (9.13), (3.39), and (9.14) apply.

$$t_e = (90)(60) = 5400 \text{ s}$$

The following properties of paraffin were obtained from Appendix A.

$$\rho = 897 \text{ kg/m}^3$$

$$S_h = 2931 \text{ J/kg} \cdot \text{K}$$

$$H_f = 147 \text{ kJ/kg}$$

---

[*]For further details, refer to *Everything You Ever Wanted to Know About Electric Heaters and Control Systems* by Watlow Electric Mfg. Co., St. Louis. MO.

The thermal capacitance of 85 kg of paraffin is

$$C_T = mS_h = (85)(2931) = 2.49 \times 10^5 \text{ J/K}$$

The heat flow rate required to raise the temperature of the paraffin is

$$Q_1 = \frac{C_T(T_{\text{final}} - T_{\text{initial}})}{t_e}$$

$$= \frac{(2.49\text{E}5)(65 - 20)}{5400}$$

$$= 2076 \text{ W}$$

The heat flow rate required to melt the paraffin is

$$Q_2 = \frac{mH_f}{t_e}$$

$$= \frac{(85)(147\text{E}3)}{5400}$$

$$= 2314 \text{ W}$$

The total heat flow required to melt the paraffin and raise the temperature is

$$Q_T = Q_1 + Q_2$$

$$= 2076 + 2314$$

$$= 4390 \text{ W} \qquad \qquad ◆$$

◆ GLOSSARY

Actuator:　A control system component that provides the power needed to carry out the control action produced by the controller. (9.3)

Actuator, hydraulic:　An actuator that uses an oil-based fluid under high pressure to carry out the control action. Hydraulic actuators are used where slow, precise positioning is required, or where heavy loads must be moved (9.3)

Actuator, pneumatic:　An actuator that uses air under moderate pressure to carry out the control action. Pneumatic actuators are used where relatively light loads are to be moved, where an air supply is available, and where fast response is required. (9.3)

Band heater:　An electric heating element that wraps around circular containers such as pipes, extruder barrels, drums, cylinders, and so on. (9.5)

Cartridge heater:　An electric heating element that consists of a nickel–chromium wire wound on a supporting core and enclosed in a metal tube sheath. (9.5)

Characteristic, equal-percentage:　A control valve characteristic that provides equal percentage changes in flow rate for equal changes in valve position. (9.4)

Characteristic, inherent:　The relationship between the position of a control valve and the flow rate through the valve when there is a constant pressure drop across the valve for all positions. (9.4)

Characteristic, installed:　The relationship between the position of a control valve and the flow rate through the valve when the valve is installed in a system and the pressure drop across the valve depends on the pressure drop in the remainder of the system. (9.4)

Characteristic, linear: A control valve characteristic that provides a linear relationship between the valve position and the flow rate through the valve. (9.4)

Characteristic, quick-opening: A control valve characteristic that provides a large change in flow rate for a small change in valve position. (9.4)

Contactor: A type of relay that has heavy-duty contacts for switching large amounts of electric power and light-duty contacts for the control circuit. *See also* Motor starter. (9.1)

Control valve: A control system component used to control the flow of a fluid by regulating the size of an opening in the flow passage. A control valve consists of a valve plus an actuator for positioning the valve stem. (9.4)

Control valve sizing: The engineering procedure of determining the correct size of a control valve for a specific installation. (9.4)

Cylinder: An actuator used to produce linear motion with a definite distance of travel. A cylinder consists of a piston and shaft enclosed in a tube that is closed at both ends, except for the shaft that protrudes through one end. The tube has ports through which hydraulic fluid or air can be passed into and out of both ends. The piston is moved by forcing fluid in one side of the tube while allowing it to exit the other side. (9.3)

Cylinder, hydraulic: A cylinder that uses hydraulic fluid to produce a linear motion. (9.3)

Cylinder, pneumatic: A cylinder that uses pressurized air to produce a linear motion. (9.3)

Diode: A two-terminal, solid-state component that allows electric current to flow in one direction, but resists flow in the other direction. (9.2)

Hydraulic motor: An actuator that uses hydraulic fluid under high pressure to produce rotary motion in a load. Types of hydraulic motors include radial piston motors, gear motors, and vane motors. (9.3)

Maintained-action switch: A type of switch in which the contacts remain in the position they are placed by a switching action until another switching action changes their position. (9.1)

Momentary-action switch: A type of switch in which the contacts are in the actuated position only while the operator holds the switch actuator in that position. As soon as the operator releases the actuator, the contacts return to their unactuated positions. (9.1)

Motor starter: A type of relay that has heavy-duty contacts for starting electric motors and light-duty contacts for the control circuit. *See also* Contactor. (9.1)

Oversize factor: A number greater than 1 that is used to increase the size of a cylinder to provide the additional force necessary to accelerate the load. An oversize factor of 1.25 is recommended for slow-moving loads, and a factor of 2.00 is recommended for fast-moving loads. (9.3)

Pneumatic motor: An actuator that uses pressurized air to produce rotary motion in a load. Types of pneumatic motors include piston motors, vane motors, and turbine motors. (9.3)

Relay: A set of switches, a coil of wire, and supporting members arranged such that the switches are actuated when electric current passes through the coil of wire. (9.1)

Silicon-controlled rectifier (SCR): A three-terminal, solid-state switching component that requires a small amount of control power to control a large amount of electric power. An SCR is both a rectifier and a latching switch that can control currents as much as 3000 times as large as the control current. (9.2)

Solenoid valve: A movable spool fitted inside a housing with one or two solenoids capable of moving the spool to different positions relative to the housing. Both the spool and the housing have fluid passages. The passages in the housing are connected together or blocked in different ways, depending on the position of the spool relative to the housing. (9.3)

Switch: One or more pairs of contacts and supporting members that are used to make or break connections in an electric circuit. (9.1)

Time-delay relay: A relay with a timing mechanism that provides a delay between the time the actuating current is applied to the coil and the time the contacts move to their actuated position. (9.1)

Transistor: A three-terminal, solid-state switching component that is used as a switch, amplifier, and oscillator. A transistor is essentially a controlled current amplifier. (9.2)

Triac: A three-terminal, solid-state switching component that requires a small amount of control power to control a large amount of electric power. A triac is similar to an SCR except that the triac can conduct in both directions, whereas the SCR can conduct in only one direction. (9.2)

Unijunction transistor: A three-terminal, solid-state component used to produce the trigger pulse for controlling SCRs and triacs. (9.2)

Valve flow coefficient (C_v):   A capacity index for control valves, which is defined as the number of U.S. gallons of water per minute that will flow through a wide-open valve with a pressure drop of 1 psi. (9.4)

Way: A passage in the housing of a solenoid valve. Each way can be connected to a supply line, a return line, or a line to a cylinder. (9.3)

Working pressure: The actual pressure required in a cylinder to move the load at the required speed. (9.3)

◆ EXERCISES

Section 9.1

9.1   A 115-V ac electric motor drives a sump pump. A normally open liquid level switch turns the motor on and off directly (no relay is used).

(a) Draw a 115-V ac ladder diagram to control the sump pump motor. Your diagram should have one rung with the level switch as the input section and the electric motor as the output element. (See Figure 9.1 for the level switch symbol. Use a circle for the motor symbol.)

(b) Electric motors have built-in thermal overload protection to prevent overheating of the motor. The thermal overload symbols are shown in Figure 9.1. Observe the heating element symbol on the left and the NC contact symbol on the right in Figure 9.1. Revise your ladder diagram to include the heating element and the NC thermal overload contact. Place the heating element between the level switch and the motor. Place the overload contact between the motor and the right vertical line (L2).

9.2   The sump pump circuit in Exercise 9.1 is simple, but it has a problem. The problem occurs when the pump flow rate is greater than the flow rate into the sump. When the level switch closes and turns on the pump motor, the level in the sump immediately begins to recede. In a very short time, the level switch will open, turning the pump off. The pump rapidly switches between on and off. This cycling is very annoying and is hard on the pump motor. We can eliminate the cycling with two level switches: a high level switch to turn the pump on and a low level switch to turn the pump off. The pump is turned on when the water level rises enough to close the high level switch. Once turned on, the pump is not turned off until the water level drops enough to open the low level switch.

(a) Draw a 115-V ac ladder diagram that uses two normally open level switches and two normally open control relay contacts to control the sump pump motor. Your diagram should have two rungs. In the top rung, place the two level switches in series. Then place a control relay contact in parallel with the high level switch. Next, place the control relay coil in the top rung as the output element. Finally, place the other control relay contact and the ac motor in the second rung.

(b) Revise your ladder diagram from part a to include the thermal overload heating element and the NC thermal overload contact. Place the overload contact in the top rung on the right side of the control relay. Place the heating element in the bottom rung between the control relay contact and the ac motor.

(c) Describe the operation of your revised circuit from part b. Begin your description with an initially empty sump, but with water flowing into the sump at a rate somewhat less than the pump flow rate. Your description should include one ON–OFF cycle of the pump.

9.3   The propane furnaces used in recreational vehicles have an interesting control circuit. You will implement some of its functions with the following ladder diagram. Draw a 12-V dc ladder diagram with the following three rungs.

(a) The top rung has a temperature switch, a time-delay relay coil (1TR), and a normally closed thermal overload contact.

(b) The middle rung has a normally open 1TR contact, a thermal overload heater, and a dc fan motor. The 1TR contact delay occurs when the coil is energized.

(c) The lower rung has a normally open flow switch in series with a normally open 1TR contact and a burner solenoid. The flow switch closes when the fan develops sufficient air flow. The burner solenoid controls the flow of propane to the burner.

(d) Describe the turn-on operation of your circuit from the closure of the temperature switch to the closure of the flow switch.

(e) Describe the turn-off operation of your circuit from the opening of the temperature switch to the opening of the flow switch.

9.4 A 240-V ac three-phase electric motor drives a shallow well pump used for irrigation. The pump is connected to a 100-gal pressure tank. The pressure in the tank turns the pump ON and OFF. When the pressure drops below 20 psi, the pump is turned ON and remains ON until the pressure reaches 50 psi. When the pressure reaches 50 psi, the pump is turned OFF and remains OFF until the pressure drops below 20 psi. A motor starter similar to Figure 9.5 is to be used to control the motor. Two NC pressure switches are to be used in place of the two pushbutton switches. Switch a is closed when the pressure is below 20 psi and open when the pressure is above 20 psi. Switch b is closed when the pressure is below 50 psi and open when the pressure is above 50 psi. Draw a diagram of the circuit for starting and stopping the pump motor. The 115-V ac control circuit should include the NC overload contact and the holding contact.

Section 9.2

9.5 Describe the four ways that an SCR can be turned ON (i.e., move from region 2 to region 3).

9.6 Describe the three ways that an SCR can be turned OFF and indicate which region the SCR is in when it is OFF.

9.7 Describe the major difference between the triac and the SCR, and explain how the triac is equivalent to two SCRs connected in parallel but in opposite directions.

9.8 A half-wave dc power supply uses one diode instead of the two diodes used in a full-wave power supply. Sketch the output voltage wave-form of the power supply in Figure 9.12 if diode D_2 is removed from the circuit.

9.9 The following voltage and current measurements were obtained from a common-emitter amplifier similar to Figure 9.14b. Draw voltage and current graphs similar to Figure 9.15 and determine the voltage gain and the current gain.

1. $V_{in} = 0.70$ V, $V_{out} = 24.2$ V,
   $i_{in} = 0$ μA, $i_{out} = 0.95$ mA
2. $V_{in} = 0.75$ V, $V_{out} = 20.0$ V,
   $i_{in} = 10.6$ μA, $i_{out} = 0.78$ mA
3. $V_{in} = 0.80$ V, $V_{out} = 15.8$ V,
   $i_{in} = 21.0$ μA, $i_{out} = 0.61$ mA
4. $V_{in} = 0.85$ V, $V_{out} = 11.6$ V,
   $i_{in} = 31.5$ μA, $i_{out} = 0.45$ mA
5. $V_{in} = 0.90$ V, $V_{out} = 7.4$ V,
   $i_{in} = 42.2$ μA, $i_{out} = 0.27$ mA
6. $V_{in} = 0.95$ V, $V_{out} = 3.4$ V,
   $i_{in} = 52.4$ μA, $i_{out} = 0.12$ mA
7. $V_{in} = 1.00$ V, $V_{out} = 0$ V,
   $i_{in} = 62.8$ μA, $i_{out} = 0$ mA

9.10 A four-stage transistor amplifier similar to Figure 9.17 has the following voltage and current gains for each stage. Determine the overall voltage gain and current gain.

Stage 1: voltage gain $= -8$, current gain $= -1.4$
Stage 2: voltage gain $= -10$, current gain $= -1.2$
Stage 3: voltage gain $= -0.95$, current gain $= 110$
Stage 4: voltage gain $= -0.96$, current gain $= 150$

Section 9.3

9.11 Determine the working pressure (WP_2) required to produce force (f_2) in each of the following hydraulic cylinder applications.

(a) 1.5-in.-diameter cylinder: $f_2 = 2450$ lb
(b) 2-in.-diameter cylinder: $f_2 = 8120$ lb
(c) 3-in.-diameter cylinder: $f_2 = 16,730$ lb
(d) 4-in.-diameter cylinder: $f_2 = 28,220$ lb
(e) 5-in.-diameter cylinder: $f_2 = 36,900$ lb
(f) 6-in.-diameter cylinder: $f_2 = 42,000$ lb
(g) 8-in.-diameter cylinder: $f_2 = 92,600$ lb
(h) 10-in.-diameter cylinder: $f_2 = 180,000$ lb

9.12 Determine the working pressure (WP_2) required to produce force (f_2) in each of the following pneumatic cylinder applications.

(a) 1.5-in.-diameter cylinder: $f_2 = 190$ lb
(b) 2-in.-diameter cylinder: $f_2 = 286$ lb
(c) 3-in.-diameter cylinder: $f_2 = 840$ lb
(d) 4-in.-diameter cylinder: $f_2 = 1060$ lb
(e) 5-in.-diameter cylinder: $f_2 = 1850$ lb
(f) 6-in.-diameter cylinder: $f_2 = 2100$ lb
(g) 8-in.-diameter cylinder: $f_2 = 4400$ lb
(h) 10-in.-diameter cylinder: $f_2 = 5500$ lb

9.13 Determine the flow rate (Q_2) required to produce speed (S_2) in each of the following hydraulic cylinder applications.

(a) 1.5-in.-diameter cylinder: $S_2 = 200$ in./min
(b) 2-in.-diameter cylinder: $S_2 = 110$ in./min
(c) 3-in.-diameter cylinder: $S_2 = 28$ in./min
(d) 4-in.-diameter cylinder: $S_2 = 16$ in./min
(e) 5-in.-diameter cylinder: $S_2 = 14$ in./min
(f) 6-in.-diameter cylinder: $S_2 = 7.8$ in./min
(g) 8-in.-diameter cylinder: $S_2 = 2.2$ in./min
(h) 10-in.-diameter cylinder: $f_2 = 6.4$ in./min

9.14 A manufacturing operation requires the movement of a workpiece a distance of 10 in. in 30 s. A force of 22,500 lb is required to move the workpiece. The available space limits the cylinder diameter to a maximum of 8 in. Leakage and noise considerations make it desirable to limit the working pressure to a maximum of 1000 psi. A hydraulic pump with a capacity of 2.5 gal/min is available in the surplus equipment yard. Select a cylinder for this application.

9.15 A manufacturing operation requires the movement of a workpiece a distance of 10 in. in 0.6 s. A force of 125 lb is required to move the workpiece. The available space limits the cylinder diameter to a maximum of 4 in. Heavy usage of the air supply sometimes limits the supply pressure to 80 psi. Select a cylinder for this application.

Section 9.4

9.16 An equal-percentage valve has a rangeability of 20. Use Equation (9.4) to calculate the value of Q/Q_{max} for values of P/P_{max} from 0 to 1 in increments of 0.1.

9.17 Determine the size of the control valve for each of the following liquid flow applications. Use a safety factor of 25%.

	Q_L (gpm)	$P_1 - P_2$ (psi)	G_L
(a)	2	80	0.80
(b)	14	46	1.00
(c)	64	33	0.74
(d)	196	52	0.88
(e)	538	49	1.00
(f)	1148	70	0.87
(g)	2376	61	1.00
(h)	4832	57	1.00

9.18 Determine the size of the control valve for each of the following gas flow applications. Use a safety factor of 25%.

	Q_g (ft^3/hr)	P_1 (psi)	P_2 (psi)	T (°F)	G_g
(a)	2,130	40	20	100	1.11
(b)	10,840	60	30	420	1.54
(c)	47,500	70	45	140	0.98
(d)	62,700	55	15	280	1.00
(e)	232,000	110	60	80	1.39
(f)	254,000	80	50	360	0.98
(g)	573,000	70	25	230	1.00
(h)	1,010,000	80	55	170	1.11

9.19 Determine the size of the control valve for each of the following steam applications. Use a safety factor of 25%.

	W (ft^3/hr)	P_1 (psi)	P_2 (psi)	V_1 (ft^3/lb)	V_2 (ft^3/lb)
(a)	190	60	40	7.2	10.5
(b)	630	45	30	9.4	14.0
(c)	1,700	50	30	8.5	14.0
(d)	3,200	60	35	7.2	12.0
(e)	7,700	55	30	7.8	14.0
(f)	10,200	45	20	9.4	21.0
(g)	24,600	55	30	7.8	14.0
(h)	46,000	55	20	7.8	21.0

Section 9.5

9.20 Determine the heat flow rate required to raise the temperature of m kilograms of each of the following solids from a temperature of $T_{initial}$ °C to a temperature of T_{final} °C in t_e minutes.

	Solid Material	m (kg)	$T_{initial}$ (°C)	T_{final} (°C)	t_e (min)
(a)	Aluminum	10	40	200	15
(b)	Brass	16	30	240	18
(c)	Gold	4	70	170	5
(d)	Steel	21	40	80	32
(e)	Pine wood	2	20	60	20

9.21 Determine the heat flow rate required to raise the temperature of V liters of each of the following liquids from a temperature of $T_{initial}$ °C to a temperature of T_{final} °C in t_e minutes.

	Liquid	V (L)	$T_{initial}$ (°C)	T_{final} (°C)	t_e (min)
(a)	Water	8	26	100	40
(b)	Oil	20	10	120	80
(c)	Turpentine	16	30	60	60
(d)	Kerosene	24	14	48	90
(e)	Glycerine	6	28	68	30

	Liquid	$T_{initial}$ (°C)	T_{final} (°C)	Q (L/min)
(a)	Water	5	55	4
(b)	Ethanol	10	50	6
(c)	Oil	25	120	2
(d)	Kerosene	16	36	9
(e)	Turpentine	30	45	7

9.22 Determine the heat flow rate required to melt m kilograms of each of the following solids in a time interval of t_e minutes.

	Solid Material	m (kg)	t_e (min)
(a)	Aluminum	10	50
(b)	Asphalt	16	75
(c)	Ice	20	60
(d)	Paraffin	8	25
(e)	Solder (50-50)	4	15

9.23 The following liquids are to be heated from a temperature of $T_{initial}$ °C to a temperature of T_{final} °C and pumped at a flow rate of Q L/min. Determine the theoretical heat flow rate required to heat the liquid.

9.24 A tank of asphalt is to be used to repair a roof. The tank contains 65 kg of asphalt that is at an early morning temperature of 15°C. Determine the heat flow rate required to melt the asphalt and raise its temperature from 15°C to a working temperature of 140°C. The desired heating time is 45 minutes. Assume the same value of heat capacity for solid and liquid asphalt.

9.25 A wave-soldering process uses a tank of 50-50 solder. The tank is 60 cm wide by 80 cm long, and the solder depth is 10 cm. The solder is at room temperature before the start of a production run. Determine the heat flow rate required to melt the solder and raise the temperature from room temperature (20°C) to the operating temperature of 250°C. The desired heating time is 2 hr. Assume the same value of heat capacity for solid and liquid solder.

◆ CHAPTER 10   **Electric Motors**

◆ OBJECTIVES

Electric motors are frequently used as the manipulative device in control systems. Stepping motors and servomotors provide the "muscle" for robotic arms and numerically controlled machine tools. Adjustable speed drives provide efficient control of pumps and blowers. Electric motors are used to sort mail, package food, make plastic sheet, and cut cereal pellets, to name a few of the host of applications.

The purpose of this chapter is to give you an entry-level ability to discuss, select, and specify various types of electric motors. After completing this chapter, you will be able to

1. Describe
   a. AC motors—induction, synchronous, and servo
   b. DC motors—series wound, shunt wound, compound wound, permanent magnet, and brushless
   c. Stepping motors—variable reluctance, permanent magnet, and hybrid
   d. AC adjustable speed drives—VVI and PWM
2. Use specifications to determine the velocity and position transfer functions of
   a. AC servomotor
   b. Voltage-driven dc motor
   c. Simplified voltage-driven dc motor
   d. Current-driven dc motor
   e. Voltage-driven dc motor, speed reducer, and load

10.1 INTRODUCTION

An electric motor is often used as the actuator in a servo control system. In Figure 2.6, such a motor is used to control the position of an antenna and in Figures 2.15, 10.19, and 10.20, electric motors are used to control the speed of a rotating load. They are used to power the six axes of movement of a robot in Figure 2.10 and to control the thickness of a sheet in Figure 2.20. They are used to drive the belt of a solids feeder in Figures 9.18 and 9.19. They are used in numerically controlled (NC) machines to position the workpiece and move the tool. Finally, variable-speed electric motors are used to drive pumps for efficient control of the flow rate of a liquid. These are just a few of the many uses of electric motors in control systems.

Figure 10.1 shows the block diagram of a position control system. In this diagram the pre-amplifier is the controller, the power amplifier and dc motor constitute the manipulating element, and the speed reducer and load are the process. Collectively, these components are the forward path, *G*, in the closed-loop servo control system in Figure 1.6.

Classification of Electric Motors

Electric motors are classified in two ways: by function and by electrical configuration. Classification by function is based on how the motor is used. Examples include servomotors, gear motors, instrument motors, pump motors, and fan motors. Classification by electrical configuration separates motors into two major categories: dc motors and ac motors. A *dc motor* consists of two parts: a rotating cylinder called the *armature,* and stationary magnetic poles called *field poles* or simply the *field.* The armature is placed in the magnetic field between the field poles. An *ac motor* also consists of two parts: a rotating cylinder called the *rotor* and a stationary part called the *stator.* The stator is a thick-walled tube that surrounds the rotor. The stator has windings that produce a rotating magnetic field in the space occupied by the rotor. The following outline summarizes the classification of electric motors by electrical configuration.

    Classification of Electric Motors

      I. AC motors

          A. Single-phase motors

              1. Induction

                  a. Squirrel cage

                  b. Wound rotor

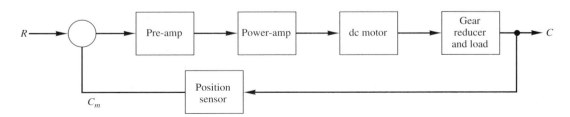

◆ **Figure 10.1** A dc motor position control system.

2. Synchronous
    a. DC excited rotor
    b. Nonexcited rotor
B. Polyphase motors
  1. Induction
    a. Squirrel cage
    b. Wound rotor
  2. Synchronous
    a. DC excited rotor
    b. Nonexcited rotor
C. Universal motors
II. DC motors
A. Wound field
  1. Series wound
  2. Shunt wound
  3. Compound wound
B. Permanent-magnet field
  1. Permanent-magnet stator, wound armature (brush type)
  2. Permanent-magnet rotor (brushless)

Force, Torque, and Induced Voltage

At the beginning of this section, you read a number of ways that electric motors are used in control. Our goal in this section is to explain the operation of these motors in simple terms without overwhelming the reader with detail. Extraneous discussion is avoided so that you can focus on the essential concepts needed to understand electric motors and their transfer functions.

Electric motors (and electric generators) are governed by two facts. The first fact describes the force that makes an electric motor go. The second fact describes the voltage produced by an electric generator. The last two sentences make it appear that the first fact applies only to electric motors and the second fact only to electric generators. Nothing could be further from the truth—both facts apply to motors and both facts apply to generators. In simple terms, the two facts are

1. A force is exerted on a conductor in a magnetic field whenever a current passes through the conductor.
2. A voltage is induced in a conductor whenever it is moved through a magnetic field.

Motion has nothing to do with the force exerted on the conductor, and current has nothing to do with the voltage induced in the conductor. In electric motors, we think of current as the cause of the force and motion as the cause of the induced voltage.

Figure 10.2 provides more detail about the force and induced voltage in an electric motor. In Figure 10.2a, the U-shaped electrical conductor represents a single *coil* in a dc motor. There are many coils in a motor, but we will focus on just this one coil. An electrical voltage source (not shown) causes electric current i to flow through the coil, entering the coil at a and leaving at b. Notice how the plus and the dot on the ends of the coil also indicate the direc-

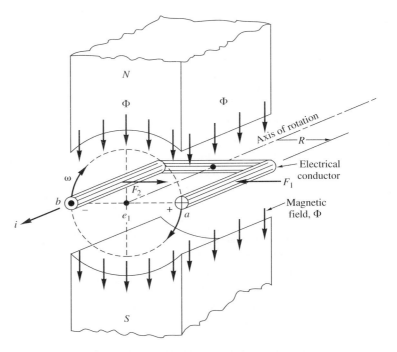

a) The U-shaped conductor rotates about its axis through the magnetic field, Φ. Current, i, passes through the conductor causing forces F_1 and F_2 to act on the conductor as shown.

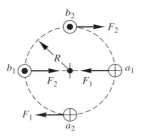

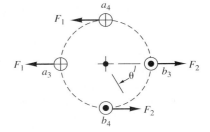

b) In position a_1–b_1, force F_1 cancels force F_2 (equal and opposite). In position a_2–b_2, force F_1 produces torque F_1R and force F_2 produces torque F_2R.
Torque = $F_1R + F_2R$

c) In position a_3–b_3, force F_1 cancels force F_2 (equal and opposite). In position a_4–b_4, force F_1 produces torque $-F_1R$ and force F_2 produces torque $-F_2R$.
Torque = $-F_1R - F_2R$

◆ **Figure 10.2** The torque exerted on the U-shaped conductor reverses direction every time the coil rotates 180°.

tion of the current. The coi! is rotating in a clockwise direction at ω rad/s. The axis of rotation is located between the two magnetic poles marked N for North and S for South. As it rotates, the coil moves through the magnetic field (Φ) between the two magnetic poles. We will use θ to indicate the angle of rotation, beginning with $\theta = 0°$ in the position shown in Figure 10.2a.

The current in the coil and the presence of the magnetic field cause forces F_1 and F_2 to be exerted on the left and right sides of the coil. These two forces are equal in magnitude and opposite in direction, as indicated by the arrows in Figure 10.2a. The forces are shown more clearly in Figure 10.2b where a_1 and b_1 mark the ends of the conductor in its original position ($\theta = 0°$). Figure 10.2b also shows the coil after it has rotated 90° to the position marked by a_2 and b_2 ($\theta = 90°$). Notice that the two forces have the same magnitude and direction as before, but instead of canceling each other, they exert a clockwise torque on the coil.

In Figure 10.2c the coil ends marked a_3 and b_3 show the coil position when $\theta = 180°$. Notice that the two forces cancel just as they did when $\theta = 0°$. The coil ends marked a_4 and b_4 show the coil position when $\theta = 270°$. Notice how the forces now exert a counterclockwise torque on the conductor. The torque on our single coil can be represented by the following equation:

$$T_1 = K_{T1}i \sin \theta \qquad (10.1)$$

where T_1 = torque on a single coil, N · m

K_{T1} = single coil torque constant, N · m/A

i = current in the coil, A

θ = angle of rotation, °

As the coil rotates, its motion through the magnetic field induces a voltage in the coil. This induced voltage is marked e_1 in Figure 10.2a. The polarity of the induced voltage is such that it opposes the external voltage that produces the current i—not enough to cancel the external voltage, but enough to reduce the current by a significant amount. The induced voltage in our single coil can be represented by the following equation:

$$e_1 = K_{E1}\omega \sin \theta \qquad (10.2)$$

where e_1 = voltage induced in a single coil, V

K_{E1} = single coil voltage constant, V · s/rad

ω = rotational velocity of the coil, rad/s

θ = angle of rotation, °

You should make three important observations about Equations 10.1 and 10.2.

1. Torque T_1 is proportional to the current.
2. Voltage e_1 is proportional to the rotational speed.
3. Torque T_1 changes direction after every half revolution ($\theta = 180°$). If we want a consistent clockwise torque on the coil, we must find a way to reverse the direction of the current after every 180° of rotation.

Figure 10.3 shows how this reversal is accomplished in a simple dc motor. Four coils are mounted in slots on a cylinder of magnetic material called the *armature*. The armature is mounted on bearings, so it is free to rotate in the magnetic field produced by the two *field*

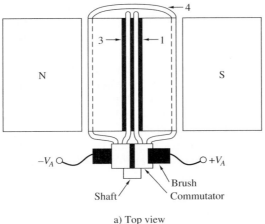

a) Top view

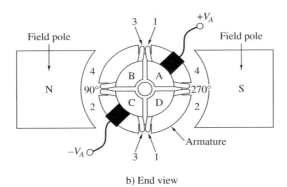

b) End view

◆ **Figure 10.3** Schematic diagram of a two-pole four-coil dc motor/generator.

poles. The ends of the coils are attached to a segmented ring called the *commutator.* The external voltage is applied to the armature coils through carbon contacts called *brushes.*

Figure 10.4 illustrates the dc motor in action. An external voltage is connected across the brushes of the dc motor, producing a current in the armature coils as shown in Figure 10.4a. This current produces a torque that causes the armature to rotate in the clockwise direction. As the armature rotates, the brushes make their way around the commutator, passing from segment to segment. When the brushes move from one segment to the next, the direction of the current in the coils is reversed. This reversal of current in the coil is called *commutation.* Notice that coils 2 and 4 are shorted by the brushes in Figure 10.4b. The reversal of the current in coils 2 and 4 can be observed by comparing the direction before commutation (Figure 10.4a) with the direction after commutation (Figure 10.4c).

In a commercial dc motor, there are many coils in slots all around the armature and more than just a single pair of field poles. The net result is a very smooth torque that depends only

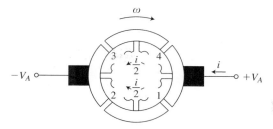

a) Before commutation

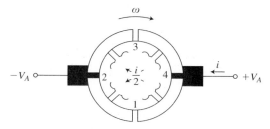

b) During commutation of coils 2 and 4

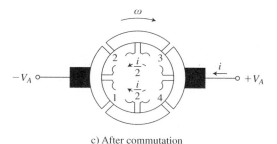

c) After commutation

◆ **Figure 10.4**   The commutator provides a switching action. During commutation (b), coils 2 and 4 are shorted by the brushes. After commutation (c), the direction of the current in coils 2 and 4 is the reverse of what it was before commutation (a).

on the current through the armature coils. This torque is equal to the product of the electric current that passes through the motor winding times a constant called the torque constant, K_T.

$$\text{Torque} = (\text{torque constant})\,(\text{armature current})$$

$$T = K_T\, i_a \tag{10.3}$$

where T = torque, N · m

   K_T = torque constant, N · m/A

   i_a = armature current, A

When the motor shaft rotates, the motion causes a voltage to be induced in the motor winding. We refer to this induced voltage as the *back emf, e_b,* to remind us that it opposes the motor current. The multiple coils and commutation also affect the back emf, resulting in a smooth back emf that depends only on the speed of the armature. The back emf is equal to the product of the rotational speed of the motor times a constant called the voltage constant, K_E.

$$\text{Back emf} = (\text{voltage constant})(\text{motor speed})$$

$$e_b = K_E\,\omega \qquad\qquad (10.4)$$

where
e_b = back emf, V

K_E = voltage constant, V · s/rad

ω = rotational speed, rad/s

Equations 10.3 and 10.4 summarize the two facts about torque and voltage in electric motors. We will use those two equations in Section 10.3 to develop the transfer function in dc motors.

10.2 AC MOTORS

AC motors dominate the application of electric motors that require a single operating speed. Single-phase motors are used for low-horsepower applications (from fractional horsepower up to about 20 hp). Polyphase motors overlap the low-horsepower range of single-phase motors and extend the range to much higher horsepower ratings.

DC motors dominate the variable-speed applications of electric motors. However, improvements in solid-state switching components and the application of large-scale integrated-circuit techniques to the complex ac drive circuits have made variable-speed ac drives more competitive with the "old standby" dc drives.

AC motors can be divided into two groups in two different ways, making four major categories of ac motors. One division is between single-phase and polyphase motors. The other division is between synchronous and induction motors. The four categories are single-phase induction motors, polyphase induction motors, single-phase synchronous motors, and polyphase synchronous motors.

Single-Phase Induction Motors

The single-phase squirrel-cage induction motor is the most common type of motor. Squirrel-cage induction motors have no brushes to generate sparks or wear out. They are very reliable, have a low initial cost, and also have a low maintenance cost. An *induction motor* consists of a *stator* with one or two windings and a *rotor* that contains the current-carrying conductors upon which the force is exerted. The rotor winding of a squirrel-cage motor consists of copper or aluminum bars that fit into slots in the rotor. The bars are connected at each end by a closed continuous ring. The assembly of conductor bars and end rings resembles a squirrel cage and gives the motor its name.

Only the stator windings of an induction motor are externally excited. The winding on the rotor is shorted and receives its energy by electromagnetic induction. The induction motor is a rotating transformer in which the stator winding is the primary and the rotor winding is the secondary.

The magnetic field produced by the stator winding does not remain fixed as it does for a dc motor. Instead, the poles of the magnetic field alternate between two positions. In effect, the magnetic field is rotating about the rotor axis. For this reason, it is called a *rotating field*. The speed at which the field of an induction motor rotates is called the *synchronous speed*. Induction motors always run at less than the synchronous speed. The difference between the synchronous speed and the actual rotor speed is called *slip*. The torque that a motor develops is proportional to the slip—if there is no slip, there will be no torque. Even at no-load, a motor needs some torque to overcome friction and wind resistance, and there will be some slip. As the torque load on the motor increases, the slip also increases to satisfy the increased torque requirement. For a motor with a two-pole stator, the synchronous speed is equal to the line frequency, 60 rps or 3600 rpm. For a four-pole motor, the synchronous speed is equal to half the line frequency, 1800 rpm.

The basic single-phase induction motor has one stator winding called the *main winding*. This type of motor is not self-starting. When the rotor is stationary, an equal torque is produced in each direction. Therefore, the net torque is zero, and the rotor remains stationary. If the rotor is started by some starting device, the motor will continue to run in the direction it was started, even if the starting device is removed. A second stator winding called the *start winding* is the most common method of starting a single-phase induction motor. The start winding is rotated 90° from the main winding as shown in Figure 10.5. A speed-sensitive switch disconnects all or part of the start circuit when the rotor reaches a preset speed. The most common methods of using a start winding are split-phase, capacitor-start, and two-capacitor.

The *split-phase motor* has only a speed switch in series with the start winding. When the motor reaches a preset speed, the switch opens, disconnecting the start winding. The motor continues to run with only the main winding excited. Figure 10.5 illustrates a split-phase motor.

The *capacitor-start motor* has a capacitor and a speed switch in series with the start winding. The speed switch disconnects the start winding and the capacitor once the motor has started. The capacitor-start motor has a larger starting torque than the split-phase motor.

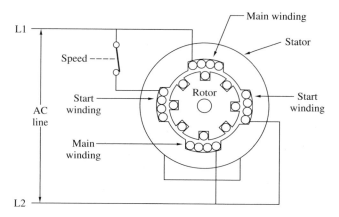

◆ **Figure 10.5**   A split-phase induction motor uses a start winding to start the motor. When the rotor reaches a set speed, the speed switch opens and the motor runs with only the main winding energized by the ac line.

The *two-capacitor motor* has two capacitors and a speed switch connected in a series/parallel combination as shown in Figure 10.6a. The start winding and one capacitor remain in the circuit after the motor has started. The speed switch removes the other capacitor from the circuit when the motor has started. The two-capacitor motor has a high starting torque and a low run current. Also, the capacitor improves the power factor, a definite advantage because induction motors are noted for their low power factor. A typical torque versus speed graph of a two-capacitor motor is shown in Figure 10.6b.

Polyphase Induction Motors

Most polyphase motors are three-phase motors. In this section we describe three-phase induction motors. There are two types of induction motors, the *squirrel-cage motor* and the *wound-rotor motor*. The rotor of a squirrel-cage motor has the same conducting bars and end rings described for the single-phase induction motor. In a wound-rotor motor the conducting bars are connected in series to form three windings. The windings are joined on one end in a wye connection. The other end of each winding is connected to a slip ring, one slip ring for each winding. Three brushes are in contact with the three slip rings so that external resistances can be connected in series with the windings. The other ends of the resistors are also connected together in a wye connection. The purpose of the external resistors is to limit the ro-

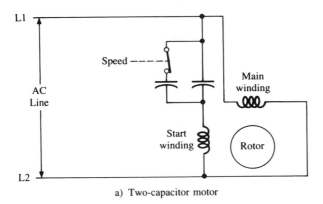

a) Two-capacitor motor

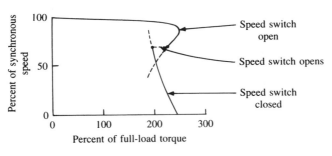

b) Speed vs. torque curve

◆ **Figure 10.6** A two-capacitor motor uses both capacitors to start the motor. When the rotor reaches about 70% of the operating speed, one of the capacitors is removed from the circuit. The other capacitor and the start winding remain in effect during normal operation of the motor.

tor current during start-up of loads with large inertia. When the external resistances are reduced to zero, the wound rotor motor is the same as a squirrel-cage motor.

A two-pole three-phase induction motor is shown in Figure 10.7. The stator has three
windings that can be connected in either a delta connection or a wye connection. The three
windings are located 120° apart and are connected to the three lines of a three-phase source.
Figure 10.7 shows the position of the magnetic field and the location of the two poles at the
beginning of a cycle of one of the three lines. During each sixth of a cycle, the poles move to
a new location by rotating 60° in the clockwise direction. In one full cycle, the poles will have
completed a 360° rotation. During that time, the north pole will have been in the six positions
numbered 1 through 6 in Figure 10.7. The synchronous speed of the two-pole motor is
3600 rpm.

The rotation of the field poles in six steps is much smoother than the two-step rotation of
the single-phase motor. One major advantage is that *three-phase induction motors are self-
starting*. They do not need an auxiliary starting device. The direction of rotation is reversed
by simply interchanging any two line connections. Interchanging two line connections also
changes the phase sequence of the motor.

A four-pole three-phase motor has six windings, located at 60° intervals. Each coil uses
stator slots that are 90° apart. The coils are paired with the coil on the opposite side of the stator. The paired coils are connected in series to form three branches. Each branch has two coils

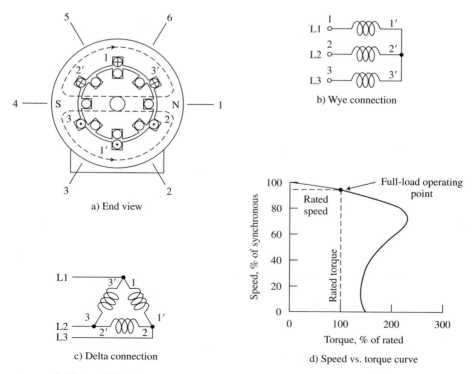

◆ **Figure 10.7**   Two-pole three-phase induction motor. The stator windings may be
connected to the three-phase ac supply in either a wye or a delta configuration.

connected in series. The three branches are connected to a three-phase line in either a delta or a wye configuration. The resulting magnetic field forms four poles, with like poles located 180° apart. The four poles rotate in six steps, in the same manner as the two poles. The synchronous speed of the four-pole motor is 1800 rpm.

Synchronous Motors

Synchronous motors normally run at synchronous speed. They are used where precise constant speed is required. Synchronous motors are made in all size ranges from fractional horsepower to several thousand horsepower. They are available in both single phase (smaller sizes) and polyphase (larger sizes). The stator windings of a synchronous motor are almost identical to the stator windings of the corresponding single-phase or polyphase induction motor. The rotor of a synchronous motor has fixed magnetic poles that lock into step with the rotating poles in the stator. The rotor may be nonexcited or direct-current excited, which gives us another way to classify synchronous motors.

Synchronous motors are not self-starting. This is true for both single-phase and polyphase synchronous motors. Two methods are used to start synchronous motors. One method uses a separate prime mover to start the synchronous motor and accelerate it to nearly synchronous speed. When the rotor locks into synchronous speed, the prime mover is removed. This method is sometimes called *prime mover starting*. The other method of starting a synchronous motor is called *induction motor starting*. A squirrel-cage winding is added to the rotor. The motor starts as an induction motor and accelerates to near-synchronous speed. When the rotor locks into synchronous speed, the current in the squirrel-cage conductor drops to zero and the motor operates as a synchronous motor. The single-phase induction-start motors use one of the starting devices for single-phase induction motors, such as split-phase, capacitor-start, shaded-pole, and so on.

The *dc excited synchronous motor* has a winding for each pole on the rotor. (Synchronous motors may have two, four, six, or more poles.) The individual field pole windings are connected in series to form one large winding that is terminated in two slip rings. The winding is excited by an external dc source through brushes that contact the slip rings. The dc excited motor is sometimes referred to as the "true synchronous motor." Large polyphase synchronous motors are dc excited motors.

Nonexcited synchronous motors include reluctance motors, hysteresis motors, and permanent-magnet motors. Reluctance-type synchronous motors use special construction to provide a variable reluctance in the rotor. This enables the rotor to establish the fixed poles without external excitation. Hysteresis motors develop fixed poles as the motor reaches synchronous speed. The shaded-pole hysteresis motor has wide application for clocks and timing devices. *Permanent-magnet motors* have permanent magnets embedded in a squirrel-cage rotor. This motor has become very popular because of its brushless construction, high power factor, and good efficiency.

AC Servomotors

AC servomotors are often used in control systems that require a low-power, variable-speed drive. The primary advantage of the ac motor over the dc motor is its ability to use the ac output of synchros, LVDTs, and other ac measuring means without demodulation of the error signal. An ac amplifier provides the gain for a proportional control mode. However, more elaborate control modes are difficult to implement with an ac signal. When additional control actions are required, the ac signal is usually demodulated, and the control action is inserted in the dc signal. The modified dc signal is then reconverted to an ac signal.

The *ac servomotor* is a two-phase, reversible induction motor with special modifications for servo operation. The schematic diagram of an ac servo is shown in Figure 10.8. The motor consists of an induction rotor and two field coils located 90° apart. One field coil serves as a fixed reference field, the other as the control field. The amplified ac error signal is applied to the control field. The signal has a variable magnitude with a phase angle of either 0° or 180°. A constant ac voltage is applied to the reference field through a 90° phase-shift network. This signal has a constant magnitude and a phase angle of −90°. The two voltages are given below.

$$e_c = V_c \cos \omega t$$

$$e_R = A \cos(\omega t - 90°) = A \sin \omega t$$

where e_c = control field voltage

e_R = reference field voltage

ω = operating frequency

V_c = variable amplitude of the control voltage

A = constant amplitude of the reference voltage

The sign and magnitude of V_c is determined by the sign and magnitude of the error signal. A negative error signal results in a negative value of V_c. This is usually interpreted as a 180° phase shift in e_c.

The linearized operating characteristics of an ac servomotor are shown in Figure 10.9. The actual operating line will depend on the speed–torque characteristics of the process. Two typical process load lines are indicated by the dashed lines. The negative values of V_c in the third quadrant simply indicate that the motor reverses direction when V_c is negative.

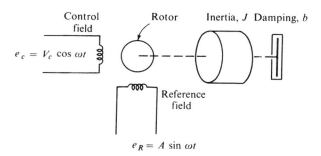

◆ **Figure 10.8** An ac servomotor is a two-phase reversible induction motor with special modifications for servo operation.

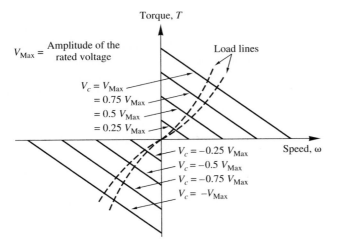

◆ **Figure 10.9** Linearized operating characteristics of an ac servomotor.

The velocity and position transfer functions are given by Equations (10.5) and (10.6).

AC SERVOMOTOR TRANSFER FUNCTIONS

Velocity Transfer Function

$$\frac{\Omega}{V_c} = \frac{K}{1 + \tau s} \text{ rad/s} \cdot \text{V} \tag{10.5}$$

Position Transfer Function

$$\frac{\Theta}{V_c} = \frac{K}{s(1 + \tau s)} \text{ rad/V} \tag{10.6}$$

where V_c = control voltage amplitude, V
   Ω = motor speed, rad/s
   Θ = motor position, rad
   $K = K_1/(b + K_2)$, rad/s · V
   $\tau = J/(b + K_2)$, s
   K_1 = stall torque/rated voltage, N · m/V
   K_2 = stall torque/no-load speed at rated voltage, N · m · s/rad
   J = moment of inertia of the load, N · m · s²/rad (or kg · m²/rad)
   b = damping resistance of the load, N · m · s/rad

**EXAMPLE
10.1**
Determine the velocity and position transfer functions of an ac servomotor with the following data:

Rated voltage: 120 V
Load inertia: 6×10^{-6} = N · m · s²/rad
Load damping: 2×10^{-5} N · m · s/rad
Stall torque: 0.04 N · m at 120 V
No-load speed: 4000 rpm at 120 V

Solution   The transfer functions are given by Equations (10.5) and (10.6).

1. Determine K_1 by dividing stall torque by rated voltage:

$$K_1 = \frac{0.04 \text{ N} \cdot \text{m}}{120 \text{ V}} = 3.33\text{E} - 4 \text{ N} \cdot \text{m/V}$$

2. Convert no-load speed to radians per second:

$$\text{No-load speed} = \frac{(4000)2\pi}{60} = 419 \text{ rad/s}$$

3. Determine K_2 by dividing stall torque by no-load speed:

$$K_2 = \frac{0.04 \text{ N} \cdot \text{m}}{410 \text{ rad/s}} = 9.55\text{E} - 5 \text{ N} \cdot \text{m} \cdot \text{s/rad}$$

4. Solve for K and τ:

$$b + K_2 = (2\text{E} - 5) + (9.55\text{E} - 5) = 1.155\text{E} - 4$$

$$K = \frac{K_1}{b + K_2} = \frac{3.33\text{E} - 4}{1.155\text{E} - 4} = 2.89 \text{ rad/s} \cdot \text{V}$$

$$\tau = \frac{J}{b + K_2} = \frac{6\text{E} - 6}{1.155\text{E} - 4} = 0.052 \text{ s}$$

5. Determine

   a. Velocity transfer function

   $$\frac{\Omega}{V_c} = \frac{2.89}{1 + 0.052s} \text{ rad/s} \cdot \text{V}$$

   b. Position transfer function

   $$\frac{\Theta}{V_c} = \frac{2.89}{s(1 + 0.052s)} \text{ rad/V}$$

   ◆

10.3 DC MOTORS

DC motors are extremely versatile drives, capable of reversible operation over a wide range of speeds, with accurate control of the speed at all times. They can be controlled smoothly from zero speed to full speed in both directions. DC motors have a high torque-to-inertia ratio that gives them quick response to control signals. DC motors are available with horsepower ratings from 1/300 to over 700.

AC motors stall at torque loads about 2 to 2.5 times their rated torque and have a starting torque of about 1.5 times their rated torque. DC motors are capable of delivering over 3 times their rated torque for a short time.

DC motors can easily accomplish dynamic braking and regenerative braking of the load. *Braking* is accomplished by momentarily turning the motor into a dc generator driven by the inertia of the load. In *dynamic braking,* the voltage from the temporary generator is applied to a bank of resistors. The resistors draw current from the generator, causing a torque that tends to slow down the generator and load (fact 1). The resistors dissipate the energy as heat. In *regenerative braking,* the current from the generator is fed back into the dc supply, thus conserving the energy that is normally lost in bringing the load to a quick stop. If the dc supply is a battery, the generator current will charge the battery.

DC servomotors have lightweight low-inertia low-inductance armatures that can respond quickly to commands for a change in position or speed. Servomotors have very low electrical and mechanical time constants. Typical electrical time constants range from 0.1 to 6 ms, and mechanical time constants range from 2.3 ms to over 40 ms. Servomotors occur in a variety of configurations and features, including permanent-magnet field poles, wirewound iron-core armatures, ironless disk armatures (pancake motors), moving-coil armatures with a stationary iron core (shell motors), and brushless motors.

Wound Field DC Motors

Wound field dc motors are classified as series, shunt, compound, and separately excited, depending on how the field winding and the armature winding are connected. The four arrangements are illustrated in Figure 10.10.

The *dc series motor* (Figure 10.10a) has the highest starting torque and the greatest no-load speed of the four types of connections. Integral horsepower series motors are always direct coupled to the load. A belt drive is never used because a broken belt would result in a runaway motor condition. A dc series motor will continue to run in the same direction when the polarity of the line voltage is reversed. The universal motor is a special type of series motor that runs equally well on direct current or alternating current.

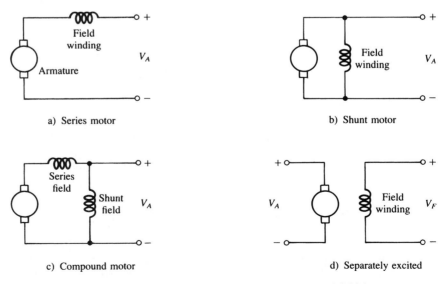

a) Series motor

b) Shunt motor

c) Compound motor

d) Separately excited

◆ **Figure 10.10** Four different methods of connecting a wound-field dc motor.

The *dc shunt motor* has a lower starting torque and a much lower no-load speed than the series motor. As one might expect, the speed and torque characteristics of the *compound motor* are between those of the series motor and the shunt motor.

The *separately excited motor* is a special case of the shunt motor, which allows separate control of the armature voltage and the field voltage. The speed of a separately excited motor can be increased by either decreasing the field voltage or increasing the armature voltage. Variable armature voltage with a fixed field voltage is the most popular method of controlling the speed of a dc motor. The term *armature-controlled dc motor* refers to this method of control.

Permanent-Magnet DC Motors

Permanent-magnet motors use permanent magnets to provide the magnetic field instead of a field winding. These motors are available in fractional and low integral horsepower ratings. The following are some of the advantages of permanent-magnet motors:

1. A simpler, more reliable motor because the field power supply is not required
2. Higher efficiency
3. Less heating, making it possible to completely enclose the motor
4. No possibility of overspeeding due to loss of field
5. A more linear torque-versus-speed curve

Alnico, ceramic, and rare-earth magnets are used in permanent-magnet (PM) motors. The use of rare-earth (samarium–cobalt) magnets has led to significant increases in the torque-to-inertia ratio of PM motors. It has also allowed improved designs such as brushless pancake motors (see the section on brushless motors).

Moving-Coil Stationary-Core DC Motors

The iron-core armature of the dc servomotor is a major obstacle to increasing the torque-to-inertia ratio of the motor. A higher ratio means that more torque is available to accelerate the load. It also means a lower mechanical time constant. In many applications, the motor inertia is 60 to 70% of the total load inertia. Thus a reduction in the motor inertia has a significant effect on the speed of the control system. The *moving-coil stationary-core motor* is one approach to overcoming this obstacle. In this case the solution is to rotate only the armature winding, leaving the core stationary. The armature winding rotates in the annular space between the field poles on the outside and the iron core on the inside. A typical moving-coil, stationary-core motor has a torque-to-inertia ratio several times greater than an iron-core armature motor.

Disk-Armature (Pancake) Motors

The disk-armature dc motor is another approach to achieving a high torque-to-inertia ratio. These motors use a large-diameter, short-length armature of nonmagnetic material. The armature operates between permanent magnetic poles mounted on two stationary disks, one on either side of the armature. The magnetic field passes from the magnetic poles on one disk,

a) Brushless servo motor

◆ **Figure 10.11**   The brushless servomotor has a permanent-magnet rotor, a three-phase stator winding, a rotor position sensor, and a solid-state circuit that controls the current in the three-phase stator winding. (Courtesy of Reliance Motion Control, Eden Prairie, Minn.)

through the armature, to the corresponding magnetic poles on the other disk. This type of construction makes use of the superior magnetic properties of rare-earth magnets. Disk armature motors are also called printed-circuit motors and *pancake* motors. The "printed circuit" name comes from the method used to place the conductors on the armature. The "pancake" name probably comes from the resemblance of the armature and the two stationary disks to a stack of pancakes.

Brushless DC Motors

The primary limitations of the dc servomotor are due to the armature winding and the brush/commutator assembly required to make electrical contact between the winding and the servo drive unit.[*] These limitations include replacement of worn brushes, arcing caused by commutation, voltage and current limitations, high rotor inertia, and a long path for dissipation of heat because most of the heat is generated in the armature. All of these limitations are eliminated by the brushless servomotor illustrated in Figure 10.11.

---

[*] Reliance Motion Control's permission to use information from their Electro-Craft *Handbook* of brushless servo systems is gratefully acknowledged.

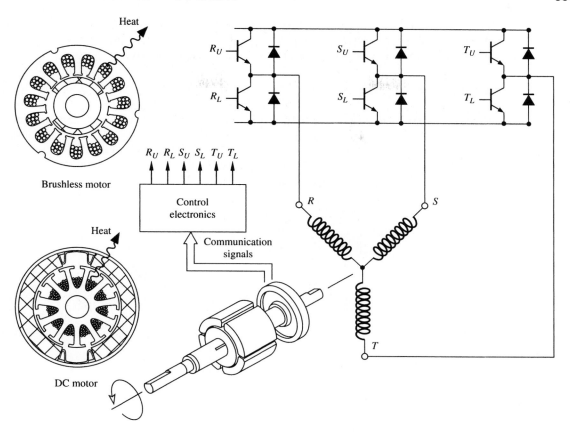

b) Construction of a brushless servo motor

◆ **Figure 10.11**   *(continued)*

The rotor of the *brushless motor* is a permanent magnet, the winding is in the stator, and a solid-state circuit replaces the brush/commutator assembly. The heat from losses is almost entirely in the stator, with a short path for dissipation to ambient. The elimination of the winding and commutator reduces the inertia of the rotor and allows higher rotor speeds. The solid-state commutation circuit eliminates brush replacement and allows higher voltages and currents in the winding.

Two types of permanent magnets are used in the rotor: rare-earth magnets (samarium–cobalt and neodymium–iron–boron) and ceramic magnets (ferrite). The ceramic magnet is low in cost and readily available but has the poorest magnetic properties. The samarium–cobalt magnet has excellent magnetic properties but is expensive and limited in supply. The neodymium–iron–boron magnet also has excellent magnetic properties and is readily available. Brushless motors with rare-earth permanent magnets have the lowest rotor inertia and the smallest motor size for a given torque rating.

The stator winding usually has three phases. The best brushless servo drives use a pulse-width-modulated (PWM) current amplifier to produce a three-phase sinusoidal current in the three stator windings. This type of drive has the smoothest operation at any speed or torque. Other drives use a simpler control circuit to produce a three-phase square-wave current, but the motor operation is not as smooth as it is with the sinusoidal drive.

Steady-State Characteristics

The steady-state operating characteristics of an armature-controlled dc motor are illustrated in Figure 10.12. The torque-versus-current graph (Figure 10.12a) shows a linear relationship between the armature current (i_a) and the motor torque (T). The slope of this line is called the torque constant (K_T). It indicates the change in torque (ΔT) produced by a change in current (Δi_a). The torque versus speed relationship is given by

$$T = K_T i_a - T_f, \text{N} \cdot \text{m} \tag{10.7}$$

$$K_T = \frac{\Delta T}{\Delta i_a}, \text{N} \cdot \text{m/A} \tag{10.8}$$

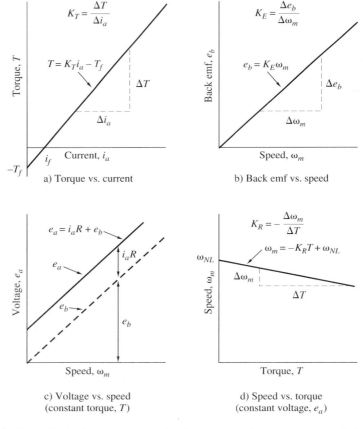

a) Torque vs. current

b) Back emf vs. speed

c) Voltage vs. speed
(constant torque, T)

d) Speed vs. torque
(constant voltage, e_a)

◆ **Figure 10.12**   Steady-state operating characteristics of the armature-controlled dc motor.

The intercept on the current axis, i_f, is the current required to overcome the friction torque, T_f.

The back emf-versus-speed graph (Figure 10.12b) shows a linear relationship between the armature speed (ω_m) and the back emf induced in the armature coil (e_b). The slope of this line is called the voltage constant (K_E). It indicates the change in back emf (Δe_b) produced by a change in armature speed ($\Delta\omega_m$). The back emf is given by

$$e_b = K_E\omega_m, \quad V \tag{10.9}$$

$$K_E = \frac{\Delta e_b}{\Delta\omega_m}, \quad V \cdot s/rad \tag{10.10}$$

The voltage-versus-speed graph (Figure 10.12c) is a graph of the armature voltage versus motor speed with constant torque. This also means a constant armature current (i_a). The armature voltage (e_a) is made up of two components: the back emf (e_b), and the voltage drop across the armature resistance (i_aR). The armature voltage (e_a) is given by

$$e_a = i_aR + e_b, \quad V \tag{10.11}$$

The speed-versus-torque graph (Figure 10.12d) is a graph of the motor torque versus motor speed with constant armature voltage. The equation for speed as a function of torque can be derived from Equations (10.7), (10.9), and (10.11).

Solve Equation (10.7) for i_a.

$$T = K_Ti_a - T_f$$

$$i_a = \frac{T + T_f}{K_T}$$

Substitute $(T + T_f)/K_T$ for i_a in Equation (10.11).

$$e_a = \frac{(T + T_f)R}{K_T} + e_b$$

Substitute $K_E\omega_m$ for e_b [Equation (10.9)], rearrange the terms, and define two constants, K_R and ω_{NL}, to get

$$\omega_m = -K_RT + \omega_{NL}, \text{rad/s} \tag{10.12}$$

$$K_R = \frac{R}{K_EK_T}, \text{rad/N} \cdot m \cdot s \tag{10.13}$$

$$\omega_{NL} = \frac{e_aK_T - RT_f}{K_EK_T}, \text{rad/s} \tag{10.14}$$

DC MOTOR STEADY-STATE CHARACTERISTICS

$$T = K_Ti_a - T_f, \text{N} \cdot m \tag{10.7}$$

$$e_b = K_E\omega_m, \text{V} \tag{10.9}$$

$$e_a = i_aR + e_b, \text{V} \tag{10.11}$$

$$\omega_m = -K_R T + \omega_{NL}, \text{rad/s} \tag{10.12}$$

$$K_R = \frac{R}{K_E K_T}, \text{rad/N} \cdot \text{m} \cdot \text{s} \tag{10.13}$$

$$\omega_{NL} = \frac{e_a K_T - R T_f}{K_E K_T}, \text{rad/s} \tag{10.14}$$

$$p = \omega_m T, \text{W} \tag{10.15}$$

where e_a = armature voltage, V

$\quad\quad e_b$ = back emf, V

$\quad\quad i_a$ = armature current, A

$\quad\quad K_E$ = voltage constant, V · s/rad

$\quad\quad K_R$ = speed regulation constant, rad/N · m · s

$\quad\quad K_T$ = torque constant, N · m/A

$\quad\quad p$ = power, W

$\quad\quad R$ = armature resistance, Ω

$\quad\quad T$ = output torque, N · m

$\quad\quad T_f$ = friction torque, N · m

$\quad\quad \omega_m$ = motor speed, rad/s

$\quad\quad \omega_{NL}$ = no-load motor speed, rad/s

Note: rotational speed is often expressed as the number of revolutions per minute (rpm) with ω replaced by N.

EXAMPLE 10.2

An armature-controlled dc motor has the following ratings:

$$T_f = 0.012 \text{ N} \cdot \text{m}$$
$$K_T = 0.06 \text{ N} \cdot \text{m/A}$$
$$I_{max} = 2 \text{ A}$$
$$K_E = 0.06 \text{ V} \cdot \text{s/rad}$$
$$\omega_{max} = 500 \text{ rad/s}$$
$$R = 1.2 \text{ }\Omega$$

Determine the following:

(a) The maximum output torque, T_{max}.
(b) The maximum power output, P_{max}.
(c) The maximum armature voltage, E_{max}.
(d) The no-load motor speed when $e = E_{max}$.

Solution

(a) The maximum output torque is obtained from Equation (10.7), when $i = I_{max}$.

$$T_{max} = K_T I_{max} - T_f = (0.06)(2) - 0.012$$
$$= 0.108 \text{ N} \cdot \text{m}$$

(b) The maximum power output is obtained from Equation (10.15) when ω and T are both a maximum.

$$P_{max} = \omega_{max} T_{max} = (500)(0.0108)$$

$$= 54 \text{ W}$$

(c) The maximum armature voltage is obtained from Equations (10.9) and (10.10) when i and ω are both a maximum.

$$e_b = K_E \omega_m$$

$$e_a = i_a R + e_b = i_a R + K_E \omega_m$$

$$E_{max} = I_{max} R + K_E \omega_{max}$$

$$= (2)(1.2) + (0.06)(500)$$

$$= 32.4 \text{ V}$$

(d) The no-load motor speed is obtained from Equation (10.14) when $e = E_{max}$.

$$\omega_{NL} = \frac{E_{max} K_T - RT_f}{K_E K_T} = \frac{(32.4)(0.06) - (1.2)(0.012)}{(0.06)(0.06)}$$

$$= 536 \text{ rad/s} \qquad \qquad \blacklozenge$$

EXAMPLE 10.3   The motor in Example 10.2 is operated at 300 rad/s with a load torque of 0.05 N · m. Determine the following:

(a) The armature voltage.

(b) The armature speed if the torque increases to 0.075 N · m and the armature voltage is not changed.

Solution   **(a)** From Equation (10.7),

$$T = K_T i - T_f$$

$$i_a = \frac{T + T_f}{K_T} = \frac{0.05 + 0.012}{0.06}$$

$$= 1.03 \text{ A}$$

From Equation (10.9),

$$e_b = K_E \omega_m = (0.06)(300) = 18 \text{ V}$$

From Equation (10.11),

$$e_a = i_a R + e_b = (1.03)(1.2) + 18$$

$$= 19.24 \text{ V}$$

(b) From Equation (10.13),

$$K_R = \frac{R}{K_E K_T} = \frac{1.2}{(0.06)(0.06)} = 333.3$$

From Equation (10.12),

$$\omega_1 = -K_R T_1 + \omega_{NL} \Rightarrow \omega_{NL} = \omega_1 + K_R T_1$$

$$\omega_2 = -K_R T_2 + \omega_{NL} = -K_R T_2 + \omega_1 + K_R T_1$$

$$= \omega_1 - K_R(T_2 - T_1)$$

$$= 300 - (333.3)(0.075 - 0.050)$$

$$= 291.7 \text{ rad/s}$$ ◆

DC Motor Transfer Functions

The schematic diagram and the block diagram of an armature-controlled dc motor are illustrated in Figure 10.13. The electrical circuit consists of the resistance (R) and inductance (L) of the armature winding plus a source to represent the back emf induced in the armature winding. Equation (10.9) defines the back emf in terms of the motor speed (ω_m).

$$e_b = K_E \omega_m$$

The corresponding s-domain (frequency domain) equation is

$$E_b(s) = K_E \Omega_m(s) \tag{10.16}$$

Equation (10.16) is represented by the feedback block (K_E) in Figure 10.13b.

A variable voltage (e_a) is applied to the armature winding. The resulting armature current (i_a) is defined by the following time-domain equation:

$$e_a = L\frac{di_a}{dt} + Ri_a + e_b$$

The corresponding s-domain equation is

$$E_a(s) = LsI_a(s) + RI_a(s) + E_b(s)$$

or

$$I_a(s) = (E_a(s) - E_b(s))\left(\frac{1}{R + Ls}\right) \tag{10.17}$$

Equation (10.17) is represented in Figure 10.13b by the summing junction and the first block on the right side of the summing junction.

The armature current (i_a) produces torque (T) as defined by Equation (10.7). In the transfer function, we neglect the friction torque and use the following equation for the motor torque:

$$T = K_T i_a$$

The corresponding s-domain equation is

$$T(s) = K_T I_a(s) \tag{10.18}$$

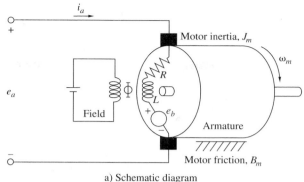

a) Schematic diagram

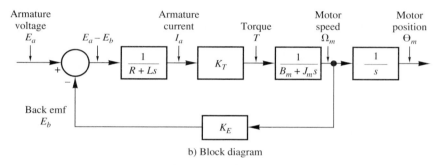

b) Block diagram

◆ **Figure 10.13**   Armature-controlled dc motor.

Equation (10.18) is represented in Figure 10.13b by the second block in the forward path, K_T.

The motor torque acts on the motor inertia (J_m) and viscous friction (B_m) as defined by the following time-domain equation:

$$T = J_m \frac{d\omega_m}{dt} + B_m\omega_m$$

The corresponding s-domain equation is

$$T(s) = J_m s\Omega_m(s) + B_m\Omega_m(s)$$

or

$$\Omega_m(s) = T(s)\left(\frac{1}{B_m + J_m s}\right) \tag{10.19}$$

Equation (10.19) is represented in Figure 10.13b by the third block in the forward path.

The motor position (θ_m) is obtained by taking the integral of the motor speed.

$$\theta_m = \int \omega_m \, dt$$

The corresponding s-domain equation is

$$\Theta_m(s) = \left(\frac{1}{s}\right)\Omega_m(s) \tag{10.20}$$

Equation (10.20) is represented in Figure 10.13b by the fourth (and last) block in the forward path.

Our objective in developing the preceding equations is to obtain a transfer function for the dc motor. In fact, we really could use two transfer functions, one for velocity (speed) control and one for position control. Proceed as follows to obtain the two transfer functions:

1. Replace $E_b(s)$ in Equation (10.17) by $K_E\Omega_m(s)$ from Equation (10.16).
2. Replace $T(s)$ in Equation (10.19) by $K_TI_a(s)$ from Equation (10.18).
3. Replace $I_a(s)$ in the new Equation (10.19) by the right-hand side of the new Equation (10.17).
4. Solve the result of step 3 for $\Omega_m(s)/E_a(s)$. This is the velocity transfer function.
5. Multiply the velocity transfer function by $1/s$ to get the position transfer function [see Equation (10.20)].

The result of step 4 is the following *velocity transfer function:*

$$\frac{\Omega_m(s)}{E_a(s)} = \frac{K_T}{RB_m + K_EK_T + (RJ_m + B_mL)s + LJ_ms^2}, \text{rad/V} \cdot \text{s}$$

or

$$\frac{\Omega_m(s)}{E_a(s)} = \frac{K_T}{RB_m + K_EK_T + RB_m(\tau_m + \tau_e)s + RB_m\tau_m\tau_es^2}, \text{rad/V} \cdot \text{s} \tag{10.21}$$

where τ_m = mechanical time constant = J_m/B_m, s

τ_e = electrical time constant = L/R, s

The electrical time constant (τ_e) is usually much smaller than the mechanical time constant (τ_m). If the armature inductance is small or the motor inertia is large, we can assume $\tau_e = 0$ to obtain the following *simplified velocity transfer function* ($\tau_e = 0$):

$$\frac{\Omega_m(s)}{E_a(s)} = \frac{K_s}{1 + \tau_ss}, \text{rad/V} \cdot \text{s} \tag{10.22}$$

where K_s = simplified gain = $\dfrac{K_T}{RB_m + K_EK_T}$, rad/V · s

τ_s = simplified time constant = $\dfrac{RJ_m}{RB_m + K_EK_T}$, s

Another variation of the velocity transfer function occurs when a current source is used to supply the armature current. Because the current source output is the armature current (i_a), we have no need for Equation (10.17). Using Equation (10.18) and Equation (10.19), we obtain the following s-domain equation:

$$\Omega_m(s) = K_TI_a(s)\left(\frac{1}{B_m + J_ms}\right)$$

The *current driven velocity transfer function* is

$$\frac{\Omega_m(s)}{I_a(s)} = \frac{K_c}{1 + \tau_m s}, \text{rad/A} \cdot \text{s} \tag{10.23}$$

where K_c = current driven gain = K_T/B_m, rad/A · s

τ_m = mechanical time constant = J_m/B_m, s

For our final version of the velocity transfer function, we consider both the motor and the load driven by the motor. In some applications, the effect of the load is slight, and the preceding transfer functions are adequate. In other applications, the effect of the load is significant and must be considered in the system design. If the motor shaft is directly coupled to a load with inertia J_L and viscous friction B_L, the transfer function is given by Equation (10.21) with the following modifications:

1. Replace B_m by $B_T = B_m + B_L$
2. Replace J_m by $J_T = J_m + J_L$
3. Replace τ_m by $\tau_T = J_T/B_T$

In some applications, the motor is connected to the load through a set of gears that cause the load to rotate at a slower speed than the motor. We call this gear assembly a speed reducer. Figure 10.14a shows a speed reducer between the motor and the load. The motor shaft is connected to the smaller gear, which has N_1 teeth. The load is connected to the larger gear, which has N_2 teeth. A speed reducer is the mechanical analog of an ideal transformer. A transformer can be used to decrease current and increase voltage. In an analogous manner, a speed reducer can be used to decrease speed and increase torque. Equations (10.24) and (10.25) define these two changes (in a speed reducer, $N_1 < N_2$).

$$\omega_L = \left(\frac{N_1}{N_2}\right)\omega_m, \text{rad/s} \tag{10.24}$$

$$T_L = \left(\frac{N_2}{N_1}\right)T_m, \text{N} \cdot \text{m} \tag{10.25}$$

The analogy does not end with decreased speed and increased torque. A transformer changes the electrical impedance such that the resistance and inductance of the load appear smaller at the input to the transformer. In an analogous manner, a speed reducer changes the mechanical "impedance" such that the inertia and friction of the load appear smaller at the input of the speed reducer. We can combine the reduced load inertia and friction with the motor inertia and friction to obtain the following total inertia, friction, and mechanical time constant.

$$J_T = J_m + \left(\frac{N_1}{N_2}\right)^2 J_L, \text{N} \cdot \text{m} \cdot \text{s}^2/\text{rad} \tag{10.26}$$

$$B_T = B_m + \left(\frac{N_1}{N_2}\right)^2 B_L, \text{N} \cdot \text{m} \cdot \text{s}/\text{rad} \tag{10.27}$$

$$\tau_T = \frac{J_T}{B_T}, \text{s} \tag{10.28}$$

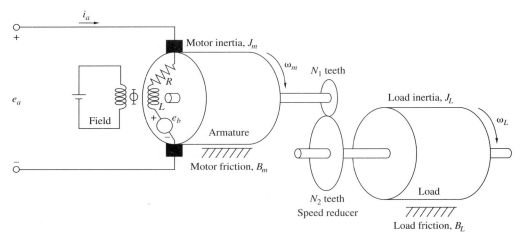

a) Schematic diagram

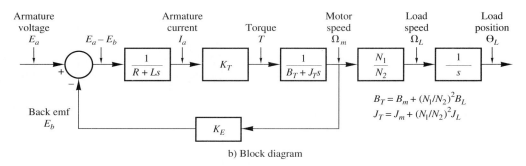

b) Block diagram

◆ **Figure 10.14** Armature-controlled dc motor with speed reducer (N_1/N_2) and load.

Figure 10.14b shows the block diagram of a dc motor, speed reducer, and load. Notice that the load inertia and friction are included with the motor inertia and friction in the motor block. It is simpler to combine the motor and the load than it is to consider them separately. To get the transfer function, we take Equation (10.21) and replace B_m by B_T, J_m by J_T, and τ_m by τ_T. Finally, we multiply by N_1/N_2 to get the following *velocity transfer function of a motor, speed reducer, and load.*

$$\frac{\Omega_L(s)}{E_a(s)} = \frac{K_T N_1/N_2}{RB_T + K_E K_T + RB_T(\tau_T + \tau_e)s + RB_T \tau_T \tau_e s^2}, \text{rad/V} \cdot \text{s} \qquad \textbf{(10.29)}$$

ARMATURE-CONTROLLED DC MOTOR TRANSFER FUNCTIONS

Note: The velocity transfer functions are given in detail. The corresponding position transfer functions are obtained by dividing the velocity transfer function by s [see Equation (10.30)].

$$\frac{\Theta_m(s)}{E_a(s)} = \frac{1}{s}\left(\frac{\Omega_m(s)}{E_a(s)}\right), \text{rad/V} \qquad \textbf{(10.30)}$$

Velocity Transfer Function of a Voltage-Driven Motor

$$\frac{\Omega_m(s)}{E_a(s)} = \frac{K_T}{RB_m + K_EK_T + RB_m(\tau_m + \tau_e)s + RB_m\tau_m\tau_es^2}, \text{rad/V} \cdot \text{s} \qquad \textbf{(10.21)}$$

Simplified Velocity Transfer Function of a Voltage-Driven Motor ($\tau_e = 0$)

$$\frac{\Omega_m(s)}{E_a(s)} = \frac{K_s}{1 + \tau_s s}, \text{rad/V} \cdot \text{s} \qquad \textbf{(10.22)}$$

where

$$K_s = \frac{K_T}{RB_m + K_EK_T}, \text{rad/V} \cdot \text{s}$$

$$\tau_s = \frac{RJ_m}{RB_m + K_EK_T}, \text{s}$$

Velocity Transfer Function of a Current-Driven Motor

$$\frac{\Omega_m(s)}{I_a(s)} = \frac{K_c}{1 + \tau_m s}, \text{rad/A} \cdot \text{s} \qquad \textbf{(10.23)}$$

where

$$K_c = K_T/B_m, \quad \text{rad/A} \cdot \text{s}$$

$$T_m = J_m/B_m, \quad \text{s}$$

Velocity Transfer Function of a Voltage-Driven Motor, Speed Reducer, and Load

$$J_T = J_m + \left(\frac{N_1}{N_2}\right)^2 J_L, \text{N} \cdot \text{m} \cdot \text{s}^2/\text{rad} \qquad \textbf{(10.26)}$$

$$B_T = B_m + \left(\frac{N_1}{N_2}\right)^2 B_L, \text{N} \cdot \text{m} \cdot \text{s}/\text{rad} \qquad \textbf{(10.27)}$$

$$\tau_T = J_T/B_T, \text{S} \qquad \textbf{(10.28)}$$

$$\frac{\Omega_L(s)}{E_a(s)} = \frac{K_TN_1/N_2}{RB_T + K_EK_T + RB_T(\tau_T + \tau_e)s + RB_T\tau_T\tau_es^2}, \text{rad/V} \cdot \text{s} \qquad \textbf{(10.29)}$$

where B_L = viscous friction of load, N $\cdot$ m $\cdot$ s/rad

B_m = viscous friction of motor, N $\cdot$ m $\cdot$ s/rad

B_T = friction of motor and load, N $\cdot$ m $\cdot$ s/rad

E_a = armature voltage, V

J_L = moment of inertia of load, N $\cdot$ m $\cdot$ s^2/rad*

J_m = moment of inertia of motor, N $\cdot$ m $\cdot$ s^2/rad

J_T = inertia of motor and load, N $\cdot$ m $\cdot$ s^2/rad

K_c = current driven gain, rad/A $\cdot$ s

K_E = voltage constant, V $\cdot$ s/rad

K_s = simplified gain, rad/V $\cdot$ s

K_T = torque constant, N $\cdot$ m/A

L = armature inductance, H

N_1 = number of teeth in the input gear

N_2 = number of teeth in the output gear

R = armature resistance, Ω

s = frequency parameter, 1/s

τ_e = electrical time constant = L/R, s

τ_m = mechanical time constant = J_m/B_m, s

τ_s = simplified time constant, s

Θ_L = position of load, rad

Θ_m = position of motor, rad

Ω_L = speed of load, rad/s

Ω_m = speed of motor, rad/s

---

*The units of J are sometimes expressed as kg · m^2/rad. We used the equivalent units, N · m · s^2/rad, because they facilitate the dimensional analysis of the torque–load equation.

$$T = J\frac{d\omega}{dt}$$

EXAMPLE 10.4

A dc motor has the following specifications.

Type: dc permanent magnet motor

Model No. ICST-5

Maximum operating speed, ω_{max} (rad/s)	500
Maximum armature current, I_{max} (A)	2.0
Voltage constant, K_E (V · s/rad)	0.06
Torque constant, K_T (N · m/A)	0.06
Friction torque, T_f (N · m)	0.012
Armature resistance, R (Ω)	1.2
Armature inductance, L (H)	0.02
Armature moment of inertia, J_m (N · m · s^2/rad)	6.2E − 4
Armature viscous friction, B_m (N · m · s/rad)	1.0E − 4

Determine the following velocity and position transfer functions.

(a) The voltage-driven motor

(b) The simplified voltage-driven motor with τ_e neglected

(c) The current-driven motor

(d) The voltage-driven motor with the following speed reducer and load specifications

$N_1 = 16$ teeth $\qquad\qquad N_2 = 32$ teeth

$J_L = 8.2E − 4$ N · m · s^2/rad $\qquad\qquad B_L = 0.15E − 5$ N · m · s/rad

Solution

(a) From Equation (10.21)

$$\tau_m = \frac{J_m}{B_m} = \frac{6.2E − 4}{1E − 4} = 6.2 \text{ s}$$

$$\tau_e = \frac{L}{R} = \frac{0.02}{1.2} = 0.0167 \text{ s}$$

$$RB_m = (1.2)(1E - 4) = 1.2E - 4$$

$$RB_m + K_E K_T = 1.2E - 4 + (0.06)(0.06) = 0.00372$$

$$RB_m(\tau_m + \tau_e) = 1.2E - 4(6.2 + 0.0167) = 7.46E - 4$$

$$RB_m \tau_m \tau_e = 1.2E - 4(6.2)(0.0167) = 1.24E - 5$$

The voltage-driven velocity transfer function is

$$\frac{\Omega_m(s)}{E_a(s)} = \frac{0.06}{0.00372 + 7.46 \times 10^{-4}s + 1.24 \times 10^{-5}s^2} \text{ rad/V} \cdot \text{s}$$

or

$$\frac{\Omega_m(s)}{E_a(s)} = \frac{16.13}{1 + 0.201s + 0.00333s^2} \text{ rad/V} \cdot \text{s}$$

The voltage-driven position transfer function is

$$\frac{\Theta_m(s)}{E_a(s)} = \frac{16.13}{s + 0.201s^2 + 0.00333s^3} \text{ rad/V}$$

(b) From Equation (10.22)

$$K_s = \frac{K_T}{RB_m + K_E K_T} = \frac{0.06}{0.00372} = 16.13 \text{ rad/V} \cdot \text{s}$$

$$\tau_s = \frac{RJ_m}{RB_m + K_E K_T} = \frac{(1.2)(6.2E - 4)}{0.00372} = 0.20 \text{ s}$$

The simplified velocity transfer function is

$$\frac{\Omega_m(s)}{E_a(s)} = \frac{16.13}{1 + 0.20s} \text{ rad/V} \cdot \text{s}$$

The simplified position transfer function is

$$\frac{\Theta_m(s)}{E_a(s)} = \frac{16.13}{s + 0.20s^2} \text{ rad/V}$$

(c) From Equation (10.23)

$$K_c = \frac{K_T}{B_m} = \frac{0.06}{1E - 4} = 600 \text{ rad/A} \cdot \text{s}$$

$$\tau_m = 6.2 \text{ s}$$

The current-driven velocity transfer function is

$$\frac{\Omega_m(s)}{I_a(s)} = \frac{600}{1 + 6.2s} \text{ rad/A} \cdot \text{s}$$

The current-driven position transfer function is

$$\frac{\Theta_m(s)}{I_a(s)} = \frac{600}{s + 6.2s^2} \text{ rad/V}$$

(d) From Equations (10.26) through (10.29)

$$\frac{K_T N_1}{N_2} = \frac{(0.06)(16)}{32} = 0.03$$

$$J_T = (16/32)^2 (8.2\text{E} - 4) + 6.2\text{E} - 4 = 8.25\text{E} - 4 \text{ N} \cdot \text{m} \cdot \text{s}^2/\text{rad}$$

$$B_T = (16/32)^2 (0.15\text{E} - 5) + 1.0\text{E} - 4 = 1.004\text{E} - 4 \text{ N} \cdot \text{m} \cdot \text{s}/\text{rad}$$

$$\tau_T = \frac{J_T}{B_T} = \frac{8.25\text{E} - 4}{1.004\text{E} - 4} = 8.22s$$

$$RB_T = (1.2)(1.004\text{E} - 4) = 1.205\text{E} - 4$$

$$RB_T + K_E K_T = 1.205\text{E} - 4 + (0.06)(0.06) = 3.72\text{E} - 3$$

$$RB_T(\tau_T + \tau_e) = 1.205\text{E} - 4(8.22 + 0.0167) = 9.925\text{E} - 4$$

$$RB_T \tau_T \tau_e = 1.205\text{E} - 4(8.22)(0.0167) = 1.654\text{E} - 5$$

The velocity transfer function is

$$\frac{\Omega_L(s)}{E_a(s)} = \frac{0.03}{0.00372 + 9.925 \times 10^{-4}s + 1.654 \times 10^{-5}s^2} \text{ rad/V} \cdot \text{s}$$

or

$$\frac{\Omega_L(s)}{E_a(s)} = \frac{8.06}{1 + 0.267s + 0.00445s^2} \text{ rad/V} \cdot \text{s}$$

The position transfer function is

$$\frac{\Theta_L(s)}{E_a(s)} = \frac{8.06}{s + 0.267s^2 + 0.00445s^3} \text{ rad/V}$$

◆

EXAMPLE 10.5

The following specifications were obtained from a manufacturer's catalog.[*]

Type: dc permanent magnet motor

Model No. E550-MG

Rated voltage, E_r (V)	30
No-load speed, $N\sigma$ (rpm)	6500
Maximum armature current, I_{max} (A)	2.2
Maximum torque, T_{max} (oz · in.)	13.6
Friction torque, T_f (oz · in.)	3.0
Voltage constant, K_E (V/krpm)	4.6
Torque constant, K_T (oz · in./A)	6.2
Armature resistance, R (Ω)	2.2
Armature inductance, L (mH)	4.6
Armature mom. of inertia, J_m (oz · in. · s²/rad)	4.4E − 3
Armature viscous friction, B_m (oz · in./krpm)	0.15

[*]*Speed and Position Control Systems*, Electro-Craft Corporation, Hopkins, Minn., undated.

Convert the motor parameters to SI units and determine the full and simplified velocity transfer functions.

Solution   The necessary conversion factors are in Appendix B.

$$T_{max} = (13.6 \text{ oz} \cdot \text{in.})(7.062\text{E} - 3) = 0.0960 \text{ N} \cdot \text{m}$$

$$T_f = (3.0 \text{ oz} \cdot \text{in.})(7.062\text{E} - 3) = 0.0212 \text{ N} \cdot \text{m}$$

$$K_E = (4.6 \text{ V/krpm})(9.5493\text{E} - 3) = 0.044 \text{ V} \cdot \text{s/rad}$$

$$K_T = (6.2 \text{ oz} \cdot \text{in./A})(7.062\text{E} - 3) = 0.044 \text{ N} \cdot \text{m/A}$$

$$J_m = (4.4\text{E} - 3 \text{ oz} \cdot \text{in.} \cdot \text{s}^2/\text{rad})(7.062\text{E} - 3) = 3.11\text{E} - 5 \text{ N} \cdot \text{m} \cdot \text{s}^2/\text{rad}$$

$$B_m = (0.15 \text{ oz} \cdot \text{in./krpm})(6.744\text{E} - 5) = 1.01\text{E} - 5 \text{ N} \cdot \text{m} \cdot \text{s/rad}$$

$$\tau_e = \frac{L}{R} = \frac{0.046}{2.2} = 2.09 \text{ ms}$$

$$\tau_m = \frac{J_m}{B_m} = \frac{3.1\text{E} - 5}{1.0\text{E} - 5} = 3.07 \text{ s}$$

From Equation (10.21)

$$RB_m = (2.2)(1.01\text{E} - 5) = 2.23\text{E} - 5$$

$$RB_m + K_E K_T = 2.23\text{E} - 5 + (0.044)(0.044) = 1.95\text{E} - 3$$

$$RB_m (\tau_m + \tau_e) = 2.23\text{E} - 5(3.07 + 0.00209) = 6.84\text{E} - 5$$

$$RB_m \tau_m \tau_e = 2.23\text{E} - 5(3.07)(0.00209) = 1.43\text{E} - 7$$

The voltage-driven velocity transfer function is

$$\frac{\Omega_m(s)}{E_a(s)} = \frac{0.044}{1.95 \times 10^{-3} + 6.84 \times 10^{-5} s + 1.43 \times 10^{-7} s^2} \text{ rad/V} \cdot \text{s}$$

or

$$\frac{\Omega_m(s)}{E_a(s)} = \frac{22.5}{1 + 0.035s + 7.35 \times 10^{-5} s^2} \text{ rad/V} \cdot \text{s}$$

From Equation 10.22

$$K_s = \frac{K_T}{RB_m + K_E K_T} = \frac{0.044}{1.95\text{E} - 3} = 22.5 \text{ rad/V} \cdot \text{s}$$

$$\tau_s = \frac{RJ_m}{RB_m + K_E K_T} = \frac{2.2(3.11\text{E} - 5)}{1.95\text{E} - 3} = 0.035 \text{ s}$$

The simplified velocity transfer function is

$$\frac{\Omega_m(s)}{E_a(s)} = \frac{22.5}{1 + 0.035 s} \text{ rad/V} \cdot \text{s}$$

◆

10.4 STEPPING MOTORS

Stepping motors are brushless dc motors designed to convert digital pulses into fixed increments of motion called steps. They provide accurate open-loop positioning of a load that can be controlled directly by computers, microprocessors, and programmable controllers. Because of their brushless construction, stepping motors are reliable, robust, and maintenance-free.

Stepping motors are used in many applications, including robots, machine tools, medical equipment, scientific instruments, plotters, and computer peripherals. Figure 10.15 shows a typical machine tool application of a stepping motor. The pulse inputs to the control unit could come from a computer or a programmable controller. The control unit converts each pulse into the appropriate switching of the drive's power transistors to move the motor shaft one step. The lead screw converts the rotary motion of the motor shaft into linear motion of the workpiece. For example, consider a stepping motor with a step size of 1.8° and a lead screw with a pitch of 0.2 in. If a programmable controller sends 200 pulses to the control unit, the motor shaft will rotate 360° (200 × 1.8° = 360°). The lead screw will also rotate 360°, and one revolution of the lead screw will move the workpiece a distance of 0.2 in. Thus each pulse moves the workpiece 0.001 in. (0.2 in./200 steps = 0.001 in./step).

Stepping motors have a wound stator and a nonexcited rotor. The stator has an even number of equally spaced poles (or teeth), each with an electrical coil. Opposing pairs of stator coils are connected in series, such that when one acts as a north pole, the other will act as a

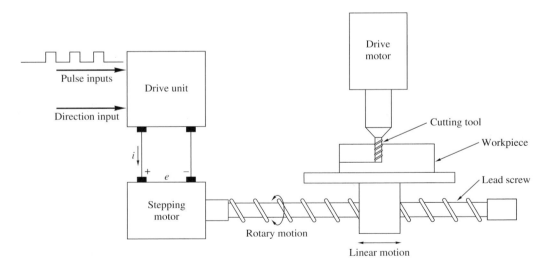

◆ **Figure 10.15** A stepping motor provides open-loop, digital control of the position of a workpiece in a numerical control machine. The drive unit receives a direction input (cw or ccw) and pulse inputs. For each pulse it receives, the drive unit manipulates the motor voltage and current, causing the motor shaft to rotate by a fixed angle (one step). The lead screw converts the rotary motion of the motor shaft into linear motion of the workpiece.

south pole. The stator may have two, three, or four independent circuits, or phases, connected to north–south pole pairs. The rotor has equally spaced external teeth with a small air gap between the rotor teeth and the stator teeth (or poles).

The number of teeth on the rotor, the number of teeth on the stator, and the number of phases on the stator determine the size of the step (called the step angle). For simple stepping motors, the following equation may be used:

$$\text{Step angle} = \frac{360°}{(\text{rotor teeth})(\text{stator phases})}$$

A stepping motor with two rotor teeth and two stator phases has a 90° step angle. A stepping motor with four rotor teeth and three stator phases has a 30° step angle. Electronic methods are used to reduce the step angle even further. Half-step operation is a method of electronically dividing each step into two half-steps. Microstep operation is a method of electronically dividing a step into 10, 16, 32, or 125 microsteps. See the sections on half-step and microstep operation for further explanation of these techniques.

Stepping motors are classified as two-, three-, or four-phase, depending on the number of windings on the stator. They are also classified as variable reluctance, permanent magnet, or hybrid, depending on the type of rotor.

Variable-reluctance (VR) stepping motors have a soft-iron rotor. When electric current is applied to a stator phase, magnetic flux is generated that causes teeth in the rotor to line up with teeth in the stator. When the current is switched to the next phase, the rotor moves a distance of one step angle to again line up the rotor and stator teeth. The VR stepping motors are used for larger step angles (e.g., 15°, 30°, and 45°).

Permanent-magnet (PM) motors have one or more permanent pole pairs in the rotor. Early PM stepping motors had only two rotor teeth and large step angles. The use of rare-earth magnets has greatly improved the design and performance of PM stepping motors.

Hybrid stepping motors have a variable-reluctance rotor with a permanent magnet in its magnetic path, usually in the rotor. The term *hybrid* refers to the use of two sources of magnetic field, the stator windings and the permanent magnet. Hybrid stepping motors are used when small step angles are required (e.g., 1.8°, 2.5°). The 1.8° stepping motor is the predominant standard for industrial automation, scientific applications, and office systems.

A typical 1.8° hybrid stepping motor has eight poles, each with 5 teeth and a simple coil. The rotor has two end caps, each with 50 teeth. The end caps are separated by a permanent magnet, so that one cap is given a north polarity, the other a south polarity. The stator coils are connected into two phases, A and B. The motor is stepped in the clockwise direction by energizing the phases in the sequence +A, −B, −A, +B, +A. The motor is stepped in the counterclockwise direction by energizing the phases in the reverse sequence, +A, +B, −A, −B, +A.

Full-Step Operation

As the name implies, *full-step operation* of a stepping motor consists of a movement of one full step for each input pulse. Full-step operation is the simplest of the three methods of stepping. An understanding of full-step operation is also a prerequisite to understanding two-step and microstep operation. Figure 10.16a shows the model of a stepping motor that will be used to explain the three methods of stepping.

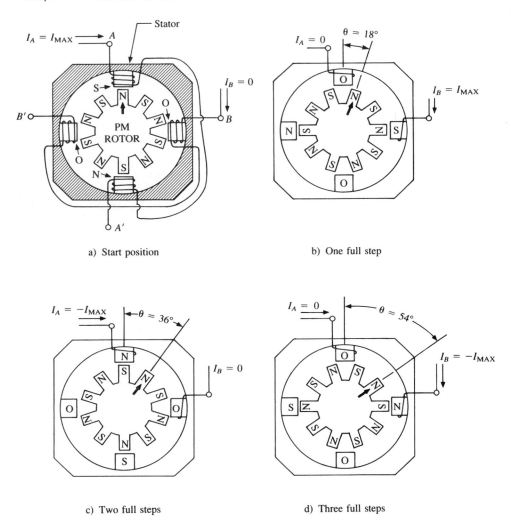

a) Start position

b) One full step

c) Two full steps

d) Three full steps

◆ **Figure 10.16** This model of a stepping motor has four teeth/pole on the stator and 10 teeth/pole on the PM rotor, which gives it 20 full steps per revolution. Each full step is $18°$ (i.e., $360°/20 = 18°$).

The model stepping motor is a two-phase motor with four wound poles in the stator. The top and bottom coils are connected in series to form phase A. The two side coils are connected in series to form phase B. The input currents are applied to the terminals marked A and B on the top and right side. The rotor is a permanent-magnet type with 10 alternating north and south poles. The resolution of the motor is equal to the product of the number of poles in the rotor (10) times the number of pole pairs (2) in the stator. Thus the model stepping motor has 20 steps per revolution and a step angle of $18°$. The top rotor pole in Figure 10.16a is marked with an arrow. The arrow will enable us to follow the movement of this "marked" pole as the rotor makes several full steps.

In Figure 10.16a, the motor is in the start position. A current of I_{max} A enters the phase A winding at terminal A. No current enters the phase B winding. The direction of current I_A is such that the top stator pole is a south pole (labeled "S") and the bottom stator pole is a north pole (labeled "N"). The absence of current in phase B results in a demagnetized condition for the two side poles (labeled "0"). The alignment of the rotor poles is such that the magnetic attraction between unlike poles will hold the rotor in the position shown.

In Figure 10.16b, a current of I_{max} enters the phase B winding at terminal B, and no current enters the phase A winding. Notice the 0 labels on the top and bottom stator poles, the S label on the right-side stator pole, and the N label on the left-side stator pole. The magnetic attraction between opposite poles and the repulsion between like poles has caused the rotor to rotate one step in the clockwise direction. Once again, the magnetic forces will hold the rotor in the new position shown in Figure 10.16b. Two more steps are shown in Figure 10.16c and d. Table 10.1 lists the angle θ and the two phase currents for five full steps.

Half-Step Operation

Half-steps are accomplished by applying partial currents to both phase windings to position the rotor halfway between two full-step positions. Figure 10.17 illustrates three half-steps of the model stepping motor. The start conditions in Figure 10.17a are identical to the conditions in Figure 10.16a. The first half-step is accomplished by reducing the phase A current to $0.707I_{max}$ A and increasing the phase B current to $0.707I_{max}$ A (see Figure 10.17b). Notice that all four stator coils are magnetized. The top and right side are south poles and the bottom and left side are north poles. The magnetic forces are such that the rotor is held in the position shown in Figure 10.17b. Notice that the position of the marked rotor pole, θ, is 9°, exactly half of a full 18° step. Two more half-steps are shown in Figure 10.17c and d. Table 10.2 lists the angle θ and the two phase currents for 10 half-steps.

Microstep Operation

In half-step operation, we saw how the rotor could be positioned halfway between two full-step positions by supplying current to both phase windings. Microstepping simply extends this technique to more than one midposition by using different values of current in each phase. The microstep sizes that are most commonly used are 1/10, 1/16, 1/32, and 1/125 of

◆ **TABLE 10.1** Sequence of Phase Currents for Five Full Steps of the Model Stepping Motor (Figure 10.16)

Step	θ (deg)	I_A/I_{max}	I_B/I_{max}
0	0	1	0
1	18	0	1
2	36	−1	0
3	54	0	−1
4	72	1	0
5	90	0	1

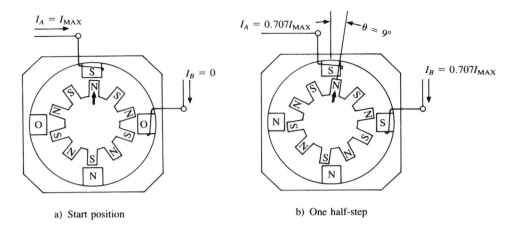

a) Start position

b) One half-step

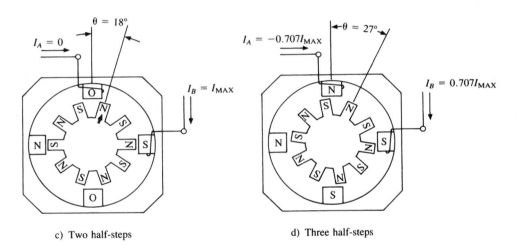

c) Two half-steps

d) Three half-steps

◆ **Figure 10.17** Half-stepping increases the resolution of the model stepping motor to 40 half-steps per revolution. In half-stepping, the driver first turns on phase A, then both A and B, then B only, then both A and B (with reversed polarity on A).

a full step. An obvious advantage of microstepping is the much finer resolution it provides. For example, when 125 microsteps are used in a stepping that has 200 full steps per revolution, the resolution is 200(125) = 25,000 microsteps per revolution.

The values of the phase currents required for a microstep are given by

$$I_A = \cos\left(\frac{90n}{s}\right)I_{\text{max}} \tag{10.31}$$

$$I_B = \sin\left(\frac{90n}{s}\right)I_{\text{max}} \tag{10.32}$$

◆ **TABLE 10.2** Sequence of Phase Currents for 10 Half-Steps of the Model Stepping Motor (Figure 10.16)

Half-Step	θ (°)	I_A/I_{max}	I_B/I_{max}
0	0	1	0
1	9	0.707	0.707
2	18	0	1
3	27	−0.707	0.707
4	36	−1	0
5	45	−0.707	−0.707
6	54	0	−1
7	63	0.707	−0.707
8	72	1	0
9	81	0.707	0.707
10	90	0	1

where I_{max} = maximum value of phase current, A

$\quad I_A$ = current in phase A winding, A

$\quad I_A$ = current in phase B winding, A

$\quad n$ = number of microsteps from start position

$\quad s$ = number of microsteps in a full step

The values of these currents may be stored in a ROM memory chip to be read when needed by the driver circuit. Table 10.3 lists the angle θ and the two phase currents for ten 1/10 microsteps for the model stepping motor (Figure 10.16).

◆ **TABLE 10.3** Sequence of Phase Currents for 10 Microsteps of the Model Stepping Motor (Figure 10.16)[a]

Microstep	θ (°)	I_A/I_{max}	I_B/I_{max}
0	0	1.000	0.000
1	1.8	0.988	0.156
2	3.6	0.951	0.309
3	5.4	0.891	0.454
4	7.2	0.809	0.588
5	9.0	0.707	0.707
6	10.8	0.588	0.809
7	12.6	0.454	0.891
8	14.4	0.309	0.951
9	16.2	0.156	0.988
10	18.0	0.000	1.000

[a]Microstep size = 0.1 (full step size).

10.5 AC ADJUSTABLE-SPEED DRIVES

AC adjustable-speed drives (both single-phase and three-phase) consist of two major parts, a converter and an inverter. The *converter* converts the ac input power to a dc voltage. The *inverter* changes the dc voltage back into an ac voltage of any desired frequency from about 3 to 60 (or 120) Hz. The output of the inverter will drive an ac synchronous motor or an ac induction motor at a speed determined by the frequency of the inverter output. For example, the speed of an 1800-rpm synchronous motor can be adjusted from 90 to 1800 rpm by a 3- to 60-Hz inverter. Figure 10.18a shows the block diagram of a three-phase inverter and the ideal waveforms of the output. The actual voltage produced by the inverter is not sinusoidal but approximates a sine wave. Two methods of approximating a sine wave, the variable-voltage inverter and the pulse-width-modulated inverter, are explained later in this section.

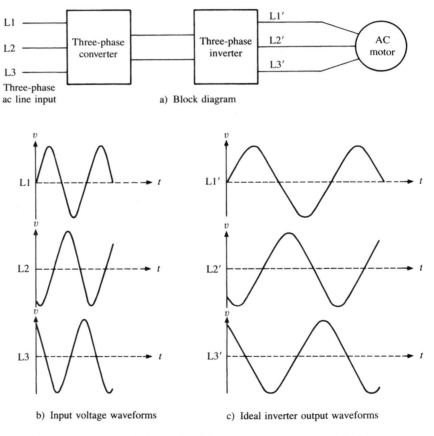

a) Block diagram

b) Input voltage waveforms

c) Ideal inverter output waveforms

◆ **Figure 10.18** A variable-speed ac drive uses a converter to convert the three-phase 60-Hz input voltage into a dc voltage source for the inverter. The inverter converts the dc voltage into a three-phase voltage with a frequency that can be varied from about 3 to 60 (or 120) Hz.

The initial use of adjustable-frequency ac drives was to control pumps, fans, and conveyors. Many applications with single-speed drives were converted to adjustable speed with the intent of reducing energy consumption or increasing manufacturing flexibility. In many of these conversions, the same ac motor was used, with the ac drive providing the variable frequency required for adjustable speed. In pumps and fans, variable speed control is more efficient than throttling to obtain lower flow rates. Energy savings of 40 to 50% have been achieved by replacing throttling controls with variable-speed controls. Another advantage of variable ac drives is the low cost and rugged construction of ac motors. AC induction motors, in particular, are less expensive, are more reliable, and require less maintenance than dc motors.

The use of adjustable-frequency ac drives in applications more demanding than pumps, fans, and conveyors was deterred by the ac drive's complicated control circuitry and poor low-speed performance. However, advances in microprocessor-based control logic and control algorithms have enabled adjustable-speed ac drives to challenge the dc drive's hold on many applications. Improved pulse-width-modulation (PWM) techniques with flux vector control have significantly improved the low-speed performance of ac drives. The improved performance includes full torque from zero speed and regenerative braking. Vector-controlled regenerative ac drives now compete with dc motors for the control of hoists, extruders, punch presses, machine tools, and other applications that require high starting torque and/or regenerative braking. Flux vector control and regenerative braking are explained later in this section.

Variable-Voltage Inverter

The purpose of the *variable-voltage inverter (VVI)* is to manufacture a sinusoidal waveform whose frequency can be varied at will over some frequency range. In the VVI method, a converter converts the ac input voltage into a variable dc voltage. The VVI inverter then inverts the dc voltage into a voltage whose amplitude and frequency are variable. The voltage is a simple approximation to a sine wave. Figure 10.19 shows the output waveforms of a three-phase variable-voltage inverter. This simple approximation of a sine wave is easy to produce, but it also results in high current spikes that generate high voltages that can break down the insulation in the motor. Another problem occurs at motor speeds below 20% of full speed. The crude approximation of the sine wave causes a jerky movement of the motor called *cogging*. The major advantage of VVI drives is a low installed cost. The major disadvantage is that they are hard on the motor owing to high voltage spikes.

SCRs were the first solid-state switching devices used in VVI drives. SCR-based VVI drives capable of driving ac motors of hundreds of horsepower have been available for many years. The latching characteristic of an SCR (an advantage in some applications) is a problem in ac inverters. Once an SCR is turned ON, it remains ON until the current is reduced to zero. This is no problem in a converter because the ac line becomes negative at the end of each positive cycle and turns the SCR OFF. In an inverter, there is nothing to turn the SCR OFF, and the control circuit must provide the means of doing so. This requires complex and expensive commutating circuitry that doubles or even triples the number of SCRs required in the inverter. The commutating circuit also requires additional capacitors and inductors, further increasing the cost of the circuit. The silicon bilateral switch (SBS) solves the problem of SCR turn-off. The SBS can be pulled OFF without the complex circuitry required by the SCR.

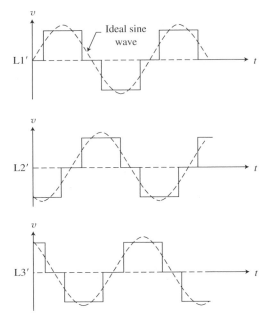

◆ **Figure 10.19** The three-phase output of a voltage-variable inverter (VVI) produces a simple approximation of a three-phase sinusoidal voltage. At low speeds, a VVI-driven motor produces a jerky motion called cogging, due to the crude approximation of a sine wave.

As higher-power transistors become available, designers began replacing SCRs with transistors. Transistors can be turned ON and OFF at will with a considerably simpler circuit and with less power than required by an SCR. Transistors also switch much faster than SCRs. This gives the transistor two more advantages over the SCR. First, the fast switching time, usually nanoseconds against microseconds for an SCR, allows very little time for heating to occur. Consequently, the transistor accomplishes the switching more efficiently than the SCR. Second, fast switching gives the transistor-based ac drive much faster response for critical control applications. The transistor's major problem is lower voltage and current ratings than SCRs. As transistors with higher ratings became available, transistor ac drives began replacing the slower, more expensive SCR drives.

Pulse-Width-Modulated Inverters

Pulse-width modulation (PWM) is another method of approximating a sine wave. The PWM inverter produces a much better approximation of a sine wave than that obtained with a VVI. Pulse-width modulation is not a new idea. It has been used in communication for many years to transmit information on a sequence of pulses called a carrier. The pulses have a constant amplitude and a variable width. A pulse-width modulator varies the width of the pulse according to an information signal.

The PWM inverter uses a sequence of pulses to approximate a sine wave with a variable amplitude and a variable frequency. Figure 10.20 illustrates how a sine wave can be approximated by pulse-width modulation. After appropriate filtering, the output of a PWM inverter is a fairly good approximation of a sine wave. One PWM method uses the magnitude of the sine function to determine the width of each pulse. This method is called *sine-coded PWM*.

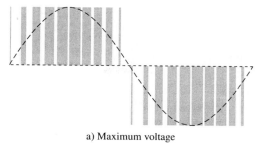

a) Maximum voltage

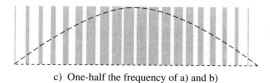

b) One-half the maximum voltage

c) One-half the frequency of a) and b)

◆ **Figure 10.20** A pulse-width-modulated inverter (PWM) produces a reasonably good approximation of a sine wave. The inverter chops the constant dc input voltage into a sequence of pulses (1 to 4 kHz) and modulates the width of the pulses to produce a sine wave. As shown here, pulse modulation can vary both the amplitude and the frequency of the inverter output voltage.

The values of the sine function are stored in a RAM memory unit for use by the controller in manufacturing a sine wave with the desired amplitude and frequency. Another advantage of PWM is that it uses a constant-amplitude dc input voltage. This means that the rectifier can be a simple diode bridge circuit. The VVI requires a variable dc voltage, which is usually provided by an SCR rectifier. The following comparison of size and cost of a 20-hp ac controller illustrates the progress in ac drives over a 10-year period.

Year	Controller Type	Relative Cost (%)	Package Size (in.)
1978	SCR—VVI	100	20 × 30 × 86
1982	Transistor—VVI	40	15 × 26 × 29
1983	Transistor—PWM	20	14 × 16 × 25

Source: Control Engineering, February 1988, p. 73.

Flux Vector Control

Flux vector control is a method of optimizing PWM inverters. It is a microprocessor-managed control algorithm that improves speed regulation and maximizes the starting torque of adjustable-speed ac induction motors. Flux vector control adjusts the frequency and phase of the voltage and current applied to the motor to maintain the rated torque capability of the motor at all speeds including zero. Our discussion of flux vector control begins with a brief review of induction motors.

A three-phase induction motor is shown in Figure 10.7. The stator of this motor has three windings that may be connected to a three-phase ac supply in either a wye or a delta configuration. The rotor has conducting bars, which are shorted at each end of the rotor. When ac power is applied to the stator windings, a rotating magnetic field is produced, current is induced in the rotor bars, and a torque is developed that causes the rotor to turn. From the conventional viewpoint, the induction motor generates torque when the rotor turns at a slightly slower speed than the rotating magnetic field produced by the stator current. This speed difference is called slip. The motor torque is directly proportional to the slip speed, ω_{slip} (rad/s).

$$\text{Torque} = (\text{torque constant}) \, \omega_{slip}$$

From a more fundamental viewpoint, the motor torque is generated by the interaction of two magnetic fields. One is the rotating magnetic field produced by the stator current. This is similar to the field of a dc motor, and we will refer to it as the *motor flux*. The other is the magnetic field produced by the induced current in the rotor. This is similar to the field produced by the armature current in a dc motor, and we will refer to it as the *torque-producing field*.

The vector control algorithm provides for separate control of motor flux and motor torque. It accomplishes this task by separating the stator current of the ac motor into two perpendicular components, i_{flux} and i_{torque}. The first component, i_{flux}, is parallel to the motor flux vector, and it is proportional to the magnitude of the motor flux as defined by the following equation.

$$|\text{Motor flux}| = (\text{flux constant}) \, i_{flux}$$

The second component, i_{torque}, is proportional to the magnitude of the motor torque as defined by the following equation:

$$|\text{Motor torque}| = (\text{torque constant}) \, |\text{Motor flux}| \, i_{torque}$$

The vector controller adds the two current components to form the stator current as shown in Figure 10.21. The flux current, i_{flux}, is essentially constant over the rated speed range of the motor. The torque current, i_{torque}, is varied in direct proportion to the applied load. By separate control of flux and torque, the vector controller provides rated torque of an induction motor from zero speed to rated speed. Flux vector control gives us the ability to use a standard induction motor and get performance comparable to dc adjustable drives in all but the most demanding applications.

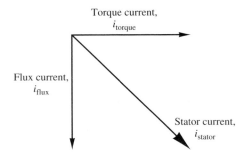

Torque current,
i_{torque}

Flux current,
i_{flux}

Stator current,
i_{stator}

◆ **Figure 10.21** The flux vector controller provides separate control of the flux current, i_{flux}, and the torque current, i_{torque}, components of the stator current, i_{stator}. *Source*: Baldor Electric Company, 5711 South 7th Street, Fort Smith, AR 72902-2400.

10.6 DC MOTOR AMPLIFIERS AND DRIVES

Amplifiers

Permanent-magnet brush-type dc servomotors are sometimes controlled with a power operational amplifier. Power op amps obey all the rules of the op-amp family and are also designed to operate at higher voltage and current levels than the lower-power members of the family. Power op amps operate at voltages above 44 V and can deliver currents above 0.1 A. Low-voltage power op amps (below 100 V) consist of an IC op-amp front end plus a high-current output stage housed in a heat-dissipating package. High-voltage power op amps (above 100 V) have a more complex circuit using many individual transistors. Construction of the high-voltage units is more labor intensive and hence more expensive. They do tend to have better accuracy due to improved control of individual components and laser trimming of the input stage. Figure 10.22 illustrates the use of a power op amp in a dc speed control system.

The transfer function of the position control system can be obtained from the block diagram in Figure 10.22. The objective is to obtain an equation for the ratio of output, Ω, over input, SP (i.e., Ω/SP). Each block in the transfer function defines an equation of the form

$$\text{Output} = (\text{input})(\text{transfer function of the block})$$

The transfer function of a block is written inside the block. For example, the output of the power op amp, E, is given by the equation

$$E = (SP - E_G)(P) \tag{10.33}$$

The dc motor is a little more complicated, due to the internal feedback loop. However, with a little algebra, we can obtain an equation for the output of the motor, Ω. Proceed as follows:

1. Determine an equation for E_i.

$$E_i = K_E\Omega \tag{10.34}$$

2. Get the output of the internal summing junction.

$$E_s = E - E_i \tag{10.35}$$

3. Get an equation for I.

$$I = E_S\left(\frac{1}{R + sL}\right) \tag{10.36}$$

4. Get an equation for Ω.

$$\Omega = I\left(\frac{K_T}{B + Js}\right) \tag{10.37}$$

5. Use Equations (10.34) through (10.37) to solve for Ω in terms of E with both E_i and I eliminated from the equation. Equation (10.38) is the final equation for the dc motor.

$$\Omega[(R + Ls)(B + Js) + K_E K_T] = K_T E \tag{10.38}$$

The last component, the tachometer, is easy.

$$E_G = K_G\Omega \tag{10.39}$$

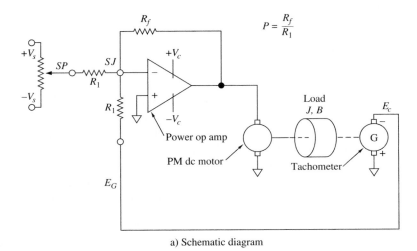

a) Schematic diagram

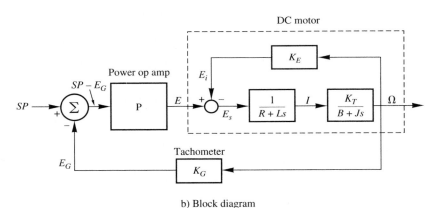

b) Block diagram

◆ **Figure 10.22** A power op amp is used to control the velocity of a small permanent-magnet brush-type dc motor.

Finally, use Equations (10.33), (10.38), and (10.39) to obtain the transfer function as given by

$$\frac{\Omega}{SP} = \frac{K_T P}{(K_E K_T + K_G K_T P + RB) + (JR + LB)s + JLs^2} \qquad \textbf{(10.40)}$$

DC Adjustable-Speed Drives

DC adjustable-speed drives use a rectifier to produce a controlled dc voltage that can be smoothly adjusted from 0 to 100% of the rated voltage. Some dc drives also provide a reverse energy path for regenerative braking. Figure 10.23 illustrates an SCR-based dc adjustable-speed drive with a built-in, computed velocity signal.

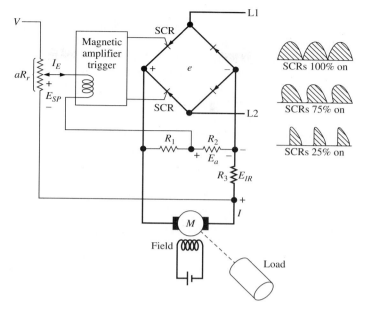

a) Schematic diagram

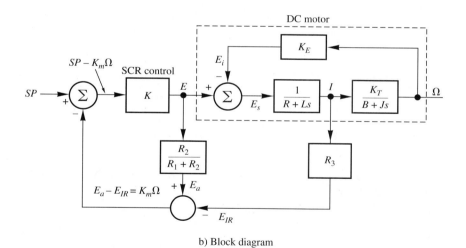

b) Block diagram

◆ **Figure 10.23**   This SCR-based dc adjustable-speed drive has a full-wave bridge rectifier and a computed-velocity feedback signal.

In Figure 10.23a, two SCRs and two diodes are connected to form a full-wave bridge rectifier. When the SCRs are turned OFF, the bridge acts as an open switch. When the SCRs are turned ON, the bridge acts as a full-wave rectifier. The output of the bridge is a pulsating voltage, which is applied to the armature of a dc motor (see the upper right corner of Figure 10.23a).

The magnetic amplifier trigger circuit produces the triggering pulses that turn the SCR ON. The SCR is said to be 100% ON if the trigger occurs at the start of each ac half-cycle. With the SCR 100% ON, the bridge acts as a full-wave rectifier. The SCR is 75% ON if the trigger occurs 45° after the start of each ac half-cycle. With the SCR 75% ON, the first 25% of each half-cycle is removed from the output waveform, as shown in Figure 10.23a. The SCR is 25% ON if the trigger occurs 135° after the start of each half-cycle. With the SCR 25% ON, the first 75% of each half-cycle is removed from the output. The inductance of the armature coil and the inertia of the motor and load help smooth out the pulsations in the power caused by the pulses in the armature voltage.

An interesting feature of the system, illustrated in Figure 10.23, is the fact that a tachometer is not used to measure the velocity of the load. Instead, the velocity is calculated from measurements of the armature voltage (E) and the armature current (I). The calculation is based on the following frequency domain version of the combination of Equations (10.9) and (10.11):

$$E - IR = K_E \Omega. \tag{10.41}$$

In Figure 10.23, E_a is the voltage drop across resistor R_2 and E_{IR} is the voltage drop across resistor R_3. Voltage E_a can be obtained by applying the voltage-divider rule to resistors R_1 and R_2.

$$E_a = \left(\frac{R_2}{R_1 + R_2} \right) E \tag{10.42}$$

Voltage E_{IR} is obtained by applying Ohm's law to R_3.

$$E_{IR} = IR_3 \tag{10.43}$$

In the design of the internal feedback, R_3 is defined as

$$R_3 = \left(\frac{R_2}{R_1 + R_2} \right) R \tag{10.44}$$

where R represents the internal resistance of the motor armature. Equations (10.42) through (10.44) can be used to form the following equation for $E_a - E_{IR}$:

$$E_a - E_{IR} = \left(\frac{R_2}{R_1 + R_2} \right) (E - IR) \tag{10.45}$$

A combination of Equations (10.41) and (10.45) produces the following result:

$$E_a - E_{IR} = K_M \Omega \tag{10.46}$$

where $K_m = \dfrac{K_E R_2}{R_1 + R_2}$, V/rad/s

Ω = armature speed, rad/s

E_a = voltage across resistor R_2, V

E_{IR} = voltage across resistor R_3, V

R_1 and R_2 are arbitrarily selected resistors

R_3 is defined by Equation (10.44)

Equation (10.46) states that the voltage difference $(E_a - E_{IR})$ is proportional to the armature speed Ω and can be used as a measurement of Ω.

◆ GLOSSARY

AC motor: An electric motor that is powered by an ac voltage. (10.2)

AC motor, induction: An ac motor that has no electrical connection to the rotor. The transformer action induces a current in current-carrying conductors in the rotor. Induction motors are very reliable and have a low initial cost and a low maintenance cost. (10.2)

AC motor, servo: A two-phase, reversible induction motor with special modifications for servo control. (10.2)

AC motor, squirrel-cage: A type of induction motor that has conducting bars and end rings on the rotor to provide an electrical path for the induced current. The name comes from the resemblance between the bars/end ring assembly and a squirrel cage. (10.2)

AC motor, synchronous: An electric motor that normally runs at synchronous speed. The stator has windings similar to an induction motor. The rotor has fixed poles that lock into step with the rotating poles in the stator. (10.2)

AC motor, wound-rotor: A type of induction motor in which the conduction bars are connected in series to form three windings. The windings are joined on one end to form a wye connection. The other end of each winding is connected to a slip ring. (10.2)

Armature: The cylindrical, rotating part of a dc motor. (10.1)

Back EMF: A voltage induced in the coils of a motor by the generator effect. This induced voltage opposes the external voltage applied to the coil and thus reduces the motor current. (10.1)

Braking: A method of slowing down a dc motor by causing the motor to act as a dc generator, thus producing a force that tends to slow down the armature. (10.3)

Braking, dynamic: A method of braking a dc motor in which the voltage produced by the generator action is applied to a bank of resistors. The resistors draw current from the generator, causing a force that tends to slow down the armature. (10.3)

Braking, regenerative: A method of braking a dc motor in which the current from the generator is fed back into the dc supply. If the dc supply is a battery, the current will charge the battery. (10.3)

Cogging: A jerky motion that occurs when an ac motor is driven at slow speeds by a variable-voltage inverter. The jerky motion is caused by the crude approximation of the sine wave. (10.5)

Commutation: The reversal of current and polarity in the coil of a dc motor as the brushes move from one segment to the next. (10.1)

Commutator: A segmented ring on the armature of a dc motor. Each segment is connected to one end of one of the armature coils. Electrical connection is made to the commutator segments (and hence the armature coils) through carbon contacts called brushes. (10.1)

DC motor: An electric motor that is powered by a dc voltage. (10.3)

DC motor, brushless: A dc motor that has a permanent-magnet armature and a wound field. There is no need for brushes and a commutator. Instead, the commutation is performed electronically on the stator winding. (10.3)

DC motor, compound: A dc motor in which part of the field winding is connected in series with the armature winding and the remainder is connected in parallel with the armature-series field windings. Compound motors have a starting torque and no-load speed that lie between the series motor and the shunt motor. (10.3)

DC motor, moving coil: A dc motor in which the armature winding rotates in an annular space between the field pole on the outside and the iron core on the inside. The result is a motor with a torque-to-inertia ratio several times greater than an iron-core armature motor. (10.3)

DC motor, pancake: A dc motor with a disk-shaped armature that achieves a high torque-to-inertia ratio. (10.3)

DC motor, permanent-magnet: A dc motor in which the field is provided by permanent magnets. The speed of the permanent-magnet motor can be controlled by the armature voltage, making this a popular motor for servo control systems. (10.3)

DC motor, separately excited: A dc motor in which the field winding and the armature winding are excited by separate sources. The speed of a separately excited motor can be increased by either decreasing the field voltage or increasing the armature voltage. (10.3)

DC motor, series: A dc motor in which the field winding and the armature winding are connected in series. Series motors produce the highest starting torque and the greatest no-load speed of the four types of dc motors. (10.3)

DC motor, shunt: A dc motor in which the field winding and the armature winding are connected in parallel. Shunt motors have a lower starting torque and a much lower no-load speed than a series motor. (10.3)

Field poles: The stationary part of a dc motor. (10.1)

Full-step operation: A movement of a stepper motor of one full step for each input pulse. *See also* Half-step and Microstep operation. (10.4)

Half-step operation: A movement of a stepper motor of one-half step for each input pulse. *See also* Full-step and Microstep operation. (10.4)

Inverter: An electronic device that converts a dc voltage into an ac voltage. The frequency of the output may be 60 Hz, or it may be variable over a specific range of frequencies such as 3 to 60 Hz. (10.5)

Inverter, pulse-width-modulated (PWM): A variable-frequency inverter that uses pulse-width modulation of a sequence of pulses to produce a fairly good approximation of a sine wave. (10.5)

Inverter, variable-voltage (VVI): A variable-frequency inverter that uses variable voltage pulses to form a stairstep approximation of a sinusoidal voltage. (10.5)

Microstep operation: A movement of a stepper motor of a fraction of a step for each input pulse. The fraction is usually 1/10, 1/16, 1/32, or 1/125. *See also* Full-step and Half-step operation. (10.4)

Rotor: The cylindrical rotating part of an ac motor. (10.2)

Slip: The difference between the synchronous speed and the actual rotor speed of an ac motor. (10.2)

Stator: The stationary part of an ac motor. (10.2)

Stepping motor: An electric motor that transforms electrical pulses into equal increments of rotary shaft motion called steps. (10.4)

Synchronous speed: The speed at which the field of an induction motor rotates. (10.2)

◆ EXERCISES

Section 10.1

10.1 Imagine you are explaining how electric motors and generators work to a friend who has little technical background. Write a paragraph that explains the two facts that are the basis of the operation of motors and generators.

10.2 Write a paragraph that explains the function of the brushes and commutator in a dc motor.

10.3 Explain what happens to the voltage at the output terminals of a dc generator if the commutator is replaced by two slip rings, one for each end of the armature winding. Assume that one brush is in contact with each slip ring.

10.4   Assume you are given the assignment of measuring the torque constant of five dc motors. Your company has a dynamometer for measuring the torque developed by a dc motor. The dynamometer is equipped with an ammeter for measuring the armature current and a tachometer for measuring the motor speed. Using this equipment, you obtained the following data. Use your data to determine the torque constant, K_T, of each motor in SI units of N · m/A.

Motor ID	Torque (N · m)	Armature Current (A)	Speed (rpm)
(a) ICST-5	0.108	2.0	4800
(b) E550-MG	0.0758	2.2	6500
(c) ICST-6	0.330	2.4	5000
(d) SD12-20	1.226	2.3	1000
(e) SX12-30	0.807	2.1	1600

10.5   Convert the torque constants obtained in Exercise 10.4 from SI units of N · m/A to English units of ozf · in./A and lbf · in./A.

10.6   Assume you are given the additional assignment of measuring the voltage constant of the five dc motors in Exercise 10.4. The company's dynamometer can also be used to drive the motor with an open armature ($i_a = 0$ A). With an open armature, the armature voltage is the back emf, e_b. Using the dynamometer in this way, you obtained the following data. Use your data to determine the voltage constant, K_E, of each motor in units of (V/krpm).

Motor ID	Armature Voltage (V)	Speed (rpm)
(a) ICST-5	30.1	4800
(b) E550-MG	29.9	6500
(c) ICST-6	78.5	5000
(d) SD12-20	65.0	1000
(e) SX12-30	78.4	1600

10.7   Convert the voltage constants obtained in Exercise 10.4 from units of (V/krpm) to units of (V · s/rad).

Section 10.2

10.8   A three-phase induction motor has a rated torque of 15 ft-lb and is connected to a load that has a starting torque of 30 ft-lb. The motor speed versus torque curve is given in Figure 10.7d. Explain what will happen when the motor is turned on.

10.9   Determine the velocity and position transfer functions of an ac servomotor with the following data:

Rated voltage: 120 V
Load inertia: 15×10^{-6} N · m · s²/rad
Load damping: 6×10^{-6} N · m · s/rad
Stall torque: 0.12 N · m
No-load speed: 5000 rpm

Section 10.3

10.10   Explain dynamic braking and regenerative braking.

10.11   Your results in Exercise 10.4 are based on the assumption that the friction torque can be neglected in the computation of the torque constant. Your follow up assignment is to measure the torque constants of the five dc motors, including the effect of the friction torque—review Figure 10.12a, Equation (10.7), and Equation (10.8). Using a dynamometer, you obtained the following data. Use your data to determine the torque constant, K_T, of each motor in SI units of N · m/A.

Motor ID	T_1 (N · m)	i_{a1} (A)	T_2 (N · m)	i_{a2} (A)	Speed (rpm)
(a) ICST-5	0.0480	1.0	0.1080	2.0	4800
(b) E550-MG	0.0274	1.1	0.0758	2.2	6500
(c) ICST-6	0.150	1.2	0.330	2.4	5000
(d) SD12-20	0.544	1.2	1.226	2.3	1000
(e) SX12-30	0.337	1.1	0.807	2.1	1600

10.12   Compare the torque constants you obtained in Exercise 10.4 with those you obtained in Exercise 10.11. Compute the percent difference as follows.

$$\% \text{ difference} = 100 \frac{K_T(\text{from Exercise 10.11}) - K_T(\text{from Exercise 10.4})}{K_T(\text{from Exercise 10.11})}$$

10.13   Convert the torque constants obtained in Exercise 10.11 from SI units of N · m/A to English units of ozf · in./A and lbf · in./A.

10.14   An armature-controlled dc motor has the following ratings:

$$T_f = 0.03 \text{ N} \cdot \text{m}$$
$$K_T = 0.15 \text{ N} \cdot \text{m/A}$$
$$I_{\max} = 2.4 \text{ A}$$
$$K_E = 0.15 \text{ V} \cdot \text{s/rad}$$

$$N_{max} = 5000 \text{ rpm}$$
$$R = 0.8 \ \Omega$$

Determine the following:
(a) The maximum output torque
(b) The maximum power output
(c) The maximum armature voltage
(d) The no-load motor speed when $e = E_{max}$

10.15 The motor in Exercise 10.7 is operated at 3000 rpm with a load torque of 0.15 N · m. Determine the following:
(a) The armature voltage
(b) The armature speed if the torque increases to 0.30 N · m and the armature voltage is not changed

10.16 A dc motor has the following specifications:

Type: dc permanent magnet motor	
Model No. ICST-6	
Maximum operating speed, ω_{max} (rad/s)	524
Maximum armature current, I_{max} (A)	2.4
Voltage constant, K_E (V · s/rad)	0.15
Torque constant, K_T (N · m/A)	0.15
Friction torque, T_f (N · m)	0.03
Armature resistance, R (Ω)	0.8
Armature inductance, L (H)	0.045
Armature moment of inertia,	
$\quad J_m$ (N · m · s²/rad)	10.2E − 4
Armature viscous friction,	
$\quad B_m$ (N · m · s/rad)	1.2E − 4

Determine the following velocity and position transfer functions.
(a) The voltage-driven motor
(b) The simplified voltage-driven motor with τ_e neglected
(c) The current-driven motor
(d) The voltage-driven motor with the following speed-reducer and load specifications:

$$N_1 = 16 \text{ teeth}$$
$$N_2 = 48 \text{ teeth}$$
$$J_L = 7.6\text{E} - 3 \text{ N} \cdot \text{m} \cdot \text{s}^2/\text{rad}$$
$$B_L = 0.20\text{E} - 4 \text{ N} \cdot \text{m} \cdot \text{s/rad}$$

10.17 The following specifications were obtained from a manufacturer's catalog:[*]

Type: dc ceramic permanent magnet motor	
Model No. SD 12-20	
Rated voltage, E_r (V)	90
No-load speed, N_o (rpm)	1386
Maximum continuous current, I_{max} (A)	2.3
Continuous stall torque, T_{max} (lb · in.)	12
Friction torque, T_f (lb · in.)	1.8
Voltage constant, K_E (V/krpm)	65
Torque constant, K_T (lb · in./A)	5.49
Armature resistance, R (Ω)	11.30
Armature inductance, L (mH)	30.0
Armature moment of inertia,	
$\quad J_m$ (lb · in. ·s²/rad)	0.007
Armature viscous friction,	
$\quad B_m$ (lb · in. · s/rad)	0.003

Convert the motor parameters to SI units and determine the full and simplified velocity transfer functions.

10.18 Repeat Exercise 10.17, with the following manufacturer's specifications:[†]

Type: dc rare-earth permanent magnet motor	
Model No. SX 12-30	
Rated voltage, E_r (V)	90
No-load speed, N_o (rpm)	1828
Maximum continuous current, I_{max} (A)	2.1
Continuous stall torque, T_{max} (lb · in.)	8.9
Friction torque, T_f (lb · in.)	1.59
Voltage constant, K_E (V/krpm)	49
Torque constant, K_T (lb · in./A)	4.16
Armature resistance, R (Ω)	5.10
Armature inductance, L (mH)	9.00
Armature moment of inertia,	
$\quad J_m$ (lb · in. · s²/rad)	0.0035
Armature viscous friction,	
$\quad B_m$ (lb · in. · s/rad)	0.004

10.19 The voltage-driven motor, speed reducer, and load are shown in Figure 10.14. The transfer function for this system is defined by Equation (10.29) which uses the total inertia (J_T) defined by Equation (10.26). The inertia of the two gears was ignored in Equation (10.26). Revise Equation (10.26) so that it includes the inertia of the gears.

---

[*]*D.C. Servomotors*, Baldor Electric Company, Fort Smith, AR, Form #DB-210-A, 1986.

[†]*Rare Earth D.C. Servomotors*, Baldor Electric Company, Fort Smith, AR, Form #DB-220-A, undated.

$J_{N1} =$ inertia of gear N_1

$J_{N2} =$ inertia of gear N_2

Section 10.4

10.20 Extend Table 10.1 to include full steps from 6 through 20.

10.21 Extend Table 10.2 to include half-steps from 11 through 40.

10.22 Extend Table 10.3 to include microsteps from 11 through 40.

Section 10.5

10.23 Explain the voltage-variable inverter and the pulse-width-modulated inverter.

Section 10.6

10.24 The SCR control in Figure 10.23 is to be used to control a dc motor with an armature resistance, R, of 0.4 Ω. The feedback voltage, E_a must be 20% of the armature voltage, E, for proper operation of the control circuit. A total value of 10 KΩ for R_1 and R_2 is required. Determine the resistance values required for R_1, R_2, and R_3.

10.25 Show that Equation (10.37) can be transformed into the transfer function given by Equation (10.21).

10.26 Use Equation (10.45) to simplify the block diagram in Figure 10.23b. In doing this, a single block with a transfer function equal to K_m replaces the two blocks with transfer functions equal to R_3 and $R_2/(R_1 + R_2)$. Derive the transfer function of the control system from your new block diagram.

CHAPTER 11

Control of Discrete Processes

◆ OBJECTIVES

A discrete process consists of a series of distinct operations with a definite condition for initiating each operation. When the series of operations has a beginning, an end, and a definite controlled form, the process is called a sequential process or a batch process. Discrete process operations can be grouped into two categories, those which are initiated by time and those which are initiated by an event. We call these categories time-driven and event-driven operations. The simplest discrete process consists of a single output that has only two possible values (e.g., ON and OFF, high and low, hot and cold, etc.). More complicated discrete processes consist of a number of operations, and each operation may consist of a number of distinct parts called steps.

The purpose of this chapter is to give you an entry-level ability to discuss, select, specify, and design discrete process control systems. After completing this chapter, you will be able to complete the following for a discrete process:

1. Statement list
2. Timing diagram
3. Sequential function chart
4. State chart
5. Ladder diagram circuit

11.1 INTRODUCTION

Discrete processes occur in a number of different places. Home appliances use discrete processes to wash dishes, to dry clothes, and to cook a frozen dinner. Manufacturing industries use a sequence of operations to produce discrete parts and assemble them into finished products. The chemical industry uses a batch process in a chemical reactor to produce a definite amount of a product. The food industry uses batch processes for operations such as sterilization, freeze drying, and extraction.

Some discrete processes are nonsequential in nature, consisting primarily of event-driven operations. These processes are controlled by programmable logic controllers or by panels of mechanical relays. However, most discrete processes are sequential, consisting of a combination of event-driven operations and time-driven operations. Depending on where they are used, sequential, discrete processes are called sequential processes or *batch processes*. In the following discussion, the term *sequential process* will include both sequential and batch processes.

A *sequential process* consists of a sequence of one or more operations (called steps) that must be performed in a defined order. The completion of this sequence of steps creates a definite amount of finished product. The product may be a discrete part, such as an automobile; an amount of liquid, such as mouthwash; or an amount of solid material, such as sugar. The sequence of operations must be repeated to produce more of the product. Sequential processes have a *set of directions* that defines each step. (In food processing the set of directions is called a recipe; in computer programming it is called an algorithm). Most sequential processes assemble specific parts or ingredients and process them according to the set of directions. The processing may be drilling, punching, painting, cooking, blending, reacting, or other operations that modify the product.

The steps in a sequential process can often be grouped into a few general operations, such as preparation, loading, processing, unloading, and postpreparation. Each general operation consists of many individual steps. Each step is a single event, such as opening a valve, starting a motor, putting a controller on automatic control, and so on. The general operations are sometimes called phases. In this book they will be called *operations,* and the individual parts of an operation will be called *steps*.

In a strictly sequential process, each step must be completed before the next step can be initiated. However, many actual processes include *parallel operations* that are not part of any other operation. For example, high- and low-level switches may start and stop a pump to maintain the level of liquid in a tank between two limits. The supply of liquid in the tank may be part of a sequential process, but the control of the level is independent of the sequential control of the process.

Auxiliary operations are another type of operation that is not included in the main sequential process. The auxiliary operation appears as one or two steps in the main process. However, it consists of a number of steps that are not included in the main process. The auxiliary operation is usually initiated and terminated by steps in the main process and runs concurrently with the main process. An example of an auxiliary operation is a continuous PID controller used to control the temperature in a batch cooker. The main process includes steps that switch the controller from manual to automatic and from automatic to manual. The main process may also include steps that adjust the setpoint of the controller to different values to follow a prescribed time-versus-temperature curve.

Methods of describing a sequential process include statement lists, flowcharts, ladder diagrams, sequential function charts, state charts, process timing diagrams, and a mathematical (Boolean) language. The *statement list* is an English-language list of actions that must be carried out for each step. A *flowchart* uses blocks to represent each step and lines to show the path from step to step. A *ladder diagram* is an electrical diagram showing the connections between various contacts, relay coils, solenoids, motors, and so on. A *state chart* is a truth table that shows the outputs produced by each step and is usually accompanied by a diagram or chart that shows the transfer paths from state to state. The *sequential function chart* uses blocks to represent each step and lines to represent the transitions from step to step. A *process timing diagram* is a graph of the outputs plotted versus time. The mathematical language is similar to a computer program. These methods will be used and explained at appropriate points in the remainder of this chapter.

11.2 TIME-DRIVEN SEQUENTIAL PROCESSES

In a *time-driven* sequential process, each step is initiated at a given time, or after a given time interval. The statement list and the process timing diagram are two methods used to describe a time-driven process. You may recognize the following statement list from a popular product found in every supermarket:

Microwave Instructions
STEP 1. Remove tray from carton and peel back film from one end to vent.
STEP 2. Place tray in microwave oven.
STEP 3. Cook 10 min on medium power (50%) or defrost.
STEP 4. Rotate $\frac{1}{2}$ turn.
STEP 5. Cook 4 min on full power.
STEP 6. Let stand for 1 min.
STEP 7. Remove film and serve.

Figure 11.1 shows an instrumentation and piping diagram for a batch blending process. This is another example of a time-driven sequential process. Two liquid ingredients are blended and heated until they are warm and thoroughly mixed. Then a solid material is added and the blending continues until the solid is completely dissolved in the liquid mixture. Finally, the blended product is pumped to a bottling line. The mixture is heated during the blending operation to facilitate the dissolving of the solid material.

The two flow controllers, FQC-100 and FQC-101, deliver measured quantities of the two liquid ingredients. Upon receiving a START signal from the sequential controller, FQC-100 and FQC-101 automatically deliver measured amounts of ingredients A and B to the mixing tank. When done, the controllers automatically shut off. The temperature controller, TC-102, controls the temperature of the contents of the blending tank. When the controller is in manual control, control valve TV-102 is turned OFF. When in automatic control, controller TC-102 manipulates the steam flow rate to maintain the temperature of the blending tank contents at the controller setpoint. The belt feeder is driven by an ac induction motor with an ac adjustable speed drive, SC-104. The speed of the belt drive is adjusted by the operator to deliver

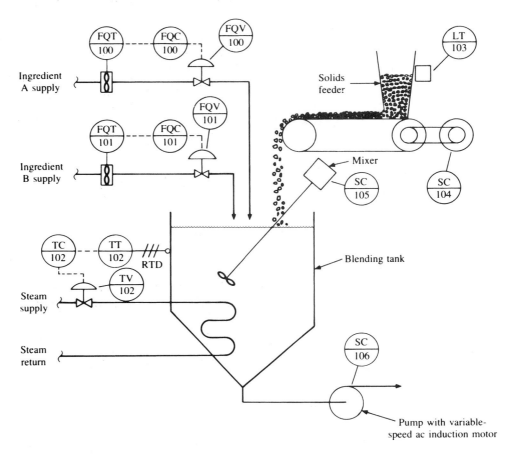

◆ **Figure 11.1** Instrumentation and piping diagram for a batch blending process. This is an example of a time-driven sequential process.

the entire amount of solid material in slightly less than 5 min. The mixer and the pump are also driven by induction motors with ac adjustable speed drives, SC-105 and SC-106. The following is a statement list of the process.

Statement List of the Batch Blending Process (Figure 11.1)

STEP 1 (4 min). Prepare the solid material and fill the solids feeder hopper with a measured amount of material.

STEP 2 (30 s). Press the START button.

STEP 3 (3 min). Fill the blending tank with a measured quantity of ingredients A and B.

STEP 4 (4 min). Heat and mix the contents of the blending tank (i.e., ingredients A and B).

STEP 5 (5 min). Gradually deliver solid material from the hopper to the blending tank. Continue to heat and mix the contents of the blending tank.

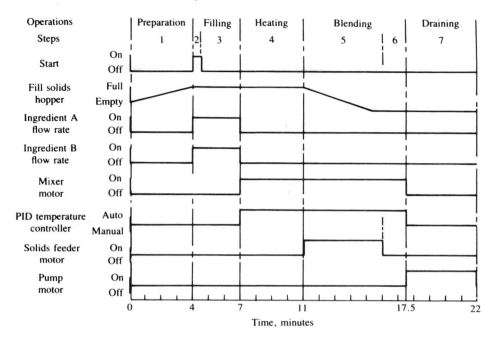

◆ **Figure 11.2** Process timing diagram for the batch blending process in Figure 11.1.

STEP 6 (1.5 min). Continue to heat and mix the contents of the blending tank.

STEP 7 (4.5 min). Pump the contents of the blending tank to the bottling line.

Figure 11.2 shows the process timing diagram for the batch blending process. Notice that the process is divided into five operations: preparation, filling, heating, blending, and draining. The filling and blending operations each have two steps. The other operations each have only one step. Batch processes often have several steps for each operation.

11.3 EVENT-DRIVEN SEQUENTIAL PROCESSES

In an *event-driven process,* each step is initiated by the occurrence of an event. The event may be a single action, such as an operator pressing a pushbutton, the closing of a limit switch, the opening of a pressure switch, or some other action that causes a switch to open or close. The event could also be a combination of several actions. For example, the event may consist of the simultaneous occurrence of several actions, with the contacts for each action connected in series. The event may also be the occurrence of any one of several actions, with the contacts for each action connected in parallel.

> The action-causing event in a sequential process is determined by an interconnection of one or more contacts. The event has occurred when there is a closed path through this interconnection of contacts.

The *ladder diagram* is a very popular graphical method of describing an event-driven process. It was developed to represent systems consisting of switches, relays, solenoids, motor starters, and other switching components used to control industrial machinery. The ladder diagram is very easy to learn and very logical in its interpretation. Only one rule is required to understand a ladder diagram: When there is a closed path through a relay coil from one side to the other side, the coil will close its associated contacts. The ladder diagram got its name from its general appearance—it looks like a ladder. A ladder diagram consists of two vertical lines and any number of horizontal circuits that connect the two vertical lines. The horizontal circuits are called the *rungs* of the ladder diagram. The ladder diagram in Figure 11.3 has three rungs.

Each rung in a ladder diagram defines one operation (or step) in the process. A rung has two parts: an *input* section and an *output* element. In the top rung in Figure 11.3, the series connection of the two limit switches (LS1 and LS2) is the input section, and relay coil CR1 is the output element. Both limit switches must be closed to energize coil CR1. When CR1 is energized, all normally open (NO) contacts labeled CR1 will close, and all normally closed (NC) contacts labeled CR1 will open. These contacts may appear in any rung in the ladder diagram. The name *CR1* is used to associate a contact with the coil that operates the contact.

In the middle rung, the parallel combination of the level switch (LV1) and the pressure switch (PS1) is the input section, and solenoid A is the output element. If either LV1 or PS1 closes, solenoid A will be energized. In the bottom rung, the input section is the parallel/series combination of contacts A, B, and CR1, and solenoid B is the output element. Solenoid B is energized when the following expression is true: (A OR B) AND NOT CR1.

Boolean equations are another method of representing the rungs in a ladder diagram. The following Boolean equations represent the three rungs in the ladder diagram in Figure 11.3.

$$CR1 = LS1 \text{ AND } LS2$$

$$\text{Sol. A} = LV1 \text{ OR } PS1$$

$$\text{Sol. B} = (A \text{ OR } B) \text{ AND NOT } CR1$$

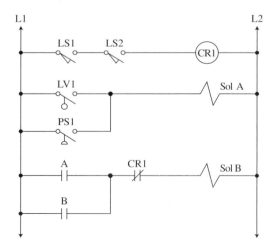

◆ **Figure 11.3** This ladder diagram has three rungs, each controlling one operation (or step) in an event-driven sequential process.

The hydraulic hoist in Figure 11.4 is an example of an event-driven control system. The schematic diagram shows the physical configuration, and the ladder diagram defines the sequential control system. The ladder diagram consists of six rungs between line L1 and line L2. Each rung consists of two or more switches or contacts on the left and a single operational element on the right. The operational element is turned on whenever there is a closed path through the rung from L1 to L2 (i.e., whenever all switches and contacts in the rung are closed).

The STOP symbol represents a *normally closed* pushbutton that is open only when someone is pressing the STOP button. The START, UP, and DOWN symbols represent *normally open* pushbutton contacts that are closed only when someone is pushing the appropriate button. The circle labeled 1M is the coil of a relay used to start the pump motor. The three contacts labeled 1M are open when coil 1M is OFF (deenergized), and they are closed when coil 1M is ON (energized). The circles labeled 1CR and 2CR are the coils of two control relays. Each control relay has one normally open contact and one normally closed contact. The normally open contacts are open when the coil is deenergized and closed when the coil is energized. The normally closed contacts are just the opposite, closed when the coil is deenergized and open when the coil is energized.

The devices labeled Sol *a* and Sol *b* in the ladder diagram are the two coils of the three-position four-way solenoid valve in the schematic diagram. The valve symbol consists of three position blocks located between the two coils. Each position block shows its method of connecting the four ways (the two lines on each side of the valve). The valve is shown in the position it occupies when both coils are deenergized. When Sol *a* is energized, the valve moves up one block, placing the bottom position block in line with the four ways. When Sol *b* is energized, the valve moves down, placing the top position block in line with the four ways. The ladder diagram includes an interlock feature that prevents simultaneous energizing of both solenoids.

The hydraulic hoist system is turned ON by pressing the START button. This completes the path through coil 1M, so coil 1M is energized and all three contacts labeled 1M are closed. When the operator releases the START button, the 1M contact directly below the START button maintains the closed path and coil 1M remains energized. A contact used in this way is called a holding contact. The second 1M contact turns ON the pump motor, and the third 1M contact activates the bottom half of L1. The operator can turn the hoist OFF at any time by pressing the STOP button. This will deenergize coil 1M and open the three 1M contacts. When the operator releases the STOP button, coil 1M remains OFF because both the START contact and the holding contact are open. This pushbutton and holding contact arrangement is a standard method of starting electric motors.

When the hoist is turned ON, the bottom half of L1 is activated, and the operator can use the UP and DOWN pushbuttons to operate the hoist. When the operator presses the UP button, relay 1CR is energized, the normally open 1CR contact closes, and Sol *a* is energized. In the schematic diagram, the solenoid valve moves up because Sol *a* is energized. This connects the pump supply line to the bottom end of the cylinder and the top end of the cylinder to the return line. The pump forces hydraulic fluid into the bottom end of the cylinder and the piston moves up, raising the platform and its load. When the operator releases the UP button, the valve returns to its normal position, blocking both sides of the cylinder and locking the cylinder in its current position. When the operator presses the DOWN button, the platform moves down in a similar manner.

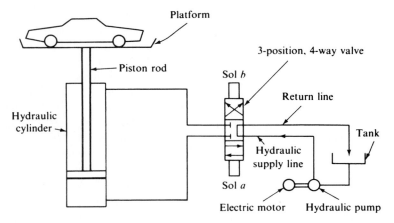

a) Schematic diagram

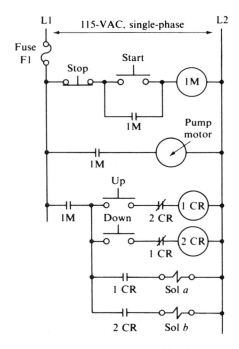

b) Ladder diagram

◆ **Figure 11.4** Event-driven sequential control system for a hydraulic hoist.

In summary: The platform goes up when the UP button is pressed, goes down when the DOWN button is pressed, and remains stationary when neither button is pressed. If the UP button is pressed when the cylinder is at the top, a relief valve (not shown) provides a direct path from the pump outlet to the tank. The same is true if the DOWN button is pressed when the cylinder is at the bottom. In addition, the system is interlocked so that both solenoids cannot be energized at the same time.

Automated Drilling Machine Example

The automated drilling machine in Figure 11.5 will be used as an example for a study of event-driven sequential processes. The drilling machine consists of an electric drill mounted on a movable platform. A hydraulic cylinder moves the platform and drill unit up and down between a *Drill Reset* position (up) and a *Hole Drilled* position (down). Limit switch LS1 is actuated when the drill platform is in the Reset position. Limit switch LS2 is actuated when

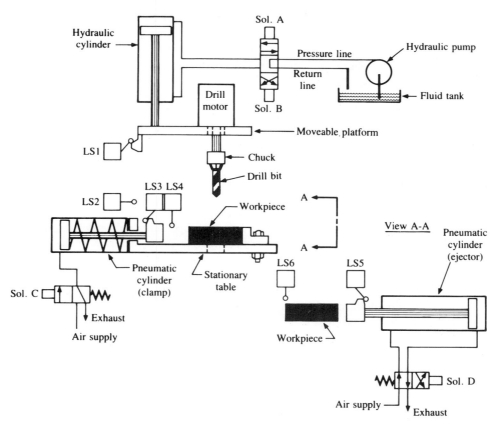

◆ **Figure 11.5**   An automated drilling machine is an example of an event-driven sequential process.

the drill platform is in the Drilled position. Notice that the term *actuated* is used instead of the term *closed*. The reason is that limit switches have both normally open and normally closed contacts. Actuating a limit switch closes the normally open contacts and opens the normally closed contacts.

A stationary table supports the part (called the workpiece) in which a hole is to be drilled. The table has a hole in line with the drill bit, so the workpiece can be drilled clean through if desired. A single-action spring-return pneumatic cylinder clamps the workpiece against a bracket that is bolted to the workpiece table. Limit switch LS3 is actuated when the clamping cylinder is in the *Unclamped* position. Limit switch LS4 is actuated when the clamping cylinder is in the *Clamped* position.

A double-acting pneumatic cylinder ejects the workpiece after the hole is drilled. The ejector cylinder is oriented 90° from the clamping cylinder, so the workpiece slides along the clamping bracket as it is ejected. Limit switch LS5 is actuated when the cylinder is in the *Ejector Retracted* position. Limit switch LS6 is actuated when the cylinder is in the *Ejected* position.

Three solenoid valves are used to control the cylinders. A hydraulic three-position four-way valve controls the hydraulic cylinder. When the pump is running and solenoid A is energized, the cylinder will move the platform down. When solenoid B is energized, the cylinder will move the platform up. When neither solenoid is energized, the valve returns the fluid to the supply tank and the platform is stationary. A two-position three-way pneumatic valve controls the clamping cylinder. When solenoid C is deenergized, the cylinder is vented and the spring returns it to the Unclamped position. When solenoid C is energized, the air supply is ported to the cylinder and it moves to the Clamped position. A two-position four-way valve controls the ejector cylinder. When solenoid D is deenergized, the ejector is forced into the Retracted position. When solenoid D is energized, the workpiece will be ejected as the cylinder moves to the Ejected position.

Sequential Function Chart

The first step in the design of a sequential control system for the drilling machine is to prepare a diagram or chart that describes the operations in the process. The sequential function chart shown in Figure 11.6 is an example of such a diagram. It uses a box to represent each step in the process. A double box is used for the first step in the cycle. The progression is from top to bottom, in order of the numbers of the steps. Arrows are not required to show the direction of the flow and are not used. A rectangular box on the right is used to describe the operation performed in each step. The condition for advancing to the next step is written next to the horizontal lines that cross the transfer path between two steps. For example, in the first step, the system waits for an operator to press the START button. The condition for advancing to step 2 is for an operator to press the START button. When this is done, we say that START is true and has a value of 1. When the START button is not pressed, we say that START is false and has a value of zero.

The sequential function chart has a provision for simultaneous sequences on parallel paths. A pair of horizontal lines just above the parallel paths indicates simultaneous activation of all the paths. Another pair of horizontal lines near the bottom indicates the simultaneous completion of all the paths. Each pair includes a wait step just above the lines, indicating si-

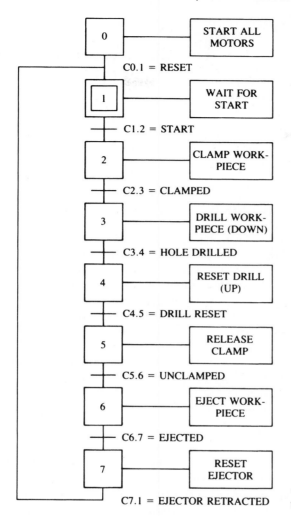

◆ **Figure 11.6**   Sequential function chart for the drilling machine in Figure 11.5.

multaneous completion. Between the simultaneous start and the simultaneous completion, each path completes its operations independently of all the other paths. Parallel operations of this type are common in manufacturing workstations.

State Chart

A state chart is a truth table showing the condition of each output at each step in the cycle. The condition of the output is indicated by an × if the output is ON or nothing if the output is OFF. (Sometimes a 1 is used in the place of the × and a 0 in place of nothing.) Table 11.1 is an example of a state chart.

◆ **TABLE 11.1** State Chart for the Drilling Machine in Figure 11.5

Step No.	Sol. A	Sol. B	Sol. C	Sol. D	Motors
0					
1					x
2			x		x
3	x		x		x
4		x	x		x
5					x
6				x	x
7					x

Process Timing Diagrams

A process timing diagram similar to Figure 11.2 can be used to describe an event-driven process. The major difference is that the timing diagram for an event-driven process uses an arbitrary time scale, whereas the timing diagram for a time-driven process uses a real-time scale. The timing diagram is a very useful tool for analyzing and designing an event-driven sequential process. It gives the designer a clear picture of the sequence in which the various inputs and outputs occur. In particular, it clearly shows the inputs and outputs whose values change during a particular step. The designer may need this information to determine whether or not a holding contact is necessary to complete an action initiated by a condition that changes before the action is completed. Figure 11.7 is the timing diagram for the drilling machine in Figure 11.5. The 0 level in the timing diagram indicates that the device is deactivated or deenergized. The 1 level indicates that the device is activated or energized.

Sequential Circuit Design

This section presents a step-by-step procedure for the design of an event-driven sequential circuit. The purpose is to illustrate the design procedure and to obtain a finished ladder diagram that is easy to follow. The design does not attempt to minimize the number of relays and contacts that are used. With microcomputers and programmable logic controllers (PLCs) providing the control function, the number of relays and contacts is not a significant cost factor. It is usually more cost-effective if the design and its documentation are easy to understand, maintain, and modify than it is to reduce the number of components in the controller. The next example does, however, illustrate a design that uses fewer components.

The basic concept of the design is the use of a control relay for each step in the process. When the process is in a given step, the control relay associated with that step is energized and all other relays are deenergized. We say that each control relay represents a "state" of the controller. There is a one-to-one correspondence between steps in the process and states of the controller, and the step numbers are used to identify the states. When the operator presses the RESET button (or the control system is automatically reset), relay CR1 is energized and all other relays are deenergized. The controller is in STEP 1, waiting for the START command. When the operator presses START, the controller moves to STEP 2 by energizing CR2 and deenergizing CR1. When STEP 2 is completed, the controller waits for the occurrence of the con-

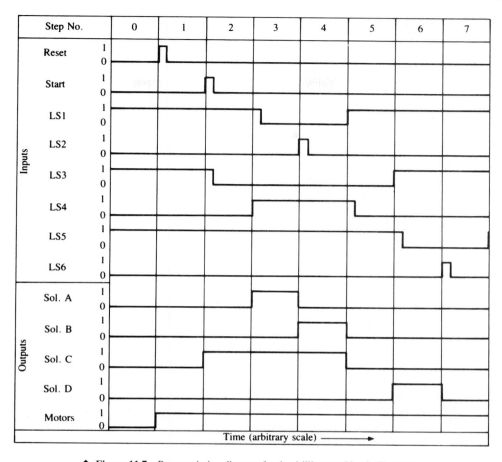

◆ **Figure 11.7**   Process timing diagram for the drilling machine in Figure 11.5.

dition for transition from STEP 2 to STEP 3. When the transition condition occurs, the controller moves to STEP 3 by energizing CR3 and deenergizing CR2. This process of moving from step to step continues until the controller is back at STEP 1 and the cycle has been completed.

Each state in the controller consists of one rung in the controller ladder diagram. The general form and one specific example of these "state" rungs are shown in Figure 11.8. Each rung begins on the left with a normally closed contact from the control relay for the next state. This NC contact opens when the advance to the next state is made. Its purpose is to turn OFF the previous state. Otherwise, the controller would end up with all states energized at the end of the cycle. Next comes a parallel section. The bottom branch of the parallel section contains the holding contact. The top branch establishes the transition condition. The transition function is of the following form:

$$\text{STATE } j = \text{STATE } i \text{ AND CONDITION } i.j$$

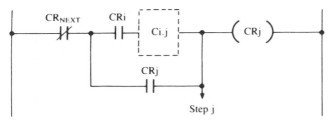

a) Step J ladder rung

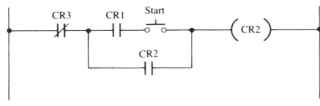

b) Step 2 ladder rung

◆ **Figure 11.8** The rungs that establish the steps in a sequential control circuit have the general form shown above.

The AND represents the logical and operator. CONDITION i.j is the condition required to move from STATE i to STATE j. In essence, the transition function states that arrival at STATE j depends on two things. First, the controller must be in STATE i (i.e., the previous state). Second, the condition for the transition from STATE i to STATE j must be true. If these two conditions are satisfied, the controller will move to STATE j, and relay CRj will be energized.

Figure 11.8b shows the ladder rung for STATE 2. The condition for moving to STATE 2 is (1) the controller is in STATE 1, and (2) the START button is pressed. We will use C1.2 to represent the condition for transfer from STATE 1 to STATE 2. It is obvious that C1.2 is described by the following expression.

$$C1.2 = \text{START}$$

The design process consists of the following steps:

1. *Define the process.* A piping and instrumentation drawing or a schematic diagram is an excellent method of graphically defining the process.
2. *Define the step (states).* The sequential function chart and the state chart are two methods of documenting the steps in the process.
3. *Define the input and output conditions.* The input and output conditions should be defined for each step in the process. Any values that change during a step should also be noted. The timing diagram is an excellent method of graphically defining the input and output conditions.
4. *Define the transition conditions.* The transition conditions are the Ci.j terms that define the condition for a transition from one state to the next state. They determine the transition portion of the circuit in the ladder rung for each state (or step). The timing diagram is an excellent tool for developing the transition conditions. The following equation defines the condition for transition from STEP 2 to STEP 3:

$$C2.3 = \text{CLAMPED} = LS2$$

5. *Define the output functions.* The output functions are Boolean equations that define each of the output elements. The following output function defines solenoid A to be energized only during STEP 3.

$$\text{Sol. A} = \text{STEP } 3$$

6. *Construct the controller ladder diagram.* The controller ladder diagram has a rung similar to those in Figure 11.8 for each step in the process.
7. *Construct the output ladder diagram.* The output ladder diagram consists of a rung for each output element in the process. The output function defines the input section of each rung.
8. *Document the design.* The design documentation is a collection of the charts and diagrams described in design steps 1 through 7.

The first three steps of the design process have already been done. Figure 11.5 and the accompanying discussion covered step 1. The sequential function chart in Figure 11.6 and the state chart in Table 11.1 are the culmination of design step 2. The timing diagram in Figure 11.7 covers step 3. The remaining design steps follow.

Design Step 4. Define the transition conditions. The timing diagram in Figure 11.7 is most useful in determining each transition condition. Look for an input signal that changes exactly at the boundary between the present state and the next state. The signals are usually from some type of switch. Usually, there is only one signal that changes at the boundary and the choice is obvious. Occasionally, there will be more than one input that changes at the boundary and an arbitrary choice must be made. The following transition conditions were determined from an examination of Figure 11.7:

$$C0.1 = \text{RESET}$$
$$C1.2 = \text{START}$$
$$C2.3 = \text{LS4}$$
$$C3.4 = \text{LS2}$$
$$C4.5 = \text{LS1}$$
$$C5.6 = \text{LS3}$$
$$C6.7 = \text{LS6}$$

The function of the limit switches is obvious from the preceding set of transition conditions. It is also obvious that all six switches are required.

Design Step 5. Define the output functions. The output functions are obtained from either the timing diagram or the state chart. Examine each output and list all the steps during which a particular output is energized. The output function for that output is the logical OR of all the steps for which it is energized. The following output conditions were determined from an examination of Figure 11.7.

$$\text{Sol. A} = \text{STEP } 3$$
$$\text{Sol. B} = \text{STEP } 4$$
$$\text{Sol. C} = \text{STEP } 2 \text{ OR STEP } 3 \text{ OR STEP } 4$$
$$\text{Sol. D} = \text{STEP } 6$$
$$\text{Motors} = 1$$

The last output requires additional explanation. The motors are started by a START MOTORS pushbutton during STEP 0. From STEP 0, a RESET signal puts the controller into STATE 1 (or STEP 1). The control circuit is not involved in the starting or stopping of the two electric motors. Figure 11.9 shows the main power and motor starting circuit. The 110-V output of the transformer provides the electrical power for the control circuit (lines 2 and 6 in Figure 11.9) and for the output circuit (lines 1 and 7 in Figure 11.9).

Design Steps 6, 7, and 8. The results of design steps 6 and 7 are shown in Figure 11.10 (the control circuit) and Figure 11.11 (the output circuit). Design step 8 consists of the following:

The schematic diagram (Figure 11.5)

The sequential function chart (Figure 11.6)

The state chart (Figure 11.1) (optional)

The timing diagram (Figure 11.7)

The transition conditions

The output functions

The main power and motor control diagram (Figure 11.9)

The control system ladder diagram (Figure 11.10)

The output ladder diagram (Figure 11.11)

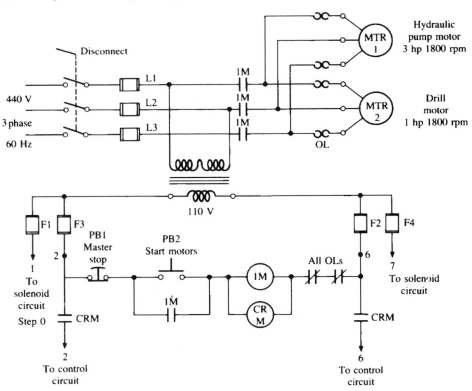

◆ **Figure 11.9** Main power and motor control circuit for the drilling machine in Figure 11.5.

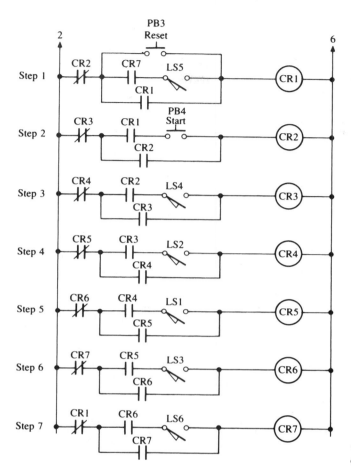

◆ **Figure 11.10**   Control circuit for the drilling machine in Figure 11.5.

An Alternative Design

Figure 11.12 shows an alternative design for the drilling machine controller. This design reduces the number of relays from seven to five (four if CR4 and TD1 can be combined into a single time-delay relay). However, the design of the controller is more empirical than the previous design. An empirical design is similar to working a puzzle. The designer keeps trying different combinations until a design that works is found. In testing different ideas, the designer develops a "feel" for the process that helps to develop a satisfactory design.

In the alternative design, the process is divided into two major operations: drilling and ejection. The drilling operation consists of three steps. New STEP 1 is a combination of STEP 1 and 2 from the original design. New STEP 2 is old STEP 3, and new STEP 3 is old STEP 4. The ejection operation is divided into two steps: the first step in ejection is old STEP 5; the second

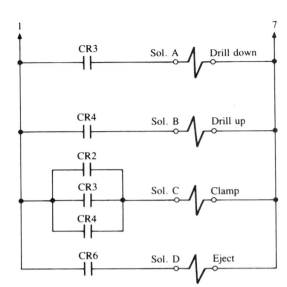

◆ **Figure 11.11** Solenoid circuit for the drilling machine in Figure 11.5.

step is a combination of old STEP 6 and 7. Actually, old STEP 7 is ignored and old STEP 6 is accomplished by a time-delay relay.

In an empirical design such as Figure 11.12, it is especially important to carefully examine a timing diagram of the control relays. Figure 11.13 shows such a diagram. A good way to check the design is to imagine that you are the controller, and walk through the operations for a complete cycle. The following is an example of a walk-through of the design in Figure 11.12 using the timing diagram in Figure 11.13.

The main power and motor control circuit is the same as in the original design. We begin in STEP 0 with the motors running and the ejector retracted (LS5 actuated). When the operator presses the START switch (Figure 11.12), CR1 is energized, and the CR1 holding contacts (in parallel with the START switch) close, and holding relay CR1 is energized. The controller is in the drilling operation. A second CR1 contact energizes solenoid C, which causes the single acting air cylinder to clamp the workpiece.

When the workpiece is clamped, LS4 is actuated and relay CR2 is energized. A CR2 contact closes to energize solenoid A, causing the hydraulic cylinder to move the drill platform down to drill a hole in the workpiece.

When the hole is drilled, switch LS2 is actuated. This opens the normally closed LS2 contact in the CR2 rung and closes the normally open LS2 contact in the CR3 rung. The result of these actions is that CR2 is deenergized and CR3 is energized. The CR3 holding contact is necessary because LS2 does not remain closed. When CR3 is energized, a CR3 contact closes to energize solenoid B. This causes the hydraulic cylinder to move the platform up, and LS2 soon becomes deactuated. The CR3 holding contact keeps the platform moving up until it reaches the drill reset position.

When the platform reaches the drill reset position, LS1 is actuated and relay CR4 is energized. The normally closed CR4 contact in the top rung opens, and coils CR1 and CR3 are deenergized. The controller has moved from the drilling operation to the ejection operation.

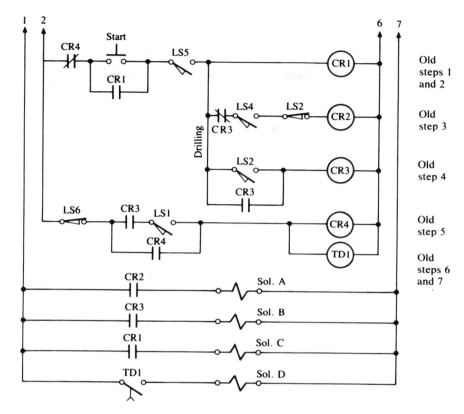

◆ **Figure 11.12**   This alternative control circuit for the drilling machine in Figure 11.5 uses fewer components than the circuit in Figure 11.10, but the design of the alternative circuit is more empirical and may have some unexpected problems.

The ejection operation begins when CR4 is energized and time-delay relay TD1 begins a time delay. After the time delay, delayed TD1 contact closes and energizes solenoid D. The time delay allows the clamp to retract before the ejector removes the workpiece.

When the workpiece is ejected, LS6 is actuated, deactivating both CR4 and TD1. This ends the ejection operation. The controller is ready for another cycle as soon as the ejector is retracted and LS5 is actuated.

11.4 TIME/EVENT-DRIVEN SEQUENTIAL PROCESSES

Many discrete processes have both time-driven and event-driven operations. Batch processes, for example, are usually time/event-driven sequential processes. The following statement list describes a reaction and blending process that will serve as an example of a time/event-driven sequential process.

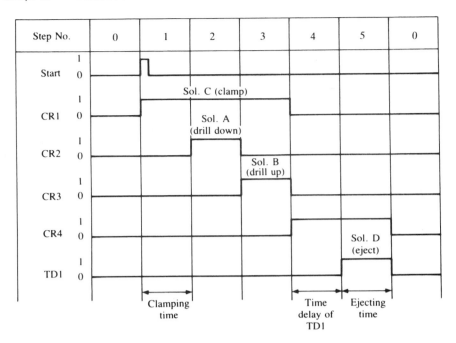

◆ **Figure 11.13** Timing diagram for the control circuit in Figure 11.12.

Statement List for a Reaction and Blending Process

STEP 1: Prepare (nominal time = 3 min)

    a. Operator prepares ingredient A

    b. Operator prepares ingredient B

    c. Wait for START PB to close

STEP 2: Charge (nominal time = 3 min)

    a. Initiated when operator presses START PB

    b. Ingredient A flow rate = 80% of FS

    c. Ingredient B flow rate = 65% of FS

    d. Mixer turned ON—stays ON until beginning of drain operation

    e. Level increases from 0 to 70%

STEP 3: Reaction (exact time = 8 minutes)

    a. Initiated when level = 70%

    b. Ingredient A flow rate = 0

    c. Ingredient B flow rate = 0

    d. Temperature controller set to AUTO—stays in AUTO until end of blend operation

STEP 4: Blend (exact time = 6 minutes)

    a. Initiated 8 min after start of reaction operation

    b. Ingredient C flow rate = 10 lb/min until total flow $Q = 40$ lb

    c. Level increases to 95%

STEP 5: Cool (nominal time = 3 min)

   a. Initiated 6 min after start of blend operation

   b. Temperature controller set to MANUAL

   c. Cooling water ON

STEP 6: Drain (normal time = 3 min)

   a. Initiated when $T = 60°$

   b. Cooling water and mixer OFF

   c. Pump ON

   d. Terminated when level = 0%

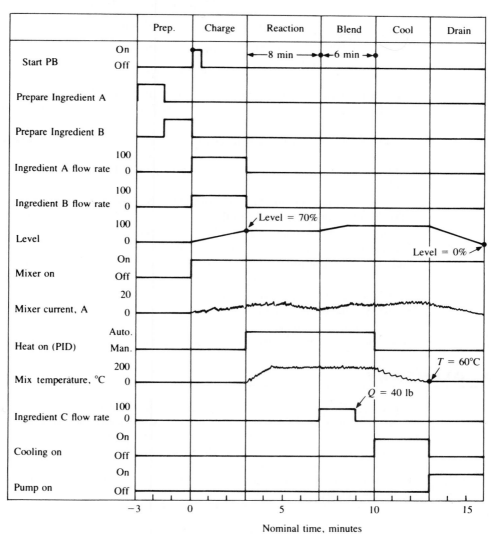

◆ **Figure 11.14**   Process timing diagram for the example reaction and blending process. This process has both time-driven and event-driven operations.

Figure 11.14 is the process timing diagram for the reaction and blending process. The transition times in the diagram are nominal times. The actual transition times may occur before or after the nominal times. A large dot marks the transition signal or condition. There are four event-driven transitions and two time-driven transitions. The START PB, two level switches, and a temperature switch provide the event-driven transitions. Two time-delay relays provide the time-driven transitions. The following is a summary of these switching devices.

$$START\ PB\ =\ normally\ open\ pushbutton\ switch$$

$$LS1\ =\ level\ switch\ set\ at\ a\ level\ of\ 70\%$$

$$LS2\ =\ level\ switch\ set\ at\ a\ level\ of\ 0\%$$

$$TS1\ =\ temperature\ switch\ set\ at\ 60°C$$

$$TD1\ =\ time\text{-}delay\ relay\ set\ for\ a\ delay\ of\ 8\ min$$

$$TD2\ =\ time\text{-}delay\ relay\ set\ for\ a\ delay\ of\ 6\ min$$

All switches have a normally open contact and a normally closed contact. The delays in TD1 and TD2 occur when the coil is energized; they also have a normally open contact and a normally closed contact.

Ingredients A and B are controlled by PI flow controllers set at 80% and 65% of full scale, respectively. The charge operation consists of filling the blending tank up to the 70% full mark. When the 70% level is reached, the PI controllers are turned off and the flow of ingredients stops.

The reaction operation has a duration of exactly 8 min, determined by time-delay relay TD1. The PID temperature controller raises the temperature of the contents of the blending tank to 200°C and holds it there until the end of the blending operation. The blending operation has a duration of exactly 6 min, determined by time-delay relay TD2. The ingredient C flow controller determines both the flow rate of ingredient C and the total amount of ingredient C that is delivered to the blending tank. The flow rate of ingredient C is 10 lb/min; the total quantity delivered is 40 lb. The ingredient C controller turns itself OFF as soon as it delivers 40 lb to the blending tank.

The cool operation sends cool water through a coil to cool the contents of the blending tank. The cool operation ends as soon as the temperature in the blending tank reaches 60°C.

◆ GLOSSARY

Batch process: *See* Discrete process; Sequential process. (11.1)

Boolean equations: A mathematical method of representing the rungs in a ladder diagram. (11.3)

Discrete process: A series of distinct operations with a definite condition for initiating each operation. If the series of operations has a beginning, an end, and a definite controlled form, the process is called a sequential process or a batch process. (11.1)

Event-driven process: A sequential process in which each step is initiated by the occurrence of an event. (11.3)

Flowchart: A diagram that uses blocks to represent the steps in a sequential process and lines to show the path from step to step. (11.1)

Ladder diagram: An electrical diagram that shows the connections between various contacts, coils, solenoids, motors, etc. in a sequential process. The ladder diagram got its name from its general appearance—it looks like a ladder. (11.1)

Operation: A major step in a sequential process such as preparation, loading, processing, etc. An operation consists of one or more parts called steps. (11.1)

Operations, auxiliary: An operation that is initiated and terminated by steps in a sequential process but operates concurrently with the main process. A PID controller is an example of an auxiliary operation. (11.1)

Operations, parallel: Operations in a sequential process that occur at the same time. (11.1)

Output functions: Boolean equations that define each output element in a sequential process. (11.3)

Process timing diagram: A graph of the outputs of a sequential process plotted versus time. (11.1)

Sequential function chart: A diagram that uses blocks to represent each step in a sequential process and lines to represent the transition from step to step. (11.1)

Sequential process: A sequence of operations that must be performed in a defined order to make a definite amount of finished product or to perform a definite amount of work. (11.1)

State chart: A truth table that shows the condition of each output in a sequential process at each step in the cycle. The state chart is usually accompanied by a diagram or chart that shows the transfer paths from state to state. (11.1)

Statement list: An English-language list of the actions that must be carried out in each step of a sequential process. (11.1)

Steps: The individual parts of a sequential operation. (11.1)

Time-driven process: A sequential process in which each step is initiated at a given time or after a given time interval. (11.2)

Transition conditions: The condition for a transition to occur from one state to another state. (11.3)

◆ EXERCISES

Section 11.1

11.1 Write a paragraph that describes something from your own experience that fits the definition of a sequential process. Some possibilities are driving to work or school, preparing a meal, taking a shower, the operations of a car wash, etc.

Section 11.2

11.2 Draw a process timing diagram similar to Figure 2.7 for the microwave instructions in Section 11.2. Estimate times for STEPS 1, 2, 4, and 7.

11.3 Find another example of a statement list similar to the microwave instructions in Section 11.2. Draw a process timing diagram similar to Figure 2.7 for your statement list.

11.4 The industrial blending process shown in Figure 9.18 could be controlled by a time-driven sequential control system described by the following statement list:

Blending Process Statement List

Initial condition: Drain valve OPEN, ready light ON. Agitator OFF, water and syrup valves OFF, solids feeder OFF.

STEP 1: (1 to 3 s) Press START switch.

STEP 2: (30 s) Drain valve CLOSED, ready light OFF, water valve ON, delay 30 s.

STEP 3: (85 s) Syrup valve ON, agitator ON, water valve remains ON, delay 85 s.

STEP 4: (175 s) Syrup valve OFF, solids feeder ON, water valve and agitator remain ON, delay 175 s.

STEP 5: (175 s) Solids feeder OFF, water valve OFF, agitator remains ON, delay 175 s.

STEP 6: (115 s) Drain valve OPEN, agitator remains ON, delay 115 s.

STEP 7: (2 s) Agitator OFF, ready light ON, water and syrup valves remain OFF. Return to initial condition.

Draw a process timing diagram from this statement list.

11.5 A batch blending process consists of a blending tank that mixes five liquid ingredients. The ingredients are designated as A, B, C, D, and E. Each ingredient is controlled by an ON/OFF solenoid valve. The following statement list could be used to control this process:

Blending Process Statement List

Initial condition: Drain valve OPEN, ready light ON, all liquid valves OFF, mixer OFF.

STEP 1: (1 to 3 s) Press START switch.

STEP 2: (20 s) Drain valve CLOSED, ready light OFF, valve A ON, delay 20 s.

STEP 3: (40 s) Valve A OFF, valves C and D ON, mixer ON, delay 40 s.

STEP 4: (60 s) Valves C and D OFF, valves B and E ON, mixer remains ON, delay 60 s.

STEP 5: (20 s) Valves B and E OFF, mixer remains ON, delay 20 s.

STEP 6: (120 s) Drain valve OPEN, mixer remains ON, delay 120 s.

STEP 7: (2 s) Mixer OFF, ready light ON, all ingredient valves remain OFF. Return to initial condition.

Draw a process timing diagram from this statement list.

Section 11.3

11.6 Write Boolean equations for the ladder rungs in Figure 11.8.

11.7 Write a Boolean equation for the ladder rung in Figure 11.9. Use the term NOT OL to account for all the overload contacts.

11.8 Write the Boolean equations for the ladder diagram in Figure 11.10. There will be seven equations, one for each rung in the ladder.

11.9 Convert the ladder diagram in Figure 11.12 into a set of Boolean equations.

11.10 Convert the following set of Boolean equations into a ladder diagram. The letters A through F represent normally open contacts.

$$CR1 = A \text{ AND } B \text{ AND } C$$
$$CR2 = D \text{ OR } E \text{ OR } F$$
$$CR3 = \text{NOT } CR1 \text{ AND NOT } CR2 \text{ AND } (LS1 \text{ OR } PS1)$$
$$SOLA = CR2$$

11.11 Convert the following set of Boolean equations into a ladder diagram.

$$IM \text{ OR } CR1 = \text{NOT STOP AND (MSTART OR } 1M) \text{ AND NOT OL}$$
$$CR2 = CR1 \text{ AND NOT } CR3 \text{ AND (START OR } CR2)$$
$$CR3 = CR1 \text{ AND NOT } CR4 \text{ AND (LS2 OR } CR3)$$
$$CR4 = CR1 \text{ AND NOT } CR2 \text{ AND (LV1 OR } CR4)$$

11.12 Sketch the circuit diagram of a sequential control system with the following features.

1. Motor starter 1M is energized by pressing a momentary contact start switch.
2. Motor starter 1M can be deenergized only by pressing a momentary contact stop switch.

11.13 Modify the circuit diagram of Exercise 11.10 to include the following additional features:

1. Limit switch 1L must be closed before the start switch can energize the motor starter 1M.
2. Either opening limit switch 1L or pressing the stop switch can deenergize motor starter 1M.

11.14 Modify the circuit diagram of Exercise 11.10 to include the following additional features:

1. Limit switch 1L must close before the start switch can energize the motor starter 1M.
2. Opening limit switch 1L cannot deenergize the motor starter. The motor starter 1M can be deenergized only by pressing a momentary contact stop switch.

11.15 Construct a timing diagram of the control relays in Figure 11.10.

11.16 A solenoid-operated lock is used on the gate of an enclosed storage area. When the solenoid is activated, the gate is unlocked, and a light goes ON. Three pushbutton switches are used in the solenoid control circuit, SW1, SW2, and SW3. The three switches are located such that a single operator can close only one switch at a time, two operators are required to close two switches, and three operators are required to close all three switches. Draw a ladder diagram for each condition listed and describe how the operators can unlock the gate:

(a) The solenoid is activated when any one of the three switches is activated.

(b) The solenoid is activated only when all three switches are activated.

(c) The solenoid is activated when any two of the three switches are activated.

11.17 The drilling machine control circuit in Figure 11.10 has a potential reset problem. Pressing the RESET pushbutton, PB3, activates the STEP 1 control relay, CR1, and deactivates the STEP 7 control relay, CR7, as it should. However, if one or more of the other control relays are activated (by Murphy's Law), pressing the RESET pushbutton will not deactivate them. Draw the ladder diagram of a revised control circuit that corrects the potential reset problem. One approach is to add another control relay, CR0, that has one NO contact and five NC contacts. Use PB3 to activate CR0, and use the CR0 contacts to activate CR1 and deactivate CR2 through CR6 (CR1 will deactivate CR7).

11.18 A reversible motor is used to open and close a gate. The motor has three terminals: L1A, L1B, and L2. When L2 is connected to ground and power is applied to L1A, the motor runs in the clockwise, CW, direction, opening the gate. When power is applied to L1B, the motor runs in the counterclockwise, CCW, direction, closing the gate. The gate has two limit switches: LS1 is activated when the gate is closed, and LS2 is activated when the gate is open. Each limit switch has one normally open, NO, contact and one normally closed, NC, contact. Two control relays, CR1 and CR2, are used in the gate control circuit. These relays have two NO contacts and two NC contacts. A NO pushbutton START switch is used to initiate the opening of the gate.

Draw a ladder diagram that will open the gate after a momentary closure of the START switch, close the gate immediately after the gate is fully open, and turn OFF the power to the motor when the gate is closed. Your circuit must include an interlock feature that will not allow simultaneous application of power on L1A and L1B.

Section 11.4

11.19 Redraw the ladder diagram in Exercise 11.15 to include a time-delay relay, TR1 (see Figure 9.3). This relay has an adjustable 10- to 200-s delay when the coil is energized, one NO contact, and one NC contact. The purpose of the time-delay relay is to provide an adjustable delay between the time the gate is fully open and the time it begins to close.

11.20 Redraw the ladder diagram in Exercise 11.16 to include a direct scan photoelectric sensor (see Figure 7.16) that senses the presence of an object in the path of the gate.

The photoelectric sensor (PS1) has one NO contact and one NC contact. When an object is detected, PS1 will prevent the motor from closing the gate, and, if the gate is partly open, it will cause the motor to open the gate. The gate will close as soon as the object is removed and the time delay has passed.

11.21 A small electric heater has two heating elements. When the heater is first turned ON, both elements are energized for quick warm-up. After a 10-min delay, the second element is turned OFF. A temperature activated switch turns OFF both elements if the heater gets too hot (see Figure 9.1). Draw the ladder diagram for the control circuit.

11.22 Complete the following for the reaction and blending process described by the statement list in Section 11.4 and the process timing diagram in Figure 11.14.

(a) Draw an instrumentation and piping diagram of the process. Assign a number for each control loop in the process. Use the following switching devices described in the text: START PB, LS1, LS2, TS1. Add any additional components you consider necessary or advantageous.

(b) Construct a sequential function chart.

(c) Construct a state chart.

(d) Determine the transition conditions.

(e) Determine the output conditions.

(f) Construct the sequential controller ladder diagram.

(g) Construct the output ladder diagram.

11.23 Assemble the documents prepared in Exercise 11.20 into a design document for the reaction and blending process. Add copies of the statement list and the process timing diagram to complete your design document.

◆ CHAPTER 12

Programmable Logic Controllers

◆ OBJECTIVES

The programmable logic controller (PLC) began as a reusable, inexpensive, flexible, reliable replacement of hard-wired relay panels. From this beginning in 1968, PLCs have developed into much more. Modern PLCs perform a host of functions such as logic, timing, counting, sequencing, PID control, and fuzzy logic. They can perform arithmetic operations, analyze data, and communicate with other PLCs and host computers.

The purpose of this chapter is to introduce you to programmable logic controllers. After completing this chapter, you will be able to

1. Describe the following parts of a PLC: power supply, processor, input module, output module, and programming unit
2. Describe the step-by-step operation of a PLC as it runs a small ladder diagram program
3. Draw a PLC input/output diagram
4. Draw a PLC ladder diagram program
5. Use timing functions in a PLC program
6. Use counter functions in a PLC program
7. Use sequencer functions in a PLC program

12.1 INTRODUCTION

A *PLC* is a digital electronic device designed to control machines and processes by performing event-driven and time-driven sequential operations. The PLC is designed for use in harsh industrial environments. It can be programmed without special programming skills, and it can be maintained by factory maintenance technicians.

The automobile industry originated the PLC to eliminate the high changeover cost of inflexible, hard-wired relay panels. The Hydramatic Division of the General Motors Corporation specified the design criteria for the first programmable logic controller in 1968. They specified a solid-state system with computer flexibility that would survive in the industrial environment, could be easily programmed and maintained by plant personnel, and would be reusable. The intent of the specification was to reduce the time and cost of model changeover. Hundreds of hard-wired relay panels were junked each time a car model was changed.

From this beginning in 1968, PLCs have developed into much more than just an inexpensive replacement for relay logic panels. The newer PLCs include modules that perform a host of functions such as logic, timing, counting, sequencing, PID control, and fuzzy logic. They can perform arithmetic operations, analyze data, and communicate with other programmable controllers and host computers. Some advantages of PLCs include

Flexibility. One PLC can handle many different operations, and software changes are easier to implement than hardware changes.

Reliability. Solid-state devices are more reliable and easier to maintain than mechanical relays and timers.

Lower cost. PLCs benefit from the ability of computer systems to deliver increasingly complex functions at diminishing cost.

Documentation. PLC programming equipment can provide an immediate printout of the current control circuit. There is no need for drawings that become outdated as the control circuit is modified.

The trend in industrial control is to use smaller PLCs that are distributed near the process rather than one large, centralized PLC with many inputs and outputs. A supervisory computer communicates with the individual PLCs on a local area network (LAN) to coordinate their activities. One reason for this trend is the tremendously complex program required to control a centralized system with one large PLC. Breaking the process into smaller, more manageable portions simplifies the program and allows the use of smaller, less expensive PLCs. Problem-solving experts will recognize this approach as an application of a proven problem-solving technique called *divide and conquer*. This method solves large, difficult problems by dividing them into a number of smaller, more easily solved problems.

Figure 12.1 shows a training and industrial work cell in which a CIM control station (see Section 2.7) is networked to two CNC-controlled machines and three PLC-controlled robots. The work cell consists of the following parts:

1. A computer-controlled conveyor
2. A CNC-controlled lathe
3. A PLC-controlled transverse robot to load and unload the lathe
4. A CNC-controlled milling machine

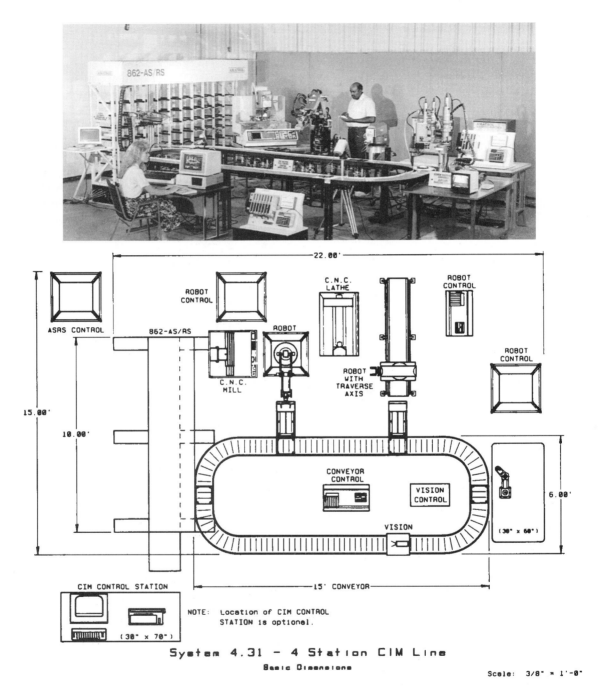

◆ **Figure 12.1** A training and industrial work cell contains two CNC-controlled machines and three PLC-controlled robots, all networked to a CIM control station. (Courtesy of Amatrol, Inc.)

5. A PLC-controlled pick-and-place robot to load and unload the mill
6. A PLC-controlled SCARA robot for drilling or small assembly on the conveyor
7. A computer-controlled storage rack (ASRS) Automatic Warehouse
8. A master computer (the CIM control station) to coordinate the entire operation

Three network configurations can be used with the work cell:

1. *A star network configuration.* All communication is between the CIM control station (master computer) and the individual units. Communication from one unit to another must pass through the CIM control station.
2. *A semi–star network configuration.* The CIM control station communicates with the PLCs, and each PLC communicates with the machine and robot under its control.
3. *A bus network configuration.* The PLCs can communicate with each other without going through the CIM control station. Figure 12.2 shows a bus network configuration.

The star configuration is the least flexible and the easiest to program. The bus configuration is the most flexible and the hardest to program.

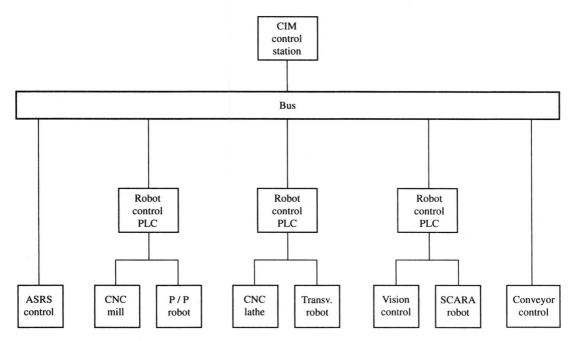

◆ **Figure 12.2**   A bus network configuration for the CIM work station. The PLCs can communicate with each other without going through the CIM control station.

12.2 PLC HARDWARE

Figure 12.3 illustrates five functional components found in all programmable logic controllers, and two others (6 and 7) found in some PLCs.

1. Power supply
2. Processor module
3. Input modules
4. Output modules
5. Programming unit
6. PID controller unit
7. Data communication unit

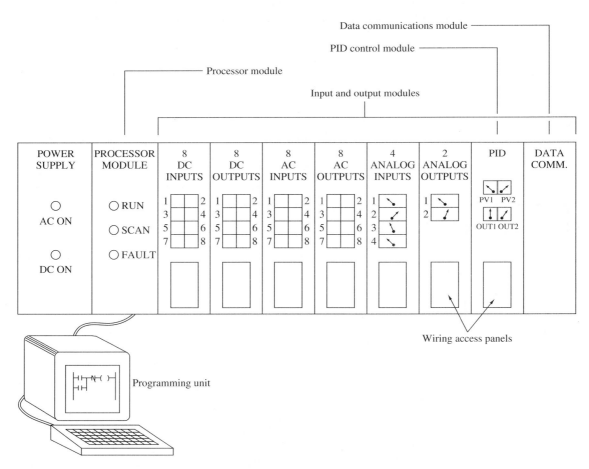

◆ **Figure 12.3** Components in a programmable controller include a power supply, a processor module, input modules, output modules, and a programming unit. The size of PLCs ranges from small units that replace eight relays to large units with thousands of inputs and outputs.

Power Supply

A PLC actually requires two power supplies. One is the external supply shown in Figure 12.3. It connects directly to a 120- or 240-V ac power line and provides ac and dc power for the input devices and the output loads. The other is an internal power supply in the processor module that powers the central processing unit (CPU) and processor logic circuitry.

Processor Module

The processor module is also referred to as the *central processing unit* (or *CPU*). It contains a microprocessor, a read-only memory unit (ROM), a random access memory unit (RAM), and an I/O interface. The information stored in ROM is permanent. It is set once, and then it cannot be changed. The information stored in RAM is temporary. It can be changed at any time, and it is lost when the processor module loses power.

The PLC operating system (OS) is stored in ROM (the OS is a program that serves the same purpose as DOS or Windows in a PC). The user program can be stored in RAM or in one of the following less volatile memory chips:

Programmable read-only memory (PROM). Can be programmed just once, then it is permanent.

Erasable programmable read-only memory (EPROM). Can be programmed many times by erasing the old program with ultraviolet light.

Electrically erasable read-only memory (EEPROM). Can be programmed many times by erasing the old program with an electrical signal.

Nonvolatile random access memory (NOVRAM). Can be programmed many times, but unlike ordinary RAM, the contents of NOVRAM are not lost when the processor loses power.

The data used by the PLC is stored in RAM. This data is organized into sections depending on the nature of the data. Each section is identified by a capital letter that is used as part of the address of locations in that section of memory. The memory sections include, but are not limited to, the following:

Input image status (I). Stores the status (1's or 0's) of the inputs from switches and ON/OFF signals from the process.

Output image status (O). Stores the binary data (1's or 0's) that will activate or deactivate ON/OFF devices in the process.

Timer status (T). Stores the time base, present value, accumulated value, and status bits of timers in the user program.

Counter status (C). Stores the present value, accumulated value, and status bits of counters in the user program.

Numerical data (N). Stores data used for number conversions, etc.

Functions (F). Stores status and data used by other functions in the user program.

The processor has two operating modes, PROGRAM and RUN. In the PROGRAM mode, the processor allows the user to make changes in the program. The processor has several status indicators that provide information to the programmer or operator. In the RUN mode, the processor repeats the following four-step cycle under the control of the operating system:

1. *Input scan.* The processor scans the inputs and stores a new image of the input conditions.
2. *Program scan.* The processor scans the program and derives a new image of the output conditions from the new image of the inputs and the old image of the outputs.
3. *Output scan.* The new image of the output conditions is transferred to the output devices.
4. *Housekeeping tasks.* Communication and other tasks are completed on a time-available basis.

The cycle may begin again immediately after completing the housekeeping tasks, or it may begin at a fixed interval.

Input Modules

There are two types of input modules, discrete and analog. The majority of inputs to a PLC are of the discrete type, bringing open and closed inputs from pushbutton contacts, limit switches, pressure switches, etc. Both ac and dc input modules are used, depending on the power source used on the input switch.

A typical input module has 4, 8, 16, or 32 input terminals, plus a common terminal (L2), and a ground terminal. One terminal of the input module is connected to one side of a contact in the process. The other side of the contact is connected to the hot side of the ac or dc power source. The common side of the power source is connected to the common terminal on the input module, and the input module provides a path to the common line through its sensing circuit.

The input module converts each input into a logic level voltage and isolates the inputs from the PLC circuitry. The input module converts an open contact into a logic 0 voltage (0 V), and it converts a closed contact into a logic 1 voltage (5 V). It is important to note that it doesn't matter if a contact is closed because it is a NO contact that is pressed or because it is a NC contact that isn't pressed. *If the contact is closed, the input module will convert it to a logic 1 voltage.*

When the processor scans the inputs, it reads the logic levels of all inputs and stores the logic levels in the input image section of RAM. Closed contacts are stored as logic 1's and open contacts as logic 0's. *The image can only indicate that a contact is open or closed; it cannot indicate whether the input contact is NO or NC.*

Although most PLC inputs are discrete, the inputs of analog signals are used for control and data acquisition purposes. Most process variables are analog in nature and must be converted to a digital form for input to a digital processor. The analog input module performs this function with an analog-to-digital converter (see Section 6.5).

Output Modules

Discrete output modules provide ON/OFF signals to operate lamps, relays, solenoids, motor starters, etc. When the processor scans the output image section of RAM, a logic 1 results in an ON signal to the output device, and a logic 0 results in an OFF signal. Triacs or relays are used to control ac output devices, and transistors or relays are used to control dc output devices. A typical output module has 4, 8, 16, or 32 output terminals, plus a hot terminal (L1), and a ground terminal.

Analog outputs require 8 to 16 bits from the output image section of memory. A digital-to-analog converter converts the binary number stored in the output image into an analog voltage (or current) between one output terminal and the common terminal.

PID and Communications Modules

The *PID module* provides three-mode control of two process variables (refer to Section 1.4 for further discussion of the three control modes). The *communications module* provides direct communication with the operator, a programming terminal, other PLCs, or a supervisory computer.

Programming Unit

The programming unit enables the operator to enter a new program, examine the program in memory, change the program in memory, monitor the status of inputs or outputs, display the contents of registers, and display timer and counter values. In addition, a password can be entered to protect the program in memory from unauthorized changes.

Programming devices include hand-held *programmers,* CRT terminals, and personal computers (with special software). The hand-held programmer resembles a pocket calculator with an LCD for displaying instructions, addresses, timer and counter values, data, and so on. It also has a keypad for entering instructions, addresses, and data. A CRT terminal looks like a personal computer, but it is specifically designed for programming a PLC. CRT terminals and personal computers allow the user to type in the program using a variety of programming languages including the Sequential Function Chart (SFC) language, the Ladder Diagram (LD) language, the Function Block Diagram (FBD) language, and the Structured Text (ST) language.

12.3 PLC PROGRAMMING AND OPERATION

This section presents a simple ladder diagram program for a PLC followed by a step-by-step analysis of the operation of the PLC as it runs the ladder diagram program.

Ladder Diagram Programming

The LD programming language uses contact and coil symbols to construct diagrams that are very similar to the ladder diagrams used for relay logic. The symbol for a normally open contact resembles the symbol for an electrical capacitor. The symbol for a normally closed contact is the open contact symbol with a diagonal line through it. The symbol for an output consists of several spaces enclosed in parentheses. Every contact and output symbol has a number written immediately below (or above) it. This number identifies the location in the processor memory (RAM) where the image of that contact or output is stored. Figure 12.4 shows a simple example of the implementation of ladder diagram logic in a programmable logic controller.

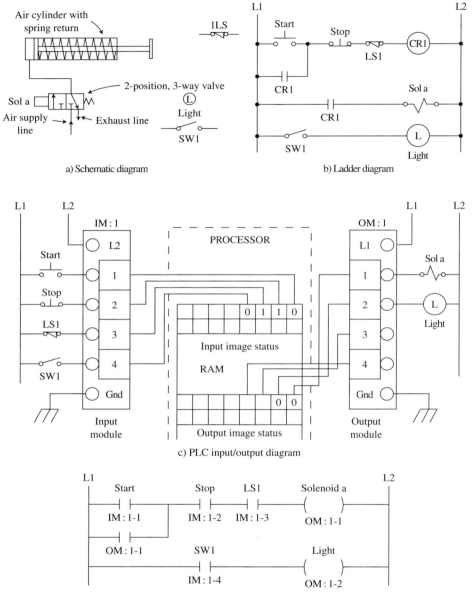

a) Schematic diagram

b) Ladder diagram

c) PLC input/output diagram

d) PLC program (ladder diagram language)

◆ **Figure 12.4** PLC implementation of a ladder diagram control circuit for a pneumatic cylinder (see Figure 2.8). Observe the use of a NC stop pushbutton in the input diagram with a NO contact symbol in the PLC program to implement the NC stop function.

Figure 12.4a shows the schematic diagram, and Figure 12.4b shows the ladder diagram of a relay logic circuit for controlling a pneumatic cylinder (see Figure 2.8). The switch-operated light was included to illustrate a second control function without unduly complicating the circuit. The first rung in the ladder diagram includes start, stop, and holding contacts in a conventional configuration plus a normally closed limit switch and a control relay. The second rung includes a normally open control relay contact and the solenoid for the pneumatic cylinder. The third rung consists of a switch and a light.

Figure 12.4c shows the input, processor, and output units of a PLC used to control the pneumatic cylinder and light. The boxes (1, 2, 3, and 4) in input module IM:1 contain the signal conditioners that convert the switch inputs into logic level voltages. Four lines connect the input boxes to the memory cells that hold the status of the inputs from those boxes. The upper right cell of the Input image memory section is connected to switch box 1, and it is identified as memory cell IM:1-1. Moving left from Memory cell IM:1-1 are cells IM:1-2, IM:1-3, and IM:1-4. The zero in memory cell IM:1-1 indicates that the START switch connected to terminal 1 is open, which it is. The 1's in memory cells IM:1-2 and IM:1-3 indicate that the STOP switch and switch LS1 are closed, which they are. The zero in cell IM:1-4 indicates that SW1 is open. Remember, the input image indicates that the contacts connected to IM:1-2 and IM:1-3 are closed, but it cannot indicate whether they are nonactivated NC contacts or activated NO contacts.

The boxes (1, 2, 3, and 4) in output module OM:1 contain the switching circuits that use the logic level voltages from output image memory cells to switch output devices ON or OFF. The line from output box 1 connects it to the upper right memory cell in the Output image memory section, that is, the OM:1-1 memory cell. Moving left from cell OM:1-1 are cells OM:1-2, OM:1-3, and OM:1-4. The zeros in memory cells OM:1-1 and OM:1-2 turn OFF solenoid a and the light.

Figure 12.4d shows the PLC program using the LD programming language. Compare the PLC program with the ladder diagram in Figure 12.4b. There are many similarities and a few differences. The ladder diagram has three rungs, but the program has only two. The second rung is missing because solenoid a is actually an output device connected to output terminal 0:1. For this reason, it does not appear in the program. Output 0:1 in the first rung of the program is actually the control relay, CR1, that activates solenoid a. The first rung in the program is equivalent to the first two rungs in the ladder diagram.

The LD language places a number of restrictions on ladder diagram programs, and the restrictions vary with different manufacturers. Typical limitations include the following:

1. The output must be at the right end of the rung.
2. Power must flow from left to right, up, or down.
3. Power must never flow from right to left.
4. The number of series contacts is limited to 11 (or some other number set by the manufacturer).
5. The number of parallel contacts is limited to 7 (or some other number set by the manufacturer).
6. A rung can have only one output.
7. The single output must be on the top line in a rung.

In a relay panel, power flow from right to left through a closed contact is called a *sneak path*. Sneak paths are not allowed in programming a PLC. Not allowing sneak paths simplifies the programming but does require some adjustment when converting a relay panel to a PLC program. The sneak paths must be converted into equivalent ladder diagram programs that do not have a sneak path.

In the LD programming language, the open and closed contact symbols are actually programming instructions. They tell the computer how to interpret the 1 or 0 in the Input image memory cell for each contact. The normally open contact symbol (⊣ ⊢) instructs the PLC to interpret a 1 as TRUE and a 0 as FALSE. If the PLC finds a path of true contacts from L1 to the output, the output is ON and a 1 is placed in its Output image memory cell. If no true path is found, the output is OFF, and a 0 is placed in its Output image memory cell.

The NO contact symbol can be used with both NO and NC physical switches, and the results are what we would expect them to be (see the two NO contact symbols used with PB1 and PB4 in Figure 12.5). A NO physical switch is evaluated as TRUE if it is actuated and FALSE if it is not. A NC physical switch is evaluated as TRUE if it is not actuated and FALSE if it is. In the PLC program in Figure 12.4d, for example, the NO contact symbol is used for all five contacts in the program. When the STOP switch is not activated, its image is a 1, and it is evaluated as TRUE, which is correct. When the STOP switch is activated, its image is a 0, and it is evaluated as FALSE, as it should be. This analysis also applies to switch LS1.

In a PLC program, the normally closed contact symbol is different, and its use can lead to a double negative effect that can cause considerable confusion (see the two NC contact

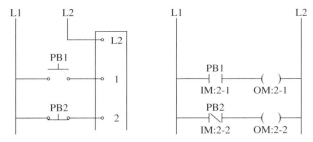

(a) Two ways to implement a NO switching function. Output OM:2-1 is OFF when PB1 is not pressed and ON when PB1 is pressed. Output OM:2-2 is OFF when PB2 is not pressed and ON when PB2 is pressed.

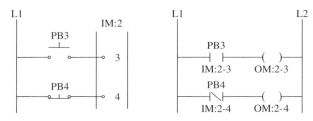

(a) Two ways to implement a NC switching function. Output OM:2-3 is ON when PB3 is not pressed and OFF when PB3 is pressed. Output OM:2-4 is ON when PB4 is not pressed and OFF when PB4 is pressed.

◆ **Figure 12.5** Two ways to implement a NO switch and two ways to implement a NC switch. Observe the double negative effect in PB2; it causes the NC switch, PB2, to behave like a NO switch. Also note that both methods of implementing a NC switch involve use of a single negative.

symbols used with PB2 and PB3 in Figure 12.5). The NC contact symbol (⊣/⊢) instructs the PLC to interpret a 1 in an Input image cell as FALSE, and a 0 as TRUE. If we use a NO physical switch with a NC contact symbol, the NO switch is evaluated as if it were a NC switch. If we use a NC physical switch with a NC contact symbol, the NC switch is evaluated as if it were a NO switch. The confusion caused by the double negative is reason enough to avoid the use of a NC switch with a NC contact symbol. For this reason, we will always use the NO switch with a NO contact symbol to implement the NO switch function. However, both methods of implementing the NC switch function use a single negative, and both methods (PB3 and PB4 in Figure 12.5) are used to implement the NC switch function.

PLC Operation

When a PLC is in the RUN mode, the processor repeats the following four-step cycle:

1. *Input scan.* Store a new image of input conditions.
2. *Program scan.* Derive the new image of output conditions.
3. *Output scan.* Transfer the new image of output conditions to output devices.
4. *Housekeeping tasks.* Communication and other tasks.

The discussion that follows will walk through four cycles of a PLC with the configuration and program shown in Figure 12.4. Our walk-through begins with the input image, the output image, and the four switches as shown in Figure 12.4c.

Cycle 1. START and SW1 are open, STOP and LS1 are closed.

Input scan. New input image is IM:1-1 = 0, IM:1-2 = 1, IM:1-3 = 1, IM:1-4 = 0.

Program scan. Rung 1: IM:1-1 = FALSE, OM:1-1 = FALSE, IM:1-2 = TRUE, IM:1-3 = TRUE, path is FALSE, so 0 is stored in cell OM:1-1.

Rung 2: IM:1-4 = FALSE so 0 is stored in cell OM:1-2.

Output scan. Both solenoid a and the light are OFF.

Cycle 2. START, SW1, STOP, and LS1 are all closed.

Input scan. New input image: IM:1-1 = 1, IM:1-2 = 1, IM:1-3 = 1, IM:1-4 = 1.

Program scan. Rung 1: IM:1-1 = TRUE, OM:1-1 = FALSE, IM:1-2 = TRUE, IM:1-3 = TRUE, path is TRUE, so 1 is stored in cell OM:1-1.

Rung 2: IM:1-4 = TRUE, so 1 is stored in cell OM:1-2.

Output scan. Both solenoid a and the light are ON.

Cycle 3. START is open, SW1, STOP and LS1 are all closed.

Input scan. New input image: IM:1-1 = 0, IM:1-2 = 1, IM:1-3 = 1, IM:1-4 = 1.

Program scan. Rung 1: IM:1-1 = FALSE, OM:1-1 = TRUE, IM:1-2 = TRUE, IM:1-3 = TRUE, path is TRUE, so 1 is stored in cell OM:1-1.

Rung 2: IM:1-4 = TRUE so 1 is stored in cell OM:1-2.

Output scan. Both solenoid a and the light are ON.

Cycle 4. START and LS1 are open, STOP and SW1 are closed.

Input scan. New input image: IM:1-1 = 0, IM:1-2 = 1, IM:1-3 = 0, IM:1-4 = 1.

Program scan. Rung 1: IM:1-1 = FALSE, OM:1-1 = TRUE, IM:1-2 = TRUE, IM:1-3 = FALSE, path is FALSE, so 0 is stored in cell OM:1-1.

Rung 2: IM:1-4 = TRUE so 1 is stored in cell OM:1-2.

Output scan. Solenoid a is OFF, the light is ON.

A couple of observations are evident from the operation of a PLC. First, the order of the operations may be important because the PLC evaluates the program one rung at a time. Second, there is a small delay between an input device action and the resulting change in the output device. This delay could be as much as the time it takes the PLC to complete one cycle.

12.4 PLC PROGRAMMING FUNCTIONS

In Section 12.3, you learned about three LD programming instructions: the NO contact instruction, the NC contact instruction, and the output instruction. These three instructions are sufficient to handle most, if not all, event-driven processes with ON/OFF inputs and ON/OFF outputs. Time-driven processes require additional instructions to handle timing and sequencing functions. Some control applications require counting functions, and others could use the ability to compare or manipulate numbers with arithmetic functions. Control of a process variable requires the PID control function. All of these advanced functions, and more, are available in programmable logic controllers.

Before launching into a study of PLC programming functions, let's examine another example of LD programming of a simple event-driven process. Our choice for this example is the event-driven sequential control system for a hydraulic hoist shown in Figure 11.4. This system uses four pushbutton switches for input. In Figure 11.4, the STOP switch was a NC pushbutton, with a NO contact instruction. In this example, we will use a NO STOP pushbutton with a NC contact instruction to provide the correct operation of the STOP switch. The outputs include a pump motor and two solenoids. The PLC input/output diagram is shown in Figure 12.6a. The PLC ladder diagram program is shown in Figure 12.6b.

Timer Function

The Timer is the most frequently used PLC function, and Time Delay On is the most common timing function. Both Time Delay On and Time Delay Off are used in PLCs. Our discussion will focus on Time Delay On. Two types of timers are used in programmable logic controllers, the one-input *nonretentive* timer and the two-input *retentive* timer.

The nonretentive timer has one input, one output, and a preset delay time. Consider a nonretentive timer with a preset delay time of 12 s. When the input is turned ON, the timer runs for 12 s before it turns the output ON. Once the output is turned ON, it remains ON until the input is turned OFF. The timer resets every time the input is turned OFF, and starting the timer again will result in a full 12-s delay.

The retentive timer has two inputs, one output, and two status bits that can be used as internal outputs. The inputs are the Enable/Reset line and the Run line. The output is the Done

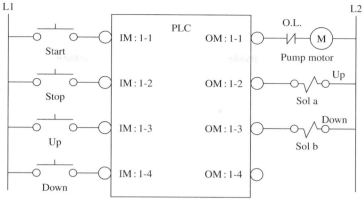

a) PLC input/output diagram

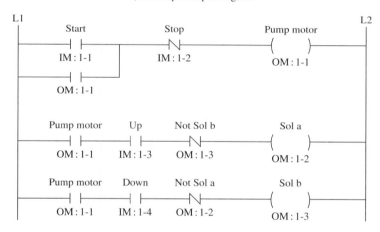

b) PLC ladder diagram program

◆ **Figure 12.6**   The input/output diagram and the LD program of a PLC used to control the hydraulic hoist shown in Figure 11.4. Observe the use of a NO stop pushbutton in the input diagram with a NC contact symbol in the PLC program to implement the NC stop function (see Figure 12.4 for the alternative implementation).

(or D) line. The status bits are the Enable and Run (or E and R) lines. When the Enable/Reset line is turned OFF, the timer is reset and cannot run. If the Enable/Reset line is ON, the timer runs whenever the Run line is ON, but it stops running when the time delay is satisfied (i.e., when output Done is ON). The E status line is ON whenever Enable/Reset is ON, but the R status line is ON only when the timer is actually running (i.e., E is ON, Run input is ON, and Done is OFF). If the Run line is turned OFF, the timer stops running, but the timer is not reset unless the Enable/Reset line is also turned OFF. If the Run and Enable/Reset lines are turned ON and OFF at the same time, the retentive timer operates as a nonretentive timer.

Consider a retentive timer with a preset delay time of 12 s and an Enable/Reset line that has just been turned ON. Assume the Run line is ON for 7 s and then goes OFF for 20 s. The timer holds the 7-s delay time while the timer is OFF. After 20 s, the Run line is turned ON

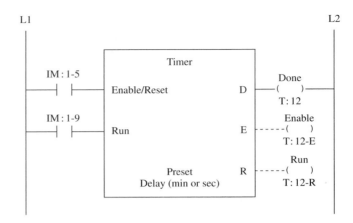

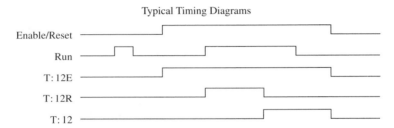

Typical Timing Diagrams

◆ **Figure 12.7** A generic, two-input, retentive timer and a typical timing diagram. The timer is shown with the output address T:12. In the timing diagram, Enable/Reset, and Run are inputs, T:12, T:12-E, and T:12-R are outputs. Observe that the timing diagrams of T:12-E and Enable/Reset match exactly, but the diagrams of T:12-R and Run do not. Output T:12-R is ON only when Enable/Reset is ON, Run input is ON, and Done is OFF, Output T:12 (Done) is ON only after the preset time delay has been reached, and it goes OFF when Enable/Reset goes OFF.

again for 200 s. Five seconds after the Run line is turned ON for the second time, the delay reaches the preset time of 12 s, and the timer output is turned ON. In this example, the actual delay time is $7 + 20 + 5 = 32$ s. After a total elapsed time of 227 s, the Run line is turned OFF. The timer output is turned OFF, but the timer is not reset (Enable/Reset is still ON). If Run is turned ON again, the output will go ON immediately, because the timer has not been reset. The timer is reset only when the Enable/Reset line is turned OFF.

Figure 12.7 shows a generic retentive timer in a simple ladder diagram program. Our intent is to illustrate the use of timers in PLC circuits in a general way without reference to specific hardware. Figure 12.7 defines the timer in sufficient detail for use in the Examples and Exercises in this text. For real applications, the reader should refer to manufacturer's reference manuals for information about the operation and programming of a specific PLC.

EXAMPLE 12.1 The batch blending process shown in Figure 12.8a is to be controlled by a Time/event-driven sequential control system.

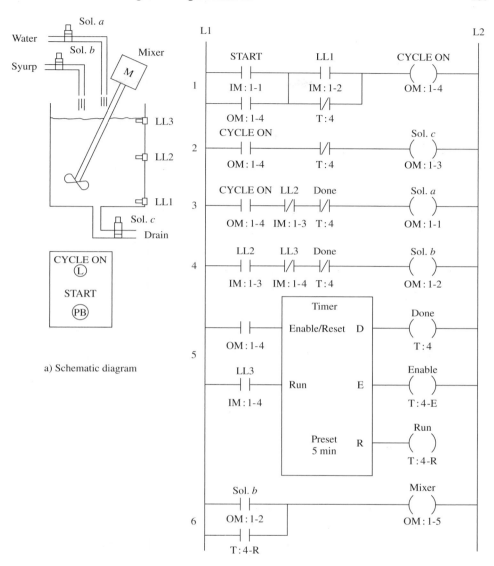

a) Schematic diagram

b) PLC ladder diagram program

◆ **Figure 12.8** The batch mixing process (a) is controlled by a PLC program (b) that uses a timer in a time/event-driven sequential control system (Example 12.1).

Initial conditions. The process begins with the drain valve open, the water and syrup valves closed, the mixer OFF, and the tank empty. The design criteria specifies the following five-step process:

STEP 1: Begins when START is pressed and ends when the liquid level switch, LL2, closes. The purpose of STEP 1 is to fill the tank with water to the level of LL2.

STEP 2: Begins when LL2 closes and ends when LL3 closes. The purpose of STEP 2 is to add syrup to the level of switch LL3 and begin mixing the two ingredients. When STEP 2 is done, the mixture in the tank will be 60% water and 40% syrup.

STEP 3: Begins when LL3 closes and ends 5 min later. The purpose of STEP 3 is to finish mixing the two ingredients.

STEP 4: Begins when STEP 3 is done and ends when switch LL1 opens. The purpose of STEP 4 is to drain the mixture from the tank.

STEP 5: Begins when LL1 opens and ends when START is pressed to begin another batch. The purpose of STEP 5 is to maintain the initial conditions listed above.

Draw a PLC ladder diagram program that satisfies the above design criteria.

Solution   The PLC program, shown in Figure 12.8b, consists of six rungs. The following is a brief explanation of each rung:

Rung 1 turns ON output OM:1-4 (CYCLE ON light) when START is pressed and keeps it ON until the tank is empty at the end of STEP 5. Initially, the true path is through the normally closed T:4 contact. In STEP 4, the true path is through the normally open LL1 contact.

Rung 2 closes the normally open drain valve when OM:1-4 is ON and before timer T:4 is done.

Rung 3 energizes solenoid a when OM:1-4 goes ON, deenergizes it when LL2 closes, and keeps it from going back ON when the level drops below LL2 in step 4.

Rung 4 energizes solenoid b when LL2 closes, deenergizes it when LL3 closes, and keeps it from going back ON when the level drops below LL3 in step 4.

Rung 5 enables the timer while the CYCLE ON light is ON, starts the timer when LL3 closes, and turns ON output T:4 when the timer is done (5-min delay).

Rung 6 turns the mixer ON during STEPS 2 and 3.   ◆

Counter Function

The PLC counter keeps track of the number of times a contact opens and ignores the times the contact closes between each pair of openings. In this way, the counter is "counting" the event that causes the contact to open. The event might be an item that is conveyed past a photodetector. It might be a vehicle that is driven over a pneumatic pressure detector. It might be a metal object on a shaft that rotates past a magnetic detector. Industry has many events to count.

Some PLC counters count from 0 up to a preset value; others count from the preset value down to 0. Most PLCs include both up and down counters. A generic PLC up counter is illustrated in Figure 12.9. You may notice a similarity between the counter and the timer in the previous section. The generic counter has two inputs, an external output, an internal output, a memory register that stores a preset value, and a memory register that stores the count.

When the Enable/Reset input line is ON, the counter adds 1 to the count register for each OFF-to-ON change on the count input line. When the Enable/Reset input line is OFF, the count is reset to zero, and the counter ignores changes on the count input line. The output of an up counter is OFF until the count reaches the preset value; then it is ON until the counter is reset. The counter continues to count after the preset value is reached. The output of a down counter is OFF until the count reaches 0, and then it is ON until the counter is reset.

Two or more counters can use the same count register. Consider, for example, two counters in an antique store—an up counter for counting shoppers entering and a down counter for counting shoppers leaving. The two counters use the same memory register for holding the count. For each person who enters, the up counter adds 1 to the count register. For each person leaving, the down counter subtracts 1 from the count register. At any time, the count

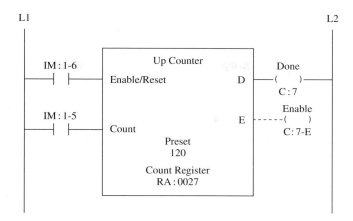

◆ **Figure 12.9**   A generic, two-input counter. The counter is shown with the output address C:7, the preset value 120, and the count register address RA:0027. When the Enable/Reset line is OFF, the counter ignores changes on the count input line, output (C:7) is OFF, and the count register (RA:0027) is reset to 0. When the Enable/Reset line is ON, the counter increments the count register by 1 each time the count input line goes from OFF to ON. The counter output (C:7) is OFF until the count in RA:0027 reaches the preset value (120). At that time, C:7 is turned ON and remains ON until Enable/Reset is turned OFF. The counter continues to increment the count register beyond the preset value, even though the output is ON, indicating it is done.

register will give the store manager an accurate count of the number of people in the store (assuming no one slips around the counter turnstiles). Figure 12.10 shows a PLC program that maintains a count of the number of shoppers in a store.

Sequencer Function

The sequencer function goes through a sequence of steps, producing specified conditions in a number of ON/OFF outputs at each step. The specified output pattern for each step is stored in one word of memory, and patterns for successive steps are stored in successive memory cells. Each bit in the word is matched with one output (bit 0 is matched with output No. 1, bit 1 is matched with output No. 2, . . . , bit 7 is matched with output No. 8). If a bit holds a logical 1, the sequencer turns the matching output ON (or leaves it ON). If a bit holds a logical 0, the sequencer turns the matching output OFF (or leaves it OFF). The number of outputs that a sequencer can control depends on the word size of the PLC. An 8-bit PLC can control up to 8 outputs, and a 16-bit PLC can control up to 16 outputs. Additional sequencer functions can be used to control additional outputs. For example, two 16-bit sequencers can control up to 32 outputs.

The generic sequencer shown in Figure 12.11 controls eight outputs in module OM:1 in a five-step sequence. The desired output conditions are stored in memory cells RA:0151 through RA:0155. The contents of these five memory cells are shown in Figure 12.11. Also shown are the lines connecting bits 0 through 7 of memory to the eight terminals in output module OM:1. The step pointer selects the memory cell whose contents are sent to OM:1. The

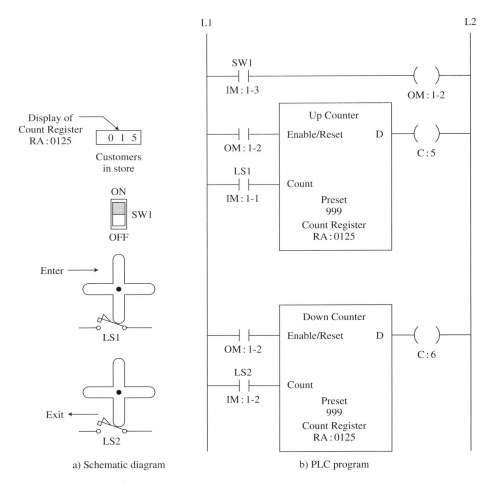

◆ **Figure 12.10** A two-counter PLC program maintains a count of the number of customers in a store.

step pointer is actually memory cell RA:0150. It contains the address of the currently selected memory cell. For example, when the sequencer is in STEP 1, RA:0150 contains the address of the starting register, RA:0151. The contents of RA:0151 are sent to module OM:1 where the zeros turn OFF all eight outputs. When the sequencer advances to STEP 2, RA:0151 contains the address of resister RA:0152. The contents of RA:0152 are sent to module OM:1 where the 1 in bit 0 turns ON output OM:1-1. When the set pointer advances to RA:0155, the sequencer output, SE:2, is turned ON. When Enable/Reset is turned OFF, output SE:2 is turned OFF, and the Step Pointer is reset to the first step (i.e., the address of the first step, RA: 0151, is stored in Memory cell RA:0150).

The generic sequencer is an event-driven function—the event is the occurrence of an OFF-to-ON change on the step input line. Combined with a timer, as shown in Figure 12.12, the se-

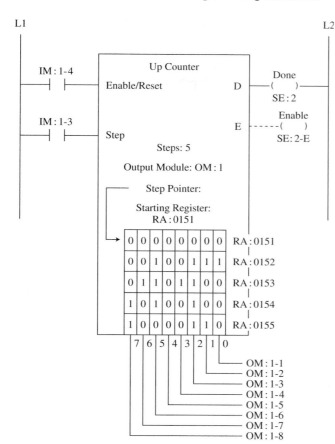

L1   L2

◆ **Figure 12.11**   A generic sequencer with the output address SE:2. When the Enable/Reset line is OFF, the sequencer is reset to STEP 1, it ignores changes on the step input line, and output SE:2 is OFF. When the Enable/Reset line is ON, the step pointer advances one step each time the step input line goes from OFF to ON. Output SE:2 goes ON when the sequencer advances to the last step. The sequencer remains in the last step and SE:2 remains ON until Enable/Reset is turned OFF.

quencer becomes a time-driven function. When OM:1-1 is ON, the sequencer Enable/Reset line is ON, and the timer output, T:8, produces an OFF-ON-OFF pulse every 30 s. The sequence of pulses on T:8 advances the sequencer once every 30 s. In this circuit, the sequencer has a constant step dwell time of 30 s. However, you can repeat a set of conditions in two or more successive memory cells to create step dwell times that are multiples of 30 s. For example, you could repeat the STEP 1 conditions three times for a dwell time of 90 s, and you could repeat the STEP 2 conditions ten times for a dwell time of 300 s.

A sequencer with multiple-step dwell times may not be good enough for some applications. You can use two sequencers driven by one timer to get variable-step dwell times. The second sequencer takes preset times from a second set of memory addresses and stores them in the memory cell that holds the preset time for the timer driving the two sequencers. In each step, the first sequencer sends the output conditions for that step to the destination output module. The second sequencer stores the dwell time for the next step in the timer's preset memory cell.

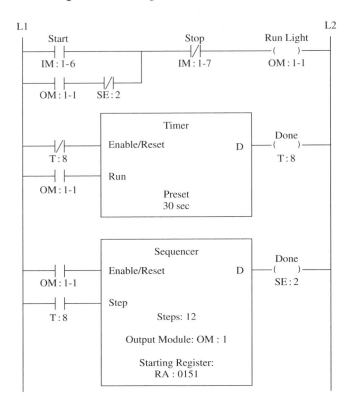

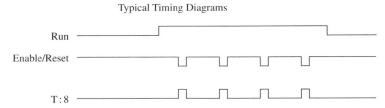

◆ **Figure 12.12** A generic sequencer with timed steps (see Example 12.2). The timer is enabled by its own output, causing it to reset immediately after it has finished a 30-s delay. After it has reset, the timer immediately begins another 30-s delay (assuming OM:1-1 is ON). The timing diagram illustrates the sequence of pulses produced by the timer in this circuit.

EXAMPLE 12.2

The industrial blending process in Figure 12.13 is to be controlled by the timed-step sequencer shown in Figure 12.12. The process consists of a blending tank with provisions to add measured amounts of water, syrup, and solids to the tank. A RUN light indicates when the process is in operation. The process has the following six steps:

1. Hold the sequencer in the reset condition with the drain valve open and everything else OFF. Delay 30 s after START is pressed.

2. Close the normally open tank drain valve, turn the water metering pump ON, turn the run light ON, and delay for 90 s.
3. Turn the syrup metering pump ON, turn the mixer ON, and delay for 60 s.
4. Turn the syrup metering pump OFF, turn the solids feeder ON, and delay for 30 s.
5. Turn the water metering pump OFF, turn the solids feeder OFF, and delay for 90 s.
6. Open the tank drain valve, turn the mixer OFF, and delay 60 s to drain the tank.

Draw a PLC input/output diagram for this system. Your diagram should show the timer preset, the sequencer parameters, and the contents of the memory addresses that hold the output conditions.

Solution   The PLC input/output diagram is shown in Figure 12.14. The input and output assignments are as follows:

Start PB	IM:1-6	RUN light	OM:1-1
Stop PB	IM:1-7	Water metering pump	OM:1-2
		Syrup metering pump	OM:1-3
		Solids feeder	OM:1-4
		Agitator	OM:1-5
		Drain valve	OM:1-6

The timer has a 30-s delay. All of the delays are multiples of 30 s, so the timed-step sequencer in Figure 12.12 will be satisfactory.

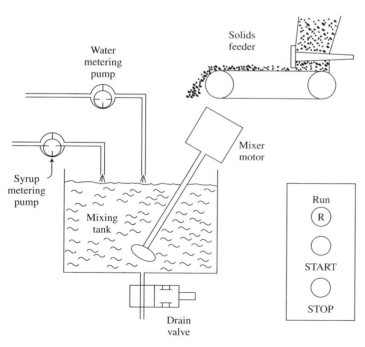

◆ **Figure 12.13**   Industrial blending process (see Example 12.2).

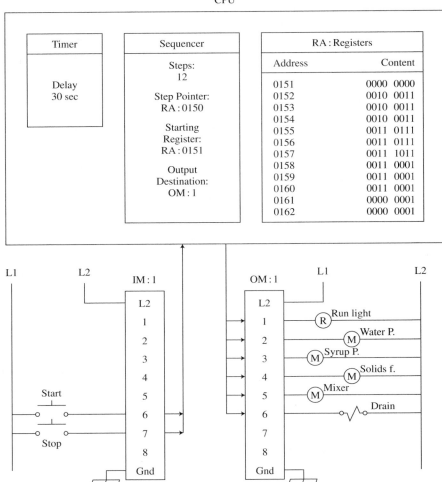

◆ **Figure 12.14** PLC input/output diagram for the industrial blending process in Example 12.2.

STEP 1 holds the sequencer in the reset condition, and only memory cell RA:0151 holds the conditions for STEP 1:

$$0 \quad 0 \quad 0 \quad 0 \quad 0 \quad 0 \quad 0 \quad 0$$

STEP 2 has a dwell of 90 s, so memory cells RA:0152, RA:0153, and RA:0154 hold the following conditions for STEP 2:

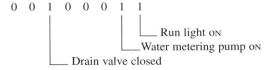

STEP 3 has a dwell of 60 s, so memory cells RA:0155 and RA:0156 hold the conditions for STEP 3:

0 0 1 1 0 1 1 1

STEP 4 has a dwell of 30 s, so only memory cell RA:0157 holds the conditions for STEP 4:

0 0 1 1 1 0 1 1

STEP 5 has a dwell of 90 s, so memory cells RA:0158, RA:0159, and RA:0160 hold the conditions for STEP 5:

0 0 1 1 0 0 0 1

STEP 6 has a dwell of 60 s, so memory cells RA:0161 and RA:0162 hold the conditions for STEP 6:

0 0 0 0 0 0 0 1   ◆

Compare Functions

Most PLCs have the ability to compare two operands. Both operands can be numbers stored in memory registers, or one operand can be a constant and the other a memory register. A compare function is programmed as a rung in the ladder diagram program with a NO enable contact, a box that specifies the operands, and an output symbol. There are six possible comparisons that a PLC could make:

1. The EQ function is ON if the two operands are equal, or OFF if they are not.
2. The NE function is ON if the operands are not equal, or OFF it they are.
3. The GE function is ON if operand 1 is greater than or equal to operand 2, or OFF if it is not.
4. The LT function is ON if operand 1 is less than operand 2, or OFF if it is not.
5. The GT function is ON if operand 1 is greater than operand 2, or OFF if it is not.
6. The LE function is ON if operand 1 is less than or equal to operand 2, or OFF if it is not.

Many PLCs have only two compare functions, EQ and GE. The other four compare functions can be easily derived from the EQ and GE functions.

Arithmetic Functions

As the name implies, arithmetic functions perform arithmetic operations on one or two operands and store the result in a destination memory register. An arithmetic function is programmed as a rung in the ladder diagram program with a NO enable contact, a specification box, and an output symbol. The following operations are performed:

1. *Addition.* The two operands are added and the sum is stored in the destination register.
2. *Subtraction.* One operand is subtracted from the other and the difference is stored in the destination register.
3. *Multiplication.* One operand is multiplied by the other and the product is stored in two consecutive registers.

4. *Division.* A double register operand is divided by a single register operand and the quotient is stored in the destination register.

5. *Square root.* The square root of an operand is computed and stored in the destination register.

Control Functions

A control function alters the way a PLC scans its program. The control function is programmed as a rung in the ladder diagram program with a NO enable contact and an output symbol. The following four control functions are used in PLCs:

1. *Skip (SK).* When its Enable contact is ON, the skip function causes a portion of a program to be passed over when the program is scanned. The number of rungs to be skipped is specified above or below the skip output symbol. For example, an enabled skip 3 function will cause the PLC to pass over the next three rungs in the program. The output of the skipped rungs will remain unchanged—outputs that were ON remain ON, and outputs that were OFF remain OFF.

2. *Master Control Relay (MCR).* The MCR function is similar to the skip function in that it affects a specified number of rungs that immediately follow the MCR rung. However, there are two differences. The skip instruction does nothing when its Enable contact is OFF; the MCR instruction does nothing when its Enable contact is ON. In other words, the MCR instruction is activated when its Enable input is OFF. When activated (enable OFF), the MCR instruction turns the specified number of outputs OFF. For example, an activated MCR 3 function will turn OFF the next three outputs in the program.

3. *Jump (JMP).* The jump function causes the PLC scan to go immediately to the next rung in the program that has a jump label (LBL) on the left side of the rung. The difference between a jump and a skip is that the skip specifies how many rungs to pass over and the jump specifies where the bypass ends.

4. *Jump to subroutine (JSR).* The JSR function causes the scan to go to another section of the program that ends with a rung that has a return label (RET). When it reaches the RET label, the PLC scan moves to the rung immediately after the JSR rung that sent it to the subroutine. This is a very powerful programming tool that allows a programmer to develop sections of program that can be used in many programs and to break a complex program into smaller, more manageable parts.

Data Move Functions

Medium and large PLCs have programming functions that copy data from one place and put them in another place. A data move function is programmed as a rung in the ladder diagram program with a NO Enable contact, a box that specifies the source and destination of the move, and an output symbol. The following are four move functions:

1. *Move (MV).* When enabled, the move function copies data from the source register into the destination register. The content of the source register is unchanged by this operation.

2. *Block Transfer (BT).* When enabled, the block transfer function copies data from a block of source registers into a block of destination registers. Both blocks consist of

consecutive memory registers. The box of the block move function specifies the number of registers in the block and the addresses of the first registers in the source and destination blocks.

3. *Table to Register (TR)*. When enabled, the TR function copies data from a block of source registers into a single destination register, one source register at a time. The program rung of a TR function is similar to that of a sequencer. The TR box has three parallel input contacts, Enable, Step, and Reset. The information in the box specifies the number of registers in the block, the address of the first register in the source block, and the address of the single destination register. When enabled, the TR function moves data from the one source register to the destination register each time the step input changes from OFF to ON. After each transfer, the TR function moves down to the next source register.

4. *Register to Table (RT)*. When enabled, the RT function copies data from a single source register into a block of destination registers, one destination register at a time. This is the reverse of the TR function.

Bit Manipulation Functions

Bit manipulation functions change the bits in a memory register in some manner. These functions fall into three general classes: changing the status of individual bits, moving bits in a register left or right, and performing logical operations on corresponding bits in two operands (stored in memory registers).

1. Changing status functions include the bit set (BS) function, the bit clear (BC) function, and the bit follow (BF) function. The names describe the operation performed by these functions. For example a BS 4, RA:0032 function, when enabled, will put a logical 1 in bit 4 of memory register RA:0032. A BC 0, RA:0125 function, when enabled, will put a logical 0 in bit 0 of memory register RA:0125. A BF 7, RA:0114 function, when enabled, will put a 1 in bit 7 of RA:0114; and when disabled, will put a 0 in that same bit.

2. Bit moving functions include shift right, shift left, rotate right, and rotate left. In some PLCs, the programmer can shift or rotate registers, several registers, a portion of a register, or a specified number of consecutive bits. We will consider only whole registers in this discussion. A rotate function moves all the bits in a register 1 bit in the designated direction (right or left). For example, in an 8-bit register, a rotate right moves bit 7 to the bit 6 position, bit 6 to the bit 5 position, . . . , bit 1 to the bit 0 position, and bit 0 to the bit 7 position. The last move is the reason for the name of the rotate function. After eight rotations to the right, the register will be back where it started. The shift also moves bits right or left, but in a shift right, bit 0 does not go to the bit 7 position. Instead, a 1 or 0 goes into the bit 7 position, and the original bit 0 value is lost.

3. Logical functions perform logical AND, OR, and EXCLUSIVE OR operations on corresponding bits of two operands and store the results in a specified destination register. For example, a logical AND function performs the following eight AND operations:

(bit 0 of op1) AND (bit 0 of op 2) → bit 0 of dest.
(bit 1 of op1) AND (bit 1 of op 2) → bit 1 of dest.
(bit 2 of op1) AND (bit 2 of op 2) → bit 2 of dest.

(bit 3 of op1) AND (bit 3 of op 2) → bit 3 of dest.
(bit 4 of op1) AND (bit 4 of op 2) → bit 4 of dest.
(bit 5 of op1) AND (bit 5 of op 2) → bit 5 of dest.
(bit 6 of op1) AND (bit 6 of op 2) → bit 6 of dest.
(bit 7 of op1) AND (bit 7 of op 2) → bit 7 of dest.

PID Control Function

In some PLCs, the PID control function is built into an I/O module. Figure 12.3 shows a PID module that provides two PID controllers capable of controlling two process variables. In other PLCs, the PID function is programmed as a rung in the ladder diagram program. Figure 12.15 shows a programmed PID control function.

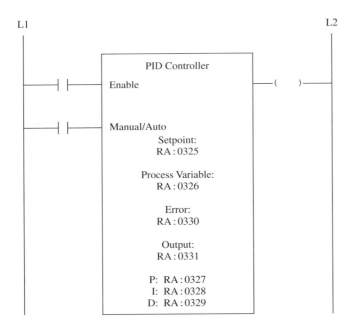

Enable: Enables the function when ON.

Manual/Auto: OFF for manual, ON for automatic control.

Setpoint: Register that holds the setpoint.

Process Variable: Register that holds the process variable.

Error: Register that holds the error.

Output: Register that holds the output.

P: Register that holds the proportional gain setting.

I: Register that holds the the integral action time constant.

D: Register that holds the derivative time setting.

◆ **Figure 12.15** A generic PID controller function.

◆ GLOSSARY

Arithmetic functions: PLC programming functions that perform an arithmetic operation on one or two operands and store the result in a memory register. The operations include addition, subtraction, multiplication, division, and square root. (12.4)

Bit manipulation functions: PLC programming functions that change the bits in memory by changing the status of individual bits, moving bits in a register right or left, or performing logical operations on corresponding bits in two operands. (12.4)

Communications module: A circuit that provides direct communication with the operator, a programming terminal, other PLCs, or a supervisory computer. (12.2)

Compare functions: PLC functions that compare two operands and report the results through an ON/OFF output. (12.4)

Control functions: PLC programming functions that alter the way a PLC scans its program through skip, master control relay, and jump operations. (12.4)

CPU: The central processing unit of a PLC. *See* processor module. (12.2)

Counter function: A PLC programming function that performs counting operations in a ladder diagram program. (12.4)

Data move functions: PLC programming functions that copy data from one place and put them in another place. (12.4)

Input image status: The section of PLC memory that stores the status (1's or 0's) of the inputs from switches and ON/OFF signals from the process. (12.2)

Input module, analog: A circuit that converts analog input signals to a digital form that can be used by a PLC. (12.2)

Input module, discrete: A circuit that converts discrete (ON/OFF) signals into logic level voltages that can be used by a PLC. Discrete input modules are designated as ac or dc, depending on the voltage used by the input device. (12.2)

Input scan: The first step in the operating cycle of a PLC, in which the processor scans the inputs and stores a new image of the input conditions. (12.2)

Ladder Diagram programming language: A language for programming PLCs that uses contact and coil symbols to construct diagrams that are similar to the ladder diagrams used for relay logic. (12.3)

Module: A circuit that mates with a standard socket in a rack that usually has a number of identical sockets. (12.2)

Output image status: The section of PLC memory that stores the binary data (1's or 0's) that will turn discrete output devices ON or OFF. (12.2)

Output module, analog: A circuit that converts a digital value in a PLC into an analog signal that can be used by continuous output devices. (12.2)

Output module, discrete: A circuit that provides ON/OFF signals to operate lamps, relays, solenoids, motors, etc. Discrete output modules are designated as ac or dc, depending on the voltage used by the output device. (12.2)

Output scan: The third step in the operating cycle of a PLC, in which the processor transfers the new image of the output conditions to the output devices. (12.2)

PID control function: A PLC programming function that provides PID control of a continuous process variable. (12.4)

PID module: A circuit that provides three-mode control of one or two process variables. (12.2)

Processor module: The PLC module that contains the microprocessor and memory units. Also called the central processing unit, or CPU. (12.2)

Programmable logic controller (PLC):   A digital, electronic device designed to control machines and processes by performing event-driven and time-driven sequential operations. (12.1)

Programmer: A hand-held device used to enter, examine, and change programs stored in a PLC. (12.2)

Program scan: The second step in the operating cycle of a PLC, in which the processor scans the program and derives a new image of the output conditions from the new image of the inputs and the old image of the outputs. (12.2)

Sequencer function: A PLC programming function that goes through a sequence of steps and produces a specified output pattern for each step. (12.4)

Sneak path: A path in a ladder diagram in which power flows from right to left through a contact. (12.3)

Timer function: A PLC programming function that produces a time delay in a ladder diagram program. (12.4)

◆ EXERCISES

Section 12.1

12.1 Write a description of a programmable logic controller for a friend who knows nothing about control.

12.2 Explain the "divide and conquer" problem-solving technique in your own words.

Section 12.2

12.3 Write a description of the five functional components found in all PLCs.

12.4 In a program scan, the processor scans which of the following?

1. Input and derives a new output from the old input and the old output
2. Input and derives a new output from the new input and the old output
3. Program and derives a new output from the old input and the old output
4. Program and derives a new output from the new input and the old output

12.5 Determine the logic level stored in an input image memory cell when the processor scans each of the following inputs.

(a) A NO switch that is not pressed.
(b) A NC switch that is not pressed.
(c) A NO switch that is pressed.
(d) A NC switch that is pressed.

Section 12.3

12.6 Determine the input image logic level for each of the following input conditions. Then determine how the PLC computer interprets that logic level (i.e., TRUE or FALSE).

	Switch Type	Switch Action	PLC Program Contact Symbol
(a)	NO	not pressed	NO
(b)	NO	not pressed	NC
(c)	NO	pressed	NO
(d)	NO	pressed	NC
(e)	NC	not pressed	NO
(f)	NC	not pressed	NC
(g)	NC	pressed	NO
(h)	NC	pressed	NC

12.7 Rung 3 in a ladder diagram program has a single NC contact symbol labeled IM:1-4 and an output symbol labeled OM:1-3. The physical switch that is connected to terminal IM:1-4 is a normally closed pushbutton switch. If output OM:1-3 is ON, is the pushbutton switch connected to IM:1-4 pressed or not pressed?

12.8 Draw a PLC input/output diagram and a ladder diagram program for the hydraulic cylinder control circuit in Figure 2.21. Refer to Figure 12.6 for an example of the I/O diagram and LD program.

12.9 Revise the hydraulic cylinder control circuit in Figure 2.21 to include a motor starter rung similar to the first rung in Figure 11.4. Then draw a PLC input/output diagram and a ladder diagram program for your revised control circuit.

Section 12.4

12.10 Draw a PLC input/output diagram and a ladder diagram program for the alternative drilling machine control shown in Figure 11.12. Include the motor control circuit in Figure 11.9 in your I/O diagram and LD program. Refer to Figure 12.4.

12.11 Draw a timing diagram of the control system in Example 12.1.

12.12 Draw a PLC input/output diagram for the control system in Example 12.1.

12.13 In Example 12.1, what is the purpose of the normally closed T:4 contact in rungs 1, 2, 3, and 4? What would happen if the T:4 contacts were removed from the four rungs?

12.14 Draw a PLC ladder diagram program that will turn a blower ON 12 s after a heater is turned ON. Reset the timer when the heater is turned OFF, even if the 12 s has not elapsed.

12.15 Draw a PLC ladder diagram program that will turn solenoid a ON immediately after switch SW1 closes and then turn solenoid b ON 25 s later.

12.16 Switch SW1 is turned ON and remains ON for the duration of an operation, but switch SW2 goes ON and OFF many times during the operation. Write a ladder diagram program that will turn ON a light when switch SW2 has accumulated 60 s of ON time.

12.17 Draw a PLC ladder diagram program that will turn ON a light when a count reaches 30 and then turn the light OFF when the count reaches 50.

12.18 Draw a PLC ladder diagram program that will turn ON a solenoid valve when count A reaches 10 or when count B reaches 15.

12.19 The batch blending process in Exercise 11.3 is to be controlled by a timed-step sequencer similar to Figure 12.11. Draw a PLC input/output diagram for this system. Your diagram should show the timer preset, the sequencer parameters, and the contents of the memory addresses that hold the output conditions.

12.20 Draw a ladder diagram program that uses two comparators, EQ and GE, to produce all six compare functions as follows:

Rung 1 is TRUE when op1 is equal to op2 (EQ)

Rung 2 is TRUE when op1 is not equal to op2 (NE).

Rung 3 is TRUE when op1 is greater than or equal to op2 (GE).

Rung 4 is TRUE when op1 is less than op2 (LT).

Rung 5 is TRUE when op1 is greater than op2 (GT).

Rung 6 is TRUE when op1 is less than or equal to op2 (LE).

CHAPTER 13

Control of Continuous Processes

◆ OBJECTIVES

A continuous process has uninterrupted inputs and outputs. The value of at least one input is changed in a manner that tends to maintain the controlled variable equal to the setpoint. The output of a continuous-process controller is determined by one or more modes of control. The most common control modes are the two-position, floating, proportional, integral, and derivative modes. Usually, the proportional mode is combined with the integral and/or derivative modes to form two- or three-mode controllers. Continuous process controllers can be grouped into two categories, those in which the setpoint is constant for long periods of time and those in which the setpoint is constantly changing. Control systems in the first category are called regulator systems, and those in the second category are called servo systems. However, control system analysis and design methods work equally well on systems in either category.

The purpose of this chapter is to give you an entry-level ability to discuss, select, and specify continuous process controllers. After completing this chapter, you will be able to

1. Describe, select, and specify the following control modes
   a. Two-position
   b. Floating
   c. Proportional (P)
   d. Proportional plus integral (PI)
   e. Proportional plus derivative (PD)
   f. Proportional plus integral plus derivative (PID)
2. Describe
   a. Data sampling
   b. Digital control algorithms
   c. Cascade control
   d. Feedforward control
   e. Adaptive control
   f. Multivariable control
   g. Fuzzy logic control

◆◆◆◆

13.1 INTRODUCTION

The block diagram of a closed-loop control system was introduced in Chapter 1 and is reproduced in Figure 13.1. Control is achieved by performing the following three operations:

Measurement. Measure the value of the controlled variable.

Decision. Compute the error (desired value minus measured value) and use the error to form a control action.

Manipulation. Use the control action to manipulate some variable in the process in a way that will tend to reduce the error.

The controller accomplishes the decision step and is the subject of this chapter. The controller consists of an error detector and a control mode unit. The error detector computes the error by subtracting the measured value (C_m) from the setpoint (SP). The control mode unit uses the error signal to produce the control action (V).

One of the most important characteristics of a controller is the way it uses the error to form the control action. The different ways the controller forms the control action are called *modes of control*. Common modes of control include (1) two-position, (2) floating, (3) proportional, (4) integral, and (5) derivative. This chapter begins with a detailed study of these five modes of control. The methods used to describe control modes include input/output graphs, time-domain equations, frequency-domain equations, transfer functions, and Bode diagrams. Table 13.1 lists some variables that are used throughout this chapter.

The controller can be implemented by pneumatic circuits, analog electronic circuits, or digital electronic circuits. Pneumatic controllers use a pneumatic equivalent of the operational

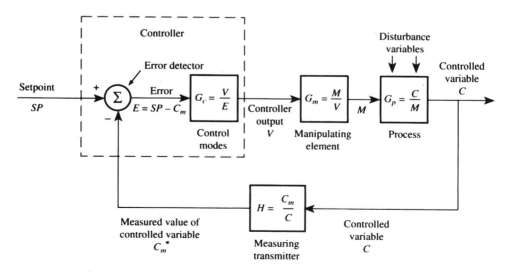

◆   **Figure 13.1**   Block diagram of a closed-loop control system. The controller consists of an error detector and a control mode unit.

◆ **TABLE 13.1** Variables Used to Define Control Modes

Symbol	Description	Units
D	Derivative action time constant	seconds
e	Time-domain error	% of F.S.[a]
E	Frequency-domain error	% of F.S.
I	Integral action rate	1/s
P	Proportion gain	[b]
v	Time-domain controller output	% of F.S.
v_0	Controller output when error $= 0$	% of F.S.
V	Frequency domain output[c]	% of F.S.
α	Derivative limiter coefficient $0 < \alpha < 1$	(none)

[a] F.S. = full scale.

[b] Gain is dimensionless when error and output are expressed in percentage of full scale.

[c] Frequency-domain equations are simplified by the assumption that v_0 is zero.

amplifier to generate the control action. Electronic analog controllers use a resistive circuit to compute the error and an operational amplifier to generate the control action. Digital controllers use a microprocessor and a control algorithm to generate the control action. Section 13.3 covers electronic analog controllers, and Section 13.4 deals with microprocessor-based digital controllers.

The chapter concludes with an overview of advanced control methods in Section 13.5. Advanced control refers to various methods of going beyond the single-loop single-variable feedback control system with three modes of control. Cascade control involves two controllers in which the output of the primary controller is the input to the secondary controller. Feedforward uses a model of the process to make changes in the controller output in response to measured changes in a major load variable without waiting for the error to occur. Adaptive control changes controller parameters to "adapt" to changes in the process. Expert system control uses artificial intelligence to incorporate the knowledge of control experts into the adaptive control system. Multivariable control uses measurements of several process load variables and may involve the manipulation of more than one process variable.

A brief review of the effect of *load changes* will be quite helpful in developing an understanding of the function of the controller. The first-order lag process illustrated in Figure 13.2 will be used to develop the concept of a process load line. The liquid process in Figure 13.2a consists of a tank with liquid level, h, input valve resistance, R_m, and output resistance, R_L. The input valve is the final control element, and the input flow rate, m, is the manipulated variable. The liquid level, h, is the controlled variable. The output resistance, R_L is the disturbance variable. Load changes are created whenever R_L changes.

The process is maintained in a balanced condition by adjusting the input flow rate, m, until it balances the output flow rate, q. Assuming laminar flow, the output flow rate, q, is given by

$$q = \left(\frac{g\rho}{R_L}\right)h \tag{13.1}$$

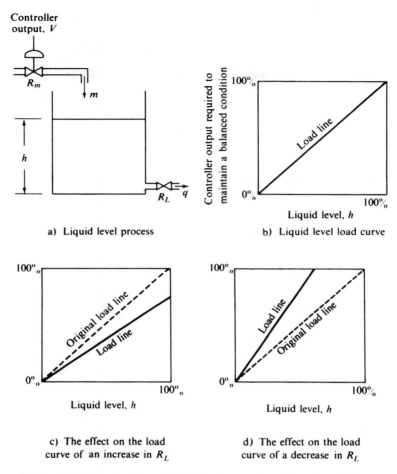

a) Liquid level process

b) Liquid level load curve

c) The effect on the load
curve of an increase in R_L

d) The effect on the load
curve of a decrease in R_L

◆ **Figure 13.2**   The load line of a liquid level process depends on the outlet flow resistance, R_L. A nominal value of R_L produces the original load line (b). A value of R_L larger than the nominal moves the load line down (c), and a smaller value moves the load line up (d).

where g = acceleration by gravity = 9.81 m/s^2
         ρ = liquid density, kg/m^3
         R_L = liquid resistance, Pa · s/m^3
         h = liquid level, m

If R_L is constant, q is proportional to h. For each level, h, there will be a specific output flow rate, q, and hence a specific input flow rate, m, necessary to maintain a balanced condition. Each input flow rate, m, requires a specific position of the final control element, R_m, which, in turn, requires a specific controller output signal, v. Eliminating the middle variables, each level requires a different controller output signal to maintain a balanced condition in the process. This concept is the basis of the process load line in Figure 13.2b.

Now let us examine what happens to the load line if R_L changes. For example, if R_L increases for some reason, the output flow rate, q, will decrease [see Equation (13.1)]. This will be reflected in a decrease in the controller output required to maintain a balanced condition, as illustrated by the load line in Figure 13.2c. On the other hand, if R_L decreases, the output flow rate, q, will increase [Equation (13.1)]. This will be reflected in an increase in the controller output, as indicated in Figure 13.2d. A load change is any condition that changes the location of the load line. After a load change, the controller must adjust its output to the value determined by the new load line. This change is accomplished by the action of the control modes in the controller.

13.2 MODES OF CONTROL

Two-Position Control Mode

The *two-position control mode* is the simplest and least expensive mode of control. The controller output has only two possible values, depending on the sign of the error. If the two positions are fully open and fully closed, the controller is called an ON/OFF *controller*. Most two-position controllers have a neutral zone to prevent chattering. The neutral zone is a range of values around zero in which no control action takes place. The error must pass through the neutral zone before any control action takes place. Figure 13.3 shows the input/output relationship of a two-position controller.

The two-position control mode supplies pulses of energy to the process, which causes a cycling of the controlled variable. The amplitude of the cycling depends on three factors: the capacitance of the process, the dead-time lag of the process, and the size of the load change the process is capable of handling. The amplitude of the oscillation is decreased by increasing the capacitance, decreasing the dead-time lag, or decreasing the size of the load change that the process can handle. Two-position control is suitable only for processes that have a large enough capacitance to counteract the combined effect of the dead-time lag and the load-change capability of the process. Two-position control is simple and inexpensive. Its use is preferred whenever the cycling can be reduced to an acceptable level.

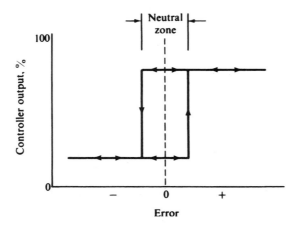

◆ **Figure 13.3** The output of a two-position controller has one of two values, depending on the value of the error signal from the error detector. The neutral zone prevents chattering (i.e., oscillation of the output between the two values when the error fluctuates around zero).

A household heating system is an example of a two-position control system. The air in the house has a relatively large thermal capacitance, and the dead-time lag is small. The rate of heat input from the furnace is just sufficient to heat the house on the most severe winter day and is small compared with the capacitance of the room. The room temperature cycles with an amplitude that is well within the acceptable limits of human comfort. This is an example of a good application of the two-position control mode. The two positions provide inputs equal to the maximum and minimum process loads (i.e., from no heat to just enough heat to handle the coldest day). A poor design is one that uses a furnace 10 times as large as required. The large rate of heat input from an oversized furnace results in a large amplitude of oscillation.

> The two-position control mode is used on processes with a capacitance large enough to reduce the cycling to an acceptable level. This implies a large capacitance process with a small dead-time lag and small load changes.

EXAMPLE 13.1

The liquid level process in Figure 13.2 is controlled by a two-position controller that opens and closes the inlet valve. The inlet flow rate, m, is 0 when the valve is closed and 0.004 m³/s when the valve is open. The oscillations in level are small enough that the outlet flow rate, q, is essentially constant at 0.002 m³/s. The tank has a cross-sectional area of 2.0 m², and the process dead-time lag is 10 s. The neutral zone of the controller is equivalent to a ±0.005-m change in level. Determine the amplitude and period of the oscillation in level, h.

Solution

(a) The rate of accumulation of liquid in the tank (a) is equal to the inflow minus the outflow.

$$a = m - q \text{ m}^3/\text{s}$$

The rate of change of the level, dh/dt, is equal to the rate of accumulation divided by the area of the tank, A.

$$\frac{dh}{dt} = \frac{a}{A} = \frac{m - q}{2.0} \text{ m/s}$$

When the valve is open,

$$\frac{dh}{dt} = \frac{0.004 - 0.002}{2.0} = 0.001 \text{ m/s}$$

When the valve is closed,

$$\frac{dh}{dt} = \frac{0 - 0.002}{2.0} = -0.001 \text{ m/s}$$

(b) A graph of the oscillation is illustrated in Figure 13.4. At point a, the valve is closed. The level is changing at a rate of -0.001 m/s (decreasing), and the level has reached the lower limit of the neutral zone. After the 10-s dead-time lag, the controller opens the valve at point b. The level has decreased an additional amount equal to (10 s) × (-0.001 m/s) = -0.01 m.

$$t_b - t_a = 10 \text{ s}$$

$$h_b - h_a = -0.01 \text{ m}$$

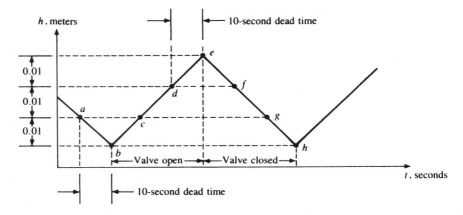

◆ **Figure 13.4** Oscillation of the level of the liquid tank in Example 13.1.

(c) The valve is open from point b to point e, and the level is increasing at a rate of 0.001 m/s. It takes 10 s to reach point c.

$$t_c - t_b = 10 \text{ s}$$

$$h_c - h_b = 0.01 \text{ m}$$

The time required to move from point c to point d is equal to the change in level (0.01 m) divided by the rate of change of level (0.001 m/s).

$$t_d - t_c = \frac{0.01}{0.001} = 10 \text{ s}$$

$$h_d - h_c = 0.01 \text{ m}$$

At point d, the level has reached the upper level of the neutral zone. After the 10-s dead-time lag, the controller closes the valve at point e.

$$t_e - t_d = 10 \text{ s}$$

$$h_e - h_d = 0.01 \text{ m}$$

(d) Because the rate of increase and the rate of decrease of the level are equal, the time from e to h is the same as the time from b to e.

$$t_h - t_e = t_e - t_b = 30 \text{ s}$$

(e) The amplitude of oscillation is equal to $h_e - h_b$, and the period is equal to $t_h - t_b$.

$$\text{Amplitude} = (h_c - h_b) + (h_d - h_c) + (h_e - h_d)$$

$$= 0.01 + 0.01 + 0.01$$

$$= 0.03 \text{ m}$$

$$\text{Period} = (t_e - t_b) + (t_h - t_e)$$

$$= 30 + 30$$

$$= 60 \text{ s}$$ ◆

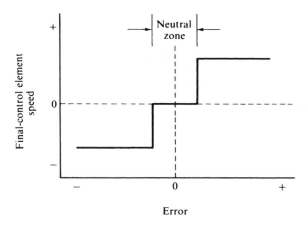

◆ **Figure 13.5** Input/output graph of the floating control mode.

Floating Control Mode

The *floating control mode* is a special application of the two-position mode in which the final control element is stationary as long as the error remains within the neutral zone. When the error is outside the neutral zone, the final control element changes at a constant rate in a direction determined by the sign of the error. The final control element continues to change until the error returns to the neutral zone, or until the final control element reaches one of its extreme positions. The input/output curve of a floating controller is illustrated in Figure 13.5.

The floating control mode has a tendency to produce cycling of the controlled variable. The amplitude of the cycling depends on the dead-time lag of the process, the capacitance of the process, and the speed at which the controller increases and decreases the final control element. The speed of the final control element determines the fastest load change that the controller can keep pace with, but not the size of the load change. The main advantage of floating control is its ability to handle load changes by gradually adjusting the final control element. As with two-position control, the amplitude is decreased by increasing the capacitance, decreasing the dead-time lag, or decreasing the speed of the final control element. Floating control is used when large load changes are anticipated, and the capacitance is large enough to counteract the effects of the dead-time lag and the speed of the final control element. Floating control is frequently used because it is inherent in the type of actuator used to drive the final control element (e.g., electric motors and hydraulic cylinders operated by ON/OFF relays or solenoid valves provide a floating control mode).

> The floating control mode is used on processes with large, slow-moving load changes and a capacitance large enough to reduce the cycling to an acceptable level. This implies a large capacitance process with a small dead-time lag. The floating control mode is inherent in some final control elements.

EXAMPLE 13.2   The solid flow rate control system in Figure 2.19 is an example of a floating mode control system. Describe the response of this system to a step increase in the density of the solid material passing over the belt. Assume that the solid flow rate is at the setpoint before the step change in density.

Solution   Before the step change, the beam is balanced and positioned at the midpoint between the two lever-actuated switches. The gate drive motor is OFF, and the gate is stationary. As the higher-density material begins to pass over the belt, the weight of solid material on the belt begins to increase at a steady rate. The weight on the belt continues to increase until the entire platform is covered with the higher-density material. As the weight on the platform increases, the added weight moves the force beam up against the top switch (the DOWN switch). Eventually, the beam moves far enough to close the DOWN switch—the beam has passed through the dead zone.

While the DOWN switch is closed, the drive motor turns the cam, lowering the gate at a steady rate. As the gate height decreases, the weight of material on the platform also decreases, but the effect is delayed by the time it takes for the change to move across the belt. As the weight on the platform decreases, the force beam moves down until, eventually, it no longer holds the DOWN switch closed. The gate drive motor stops, and the gate position is again stationary.

The time it takes for changes to move across the platform presents a potential problem. When the beam finally senses the correct weight on the platform, the gate has continued to move down as the correct weight moves across the platform. Thus when the gate finally stops, it has moved too far, and the weight of solid material is less than it should be. If the change in density is small, the beam may not move far enough to close the UP switch. The flow rate will be less than the setpoint, but it will be within the dead zone. The gate will remain stationary.

If, however, the step change is large, the force beam will pass through the dead zone, closing the UP switch. As the gate moves up, the weight on the belt increases, but the increase is delayed by the belt travel time. Floating control produces an inherent cycling of the solid flow rate as the system "searches" for the correct gate position. We say that the control system "floats" about the desired setpoint, hence the name *floating control mode*.   ◆

Proportional Control Mode

The *proportional control mode* produces a change in the controller output proportional to the error signal. There is a fixed linear relationship between the value of the controlled variable and the position of the final control element. A simple example of a proportional controller is shown in Figure 13.6. The controlled variable is the liquid level in the tank. The float is the measuring instrument, the valve is the manipulating element, and the lever provides the control action. Notice that there is a different valve position for each level. The desired level is h_0, and v_0 is the valve position corresponding to h_0 (v_0 is the position of the valve when the error is zero). The valve position, v, is given by the following equation:

$$\frac{v - v_0}{a} = \frac{h_0 - h}{b}$$

$$\text{Error signal} = e = h_0 - h$$

$$v = \left(\frac{a}{b}\right)e + v_0 \qquad \textbf{(13.2)}$$

where v = valve position, m
         v_0 = valve position with zero error, m
         e = error signal, m

Equation (13.2) is one version of the defining equation of the proportional control mode.

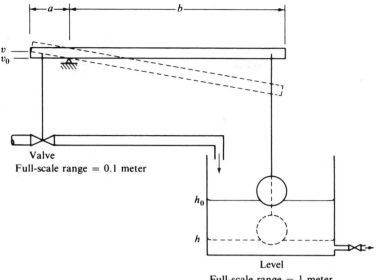

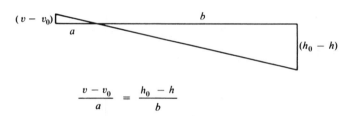

$$\frac{v - v_0}{a} = \frac{h_0 - h}{b}$$

◆ **Figure 13.6**   Simple proportional mode controller. The lever establishes a fixed linear relationship between the level (h) and the valve stem position (v).

The gain (P) of the proportional controller in Figure 13.6 is the change in valve position divided by the corresponding change in level. Both are expressed in percentage of the full-scale range.

$$\text{Percentage change in valve position} = \frac{100(v - v_0)}{0.1}$$

$$= 1000(v - v_0)$$

$$\text{Percentage change in level} = \frac{100(h_0 - h)}{1}$$

$$= 100(h_0 - h)$$

$$\text{Gain, } P = \frac{1000(v - v_0)}{100(h_0 - h)} = 10\left(\frac{v - v_0}{h_0 - h}\right) = 10\left(\frac{a}{b}\right)$$

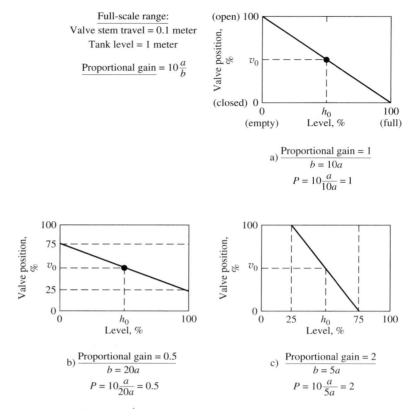

Full-scale range:
Valve stem travel = 0.1 meter
Tank level = 1 meter

Proportional gain = $10\dfrac{a}{b}$

a) $\dfrac{\text{Proportional gain} = 1}{b = 10a}$

$$P = 10\dfrac{a}{10a} = 1$$

b) $\dfrac{\text{Proportional gain} = 0.5}{b = 20a}$

$$P = 10\dfrac{a}{20a} = 0.5$$

c) $\dfrac{\text{Proportional gain} = 2}{b = 5a}$

$$P = 10\dfrac{a}{5a} = 2$$

◆ **Figure 13.7** Input/output graphs for the controller in Figure 13.6 with gains of 1, 0.5, and 2. The gain is changed by moving the pivot point on the lever to change the ratio of distance a over distance b.

Figure 13.7 includes input/output graphs of the proportional controller in Figure 13.6 with gains of 0.5, 1, and 2. In this illustration, v_0 is set at 50% open when h_0 is at 50% full. Observe that the valve stem moves a distance of 0.1 m between open and closed, and the level sensor moves a distance of 1 m between empty and full.

A proportional gain of 1 is achieved by setting the fulcrum such that distance b is 10 times distance a.

$$P = 10\frac{a}{b} = 10\frac{a}{10a} = 1$$

Figure 13.7a shows the input/output graph for a gain of 1. A level of 0% full (i.e., empty) yields a valve position of 100% open, a level of 50% full yields a valve position of 50% open, and a level of 100% full yields a valve position of 0% open. With a gain of 1, a given percentage increase in level produces the same percentage decrease in valve position (and vice versa).

A proportional gain of 0.5 is achieved by setting the fulcrum such that distance b is 20 times distance a.

$$P = 10\,\frac{a}{b} = 10\,\frac{a}{20a} = 0.5$$

Figure 13.7b shows the input/output graph for a gain of 0.5. A level of 0% full yields a valve position of 75% open, a level of 50% full yields a valve position of 50% open, and a level of 100% full yields a valve position of 25% open. With a gain of 0.5, a given percentage increase in level produces only half that percentage decrease in valve position.

A proportional gain of 2 is achieved by setting the fulcrum such that distance b is 5 times distance a.

$$P = 10\frac{a}{b} = 10\frac{a}{5a} = 2$$

Figure 13.7c shows the input/output graph for a gain of 2. A level of 25% full yields a valve position of 100% open, a level of 50% full yields a valve position of 50% open, and a level of 75% full yields a valve position of 0% open. With a gain of 2, a given percentage change in level produces twice that percentage change in valve position.

In general, an increase in the gain reduces the size of the error required to produce a 100% change in the valve position. In other words, a high gain requires a small error to produce the change in valve position necessary to balance the process. Although this seems to imply that the gain should be as high as possible, unfortunately, increasing the gain increases the tendency for oscillation of the controlled variable. A compromise is necessary in which the gain is made as large as possible without producing unacceptable oscillations.

One problem with the proportional control mode is that it cannot completely eliminate the error caused by a load change. A residual error is always required to maintain the valve at some position other than v_0. This is obvious in Equation (13.2), and it is equally obvious in the simple system illustrated in Figure 13.6. A load change means that a different valve position is required to maintain a balanced condition in the process. With a proportional control mode, a change in level is the only way that the valve position can be changed. This change or *residual error* is called the *proportional offset*. The size of the offset is directly proportional to the size of the load changes and inversely proportional to the gain, P. The proportional control mode is used when the gain can be made large enough to reduce the proportional offset to an acceptable level for the largest expected load change.

The response of the proportional mode control action is instantaneous. There is no delay between a change in level and the corresponding change in valve position. The Bode diagram in Figure 13.8 is another way of looking at the response of the proportional control mode. Notice that the phase angle is 0° for all values of frequency. The absence of any phase lag is another indication of the speed of response of the proportional control mode. The gain is also constant for all values of frequency, with the decibel level determined by the value of the gain P.

The proportional control mode is used on processes with a small capacitance and fast-moving load changes when the gain can be made large enough to reduce the offset to an acceptable level. This implies a process with a capacitance that is too small to permit the use of two-position or floating control.

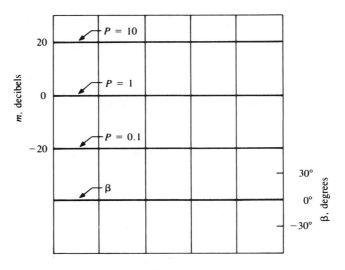

◆ **Figure 13.8** Bode diagram of proportional control modes with gains of 10, 1, and 0.1. The phase angle, β, is $0°$ and the magnitude, m, is constant for all values of radian frequency, ω.

PROPORTIONAL CONTROL MODE

Time-Domain Equation

$$v = Pe + v_0 \qquad\qquad (13.3)$$

Frequency-Domain Equation

$$V = PE \qquad\qquad (13.4)$$

Transfer Function

$$\frac{V}{E} = P \qquad\qquad (13.5)$$

(See Table 13.1 for definitions of the variables.)
Note: In the frequency domain, v_0 is assumed to be zero to satisfy the condition of zero initial conditions for transfer functions.

EXAMPLE 13.3 A proportional controller has a gain of 4. Determine the proportional offset required to maintain $v - v_0$ at 20%.

Solution

$$\text{Gain} = \frac{v - v_0}{E}$$

$$4 = \frac{20}{E}$$

$$E = 5\%$$

◆

EXAMPLE
13.4

The dc motor speed control system in Figure 2.15 is an example of a proportional mode control system. Actually, the dc motor works fairly well as an open-loop speed control system due to the feedback effect of the motor's back emf, e_b. This was demonstrated in Example 10.3, where a dc motor underwent a load change. The motor was running at 300 rad/s with a load torque of 0.05 N · m. The load torque was increased to 0.075 N · m, causing the motor speed to drop to 291.7 rad/s, a decrease of only 8.3 rad/s, or 2.77%. Here is a summary of the motor conditions before and after the load change.

DC Motor Open-Loop Control

	Before Load Change	After Load Change
Torque, T (N · m)	0.05	0.075
Voltage, e_a (v)	19.24	19.24
Current, i_a (A)	1.033	1.45
Speed, ω (rad/s)	300	291.7

Let us see how the proportional control mode in Figure 2.15 improves the speed control of this motor. The motor parameters from Example 10.3 are given in the following list. Included are the tachometer gain, K_{tac}, and the overall gain, G_{amp}, of the op amp and the power amp in Figure 2.15.

$$T_f = 0.012 \text{ N} \cdot \text{m} \qquad K_T = 0.06 \text{ N} \cdot \text{m/A}$$

$$T_1 = 0.05 \text{ N} \cdot \text{m} \qquad K_E = 0.06 \text{ V} \cdot \text{s/rad}$$

$$T_2 = 0.075 \text{ N} \cdot \text{m} \qquad K_{tac} = 0.11 \text{ V} \cdot \text{s/rad}$$

$$R = 1.2 \, \Omega \qquad G_{amp} = 10.0 \text{ V/V}$$

Determine the "before" and "after" conditions for the dc motor closed-loop control system. Include the proportional offset in the "after" conditions.

Solution

The following dc steady-state equations are used in the solution:

$$T = K_T i_a - T_f, \text{ N} \cdot \text{m} \qquad\qquad \textbf{(10.7)}$$

$$e_b = K_E \omega, \text{ V} \qquad\qquad \textbf{(10.9)}$$

$$e_a = i_a R + e_b, \text{ V} \qquad\qquad \textbf{(10.11)}$$

or

$$e_a = i_a R + K_E \omega, \text{ V}$$

Additional subscripts are used to differentiate between the two load conditions. Thus T_1, i_{a1}, e_{a1}, ω_1, and c_{m1} indicate conditions when the load torque is 0.5 N · m, and T_2, I_{a2}, e_{a2}, ω_2, and c_{m2} indicate conditions when the load torque is 0.075 N · m.

(a) In the closed-loop control system, the input to the op amp is equal to the setpoint, *sp*, minus the feedback voltage, c_m. The armature voltage, e_a, is equal to the input to the op amp multiplied by the overall gain of the two amplifiers. We compute the setpoint as follows:

$$c_{m1} = K_{tac} \omega_1 = (0.11)(300) = 33 \text{ V}$$

$$e_{a1} = (sp - c_{m1}) G_{amp}$$

$$sp = \frac{e_{a1}}{G_{\text{amp}}} + c_{m1}$$

$$sp = \frac{19.24}{10} + 33 = 34.924 \text{ V}$$

(b) The setpoint of 34.924 V produces a motor speed of 300 rad/s with a load torque of 0.05 N · m. Our task is to determine the motor speed when the torque increases to 0.075 N · m with the setpoint still at 34.924 V. To do this, we combine the equation for c_{m1} with the equation for e_{a1} and change the subscripts as follows:

$$e_{a2} = (sp - K_{\text{tac}}\omega_2)G_{\text{amp}}$$

$$e_{a2} = (34.924 - 0.11\omega_2)10.0$$

The preceding equation has two unknowns, e_{a2} and ω_2. We obtain a second equation from the second version of Equation (10.11) as follows.

$$e_{a2} = i_{a2}R + K_E\omega_2$$

$$e_{a2} = (1.45)(1.2) + 0.06\omega_2$$

Solving the two equations for ω_2 gives us the following:

$$\omega_2 = \frac{347.5}{1.16} = 299.57 \text{ rad/s}$$

Comparison of Open-Loop and Closed-Loop Control

	Open-Loop	Closed-Loop
i_{a1} (A)	1.033	1.033
i_{a2} (A)	1.45	1.45
ω_1 (rad/s)	300	300
ω_2 (rad/s)	291.7	299.57
Residual error (rad/s)	8.3	0.43
Residual error (%)	2.77	0.143

The closed-loop proportional offset (residual error) is only 0.43 rad/s. This small offset error is due to the high gain of the proportional mode and the inherent feedback of the dc motor. ◆

Integral Control Mode

The *integral control mode* changes the output of the controller by an amount proportional to the integral of the error signal. As long as there is an error, the integral control mode will change the output at a rate proportional to the size of the error. Figure 13.9 illustrates the relationship between the error signal and the controller output. Notice that the rate of change of the controller output is proportional to the error signal (the rate of change is equal to the slope of the graph).

The Bode diagram of the integral control mode is shown in Figure 13.10. The gain decreases at a rate of 20 dB per decade increase in frequency and passes through 0 dB at a ra-

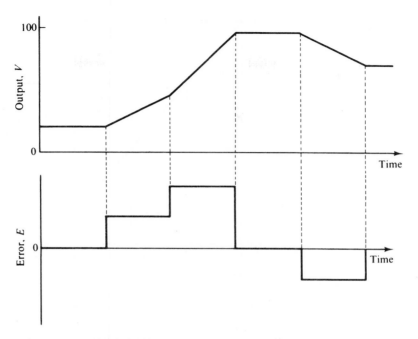

◆ **Figure 13.9**   The integral control mode responds to an error signal by changing the controller output at a rate proportional to the size of the error.

dian frequency equal to I, where I is the integral action rate. The phase angle is a constant $-90°$ for all frequency values. The integral control mode is almost always used with the proportional control mode.

INTEGRAL CONTROL MODE

Time-Domain Equation

$$v = I \int_0^t e\, dt + v_0 \tag{13.6}$$

Frequency-Domain Equation

$$V = \left(\frac{I}{s}\right) E \tag{13.7}$$

Transfer Function

$$\frac{V}{E} = \frac{I}{s} \tag{13.8}$$

(See Table 13.1 for definitions of the variables.)

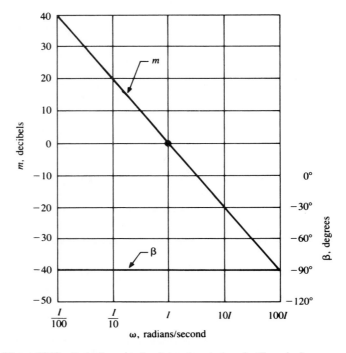

◆ **Figure 13.10** Bode diagram of an integral control mode. The gain decreases as frequency increases and the phase angle is a constant $-90°$.

EXAMPLE 13.5 Determine the equation of the straight line segments in the output graph in Figure 13.9, given the following values of the integral action rate (I), error (e), and initial output ($v(0)$).

$$I = 1 \text{ s}^{-1}, \quad v = 1.5, \quad t = 0$$

$$e = 0, \quad 0 \leq t \leq 1 \text{ s}$$

$$e = 2, \quad 1 \leq t \leq 2 \text{ s}$$

$$e = 4, \quad 2 \leq t \leq 3 \text{ s}$$

$$e = 0, \quad 3 \leq t \leq 4 \text{ s}$$

$$e = -2, \quad 4 \leq t \leq 5 \text{ s}$$

Solution

(a) For $0 \leq t \leq 1$ s, $e = 0$, $v(0) = 1.5$

$$v(t) = I \int_0^t e \, dx + v(0) = (1) \int_0^t (0) \, dx + 1.5 = 0 + 1.5$$

$$v(t) = 1.5, \quad 0 \leq t < 1 \text{ s}$$

(b) For $1 \leq t \leq 2$ s, $e = 2$, $v(1) = 1.5$

$$v(t) = I \int_1^t e \, dx + v(1) = (1) \int_1^t (2) \, dx + 1.5$$

$$v(t) = 2x|_1^t + 1.5 = 2(t - 1) + 1.5 = 2t - 0.5$$

$$v(t) = 2t - 0.5, \quad 1 \le t < 2 \text{ s}$$

$$v(2) = (2)(2) - 0.5 = 3.5$$

(c) $2 \le t \le 3$ s, $e = 4$, $v(2) = 3.5$

$$v(t) = (1) \int_2^t (4) \, dx + 3.5 = 4(t - 2) + 3.5$$

$$v(t) = 4t - 4.5, \quad 2 \le t \le 3 \text{ s}$$

$$v(3) = (4)(3) - 4.5 = 7.5$$

(d) $3 \le t < 4$ s, $e = 0$, $v(3) = 7.5$

$$v(t) = 7.5, \quad 3 \le t \le 4 \text{ s}$$

$$v(4) = 7.5$$

(e) $4 \le t < 5$ s, $e = -2$, $v(4) = 7.5$

$$v(t) = (1) \int_4^t (-2) \, dx + 7.5 = -2(t - 4) + 7.5$$

$$v(t) = -2t + 15.5, \quad 4 \le t \le 5 \text{ s}$$ ◆

PI Control Mode

The integral mode is frequently combined with the proportional mode to provide an automatic reset action that eliminates the proportional offset. The combination is referred to as the *proportional plus integral (PI) control mode*. The integral mode provides the reset action by constantly changing the controller output until the error is reduced to zero. Figure 13.11 illustrates the step response of a proportional plus integral controller. The proportional mode provides a change in the controller output that is proportional to the error signal. The integral mode provides an additional change in the output that is proportional to the integral of the error signal. The reciprocal of integral action rate (I) is the time required for the integral mode to match the change in output produced by the proportional mode.

References to Figures 13.6 and 13.7 will help explain the effect of the integral mode portion of the PI controller. In Figure 13.6, the proportional mode is implemented by the lever and the proportional gain is determined by the distances a and b. More specifically, the proportional gain is equal to 10 times distance a divided by distance b. The effect of the integral mode is to change the length of the link from the left end of the lever down to the control valve stem. The integral mode changes the length of this link at a rate that is proportional to the error signal. By changing the length of the link, the integral mode changes the position of the valve stem for a given level in the tank. For example, assume that the level is at 50%. The integral mode can change the length of the link such that the valve could be at any position from 0% to 100% open.

Figure 13.7 gives us another view of the integral control mode. The three graphs in Figure 13.7 show diagonal I/O lines with different slopes. The proportional mode determines the slope of the I/O lines, as explained in Figure 13.7. The integral mode raises or lowers the

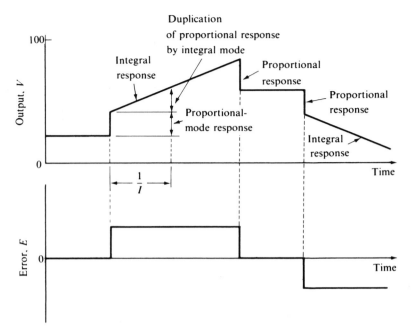

◆ **Figure 13.11** Step response of a proportional plus integral (PI) control mode.

I/O line without changing its slope. By raising or lowering the I/O line, the integral mode automatically adjusts it to accommodate changes in the load on the process.

One problem with the integral mode is that it increases the tendency for oscillation of the controlled variable. The gain of the proportional controller must be reduced when it is combined with the integral mode. This reduces the ability of the controller to respond to rapid load changes. If the process has a large dead-time lag, the error signal will not immediately reflect the actual error in the process. This delay often results in overcorrection by the integral mode—that is, the integral mode continues to change the controller output after the error is actually reduced to zero, because it is acting on an "old" signal.

The Bode diagram of a PI control mode is shown in Figure 13.12. The diagram is divided into two halves by the integral action break-point frequency, which is equal to the integral action rate.

$$\omega_i = I$$

On the left side of the diagram, $\omega < \omega_i$, the integral action dominates with the gain decreasing at 20 dB per decade and the phase angle equal to $-90°$. On the right side of the diagram, $\omega > \omega_i$, the proportional action dominates with a phase angle of $0°$ and a magnitude determined by the proportional gain, P. The region between $0.1\omega_i$ and $10\omega_i$ is a transition zone between the two sides of the diagram. In Figure 13.12, the proportional gain, P, is equal to 1, which gives a magnitude of 0 dB on the Bode diagram. The effect of a proportional gain, P, other than 1 is to raise or lower the gain curve without affecting the phase curve. A gain of $P = 10$, for example, would raise the entire gain curve 20 dB. A gain of 0.1 would lower the entire gain curve by 20 dB.

The proportional plus integral control mode is used on processes with large load changes when the proportional mode alone is not capable of reducing the offset to an acceptable level. The integral mode provides a reset action that eliminates the proportional offset.

PROPORTIONAL PLUS INTEGRAL (PI) CONTROL MODE

Time-Domain Equation

$$v = Pe + PI \int_0^t e \, dt + v_0 \qquad (13.9)$$

Frequency-Domain Equation

$$V = PE + P\left(\frac{I}{s}\right)E \qquad (13.10)$$

Transfer Function

$$\frac{V}{E} = P\left(\frac{I + s}{s}\right) \qquad (13.11)$$

(See Table 13.1 for definitions of the variables.)

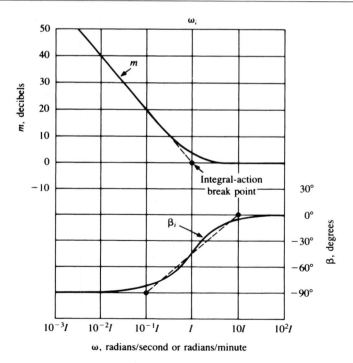

◆ **Figure 13.12**   Bode diagram of a proportional plus integral control mode. The integral mode dominates at frequencies below ω_i. The proportional mode dominates at frequencies above ω_i.

EXAMPLE 13.6

A PI controller has a gain, P, of 2 and an integral action rate, I, of 0.02 s^{-1}. The value of v_0 is 32% (at $t = 0$). The graph of the error signal is given in Figure 13.13a. Determine the value of the controller output at the following times: $t =$ (a) 0, (b) 10, (c) 50, (d) 75, and (e) 100 s.

Solution

The controller output is given by Equation (13.9). Substituting the values of v_0, P, and I, the equation of the controller is

$$v = 2e + (2)(0.02) \int_0^t e \, dt + 32$$

The term $\int_0^t e \, dt$ is equal to the net area under the error curve between 0 and t s.

(a) At $t = 0$ s, $e = 0$ and the net area $= 0$;

$$v = (2)(0) + \left(\frac{1}{25}\right)(0) + 32 = 32\%$$

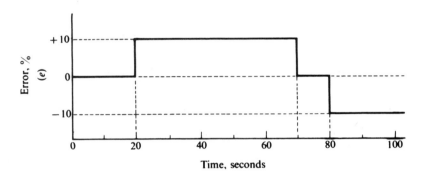

a) Graph of the error signal, e

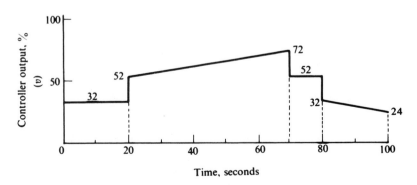

b) Graph of the controller output signal,

◆ **Figure 13.13** Error and controller output graphs for Example 13.6.

(b) At $t = 10$ s, $e = 0$ and net area $= 0$:

$$v = (2)(0) + \left(\frac{1}{25}\right)(0) + 32 = 32\%$$

(c) At $t = 50$ s, $e = 10\%$ and net area $= (10)(50 - 20) = 300\%$:

$$v = (2)(10) + \left(\frac{1}{25}\right)(300) + 32 = 64\%$$

(d) At $t = 75$ s, $e = 0$ and net area $= (10)(70 - 20) = 500$:

$$v = (2)(0) + \left(\frac{1}{25}\right)(500) + 32 = 52\%$$

(e) At $t = 100$ s, $e = -10\%$ and net area $= (10)(70 - 20) + (-10)(100 - 80) = 300$:

$$v = (2)(-10) + \left(\frac{1}{25}\right)(300) + 32 = 24\%$$

A graph of the controller output is shown in Figure 13.13b.   ◆

Derivative Control Mode

The *derivative control mode* changes the output of the controller proportionally to the rate of change of the error signal. This change may be caused by a variation in the measured variable, the setpoint, or both. The derivative mode is an attempt to anticipate an error by observing how fast the error is changing and using the rate of change to produce a control action that will reduce the expected error. The derivative mode contributes to the output of the controller only while the error is changing. For this reason, the derivative control mode is always used in combination with the proportional or proportional plus integral control modes.

> The derivative control mode is never used alone. It is always used in combination with the proportional or proportional plus integral modes.

The step and ramp responses of the ideal derivative control mode are given in Figure 13.14. At every instant, the output of the derivative control mode is proportional to the slope or rate of change of the error signal. The step response indicates the reason that the ideal derivative control mode is never used in practical controllers. The error curve has an infinite slope when the step change occurs. The ideal derivative mode must respond with an infinite change in the controller output. In practical controllers, the response of the derivative action to rapidly changing signals is limited. This greatly reduces the sensitivity of the controller to the unwanted noise spikes that frequently occur in practice.

The Bode diagram of the ideal derivative mode (not shown) is the opposite of the integral mode diagram shown in Figure 13.10. The gain increases at a rate of 20 dB per decade increase in frequency, and passes through 0 dB at a radian frequency equal to $1/D$. The phase angle is a constant $+90°$ for all frequency values. The equations of the ideal derivative control mode are given below. The practical derivative mode is covered in the next section.

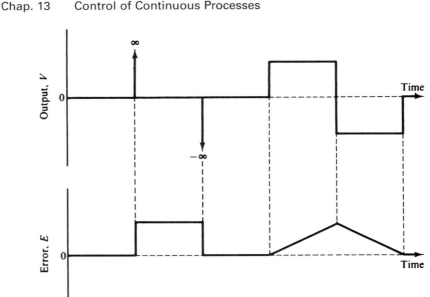

◆ **Figure 13.14** Step and ramp response of the ideal derivative control mode.

IDEAL DERIVATIVE CONTROL MODE

Time-Domain Equation

$$v = D\frac{de}{dt} \tag{13.12}$$

Frequency-Domain Equation

$$V = DsE \tag{13.13}$$

Transfer Function

$$\frac{V}{E} = Ds \tag{13.14}$$

(See Table 13.1 for definitions of the variables.)

PD Control Mode

The derivative control mode is sometimes used with the proportional mode to reduce the tendency for oscillations and allow a higher proportional gain setting. The combination of proportional and derivative modes is referred to as the *PD control mode*. The proportional mode provides a change in the controller output that is proportional to the error signal. The derivative mode provides an additional change in the controller output that is proportional to the rate of change of the error signal. The derivative mode anticipates the future value of the error signal and changes the controller output accordingly. This anticipatory action makes the derivative mode useful in controlling processes with sudden load changes. For this reason, the derivative mode is usually used with proportional or proportional plus integral control when the sudden load changes produce excessive errors. The derivative mode control action opposes the change of a controlled variable, which helps damp out oscillations of the controlled variable.

The proportional plus derivative control mode is used on processes with sudden load changes when the proportional mode alone is not capable of keeping the error within an acceptable level. The derivative mode provides an anticipatory action that reduces the maximum error produced by sudden load changes. It also allows a higher gain setting, which helps reduce the proportional offset.

Equation (13.15) is the time-domain equation of a practical proportional plus derivative control mode. The *Pe* term is the proportional mode action. The *PD de/dt* term is the ideal derivative mode action, and $\alpha D\, dv/dt$ is the term that limits the response produced by rapidly changing signals. Equation (13.16) is the frequency-domain equation, and Equation (13.17) is the transfer function.

PROPORTIONAL PLUS DERIVATIVE (PD) CONTROL MODE

Time-Domain Equation

$$v = Pe + PD\frac{de}{dt} - \alpha D\frac{dv}{dt} + v_0 \tag{13.15}$$

Frequency-Domain Equation

$$V = PE + PDsE - \alpha DsV \tag{13.16}$$

Transfer Function

$$\frac{V}{E} = P\left(\frac{1 + Ds}{1 + \alpha Ds}\right) \tag{13.17}$$

(See Table 13.1 for definitions of the variables.)

The Bode diagram of a PD control mode is shown in Figure 13.15. The proportional mode dominates the left side of the diagram (where $\omega < \omega_d = 1/D$). The proportional gain raises or lowers the entire gain curve, just as it did in the PI control mode. The derivative mode causes the gain curve to slope up at 20 dB per decade at the derivative-action break point. The derivative limiter causes the gain to return to horizontal at the limiter break point. The diagram clearly shows how the derivative mode amplifies high-frequency signals, and how the derivative limiter reduces the amplification of high-frequency signals. Notice also that the limiter causes the phase angle to return to 0° at the higher frequencies. In effect, the PD control mode provides a phase lead over a band of frequencies. Controller design involves the placement of this phase lead where it will do the most good.

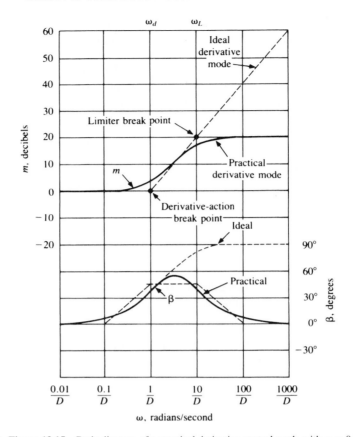

◆ **Figure 13.15** Bode diagram of a practical derivative control mode with $\alpha = 0.1$.

EXAMPLE 13.7

A PD controller has a gain (P) of 0.8, a derivative action time constant (D) of 1 s, an initial output (v_0) of 40%, and a derivative limiter coefficient (α) of 0.1. The graph of the error signal is given in Figure 13.16. Determine the value of the controller output (v) at the following times: (a) $t = 0$ s, (b) $t = 10-$, an instant before $t = 10$ s, (c) $t = 10+$, an instant after $t = 10$ s, (d) $t = 20$ s, (e) $t = 40$ s, and (f) $t = 60$ s. Assume that the derivative limiter term is negligible.

Solution

The controller output is given by Equation (13.15). Using the values and assumptions given above, the equation for the controller is

$$v = 0.8e + 0.8\frac{de}{dt} + 40$$

The term de/dt is equal to the slope of the curve at any given instant of time.

(a) At $t = 0$, $e = 0$, and the slope $= de/dt = 0$.

$$v = (0.8)(0) + (0.8)(0) + 40 = 40\%$$

(b) At $t = 10-$, an instant before $t = 10$ s, $e = 0$, and the slope $= de/dt = 0$.

$$v = (0.8)(0) + (0.8)(0) + 40 = 40\%$$

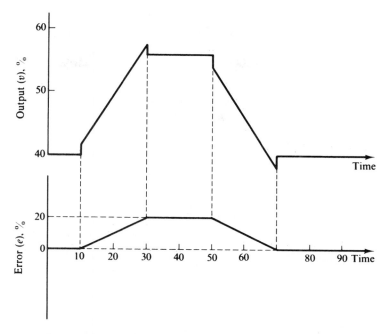

◆ **Figure 13.16**   Error and controller output graphs for Example 13.7.

(c) At $t = 10+$, an instant after $t = 10$ s, $e = 0$, and the slope $= de/dt = 20/20 = 1\%/s$.

$$v = (0.8)(0) + (0.8)(1) + 40 = 40.8\%$$

(Notice the increase in v from 40.0% to 40.8% at t passes 10 s. We say there is a step change in v of 0.8% at $t = 10$ s.)

(d) At $t = 20$ s, $e = 10\%$ and $de/dt = 20/20 = 1\%/s$.

$$v = (0.8)(10) + (0.8)(1) + 40 = 48.8\%$$

(e) At $t = 40$ s, $e = 20\%$ and $de/dt = 0$.

$$v = (0.8)(20) + (0.8)(0) + 40 = 56\%$$

(f) At $t = 60$ s, $e = 10\%$ and $de/dt = -20/20 = -1\%/s$.

$$v = (0.8)(10) + (0.8)(-1) + 40 = 47.2\%$$

The diagram of the output, v, is shown in Figure 13.16.   ◆

PID Control Mode

The *PID control mode* is a combination of the proportional, integral, and derivative control modes. A PID controller is also referred to as a three-mode controller. The integral mode is used to eliminate the proportional offset caused by large load changes. The derivative mode reduces the tendency toward oscillations and provides a control action that anticipates changes in the error signal. The derivative mode is especially useful when the process has sudden load changes.

> The proportional plus integral plus derivative control mode is used on processes with sudden, large load changes when one- or two-mode control is not capable of keeping the error within acceptable limits. The derivative mode produces an anticipatory action that reduces the maximum error produced by sudden load changes. The integral mode provides a reset action that eliminates the proportional offset.

Equation (13.18) is the defining equation for an ideal three-mode controller.

$$v = Pe + PI \int_0^t e\, dt + PD \frac{de}{dt} + v_0 \tag{13.18}$$

The practical PID controller includes the derivative limiter term introduced in the section on the PD control mode. Equations (13.19), (13.20), and (13.21) define the practical PID controller.

PROPORTIONAL PLUS INTEGRAL PLUS DERIVATIVE (PID) CONTROL MODE

Time-Domain Equation

$$v = Pe + PI \int_0^t e\, dt + PD \frac{de}{dt} - \alpha D \frac{dv}{dt} + v_0 \tag{13.19}$$

Frequency-Domain Equation

$$V = PE + P\left(\frac{I}{s}\right)E + PDsE - \alpha DsV \tag{13.20}$$

Transfer Function

$$\frac{V}{E} = P\left(\frac{I + s + Ds^2}{s + \alpha Ds^2}\right) \tag{13.21}$$

(See Table 13.1 for definitions of the variables.)

EXAMPLE 13.8

A PID controller has the following parameters: $P = 4.3$, $I = (1/7)$ s^{-1}, $D = 0.5$ s, $v_0 = 10\%$. The graph of the error signal is given in Figure 13.17. Determine the output of the controller at $t =$ (a) 5, (b) 10, (c) 15, and (d) 25 s. Assume that the derivative limiter term is negligible.

Solution

The controller output is given by Equation (13.19). Using the values and assumptions given above, the equation for the controller is

$$v = 4.3e + 0.614 \int_0^t e\, dt + 2.15\left(\frac{de}{dt}\right) + 10$$

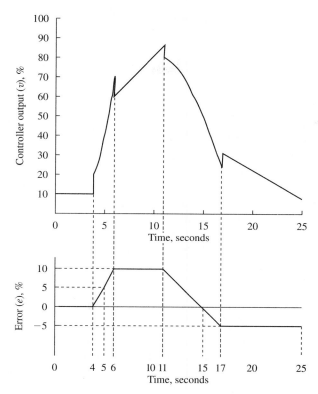

◆ **Figure 13.17** Error and controller output graphs for Example 13.8.

The term $\int_0^t e\, dt$ is equal to the net area under the error curve between 0 and t s. The term de/dt is equal to the slope of the curve at any given instant of time.

(a) At $t = 5$ s, $e = 5\%$, net area $= (0.5)(1)(5) = 2.5\%/s$ and slope $= 10/(6 - 4) = 5\%/s$:

$$v = (4.3)(5) + (0.614)(2.5) + (2.15)(5) + 10$$

$$= 44\%$$

(b) At $t = 10$ s, $e = 10\%$, net area $= (0.5)(2)(10) + (4)(10) = 50$, and slope $= 0$:

$$v = (4.3)(10) + (0.614)(50) + (2.15)(0) + 10$$

$$= 84\%$$

(c) At $t = 15$ s, $e = 0$, net area $= (0.5)(2)(10) + (5)(10) + (0.5)(4)(10) = 80$, and slope $= -15/(17 - 11) = -2.5\%/5$:

$$v = (4.3)(0) + (0.614)(80) + (2.15)(-2.5) + 10$$

$$= 54\%$$

(d) At $t = 25$ s, $e = -5\%$, net area $= 80 - (0.5)(2)(5) - (8)(5) = 35$, and slope $= 0$:

$$v = (4.3)(-5) + (0.614)(35) + (2.15)(0) + 10$$

$$= 10\%$$

A graph of the controller output is included in Figure 13.17. ◆

13.3 ELECTRONIC ANALOG CONTROLLERS

An electronic *analog controller* has two main parts: the error detector and the control mode unit. An example of an electrical error detector is illustrated in Figure 13.18. The output of the measuring transmitter is a 4- to 20-mA electric current signal. Each value of the current represents a unique value of the controlled variable, c. The 4-mA signal represents the minimum value of c, and the 20-mA signal represents the maximum value. The current signal is applied to a 62.5-Ω resistor, resulting in a 0.25- to 1.25-V signal across the resistor. The setpoint signal is produced by a potentiometer with a 0.25- to 1.25-V output range. The two voltage signals are connected in opposition so that the voltage between points a and b is equal to the setpoint signal minus the measured value signal.

$$e = sp - c_m$$

The control mode unit is sometimes called "the controller," although it is actually one part of the unit that is usually called by that name. The electronic analog controller uses a single operational amplifier and some resistors and capacitors to form the control mode unit. The operational amplifier is used as a function generator, and the resistors and capacitors are arranged to implement the transfer function of the desired control mode or combination of modes.

The analog proportional controller uses three resistors to form an inverting amplifier (see Figure 13.19). The circuit has two inputs, the error, e, and the output offset, v_0. The proportional gain, P, is equal to the feedback resistor, R_f, divided by the error input resistor, R_1. The offset resistor, R_o, must be equal to the feedback resistor, R_f, to satisfy the time-domain equa-

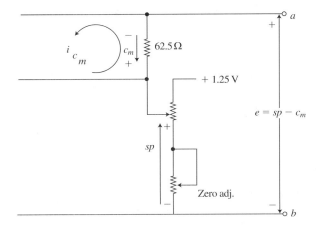

◆ **Figure 13.18** Typical error detector in an electronic analog controller.

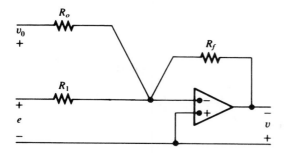

Time-domain equation: $v = Pe + v_0$

Transfer function $= \dfrac{V}{E} = P$

$P = \dfrac{R_f}{R_1}$

$R_o = R_f$

◆ **Figure 13.19**   An analog proportional controller is essentially an op-amp inverting amplifier.

tion. The output lines can be reversed to make the output either positive or negative with respect to the sign of the error. Some applications of the controller will require a positive output for a positive error, and other applications will require a negative output for a positive error.

   The proportional plus integral controller uses two resistors and a capacitor to implement the PI transfer function (see Figure 13.20). The capacitor, C_i, is placed in series with the feedback resistor, R_i. The gain, P, is equal to the feedback resistor, R_i, divided by the input resistor, R_1. The integral action rate is equal to the reciprocal of the product of the input resistor, R_i, and the capacitor, C_i.

   The proportional plus derivative controller uses four resistors and a capacitor to implement the PD mode (Figure 13.21). The circuit is a proportional controller with a parallel combination of resistor, R_d, and capacitor, C_d, placed in series with the input resistor, R_1. The equations for the gain, derivative action time constant, and derivative limiter coefficient are given in Figure 13.21. A typical value of α is 0.1.

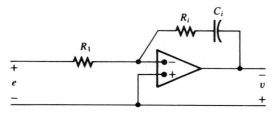

Transfer function $= V/E = P\left[\dfrac{I + S}{S}\right]$

$I = \dfrac{1}{R_iC_i} =$ integral action rate, second^{-1}

$P = R_i/R_1 =$ gain

◆ **Figure 13.20**   Analog proportional plus integral (PI) controller.

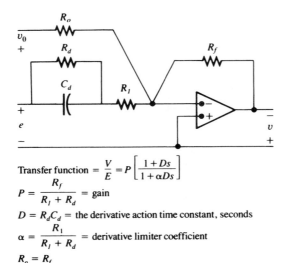

Transfer function $= \dfrac{V}{E} = P \left[\dfrac{1+Ds}{1+\alpha Ds} \right]$

$P = \dfrac{R_f}{R_1 + R_d} = \text{gain}$

$D = R_d C_d = \text{the derivative action time constant, seconds}$

$\alpha = \dfrac{R_1}{R_1 + R_d} = \text{derivative limiter coefficient}$

$R_o = R_f$

◆ **Figure 13.21** Analog proportional plus derivative (PD) controller.

Two versions of the analog PID controller are shown in Figure 13.22. One version (Figure 13.22a) forms the derivative action on the input side and the integral action on the output side. The other version (Figure 13.22b) does just the opposite and forms the integral action on the input side and the derivative action on the output side. The equations for the controller parameters for each version are given in Figure 13.22.

The transfer function for the analog PID controller is a modified version of Equation (13.21). The modification is done for reasons of economy and consists of two first-order networks in series. The implementation of Equation (13.21) in its exact form requires three operational amplifiers. The derivative and integral terms must be formed in parallel and then summed with a summing amplifier. The modification consists of inserting an interaction term (*PIDe*) in the time-domain equation, as shown below:

$$v = Pe + PIDe + PI \int_0^t e \, dt + PD \frac{de}{dt} - \alpha D \frac{dv}{dt} + v_0$$

A Laplace transformation of the equation above with $v_0 = 0$ gives the following frequency-domain equation:

$$V = PE + PIDE + P\left(\frac{I}{s}\right)E + PDsE - \alpha DsV$$

Solving for the ratio V/E gives the following transfer function:

$$\frac{V}{E} = \frac{P + PID + PI/s + PDs}{1 + \alpha Ds} \tag{13.22}$$

or

$$\frac{V}{E} = P\left(\frac{I + (1 + ID)s + Ds^2}{s + \alpha Ds^2}\right) = P\left(\frac{I + s}{s}\right)\left(\frac{1 + Ds}{1 + \alpha Ds}\right) \tag{13.23}$$

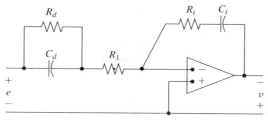

a) Derivative input and integral output

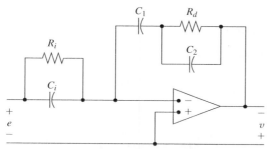

b) Integral input and derivative output

$$\text{Transfer function} = \frac{V}{E} = P\left(\frac{I + s}{s}\right)\left(\frac{1 + Ds}{1 + \alpha Ds}\right)$$

Figure 13.22a   Figure 13.22b

$$P = \frac{R_i}{R_1 + R_d}$$ $$P = (C_i/C_1)$$

$$\alpha = \frac{R_1}{R_1 + R_d}$$ $$\alpha = \frac{C_2}{C_1 + C_2}$$

$$I = 1/(R_iC_i)$$ $$I = 1/(R_iC_i)$$

$$D = R_dC_d$$ $$D = R_d(C_1 + C_2)$$

Figure 13.22   Two versions of the analog proportional plus integral plus derivative (PID) controller.

EXAMPLE 13.9   Determine the values of R_1 and R_i for an electronic proportional plus integral controller with a gain, P, of 2 and an integral action rate, I, of 0.02 s^{-1}. Use a 10-μF capacitor for C_i. Determine the transfer function.

Solution   The equations are given in Figure 13.20.

   (a) $I = 1/(R_iC_i)$; therefore, $R_i = 1/(IC_i)$:

$$R_i = \frac{50}{10^{-5}} = 5\ \text{M}\Omega$$

   (b) $P = R_i/R_1$; therefore, $R_1 = R_i/P$:

$$R_1 = \frac{5 \times 10^6}{2} = 2.5\ \text{M}\Omega$$

The transfer function is

$$\frac{V}{E} = 2\left(\frac{1 + 50s}{50s}\right)$$

◆

EXAMPLE 13.10

Determine the value of R_1, R_i, R_d, and C_d for the analog PID controller in Figure 13.22a. The controller has a gain, P, of 4, an integral action rate, I, of $\frac{1}{7}$ s^{-1}, a derivative action time constant, D, of 0.5 s, and a derivative limiter coefficient, α, of 0.1. Use a 10-μF capacitor for C_i. Also determine the transfer function.

Solution

The equations are given in Figure 13.22a.

(a) $I = 1/(R_iC_i)$; therefore, $R_i = 1/(IC_i)$:

$$R_i = \frac{7}{10^{-5}} = 700 \text{ k}\Omega$$

(b) $\alpha = R_1/(R_1 + R_d)$; therefore,

$$R_d = \left(\frac{1 - \alpha}{\alpha}\right)R_1 = \left(\frac{1 - 0.1}{0.1}\right)R_1 = 9R_1$$

(c) $P = R_i/(R_1 + R_d) = R_i/(10R_1)$; therefore,

$$R_1 = \frac{R_i}{10P} = \frac{700 \text{ k}\Omega}{40} = 17.5 \text{ k}\Omega$$

(d) $R_d = 9R_1 = (9)(17.5 \text{ k}\Omega) = 157.5 \text{ k}\Omega$

(e) $D = R_dC_d$; therefore, $C_d = D/R_d$:

$$C_d = \frac{0.5}{157.5 \text{ k}\Omega} = 3.2 \text{ }\mu\text{F}$$

The transfer function is

$$\frac{V}{E} = 4\left[\frac{\frac{1}{7} + (1 + 0.5/7)s + 0.5s^2}{s + 0.05s^2}\right] = \frac{1 + 7.5s + 3.5s^2}{1.75s + 0.0875s^2}$$

◆

13.4 DIGITAL CONTROLLERS

Microprocessor-based *digital controllers* are now very commonplace in industrial control systems. There are many reasons for the popularity of digital controllers. The power of the microprocessor provides advanced features such as adaptive self-tuning, multivariable control, and expert systems. The ability of the microprocessor to communicate over a field bus or local area network is another reason for the wide acceptance of the digital controller. Digital controllers used for closed-loop control generally implement the PI, PD, or PID control modes. In this section we examine the PID digital controller.

Sampling

A digital controller measures the controlled variable at specific times, which are separated by a time interval called the sampling time, Δt. Each sample (or measurement) of the controlled variable is converted to a binary number for input to a digital computer or microcomputer. The computer subtracts each sample of the measured variable from the setpoint to determine a set of error samples.

$$e_1 = sp - c_{m1} = \text{first error sample}$$

$$e_2 = sp - c_{m2} = \text{second error sample}$$

$$e_3 = sp - c_{m3} = \text{third error sample}$$

$$\vdots$$

$$e_n = sp - c_{mn} = \text{present error sample}$$

Control Algorithms

After computing each error sample, a digital PID controller follows a procedure called the PID algorithm to calculate the controller output based on the error samples: $e_1, e_2, e_3, \ldots, e_n$. The PID algorithm has two versions, the positional version and the incremental version.

The *positional PID algorithm* determines the valve position, v_n, based on the error signals. Equation (13.24) is a simplified version of the positional algorithm.

$$v_n = Pe_n + PI\,\Delta t \sum_{j=1}^{j=n} e_j + PD\,\frac{\Delta e_n}{\Delta t} \qquad \textbf{(13.24)}$$

where v_n = present valve position, percentage of full scale

$\quad\quad\quad P$ = controller gain

$\quad\quad\quad e_n$ = present error sample, percentage of full scale

$\quad\quad\quad \Delta t$ = sample time, s

$\quad\quad\quad I$ = integral action rate, s^{-1}

$\quad\quad\quad D$ = derivative action time constant, s

$\quad\quad \Delta e_n = e_n - e_{n-1}$ = change in the error signal

A flow diagram of a positional PID algorithm is shown in Figure 13.23.

The *incremental PID algorithm* determines the change in the valve position, $\Delta v_n = v_n - v_{n-1}$, based on the error samples. The incremental algorithm can be determined by using Equation (13.24) to determine v_n and v_{n-1} and then subtracting to obtain Equation (13.25) as follows:

$$v_{n-1} = Pe_{n-1} + PI\,\Delta t \sum_{j=1}^{j=n-1} e_j + PD\,\frac{\Delta e_{n-1}}{\Delta t}$$

$$\Delta v_n = P\,\Delta e_n + PI\,\Delta t\,e_n + PD\left(\frac{\Delta e_n - \Delta e_{n-1}}{\Delta t}\right) \qquad \textbf{(13.25)}$$

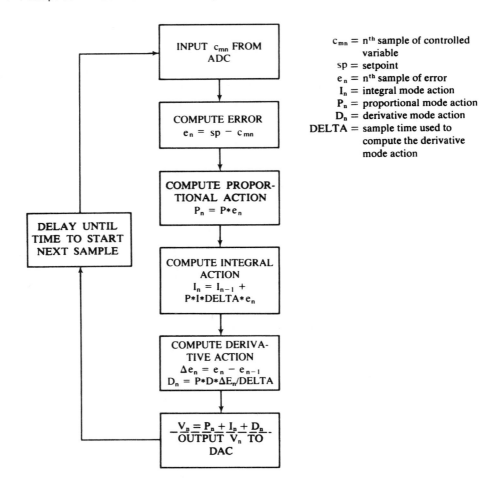

c_{mn} = n^{th} sample of controlled variable
sp = setpoint
e_n = n^{th} sample of error
I_n = integral mode action
P_n = proportional mode action
D_n = derivative mode action
DELTA = sample time used to compute the derivative mode action

◆ **Figure 13.23** Flow diagram of a positional PID control algorithm.

where
$$\Delta v_n = v_n - v_{n-1}$$
$$\Delta e_n = e_n - e_{n-1}$$
$$\Delta e_n - \Delta e_{n-1} = e_n - 2e_{n-1} + e_{n-2}$$

The incremental algorithm is especially suited to incremental output devices such as stepper motors. The positional algorithm is more natural and has the advantage that the controller "remembers" the valve position. If the sample time, Δt, is much shorter than the integral action time constant, $T_i = 1/I$, the positional algorithm will produce a behavior similar to an analog controller.

The Integral Mode

The integral mode in Equation (13.24) presents computational problems that can produce unsatisfactory results. The integral mode is given by the following term in Equation (13.24).

$$\text{Integral term} = PI\,\Delta t \sum_{j=1}^{j=n} e_j \tag{13.26}$$

For each sample, the integral mode must produce a change given by

$$\text{Integral mode change} = PI\,\Delta t\, e_j \tag{13.27}$$

When the value of $PI\,\Delta t$ is less than 1, it is more convenient to work with the reciprocal of $PI\,\Delta t$, which could be stored in the computer as an integer. In this case, Equation (13.27) would be revised as follows:

$$I_{\text{DIV}} = \frac{1}{PI\,\Delta t}$$
$$\text{Integral mode change} = \frac{e_j}{I_{\text{DIV}}} \tag{13.28}$$

If $PI\,\Delta t$ is very small, the computer may ignore relatively large errors because of insufficient resolution. For example, consider a digital controller with a 12-bit word length. The resolution of a 12-bit binary number is 1 part in 4096. To illustrate a point, let's assume that a 12-bit binary number is used to represent a range of errors from -2048 to $+2047$. If $P = 0.5$, $\Delta t = 1$ s, and $I = 0.002$ s^{-1}, then

$$PI\,\Delta t = (0.5)(0.002)(1) = 0.001$$
$$I_{\text{DIV}} = 1000$$

Any value of error greater than -1000 and less than $+1000$ (i.e., 48% of the full-scale range) would result in an integral mode change less than 1, which would be ignored. This small change would then be lost unless a special provision is made to include the change in the calculations for the next sample. The end result is a permanent offset error that the integral mode is unable to eliminate.

One solution to this integral offset problem is to increase the precision by increasing the word length in the computer. A 16-bit word length has a precision of 1 part in 65,536, which could represent a range of errors from $-32,768$ to $+32,768$. This would reduce the offset error to about 3% of the full-scale range.

Another solution is to add the unused portion of the sum of the error samples to the current error sample, e_n, before computing the integral mode change. In the preceding example, an error of 900 in each of two successive samples would not produce an integral mode change because each sample is less than 1000. However, if the first sample is retained, the sum of 1800 would produce an integral mode change of $1800/1000 = 1$ with a remainder of 800. The remainder of 800 would be retained to be added to the next error sample. Every time the accumulated remainder plus the current error is greater than 1000, another increment will be added to the integral mode change.

Derivative Mode

The derivative mode in Equation (13.24) also presents computational problems that can produce unsatisfactory results. A slowly changing signal, for example, results in a "jumpy" derivative mode action. Let's examine how this can occur and what can be done to smooth out the derivative action.

The derivative mode is given by the following term in Equation (13.24)

$$\text{Derivative term} = PD\left(\frac{e_n - e_{n-1}}{\Delta t}\right) \tag{13.29}$$

The term $(e_n - e_{n-1})/\Delta t$ is actually an *estimate* of the rate of change of the error, de/dt. Because Δt is fixed by the sampling rate, our attention will focus on the term $(e_n - e_{n-1})$, which we will represent as est_1. The derivative term produced by est_1 will be called D_1.

$$\text{est}_1 = e_n - e_{n-1} \tag{13.30}$$

$$D_1 = PD\frac{\text{est}_1}{\Delta t} \tag{13.31}$$

If $P = 6$, $\Delta t = 1$ s, and $D = 100$ s, then

$$D_1 = (6)\left(\frac{100}{1}\right)(\text{est}_1) = (600)(\text{est}_1)$$

Table 13.2 shows the derivative term that is produced by a controlled variable that decreases at the rate of 0.5% per second. Notice that the derivative term, D_1, jumps back and forth between 0 and 600 because the estimate, est_1, oscillates between 0 and 1.

What is needed is a better estimate of Δe. The theory of digital estimators is beyond the scope of this book. However, a simple example will show how a good estimator can smooth out the derivative term. The idea of an estimator is to use previous samples to improve the estimate. For our example, we use an estimator that uses the last four samples to estimate Δe. We will call this estimate est_2 and the derivative term it produces, D_2.

$$\text{est}_2 = (e_n + e_{n-1}) - (e_{n-2} + e_{n-3}) \tag{13.32}$$

$$D_2 = PD\frac{\text{est}_2}{2^2\,\Delta t} \tag{13.33}$$

If $P = 6$, $\Delta t = 1$ s, and $D = 100$ s, then

$$D_2 = 6\left(\frac{100}{4}\right)\text{est}_2 = 150\text{est}_2$$

Table 13.2 shows how our simple estimator has smoothed out the derivative term. The est_2 estimator has an effective sample period of 2 s. It used two samples to estimate e_n and two more samples to estimate e_{n-2}. The 2^2 term in the equation for D_2 accounts for the doubling of the sample period and the use of two samples to determine an average. The idea of the est_2 estimator can be extended to include more previous samples. An est_5 estimator would increase

◆ **TABLE 13.2**   Derivative Action Produced by Two Δe Estimators (setpoint, $sp = 9\%$)

n	c	c_m	e	est_1	D_1	est_2	D_2
1	9.5	9	0	0	0	0	0
2	9.0	9	0	0	0	0	0
3	8.5	8	1	1	600	1	150
4	8.0	8	1	0	0	2	300
5	7.5	7	2	1	600	2	300
6	7.0	7	2	0	0	2	300
7	6.5	6	3	1	600	2	300
8	6.0	6	3	0	0	2	300
9	5.5	5	4	1	600	2	300
10	5.0	5	4	0	0	2	300

the effective sample time to 5 s and use five samples to estimate e_n and five samples to estimate e_{n-5}.

$$est_5 = (e_n + e_{n-1} + e_{n-2} + e_{n-3} + e_{n-4})$$
$$-(e_{n-5} + e_{n-6} + e_{n-7} + e_{n-8} + e_{n-9})$$

$$D_5 = PD\frac{est_5}{5^2\,\Delta t}$$

13.5 ADVANCED CONTROL

Advanced control refers to various methods of going beyond the single-loop single-variable feedback control system with three modes of control. Topics in this section include cascade control, feedforward control, adaptive self-tuning controllers, and multivariable control systems.

Cascade Control

Cascade control involves two controllers with the output of the primary controller providing the setpoint of the secondary controller. The level control loop in Figure 2.14 provides an excellent application of cascade control. The changes in level occur slowly owing to the capacitance of the tank. In contrast, changes in flow occur very quickly. When a disturbance causes a change in the input flow rate, there is a considerable delay before the level changes enough to correct the disturbance. The disturbance often changes before the correction is made. The slow-moving correction of disturbances results in fluctuations of the level. Figure 13.24 shows how cascade control is used to improve the level control system.

In Figure 13.24, a flow transmitter and a secondary controller are used to form a flow control loop within the level control loop. The output of the level controller is the remote setpoint of the flow controller. The flow control loop responds quickly to flow disturbances, virtually eliminating the level fluctuations they caused in a simple level control loop. Industrial processes have many applications of cascade control.

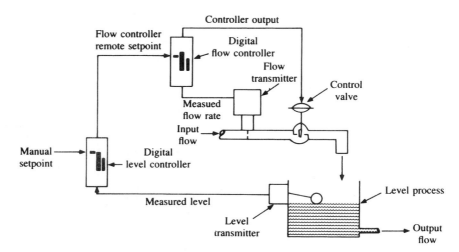

◆ **Figure 13.24** A cascade level control system uses a flow control loop inside the level control loop for improved response to flow disturbances.

Feedforward Control

Feedforward control uses a model of the process to make changes in the controller output in response to measured changes in a major load variable without waiting for the error to occur. The tubular heat exchanger control loop in Figure 2.3 is a prime candidate for feedforward control. The product flow rate is the major load on the process. An increase in the product flow rate requires an increase in the flow rate of the heating fluid to maintain the product temperature at the setpoint. Figure 13.25 shows the application of feedforward control to the tubular heat exchanger.

In Figure 13.25, a flow transmitter measures the product flow rate and sends the signal to a load compensator. The load compensator computes the correction necessary to adjust for the product flow rate. The output of the compensator is added to the output of the temperature controller. There is no delay in the compensator signal; the correction is made as soon as the change in product flow rate is measured. The term *feedforward* comes from the fact that the compensation signal travels in the same direction as the product. This is in contrast to the measured temperature signal, which travels in the opposite direction as the product; hence the name *feedback* for the primary loop.

If the feedforward compensation is perfect and there are no other disturbance variables in the process, the feedback loop could be eliminated. These ideal conditions never occur in practice, so feedforward control systems invariably include a feedback loop to make the final adjustments.

Adaptive Controllers

Adaptive controllers change the controller parameters to "adapt" to changes in the process. For example, a change in the product flow rate in the temperature control system in Figure 13.25 will change the dead time of the process. A change in the process dead time means a

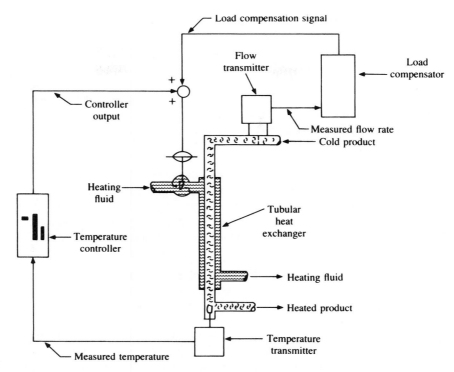

◆ **Figure 13.25** A feedforward signal compensates for changes in product flow rate in this temperature control system.

change in controller parameters is necessary to "tune" the controller to the process. An adaptive controller determines the values of P, I, and D necessary to adapt to the new process conditions and makes the necessary changes. Many different techniques are used to "adapt" the controller to changes in the process. Self-tuning controllers fall into two general categories: those that use a model of the process as the basis of the tuning, and those that use pattern recognition and stored knowledge as the basis.

A typical model-based adaptive controller introduces a step change in the setpoint and observes the resulting process response. The controller then forms a model of the process based on the response to the step change. This "self-learning" operation is repeated and the model and tuning parameters are adjusted until they match the actual process.

A pattern-recognition approach to adaptive control uses a graph of error versus time. The controller constantly examines the response to natural disturbances, looking for the presence or absence of peak heights, the time between peaks, and the proportional offset. Following a disturbance, the controller automatically computes P, I, and D based on the observed response pattern and knowledge stored in the controller's memory.

Multivariable Control

Multivariable control uses measurements of several process load variables and may involve the manipulation of more than one process variable. The computer control systems used to control the fuel injectors and spark timing of automobiles are excellent examples of multivariable control systems. Figure 13.26 illustrates a representative automotive control system.

The following is a general description of automotive control systems for the purpose of illustrating a multivariable control system. The discussion is intended to be generic, and no attempt is made to give a complete description of a particular system.

The purpose of the system shown in Figure 13.26 is to control the fuel injector flow rate, the ignition spark timing, and the idle speed. The inputs to the controller include engine coolant temperature, manifold air temperature, manifold vacuum, barometric pressure, throttle position, engine speed, fuel pressure, and the oxygen content in the exhaust gas. A single computer controls all three output variables. The control system has eight inputs and three outputs, making it a multivariable control system.

The major operating modes of the control system are

1. *Starting.* The controller varies the amount of fuel sprayed into the intake manifold according to the engine coolant temperature. A cold engine receives more fuel than a warm engine. The ignition system generates the spark-timing signals internally and ignores the timing signals from the computer.
2. *Normal running.* The computer uses four input signals to maintain a nearly ideal air/fuel ratio (about 14.7 : 1). The four input variables are manifold air temperature, manifold vacuum, fuel pressure, and the oxygen content. The computer also modifies the ignition timing based on engine speed, manifold vacuum, engine coolant temperature, and barometric pressure.

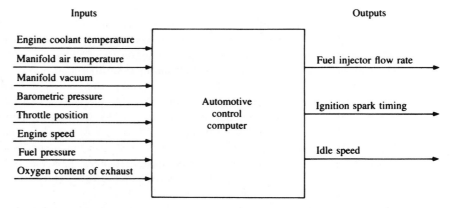

◆ **Figure 13.26** An automotive control computer is an example of a multivariable control system. The generic system shown here is intended only to illustrate multivariable control.

3. *Cold running.* The computer provides extra fuel when the engine coolant temperature is below a predetermined value.
4. *Acceleration.* The computer provides extra fuel during acceleration.
5. *Deceleration.* The computer reduces the amount of fuel during deceleration to reduce the pollution produced by the engine.
6. *Idle.* The idle speed is increased when the engine coolant temperature is below a predefined value. Idle speed is also increased when the battery voltage is low, when the transmission is shifted into drive or reverse, and when the air-conditioning compressor comes on.

13.6 FUZZY LOGIC CONTROLLERS

Introduction

Fuzzy logic is a relatively new methodology that uses language and reasoning principles similar to the way humans solve problems. An objective of fuzzy logic is to make computers "think" like we do. The computer's world is black and white; our world has many shades of gray. Consider the simple statement "It is warm outside today." What does "warm" mean? It could mean 75, 80, or 85°F. In northern Minnesota, where I was raised, the term *warm* in January meant 0°F. After two weeks when the thermometer remained at −30 to −40°F, we called it warm outside when the thermometer got all the way up to 0°F. The point is that there is a vagueness in the words we use to describe conditions such as cold, cool, warm, and hot. When we say it is raining, is it a mist, a light rain, a moderate rain, or a heavy rain? Give it some thought, and you will agree that we deal with vagueness every day, and yet we are able to produce crisp actions based on a somewhat vague (or fuzzy) knowledge of conditions. Fuzzy logic is a method that mimics our ability to produce a crisp action from fuzzy or incomplete inputs.

Fuzzy logic began in the 1960s when Professor Lotfi Zadeh, University of California at Berkeley, proposed a mathematical way of looking at the intrinsic vagueness of human language. He named this new method *fuzzy logic.* Observing that human reasoning often uses variables whose values are vague (or fuzzy), he introduced the concept of linguistic variables—variables whose values are words that describe a condition, such as HOT, NORMAL, or COLD.

A linguistic value is not a single entity; rather, it is a set of elements that have different degrees of membership in the set. This set of elements, called a fuzzy set, differs from conventional sets in which an element is either in the set or it is not. In a fuzzy set, an element can be entirely in the set, partially in the set, or not in the set. Consider the linguistic variable ENGINE__TEMPERATURE with linguistic values of HOT, NORMAL, and COLD. The value NORMAL can be represented by the fuzzy set illustrated in Figure 13.27. The value NORMAL names a fuzzy set of normal operating temperatures for an automobile engine. The temperatures on the horizontal axis are possible members of the set of normal operating temperatures. The vertical axis is the degree of membership in the set $\mu(x)$. Notice that the degree of membership ranges from 0 to 1. A membership value of 0 means the element is not in the set, a value of 0.5 means the element is 50% in the set, and a value of 1.0 means the element is 100% in the set. The fuzzy set allows us to precisely define variables that are inherently vague.

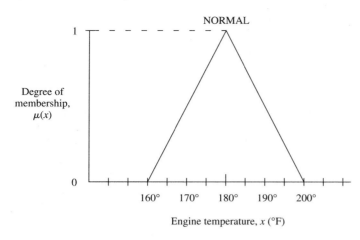

◆ **Figure 13.27** The linguistic value NORMAL names a fuzzy set of temperature elements. The element $x = 180°F$ has a membership in the set of 1.0 (100%). Elements $x = 170°F$ and $x = 190°F$ have memberships of 0.5 (50%). Elements with values of 160°F or less are not in the set, and elements with values of 200°F or greater are also not in the set.

A fuzzy set consists of three components:

1. A horizontal axis of values of x, the elements in the set
2. A vertical axis of values of $\mu(x)$, the degree of membership in the set
3. A line that defines matching pairs of x and $\mu(x)$ values

For example, the fuzzy set in Figure 13.27 defines the following pairs of x and $\mu(x)$ values:

$$(160, 0) \quad (170, 0.5) \quad (180, 1.0) \quad (190, 0.5) \quad (200, 0)$$

The shape of the fuzzy set is not inherently triangular. Trapezoids and parabolic shapes are also used. The triangle is often chosen because it is easy to implement in computer software, and it provides a natural decrease as the value of the element moves away from the full membership value.

Fuzzification

The first step in the design of a fuzzy logic control system is the *fuzzification* of the input and output variables. Figure 13.28 shows the fuzzification of the input and output variables of a control system that uses a heating element to control the temperature of an extruder. The input to the controller is the extruder temperature. The output from the controller is a change in the heating element current. The linguistic name TEMPERATURE is given to the input variable, and it is assigned five linguistic values: COLD, COOL, NORMAL, WARM, and HOT. The name CURRENT__CHANGE is given to the output variable, and it is also assigned five values: PM (positive medium), PS (positive small), ZERO, NS (negative small), and NM (negative medium). Triangles are used to define the fuzzy sets for all values with the exception of the two end values of TEMPERATURE. Trapezoids are used for COLD and HOT to extend their membership to the extreme ends. The fuzzy sets overlap to provide a smooth transition from one linguistic value to the next. As temperature increases, COLD gradually gives way to COOL, then COOL gradually gives way to NORMAL, and so on.

An odd number of linguistic values is always used to fuzzify a process variable, with five and seven being the most common numbers. The upper limit is eleven linguistic values, and, of course, the lower limit is three.

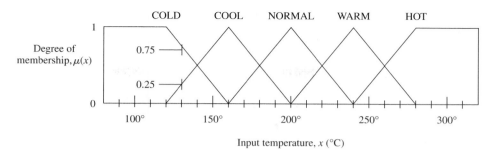

a) Fuzzification of input temperature

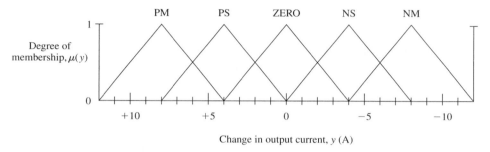

b) Fuzzification of output current change

◆ **Figure 13.28**   Fuzzification of the input and output variables of a temperature control system. The input temperature is represented by five overlapping linguistic values: COLD, COOL, NORMAL, WARM, and HOT. The change in output current is also represented by five overlapping linguistic values: PM (positive medium), PS (positive small), ZERO, NS (negative small), and NM (negative medium).

The Fuzzy Rule Base

The second step in the design of a fuzzy logic control system is to convert the designer's knowledge of the process into a set of if . . . then rules that relate the input to the output. The rules consist of a number of if . . . then statements that take the following form:

$$\text{If } X \text{ is } x \text{ then } Y \text{ is } y$$

where X and Y are linguistic variables and x and y are their linguistic values. The rule base defines a set of imprecise dependencies between the two linguistic variables. The following is a rule base for the linguistic variables in Figure 13.28:

1. If TEMPERATURE is COLD   then CURRENT__CHANGE is PM
2. If TEMPERATURE is COOL   then CURRENT__CHANGE is PS
3. If TEMPERATURE is NORMAL   then CURRENT__CHANGE is ZERO
4. If TEMPERATURE is WARM   then CURRENT__CHANGE is NS
5. If TEMPERATURE is HOT   then CURRENT__CHANGE is NM

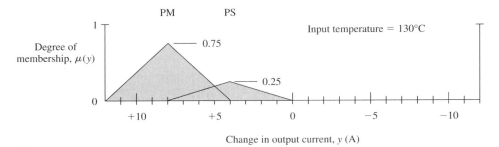

◆ **Figure 13.29** Graphic representation of the fuzzy output current change that is produced by an input temperature of 130°C. The output consists of two weighted fuzzy sets produced by a 75% application of rule No. 1 and a 25% application of rule No. 2. The centroid method of defuzzification produces a crisp output that is the centroid of the two triangles.

Defuzzification

The third and last step in the design of a fuzzy logic control system is to defuzzify the results of the rule base to produce a crisp control action. At any given input value, some rules apply, and the others do not. The objective of defuzzification is to find the output action that best represents the information contained in the rules activated by the current input. In other words, how can we use the active rules to produce a crisp control action?

Consider an input temperature of 130°C in Figure 13.28. This temperature has a 0.75 membership in the fuzzy set COLD and a 0.25 membership in the fuzzy set COOL. This means that rule No. 1 applies at a 75% level and rule No. 2 applies at a 25% level. Figure 13.29 shows a graphic representation of a 75% application of rule No. 1 and a 25% application of rule No. 2. Where would you place the output action? Common sense dictates an action between the two peaks and nearer to the larger peak. In reaching this conclusion, you have intuitively applied the centroid, or center-of-gravity, method of defuzzification. The centroid method is usually used in control applications because it provides smooth changes in output as the process changes.

In centroid defuzzification, a crisp output is determined by the weighted sum of the centroids of the *active* fuzzy sets. For a temperature of 130°C, the centroid value of the current change is determined as follows:

$$\text{CURRENT\_CHANGE}\,(130°C) = \frac{y_1\mu(y_1) + y_2\mu(y_2)}{\mu(y_1) + \mu(y_2)}$$

$$= \frac{8(0.75) + 4(0.25)}{0.75 + 0.25} = \frac{6 + 1}{1} = 7\,\text{A}$$

In general, the centroid output action is given by the following formula:

$$\text{Control action} = \frac{\displaystyle\sum_{i=1}^{n} y_i\,\mu(y_i)}{\displaystyle\sum_{i=1}^{n} \mu(y_i)} \qquad \textbf{(13.34)}$$

where n = the number of if . . . then rules

y_i = centroid of the ith fuzzy set

$\mu(y_i)$ = weighted membership in the ith fuzzy set

$\mu(y_i)$ = 0 for all inactivated fuzzy sets

EXAMPLE 13.11

Determine the output action of the fuzzy logic controller (FLC) described above for extruder temperatures of 140 and 150°C.

Solution

From Figure 13.28, we determine that 140°C is 0.5 COLD and 0.5 COOL. Using Equation 13.34, we get the following control action:

$$\text{CURRENT\_CHANGE } (140°C) = \frac{8(0.5) + 4(0.5)}{0.5 + 0.5} = \frac{4 + 2}{1} = 6 \text{ A}$$

For a temperature of 150°C, we get the following control action:

$$\text{CURRENT\_CHANGE } (150°C) = \frac{8(0.25) + 4(0.75)}{0.25 + 0.75} = \frac{2 + 3}{1} = 5 \text{ A}$$ ◆

FLC Configurations

The operations of fuzzification, application of the fuzzy rule base, and defuzzification are digital operations accomplished by nested loops in a computer program. Figure 13.30a shows a block diagram of a fuzzy logic control module. There are two ways a FLC module can be used. Figure 13.30b shows a possible configuration in which the fuzzy logic control module replaces the conventional PID module. In this configuration the input to the FLC is the error signal, and the output of the FLC goes to the manipulating element. A more common configuration is shown in Figure 13.30c. In this configuration, the FLC is used to feed artificial setpoints to the PID controller to eliminate overshoot and improve response to process upsets. This emulates techniques an expert operator uses to improve the performance of a PID controller.

The current trend is to blend FLCs with conventional PIDs rather than to replace PIDs with FLCs. Instead of replacing existing methods of control, fuzzy logic provides additional methods that increase the range of problems that can be solved. Fuzzy logic has much to offer in supervisory control problems such as traffic control, quality control, transportation systems, and so forth.

Final Observations

The fuzzification example in Figure 13.28 could lead you to the conclusion that the memberships in fuzzy sets always add up to 1. There is no such requirement in fuzzy sets, and memberships can add up to more than 1, less than 1, or equal to 1. Simply reducing the overlap in Figure 13.28, for example, would result in memberships that add up to less than 1, as illustrated in Figure 13.31. Increasing the overlap in Figure 13.28 would result in memberships that total more than 1.

Fuzzy logic control modules are not limited to one input and one output. The rate of change of the input variable is a common second input. The extruder temperature control system in our example could have the rate of change of temperature as a second input. A fuzzy

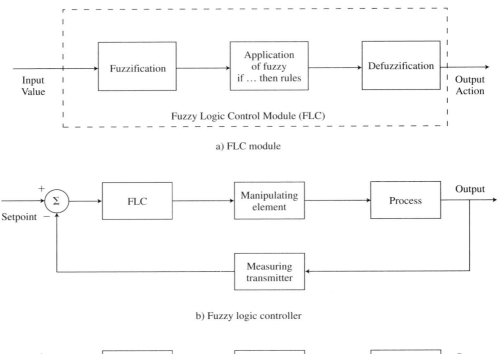

a) FLC module

b) Fuzzy logic controller

c) Blended FLC/PID controller

◆ **Figure 13.30**   A fuzzy logic control module and two configurations of fuzzy logic controllers. The FLC module (a) is a software module that uses nested loops to accomplish the three FLC operations. The fuzzy logic controller (b) replaces the conventional PID module with an FLC module. The blended controller (c) combines the FLC module with the PID module. The FLC module feeds artificial setpoints to the PID module to reduce fluctuations caused by process upsets.

logic controller of a position control system could have both position and velocity as inputs. When two inputs are used, an AND or OR operator is introduced in the if . . . then statements as follows:

If POSITION is POSITIVE__SMALL AND VELOCITY is ZERO then
MOTOR__CURRENT is NEGATIVE__SMALL

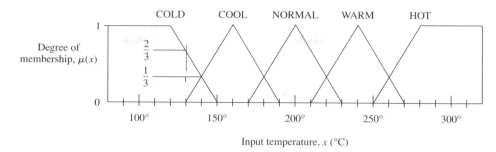

◆ **Figure 13.31** Reducing the overlap in the fuzzy sets results in memberships that add up to less than 1. A temperature of 130°C has a membership of two thirds in COLD and 0 in COOL. A temperature of 140°C has a membership of one third in COLD and one third in COOL. Both temperatures have total memberships of two thirds.

The extruder temperature controller used as the example in Figure 13.28 is essentially an integral mode controller. The cycle time of the control loop must allow sufficient time for the extruder temperature to adjust to a change in current before the next correction is applied. With improper timing, the FLC will produce large swings in the extruder temperature. In other words, the FLC must be tuned to the process to obtain satisfactory performance.

The design of a fuzzy logic controller can be a very time-consuming process. Specifying the linguistic variables, the fuzzy membership functions, and the if . . . then rule base requires extensive knowledge about the process. The designer must study the process, seek the experience of skilled operators, and fine-tune the results to obtain a successful design. This hard work is not without reward, however, because fuzzy logic applications have performed very well. In some applications, fuzzy logic has been able to reduce overshoot in response to disturbances without increasing the response time. In one example, the insertion of a fuzzy logic control module in a conventional PID control loop reduced temperature fluctuations from $\pm 15°F$ to less than $\pm 3°F$.

◆ GLOSSARY

Adaptive controller: A control system that changes the control mode settings in response to changes in the process. (13.5)

Analog controller: A controller producing a controller output that can have any value between 0 and 100% of the full-scale output. (13.3)

Cascade control: A system that uses two controllers, with the output of the primary controller providing the setpoint of the secondary controller. (13.5)

Control mode: Any of several different ways the controller forms the control action. Common control modes include two-position, floating, proportional, integral, and derivative. (13.1)

Control mode, derivative (D): The derivative control mode changes the output of the controller by an amount proportional to the rate of change of the error signal. (13.2)

Control mode, floating: A variation of the two-position mode in which the controller output is constant as long as the error remains within a small band around zero (called the neutral zone). When the error is outside the neutral zone, the controller output changes at a fixed rate until the error returns to the neutral zone, or the output reaches one of its extreme positions (0 or 100%). (13.2)

Control mode, integral (I): The integral control mode changes the output of the controller by an amount proportional to the accumulation (integral) of the error signal. (13.2)

Control mode, PD: The combination of the proportional and derivative control modes. (13.2)

Control mode, PI: The combination of the proportional and integral control modes. (13.2)

Control mode, PID: The combination of the proportional, integral, and derivative control modes. (13.2)

Control mode, proportional (P): The proportional mode produces a change in the controller output that is proportional to the error signal. There is a fixed linear relationship between the error signal and the output of the controller, *See also* Proportional offset. (13.2)

Control mode, two-position: The two-position mode produces only two possible controller output values, depending on the sign of the error. If the two positions are 0 and 100%, the controller is called an ON/OFF controller. (13.2)

Defuzzification: The process of finding the best crisp representation of a fuzzy value of a linguistic variable. (13.6)

Digital controller: A controller producing a controller output that can have only a definite number of discrete values between 0 and 100% of the full-scale output. (13.4)

Feedforward control: A control system that uses a model of the process to change the controller output in response to measured changes in a major load variable without waiting for the error to occur. (13.5)

Fuzzification: The process of turning a crisp value of a variable into a fuzzy value of a linguistic variable. (13.6)

Fuzzy set: A function that maps a value that might be in the set to a number between 0 and 1 that defines its degree of membership in the set. A degree of 0 means the element is not in the set, a degree of 0.5 means the element is 50% in the set, and a degree of 1.0 means the element is 100% in the set. (13.6)

Linguistic variable: A variable whose values are words that describe its condition, such as HOT, NORMAL, or COLD. The value HOT names a fuzzy set that specifies the elements that constitute HOT. The value NORMAL names a fuzzy set that specifies the elements in NORMAL, and the value COLD names a fuzzy set that specifies the elements in NORMAL. (13.6)

Load change: Any condition in a control system that changes the location of the load line. (13.1)

Multivariable control: A control system that uses measurements of several process load variables to manipulate several other process variables in order to maintain a specified control objective. (13.5)

PID algorithm, incremental: A procedure used by a digital controller to calculate the change in the value of the controller output based on the error samples. (13.4)

PID algorithm, positional: A procedure used by a digital controller to calculate the value of the controller output based on the error samples. (13.4)

Proportional offset: The error required by a proportional control mode to hold the controller output at some value other than the value it has when the error is zero. *See also* Residual error. (13.2)

Residual error: Another name for the proportional offset. (13.2)

◆ EXERCISES

Section 13.1

13.1 Describe the three operations performed by a feedback control system.

13.2 Name the five common modes of control.

13.3 Name the two parts of a controller.

13.4 Name three methods of implementing a controller.

13.5 Describe how you would determine if there has been a load change in a control system that is maintaining the controlled variable at a constant set point value (e.g., a regulator system).

Section 13.2

13.6 Describe the operation of each of the five modes of control.

13.7 Describe three ways to reduce the amplitude of the oscillation in a two-position control system.

13.8 Describe the conditions for which you would select a two-position control mode.

13.9 The two-position controller in Example 13.1 is modified so that the valve moves between two partially open positions instead of between ON and OFF. The inlet flow rate, m, is 0.001 m^3/s when the valve is in the minimum flow position and 0.003 m^3/s when the valve is in the maximum flow position. The rest of the conditions are the same as in Example 13.1. Determine the amplitude and the period of oscillation of the level, h. Compare your results with the results in Example 13.1. What was given up to reduce the amplitude of the oscillation?

13.10 Describe the conditions for which you would select a floating control mode.

13.11 The solid flow rate control system shown in Figure 2.19 uses a gate to control the level of material on the belt. A single-speed reversible motor is used to drive the cam that positions the gate. If the solid feed rate is below a predetermined value, the controller drives the gate up at a constant rate. If the feed rate is above a second predetermined value, the controller drives the gate down at a constant rate. Between the two predetermined values, the gate is motionless. Identify the mode of control used in this system.

13.12 Sketch the input/output graph for each of the following proportional controller conditions:
(a) $P = 8$, $v_0 = 40\%$ **(c)** $P = 0.5$, $v_0 = 25\%$
(b) $P = 0.25$, $v_0 = 55\%$

13.13 Describe the conditions for which you would select a proportional control mode.

13.14 Determine the proportional offset required to maintain $v - v_0$ at 12% for proportional controllers with each of the following gain values:
(a) $P = 0.2$
(b) $P = 0.6$
(c) $P = 1.2$

13.15 Describe the conditions for which you would select a PI control mode.

13.16 A PI controller has a gain of 0.5, an integral action rate of 0.0125 s^{-1}, and a value of v_0 of 25%. The graph of the error signal is given in Figure 13.32. Determine the value of the controller output at $t = 15$ s and $t = 30$ s.

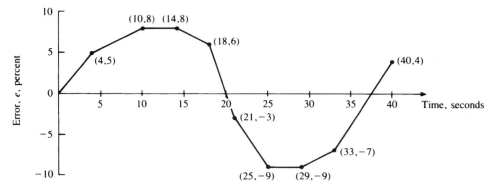

◆ **Figure 13.32**   Error graph used in Exercises 13.16 through 13.8.

13.17 Describe the conditions for which you would select a PD control mode.

13.18 A PD controller has the following parameter values: $P = 0.5$, $D = 12$ s, $\alpha = 0$, and $v_0 = 42\%$. The graph of the error signal is given in Figure 13.32. Determine the value of the controller output at **(a)** $t = 15$ s and **(b)** $t = 30$ s.

13.19 Describe the conditions for which you would select a PID control mode.

13.20 A PID controller has the following parameter values: $P = 0.5$, $I = 0.0125$ s^{-1}, $D = 12$ s, $\alpha = 0$, and $v_0 = 55\%$. The graph of the error signal is given in Figure 13.32. Determine the value of the controller output at **(a)** $t = 15$ s and **(b)** $t = 30$ s.

13.21 Equation (13.21) defines the transfer function for a PID controller. Use program BODE to print a Bode Data Table for each of the following sets of controller parameters. Parameter set a sets a basis for comparison, and parameter sets b, c, and d each reduce one control mode by 50%. Highlight the gain and phase values in the Bode Data Tables for parameter sets b, c, and d that differ from the corresponding gain and phase values in the table for parameter set a. The highlighted values mark the values changed by the reductions in the P, I, and D control modes.

(a) $P = 4.0$, $I = 0.14$ s^{-1}, $D = 0.50$ s, $\alpha = 0.1$
(b) $P = 2.0$, $I = 0.14$ s^{-1}, $D = 0.50$ s, $\alpha = 0.1$
(c) $P = 4.0$, $I = 0.07$ s^{-1}, $D = 0.50$ s, $\alpha = 0.1$
(d) $P = 4.0$, $I = 0.14$ s^{-1}, $D = 0.25$ s, $\alpha = 0.1$

13.22 The conditions in Exercise 13.21 were selected to show how each control mode (P, I, D) affects the frequency response of the controller.

(a) Review your Bode Data Table for parameter set b, and comment on the effect of the 50% reduction in the P control mode.

(b) Review your Bode Data Table for parameter set c, and comment on the effect of the 50% reduction in the I control mode.

(c) Review your Bode Data Table for parameter set d, and comment on the effect of the 50% reduction in the D control mode.

(d) Draw a Bode diagram and plot the gain and phase angle for parameter set b, using solid lines. Then plot the gain and phase angle for parameter set a, using dashed lines where the values differ from set b.

(e) Repeat Exercise 13.22d with parameter set b replaced by parameter set c.

(f) Repeat Exercise 13.22d with parameter set b replaced by parameter set d.

13.23 A certain process has a small capacitance. Sudden, moderate load changes are expected, and a small offset error can be tolerated. Recommend the control mode or combination of modes most suitable for controlling this process.

13.24 A process has a large capacitance and no dead-time lag. The anticipated load changes are relatively small. Recommend the control mode or combination of modes most suitable for controlling this process.

13.25 A dc motor is used to control the speed of a pump. A tachometer–generator is used as the speed sensor. The load changes are insignificant, and there is no dead-time lag in the process. Recommend the control mode or combination of modes most suitable for controlling the motor speed.

13.26 A liquid flow process is fast and the flow rate signal has many "noise spikes." Large load changes are quite common, and there is very little dead-time lag. Recommend the control mode or combination of modes most suitable for controlling the flow rate. Why is the derivative mode usually avoided in liquid flow controllers?

13.27 An electric heater is used to control the temperature of a plastic extruder. The process is a first-order lag with almost no dead-time lag. The time constant is very large. Under steady operation, the load changes are insignificant. Recommend the control mode or combination of modes most suitable for controlling the temperature of the extruder.

13.28 A dryer is a slow process with a very large dead-time lag. Sudden load changes are common, and proportional offset is undesirable. Recommend the control mode or combination of modes most suitable for controlling the dryer.

13.29 A PI controller defined by Equation 13.11 has a gain that increases as frequency decreases. At very low frequencies, the gain becomes very large. In practical PI controllers, the maximum gain is limited. In the following transfer function of a PI controller, the term b in the denominator limits the gain to a maximum value of $1/b$.

$$\frac{V}{E} = P\left(\frac{I + s}{bI + s}\right)$$

Use the program BODE to generate two sets of frequency data for parameter sets 1 and 2 below (see Example 4.19). Plot both sets of data on a single Bode diagram and comment on the comparison of the effect of term b.

Set 1: $P = 1$, $I = 0.1$ s^{-1}, $b = 0$
Set 2: $P = 1$, $I = 0.1$ s^{-1}, $b = 0.01$

Section 13.3

13.30 An analog proportional mode controller (Figure 13.19) uses a value of 100 kΩ for R_f. Determine the value of R_1 for each of the following controller gains:

(a) $P = 0.25$

(b) $P = 1.75$

(c) $P = 8.6$

13.31 Determine the values of R_1 and R_i for an analog PI controller (Figure 13.20) with $P = 6.4$ and $I = 0.00541$ s^{-1}. Use a 1000-μF capacitor for C_i.

13.32 Determine the value of R_1, R_d, and R_f for an analog PD controller (Figure 13.21) with $P = 3.6$, $D = 12$ s, and $\alpha = 0.1$. Use a 100-μF capacitor for C_d.

13.33 Determine the values of R_1, R_i, R_d, and C_i for an analog PID controller, as shown in Figure 13.22a. The controller parameters are $P = 0.65$, $I = 1/70$ s^{-1}, $D = 0.32$ s, $\alpha = 0.1$. Use a 1-μF capacitor for C_d.

13.34 Use the program BODE to generate frequncy-response data for (a) the noninteracting PID controller [Equation (13.21)], and (b) the interacting PID controller [Equation (13.23)]. Plot Bode diagrams for both controllers. Comment on the difference between the two Bode diagrams. The controller has the following parameters: $P = 4.0$, $I = 1/7$, $D = 0.5$, and $\alpha = 0.1$.

Section 13.4

13.35 A digital controller is started up with the setpoint (SP) at 20%, and the controlled variable at 0%. The first 24 samples are listed below. For each sample, compute the value of the error, e_n, the summation of errors, S_n, and the change in the error signal, Δe_n, as defined by the following equations.

$$e_n = SP - C_{mn} \qquad S_n = \sum_{j=1}^{j=n} e_j \qquad \Delta e_n = e_n - e_{n-1}$$

Put your result in a table with five columns and the following heading:

n	$C_{mn}(\%)$	$e_n(\%)$	$S_n(\%)$	$\Delta e_n(\%)$

First 24 Samples of the Digital Controller

n	$C_{mn}(\%)$	n	$C_{mn}(\%)$	n	$C_{mn}(\%)$
1	0.0	9	18.5	17	19.8
2	3.1	10	20.0	18	20.0
3	5.9	11	21.4	19	20.1
4	8.5	12	22.0	20	20.2
5	10.9	13	21.2	21	20.1
6	13.1	14	20.0	22	20.0
7	15.1	15	19.8	23	20.0
8	16.9	16	19.5	24	20.0

13.36 The digital controller in Exercise 13.35 uses the positional algorithm defined by Equation (13.24).

$$v_n = P_{mode} + I_{mode} + D_{mode}$$

where

$P_{mode} = Pe_n$

$I_{mode} = PI\Delta t S_n$

$D_{mode} = PD\Delta e_n/\Delta t$

The PID modes and sample time are set to the following values:

$$P = 2 \qquad I = 0.1 \text{ s}^{-1} \qquad D = 1 \text{ s} \qquad \Delta t = 2 \text{ s}$$

Compute P_{mode}, I_{mode}, D_{mode}, and V_n for each sample in Exercise 13.36. Put your results in a table with five columns and the following heading.

n	P_{mode} (%)	I_{mode} (%)	D_{mode} (%)	V_n (%)

13.37 Construct the following graphs of your results from Exercise 13.36: C_{mn} vs. n, P_{mode} vs. n, I_{mode} vs. n, and D_{mode} vs. n.

13.38 Draw a PID flow diagram similar to Figure 13.23 with the following change: Replace the integral mode operation with the following operations:

(a) Add the nth sample error, e_n, to a new variable called the unused sum, S:

$$S = S + e_n$$

(b) Divide S by I_{DIV} to obtain a quotient, Q, and a remainder, R.

$$I_{DIV} = \frac{1}{PI \, \Delta t}$$

$$Q = S \text{ div } I_{DIV}$$

$$R = S \text{ mod } I_{DIV}$$

(c) Add Q to the past integral mode action, I_{n-1}, to form the current integral mode action, I_n.

$$I_n = I_{n-1} + Q$$

(d) Set the unused sum, S, equal to the remainder, R:

$$S = R$$

Explain the purpose of these operations.

13.39 Draw a PID flow diagram similar to Figure 13.23 with the following change: Replace the derivative mode operation in the PID flow diagram (Figure 13.23) with a new derivative mode operation that uses an improved esti-mate of the rate of change of the error, de/dt. Use the esti-mate defined by Equation (13.32) and use Equation (13.33) to compute the derivative term.

Section 13.5

13.40 Select which of the following four types of ad-vanced control is described by each statement.

1. Adaptive control
2. Cascade control
3. Feedforward control
4. Multivariable control

(a) An automative control computer.

(b) Keeps the controller "in tune" when the product flow rate is changed.

(c) Output of the first controller is the setpoint of the sec-ond controller.

(d) Under ideal conditions, the feedback loop could be eliminated.

(e) Improves the response to flow disturbances in a level control system.

(f) Uses a "self-learning" operation to update a model of the process.

(g) Changes the controller output in response to measured changes in a process variable.

(h) Uses stored knowledge about response patterns to compute new PID values.

(i) Uses measurements of more than one process variable to manipulate one or more process variables.

(j) Changes the controller parameters (P, I, D) in re-sponse to a measured change in a process variable.

Section 13.6

13.41 Determine the output action of the fuzzy logic controller in Example 13.11 for the following extruder temperatures:

(a) 100°C	**(d)** 160°C	**(g)** 220°C	**(j)** 280°C
(b) 120°C	**(e)** 180°C	**(h)** 240°C	**(k)** 300°C
(c) 140°C	**(f)** 200°C	**(i)** 260°C	

13.42 Draw a graph of your results in Exercise 13.41. Place extruder temperature on the x-axis (input) and cur-rent change on the y-axis (output).

13.43 A fuzzy logic controller is used to control a dc mo-tor position control system. An encoder measures the angu-lar position, and a tachometer measures angular velocity. The encoder output is mapped such that one rotation pro-duces a position signal that ranges from -255 to $+255$. The tachometer output is mapped such that full speed in one direction produces a signal of -255 and full speed in the other direction produces a signal of $+255$. The output of the controller is the current applied to the motor arma-ture, and it ranges from -3 to $+3$ A. Draw fuzzification graphs for position, velocity, and current. Use the five lin-guistic values in Figure 13.28b for each variable, and use five triangular fuzzy sets with 50% overlap as in Figure 13.28.

13.44 Develop a set of 9 if . . . then rules for the fuzzy logic controller in Exercise 13.43 using only the inner three linguistic values for each input variable (i.e., NS, ZERO, and PS). Use the following if . . . then rule as a guide:

If POSITION is NS AND VELOCITY is ZERO then CURRENT is PS

PART FIVE ◆ ANALYSIS AND DESIGN

◆ **CHAPTER 14**

Process Characteristics

◆ OBJECTIVES

The objective of a control system is to maintain a desired value of a "controlled variable" in a process. To do this, the controller adjusts the value of another variable in the process, the "manipulated variable." The characteristics of a process are an expression of the relationship between the manipulated variable (input) and the controlled variable (output). The Bode diagram design method covered in Chapter 16 requires a knowledge of the characteristics of a process. Process characteristics can be expressed in several ways. In this chapter we use the following four methods: step response graph, time-domain equation, transfer function, and Bode diagram.

The purpose of this chapter is to provide you with the means of determining the characteristics of the following five types of processes: **integral, first-order lag, second-order lag, dead time,** and **first-order lag plus dead time.** After completing this chapter, you will be able to sketch or determine the following for electrical, thermal, liquid flow, gas flow, and mechanical examples of the five types of processes:

1. Characteristic parameters
   a. Steady-state gain (all processes)
   b. Time constant (first-order lag processes)
   c. Integral action time constant (integral processes)
   d. Dead-time lag (dead-time processes)
   e. Resonant frequency (second-order lag processes)
   f. Damping ratio (second-order lag processes)
2. Step response graph
3. Time-domain equation
4. Transfer function
5. Bode diagram

14.1 INTRODUCTION

A process or component is characterized by the relationship between the input signal and the output signal. It is this input/output (I/O) relationship that determines the design requirements of the controller. If the I/O relationship of the process is completely defined, the designer can specify the optimum controller parameters. If the I/O relationship is poorly defined, the designer must provide a large adjustment of the controller parameters so that the optimum settings can be determined during start-up of the system. This chapter provides the information necessary to determine the I/O relationship of the following types of processes:

1. The integral or ramp process (Section 14.2).
2. The first-order lag process (Section 14.3).
3. The second-order lag process (Section 14.4).
4. The dead-time process (Section 14.5).
5. The first-order lag plus dead-time process (Section 14.6).

The I/O relationship of a process may be defined by any or all of the following:

1. The step response graph
2. The time-domain equation
3. The transfer function
4. The frequency response graph

The *step response graph* is the time graph of the output signal following a step change in the input signal from one value to another. The *time-domain equation* defines the *size* versus *time* relationship between the input signal and the output signal. Time-domain equations are expressed in terms of the basic elements defined in Chapter 3 and frequently contain integral or derivative terms. The transfer function of a process is obtained by transforming the time-domain equation into a frequency-domain algebraic equation and then solving for the ratio of output over input. The *transfer function* defines the *gain and phase difference* versus *frequency* relationship between the input and output signals. The *frequency response graph* (or Bode diagram) is a dual plot of gain versus frequency and phase difference versus frequency.

The characteristics of a process depend on its basic elements (resistance, capacitance, inertance, dead time), not on the type of system (thermal, electrical, mechanical, etc.). Two different systems may have the same process characteristics. That is, they may have the same time-domain equation, transfer function, step response, and Bode diagram. This means that we can extend our knowledge of one system to all other systems that have the same process characteristics. For example, electrical, liquid flow, gas flow, and thermal first-order lag processes are all characterized by a time constant that is equal to the product of resistance times capacitance ($\tau = RC$).

Each of the following sections presents the step response graph, time-domain equation, transfer function, and Bode diagram that characterize one type of process. The equations are included in a box with all terms and their units. Each section concludes with one or more examples for use as guides in calculating the parameters of different systems with the same process characteristics.

14.2 THE INTEGRAL OR RAMP PROCESS

The *integral* or *ramp process* consists of a single capacitive element configured such that the outflow of material or energy is independent of the amount of material or energy stored in the capacitive element. The quantity of stored material or energy remains constant only if the inflow rate is exactly equal to the outflow rate. If the inflow rate is greater than the outflow rate, the quantity stored will increase at a rate proportional to the difference. If the inflow rate is less than the outflow rate, the quantity stored will decrease at a rate proportional to the difference. The input flow rate is the input signal to the integral process, but the output flow rate is not the output signal. The output signal of the integral process is a variable, such as liquid level, that is a measure of the amount of material or energy stored in the capacitive element.

The *step response* of an integral process is illustrated in Figure 14.1. Before the step change, the input flow rate is equal to the output flow rate and the level is maintained constant. After the step change, the input flow rate is greater than the output flow rate, and the level increases at a constant rate. The term *ramp process* is derived from the ramplike shape of the output response graph. The step response of an integral process is measured by the *integral action time constant,* which is the number of seconds (or minutes) required for the output to reach the same percentage change as the input (see Figure 14.1).

Two integral processes are illustrated in Figure 14.2. In the liquid level integral process, the input signal is the input flow rate and the output signal is the level of liquid in the tank. In the sheet loop integral process, the input signal is the input sheet velocity and the output signal is the distance from the belt to the bottom of the sheet loop. The input and output signals are expressed as a percentage of the full-scale range. In both integral processes, the change in the output signal during the time interval from t_0 to t_1 is equal to the integral of the difference between the input flow rate and the output flow rate divided by the *integral action time constant* (T_i). Equations (14.1) and (14.2) are the time-domain equation and transfer function of the liquid flow integral process. The equations for the sheet loop integral process are obtained from Equations (14.1) and (14.2) by replacing q_{in} and q_{out} by v_{in} and v_{out}. See the development of Equation 4.7 in Section 4.2 for further details concerning Equation (14.1). See also Example 4.12 for the development of Equation (14.2).

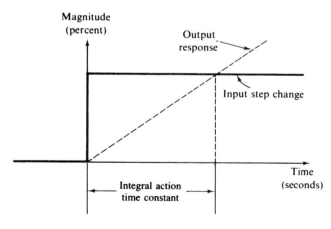

Magnitude (percent)

Output response

Input step change

Time (seconds)

Integral action time constant

◆ **Figure 14.1** Response of an integral process to a step change in the input signal. Before the step change, the input signal (inflow minus outflow) was zero and the output signal (level) remained constant. After the step change, the inflow was greater than the outflow, and the level increased at a constant rate. The integral action time constant, T_i, is the time required for the output signal to change by an amount equal to the input step change.

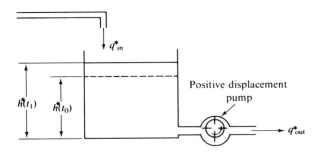

a) A liquid level integral process

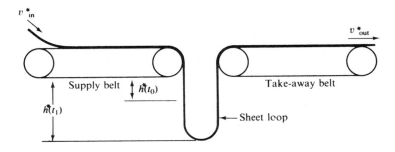

b) A sheet loop integral process

◆ **Figure 14.2** Examples of integral processes.

INTEGRAL PROCESS

Time-Domain Equation

$$h^*(t_1) - h^*(t_0) = \frac{1}{T_i} \int_{t_0}^{t_1} (q_{in}^* - q_{out}^*) \, dt \qquad (14.1)$$

Transfer Function

$$\frac{H^*(s)}{Q_{in}^*(s)} = \frac{1}{T_i s} \qquad (14.2)$$

where $h^*(t_0)$ = normalized output at time t_0, percentage of FS_{out}
$h^*(t_1)$ = normalized output at time t_1, percentage of FS_{out}
FS_{in} = full-scale range of the input
FS_{out} = full-scale range of the output
q_{in}^* = normalized input flow rate, percentage of FS_{in}
q_{out}^* = normalized output flow rate, percentage of FS_{in}
t = time, s
T_i = integral action time constant, s

Liquid Flow Integral Process

$$T_i = A \left(\frac{FS_{out}}{FS_{in}} \right)$$

where A = cross-sectional area of the tank at the liquid surface, m^2

FS_{in} = full-scale input range, m^3/s

FS_{out} = full-scale output range, m

Sheet Loop Integral Process

Replace q_{in}^* and q_{out}^* by v_{in}^* and v_{out}^* in Equations (14.1) and (14.2).

v_{in}^* = input sheet velocity, percentage of FS_{in}

v_{out}^* = output sheet velocity, percentage of FS_{in}

$$T_i = 2 \left(\frac{FS_{out}}{FS_{in}} \right)$$

The Bode diagram of an integral process (Figure 14.3) consists of straight-line gain and phase graphs. The gain line has a slope of -20 dB per decade increase in frequency. The phase line is horizontal at $-90°$. A gain line with a slope of -20 dB per decade increase in frequency means that the gain decreases by a factor of $1/10$ when the frequency increases by a factor of 10, or vice versa. It also means that the gain decreases by a factor of $1/2$ when the frequency increases by a factor of 2. In general, if the frequency changes by the factor a, the gain changes by the factor $1/a$ (i.e., if $\omega_2 = a\omega_1$, then $g_2 = (1/a)g_1$. With a little algebraic manipulation, we can eliminate the factor a with the following simpler result for a gain line with a slope of -20 dB per decade increase in frequency.

$$g_2 = g_1 \left(\omega_1 / \omega_2 \right)$$

Notice that the integral process Bode diagram is identical to the Bode diagram of the integral control mode. An integral process has a built-in integral control mode action.

EXAMPLE
14.1

A liquid level integral process has the following parameters:

Tank height: 4 m

Tank diameter: 1.5 m

$FS_{in} = 0.01$ m^3/s

$FS_{out} = 4$ m

$h^*(t_0) = 22.5\%$ of FS (0.9 m)

$q_{out}^* = 60\%$ of FS (0.006 m^3/s)

$q_{in}^* = 80\%$ of FS (0.008 m^3/s)

Determine the time-domain equation, the transfer function, the integral action time constant, and the level at time $t_0 + 100$ s.

Solution

$$A = \text{area} = \frac{\pi d^2}{4} = \frac{\pi (1.5)^2}{4} = 1.77 \text{ m}^2$$

$$T_i = A \frac{FS_{out}}{FS_{in}} = \frac{(1.76)(4)}{0.01} = 703 \text{ s}$$

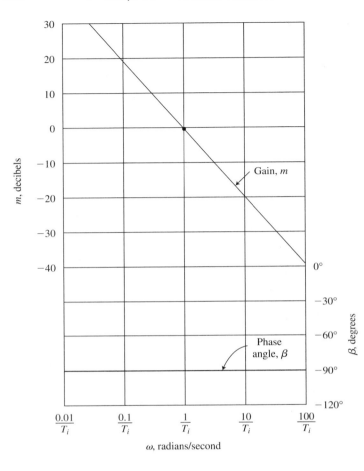

◆ **Figure 14.3** The integral process has the same Bode diagram as the integral control mode. For a liquid level integral process, $T_i = A(\text{FS}_{out})/(\text{FS}_{in})$. For the sheet loop integral process, $T_i = 2(\text{FS}_{out})/(\text{FS}_{in})$.

The time-domain equation is

$$h^*(t_1) = \frac{1}{703} \int_{t_0}^{t_1} (q_{in}^* - q_{out}^*)\, dt + h^*(t_0)$$

The transfer function is

$$\frac{H^*(s)}{Q_{in}^*(s)} = \frac{1}{703\, s}$$

The change in level from t_0 to $t_0 + 100 = (80 - 60)(100)/703 = 2.83\%$ of FS.

$$h^*(t_0 + 100) = h^*(t_0) + \text{change in level}$$

$$= 22.5 + 2.83 = 25.33\% \text{ of FS}$$

$$= \frac{(25.33)(4)}{100} = 1.013 \text{ m}$$

$$= 25.33\% \text{ of FS} = 1.013 \text{ m} \qquad ◆$$

14.3 THE FIRST-ORDER LAG PROCESS

The *first-order lag process* consists of a single capacitive element configured such that the outflow of material or energy is proportional to the amount of material or energy stored in the capacitive element. For each input flow rate, there is a corresponding amount of stored material or energy that will produce an output flow rate equal to the input. The first-order lag process is a self-regulating process because it automatically produces an output flow rate to match each input rate. In contrast, the integral process is a non-self-regulating process.

The *step response* of the first-order lag process is shown in Figure 14.4. Before the step change, the input flow rate is equal to the output flow rate and the level is maintained constant. The step change consists of increasing the input flow rate. Let M represent the percentage increase in the input flow rate. The output flow rate is proportional to the level, which does not change immediately. The input is greater than the output, so the level will increase at a rate proportional to the difference. As the level increases, the difference between the input and the output decreases. This in turn reduces the rate at which the level increases. The result is the output response curve of Figure 14.4.

The step response of a first-order lag is measured by the *time constant,* which is the number of seconds (or minutes) required for the output to reach 63.2% of the total change. During each additional interval equal to the time constant, the output will reach 63.2% of the remaining change. In Figure 14.4, for example, the input step change is equal to M. During the first time constant interval, the output changed by an amount equal to $0.632M$, and the remaining change was $M - 0.632M = 0.368M$. During the second time interval, the output increased by $0.632(0.368M) = 0.232M$. Thus after two time constants, the output will reach $0.632M + 0.232M = 0.864M$. After five constants, the output will reach $0.993M$.

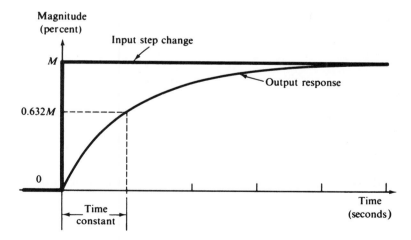

◆ **Figure 14.4**   The step response of a first-order lag process is characterized by the time constant τ. From any point on the graph, the output will cover 63.2% of the remaining distance to the final value in one time constant.

A liquid level, first-order lag, and three electrical first-order lag processes are illustrated in Figure 14.5. The *liquid level process* in Figure 14.5a was analyzed in Chapter 4 [see the development of Equation (4.6)]. The input of the liquid level process is the input flow rate, q_{in}, and the output is the level, h, of the liquid in the tank.

The *series RC circuit* in Figure 14.5b was also analyzed in Chapter 4 [see the development of Equation 4.10)]. In this circuit the input is the source voltage, e_{in}, and the output is the voltage across the capacitor, e_{out}.

In the *parallel RC circuit* in Figure 14.5c, the input is the source current, i_{in}, and the output is the voltage across the capacitor, e_{out}. The analysis of the parallel RC circuit begins by applying Kirchhoff's current law to the top node of the circuit.

$$i_{in} = i_R + i_C$$

But $i_R = e_{out}/R$ and $i_C = C\, de_{out}/dt$:

$$i_{in} = \frac{e_{out}}{R} + C\frac{de_{out}}{dt}$$

$$RC\frac{de_{out}}{dt} + e_{out} = Ri_{in} \tag{14.3}$$

Equation (14.3) is the equation of a first-order lag process with a time constant equal to RC and a gain equal to R.

The third electrical circuit is the *series RL circuit* shown in Figure 14.5d. In this circuit, the input is the source voltage, e_{in}, and the output is the voltage across the resistor, e_{out}. The analysis begins with Kirchhoff's voltage law.

$$e_{in} = e_L + e_R$$

But

$$e_L = L\frac{di}{dt}$$

and

$$i = \frac{e_{out}}{R}$$

Therefore,

$$\frac{di}{dt} = \frac{1}{R}\frac{de_{out}}{dt}$$

Substituting and rearranging gives the following equation for the *RL* circuit:

$$\frac{L}{R}\frac{de_{out}}{dt} + e_{out} = e_{in} \tag{14.4}$$

Equation (14.4) is the equation of a first-order lag process with a time constant equal to L/R and a steady-state gain equal to 1.

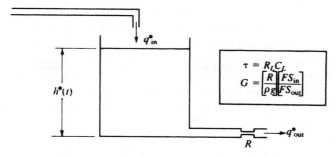

a) A liquid level, first-order lag process

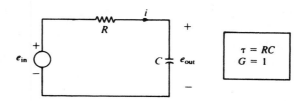

b) A series RC, first-order lag component

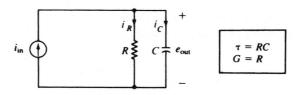

c) A parallel RC, first-order lag component

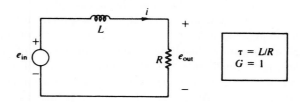

d) A series RL, first-order lag component

◆ **Figure 14.5** Liquid flow and electrical first-order lag processes.

A *thermal, first-order lag process* is shown in Figure 14.6a. A jacketed kettle is used to heat a liquid. The mixer maintains a uniform temperature throughout the liquid. The input to the process is the jacket temperature, θ_j, and the output is the liquid temperature, θ_L. The increase in the liquid temperature, $\Delta\theta_L$ is equal to the amount of heat transferred to the liquid divided by the thermal capacitance, C_T, of the liquid in the tank. The amount of heat, Δ_q, transferred to the liquid depends on the thermal resistance, R_T, of the wall between the steam and the liquid, the temperature difference $(\theta_j - \theta_L)$ between the jacket and the liquid, and the time interval, Δt.

$$\Delta\theta_L = \frac{\Delta q}{C_T} = \frac{1}{R_T C_T}(\theta_j - \theta_L)\,\Delta t$$

$$R_T C_T \frac{\Delta\theta_L}{\Delta t} + \theta_L = \theta_j$$

or as $\Delta t \to 0$,

$$R_T C_T \frac{d\theta_L}{dt} + \theta_L = \theta_j \tag{14.5}$$

Equation (14.5) is the equation of a first-order lag process with a time constant equal to $R_T C_T$ and a steady-state gain equal to 1.

A *gas-pressure first-order lag process* is shown in Figure 14.6b. In this process the input is the mass flow rate of gas entering the tank, w_{in}, and the output is the pressure, p, of the gas in the tank. The gas process is described by the following equation, in which R_g is the gas flow resistance of the outlet, and C_g is the capacitance of the tank:

$$R_g C_g \frac{dp}{dt} + p = R w_{in} \tag{14.6}$$

Equation (14.6) is the equation of a first-order lag process with a time constant equal to $R_g C_g$ and a steady-state gain equal to 1.

A *blending, first-order lag process* is illustrated in Figure 14.6c. A constant flow rate of q m³/s passes through the tank. The input of the process is the concentration, c_i, of component A in the incoming fluid. The output of the process is the concentration, c_o, of component A in the fluid in the tank (and the outgoing fluid). A material balance is used to determine the time-domain equation. In a time interval Δt, an amount of liquid equal to $q(\Delta t)$ is added at the inlet and an equal amount is removed at the outlet. The amount of component A added to the tank in the incoming fluid is equal to $q(\Delta t)c_i$. The amount of component A removed in the outgoing fluid is $q(\Delta t)c_o$. The difference is the increase of component A in the tank, which is equal to $V(\Delta c_o)$.

Amount of buildup = amount inputed − amount removed

$$V(\Delta c_o) = q(\Delta t)c_i - q(\Delta t)c_o$$

$$\frac{V}{q}\frac{\Delta c_o}{\Delta t} + c_o = c_i$$

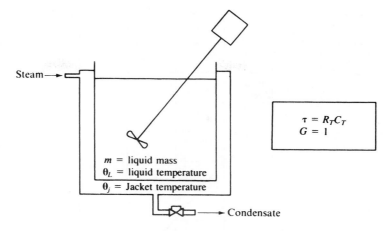

$$\tau = R_T C_T$$
$$G = 1$$

m = liquid mass
θ_L = liquid temperature
θ_j = Jacket temperature

a) A thermal, first-order lag process

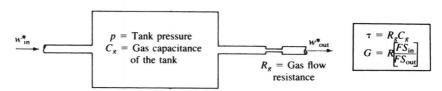

w^*_{in}

p = Tank pressure
C_g = Gas capacitance
of the tank

w^*_{out}

R_g = Gas flow
resistance

$$\tau = R_g C_g$$
$$G = R\left[\frac{FS_{in}}{FS_{out}}\right]$$

b) A gas pressure, first-order lag process

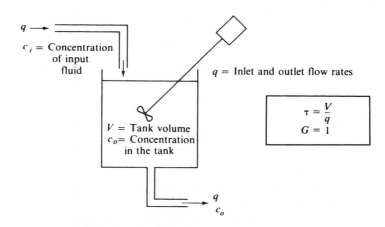

q

c_i = Concentration
of input
fluid

q = Inlet and outlet flow rates

V = Tank volume
c_o = Concentration
in the tank

$$\tau = \frac{V}{q}$$
$$G = 1$$

q
c_o

c) A blending, first-order lag process

◆ **Figure 14.6** Thermal, gas pressure, and blending first-order lag processes.

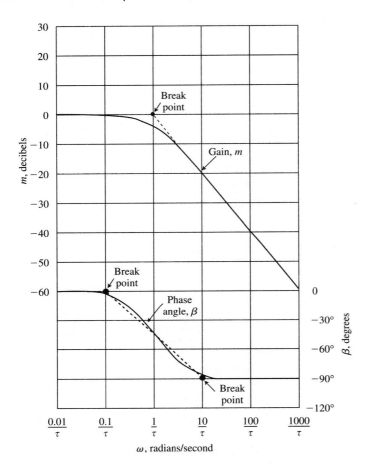

◆ **Figure 14.7** Bode diagram of a first-order lag process with a steady-state gain of 1.

or as $\Delta_t \to 0$,

$$\frac{V}{q}\frac{dc_o}{dt} + c_o = c_i \qquad (14.7)$$

Equation (14.7) is the equation of a first-order lag process with a time constant equal to V/q and a steady-state gain equal to 1.

The *Bode diagram* of a first-order lag process with a steady-state gain of 1 is shown in Figure 14.7. The gain line has a break at a radian frequency equal to the reciprocal of the time constant. This frequency is called the break-point frequency, ω_b.

$$\omega_b = \frac{1}{\tau} \qquad (14.8)$$

The phase line has two breaks, one on each side of the break-point frequency, ω_b. A gain different from one raises or lowers the gain line, but does not change the shape of the line. The phase line is unaffected by the gain of the process.

FIRST-ORDER LAG PROCESS

Time-Domain Equation

$$\tau \frac{dy}{dt} + y = Gx \qquad\qquad (14.9)$$

Transfer Function

$$\frac{Y}{X} = \frac{G}{1 + \tau s} \qquad\qquad (14.10)$$

where G = steady-state gain of the process

t = time, s

x = input to the process*

y = output of the process†

τ = time constant, s

Liquid Level First-Order Lag Process (Figure 14.5a.)

Process input: q_{in}^{*} = input flow rate, percentage of FS_{in}

(FS_{in} = input range, m³/s)

Process output: h^{*} = level, percentage of FS_{out}

FS_{out} = output range, m)   (14.11)

$$\tau = R_L C_L$$

$$G = \left(\frac{R_L}{\rho g}\right)\left(\frac{FS_{in}}{FS_{out}}\right) \qquad\qquad (14.12)$$

R_L = liquid resistance, Pa · s/m³

C_L = liquid capacitance, m³/Pa

ρ = liquid density, kg/m³

g = 9.81 m/s²

Electrical First-Order Lag Processes (Figure 14.5)

Process input: e_{in} = input voltage (Figure 14.5b, d)

i_{in} = input current (Figure 14.5c)

Process output: e_{out} = output voltage

$$\tau = RC \text{ (Figure 14.5b, c)} \qquad\qquad (14.13)$$

$$\tau = \frac{L}{R} \text{ (Figure 14.5d)}$$

---

*The input may be expressed with units or as a percentage of FS_{in}.
FS_{in} = full-scale range of the input.
†The output may be expressed with units or as a percentage of FS_{out}.
FS_{out} = full-scale range of the output.

$$G = 1 \text{ (Figure 14.5b, d)} \tag{14.14}$$

$$G = R \text{ (Figure 14.5c)} \tag{14.15}$$

$$R = \text{electrical resistance, } \Omega$$

$$C = \text{electrical capacitance, F}$$

Thermal First-Order Lag Process (Figure 14.6a)

Process input: θ_j = jacket temperature, °C

Process output: θ_L = liquid temperature, °C

$$\tau = R_T C_T \tag{14.16}$$

$$G = 1 \tag{14.14}$$

$$R_T = \text{thermal resistance, K/W}$$

$$C_T = \text{thermal capacitance, J/K}$$

Gas Pressure First-Order Lag Process (Figure 14.6b)

Process input: w_{in}^* = input flow rate, percentage of FS_{in}

(FS_{in} = input range, kg/s)

Process output: p^* = pressure in tank, percentage of FS_{out}

(FS_{out} = output range, Pa)

$$\tau = R_g C_g \tag{14.17}$$

$$G = R\left(\frac{\text{FS}_{\text{in}}}{\text{FS}_{\text{out}}}\right) \tag{14.18}$$

$$R_g = \text{gas flow resistance, Pa} \cdot \text{s/kg}$$

$$C_g = \text{gas flow capacitance, kg/Pa}$$

Blending First-Order Lag Process (Figure 14.6c)

Process input: c_i = input concentration, %

Process output: c_o = output concentration, %

$$\tau = \frac{V}{q} \tag{14.19}$$

$$G = 1 \tag{14.14}$$

$$q = \text{flow rate of the liquid, m}^3\text{/s}$$

$$V = \text{volume of the tank, m}^3$$

EXAMPLE
14.2

An oil tank similar to Figure 14.5a has a diameter of 1.25 m and a height of 2.8 m. The outlet at the bottom is smooth tubing with a length of 5 m and a diameter of 2.85 cm. The oil temperature is 15°C. The full-scale ranges are $FS_{in} = 4.0 \times 10^{-4}$ m^3/s (24 L/min) and $FS_{out} = 2.8$ m. Determine each of the following:

(a) The capacitance of the tank.
(b) The resistance of the outlet.
(c) The time constant of the process.
(d) The gain of the process.
(e) The time-domain equation.
(f) The transfer function.

Solution

(a) Equation (3.20) in Chapter 3 may be used to compute the tank capacitance.

$$C_L = \frac{A}{\rho g}$$

$$A = \frac{\pi D^2}{4} = \frac{\pi (1.25)^2}{4} = 1.23 \text{ m}^2$$

$$\rho = 880 \text{ kg/m}^3 \text{ (Appendix A)}$$

$$C = \frac{1.23}{880 \times 9.81}$$

$$= 1.4215\text{E} - 4 \text{ cm/Pa}$$

(b) The program LIQRESIS may be used to compute the liquid flow resistance. From Appendix A, the absolute viscosity of oil is 0.160 Pa · s, and the density is 880 kg/m^3. Using a flow rate of 24 LPM, LIQRESIS gives the following results:

Reynolds number $= 98$
Flow is laminar
$R_L = 4.941\text{E} + 7$ Pa · s/m^3
$p = 19.8$ kPa

(c) The time constant, $\tau = R_L C_L$

$$\tau = (4.941\text{E} + 7)(1.4215\text{E} - 4)$$

$$= 7024 \text{ s (or 117 min)}$$

(d) The process gain,

$$G = \left(\frac{R_L}{\rho g}\right)\left(\frac{FS_{in}}{FS_{out}}\right)$$

$$= \left(\frac{4.941\text{E} + 7}{880 \times 9.81}\right)\left(\frac{4.0\text{E} - 4}{2.8}\right) = 0.8176$$

(e) Equation (14.9) gives the time-domain equation.

$$7024\frac{dh^*}{dt} + h^* = 0.818 q_{in}^*$$

(f) Equation (14.10) gives the transfer function.

$$\frac{H^*(s)}{Q_{in}^*(s)} = \frac{0.818}{1 + 7024s}$$

◆

EXAMPLE 14.3

An electrical circuit similar to Figure 14.5b has an 8.2-kΩ resistance value and a 60-μF capacitance value. Determine the following:

(a) The time constant.
(b) The transfer function.

Solution

(a) Time constant,

$$\tau = RC$$

$$= (8.2E + 3)(60E - 6) = 0.492 \text{ s}$$

(b) The transfer function is

$$\frac{E_{out}(s)}{E_{in}(s)} = \frac{1}{1 + 0.492s}$$

◆

EXAMPLE 14.4

An oil-bath thermal process similar to Figure 14.6a has an inside diameter of 1 m and a height of 1.2 m. The inside film coefficient is 62 W/m² · K, and the outside film coefficient is 310 W/m² · K. The tank wall is a single layer of steel, 1.2 cm thick. Determine each of the following:

(a) The thermal resistance.
(b) The thermal capacitance.
(c) The time constant.
(d) The time-domain equation.
(e) The transfer function.

Solution

(a) Determine the thermal resistance, R_T.

From Appendix A, $K = 45$ W/m · K for steel

Thickness of the steel wall = 0.012 m

$$A = \frac{\pi D^2}{4} + \pi D h = \pi D \left(\frac{D}{4} + h \right)$$

$$= \pi(1) \left(\frac{1}{4} + 1.2 \right) = 4.5553 \text{ m}^2$$

From Equation (3.38) in Chapter 3.

$$R_L = \frac{(1/h_i + x/k + 1/h_0)}{A}$$

$$= \frac{1/62 + 0.012/45 + 1/310)}{4.5553} = 4.307E - 3 \text{ K/W}$$

(b) Equation (3.39) in Chapter 3 may be used to compute the thermal capacitance.

$$C = mS_k$$

$$S_h = 2180 \text{ J/kg} \cdot \text{K (Appendix A)}$$

$$\rho = 880 \text{ kg/m}^3 \text{ (Appendix A)}$$

$$m = \frac{\rho h \pi D^2}{4} = \frac{(800)(1.2)(\pi)(1)^2}{4} = 829.4 \text{ kg}$$

$$C = (829.4)(2180) = 1.808E + 6 \text{ J/K}$$

(c) Time constant $= \tau = R_T C_T$.

$$\tau = (4.30E - 3)(1.808E + 6)$$

$$= 7788 \text{ s (or 129.8 min)}$$

(d) The time-domain equation is

$$7788 \frac{d\theta_L}{dt} + \theta_L = \theta_j$$

(e) The transfer function is

$$\frac{\Theta_L(s)}{\Theta_j(s)} = \frac{1}{1 + 7788s}$$

◆

EXAMPLE 14.5

A carbon dioxide (CO_2) gas pressure process similar to Figure 14.6b has the following parameters:

Pressure vessel volume: 1.4 m^3
Temperature: 530 K
Gas flow resistance: $2 \times 10^5 \text{ Pa} \cdot \text{s/kg}$
$FS_{in} = 2 \text{ kg/s}$
$FS_{out} = 200 \text{ kPa}$

Determine each of the following.

(a) The capacitance of the pressure vessel.
(b) The time constant.
(c) The process gain.
(d) The time-domain equation.
(e) The transfer function.

Solution

(a) Equation (3.31) in Chapter 3 may be used to determine the capacitance of the pressure vessel.

$$C = \frac{1.2E - 4 \, MV}{T}$$

$$M = 44 \text{ for } CO_2 \text{ (Appendix A)}$$

$$V = 1.4 \text{ m}^3$$

$$T = 530 \text{ K}$$

$$C = \frac{(1.2E - 4)(44)(1.4)}{530}$$

$$= 1.3947E - 5 \text{ kg/Pa}$$

(b) The time constant, $\tau = R_g C_g$

$$\tau = (2E + 5)(1.3947E - 5) = 2.79 \text{ S}$$

(c)
$$G = R\left(\frac{FS_{in}}{FS_{out}}\right) = (2E + 5)\left(\frac{2}{2E + 5}\right)$$
$$= 2$$

(d) The time-domain equation is

$$2.79\frac{dp^*}{dt} + p^* = 2w^*$$

(e) The transfer function is

$$\frac{P^*(s)}{W^*(s)} = \frac{2}{1 + 2.79s}$$ ◆

EXAMPLE 14.6   A blending tank similar to Figure 14.6c has a tank volume of 3.1 m^3 and a flow rate of 0.0031 m^3/s. Determine the following:

(a) The time constant.
(b) The time-domain equation.
(c) The transfer function.

Solution   (a) The time constant,

$$\tau = \frac{V}{q}$$

$$= \frac{3.1}{0.0031} = 1000 \text{ s}$$

(b) The time-domain equation is

$$1000\left(\frac{dc_o}{dt}\right) + c_o = c_i$$

(c) The transfer function is

$$\frac{C_o(s)}{C_i(s)} = \frac{1}{1 + 1000s}$$ ◆

14.4 THE SECOND-ORDER LAG PROCESS

A *second-order lag process* has two capacitance elements, one capacitance and one inertial element, or two inertial elements. (The inertial elements are mass, inductance, and inertance). Three parameters characterize the response of a second-order system. The first parameter is the resonant frequency, denoted by ω_0. The second parameter is the amount of damping in the process, expressed by either the damping coefficient, α, or the damping ratio, ζ. The damping ratio is simply the damping coefficient divided by the resonant frequency ($\zeta = \alpha/\omega_0$). The third parameter is the steady-state gain, G.

The step response of a second-order system is divided into three types depending on the value of the damping ratio, ζ. If the damping ratio is greater than 1, the response is overdamped as shown in Figure 14.8a. If the damping ratio is less than 1, the response is underdamped as shown in Figure 14.8c. If the damping ratio is equal to 1, the response is critically damped as shown in Figure 14.8b. Notice the fast rise and the oscillatory nature of the underdamped response contrasted with the slow rise and the lack of oscillation of the overdamped response. The critically damped response has the fastest rise possible with no oscillation.

   A *mechanical spring-mass-damping second-order process* is shown in Figure 14.9a. In this process the applied force, f, is the input signal, and the position of the moving body, h, is the output signal. Three forces act on the moving body: the force exerted by the spring, f_s, the

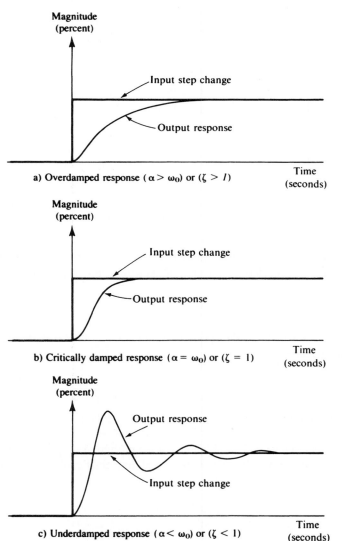

a) Overdamped response ($\alpha > \omega_0$) or ($\zeta > 1$)

b) Critically damped response ($\alpha = \omega_0$) or ($\zeta = 1$)

c) Underdamped response ($\alpha < \omega_0$) or ($\zeta < 1$)

◆ **Figure 14.8** The step response of a second-order process is overdamped, critically damped, or underdamped, depending on the damping ratio, $\zeta = \alpha/\omega_0$.

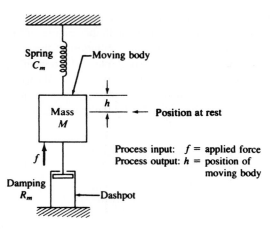

a) A spring-mass-damping second-order process

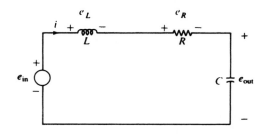

b) A series *RLC* second-order process

◆ **Figure 14.9** Mechanical and electrical second-order processes.

force exerted by the damping piston, f_r, and the externally applied force, f. The sum of these three forces will cause the mass to accelerate according to Newton's law of motion.

$$f + f_s + f_r = M \frac{d^2 h}{dt^2}$$

The spring force, f_s, is equal to the distance the spring is compressed or extended, divided by the mechanical capacitance of the spring.

$$f_s = -\frac{h}{C_m}$$

The negative sign in the equation for f_s indicates that the direction of the force is down when h is positive (up) and up when h is negative (down).

The damping force is equal to mechanical resistance of the dashpot multiplied by the velocity of the dashpot piston. The piston velocity is represented mathematically by the derivative of h with respect to time (i.e., dh/dt).

$$f_r = -R_m \frac{dh}{dt}$$

The negative sign in the equation for f_r indicates that the force is down when the piston velocity is positive (moving up) and the force is up when the piston velocity is negative (moving down).

Substituting the last two equations into the first equation, and rearranging the terms gives the following time-domain equation for a spring-mass-damping process.

$$MC_m \frac{d^2h}{dt^2} + R_m C_m \frac{dh}{dt} + h = C_m f \qquad (14.20)$$

where C_m = capacitance of the spring, m/N
         f = externally applied force, N
         h = position of the moving body, m
         M = mass of the moving body, kg
         R_m = dashpot resistance, N · s/m

The circuit shown in Figure 14.9b is the electrical analog of the spring-mass-damping process. This can be easily seen by comparing Equation (14.20) with the following equation for the series *RLC* circuit:

$$LC \frac{d^2 e_{out}}{dt^2} + RC \frac{de_{out}}{dt} + e_{out} = e_{in} \qquad (14.21)$$

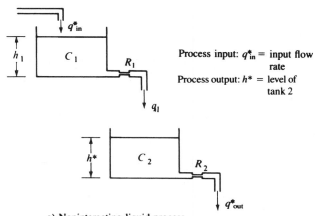

Process input: q^*_{in} = input flow rate
Process output: h^* = level of tank 2

a) Noninteracting liquid process

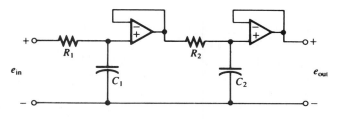

b) Noninteracting electrical process

◆ **Figure 14.10**   Two-capacitance noninteracting second-order processes.

A *noninteracting second-order process* has two capacitive elements configured such that the second capacitive element has no effect on the first capacitive element. Equation (14.22) is the time-domain equation of a noninteracting process with input signal, x, and output signal, y.

$$\tau_1\tau_2 \frac{d^2y}{dt^2} + (\tau_1 + \tau_2)\frac{dy}{dt} + y = Gx \tag{14.22}$$

where

$$\tau_1 = R_1C_1 \quad \text{and} \quad \tau_2 = R_2C_2$$

Figure 14.10 shows examples of liquid and electrical noninteracting second-order processes. The input signal of the liquid process is the input flow rate to the first tank, q_{in}, and the output signal is the level of the second tank, h. The input signal of the electrical circuit is the input voltage, e_{in}, and the output signal is the output voltage, e_{out}. The time-domain equations and the transfer functions of the liquid and electrical noninteracting processes are in the boxed summary just before the examples at the end of this section.

An *interacting second-order process* has two capacitive elements configured such that the second capacitive element does have an interactive effect on the first capacitive element. Equation (14.23) is the time-domain equation of an interacting, second-order process with input signal, x, and output signal, y.

$$A_2 \frac{d^2y}{dt^2} + A_1 \frac{dy}{dt} + y = Gx \tag{14.23}$$

$$G = \frac{R_L}{R_1 + R_2 + R_L}$$

$$A_1 = G\left[(\tau_1 + \tau_2) + \tau_1 \frac{R_2}{R_L} + \tau_2 \frac{R_1}{R_2}\right]$$

$$A_2 = G\tau_1\tau_2$$

Figure 14.11 shows examples of liquid and electrical interacting second-order processes. The input signal of the liquid process is the inlet pressure, p_{in}, and the output signal is the outlet pressure in the second tank, p_{out}. The input signal of the electrical circuit is the input voltage, e_{in}, and the output signal is the output voltage, e_{out}. The time-domain equations and the transfer functions of the liquid and electrical interacting, second-order processes are also in the boxed summary just before the examples at the end of this section.

As a final illustration of a second-order process, we will examine the dc motor–driven load shown in Figure 14.12. The input signal is the voltage, e, applied to the armature of the dc motor. The output signal is the rotational velocity of the load, ω. Three equations define this process. The first equation is the electrical equation of the armature circuit.

$$L\frac{di}{dt} + R_i + K_e\omega = e \tag{14.24}$$

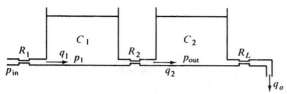

Process input: p_{in} = inlet pressure
Process output: p_{out} = pressure in tank 2

a) An interacting liquid process

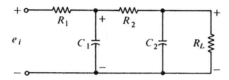

◆ **Figure 14.11**   Two-capacitance interacting
second-order processes.

b) An interacting electrical process

The second equation is the current-versus-torque equation of the dc motor.

$$\text{Torque} = K_t i$$

The third equation is the mechanical equation of the load.

$$\text{Torque} = J\frac{d\omega}{dt} + B\omega$$

Combining the last two equations gives us the following:

$$i = \frac{J(d\omega\ dt) + B\omega}{K_t}$$

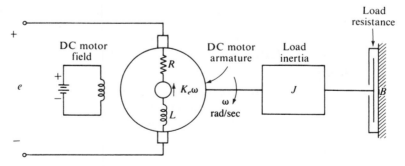

Process input: e = armature voltage, volts
Process output: ω = load speed, radian/second

◆ **Figure 14.12**   A dc motor–driven load is a second-order process.

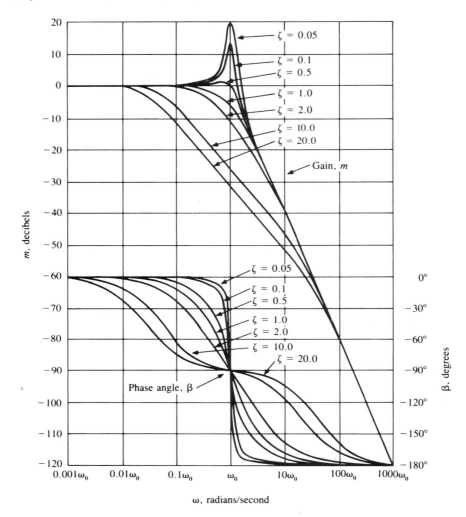

◆ **Figure 14.13** The Bode diagram of a second-order process varies considerably for different values of the damping ratio, $\zeta = \alpha/\omega_0$.

The time-domain equation of the process is obtained by substituting the right-hand side of the last equation for i in the first equation. The result can be reduced to Equation (14.25), the time-domain equation of the dc motor-driven load.

$$A_2 \frac{d^2\omega}{dt^2} + A_1 \frac{d\omega}{dt} + \omega = Ge \qquad (14.25)$$

where $\tau_m = \dfrac{J}{B} = $ mechanical time constant

$\tau_e = \dfrac{L}{R} = $ electrical time constant

$$A_2 = \frac{\tau_m \tau_e RB}{K_e K_t + RB}$$

$$A_1 = \frac{(\tau_m + \tau_e)RB}{K_e K_t + RB}$$

$$G = \frac{K_t}{K_e K_t + RB}$$

J = moment of inertia of the load, kg · m^2

B = damping resistance, N · m · s/rad

L = armature inductance, H

R = armature resistance, Ω

K_t = torque constant, N · m/A

e = armature voltage, V

ω = motor speed, rad/s

K_e = EMF constant, V · s/rad

The *Bode diagram* of a second-order lag process is shown in Figure 14.13. The frequency scale is normalized in terms of the resonant frequency, and the gain of the process is 1. A range of values of damping ratio from 0.05 to 20 illustrates the pronounced effect that damping has on both the gain and phase angle, especially at the resonant frequency. Notice the amplification at the resonant frequency when the damping ratio is 0.1 and 0.05. The process with $\zeta = 0.05$ has an output at the resonant frequency that is 10 times as large as the input. The effect of the steady-state gain, G, is to raise or lower the gain line without changing the shape of the line. The steady-state gain has no effect on the phase angle line.

SECOND-ORDER LAG PROCESS

Time-Domain Equation

$$A_2 \frac{d^2 y}{dt^2} + A_1 \frac{dy}{dt} + y = Gx \qquad\qquad \textbf{(14.26)}$$

Transfer Function

$$\frac{Y(s)}{X(s)} = \frac{G}{1 + A_1 s + A_2 s^2} \qquad\qquad \textbf{(14.27)}$$

Parameters

$$\omega_0 = \sqrt{\frac{1}{A_2}} \qquad\qquad \textbf{(14.28)}$$

$$\alpha = \frac{A_1}{2A_2} \qquad\qquad \textbf{(14.29)}$$

$$\zeta = \frac{\alpha}{\omega_0} = \frac{A_1}{2\sqrt{A_2}} = \frac{A_1 \omega_0}{2} \qquad\qquad \textbf{(14.30)}$$

$$A_2 = \frac{1}{\omega_0^2} \tag{14.31}$$

$$A_1 = \frac{2\zeta}{\omega_0} = \frac{2\alpha}{\omega_0^2} \tag{14.32}$$

where G = steady-state gain
  x = input signal
  y = output signal
  α = damping coefficient, s^{-1}
  ω_0 = resonant frequency, rad/s
  ζ = damping ratio

Mechanical Second-Order Process (Figure 14.9a)

Process input: f = applied force, N

Process output: h = position of mass, m

$$A_2 = MC_m \tag{14.33}$$

$$A_1 = RC_m \tag{14.34}$$

$$G = C_m \tag{14.35}$$

M = mass of the moving body, kg

R = damping resistance, N · s/m

C_m = capacitance of the spring, m/N

Electrical Second-Order Process (Figure 14.9b)

Process input: e_{in} = input voltage, V

Process output: e_{out} = capacitor voltage, V

$$A_2 = LC \tag{14.36}$$

$$A_1 = RC \tag{14.37}$$

$$G = 1 \tag{14.38}$$

L = electrical inductance, H

C = electrical capacitance, F

R = electrical resistance, Ω

Noninteracting Second-Order Process (Figure 14.10)

$$\tau_1 = R_1 C_1 \tag{14.39}$$

$$\tau_2 = R_2 C_2 \tag{14.40}$$

$$A_2 = \tau_1 \tau_2 \tag{14.41}$$

$$A_1 = \tau_1 + \tau_2 \tag{14.42}$$

1. Liquid level noninteracting process

> Process input: q_{in}^* = input flow rate, percentage of FS_{in}
>
> Process output: h^* = level in tank 2, percentage of FS_{out}

$$G = \left(\frac{R_2}{\rho g}\right)\left(\frac{FS_{in}}{FS_{out}}\right) \tag{14.43}$$

> R_1, R_2 = liquid resistance, Pa · s/m³
>
> C_1, C_2 = liquid capacitance, m³/Pa

2. Electrical noninteracting process

> Process input: e_{in} = input voltage, V
>
> Process output: e_{out} = voltage across C_2, V

$$G = 1 \tag{14.44}$$

> R_1, R_2 = electrical resistance, Ω
>
> C_1, C_2 = electrical capacitance, Ω

Interacting Second-Order Process (Figure 14.11)

$$G = \frac{R_L}{R_1 + R_2 + R_L} \tag{14.45}$$

$$A_2 = \tau_1\tau_2 G \tag{14.46}$$

$$A_1 = \left(\tau_1 + \tau_2 + \tau_1\frac{R_2}{R_L} + \tau_2\frac{R_1}{R_2}\right)G \tag{14.47}$$

1. Liquid level interacting process

> Process input: p_{in} = inlet pressure, Pa
>
> Process output: p_{out} = pressure in tank 2, Pa

2. Electrical interacting process

> Process input: e_{in} = input voltage, V
>
> Process output: e_{out} = voltage across R_L, V

DC Motor Second-Order Process (Figure 14.12)

> Process input: e = armature voltage, V
>
> Process output: ω = rotational velocity of the load, rad/s

$$A_2 = \frac{\tau_m\tau_e RB}{RB + K_e K_t} \tag{14.48}$$

$$A_1 = \frac{(\tau_m + \tau_e)RB}{RB + K_e K_t} \tag{14.49}$$

$$G = \frac{K_t}{RB + K_eK_t} \tag{14.50}$$

where $\tau_m = \dfrac{J}{B}$, mechanical time constant, s

$\tau_e = \dfrac{L}{R}$, electrical time constant, s

$B =$ damping resistance, N · m · s/rad

$e =$ armature voltage, V

$J =$ moment of inertia of the load, kg · m^2

$K_e =$ EMF constant, V · s/rad

$K_t =$ torque constant, N · m/A

$L =$ armature inductance, H

$R =$ armature resistance, Ω

$\omega =$ motor speed, rad/s

EXAMPLE 14.7   A spring-mass-damping process consists of a 10-kg mass, a spring capacitance of 0.001 m/N, and a damping resistance of 20 N · s/m. Determine the following:

(a) The time-domain equation.
(b) The transfer function.
(c) The resonant frequency (ω_0).
(d) The damping ratio (ζ).
(e) Whether the process is overdamped, underdamped, or critically damped.

Solution   (a) The time domain is given by Equation (14.26) and Equations (14.33) through (14.35).

$$A_2 = MC_m = (10)(0.001) = 0.01$$

$$A_1 = RC_m = (20)(0.001) = 0.02$$

$$G = C_m = 0.001$$

$$0.01\frac{d^2h}{dt^2} + 0.02\frac{dh}{dt} + h = 0.001f$$

(b) Equation (14.27) gives the transfer function.

$$\frac{H(S)}{F(S)} = \frac{0.001}{1 + 0.02s + 0.01s^2}$$

(c) From Equation (14.28),

$$\omega_0 = \sqrt{\frac{1}{0.01}} = 10 \text{ rad/s}$$

(d) From Equation (14.30),

$$\zeta = \frac{0.02}{2\sqrt{0.01}} = 0.1$$

(e) The damping ratio is less than 1; therefore, the process is underdamped.   ◆

EXAMPLE 14.8

An electrical series RLC circuit consists of a 0.022-H inductor, a 10-μF capacitor, and a 200-Ω resistance. Determine the following:

(a) The time-domain equation.
(b) The transfer function.
(c) The resonant frequency.
(d) The damping ratio.
(e) The type of damping.

Solution

(a) The time-domain equation is given by Equation (14.26) and Equations (14.36) through (14.38).

$$A_2 = LC = (0.022)(10E - 6) = 2.2E - 7$$

$$A_1 = RC = (200)(10E - 6) = 0.002$$

$$2.2 \times 10^{-7}\left(\frac{d^2 e_{out}}{dt^2}\right) + 0.002\left(\frac{de_{out}}{dt}\right) + e_{out} = e_{in}$$

(b) The transfer function is given by Equation (14.27).

$$\frac{E_{out}(s)}{E_{in}(s)} = \frac{1}{1 + 0.002s + 2.2 \times 10^{-7} s^2}$$

(c) From Equation (14.28),

$$\omega_0 = \sqrt{\frac{1}{2.2E - 7}} = 2.13 \times 10^3 \text{ rad/s}$$

(d) From Equation (14.30),

$$\zeta = \frac{(2.0E - 3)(2.13E + 3)}{2} = 2.13$$

(e) The damping ratio is greater than 1; therefore, the process is overdamped.   ◆

EXAMPLE 14.9

A liquid, noninteracting, two-capacity system has time constants τ_1 and τ_2 of 520 s and 960 s. The liquid is water and the value of R_2 is 1.6×10^{-6} Pa · s/m^3. The full-scale ranges are $FS_{in} = 5.0 \times 10^{-3}$ m^3/s and $FS_{out} = 2.0$ m. Determine the following:

(a) The time-domain equation.
(b) The transfer function.

Solution

(a) The time-domain equation is given by Equation (14.26) and Equations (14.41) through (14.43).

$$A_2 = \tau_1 \tau_2 = (520)(960) = 5E + 5 \text{ s}^2$$

$$A_1 = \tau_1 + \tau_2 = 520 + 960 = 1480 \text{ s}$$

$$\rho = 1000 \text{ kg/m}^3 \text{ (Appendix A)}$$

$$G = \left[\frac{1.6E + 6}{(1000)(9.81)}\right]\left(\frac{5.0E - 3}{2}\right) = 0.41$$

$$500{,}000\,\frac{d^2 h^*}{dt^2} + 1480\,\frac{dh^*}{dt} + h^* = 0.41 q^*_{in}$$

(b) The transfer function is given by Equation (14.27).

$$\frac{H^*(s)}{Q^*_{in}(s)} = \frac{0.41}{1 + 1480s + 500{,}000s^2}$$

◆

EXAMPLE 14.10

An electrical, interacting, two-capacity process (Figure 14.11b) has the following component values:

$$R_1 = 1000 \ \Omega$$
$$R_2 = 800 \ \Omega$$
$$R_L = 400 \ \Omega$$
$$C_1 = 10 \ \mu F$$
$$C_2 = 50 \ \mu F$$

Determine the following:

(a) The time-domain equation.
(b) The transfer function.

Solution

(a) The time-domain equation is given by Equations (14.26), (14.39), (14.40), (14.45), (14.46), and (14.47).

$$\tau_1 = (1000)(10E - 6) = 0.01 \ s$$

$$\tau_2 = (800)(50E - 6) = 0.04 \ s$$

$$G = \frac{400}{1000 + 800 + 400} = 0.1818$$

$$A_2 = (0.01)(0.04)(0.182) = 7.27 \times 10^{-5}$$

$$A_1 = \left[0.01 + 0.04 + 0.01\left(\frac{800}{400}\right) + 0.04\left(\frac{1000}{800}\right)\right] 0.1818$$

$$= 0.0218$$

The time-domain equation is

$$7.27 \times 10^{-5} \frac{d^2 e_{out}}{dt^2} + 0.0218 \frac{de_{out}}{dt} + e_{out} = 0.182 e_{in}$$

(b) The transfer function is

$$\frac{E_{out}(s)}{E_{in}(s)} = \frac{0.182}{1 + 0.0218s + 7.27 \times 10^{-5}s^2}$$

◆

EXAMPLE 14.11

An armature-controlled dc motor has the following characteristics:

$$J = 6.2 \times 10^{-4} \ kg \cdot m^2$$
$$B = 1.0 \times 10^{-4} \ N \cdot m \cdot s/rad$$
$$L = 0.020 \ H$$
$$R = 1.2 \ \Omega$$
$$K_e = 0.043 \ V \cdot s/rad$$
$$K_t = 0.043 \ N \cdot m/A$$

Determine the following:

(a) The mechanical time constant.
(b) The electrical time constant.
(c) The time-domain equation.
(d) The transfer function.

Solution

(a) $\tau_m = J/B = 6.2\text{E} - 4/1.0\text{E} - 4 = 6.2$ s.

(b) $\tau_e = L/R = 0.020/1.2 = 0.0167$ s.

(c) The time-domain equation is given by Equations (14.26), (14.48), (14.49), and (14.50).

$$A_2 = \frac{(6.2)(0.0167)(1.2)(1.0\text{E} - 4)}{(1.2)(1.0\text{E} - 4) + (0.043)(0.043)} = 0.00630$$

$$A_1 = \frac{(6.2 + 0.0167)(1.2)(1.0\text{E} - 4)}{(1.2)(1.0\text{E} - 4) + (0.043)(0.043)} = 0.379$$

$$G = \frac{0.043}{(1.2)(1.0\text{E} - 4) + (0.043)(0.043)} = 21.8$$

The time-domain equation is

$$0.00630 \frac{d^2\omega}{dt^2} + 0.379 \frac{d\omega}{dt} + \omega = 21.8e$$

(d) The transfer function is

$$\frac{\Omega(s)}{E(s)} = \frac{21.8}{1 + 0.379s + 0.00630s^2} \qquad ◆$$

14.5 THE DEAD-TIME PROCESS

A *dead-time process* is one in which mass or energy is transported from one point to another. The output signal is identical to the input signal except for a time delay. The time delay is called the dead-time lag and is denoted by t_d. The dead-time lag is the time required for the signal to travel from the input location to the output location. Dead time was included as one of the common elements in Chapter 3 (see Sections 3.1, 3.2, 3.3, and 3.6 for further discussion and examples of dead-time elements).

The step response of a dead-time process is shown in Figure 14.14. Before the step change, the output signal and the input signal are equal. The step change increases the input signal to a new value at time $t = 0$ s. The output signal remains at the original value until time $t = t_d$, when it also increases to the new value. The graph of the output signal is a duplication of the input graph moved to the right by t_d seconds.

The response of a dead-time process is characterized by the dead-time lag—the number of seconds (or minutes) that elapse between an input change and the corresponding output change. The time-domain equation of the dead-time process was developed in Section 3.1. Figure 14.15 illustrates two examples of dead-time processes.

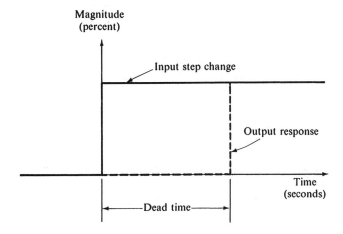

◆ **Figure 14.14**   The step response of a dead-time process is identical to the input but is delayed by the amount of dead-time lag, t_d.

The *Bode diagram* of a dead-time process is shown in Figure 14.16. The dead-time process delays a signal but does not change the size of the signal. The gain is 1 at all frequencies, and the Bode diagram gain line is on the 0-dB line. The phase angle, β, is given by the following equation:

$$\beta = -57.3\omega t_d \tag{14.51}$$

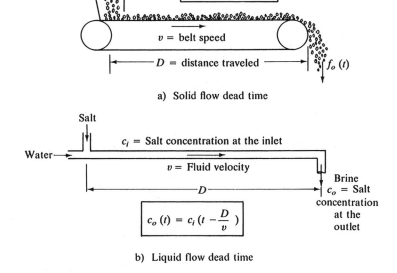

◆ **Figure 14.15**   Solid flow and liquid flow dead-time processes.

DEAD-TIME PROCESS

Time-Domain Equation

$$f_o(t) = f_i(t - t_d) \tag{14.52}$$

$$t_d = \frac{D}{v} \tag{3.1}$$

Transfer Function

$$\frac{F_o(s)}{F_i(s)} = e^{-t_d s} \tag{14.53}$$

where D = distance from input to output, m

          $f_i(s)$ = input signal

          $f_o(s)$ = output signal

          t = time, s or min

         t_d = dead-time lag, s or min

         v = velocity of signal travel, m/s

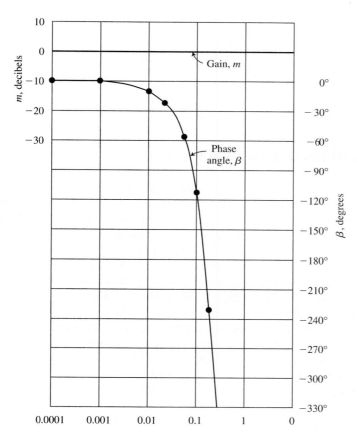

◆ **Figure 14.16** Bode diagram of a dead-time process with a dead-time lag of 20 s (Example 14.12). The gain is 1 at all frequencies, but the phase lag becomes very large as frequency increases.

**EXAMPLE
14.12**

A dead-time process similar to Figure 14.15a consists of a 12-m-long belt conveyor with a belt velocity of 0.6 m/s. Determine the dead-time lag (t_d), the time-domain equation, the transfer function, and the Bode diagram.

Solution

$$t_d = \frac{D}{v} = \frac{12}{0.6} = 20 \text{ s}$$

The time-domain equation is given by Equation (14.52).

$$f_o(t) = f_i(t - 20)$$

The transfer function is given by Equation (14.53).

$$\frac{F_o(s)}{F_i(s)} = e^{-20s}$$

The Bode diagram gain line is a constant 0 dB. The phase line is given by Equation (14.51).

$$\beta = -57.3\omega t_d = -57.3\omega(20) = -1146\omega$$

The following table of values was used to construct the phase angle line in the Bode diagram in Figure 14.16.

ω (rad/s)	β (deg)
0.001	1.15
0.01	11.5
0.02	22.9
0.05	57.3
0.1	114.6
0.2	229.2
0.4	458.4

14.6 THE FIRST-ORDER LAG PLUS DEAD-TIME PROCESS

The first-order lag plus dead-time process is a series combination of a first-order lag process and a dead-time element. The step response, illustrated in Figure 14.17, is characterized by the first-order lag time constant (τ) and the dead-time lag (t_d). The first-order lag plus dead-time characteristic is frequently used as a first approximation of the model of more complex processes for the purpose of analysis and design. The values of t_d and τ can be determined from a step response test of the process as indicated in Figure 14.17. An example of a first-order lag plus dead-time salt brine process is shown in Figure 14.18. A mixing tank is used to blend salt and water continuously to form a brine solution. The inlet water flow rate is regulated to match the outlet flow of brine solution. The salt flow rate is regulated to maintain the desired salt concentration in the mixing tank. The salt flow rate, $f_i(t)$, is the input signal of the process. The salt concentration in the tank, c_o, is the output signal of the process. The belt conveyor is the dead-time element and the mixing tank provides the first-order lag characteristic. In this example, the dead-time element preceded the first-order lag element. However, the response is the same for processes in which the first-order lag characteristic precedes the dead-time element.

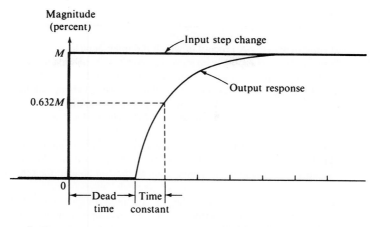

◆ **Figure 14.17** Step response of a first-order lag plus dead-time process.

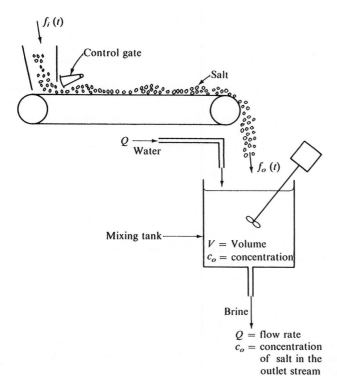

◆ **Figure 14.18** First-order lag plus dead-time blending process.

◆ GLOSSARY

Dead-time process: A process in which the input appears at the output after a time delay of t_d. The time delay, t_d, completely characterizes the dead-time process. This process usually involves transportation of mass or energy from one point to another. (14.5)

First-order lag process: A process in which the rate of change of the output is proportional to the difference between the input and the output. The first-order lag process is completely characterized by its time constant, τ, and its steady-state gain, G. First-order lag processes consist of a single resistance element and a single storage element. (14.3)

Frequency response graph: Two graphs that display the gain and the phase angle of a component plotted versus frequency. The two graphs share a common log frequency scale on the x-axis. The y-axis of one graph is the decibel gain of the component; the other is the phase angle of the component. (14.1)

Integral action time constant: The time required for the output of an integral process to change by an amount equal to the difference between the input and the nominal value that will maintain the output constant. (14.2)

Integral process: A process in which the output changes at a rate proportional to the difference between the input and a nominal value that will maintain the output constant. The integral process is characterized by the integral action time constant as defined above. (14.2)

Interacting second-order process: A second-order process configured such that the second storage element has an effect on the first storage element. (14.4)

Noninteracting second-order process: A second-order process configured such that the second storage element has no effect on the first storage element. (14.4)

Ramp process: Another name for the integral process. (14.2)

Second-order lag process: A process that has two storage elements and at least one resistive element. The second-order lag process is completely characterized by its resonant frequency, ω_0, damping ratio, ζ, and steady-state gain, G. The response of the second-order lag process is divided into three regions: underdamped, critically damped, and overdamped, depending on whether the damping ratio is less than 1, equal to 1, or greater than 1. (14.4)

Step response graph: A graph of the output of a process following a step change in the input to the process. (14.1)

Time-domain equation: An equation that defines the size-versus-time relationship between the input and the output of a process. (14.1)

Transfer function: The frequency domain ratio of the output of a process over its input with all initial conditions set to zero. The transfer function defines the gain and phase difference versus frequency relationship between the input and the output of the process. (14.1)

◆ EXERCISES

Section 14.1

14.1   Which of the following methods define the relationship between the size of the input to a process and the size of the output from the process at different times?
(a) Step response graph and frequency-response graph
(b) Time-domain equation and frequency-response graph
(c) Step response graph and transfer function
(d) Transfer function and frequency-response graph
(e) Step response graph and time-domain equation
(f) Time-domain equation and frequency-response graph.

14.2   Which methods in Exercise 14.1 a–f define the size and phase angle relationship between the input to a process and the output from the process at different frequencies.

Section 14.2

14.3   Choose all of the following conditions for which the output of an integral process will remain constant.
(a) The input flow rate is constant, and the output flow rate varies.

(b) The input flow rate varies and the output flow rate is constant.

(c) Both the input and output flow rates are constant, but they are not equal.

(d) The input flow rate is equal to the output flow rate.

(e) The amount of stored material or energy is constant.

14.4   The input signal (inflow minus outflow) to an integral process is suddenly changed from 0% to 10% and then held at 10%. The process output was at 20% when the step change occurred. Determine the output at the following times (measured from the instant the step change occurred). The process has an integral action time constant of 40 s.

(a) 20 s

(b) 40 s

(c) 80 s

(d) 120 s

14.5   The input signal (inflow minus outflow) to an integral process varies sinusoidally at a radian frequency, ω_1, of 0.025 rad/s. The output signal varies sinusoidally at the same frequency with an amplitude of 10%. Determine the amplitude of the output for each of the following input signal radian frequencies, assuming the input signal amplitude is unchanged.

(a) $\omega_2 = 0.25$ rad/s

(b) $\omega_2 = 0.10$ rad/s

(c) $\omega_2 = 0.010$ rad/s

(d) $\omega_2 = 0.005$ rad/s

14.6   A liquid level integral process (Figure 14.2a) has the following parameters and conditions:

Tank height = 5.2 m
Tank diameter = 1.8 m
FS_{in} = 0.02 m3/s
FS_{out} = 5 m
$h^*(t_o)$ = 1.7 m = 34% of FS
q^*_{out} = 0.009 m^3/s = 45% of FS
q^*_{in} = 0.011 m^3/s = 55% of FS

Determine the time-domain equation, the transfer function, the integral action time constant, and the level at time $t_o + 50$ s.

14.7   Construct a Bode diagram of the integral process in Exercise 14.1. The input signal is expressed as a percentage of FS_{in}, and the output signal is expressed as a percentage of FS_{out}. Use the Bode diagram to determine the output amplitude produced by a sinusoidal input at each frequency given below. The input signal has an amplitude of 10% of FS_{in}. Express the output amplitude both as a percentage of FS_{out}, and in meters. Indicate any output that is limited by FS_{out}.

Frequency (rad/s)

1.57×10^{-5}

1.57×10^{-4}

1.57×10^{-3}

1.57×10^{-2}

1.57×10^{-1}

1.57×10^{0}

Section 14.3

14.8   A liquid level first-order lag process has an output flow rate, q_{out}, of 2 gpm when the tank level, $h(t)$, is at 40% of full scale. Determine the output flow rate for each of the following values of the level, $h(t)$.

(a) 20%

(b) 80%

(c) 10%

(d) 100%

14.9   The input signal of a first-order lag process is suddenly changed from 10% to 30% of full scale. The process output was at 10% of full scale when the step change occurred. The process has a time constant of 20 seconds. Two hundred seconds after the step change, the output was 30% of full scale. Determine the output at each of the following times after the step change.

(a) 20 s

(b) 40 s

(c) 60 s

(d) 80 s

(e) 100 s

14.10   A first-order lag process has a time constant, τ, of 10 s, and a steady-state gain, G, of 1. Use the Bode diagram in Figure 14.7 to determine the gain, m, in decibels, and the phase angle, β, in degrees for each of the following values of the radian frequency.

(a) 0.001 rad/s

(b) 0.01 rad/s

(c) 0.1 rad/s

(d) 1.0 rad/s

(e) 10.0 rad/s

14.11   The gain values in Exercise 14.10 are expressed in decibels. If g is the ordinary gain value, and m is the decibel gain value, then the following equations convert between dB and ordinary gain.

$$m = 10 \log_{10}g \qquad g = 10^{m/20}$$

Convert the decibel gain values, m, that you obtained in Exercise 14.10 into ordinary gain values, g.

14.12 The transfer function of the first-order lag process, given by Equation (14.10), can be used to compute the gain and phase angle of the process at any given frequency. Refer to Example 4.18 for an example of the use of the transfer function to determine the gain and phase angle at a given radian frequency. Use values of 1 for G and 10 for τ in Equation (14.10), and determine the gain, g, and phase angle, β, for each of the following radian frequencies. Compare your results with the gain and phase angle values you obtained in Exercises 14.10 and 14.11.

(a) $\omega = 0.001$ rad/s
(b) $\omega = 0.01$ rad/s
(c) $\omega = 0.1$ rad/s
(d) $\omega = 1.0$ rad/s
(e) $\omega = 10.0$ rad/s

14.13 An oil tank similar to Figure 14.5a has a diameter of 1.658 m and a height of 3.6 m. The outlet at the bottom is a 4-m-long smooth tube with a diameter of 2.5 cm. The oil temperature is 15°C. The full-scale ranges are: $FS_{in} = 2.5 \times 10^{-4}$ m³/s and $FS_{out} = 3.2$ m. Determine each of the following:

(a) The capacitance of the tank.
(b) The resistance of the outlet.
(c) The time constant of the process.
(d) The gain of the process.
(e) The time-domain equation.
(f) The transfer function.

14.14 An electrical circuit similar to Figure 14.5b has a capacitance value of 10 μF. Determine the resistance value that will result in a time constant equal to 0.083 s.

14.15 Construct the Bode diagram of the first-order electrical circuit in Example 14.3. Determine the amplitude and phase of the output signal corresponding to each of the following input signals:

(a) $e_{in} = 0.25 \cos(0.2t + 45°)$ V
(b) $e_{in} = 0.5 \cos(2t + 45°)$ V
(c) $e_{in} = 1.5 \cos(20t + 45°)$ V

14.16 An oil-bath thermal system similar to Figure 14.6a has an inside diameter of 0.75 m and a height of 0.6 m. The inside film coefficient is 58 W/m² · K, and the outside film coefficient is 350 W/m² · K. The tank wall is a single layer of aluminum 2.1 cm thick. Determine each of the following:

(a) The thermal resistance.
(b) The thermal capacitance.
(c) The time constant.
(d) The time-domain equation.
(e) The transfer function.

14.17 Construct the Bode diagram of the oil-bath thermal process in Example 14.4.

14.18 A nitrogen gas process similar to Figure 14.6b has the following parameters:

Pressure vessel volume: 1.1 m³
Temperature: 540 K
Gas flow resistance: 1.8×10^5 Pa · s/kg
FS_{in}: 2.5 kg/s
FS_{out}: 100 kPa

Determine each of the following:
(a) The capacitance of the pressure vessel.
(b) The time constant.
(c) The process gain.
(d) The time-domain equation.
(e) The transfer function.

14.19 A blending system similar to Figure 14.6c has a flow rate of 0.006 m³/s. Determine the tank volume that will result in a time constant of 10 min. Then determine the time-domain equation and transfer function using the 10-min time constant.

Section 14.4

14.20 Name the three parameters that characterize a second-order lag process.

14.21 Name the three types of step response of a second-order lag process. Describe each type of response and the condition that causes this type of response.

14.22 Describe the difference between noninteracting and interacting second-order lag processes.

14.23 A second-order lag process has a resonant frequency, ω_0, of 10 rad/s, a damping ratio of 0.1, and a steady state gain, G, of 1. Use the Bode diagram in Figure 14.13 to determine the gain, m, in decibels, and the phase angle, β, in degrees for each of the following values of the radian frequency. Convert your decibel gain values, m, to ordinary gain values, g.

(a) 0.1 rad/s
(b) 1.0 rad/s
(c) 10.0 rad/s
(d) 100 rad/s
(e) 1000 rad/s

14.24 Repeat Exercise 14.23 with a damping ratio of 10, a resonant frequency of 10 rad/s, and a steady-state gain of 1.

14.25 Equation (14.26) defines the transfer function for a second-order lag process. Use the transfer function and program BODE to print two Bode Data Tables to check the graphical results you obtained in Exercises 14.23 and 14.24. Comment on the comparison between the values you obtained from the graph in Figure 14.13 and the values you obtained from program BODE.

14.26 A spring-mass-damping system consists of a 25-kg mass, a spring capacitance of 6.9×10^{-4} m/N, and a damping resistance of 42 N · s/m. Determine the time-domain equation, the transfer function, the reso-nant frequency, ω_0, the damping ratio, ζ, and whether the process is overdamped, underdamped, or critically damped.

14.27 Construct the Bode diagram of the spring-mass-damping system in Example 14.7. Determine the amplitude and phase angle of the output produced by each of the following input signals:
(a) $f = 10 \cos(t + 0°)$N
(b) $f = 10 \cos(10t + 0°)$N
(c) $f = 10 \cos(100t + 0°)$N

14.28 An electrical series *RLC* circuit consists of a 0.04-H inductor, a 4-μF capacitor, and a 100-Ω resistor. Verify that the system is underdamped. The damping of the circuit may be increased by adding a second resistor in series with the 100-Ω resistor. Determine the value of the second resistor that will result in a critically damped circuit. Determine the time-domain equation and the transfer function of the critically damped circuit.

14.29 A liquid noninteracting two-capacity system has time constants τ_1 and τ_2 of 300 s and 1200 s. The liquid is water, and the value of R_2 is 2.2×10^6 Pa · s/m^3. The full-scale ranges are $\mathrm{FS_{in}} = 1.0 \times 10^{-2}$ m^3/s and $\mathrm{FS_{out}} = 4$ m. Determine the time-domain equation and transfer function.

14.30 An electrical interacting second-order circuit has the following component values:

$$R_1 = 100 \ \Omega$$
$$R_2 = 300 \ \Omega$$
$$R_L = 50 \ \Omega$$
$$C_1 = 0.1 \ \mu F$$
$$C_2 = 0.8 \ \mu F$$

Determine τ_1, τ_2, the time-domain equation, and the transfer function.

14.31 An armature-controlled dc motor has the following characteristics:

$$B = 2.0 \times 10^{-3} \ \text{N} \cdot \text{m} \cdot \text{s/rad}$$
$$J = 3.2 \times 10^{-3} \ \text{kg} \cdot \text{m}^2$$

$$K_e = 0.22 \ \text{V} \cdot \text{s/rad}$$
$$K_t = 0.22 \ \text{N} \cdot \text{m/A}$$
$$L = 0.075 \ \text{H}$$
$$R = 1.2 \ \Omega$$

Determine the mechanical time constant, the electrical time constant, the time-domain equation, the transfer function, the resonant frequency, ω_0, the damping ratio, ζ, and the type of damping (i.e., overdamped, critically damped, or underdamped).

14.32 Construct the Bode diagram of the dc motor second-order system in Example 14.11.

14.33 The response of a control system is very much like the response of a second-order system. A closed-loop system may be overdamped, critically damped, or underdamped. The damping coefficient, α, determines the type of damping present in a system. In Equation (14.26), the equation of the second-order system is expressed in terms of A_1 and A_2. The equation can also be expressed in terms of ω_0 and α.

$$\frac{1}{\omega_0^2} \frac{d^2 y}{dt^2} + 2\left(\frac{\alpha}{\omega_0^2}\right)\frac{dy}{dt} + y = Gx$$

Notice that α is part of the coefficient of the first derivative of the output (dy/dt). The first derivative is the rate of change (or velocity) of the output. Consider a dc motor position control system that is underdamped. A speed sensor measures the rate of change of the output (i.e., dy/dt). Explain how the signal from the speed sensor could be used to increase the damping coefficient of the system.

Section 14.5

14.34 Determine the dead-time lag for each of the following distance and velocity parameters.
(a) D = 15 ft, v = 4 ft/s
(b) D = 34 ft, v = 5.5 ft/s
(c) D = 4.6 m, v = 1.2 m/s
(d) D = 9.8 m, v = 2.4 m/s

14.35 A dead-time process similar to Figure 14.15a consists of an 8.2-m-long belt conveyor with a belt velocity of 0.44 m/s. Determine the dead-time lag, the time-domain equation, and the transfer function.

14.36 Construct a Bode diagram for a dead-time lag process with a dead time of 250 s.

Section 14.6

14.37 Name and describe the two parameters that characterize a first-order lag plus dead-time lag process.

14.38 A first-order lag plus dead-time process similar to Figure 14.18 has the following parameter values:

Distance traveled on the belt: 7.6 m

Belt speed: 1.1 m/s

Tank volume: 12.2 m^3

Water flow rate: 0.01 m^3/s

Determine the dead-time lag (t_d) and the first-order time constant (τ).

14.39 A blending and heating system is illustrated in Figure 1.10. The thermal system is a first-order lag and the blending system is a first-order lag plus dead time. Determine the time constant and dead-time lag of the blending system and the time constant of the thermal system. Also determine the transfer function of the thermal system. The system parameters are as follows:

Production rate: 1.8×10^{-4} m^3/s

Mixing tank diameter: 0.55 m

Height of liquid in the mixing tank: 0.65 m

Inside thermal film coefficient:
$$h_i = 810 \text{ W/m}^2 \cdot \text{K}$$

Outside thermal film coefficient:
$$h_o = 1200 \text{ W/m}^2 \cdot \text{K}$$

Tank wall thickness: 0.7 cm

Tank wall material: steel

Density of liquid in the tank: 1008 kg/m^3

Specific heat of the liquid: 4060 J/kg · K

Diameter of the mixing tank outlet pipe: 2 cm

Distance from tank outlet to the analyzer feed pipe inlet: 1.5 m

Concentration analyzer feed pipe flow velocity: 0.2 m/s

Length of the concentration analyzer feed pipe: 2 m

◆ CHAPTER 15   **Methods of Analysis**

◆ OBJECTIVES

The analysis of a control system centers on answering the question: Is the system stable? A stable control system responds to any reasonable input with an error that diminishes with time. The error of a stable system may or may not oscillate, and it may or may not diminish to zero. If the error oscillates, the amplitude of the oscillation diminishes exponentially with time. If the error does not diminish to zero, it eventually reaches a small, steady value. In contrast, an unstable control system responds to at least some reasonable inputs with an error that increases with time. The error usually oscillates. For continuous, linear systems, an error that oscillates with constant amplitude is considered to be unstable. However, some discontinuous systems are inherently oscillatory, and an oscillating error is considered stable as long as the amplitude does not increase with time.

A second concern of control system analysis is the question: How quickly does the error diminish? Control system engineers must answer both questions in the process of designing a control system that is stable and minimizes errors resulting from disturbances, changes in setpoint, or changes in load.

The purpose of this chapter is to introduce you to three methods the control engineer uses to analyze control systems: Bode diagrams, Nyquist diagrams, and root locus. After completing this chapter, you will be able to

1. Use a computer-aided, graphical method (the program named DESIGN) to construct or determine the following for a control system
   a. Open-loop Bode diagram
   b. Closed-loop Bode diagram
   c. Error-ratio graph
   d. Stability (from an open-loop Bode diagram)
   e. Gain and phase margin (from an open-loop Bode diagram)
   f. Nyquist diagram
   g. Stability (from a Nyquist diagram)
2. Construct a root-locus diagram of a control system and determine
   a. Values of gain, K, for which the system is unstable
   b. Value of gain, K, for a given damping ratio

15.1 INTRODUCTION

In this chapter you will study a number of methods used to analyze control systems. The analysis of a control system involves the determination of the response of the system to various disturbances and the answer to the question: Is the system stable? A closed-loop control system consists of several components connected in series to form a closed loop. These components include the controller, manipulating element, process, and measuring transmitter (see Figure 1.7). Thus our study begins with the overall transfer function and frequency response of several components connected in series.

There are actually two overall transfer functions of a closed-loop control system. One transfer function is obtained with the feedback line disconnected at the error detector. We call this the *open-loop transfer function* of the control system. The second transfer function is obtained with the feedback line connected. We call this the *closed-loop transfer function* of the system. Both transfer functions are expressed as the ratio C_m/SP.

The *error ratio* and the *deviation ratio* give additional insight into the response of a closed-loop control system to disturbances of various frequencies. The response of a control system can be divided into three frequency zones. Zone 1 is the range of frequencies below the "frequency limit" of the control system. Zone 2 is the middle range of frequencies between the "frequency limit" and a second, higher-frequency value. Zone 3 is the range of frequencies above the second frequency value. In zone 1, feedback reduces the error caused by a disturbance or setpoint change. In zone 2, feedback actually increases the error caused by a disturbance or setpoint change. In zone 3, feedback has no effect on the error. The importance of knowing the frequency limit of a control system is obvious. A control system cannot handle disturbances or setpoint changes that have a frequency greater than its frequency limit.

The *frequency response* of a control system is usually displayed as a pair of graphs called Bode diagrams. One graph shows the overall gain (amplitude change from SP to C_m) versus frequency. The second graph shows the overall phase angle (phase change from SP to C_m) versus frequency. The same frequency scale is used on both graphs, and they are plotted on one diagram, thereby making it easy to read the gain and phase angle at any given frequency. As with transfer functions, two frequency responses are useful in control system analysis. The *open-loop frequency response* is obtained with the feedback line disconnected at the error detector. The *closed-loop frequency response* is obtained with the feedback line connected.

Both graphical and computer-aided methods are used to obtain the open-loop and closed-loop Bode diagrams of a control system. A BASIC program for computer-aided analysis and design is included on a disk that comes with this book. DESIGN produces both open-loop and closed-loop Bode diagrams of a control system. It also produces a Nyquist diagram and an error-ratio graph of the system. In Chapter 16, DESIGN is used for computer-aided design of PID controllers and compensation networks. With DESIGN, a designer can try a number of different designs and observe the change in the frequency response on a graphic screen. The designer can use a "what if" approach to the analysis and design of a control system. However, a certain amount of hand plotting of graphical data will help students acquire the understanding and judgment required to make full use of powerful computer-aided methods such as DESIGN.

Stability is a major concern in the design of a control system. Control system designers use two safety margins, called the *gain margin* and *phase margin,* to assure the stability of the control system. The designer reads these two margins directly from the open-loop Bode diagram, and uses them as a guide while shaping the Bode diagram to obtain the optimum control system design.

The Bode diagram does not apply to some complex types of systems. For these systems, the designer may use the *Nyquist stability criterion,* which applies to a wider range of systems than the Bode diagram. The Nyquist criterion uses a graphic technique to determine the stability or instability of the control system. Although the Nyquist criterion answers the question "Is the system stable?", it is not as useful as Bode diagrams and root locus in the actual design of the controller.

The *root-locus* technique is another graphical method of determining the stability or instability of a control system. The root locus consists of a plot of the roots of a certain characteristic equation of the control system as the gain varies from zero to infinity. From the root-locus plot, the designer can observe the relationship between the system gain and the stability of the system. The root locus is particularly useful to the designer in determining the effect of gain changes on the stability and performance of the control system.

15.2 OVERALL BODE DIAGRAM OF SEVERAL COMPONENTS

A control system consists of several components connected in series to form a closed loop. The design and analysis of a control system require a knowledge of the transfer function and the frequency response of this group of components considered as a unit. In other words, the overall transfer function and frequency response must be known.

Consider the three components in Figure 15.1 with transfer functions TF_1, TF_2, and TF_3. Signal S_1 is the input signal of component 1. Signal S_2 is the output signal of component 1 and the input signal of component 2. Signal S_3 is the output signal of component 2 and the input signal of component 3. Signal S_4 is the output signal of component 3. Also, signal S_1 is the input signal and S_4 is the output signal of the three components as a unit. By defini-

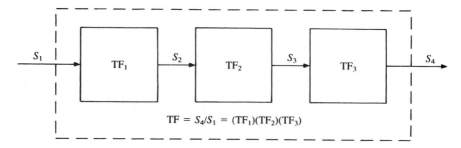

◆ **Figure 15.1**   The transfer function of three components connected in series is equal to the product of the transfer functions of the individual components.

tion, the transfer function of a a component or group of components is the ratio of the output signal over the input signal. Thus

$$\text{Transfer function of component 1} = \text{TF}_1 = \frac{S_2}{S_1}$$

$$\text{Transfer function of component 2} = \text{TF}_2 = \frac{S_3}{S_2}$$

$$\text{Transfer function of component 3} = \text{TF}_3 = \frac{S_4}{S_3}$$

$$\text{Transfer function of the group of components} = \text{TF} = \frac{S_4}{S_1}$$

But

$$\frac{S_4}{S_1} = \left(\frac{S_2}{S_1}\right)\left(\frac{S_3}{S_2}\right)\left(\frac{S_4}{S_3}\right)$$

and

$$\text{TF} = (\text{TF}_1)(\text{TF}_2)(\text{TF}_3) \tag{15.1}$$

Equation (15.1) states that the overall transfer function of several components in series is equal to the product of the transfer functions of the individual components.

The frequency response of a component or group of components is obtained by substituting $j\omega$ for s in the transfer function. The transfer function can then be reduced to a complex number in the polar form. The magnitude of the complex number is the frequency response gain, g. The angle of the complex number is the frequency response phase angle, β. Let the frequency response be represented by the following terms:

$g_1\underline{/\beta_1}$ = frequency response of component 1

$g_2\underline{/\beta_2}$ = frequency response of component 2

$g_3\underline{/\beta_3}$ = frequency response of component 3

$g\underline{/\beta}$ = overall frequency response of the three components as a unit

Then, from Equation (15.1)

$$g\underline{/\beta} = (g_1\underline{/\beta_1})(g_2\underline{/\beta_2})(g_3\underline{/\beta_3}) \tag{15.2}$$

However, the product of three complex numbers in polar form is a complex number whose magnitude is the product of the magnitudes of the three complex numbers, and whose angle is the sum of the angles of the three complex numbers.

$$g\underline{/\beta} = g_1\, g_2\, g_3\underline{/\beta_1 + \beta_2 + \beta_3} \tag{15.3}$$

Equation (15.3) states that the *overall gain* of several components in series is equal to the product of the gains of the individual components. It further states that the *overall phase angle* of several components in series is equal to the sum of the phase angles of the individual components. In other words, the gains are multiplied and the phase angles are added.

The Bode diagram uses a logarithmic gain scale so that multiplication of gain terms can be accomplished by graphically adding the logarithms of the individual gain terms; that is,

$$\log g = \log (g_1 \cdot g_2 \cdot g_3) = \log g_1 + \log g_2 + \log g_3$$

The decibel scale is a logarithmic scale, and the addition of decibel values is equivalent to multiplying the corresponding gain values.

$$m(\text{dB}) = m_1(\text{dB}) + m_2(\text{dB}) + m_3(\text{dB}) \tag{15.4}$$

> The overall transfer function of several components in series is equal to the product of the transfer functions of the individual components.
>
> The overall gain of several components in series is equal to the product of the gains of the individual components.
>
> The overall phase angle of several components in series is equal to the sum of the phase angles of the individual components.
>
> The overall frequency response of several components can be determined on a Bode diagram graphically by adding the decibel gains and by adding the phase angles of the individual components.

EXAMPLE 15.1

A thermal process and a temperature measuring means are connected in series in a control system. Determine the overall transfer function from the following individual transfer functions:

$$\text{Thermal process transfer function} = \frac{1}{1 + 6420s}$$

$$\text{Measuring means transfer function} = \frac{1}{1 + 78s}$$

Solution

The overall transfer function is the product of the individual transfer functions:

$$\text{Overall transfer function} = \left(\frac{1}{1 + 6420s}\right)\left(\frac{1}{1 + 78s}\right)$$

$$= \frac{1}{(1 + 6420s)(1 + 78s)} \qquad ◆$$

EXAMPLE 15.2

Three components are connected in series. The transfer functions are given below. Determine the overall transfer function, TF.

$$\text{TF}_1 = \frac{1 + \tau_1 s}{1 + \tau_2 s}$$

$$\text{TF}_2 = \frac{1}{1 + A_1 s + A_2 s^2}$$

$$\text{TF}_3 = \frac{1}{1 + \tau_3 s}$$

Solution   The overall transfer function is the product of the individual transfer functions:

$$TF = (TF_1)(TF_2)(TF_3)$$

$$= \left(\frac{1 + \tau_1 s}{1 + \tau_2 s}\right)\left(\frac{1}{1 + A_1 s + A_2 s^2}\right)\left(\frac{1}{1 + \tau_3 s}\right)$$

$$= \frac{1 + \tau_1 s}{(1 + \tau_2 s)(1 + A_1 s + A_2 s^2)(1 + \tau_3 s)}$$ ◆

EXAMPLE 15.3   The following gains and phase angles were measured at a frequency of 1 cycle per second. Determine the overall gain, g, and phase angle, β, if the components are connected in series.

	Gain	Phase Angle (deg)
Final control element	1.5	−5
Process	2.0	−170
Measuring means	0.9	−15

Solution   The gain, g, is the product of the individual gains and the phase angle, β, is the sum of the individual phase angles.

$$g = (1.5)(2.0)(0.9) = 2.7$$

$$\beta = (-5°) + (-170°) + (-15°) = -190°$$ ◆

15.3 OPEN-LOOP BODE DIAGRAMS

The *open-loop Bode diagram* expresses the *frequency response* of the control system when the measuring transmitter output is disconnected from the error detector. The Bode diagram data can be obtained from the transfer function of the "open-loop control system" shown in Figure 15.2. The first block represents the forward path components (controller, manipulating element, and process), and the symbol $G(s)$ is used to represent the transfer function of these components. The second block represents the feedback path component (measuring transmitter). Because the measurement, C_m, is not used, the output of the error detector is equal to the setpoint signal (i.e., $E = SP$).

The open-loop transfer function is obtained by substituting $SP[G(s)]$ for C in the equation for C_m (Figure 15.2) and then dividing the resulting equation by SP.

OPEN-LOOP RESPONSE

$$\text{Open-loop transfer function} = \frac{C_m}{SP} = G(s)H(s) \qquad \textbf{(15.5)}$$

The open-loop Bode diagram is constructed with the aid of a computer by multiplying the gains and adding the phase angles of the controller, manipulating element, process, and measuring transmitter.

The open-loop Bode diagram is constructed graphically by adding the decibel gain and phase graphs of the controller, manipulating element, process, and measuring transmitter.

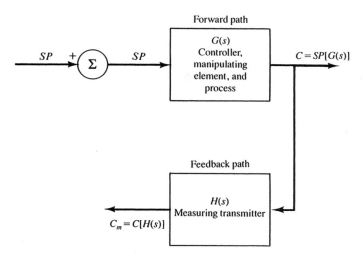

$$\frac{C_m}{SP} = [G(s)][H(s)]$$

◆ **Figure 15.2** The open-loop transfer function is obtained by disconnecting the feedback line from the error detector and solving for C_m/SP.

15.4 CLOSED-LOOP BODE DIAGRAMS

The *closed-loop Bode diagram* expresses the *frequency response* of the control system when the measuring transmitter output is connected to the error detector. The Bode diagram data can be obtained from the transfer function of the closed-loop control system shown in Figure 15.3.

In Figure 15.3, the loop is closed, making the output of the error detector equal to the set-point minus the measuring transmitter output (i.e., $E = SP - C_m$). The output of the forward path components, C, is equal to the input, E, times the transfer function $G(s)$.

$$C = E[G(s)] = (SP - C_m)G(s)$$

The output of the feedback path component, C_m, is equal to the input, C, times the transfer function $H(s)$.

$$C_m = C[H(s)] = (SP - C_m)G(s)H(s)$$

The preceding equation may be solved for the closed-loop transfer function C_m/SP.

$$C_m = SP[G(s)H(s)] - C_m[G(s)H(s)]$$

$$C_m + C_m[G(s)H(s)] = SP[G(s)H(s)]$$

$$C_m[1 + G(s)H(s)] = SP[G(s)H(s)]$$

$$\frac{C_m}{SP} = \frac{G(s)H(s)}{1 + G(s)H(s)} = \text{closed-loop transfer function}$$

where $G(s)H(s)$ is the open-loop transfer function. In other words, the closed-loop response is equal to the open-loop response divided by 1 plus the open-loop response.

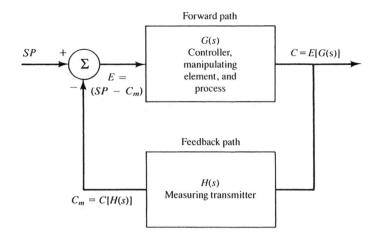

$$\frac{C_m}{SP} = \frac{[G(s)][H(s)]}{[G(s)][H(s)] + 1}$$

◆ **Figure 15.3** The closed-loop transfer function is obtained by connecting the feedback line to the error detector and solving for C_m/SP.

CLOSED-LOOP RESPONSE

$$\text{Closed-loop transfer function} = \frac{C_m}{SP} = \frac{G(s)H(s)}{1 + G(s)H(s)} \tag{15.6}$$

The closed-loop Bode diagram is constructed with the aid of a computer by dividing the open-loop response by 1 plus the open-loop response.

The closed-loop Bode diagram is constructed graphically by converting the open-loop gain and phase values to the corresponding closed-loop values. A Nichols chart is used to convert each set of open-loop gain and phase values into the corresponding closed-loop gain and phase values.

Figure 15.4 shows the open-loop and closed-loop Bode diagrams of a control system that consists of a proportional controller and a first-order lag plus dead-time process. The controller gain is 40. The process time constant is 100 s, and the dead-time lag is 2 s. The open-loop transfer function is

$$G(s)H(s) = 40\left(\frac{e^{-2s}}{1 + 100s}\right)$$

The closed-loop Bode diagram may be determined by using program DESIGN, by using a Nichols chart, or by using a calculator. The use of program DESIGN is covered in Section 15.6. The use of the Nichols chart and the calculator are covered in the following paragraphs.

Figure 15.5 illustrates the use of a Nichols chart to graphically obtain the closed-loop response from the open-loop response. The open-loop data points from Figure 15.4 are plotted on the Nichols chart using the rectangular, open-loop coordinates. (The open-loop phase an-

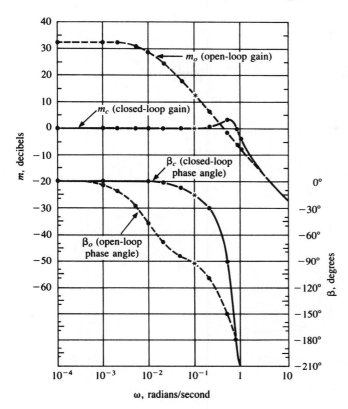

◆ **Figure 15.4** Open-loop and closed-loop Bode diagrams of a control system consisting of a proportional controller and a first-order lag plus dead-time process.

gle is on the horizontal axis, and the open-loop decibel gain is on the vertical axis.) The closed-loop gain and phase angle are read from the same plotted points using the curved, closed-loop coordinates. (The closed-loop gain coordinates are the lines with the short dashes. The closed-loop phase-angle coordinates are the lines with the long and short dashes.)

Notice the points that are marked with an asterisk on Figures 15.4 and 15.5. We will use these marked points to trace the plotting of one open-loop point on the Nichols chart and the reading of the corresponding closed-loop gain and phase angle. The trace begins by reading the open-loop gain and phase angle.

From Figure 15.4 at $\omega = 0.1$ rad/s:

$$\text{Open-loop gain} = 12 \text{ dB}$$

$$\text{Open-loop phase angle} = -95°$$

This point is plotted in the Nichols chart (Figure 15.5) and is marked with an asterisk. The frequency, $\omega = 0.1$, is written beside the point. Notice that the marked point is on the 0-dB closed-loop gain line, so we read the closed-loop gain as 0 dB. Notice also that the marked point is midway between the $-10°$ and the $-20°$ closed-loop phase-angle lines, so we read the closed-loop phase angle as $-15°$.

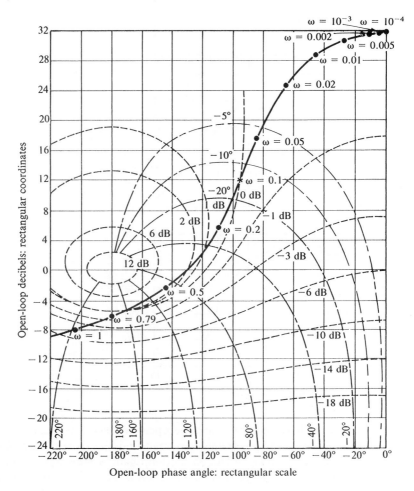

◆ **Figure 15.5** The open-loop data from Figure 15.4 is plotted in the Nichols chart using the rectangular open-loop coordinates. The closed-loop data are then read from the same plotted points using the curved closed-loop coordinates. [Based on information from H. M. James, N. B. Nichols, and R. S. Phillips, *Theory of Servomechanisms* (New York: McGraw-Hill Book Company, 1947), p. 179.]

From Figure 15.5 at $\omega = 0.1$ rad/s:

$$\text{Closed-loop gain} = 0 \text{ dB}$$

$$\text{Closed-loop phase angle} = -15°$$

These two closed-loop points are plotted in Figure 15.4 and are also marked with asterisks.

The use of a calculator to convert open-loop points to closed-loop points provides a more accurate result than the Nichols chart method. The calculator method uses Equation (15.6) with the open- and closed-loop points expressed as complex numbers in polar form (the magnitude of the complex number is the ordinary gain, g, and the angle of the complex number is the phase angle, β). In these terms, Equation (15.6) can be restated as follows:

$$g\underline{/\beta} \text{ (closed-loop)} = \frac{g\underline{/\beta} \text{ (open-loop)}}{1 + g\underline{/\beta} \text{ (open-loop)}}$$

We begin the conversion of the previous open-loop point by converting the 12-dB gain from decibels to ordinary gain.

$$\text{Open-loop gain, } g = 10^{12/20} = 3.98$$

$$g\underline{/\beta} \text{ (open-loop)} = 3.98 \underline{/-95°}$$

$$1 + g\underline{/\beta} \text{ (open-loop)} = 1 + 3.98\underline{/-95°}$$

$$= 1 - 0.35 - j3.97$$

$$= 0.65 - j3.97 = 4.02\underline{/-80.6°}$$

$$g\underline{/\beta} \text{ (closed-loop)} = \frac{3.98\underline{/-95°}}{4.02\underline{/-80.6°}} = 0.99 \underline{/-14.4°}$$

15.5 ERROR RATIO AND DEVIATION RATIO

Every closed-loop control system has a maximum frequency limit. A control system cannot respond accurately to setpoint changes above its frequency limit. Disturbances with frequencies above the frequency limit are especially troublesome. The controller cannot reduce the error produced by the disturbance and may actually increase the error. The maximum frequency limit can be determined by comparing the error when feedback is used with the error when feedback is not used. A convenient means of comparison is the *error ratio,* ratio of the closed-loop error over the open-loop error.

A typical error-ratio graph is illustrated in Figure 15.6. The graph is divided into three frequency *zones:* 1, 2, and 3. In zone 1, the error ratio is less than 1 (0 dB). This means that the closed-loop error is less than the open-loop error. The controller will reduce errors occurring in zone 1. In zone 2, the error ratio is greater than 1 (0 dB). The closed-loop error is actually greater than the open-loop error, and the controller is doing more harm than good. In zone 3, the error ratio is equal to 1 (0 dB). The closed-loop error is equal to the open-loop error, and the presence of the controller neither increases or decreases the error. The *maximum frequency limit* is the frequency that divides zone 1 and zone 2.

The equation for the error ratio is derived from the open-loop and the closed-loop transfer functions. The closed-loop error is the error signal, E, in the closed-loop control system illustrated in Figure 15.3. The open-loop error is the setpoint signal SP, minus the measured value signal, C_m, in the open-loop control system shown in Figure 15.2.

$$\text{Error ratio} = \frac{\text{closed-loop error magnitude}}{\text{open-loop error magnitude}}$$

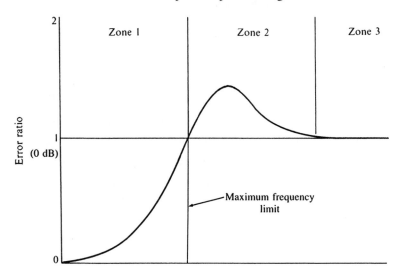

◆ **Figure 15.6**   A graph of the error ratio may be divided into three distinct zones based on a comparison of the error with the loop closed and the error with the loop open. In zone 1, closed-loop control reduces the error. In zone 2, closed-loop control increases the error. In zone 3, closed-loop control neither increases nor decreases the error.

$$\text{Error ratio} = \frac{\text{closed-loop error magnitude}}{\text{open-loop error magnitude}}$$

$$\text{Closed-loop error} = (SP - C_m) \text{ with the loop closed}$$

$$= SP - SP\left(\frac{C_m}{SP}\right) \text{ with the loop closed}$$

$$= SP - SP\left[\frac{G(s)H(s)}{1 + G(s)H(s)}\right]$$

$$= SP\left[1 - \frac{G(s)H(s)}{1 + G(s)H(s)}\right]$$

$$= SP\left[\frac{1 + G(s)H(s) - G(s)H(s)}{1 + G(s)H(s)}\right]$$

$$= SP\left[\frac{1}{1 + G(s)H(s)}\right] \tag{15.7}$$

$$\text{Open-loop error} = (SP - C_m) \text{ with the loop open}$$

$$= SP - SP\left(\frac{C_m}{SP}\right) \text{ with the loop open}$$

◆ **TABLE 15.1**   Error Ratio Versus Gain When $\beta = -180°$

$g_{-180°}$	$1 + g_{-180°}$	$1 - g_{-180°}$	Error Ratio	
			Ratio	dB
1.0	2.0	0.0	infinite	infinite
0.9	1.9	0.1	5.26	14.4
0.8	1.8	0.2	2.78	8.9
0.7	1.7	0.3	1.96	5.85
0.6	1.6	0.4	1.56	3.85
0.5	1.5	0.5	1.33	2.50
0.4	1.4	0.6	1.19	1.50
0.3	1.3	0.7	1.10	0.80
0.2	1.2	0.8	1.04	0.35
0.1	1.1	0.9	1.01	0.10
0.0	1.0	1.0	1.00	0.00

$$= SP - SP[G(s)H(s)]$$
$$= SP[1 - G(s)H(s)] \tag{15.8}$$

The error ratio is equal to the magnitude of the right-hand side of Equation (15.7) divided by the magnitude of the right-hand side of Equation (15.8).

$$\text{Error ratio} = \left| \frac{1}{[1 + G(s)H(s)][1 - G(s)H(s)]} \right| \tag{15.9}$$

The right-hand expression is a complex number. The parallel lines indicate that the error ratio is equal to the magnitude of this complex number, and the angle is ignored.

Every closed-loop control system has a zone 2: a range of frequencies for which the controller increases the error. The results in Table 15.1 clearly illustrate one frequency that is always in zone 2—the frequency for which the open-loop phase angle is $-180°$. This frequency is designated $\omega_{-180°}$. The open-loop gain at $\omega_{-180°}$ is designated by $g_{-180°}$.

$$g_{-180°} \underline{/-180°} = -g_{-180°} + j0 = -g_{-180°}$$

At $\omega_{-180°}$, the error ratio is given by

$$\text{Error ratio} = \frac{1}{(1 - g_{-180°})(1 + g_{-180°})}$$

The *deviation ratio** is equal to the closed-loop error magnitude over the setpoint magnitude. It is an indication of how accurately a control system can follow a change in setpoint and is used to evaluate follow-up control systems. The deviation ratio is obtained directly from Equation (15.7).

$$\text{Closed-loop error} = SP \left[\frac{1}{1 + G(s)H(s)} \right]$$

$$\text{Deviation ratio} = \left| \frac{\text{closed-loop error}}{\text{setpoint}} \right|$$

$$= \left| \frac{1}{1 + G(s)H(s)} \right| \tag{15.10}$$

---

*G. K. Tucker and D. M. Wills, *A Simplified Technique of Control System Engineering* (Philadelphia: Minneapolis–Honeywell Regulator Company, 1962), pp. 91–93.

Figure 15.7 compares the graphs of the error ratio and the deviation ratio for the control system with the Bode diagrams shown in Figure 15.4. The control system consists of a proportional controller with a gain of 40 and a process with a time constant of 100 s and a dead-time lag of 2 s.

ERROR RATIO

The error ratio gives the relative size of the error produced by a disturbance when the loop is closed compared with the error when the loop is open.

$$\text{Error ratio} = \frac{1}{(1 + g_o \underline{/\beta_o})(1 - g_o \underline{/\beta_o})} \tag{15.11}$$

where g_o = open-loop gain

β_o = open-loop phase angle

DEVIATION RATIO

The deviation ratio gives the relative size of the closed-loop error produced by a change in setpoint compared with the size of the change in the setpoint.

$$\text{Deviation ratio} = \frac{1}{1 + g_o \underline{/\beta_o}} \tag{15.12}$$

MAXIMUM FREQUENCY LIMIT

The maximum frequency limit is the frequency above which a control system cannot respond accurately to changes in setpoint or disturbance. The controller cannot reduce the error and may actually increase the error. The maximum frequency limit is the frequency that separates control zone 1 from control zone 2.

Three Zones of Control

  Zone 1: error or deviation ratio < 0 dB, good control
  Zone 2: error or deviation ratio > 0 dB, poor control
  Zone 3: error or deviation ratio = 0 dB, no control

15.6 COMPUTER-AIDED BODE PLOTS

The determinations of the open-loop Bode diagram, the closed-loop Bode diagram, and the error ratio of a control system are tedious, time-consuming chores that are well suited to computer analysis. DESIGN is a BASIC program for computer-aided analysis and design of closed-loop control systems. In Chapter 16 DESIGN is used to design PID controllers and control system compensation networks. In this chapter, it is used to produce Bode diagrams, error-ratio graphs, and Nyquist diagrams on the computer screen and a data table on the printer.

DESIGN is included on a disk that comes with this book. The program is not copy protected, and distribution of the program is permitted and encouraged.

DESIGN begins with the input of the transfer functions of the components that make up the control system. The system can have up to 10 polynomial transfer functions and up to 10

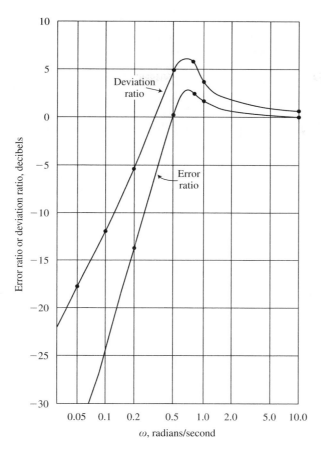

◆ **Figure 15.7**   Graph of the error ratio and the deviation ratio for the control system described by the Bode diagrams in Figure 15.4.

dead-time delays. All components are assumed to be connected in series to form a closed-loop system. Each transfer function is a polynomial of the following form:

$$TF = \frac{A(0) + A(1)s + A(2)s^2 + A(3)}{B(0) + B(1)s + B(2)s^2 + B(3)}$$

Upon completion of the input of the transfer functions, DESIGN computes and stores analysis data for a frequency range from 1×10^{-6} to 5.6×10^5 rad/s. The analysis data include the data necessary to plot the open-loop Bode graph, the closed-loop Bode graph, the error-ratio graph, and the Nyquist diagram of the control system.

When the analysis data have been stored, the program switches to the high-resolution graphics mode and puts the open-loop Bode diagram on the screen. The top line of the screen presents controller design parameters that are used in Chapter 16 and can be ignored in this chapter. The second line is the command line. The commands (Dmode), (Imode), and (Pmode) are also used only in Chapter 16. In this chapter, only the (Analysis), (Zoom), (Unzoom), and (Quit) commands are used. To execute a command, simply press the first letter of the command. For example, press the A-key to execute the (Analysis) command.

As the name implies, the (Zoom) command provides a close-up look at the graph on the screen. The (Unzoom) command returns the screen to its original scale. The (Analysis) command brings up a new command line with the commands (Closed-loop), (Nyquist), (Error-ratio), and (Open-loop). The (Closed-loop) command puts the closed-loop Bode diagram on the screen. The (Nyquist) command puts the Nyquist diagram on the screen, and the (Error-ratio) command puts the error-ratio graph on the screen.

Use the (Quit) command to exit the program. Before the program quits, it queries the user for the option to print a data table. A Yes response will result in a printed copy of the analysis data. For convenience, the analysis data table is terminated when the phase angle falls below $-999°$.

Example 15.4 illustrates the use of DESIGN to analyze a closed-loop control system.

EXAMPLE 15.4

A control system consists of a proportional controller and a first-order lag plus dead-time process. The following values were obtained from a step response test of the process:

$$\text{Dead-time} = t_d = 2 \text{ s}$$

$$\text{Time constant} = \tau = 100 \text{ s}$$

Therefore,

$$H(s) = e^{-2s}\left(\frac{1}{1 + 100s}\right)$$

The controller gain is set at 40, so $G(s) = [40]$. Combining $G(s)$ and $H(s)$ gives the following open-loop transfer function:

$$\text{TF(open-loop)} = G(s)H(s) = \left(\frac{40}{1 + 100s}\right)e^{-2s}$$

Use DESIGN to produce the following:

1. An open-loop Bode graph
2. A closed-loop Bode graph
3. An error-ratio graph
4. A printed Bode data table for open-loop, closed-loop, and error-ratio graphs

Solution

The following inputs were used in a run of DESIGN

```
Transfer function coefficients of component number 1:
        A(0) = 40,    A(1) = 0,    A(2) = 0,    A(3) = 0
        B(0) = 1,    B(1) = 100,    B(3) = 0,    B(3) = 0
Dead time delay number 1 = 2
```

The results of the run are as follows:

1. The open-loop Bode graph: Figure 15.8
2. The closed-loop Bode graph: Figure 15.9
3. The error-ratio graph: Figure 15.10
4. The Bode data: Table 15.2

◆

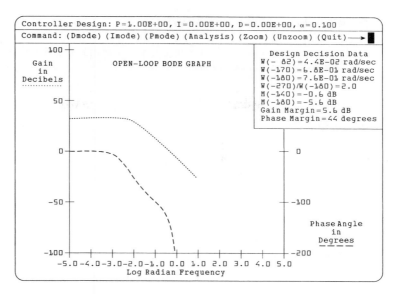

◆ **Figure 15.8** Open-loop Bode graph for the control system in Example 15.4 as displayed by DESIGN. The dotted line on the top is the open-loop gain; the dashed line on the bottom is the open-loop phase angle. The Controller Design line on the top and the Design Data table are used in Chapter 16, and can be ignored in Chapter 15.

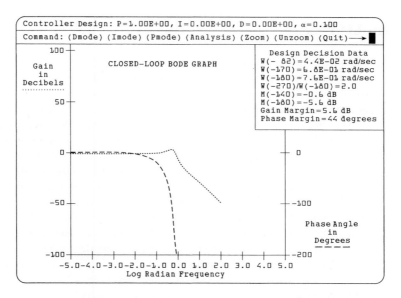

◆ **Figure 15.9** Closed-loop Bode graph for the control system in Example 15.4 as displayed by DESIGN. The dotted line is the closed-loop gain; the dashed line is the closed-loop phase angle.

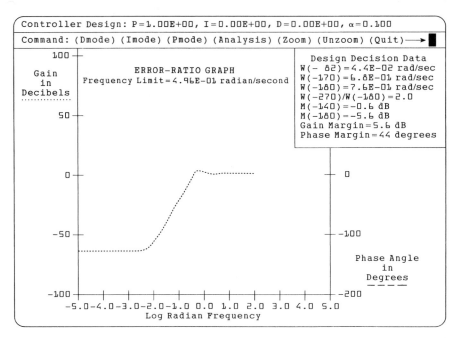

◆ **Figure 15.10** Error-ratio graph for the control system in Example 15.4 as displayed by DESIGN.

15.7 STABILITY

The possibility of sustained oscillations always exists in a closed-loop control system. When a system goes into a continuous oscillation, it is said to be unstable. *Stability* refers to the ability of a control system to dampen out any oscillations that result from an upset. The analysis of a control system seeks the answer to the question: Is the control system stable? The design of a control system involves the determination of controller parameters that will minimize the error produced by a disturbance without making the system unstable.

Every closed-loop control system has a tendency to produce an oscillation in the controlled variable. This tendency for self-oscillation is caused by the presence of the feedback signal and is directly related to $\omega_{-180°}$, the frequency at which the open-loop phase angle is equal to $-180°$. If the open-loop gain at $\omega_{-180°}$ is less than 1, the oscillations will diminish in size with each successive cycle. If the open-loop gain at $\omega_{-180°}$ is equal to 1, the oscillation will continue at a constant amplitude. If the open-loop gain at $\omega_{-180°}$ is greater than 1, the oscillation will increase in size with each successive cycle. If the oscillations diminish, the control system is said to be stable. If not, the system is said to be unstable. Therefore, *a stable control system is one in which the open-loop gain is less than 1 when the open-loop phase angle is $-180°$.*

Figure 15.11 illustrates a control system with a sustained oscillation. Graphs of the setpoint, *SP*, error *E*, and controlled variable, *C*, are located directly above the corresponding signal lines. Figure 15.11 traces a single half-wave, sinusoidal pulse as it travels around the

◆ **TABLE 15.2** Bode Data for Example 15.4

Frequency (rad/ sec)	Open-Loop Response		Closed-Loop Response		Error ratio (dB)
	Gain (dB)	Angle (°)	Gain (dB)	Angle (°)	
1.0E − 06	32.0	−0.0	−0.2	−0.0	−64.1
1.8E − 06	32.0	−0.0	−0.2	−0.0	−64.1
3.2E − 06	32.0	−0.0	−0.2	−0.0	−64.1
5.6E − 06	32.0	−0.0	−0.2	−0.0	−64.1
1.0E − 05	32.0	−0.1	−0.2	−0.0	−64.1
1.8E − 05	32.0	−0.1	−0.2	−0.0	−64.1
3.2E − 05	32.0	−0.2	−0.2	−0.0	−64.1
5.6E − 05	32.0	−0.3	−0.2	−0.0	−64.1
1.0E − 04	32.0	−0.6	−0.2	−0.0	−64.1
1.8E − 04	32.0	−1.0	−0.2	−0.0	−64.1
3.2E − 04	32.0	−1.8	−0.2	−0.0	−64.1
5.6E − 04	32.0	−3.3	−0.2	−0.1	−64.0
1.0E − 03	32.0	−5.8	−0.2	−0.1	−64.0
1.8E − 03	31.9	−10.3	−0.2	−0.3	−63.8
3.2E − 03	31.6	−17.9	−0.2	−0.5	−63.2
5.6E − 03	30.8	−30.0	−0.2	−0.8	−61.7
1.0E − 02	29.0	−46.1	−0.2	−1.4	−58.1
1.8E − 02	25.8	−62.7	−0.2	−2.5	−51.7
3.2E − 02	21.6	−76.1	−0.2	−4.5	−43.3
5.6E − 02	16.9	−86.4	−0.2	−8.0	−34.0
1.0E − 01	12.0	−95.7	−0.1	−14.4	−24.5
1.8E − 01	7.0	−107.2	0.3	−26.1	−15.4
3.2E − 01	2.0	−124.4	1.4	−49.7	−6.7
5.6E − 01	−3.0	−153.4	3.4	−112.3	1.9
1.0E + 00	−8.0	−204.0	−4.3	−218.4	0.9
1.8E + 00	−13.0	−293.5	−13.9	−304.2	−0.3
3.2E + 00	−18.0	−452.2	−18.0	−445.0	−0.1
5.6E + 00	−23.0	−734.3	−23.5	−733.4	0.0

closed loop. The system has an open-loop gain of 1 and an open-loop phase angle of $-180°$. In Figure 15.11a, the half-wave pulse is introduced at the setpoint. This is the only signal that appears at the setpoint, and the setpoint is zero for the remainder of the trace. The half-wave input pulse appears immediately on the error signal graph, but does not appear on the controlled variable graph until the error signal line has seen the entire 180° half-wave input pulse.

In Figure 15.11b, the half-wave pulse has moved through the process block and now appears as pulse 3 on the controlled variable graph. There is no time delay through the error detector, and pulse 4 appears on the error graph as the inversion of pulse 3. Notice how pulse 4 completes one cycle of the sine wave on the error graph. The dashed lines for pulses 1 and 2 indicate that those pulses occurred at an earlier time than the current pulses (3 and 4).

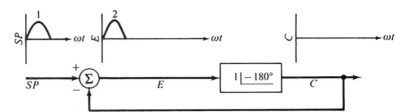

a) A single half-wave pulse (1) is introduced at the setpoint (*SP*). Since
$E = SP - C$ and $C = 0$, the error (2) is identical to the setpoint (1).

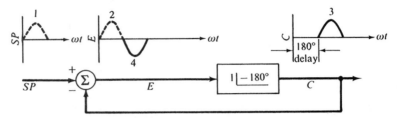

b) The system delays pulse 2 by 180°, forming pulse 3 at output (*C*).
The error detector inverts pulse 3, forming error pulse 4. Notice how
pulse 4 forms the next half of a sine wave.

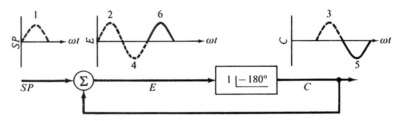

c) The system delays pulse 4 by 180°, forming pulse 5 at the output
(*C*). The error detector inverts pulse 5, forming error pulse 6. Notice
how pulse 6 continues the formation of a sine wave.

◆ **Figure 15.11**   Sustained oscillations in a closed-loop system.

In Figure 15.11c, the half-wave pulse has again moved through the process block to form
pulse 5 on the controlled variable graph and inverted pulse 6 on the error graph. Notice how
pulse 6 continues the formation of the sine wave on the error graph.

> A control system will oscillate with a constant amplitude if the open-loop gain is 1 at the
> frequency for which the open-loop phase angle is $-180°$.

Figure 15.12 illustrates a control system with a diminishing oscillation. Figure 15.12
traces a single half-wave pulse as it travels around the closed loop, just as Figure 15.11 did.
This time, however, the open-loop gain is 0.5 instead of 1.0. The open-loop phase angle re-

mains at $-180°$. Notice how the gain of 0.5 reduces the amplitude of the sine wave as it is formed by the cycling pulse.

> A control system will oscillate with a diminishing amplitude if the open-loop gain is less than 1 at the frequency for which the open-loop phase angle is $-180°$.

Figure 15.13 illustrates a control system with an increasing oscillation. Figure 15.13 traces a single half-wave pulse as it travels around the closed loop. Here, the open-loop gain

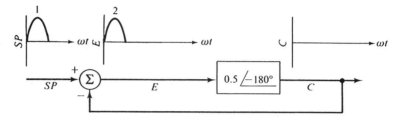

a) A single half-wave pulse (1) is introduced at the setpoint (*SP*). The error (2) is identical to the setpoint (1).

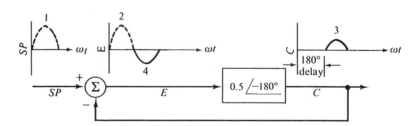

b) The system delays pulse 2 by 180° and halves the amplitude, forming pulse 3 at output (*C*). The error detector inverts pulse 3, forming error pulse 4. Notice how pulse 4 forms the next half of a diminishing sine wave.

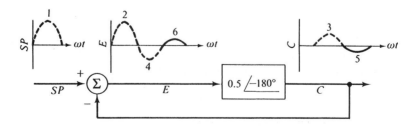

c) The system delays and halves pulse 4, forming pulse 5 at the output (*C*). The error detector inverts pulse 5, forming error pulse 6. Notice how pulse 6 continues the formation of a diminishing sine wave.

◆ **Figure 15.12**   Diminishing oscillations in a closed-loop control system.

is 2 while the open-loop phase angle remains at $-180°$. Notice how the gain of 2 increases the amplitude of the sine wave as it is formed by the cycling pulse.

> A control system will oscillate with an increasing amplitude if the open-loop gain is greater than 1 at the frequency for which the open-loop phase angle is $-180°$.

A complete discussion of stability requires a level of mathematics that is beyond the scope of this book. However, we can expand on the intuitive concept of stability developed in

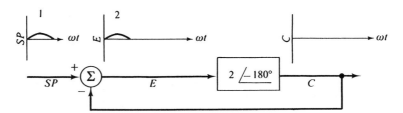

a) A single half-wave pulse (1) is introduced at the setpoint (*SP*). The error (2) is identical to the setpoint (1).

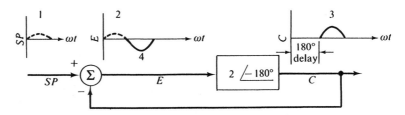

b) The system delays pulse 2 by 180° and doubles the amplitude, forming pulse 3 at the output (*C*). The error detector inverts pulse 3, forming error pulse 4. Notice how pulse 4 forms the next half of an increasing sine wave.

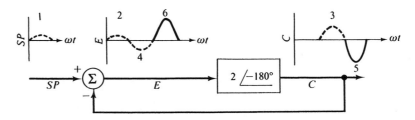

c) The system delays and doubles pulse 4, forming pulse 5 at the output (*C*). The error detector inverts pulse 5, forming error pulse 6. Notice how pulse 6 continues the formation of an increasing sine wave.

◆ **Figure 15.13** Increasing oscillations in a closed-loop control system.

the preceding discussion. Another view of stability can be obtained from an examination of the open- and closed-loop transfer functions given by Equations (15.5) and (15.6).

$$\text{Open-loop TF} = \frac{C_m}{SP} = G(s)H(s)$$

$$\text{Closed-loop TF} = \frac{C_m}{SP} = \frac{G(s)H(s)}{1 + G(s)H(s)}$$

A control system is clearly unstable if the denominator of the closed-loop transfer function is equal to zero. This condition is satisfied when the open-loop transfer function is equal to -1, which occurs when the open-loop gain is 1 and the open-loop phase angle is $-180°$. The same conclusion was reached in the discussion of Figure 15.11.

A control system is unstable if the open-loop gain is equal to or greater than 1 at a frequency where the open-loop phase angle is $-180°$.

Our final view of stability involves the use of the inverse Laplace transformation to obtain the time-domain impulse response of a control system. We begin by assuming a polynomial, open-loop transfer function of the following general form:

$$G(s)H(s) = \frac{A_0 + A_1 s + A_2 s^2 + A_3 s^3}{B_0 + B_1 s + B_2 s^2 + B_3 s^3}$$

The denominator of the closed-loop transfer function is

$$1 + G(s)H(s) = 1 + \frac{A_0 + A_1 s + A_2 s^2 + A_3 s^3}{B_0 + B_1 s + B_2 s^2 + B_3 s^3}$$

$$= \frac{(A_0 + B_0) + (A_1 + B_1)s + (A_2 + B_2)s^2 + (A_3 + B_3)s^3}{B_0 + B_1 s + B_2 s^2 + B_3 s^3}$$

The closed-loop transfer function is

$$\frac{C_m}{SP} = \frac{G(s)H(s)}{1 + G(s)H(s)} = \frac{A_0 + A_1 s + A_2 s^2 + A_3 s^3}{(A_0 + B_0) + (A_1 + B_1)s + (A_2 + B_2)s^2 + (A_3 + B_3)s^3}$$

Assume that p_1, p_2, and p_3 are the roots of the denominator, and z_1, z_2, and z_3 are the roots of the numerator. The closed-loop transfer function can then be written as follows:

$$\frac{C_m}{SP} = \frac{(s - z_1)(s - z_2)(s - z_3)}{(s - p_1)(s - p_2)(s - p_3)}$$

Next, we apply a unit impulse at the setpoint. The unit impulse is chosen because the Laplace transformation of the unit impulse is 1, giving us the simplest possible function to work with.

$$C_m = \frac{C_m}{SP}[SP] = \frac{(s - z_1)(s - z_2)(s - z_3)}{(s - p_1)(s - p_2)(s - p_3)}[1]$$

$$C_m(s) = \frac{(s - z_1)(s - z_2)(s - z_3)}{(s - p_1)(s - p_2)(s - p_3)}$$

Finally, we do a partial fraction expansion to get the inverse Laplace transformation.

$$C_m(s) = \frac{K_1}{s - p_1} + \frac{K_2}{s - p_2} + \frac{K_3}{s - p_3}$$

By inverse Laplace transformation,

$$C_m(t) = K_1 e^{p_1 t} + K_2 e^{p_2 t} + K_3 e^{p_3 t}$$

The equation for $C_m(t)$ shows that p_1, p_2, and p_3 must all be negative for $C_m(t)$ to remain bounded as t increases. For example, if $p_1 = 2$, the term $K_1 e^{p_1 t} = K_1 e^{2t}$ will increase exponentially as t increases.

A control system is stable if all the roots of $1 + G(s)H(s) = 0$ have negative real parts.

15.8 GAIN AND PHASE MARGIN

The open-loop Bode diagram of a control system provides the information necessary to determine the stability of the system. It also gives the designer an idea of the stability safety margin, that is, how much the gain or phase angle can change before the system will become unstable. The designer also uses the open-loop Bode diagram as a design tool—a topic covered in Chapter 16.

The stability of a closed-loop control system is defined by either of the following stability conditions:

Stability Conditions

1. A control system is stable only if the open-loop gain is less than 1 (or 0 dB) at the frequency for which the phase angle is $-180°$.
2. A control system is stable only if the phase angle is greater than $-180°$ (i.e., the phase lag is less than 180°) at the frequency for which the gain is 1 (or 0 dB).

The two stability conditions suggest two margins of safety for the stability of a closed-loop control system. First, the gain at the $-180°$ frequency must be a safe amount less than 1 (or 0 dB). We call this the *gain margin*. Second, the phase angle at the 0 dB frequency must be a safe amount above $-180°$. We call this the *phase margin*. Standard practice is a gain margin of 6 dB and a phase margin of 40°.

The designer considers a closed-loop control system to be *stable* only if it satisfies both the gain margin and the phase margin. In other words, a control system is considered to be stable only if the open-loop gain is less than 0.5 (or -6 dB) at the $-180°$ frequency, and the phase angle is greater than $-140°$ at the 0 dB frequency.

The designer considers a control system to be *marginally stable* if it satisfies the stability conditions but does not satisfy both the gain margin and the phase margin criteria.

> 1. A closed-loop control system is *stable* if it has a gain margin of at least 6 dB and a phase margin of at least 40°.
> 2. A closed-loop control system is *marginally stable* if it satisfies the stability conditions, but its gain margin is less than 6 dB or its phase margin is less than 40°, or both.
> 3. A closed-loop control system is *unstable* if its open-loop gain is 1 or greater at the −180° frequency.

The designer can read the gain and phase margins directly from the open-loop Bode diagram of the control system. In doing this, the designer locates two frequencies on the Bode diagram. One is the frequency at which the open-loop gain is 0 dB. This frequency is called $\omega_{0\ dB}$. The other is the frequency at which the phase angle is −180°. This frequency is called $\omega_{-180°}$. The designer uses $\omega_{0\ dB}$ and $\omega_{-180°}$ to read the gain and phase margins as illustrated in Figure 15.14. The gain margin is obtained by reading the open-loop gain at $\omega_{-180°}$. We will

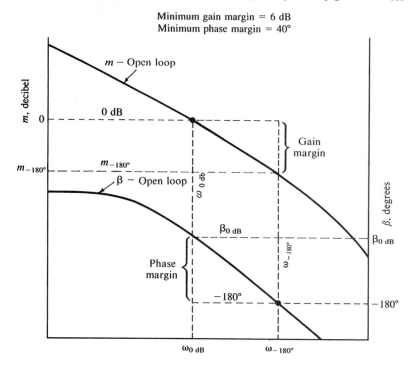

◆ **Figure 15.14**   The control system designer uses $m_{-180°}$ (the gain at the −180° frequency) and $\beta_{0\ dB}$ (the phase angle at the 0 dB frequency) to determine the gain margin and the phase margin. The gain margin is equal to $-m_{-180°}$, and the phase margin is equal to $180 + \beta_{0\ dB}$.

call this gain value $m_{-180°}$. The gain margin is equal to $-m_{-180°}$. The phase margin is obtained by reading the open-loop phase angle at ω_0 dB. We will call this phase angle value β_0 dB. The phase margin is equal to $180 + \beta_0$ dB.

Gain and Phase Margins

      Gain margin $= -m_{180°}$

      Phase margin $= 180 + \beta_0$ dB

EXAMPLE 15.5

Plot the open-loop Bode diagram for the control system in Example 15.4 and determine the gain and phase margins.

Solution

The easiest way to plot the open-loop Bode diagram is to use the Bode data table produced by DESIGN (see Table 15.2). However, a certain amount of calculator computation and hand plotting of Bode data can enhance the learning process. With that in mind, we proceed to compute four Bode data points.

    The first step in computing Bode data is to substitute $j\omega$ for s in the open-loop transfer function. Now for any given frequency (ω_i), the transfer function can be reduced to a single complex number in polar form. The magnitude and angle of this complex number are the open-loop gain and the open-loop phase angle at the given frequency (ω_i). From Example 15.4, we have the following open-loop transfer function:

$$\text{TF(open-loop)} = (40) \left(\frac{1}{1 + 100\ s} \right) e^{-2s}$$

Replacing s by $j\omega$, we obtain the following:

$$\text{Open-loop gain and phase angle} = (40) \left(\frac{1}{1 + j100\omega} \right) e^{-j2\omega}$$

(*Note:* The dead-time delay term, $e^{-2\omega}$, defines a complex number with a magnitude of 1 and an angle of 2ω rad. We must multiply this angle by 57.3 to convert it to degrees.)

    (a) $\omega = 0.001$ rad/s

$$\text{Gain and phase angle} = (40) \left(\frac{1}{1 + j0.1} \right) e^{-j0.002(57.3)}$$

$$= (40)(0.9950\ \underline{/-5.71°})(1\ \underline{/-0.11°})$$

$$= 39.8\ \underline{/-5.8°}$$

The decibel gain $= 20[\log_{10}(39.8)] = 32.0$ dB

At $\omega = 0.001$ rad/s:

      open-loop gain $= 32.0$ dB

    open-loop phase angle $= -5.8°$

    (b) $\omega = 0.01$ rad/s

$$\text{Gain and phase angle} = (40) \left(\frac{1}{1 + j1} \right) e^{-j0.02(57.3)}$$

$$= (40)(0.707) \ \underline{/-45.0°} \ (1 \ \underline{/-1.1°})$$

$$= 28.28 \ \underline{/-46.1°}$$

The decibel gain $= 20[\log_{10}(28.28)] = 29.0$ dB

At $\omega = 0.001$ rad/s:

$$\text{open-loop gain} = 29.0 \text{ dB}$$

$$\text{open-loop phase angle} = -46.1°$$

(c) $\omega = 0.1$ rad/s

$$\text{Gain and phase angle} = (40)\left(\frac{1}{1 + j10}\right)e^{-j0.2(57.3)}$$

$$= (40)(0.0995 \ \underline{/-84.3°})(1 \ \underline{/-11.5°})$$

$$= 3.98 \ \underline{/-95.7°}$$

The decibel gain $= 20[\log_{10}(3.98)] = 12.0$ dB

At $\omega = 0.001$ rad/s:

$$\text{open-loop gain} = 12.0 \text{ dB}$$

$$\text{open-loop phase angle} = -95.7°$$

(d) $\omega = 1$ rad/s

$$\text{Gain and phase angle} = (40)\left(\frac{1}{1 + j100}\right)e^{-j2(57.3)}$$

$$= (40)(0.01\underline{/-89.4°})(1\underline{/-114.6°})$$

$$= 0.40\underline{/-204.0°}$$

The decibel gain $= 20[\log_{10}(0.40)] = -8.0$ dB

At $\omega = 1$ rad/s:

$$\text{open-loop gain} = -8.0 \text{ dB}$$

$$\text{open-loop phase angle} = -204.0°$$

Examination of Table 15.2 reveals complete agreement between the calculator results and DESIGN results. Refer to Figures 14.7 and 14.16 for further insight concerning the frequency response of first-order lags and dead-time delays.

The Bode diagram is shown in Figure 15.15. The following approximate values are obtained from the Bode diagram:

$$m_{-180°} = -6 \text{ dB}$$

$$\beta_{0 \text{ dB}} = -135°$$

The gain margin $= -m_{-180°} = 6$ dB

The phase margin $= 180° + \beta_{0 \text{ dB}} = 180° + (-135°) = 45°$

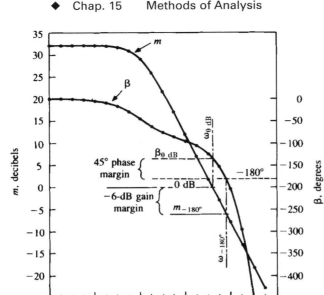

◆ **Figure 15.15** The open-loop Bode diagram is used to determine the gain margin and phase margin of a control system (see Example 15.5).

Exact values from DESIGN are a gain margin of 5.6 dB and a phase margin of 44°. Therefore, we conclude that the control system is, in the strictest sense, marginally stable. However, if the gain margin is rounded to the nearest integer, 6, and we accept the rounded value as the gain margin, we could conclude that the control system is stable. ◆

15.9 NYQUIST STABILITY CRITERION

The *Nyquist stability criterion* is a graphic method of determining if the function $1 + G(s)H(s)$ has any positive roots. A Nyquist diagram is a plot, in a complex plane, of the open-loop gain and phase angle as the frequency, ω, is varied from 0 to infinity. In plotting a Nyquist diagram, the gain and phase angle are treated as the magnitude and angle of a complex number in polar form. Nyquist diagrams are presented as a polar plot; however, the gain and phase angle could be converted to rectangular form and plotted on rectangular coordinates with the same result. *In plotting a Nyquist diagram, decibel values must be converted to gain values.* The following equation can also be used to convert from decibel, m, to gain, g:

$$g = 10^{m/20} \tag{15.13}$$

If the control system under consideration is open-loop stable, the Nyquist criterion reduces to the observation of whether or not the Nyquist plot encloses the point $-1 + j0$. If it does, then $1 + G(s)H(s)$ has positive roots and the system is unstable. If the Nyquist plot does not enclose the point $-1 + j0$, then $1 + G(s)H(s)$ has only negative roots and the system is stable or marginally stable.

Figure 15.16 shows the Nyquist plot of a stable closed-loop control system with a close-up view showing the three points used to determine the stability of the system. The first point is the *stability point,* which has the rectangular coordinates $-1 + j0$ or polar coordinates $1 \; \underline{/-180°}$.

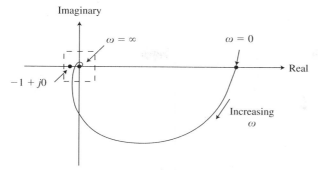

a) Nyquist diagram of a stable control system

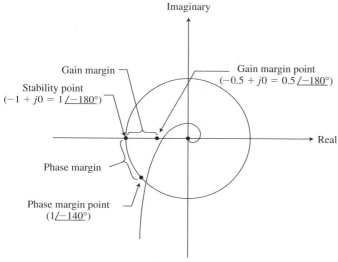

b) Close-up view of the boxed region in part a

◆ **Figure 15.16** The Nyquist diagram of a stable closed-loop control system and a close-up view showing the gain margin and phase margin. Observe how the gain and phase margins combine to keep the Nyquist plot a "safe" distance from the stability point.

The second point is the *gain margin point,* with coordinates $-0.5 + j0$ rectangular, or $0.5 \angle{-180°}$ polar. The third point is the *phase margin point,* with polar coordinates $1\angle{-140°}$. The following Nyquist criterion applies to all control systems that are open-loop stable (all systems in this book are open-loop stable).

Nyquist Stability Criterion

> A closed-loop control system that is open-loop stable is
>
> 1. *Unstable* if the Nyquist plot encircles the stability point ($-1 + j0 = 1 \angle{-180°}$) (see Figure 15.16).
>
> 2. *Marginally stable* if the Nyquist plot does not encircle the stability point ($-1 + j0$), but does encircle the gain margin point ($-0.5 + 0$), or the phase margin point ($1 \angle{-140°}$), or both the gain margin and phase margin points.
>
> 3. *Stable* if the Nyquist plot does not encircle the gain margin point ($-1 + j0$) and also does not encircle the phase margin point ($1 \angle{-140°}$).

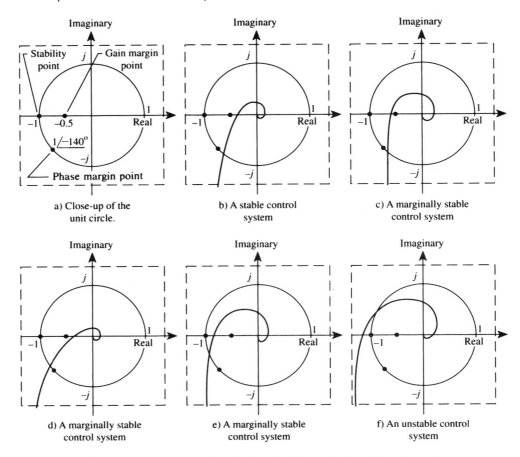

◆ **Figure 15.17** Close-up Nyquist plots of stable, marginally stable, and unstable
control systems. The control system in (c) does not meet the gain margin criterion;
the system in (d) does not meet the phase margin criterion; and the system in (e)
does not meet either the gain or phase margin criterion.

Figure 15.17 shows Nyquist plots of stable, marginally stable, and unstable control systems. The marginally stable systems fall into one of three categories: systems with an acceptable phase margin but an unacceptable gain margin, systems with an acceptable gain mar-
gin but an unacceptable phase margin, and systems with unacceptable gain and phase margins.

EXAMPLE 15.6   Plot a Nyquist diagram for the control system in Example 15.4 and determine if the system is stable or unstable.

Solution   The open-loop data from Example 15.4 will be used to construct the Nyquist diagram. The following table of values was taken from the data table in Example 15.4. Equation (15.13) was used to convert the decibel values to gain values.

ω (rad/s)	m (dB)	g	β (°)
0.001	32.0	39.8	−5.8
0.010	29.0	28.2	−46.1
0.018	25.8	19.5	−63.0
0.032	21.5	11.9	−76.3
0.056	16.9	7.00	−86.3
0.10	12.0	3.98	−95.7
0.18	6.9	2.21	−107.4
0.32	1.9	1.24	−124.9
0.56	−2.9	0.72	−153.1
1.0	−8.0	0.40	−204.0
1.8	−13.1	0.22	−295.9
3.2	−18.1	0.12	−456.5
5.6	−22.9	0.07	−731.6

The Nyquist plot is shown in Figure 15.18. The plot does not enclose the point $-1 + j0$, so the control system is stable.   ◆

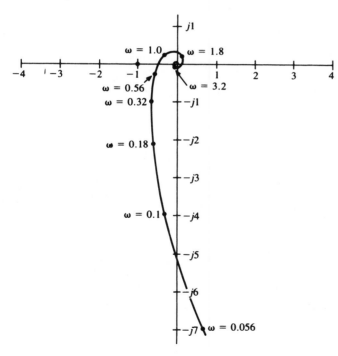

◆ **Figure 15.18** The Nyquist diagram is used to determine the stability of a control system (see Example 15.6).

15.10 ROOT LOCUS

One condition for stability of a closed-loop control system is that all roots of the following equation must have negative real parts:

$$1 + G(s)H(s) = 0 \qquad (15.14)$$

Equation (15.14) is referred to as the *closed-loop characteristic equation*. Recall that $G(s)H(s)$ is the open-loop transfer function and Equation (15.14) is the denominator of the closed-loop transfer function as defined by Equation (15.6).

The *root locus* is a graphical method of analysis in which all possible roots of the closed-loop characteristic equation are plotted as the controller gain, K, is varied from 0 to infinity. This graph of all possible roots gives the designer considerable insight into the relationship between gain, K, and the stability of the system. The root-locus method not only answers the question of stability, but also gives the designer information such as the damping ratio or damping constant of the system. The root-locus method was developed by W. R. Evans, who invented a device called a spirule that simplified the construction of root-locus plots. Computer software that produces printed root-locus plots is also available.

We begin with three simple examples to show how the roots of the characteristic equation change with changes in gain, K. We will use s_o to represent any value of s that satisfies the characteristic equation. Our first example has the following open-loop transfer function:

$$G(s)H(s) = \frac{K}{s + 2}$$

The characteristic equation is

$$1 + \frac{K}{s_o + 2} = 0$$

Solving for s_o:

$$s_o = -(K + 2)$$

Thus the root locus begins at $s_o = -2$ when $K = 0$ and moves to the left on the real axis as K increases. Figure 15.19a shows the root locus of example 1.

Our second example has the following open)^2-loop transfer function, characteristic equation, and solution:

$$G(s)H(s) = \frac{K}{(s + 2)^2}$$

$$1 + \frac{K}{(s_o + 2)^2} = 0$$

$$s_o = -2 \pm j \sqrt{K}$$

Thus the root locus begins at $s_o = -2$ when $K = 0$ and moves both up and down along a vertical line as K increases. Figure 15.19b shows the root locus of example 2.

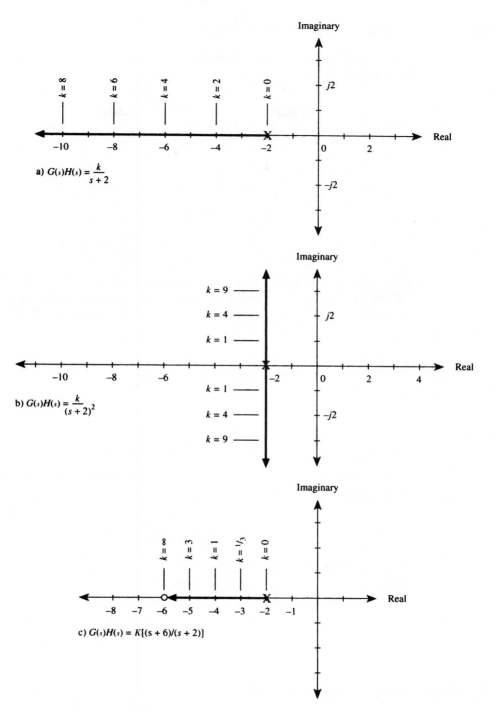

Figure 15.19 Root locus of $1 + G(s)H(s) = 0$ for three simple systems.

Our third example has the following open-loop transfer function, characteristic equation, and solution:

$$G(s)H(s) = \frac{K(s + 6)}{(s + 2)}$$

$$1 + \frac{K(s_o + 6)}{(s_o + 2)} = 0$$

$$s_o + 2 + K(s_o + 6) = 0$$

$$s_o(K + 1) + 6K + 2 = 0$$

$$s_o = -\left(\frac{6K + 2}{K + 1}\right)$$

The root locus is the plot of s_o as K varies from 0 to infinity. Our first observation is that s_o will always be a negative real number, and the magnitude of s_o will increase as K increases. We really only need the two endpoints of the locus, the values of s_o when $K = 0$ and $K = \infty$.

When $K = 0$,

$$s_o = -\left(\frac{6K + 2}{K + 1}\right)\bigg|_{K=0} = -2$$

When $K = \infty$,

$$s_o = -\left(\frac{6K + 2}{K + 1}\right)\bigg|_{K=\infty}$$

The last expression will be easier to evaluate if we divide the numerator and denominator of the fraction by K.

$$s_o = -\left(\frac{6 + 2/K}{1 + 1/K}\right)\bigg|_{K=\infty} = -6$$

Figure 15.19c shows the root-locus diagram of this system. The following table of additional values of s_o shows how s_o varies as K increases:

K	$\frac{1}{3}$	1	3	39	399	3999
s_o	3	−4	−5	−5.9	−5.99	−5.999

EXAMPLE 15.7 Determine the root locus of a control system with the following open-loop transfer function:

$$G(s)H(s) = \frac{K}{s(s + 10)}$$

Solution

$$1 + G(s_o)H(s_o) = 1 + \frac{K}{s_o(s_o + 10)} = \frac{s_o(s_o + 10) + K}{s_o(s_o + 10)}$$

The characteristic equation is

$$s_o(s_o + 10) + K = s_o^2 + 10s_o + K = 0$$

The roots are given by the quadratic formula:

$$s_1 = -5 + \sqrt{25 - K}$$

$$s_2 = -5 - \sqrt{25 - K}$$

The following table of values shows how s varies as K increases:

K	s_1	s_2
0	0	-10
9	-1	-9
16	-2	-8
21	-3	-7
24	-4	-6
25	-5	-5
50	$-5 + j5$	$-5 - j5$
125	$-5 + j10$	$-5 - j10$

Figure 15.20 shows the root-locus diagram of this system.   ◆

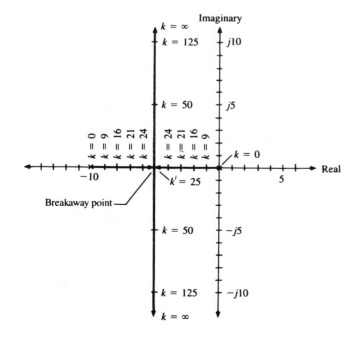

◆ **Figure 15.20**   Root locus of $1 + G(s)H(s)$ $= 0$ for the system with $G(s)H(s) = k/[s(s + 10)]$.

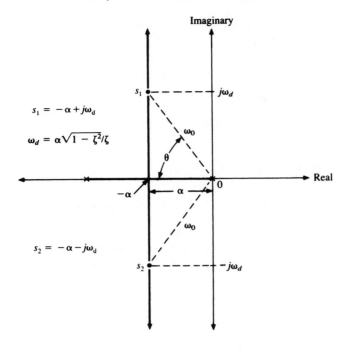

$s_1 = -\alpha + j\omega_d$

$\omega_d = \alpha\sqrt{1 - \zeta^2}/\zeta$

$s_2 = -\alpha - j\omega_d$

◆ **Figure 15.21** Root-locus diagram showing the relationship between the roots of the characteristic equation, the damping ratio ζ, and the damping coefficient α.

Figure 15.21 illustrates the relationship between the roots of the characteristic equation, the damping ratio, ζ, and the damping coefficient, α. The following symbols and terminology will be used in the following discussion:

$$\alpha = \text{damping coefficient, s}^{-1}$$

$$\omega_0 = \text{resonant frequency, rad/s}$$

$$\omega_d = \text{damped resonant frequency, rad/s}$$

$$\zeta = \text{damping ratio}$$

$$\theta = \text{operating point angle}$$

When the roots of the characteristic equation are imaginary, they always come in conjugate pairs given by the following equations:

$$s_1 = -\alpha + j\omega_d$$

$$s_2 = -\alpha - j\omega_d$$

Notice in Figure 15.21 that ω_0 is the hypotenuse of a right triangle, and ω_d and α are the two legs of the right triangle. Thus, by the Pythagorean theorem,

$$\omega_0 = \sqrt{\alpha^2 + \omega_d^2} \tag{15.15}$$

The damping ratio is given by

$$\zeta = \frac{\alpha}{\omega_0}$$

Therefore

$$\theta = \cos^{-1}\zeta \qquad\qquad\text{(15.16)}$$

and

$$\alpha = \zeta\sqrt{\alpha^2 + \omega_d^2}$$
$$\alpha^2 = \zeta^2(\alpha^2 + \omega_d^2)$$
$$\omega_d^2 = \frac{\alpha^2(1 - \zeta^2)}{\zeta^2}$$
$$\omega_d = \alpha\,\frac{\sqrt{1 - \zeta^2}}{\zeta} \qquad\qquad\text{(15.17)}$$

Equation (15.17) can be used to determine the value of K that will result in a specified damping ratio as illustrated in Example 15.8.

EXAMPLE 15.8   Find the value of gain, K, that will give a damping ratio of 0.5 in the control system in Example 15.7.

Solution   Equation (15.17) gives the required value of ω_d. From Example (15.7), $\alpha = 5$.

$$\omega_d = \frac{\alpha\sqrt{1 - \zeta^2}}{\zeta}$$
$$= \frac{5\sqrt{1 - 0.5^2}}{0.5}$$
$$= 8.66$$

In Example 15.7, when $(25 - K)$ is negative, the roots are imaginary and $\omega_d = \sqrt{K - 25}$.

$$\omega_d = \sqrt{K - 25} = 8.66$$
$$K - 25 = 8.66^2$$
$$K = 8.66^2 + 25 = 100$$

Thus the control system will have a damping ratio of 0.5 when the gain K is equal to 100.   ◆

The methods used in the previous examples are not adequate for the more realistic example that follows. Let us now consider the following general form of the open-loop transfer function:

$$G(s)H(s) = K\left[\frac{(s - z_1)(s - z_2)\dots(s - z_m)}{(s - p_1)(s - p_2)\dots(s - p_n)}\right] \qquad\text{(15.18)}$$

where $K > 0$

$z_1, z_2, \dots, z_m$ are the *zeros* of $G(s)H(s)$ (zeros are values of s for which $G(s)H(s) = 0$)

$p_1, p_2, \dots, p_n$ are the *poles* of $G(s)H(s)$ (poles are values of s for which $G(s)H(s) = \infty$)

Note that m is the number of zeros in $G(s)H(s)$ and n is the number of poles. We will be using those two numbers to compute certain characteristics of the root-locus plot. Consider also the following form of the closed-loop characteristic equation:

$$G(s)H(s) = -1 = 1\underline{/-180°} \qquad\qquad\text{(15.19)}$$

Any point s_o that makes the angle of $G(s_o)H(s_o)$ equal to an odd multiple of $\pm 180°$ will satisfy Equation (15.19). This is commonly called the *angle condition* of the closed-loop characteristic equation. All points that satisfy the following angle condition also satisfy the closed-loop characteristic equation and lie on the root locus:

Angle Condition of the Closed-Loop Characteristic Equation

$$\text{Angle of } [G(s_o)H(s_o)] = \pm N(180) \tag{15.20}$$

where $N = 1, 3, 5, 7, 9, \ldots = \{\text{set of odd integers}\}$

If s_o is a point on the root locus, then the value of K can be computed by the following *magnitude condition:*

Magnitude Condition of the Closed-Loop Characteristic Equation

$$K = \text{magnitude of } \left| \frac{(s_o - p_1)(s_o - p_2) \ldots (s_o - p_n)}{(s_o - z_1)(s_o - z_2) \ldots (s_o - z_m)} \right| \tag{15.21}$$

A detailed study of the angle condition has resulted in the development of the following rules for construction of root locus plots.*

Root-Locus Rules

$$G(s)H(s) = K \left[\frac{(s - z_1)(s - z_2) \ldots (s - z_m)}{(s - p_1)(s - p_2) \ldots (s - p_n)} \right] \tag{15.18}$$

$m = $ the number of zeros

$n = $ the number of poles

Begin the root-locus plot by drawing an s-plane diagram with equal scales for the real and imaginary axes. Then plot each zero and pole of $G(s)H(s)$ on the s-plane diagram. Use an X to mark the location of each pole and a 0 to mark the location of each zero. Then apply the following rules to plot the root locus of $1 + G(s)H(s)$:

1. The root locus will have n branches that begin at the n poles with the value $K = 0$.
2. The root locus will have m branches that will terminate on the m finite zeros as K approaches infinity.
3. The remaining $n - m$ branches will go to infinity along asymptotes as K approaches infinity.
4. The asymptotes meet at a point called the *centroid, C*, which is computed as follows:

$$C = \frac{\Sigma \text{ poles} - \Sigma \text{ zeros}}{(\text{number of poles}) - (\text{number of zeros})} \tag{15.22}$$

$$C = \frac{(p_1 + p_2 + \ldots + p_n) - (z_1 + z_2 + \ldots + z_m)}{(\text{number of poles}) - (\text{number of zeros})}$$

---

*Chester L. Nachtigal, *Instrumentation and Control Fundamentals and Applications* (New York: Wiley, 1990) pp. 630–31.

The $n - m$ asymptote angles are given by

$$\theta_N = \pm \frac{N(180°)}{n - m}; \quad N = 1, 3, 5, 7, 9, \ldots \tag{15.23}$$

For each N, 2 angles are computed. Use enough values of N to compute the required $n - m$ angles.

5. Begin at the right and number the zeros and poles that lie on the real axis without regard to whether the points are a zero or a pole. A typical numbering will appear as follows:

If there is an even number of points on the real axis, then the root locus will lie on the part of the real axis between each odd point and the following even point as illustrated below:

If there is an odd number of points on the real axis, then the root locus will extend to infinity from the highest numbered point as illustrated below:

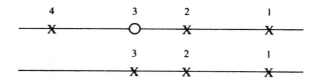

6. If two zeros are connected, there must be a breakin point between them.

 If two poles are connected, there must be a breakout point between them.

 If a pole and a zero are connected, they usually are a full branch, starting at the pole with $K = 0$ and ending at the zero as $K \to \infty$.

 Use rule 6 to draw arrows and break points as shown below:

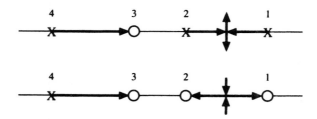

7. The break points occur where the derivative of K with respect to s is equal to zero. Computation of the break points is demonstrated in Example 15.9.

8. The points where the branches cross the imaginary axis are determined by replacing s by $j\omega$ in the characteristic equation and then solving for ω.

9. The angle condition may be used to determine the angle of departure from complex poles or the angle of arrival at complex zeros. Draw a line from each of the other zeros or poles to the zero or pole in question. Measure the angle formed by the positive real axis and the line to each of the other zeros and poles. The departure or arrival angle is equal to 180° plus the sum of all the zero angles minus the sum of all the pole angles.

10. The operating point for a given damping ratio, zeta (ζ), can be determined as follows. First, determine the operating point angle, $\theta = \cos^{-1}(\zeta)$. Then draw a line from the origin of the s-plane that forms an angle of θ with the negative real axis. This line intersects the root locus at the operating point. The length of the line from the origin to the operating point represents the natural frequency, ω_o.

11. The value of K at the operating point can be determined as follows. First, draw lines from each pole and each zero to the operating point. Measure the length of each line and determine the product of the lengths of the lines from the poles and the product of the lengths of the lines from the zeros. If there are no zeros (or poles), then use 1 as the product. The value of K at the operating point is equal to the poles product divided by the zeros product.

EXAMPLE 15.9

Determine the root locus of a control system with the following open-loop transfer function:

$$G(s)H(s) = \frac{K}{s(s + 3)(s + 9)}$$

Then determine the operating point, s_o, for a damping ratio of 0.5 ($\zeta = 0.5$). Also find the gain, K, and the resonant frequency, ω_0, at the operating point.

Solution

(a) The open-loop transfer function has no zeros and three poles at $s = 0$, $s = -3$, and $s = -9$. The root locus will have three branches that will all go to infinity along 3 asymptotes (Rules 1, 2, and 3). The poles are shown in Figure 15.22.

(b) The asymptotes meet at the centroid, C, determined by Equation (15.22) (Rule 4):

$$C = \frac{(\Sigma \text{ poles}) - (\Sigma \text{ zeros})}{(\text{number of poles}) - (\text{number of zeros})}$$

$$C = \frac{[0 + (-3) + (-9)] - [0]}{3 - 6} = -4$$

The asymptote angles are determined by Equation (15.23):

$$\theta_1 = \pm \frac{1(180)}{3 - 0} = \pm 60°$$

$$\theta_3 = \pm \frac{3(180)}{3 - 0} = \pm 180°$$

Therefore, the asymptote angles are $+60°$, $-60°$, and $+180°$.

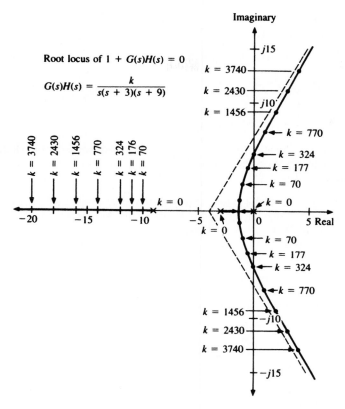

◆ **Figure 15.22**   Root locus of $1 + G(s)H(s)$ $= 0$ for the system with $G(s)H(s) = k/[s(s + 3)(s + 9)]$.

(c) The root locus lies on the portion of the real axis between $s = 0$ and $s = -3$, and the portion from $s = -9$ to $s = -\infty$ (Rule 5).

(d) There is a breakout point on the real axis between $s = 0$ and $s = -3$ (Rule 6).

(e) The breakout point occurs where the derivative of K with respect to s is equal to zero $(dK/ds = 0)$. To obtain this derivative, we first solve the characteristic equation for K:

$$1 + \frac{K}{s(s + 3)(s + 9)} = 0$$

$$K = -(s^3 + 12s^2 + 27s)$$

The power rule may be used to obtain dK/ds.

$$\frac{dK}{ds} = -(3s^2 + 24s + 27)$$

Now set $dK/ds = 0$ to obtain the breakout point.

$$\frac{dK}{ds} = -(3s^2 + 24s + 27) = 0$$

Dividing both sides of the equation by -3 and then applying the quadratic formula gives us two candidates for the desired breakout point.

$$s^2 + 8s + 9 = 0$$

$$s = \frac{-8 \pm \sqrt{64 - 36}}{2}$$

$$s = -1.35425 \text{ and } -6.64575$$

The breakout point is between 0 and -3, so $s = -1.35425$ is the desired breakout point (Rule 7).

(f) The points where the branches cross the imaginary axis are obtained from the characteristic equation with s replaced by $j\omega$.

$$s^3 + 12s^2 + 27s + K = 0$$

$$(j\omega)^3 + 12(j\omega)^2 + 27(j\omega) + K = 0$$

$$-j\omega^3 - 12\omega^2 + 27j\omega + K = 0$$

$$(K - 12\omega^2) + j(27\omega - \omega^3) = 0$$

Set the imaginary part equal to zero and solve for ω.

$$27\omega - \omega^3 = 0$$

$$\omega^2 = 27$$

$$\omega = \pm\sqrt{27} = \pm3\sqrt{3}$$

Set the real part equal to zero and solve for K.

$$K = 12\omega^2 = 12(27) = 324$$

Therefore, the branches cross the imaginary axis at $\omega = \pm3\sqrt{3}$ with $K = 324$ (Rule 8).

(g) The root-locus plot is shown in Figure 15.22. The characteristic equation, $s^3 + 2s^2 + 27s + K = 0$, involves the solution of a cubic equation, which is considerably more involved than a quadratic equation. However, the roots when $K = 0$ are easy to find. They are 0, -3, and -9. The next step is to use the fact that one of the three roots will always be real. Assume that the real root is $-a$, and divide the characteristic equation by the factor $(s + a)$ to get a general equation for the remaining two roots.

$$\frac{s^3 + 12s^2 + 27s + K}{s + a} = s^2 + (12 - a)s + 27 - a(12 - a)$$

The remainder of the division is $K - 27a + 12a^2 - a^3$. We can now express the characteristic equation as follows:

$$(s + a)[s^2 + (12 - a)s + 27 - a(12 - a)] = 0 \qquad \textbf{(15.24)}$$

The value of K must be such that the remainder is 0.

$$K = 27a - 12a^2 + a^3 \qquad \textbf{(15.25)}$$

A calculator or computer program may be used to obtain exact points on the root locus using Equations (15.24) and (15.25) and the quadratic formula. A trial-and-error procedure may also be employed to obtain the exact operating point. However, the following graphic analysis provided quick results with sufficient accuracy for most applications.

(h) The expanded view of the root locus in Figure 15.23 shows the determination of the operating point for a damping ratio of 0.5 (Rule 10). The line of constant damping ratio makes an angle of θ as given by Equation (15.16).

$$\theta = \cos^{-1} 0.5 = 60°$$

The operating point is at the intersection of the damping ratio, ζ, = 0.5 line and the root locus. From the graph, we obtain the approximate value $-1.12 + j\,1.95$ for the coordinates of the operating point. Using a calculator, the more exact value of the coordinates is $-1.125 + j1.9486$.

$$\text{Operating point} = -1.125 + j1.9486 = 2.25\,\underline{/60°}$$

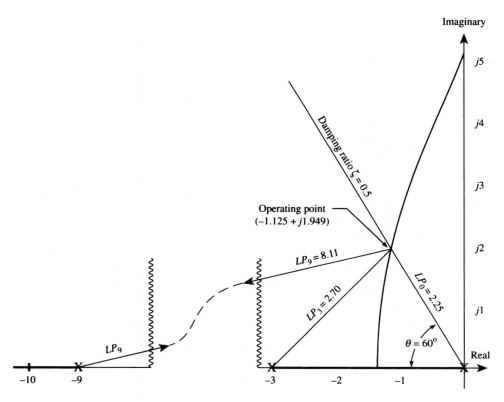

◆ **Figure 15.23**   Expanded view of the operating point region of the root locus shown in Figure 15.22. At the operating point the system has a damping ratio of 0.5 and a resonant frequency of 2.25 rad/s.

The resonant frequency is equal to the length of the line from the origin to the operating point. From the graph, this distance is estimated to be 2.25 (the same as the calculated value).

$$\text{Resonant frequency} = 2.25 \text{ rad/s}$$

(i) The value of K at the operating point is determined by the poles product divided by the zeros product (Rule 11). From the graph we obtain the following lengths:

$$LP_0 = \text{length from the pole at } (0,0) = 2.25$$

$$LP_3 = \text{length from the pole at } (-3,0) = 2.70$$

$$LP_9 = \text{length from the pole at } (-9,0) = 8.11$$

$$\text{Poles product} = (2.25)(2.70)(8.11) = 49.3$$

$$\text{Zeros product} = 1 \text{ (There are no zeros.)}$$

$$K = \frac{49.3}{1} = 49.3$$

The more exact calculated value is $K = 49.36$. ◆

◆ GLOSSARY

Angle condition: A condition used to locate points on the root locus of a control system. If s_o is on the root locus, then it will satisfy the closed-loop characteristic equation (i.e., $1 + G(s_o)H(s_o = 0)$), and the angle of $G(s_o)H(s_o)$ will equal an odd multiple of $\pm 180°$. (15.10)

Centroid: The point where the asymptotes of a root locus meet. (15.10)

Characteristic equation, closed-loop: The equation $(1 + G(s)H(s) = 0)$ that defines the poles of the closed-loop transfer function and is used to construct the root-locus plot. A necessary condition of stability is that all of the roots of this equation must have negative real parts. (15.10)

Closed-loop frequency response (Bode diagram): Graphs of frequency versus the gain and phase angle change from the setpoint to the measuring transmitter output with the transmitter output connected to the error detector. (15.4)

Closed-loop transfer function: The C_m/SP transfer function of a control system with the measuring transmitter output connected to the error detector. (15.4)

Deviation ratio: The ratio of the closed-loop error magnitude over the setpoint magnitude and a measure of how accurately a control system can follow a change in the setpoint (15.5)

Error ratio: The ratio of the closed-loop error magnitude over the open-loop error magnitude and a measure of the accuracy of a control system. (15.5)

Frequency limit: The maximum frequency of setpoint changes or disturbances that a control system can handle. A control system cannot follow setpoint changes or reduce the error caused by disturbances that are above its frequency limit. (15.5)

Gain margin: A safety factor for the stability of a closed-loop control system that specifies how much additional gain is required to make the system unstable. (15.8)

Gain margin point: A point on the Nyquist diagram, located at $-0.5 + j0$, that is used to determine if a closed-loop control system meets the gain margin criterion. (15.9)

Magnitude condition: A condition used to determine K for a point on the root locus of a control system. If s_o is on the root locus, then the magnitude of $G(s_o)H(s_o)$ will be equal to 1, and K can be determined accordingly. (15.10)

Marginally stable: A control system is marginally stable if it satisfies the stability conditions, but does not meet the gain margin, the phase margin, or both the gain and phase margin criteria. (15.8)

Nyquist stability criterion: A closed-loop control system is stable if the open-loop system is stable and the Nyquist plot does not encircle the stability point, $-1 + j0$. (15.9)

Open-loop frequency response (Bode diagram): Graphs of frequency versus the gain and phase angle change from the setpoint to the measuring transmitter output with the transmitter output disconnected from the error detector. (15.3)

Open-loop transfer function: The C_m/SP transfer function of a control system with the measuring transmitter output disconnected from the error detector. (15.3)

Overall gain: The amplitude of the output of the last of several components connected in series, divided by the amplitude of the input to the first component. The overall gain is equal to the product of the gains of the individual components. (15.2)

Overall phase angle: The phase angle of the output of the last of several components connected in series, minus the phase angle of the input to the first component. The overall phase angle is equal to the sum of the phase angles of the individual components. (15.2)

Phase margin: A safety factor for the stability of a closed-loop control system that specifies how much additional phase lag is required to make the system unstable. (15.8)

Phase margin point: A point on the Nyquist diagram, located at $1 \;/\!-140°$, that is used to determine if a closed-loop control system meets the phase margin criterion. (15.9)

Root locus: The plot of all possible roots of the closed-loop characteristic equation as the controller gain, K, is varied from 0 to infinity. (15.10)

Stability conditions: Any condition that assures that a closed-loop control system will be stable.

1. A closed-loop control system is stable if the open-loop gain is less than 1 at the frequency for which the open-loop phase angle is $-180°$.
2. A closed-loop control system is stable if all the roots of the closed-loop characteristic equation, $1 + G(s)H(s) = 0$, have negative real parts. (15.8)
3. *See* Nyquist stability criterion.

Stability point: A point on the Nyquist diagram, located at $-1 + j0$, that is used to determine if a closed-loop control system meets the stability conditions defined above. (15.9)

Three zones of control: The graph of the error ratio versus frequency is divided into three distinct zones. Zone 1 covers the low frequencies; zone 2, the midrange frequencies; and zone 3, the high-range frequencies. In zone 1, the closed-loop control system reduces the error. In zone 2, closed-loop control increases the error. In zone 3, closed-loop control neither increases nor decreases the error. The frequency limit of the control system is the frequency on the border between zone 1 and zone 2. (15.5)

Unstable: A control system is unstable if it does not meet the stability conditions defined above. (15.8)

◆ EXERCISES

Section 15.2

15.1 Determine the overall transfer function of the following three components when they are connected in series.

$$\text{Final control element TF} = \frac{31.25}{1 + 0.75s + 0.1736s^2}$$

$$\text{Blending process TF} = \frac{1}{1 + 1325s}$$

$$\text{Measuring transmitter TF} = \frac{e^{-15s}}{1 + 200s}$$

15.2 Determine the overall gain, g, and phase angle, β, for each of the following sets of components (connected in series).

(a) Final control element: $g = 2.2, \quad \beta = -8°$
   Process: $g = 12.0, \quad \beta = -45°$
   Measuring transmitter: $g = 0.8, \quad \beta = -16°$

(b) Final control element: $g = 12.0, \quad \beta = -25°$
      Process: $g = 8.0, \quad \beta = -68°$
      Measuring transmitter: $g = 2.4, \quad \beta = -12°$
(c) Final control element: $g = 0.6, \quad \beta = -5°$
      Process: $g = 1.4, \quad \beta = -120°$
      Measuring transmitter: $g = 3.6, \quad \beta = -25°$
(d) Final control element: $g = 18.0, \quad \beta = -35°$
      Process: $g = 22.0, \quad \beta = -180°$
      Measuring transmitter: $g = 15.0, \quad \beta = -62°$

Sections 15.3 and 15.4

15.3 Use the Nichols chart (Figure 15.5) to convert each of the following open-loop points (gain and phase angle) to the corresponding closed-loop point (gain and phase angle.

	m (dB)	β (°)		m (dB)	β (°)
(a)	0	−160°	**(f)**	−16	−180°
(b)	8	−40°	**(g)**	−8	−60°
(c)	−12	−100°	**(h)**	0	−40°
(d)	16	−140°	**(i)**	−4	−120°
(e)	4	−20°	**(j)**	2	−80°

15.4 Use a calculator to convert the open-loop points in Exercise 15.3 to the corresponding closed-loop points. Compare your calculated results with your graphical results in Exercise 15.3.

Section 15.5

15.5 There are many situations where people act as the controller in a control system situation. Have you ever heard the expression "I zigged when I should have zagged"? This accurately describes a human controller operating in zone 2, above its maximum frequency limit. Operation in zone 2 is a matter of bad timing.

When a disturbance changes faster than the controller can respond, the control action is delayed and, rather than reducing the error, the controller actually increases the error.

Describe, from your own experience or knowledge, an example of a person-controller operating beyond their maximum frequency limit.

Section 15.6

15.6 A proportional controller is used to control the pressure in a first-order lag plus dead-time process. The control system has the following transfer functions:

$$G(s) = [5]$$

$$H(s) = \left(\frac{2}{1 + 0.5s}\right) e^{-0.05s}$$

(a) Determine the open-loop transfer function and use program BODE to make a printed copy of the Bode Data Table.

(b) Draw an open-loop Bode diagram and an error ratio graph.

(c) Obtain the maximum frequency limit of the system from the Design Summary in the output printed by Program Bode. Convert the maximum frequency limit from radians/second, ω, to hertz, f.

15.7 A thermal control system has a proportional controller with a gain of 15. The first-order lag plus dead-time process has the following transfer function:

$$H(s) = \left(\frac{1}{1 + 1000s}\right) e^{-200s}$$

Repeat assignments a, b, and c from Exercise 15.6.

15.8 A PI controller is used to control the level in a first-order lag plus dead-time process. The control system has the following transfer functions:

$$G(s) = \frac{0.628 + 9.1s}{s}$$

$$H(s) = \left(\frac{1}{1 + 50s}\right) e^{-2s}$$

Repeat assignments a, b, and c from Exercise 15.6.

15.9 An armature-controlled dc motor is used in a position control system. The controller and process transfer functions are given below:

$$G(s) = \frac{3.2 + 0.48s}{1 + 0.015s}$$

$$H(s) = \frac{19}{s + 0.329s^2 + 0.00545s^3}$$

Repeat assignments a, b, and c from Exercise 15.6.

15.10 A solid flow control system uses a PI controller. The controller and process transfer functions are given below:

$$G(s) = \frac{0.4 + 14.4s}{18s}$$

$$H(s) = e^{-20s}$$

Repeat assignments a, b, and c from Exercise 15.6.

15.11 A first-order lag plus dead-time blending process has a time constant of 1000 s and a dead-time lag of 20 s. The PI controller used to control the process has a gain of 44 and an integral action rate of $\frac{1}{230}$ s^{-1}. Repeat assignments a, b, and c from Exercise 15.6.

15.12 A PID controller is used to control a second-order lag plus dead-time process. The transfer functions of the controller and process are given below:

$$G(s) = \frac{42.86 + 12s + 2.16s^2}{s + 0.018s^2}$$

$$H(s) = \left(\frac{0.1}{1 + 0.02s + 0.01s^2}\right) e^{-0.05s}$$

Repeat assignments a, b, and c from Exercise 15.6.

15.13 A pressure system consists of a measuring transmitter with a lag coefficient of 0.2 s, a critically damped control valve, and a pressure tank with a time constant of 1 s. The transfer functions of the components are given below:

$$\text{Controller TF} = G(s) = \frac{1.8 + 0.9s + 0.324s^2}{s + 0.036s^2}$$

$$\text{Measuring transmitter TF} = \frac{1}{1 + 0.2s}$$

$$\text{Control valve TF} = \frac{5}{1 + 0.2s + 0.01s^2}$$

$$\text{Process TF} = \frac{2}{1 + s}$$

Repeat assignments a, b, and c from Exercise 15.6.

15.14 An amplidyne position control system consists of a PD controller, an amplidyne unit, a dc motor, and a load. The transfer functions are given below:

$$\text{Controller TF} = G(s) = \frac{0.01 + 0.00067s}{1 + 0.0067s}$$

$$\text{Amplidyne TF} = \frac{8500}{1 + 0.025s + 0.0001s^2}$$

$$\text{Motor and load TF} = \frac{1}{s + 0.35s^2 + 0.015s^3}$$

Repeat assignments a, b, and c from Exercise 15.6.

15.15 The AJ Food Company uses a PID controller to control a jacketed kettle similar to Figure 14.6a. Heated liquid from the kettle flows through a pipeline to the bot-

tling machine, where it is placed in 1-L bottles. A temperature sensor measures the temperature of the liquid product as it leaves the jacketed kettle. The transfer functions of the system components are given below:

$$\text{Controller TF} = G(s) = \frac{0.0075 + 3s + 600s^2}{s + 20s^2}$$

$$\text{Measuring transmitter TF} = \frac{1}{1 + 290s + 12,000s^2}$$

$$\text{Process TF} = \left(\frac{2}{1 + 1070s}\right) e^{-1.1s}$$

$$\text{Control valve} = \frac{4}{1 + 0.074s + 0.0021s^2}$$

Repeat assignments a, b, and c from Exercise 15.6.

Section 15.7

15.16 The stability of a control system can be determined from the open-loop Bode diagram of the system by observing the open-loop gain at the frequency where the open-loop phase angle is $-180°$ (this gain is designated as $M(-180)$. Examine the printed copies of the Bode data tables you obtained in Exercises 15.6 through 15.15. The value of $M(-180)$ is listed in the design summary that follows the Bode data table. Based on the value of $M(-180)$, determine if the system is "stable" or "unstable" and write a sentence that expresses your conclusion about the stability of the system.

Section 15.8

15.17 Examine the printed copies of the Bode data tables you obtained in Exercises 15.6 through 15.15, and observe the following two values in the design summary section: $M(-180)$, the gain at the frequency where the open-loop phase angle is $-180°$, and ANGLE (0DB), the phase angle at the frequency where the open-loop phase angle is 0 dB.

Use the values of $M(-180)$ and ANGLE (0DB) to determine the gain margin and the phase margin of each control system. Then check your results with the gain and phase margin printed in the design summary.

Section 15.9

15.18 Use the printed copies of the Bode Data Table you obtained in' Exercises 15.6 through 15.15 to sketch a Nyquist diagram for each control system (refer to Figure 15.6). Determine the stability of each control system from your Nyquist diagram.

Section 15.10

15.19 Construct a root-locus diagram of the control system with the following open-loop transfer function:

$$G(s)H(s) = K\left(\frac{s + 12}{s + 4}\right)$$

15.20 Construct a root-locus diagram of the control system with the following open-loop transfer functions:

$$G(s)H(s) = \frac{4K}{s(s + 16)}$$

Determine the value of K that will produce a damping ratio of 0.65.

15.21 Construct a root-locus diagram of the control system with the following open-loop transfer function:

$$G(s)H(s) = \frac{10K}{s^2 + 20s + 16}$$

Determine the value of K that will produce a damping ratio of 0.8.

15.22 Construct a root-locus diagram of the control system with the following open-loop transfer function:

$$G(s)H(s) = \frac{K}{s(s + 2)(s + 12)}$$

15.23 Construct a root-locus diagram of the control system with the following open-loop transfer function:

$$G(s)H(s) = \frac{K(s + 5)}{s^2}$$

15.24 Construct a root-locus diagram of the control system with the following open-loop transfer function:

$$G(s)H(s) = \frac{K(s + 5)}{s^2(s + 80)}$$

15.25 Construct a root-locus diagram of the control system with the following open-loop transfer function:

$$G(s)H(s) = \frac{K(s + 10)}{s(s + 1)(s + 50)}$$

15.26 Construct a root-locus diagram of the control system with the following open-loop transfer function:

$$G(s)H(s) = \frac{K}{s(s + 5)(s + 15)}$$

Then determine the operating point, s_o, for a damping ratio of 0.707. Also find the gain, K, and the resonant frequency, ω_o, at the operating point.

15.27 Construct a root-locus diagram of the control system with the following open-loop transfer function:

$$G(s)H(s) = \frac{K}{(s + 3)(s^2 + 8s + 32)}$$

Then determine the operating point, s_o, for a damping ratio of 0.707. Also find the gain, K, and the resonant frequency, ω_o, at the operating point.

CHAPTER 16

Controller Design

Controller design involves the selection of the control modes to be used and the determination of the value of each mode. For continuous systems, the proportional mode is usually combined with the integral and/or derivative modes to form a PI, PD, or PID controller. The parameters for these three modes are the proportional gain (P), the integral action rate (I), and the derivative action time constant (D).

The simplest design method is to specify a PID controller and then determine the control mode parameters in the field during plant start-up. This field adjustment is called "tuning the controller," and a procedure called the ultimate cycle method is used to determine the mode parameters. Some controllers are designed to perform this field adjustment automatically or semiautomatically.

A more formal design method uses Bode diagrams and a computer-aided graphic technique to design the controller. Using the Bode diagram method, the designer can determine the control mode parameters before the plant is constructed. This speeds up the plant start-up considerably, although some fine tuning of the controller is usually required in the field. A major difficulty with this method is the requirement of an open-loop Bode diagram of the process. Because the plant is not yet constructed, the designer must construct a model of the process to obtain the necessary open-loop Bode diagram.

The purpose of this chapter is to give you an entry-level ability to discuss controller design, tune a controller, complete a computer-aided Bode design of a controller, and design simple compensation networks. After completing this chapter, you will be able to

1. Describe the ultimate cycle method for tuning a controller
2. Describe the process reaction method for tuning a controller
3. Discuss self-tuning adaptive controllers
4. Complete a computer-aided Bode design of a PID controller
5. Specify compensation networks that will improve the response of the control system

16.1 INTRODUCTION

Controller design consists of the selection of the control modes and control mode settings and/or compensation networks that will result in a stable system that meets the control objectives. The control objectives may specify some or all of the following characteristics:

1. The accuracy and speed of response of the measuring transmitter (refer to Chapter 5).
2. The residual error allowed by the controller after a load change (refer to Chapter 13).
3. The response of the control system to a step change in load or setpoint. Quarter amplitude decay or critical damping may be specified (refer to Chapter 1).
4. The maximum frequency limit. The control system is able to follow setpoint changes and minimize disturbances with frequencies less than the maximum limit (refer to Chapter 15).
5. The response time, the rise time, and the settling time of the control system.
6. Minimum cost. The initial cost of the hardware is only one part of this objective. Maintenance costs, operating costs, reject product costs, and environmental costs are examples of the other costs that may be a factor. This objective requires a judgment based on economic and social conditions.

Two different approaches are used to design a control system. This does not mean that all design methods use only one approach or the other. Often parts of each approach are found in the design of a control system.

One approach to controller design uses a standard P, PI, PD, or PID controller. In this method, a major design decision is the selection of which control modes to include in the controller. Then start-up control mode settings are chosen. The actual mode settings are determined during start-up by a process called *"tuning the controller."* The tuning procedure consists of deliberately inducing a small oscillation in the system so that the period of the oscillation can be measured. (Controller adjustment formulas give the optimum control mode values based on the period of oscillation of the system.) An advantage of this method is that accurate models of the process are not required (accurate transfer functions of many industrial processes are very difficult to determine). This method is used when the cost of determining the controller adjustments during start-up is less than the cost of the analysis and design required to define the control system before start-up. Even the most accurate process control system design may require final adjustments in the field.

The second approach uses PID modes or compensation networks designed for a specific application. (The term *compensation network* often includes the I and D control modes under the names integration and lead-lag networks.) The second method is used when the transfer function of the process is known. (Accurate transfer functions of most electromechanical systems are relatively easy to determine.) The design procedure consists of the use of compensation networks to modify the open-loop frequency response such that a simple loop gain adjustment will result in a stable system that meets the performance objectives. This method is used when the cost of determining the controller adjustments during start-up is greater than the cost of the analysis and design required to define the control system before start-up.

The frequency response of most industrial processes is difficult to determine or measure. For this reason, process control design is based more on the first approach. However, frequency response methods are sometimes used to select the control modes and determine approximate mode settings. Field adjustments are almost always required to determine the exact mode settings.

The second approach has been used very successfully to design servomechanisms. The frequency response of most electromechanical components is relatively easy to determine and measure. Dead time is usually negligible. The designer has the option of using the root-locus method or the Bode method to design compensation networks that are inserted into the control loop. Field adjustments are often unnecessary, but some loop gain adjustment may be necessary to fine tune the system in the field.

16.2 THE ULTIMATE CYCLE METHOD

The *ultimate cycle method* uses controller adjustment formulas to determine the controller settings. The formulas require two measurements from the process: the minimum controller gain that causes the control system to oscillate (ultimate gain, G_u) and the period of the oscillation (ultimate period, P_u). To use the ultimate cycle method, the controller must be installed and the system ready to operate.

The following procedure is used to determine the ultimate gain and ultimate period. First, set the integral and derivative modes to the least effective setting. Then, start up the process, with the controller gain at a low value. Increase the gain setting until the controlled variable begins to oscillate. The last gain setting is the ultimate gain (G_u). The period of the oscillation is the ultimate period (P_u). The controller settings are determined from Table 16.1.

The original ultimate cycle method was developed by Nichols and Ziegler[*]. The problem with the original method is that the formulas do not guarantee an optimum response. The modified ultimate cycle method was developed to solve this problem in a very direct way. The gain setting is adjusted in a trial-and-error procedure until the desired response is obtained. The most common "desired response" is quarter amplitude decay (see Section 1.11). This criterion is illustrated in Figure 16.1. As the name suggests, quarter amplitude decay is an underdamped response in which each successive positive peak is one fourth as large as the preceding positive peak.

The modified method requires only the measurement of the ultimate period, P_u, which is used to compute the integral and derivative mode settings. After the integral and derivative modes have been set, the process is disturbed with a small change in setpoint, the response of the controlled variable is observed, and the gain is adjusted. The gain adjustment is simple—if the response is overdamped, the gain is increased; if the response is underdamped, the gain is decreased. The sequence of setpoint change, observation of the response, and adjustment of the gain is repeated until a quarter amplitude decay response is obtained. The same procedure can be used to tune a controller to obtain a critically damped response (Section 1.11). The general rule on gain adjustment is that increasing the gain will decrease the damping and vice versa.

The modified ultimate cycle method is very reliable and works on all types of processes. It is particularly effective on relatively fast processes where the waiting time is short. However, it can be very time-consuming on a slow process. The author once spent most of a day tuning a process with a cycle time of about 1 h. In a slow process, a process model can be very helpful in speeding up the tuning procedure. With a model of the process, the designer has the option of using the process reaction method (Section 16.3) or the Bode design method (Section 16.5).

[*]N. B. Nichols and J. G. Ziegler, " Optimum Settings for Automatic Controllers," *ASME Transactions,* Vol. 64, No. 8, 1942, pp. 759–768.

◆ **TABLE 16.1** Ultimate Cycle Method

Control Modes	Original Method	Modified Method
Proportional control (P)	$P = 0.5G_u$	Adjust the gain to obtain quarter-amplitude decay response to a step change in setpoint (see Figure 16.1)
Proportional plus integral control (PI)	$I = \dfrac{1.2}{P_u} (\text{min}^{-1})$ $P = 0.45G_u$	$I = \dfrac{1}{P_u} (\text{min}^{-1})$ Adjust the gain to obtain quarter amplitude decay response to a change in setpoint
Proportional plus integral plus derivative control (PID)	$I = \dfrac{2.0}{P_u} (\text{min}^{-1})$ $D = \dfrac{P_u}{8} (\text{min})$ $P = 0.6G_u$	$I = \dfrac{1.5}{P_u} (\text{min}^{-1})$ $D = \dfrac{P_u}{6} (\text{min})$ Adjust the gain to obtain quarter amplitude decay response to a change in setpoint

EXAMPLE 16.1

A process control system is tested at start-up. The derivative mode is turned OFF, and the integral mode is set at the lowest setting. The gain is gradually increased until the controlled variable starts to oscillate. The gain setting is 2.2 and the period of oscillation is 12 min. Use the original ultimate cycle method to determine the PID controller settings.

Solution

$$I = \frac{2.0}{P_u} = \frac{2.0}{12} = 0.167 \text{ min}^{-1}$$

$$D = \frac{P_u}{8} = \frac{12}{8} = 1.5 \text{ min}$$

$$P = 0.6G_u = (0.6)(2.2) = 1.32 \qquad ◆$$

16.3 THE PROCESS REACTION METHOD

The *process reaction method*[*] uses controller adjustment formulas to determine the controller settings. This method assumes that the process can be approximated by a first-order lag plus dead-time model. The formulas use three parameters obtained from an open-loop step response test of the process. The test begins by operating the process on manual control (open loop) until the measured variable remains constant. Then a step change in the manipulating element is

---
[*] Ibid.

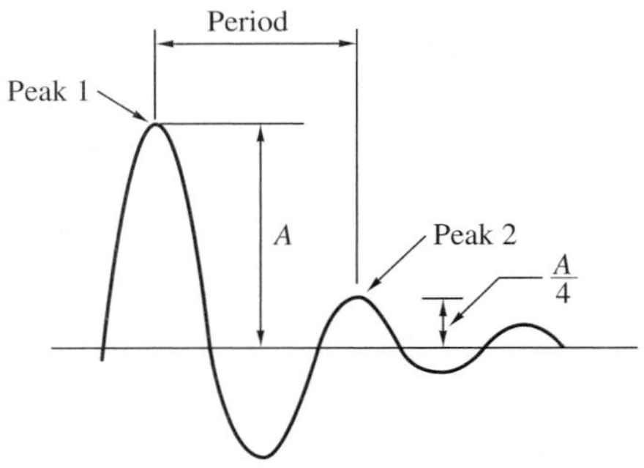

◆ **Figure 16.1** The quarter amplitude decay response is a common control system design criterion.

made and the response of the controlled variable is recorded. This record of the controlled variable is called the *process reaction graph*. A typical process reaction graph is shown in Figure 16.2.

The process reaction formulas require three variables from the step response test. Two variables are obtained from the process reaction graph. The third variable is Δp, the percentage change in the manipulating element output (e.g., percentage change in the control valve stem position). The two variables from the reaction graph are the effective delay, L, and the slope of the tangent line, N. The controller settings are determined from the following formulas.

Process Reaction Method

$$L = \text{effective delay, min}$$

$$N = \frac{\Delta C_m}{\Delta t} \, \%/\text{min}$$

$$\Delta P = \text{change in the manipulating element, } \%$$

1. Proportional control (P)

$$P = \frac{\Delta P}{NL}$$

2. Proportional plus integral control (PI)

$$P = 0.9 \left(\frac{\Delta P}{NL} \right)$$

$$I = \frac{0.3}{L} \, \text{min}^{-1}$$

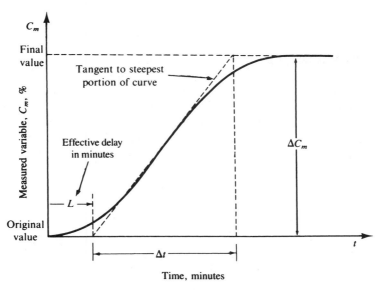

◆ **Figure 16.2**   A process reaction graph is a plot of the controlled variable after a step change in the output of the manipulating element. A first-order lag plus dead-time model of the process is derived from the graph with the dead time equal to L and the first-order time constant equal to Δt.

3. Proportional plus integral plus derivative control (PID)

$$P = 1.2\left(\frac{\Delta P}{NL}\right)$$

$$I = \frac{0.5}{L} \text{ min}^{-1}$$

$$D = 0.5L \text{ min}$$

A major advantage of the process reaction method over the modified ultimate cycle method is that once the process model is obtained, the control modes can be determined and implemented immediately. This is particularly useful on a very slow process where considerable elapsed time is required to complete the ultimate cycle method. The disadvantage of the process reaction method is that most processes are more complex than a simple first-order lag plus dead-time model. This usually means that some final adjustment of the gain is still required to obtain the desired response (e.g., quarter amplitude decay or critical damping). The process reaction method does provide an initial setting of the control modes, and these initial settings are usually close to the final values. However, a realistic expectation is that some tweaking of the gain will be required to obtain the desired response.

EXAMPLE 16.2

During start-up, the control valve of a process control system was maintained constant until the controlled variable stopped changing and reached a steady value. Then the control valve position was changed by 10%. A response graph similar to Figure 16.2 was obtained. An analysis of the reaction graph produced the following values:

$L = 5$ min
$\Delta t = 10$ min
Initial value of C_m: 40%
Final value of C_m: 48%

Use the process reaction method to determine the settings of a three-mode controller.

Solution

$$N = \frac{\Delta C_m}{\Delta t} = \frac{48 - 40}{10} = 0.8\%/\text{min}$$

$$\Delta P = 10\%$$

$$L = 5 \text{ min}$$

$$P = 1.2\left(\frac{\Delta P}{NL}\right) = 1.2\left(\frac{(10)}{(0.8)(5)}\right) = 3$$

$$I = \frac{0.5}{L} = \frac{0.5}{5} = 0.1 \text{ min}^{-1}$$

$$D = 0.5L = (0.5)(5) = 2.5 \text{ min}$$ ◆

16.4 SELF-TUNING ADAPTIVE CONTROLLERS

A major problem in designing or tuning process control systems is that the process models are often very complex, difficult to obtain, and inaccurate. Most models assume lumped elements (e.g., resistance, capacitance, inertance, inductance, inertia, dead time). In many processes, these elements are distributed in much the same way as they are in an electrical transmission line. Often the elements in the process change for various reasons. A change in production rate almost always changes the dead-time delays in a process. In one plant start-up, the author used the modified ultimate cycle method to tune a temperature control loop. The system responded to small step changes in setpoint with almost perfect quarter amplitude decay response. The next time the author checked, the response was unstable, oscillating with an unacceptably large amplitude. The problem was caused by a reduction in the production rate. A repeat of the modified ultimate cycle method resulted in new settings that returned the system to quarter amplitude decay response. However, when the production rate was returned to the original rate, the system had an overdamped response.

The control engineer has three choices in handling the control of a process whose model changes with changes in production rate (or other reasons). The first choice is to tune the controller for the worst case and accept a sluggish response for other conditions. The second choice is to change the controller mode settings every time the process model changes. (This choice assumes we know what causes the change in the process model—this is not always the case.) The third choice is to design a controller that will change itself every time the process changes. The self-tuning adaptive controller is an implementation of the third choice.

The objective of *self-tuning adaptive controllers* is to maintain the desired control criteria when the load or production rate changes from one value to another. The adaptive controller accomplishes this objective by observing the response of the control system and automatically adjusting the PID values accordingly.

Adaptive controllers use a conventional PID algorithm within an outer loop that handles the adaptive algorithm. The outer loop is called a shell because it encloses the PID algorithm and determines the values of P, I, and D. In other words, the adaptive shell establishes the "PID environment" used by the PID algorithm. The adaptive shell is also a feedback loop. It observes the control system and uses these observations to determine the values of P, I, and D that will achieve optimum performance.

Many different techniques are used to "adapt" a controller to changes in the process. Adaptive controllers fall into one of three categories: those that use programmed adjustment of the controller gain, those that use a model of the process to determine the PID values, and those that use pattern recognition to determine the PID values. The model-based adaptive controllers use a variation of the process reaction method for computing the PID values. The pattern-recognition adaptive controllers use a variation of the ultimate cycle method for determining the PID values.

Programmed Adaptive Controllers

A *programmed adaptive controller* automatically adjusts the proportional gain (P) as a function of a number of process-related variables. The proportional gain may be programmed as a function of the controlled variable, the setpoint, the error, the controller output, or a remote input variable. The adaptive algorithm can be set for independent gain adjustment using any combination of the process-related variables. For example, the remote input could be the production rate. The controller could adjust the controller gain according to the production rate to maintain the quarter amplitude decay criteria for all production rates. The controller could also make an independent adjustment in the gain based on the value of the setpoint.

Model-Based Adaptive Controllers

A *model-based adaptive controller* uses an internal model of the process to determine the optimum PID values. The adaptive algorithm uses an identification or "self-learning" mode to determine the model of the process. In this mode, the controller introduces step changes above and below the setpoint and observes the reaction of the process to these changes. The setpoint changes and observations are repeated until sufficient data have been obtained to establish a process reaction curve similar to Figure 16.2. The process model is obtained from the process reaction curve, and the PID values are determined by formulas similar to the process reaction formulas in Section 16.3.

Once the model has been established, the adaptive algorithm uses an operating mode in which the PID values remain fixed at the values established in the identification mode. Some algorithms also provide a continuous update operating mode. In this mode, the process model is continuously updated in the same manner as it was in the identification mode. If the process changes, the continuous update mode will adjust the process model and the PID values accordingly.

Pattern-Recognition Adaptive Controllers

A *pattern-recognition adaptive controller* uses a graph of error versus time similar to Figure 16.1 to obtain the optimum PID values. The adaptive algorithm constantly examines the response of the control system to naturally occurring disturbances caused by changes in load or setpoint. Whenever the magnitude of the error exceeds a threshold value, the adaptive algorithm searches for peaks in the magnitude of the error (see peaks 1, 2, etc., in Figure 16.1). When the algorithm finds two or three peaks, it uses the time between peaks and the magnitudes of the peaks to determine the amplitude decay ratio and the period of the oscillation. Formulas similar to the ultimate cycle formulas in Section 16.2 are then used to compute the optimum PID values.

16.5 COMPUTER-AIDED PID CONTROLLER DESIGN

The design procedure presented in this section uses the open-loop Bode diagram of the control system as the major design tool. For this reason the design procedure is known as the *Bode design method*. The objective of the design procedure is to obtain a stable control system that provides accurate control for a wide range of frequencies. Stable control means that the final design satisfies the gain margin and phase margin criteria covered in Chapter 15. Control over a wide range of frequencies means that the maximum frequency limit of the control system is as high as possible. Accurate control means that the open-loop gain is as large as possible at frequencies below the $-180°$ phase-angle frequency. The design procedure consists of using the PID control modes to modify the open-loop Bode diagram in such a way that the control objectives are satisfied. The Bode design of a controller is accomplished in four steps.

1. The open-loop Bode diagram of the process is used to determine the usefulness of the derivative control mode. If the derivative mode is found to be useful, the Bode diagram is used to determine the value of the derivative action time constant, D. If the derivative mode is not useful, D is given a value of 0.
2. The open-loop Bode diagram of the process plus the derivative control mode is used to determine the usefulness of the integral control mode. If the integral mode is found to be useful, the Bode diagram is used to determine the value of the integral action rate, I. If the integral mode is not useful, I is given a value of 0.
3. The open-loop Bode diagram of the process plus the integral and derivative control modes is used to determine the proportional mode gain, P.
4. Three diagrams of the process and PID controller are used to evaluate the final design of the control system: (a) The open-loop Bode diagram is used to determine the gain margin and the phase margin of the control system. (b) The error ratio graph is used to determine the maximum frequency limit of the control system. (c) The closed-loop Bode diagram presents the frequency response of the closed-loop control system.

   The Bode design method is a graphical method that helps the designer visualize the effect of each control mode on the frequency response of the control system. The designer uses straight-line approximations on semilog graph paper or a computer program to construct the open-loop Bode diagrams used in each step of the process. In step 4, the designer uses a Nichols chart or a computer program to translate the open-loop Bode diagram into the closed-loop Bode diagram. A major advantage of the Bode design method is the excellent visual aid provided by the graphical approach. The designer can observe exactly how each control mode reshapes the open-loop frequency response of the system.

   In this section we present a BASIC computer program to facilitate the Bode design method. The program is named CONTROL SYSTEM DESIGN and the file name is DESIGN. The designer may use DESIGN to complete the four-step Bode-method design of a PID controller. DESIGN can handle a process with up to nine components (first, second, or third order) and ten dead-time delays. It uses the bottom 23 lines of the computer screen to display one of the following graphs as selected by the designer:

1. Open-loop Bode diagram
2. Closed-loop Bode diagram
3. Nyquist diagram
4. Error-ratio graph

A table of "Design Decision Data" is displayed in the upper right corner of the graph. The designer uses data from this table to determine values of P, I, D, and α. The top line of the computer screen is the *Status Line*. It usually displays the current values of P, I, D, and α, but may display other information pertinent to the completion of a command.

The second line of the screen is the *Command Line*. It usually prompts the user for one of the following commands: *Dmode, Imode, Pmode, Analysis, Zoom, Unzoom,* and *Quit*. A command is given by simply pressing the first letter of the command (i.e., the capitalized letter). When a P, I, or D command is given, the command Line prompts the designer for the appropriate control mode parameter(s). When the Analysis command is given, the Command Line prompts the designer for the choice of Closed-loop, Nyquist, Error-ratio, or Open-loop. The Zoom command provides a close-up view of the graph in the region of 0 dB and $-180°$ phase angle. The Unzoom command returns the graph to normal size.

DESIGN begins with $P = 1$, $I = 0$, $D = 0$, and $\alpha = 0.1$. As the design steps unfold, the designer enters new values of D, α, I, and P in that order. After each mode is entered, the program displays both the old and the new gain and phase angle graphs. The designer can observe exactly how the mode just entered has changed the open-loop gain and phase angle graphs. After viewing the changes just made, the designer presses any key to move to the next step in the design procedure.

A major feature of DESIGN is the ease with which the designer can go back and change any control mode. This enables the designer to use a "what-if" analysis in a "trial-and-examine" procedure to search for the best possible control system design. When the Quit command is given, the designer has the option of printing a final design report, which includes a design summary and a table of open-loop, closed-loop, and error-ratio data.

The essence of the Bode design method is three decisions the designer makes to change the open-loop gain and phase graphs in the Bode diagram. The three decisions are the values of the D, I, and P control modes. To determine the three control modes, the designer uses data from open-loop Bode diagrams in formulas based on analysis and experience. As with many such formulas, there are exceptions. Usually the formulas result in a good design, but sometimes they do not. The designer must examine the Bode diagrams, recognize a poor design when it occurs, and then modify the control modes to obtain a good design. If the designer uses the traditional graphical design method, the construction of the necessary Bode diagrams can be a tedious and time-consuming task. Using DESIGN, the designer is relieved of the burden of constructing Bode diagrams and can quickly analyze the effect of different values for the three control modes.

Our study of the Bode design method begins with an example in which the control mode formulas result in a good design. The design formulas are built into DESIGN, and with a little coaching, you can complete the computer-aided design procedure. Example 16.3 walks you through the design of a PID controller for a blending process. Example 16.3 is followed by a detailed discussion of the four design steps and the formulas used to determine the three control modes.

EXAMPLE 16.3

Use DESIGN to design a PID controller for a blending system with the following overall transfer function (measuring transmitter, process, and control valve).

$$TF = \left(\frac{e^{-15s}}{1 + 200s} \right)\left(\frac{1}{1 + 1325s} \right)\left(\frac{31.25}{1 + 0.75s + 0.1736s^2} \right)$$

Note: Example 16.3 walks through a run of DESIGN. A good way to study a walk-through is to read the entire example once, then run DESIGN and duplicate the example on your computer.

Solution   Enter the following inputs in a run of DESIGN to initiate the design of the concentration control
system:

```
Transfer function coefficients of component number 1:
A (0) = 1,      A (1) =   0,     A (2) = 0,     A (3) = 0
B (0) = 1,      B (1) = 200,     B (2) = 0,     B (3) = 0

Transfer function coefficients of component number 2:
A (0) = 1,      A (1) =    0,    A (2) = 0,     A (3) = 0
B (0) = 1,      B (1) = 1325,    B (2) = 0,     B (3) = 0

Transfer function coefficients of component number 3:
A (0) = 31.25,     A (1) = 0,    A (2) = 0,        A (3) = 0
B (0) = 1,         B (1) = 0.75, B (2) = 0.1736,   B (3) = 0

Dead time delay number 1 = 15
```

(a) Design Step 1

Figure 16.3 is the initial graphic display of DESIGN after entering the preceding inputs. The
graph shows the open-loop Bode diagram of the process and the Design Decision Data obtained from the open-loop Bode graph.

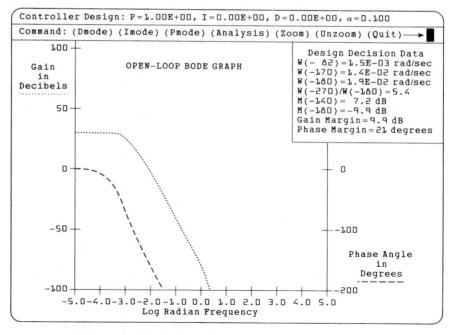

◆ **Figure 16.3**   The initial graphic display of DESIGN for Example 16.3. Line 2
lists the available commands. The designer initiates design step 1 by pressing the
"D" key to invoke the Dmode command.

We now press the "D" key to initiate the derivative mode design. The top two lines of the screen appear as follows:

```
Derivative mode design: Dmode is definitely useful.
```

```
Suggested D = 1.06E+02, Current D = 0.00E+00, Enter D → ■
```

The first line is called the Status Line. It displays various information about the control modes. The second line is enclosed in a box that we will call the Command Line Window. Notice that the Command Line Window includes a suggested value for D, the current value for D, and a prompt for entry of the design value of D. We choose the suggested value of 106 for D and enter that value at the D → ■ prompt in the Command Line Window.

The Command Line Window now appears as follows:

```
Current α = 0.10 Enter desired α → ■
```

We enter the value of 0.1 for α. As the program computes the new open-loop Bode data, it scrolls the data through the Command Line Window. When the computations are completed, the screen appears as shown in Figure 16.4.

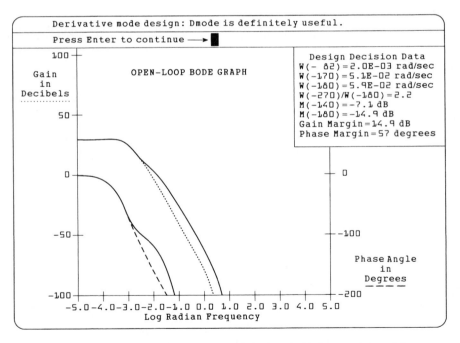

◆ **Figure 16.4** The graphic display of DESIGN is on hold after completion of design step 1 for Example 16.3. The dashed lines are the original gain and angle graphs. The solid lines show how the derivative mode has changed the gain and angle graphs.

The screen is on hold and we press the Enter key to conclude the design step 1. The graph on the screen now displays the open-loop Bode graph including the derivative control mode. The Status and Command Lines appear as follows:

```
Controller Design: P = 1.00E+00, I = 0.00E+00, D = 1.06E+02, α = 0.100
```

```
Command: (Dmode) (Imode) (Pmode) (Analysis) (Zoom) (Unzoom) (Quit) → ■
```

(b) Design Step 2

The graph in Figure 16.4 shows the open-loop Bode graph including the derivative mode We now press the "I" key to initiate the integral mode design. The Command Line Window appears as follows:

```
Suggested I = 2.00E−03, Current I = 0.00E+00, Enter I → ■
```

DESIGN suggested a value of 0.002 for the integral control mode. We choose the suggested value of 0.002 for I, and enter that value at the I → ■ prompt in the Command Line Window. As before, the new open-loop Bode data scrolls through the Command Line Window. When the computations are completed, the screen appears as shown in Figure 16.5.

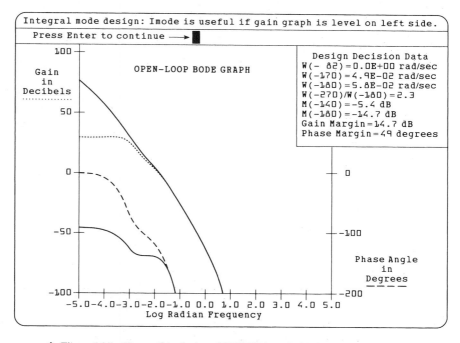

◆ **Figure 16.5** The graphic display of DESIGN is on hold after completion of design step 2 for Example 16.3. The dashed lines are the gain and angle graphs before adding the integral mode. The solid lines show how the integral mode has changed the gain and angle graphs.

The screen is on hold and we press the Enter key to conclude design step 2. The Status and Command Lines appear as follows:

```
Controller Design: P = 1.00E+00, I = 2.00E-03, D = 1.06E+02, α = 0.100
```

```
Command: (Dmode) (Imode) (Pmode) (Analysis) (Zoom) (Unzoom) (Quit) → ■
```

(c) Design Step 3

The graph in Figure 16.5 shows the open-loop Bode graph including the integral and derivative modes. We now press the "P" key to initiate the proportional mode design. The Command Line Window appears as follows:

```
Suggested P = 1.85E+00, Current P = 1.00E+00, Enter P → ■
```

Once again we choose the suggested value and enter 1.85 at the P → ■ prompt in the Command Line. The new open-loop Bode data again scrolls through the Command Line Window. When the computations are completed, the screen appears as shown in Figure 16.6. *This is the final design open-loop Bode graph.* The Design Decision Data in Figure 16.6 are the final design data.

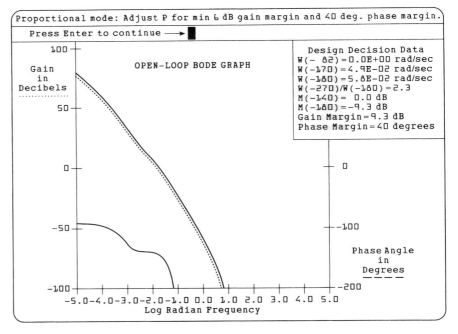

◆ **Figure 16.6** The graphic display of DESIGN is on hold after completion of design step 3 for Example 16.3. The dashed lines are the gain and angle graphs before adding the proportional mode. The solid line shows how the proportional mode has changed the gain graph. Notice that the angle graph is unchanged.

The screen is on hold, and we press the Enter key to conclude design step 3.

The graph on the screen now displays the open-loop Bode graph including the PID control modes. The Status and Command Lines appear as shown below.

```
Controller Design: P = 1.85E+00, I = 2.00E-03, D = 1.06E+02, α = 0.100
```

```
Command: (Dmode) (Imode) (Pmode) (Analysis) (Zoom) (Unzoom) (Quit) → ■
```

(d) Design Analysis

The (Analysis) command may be used to observe the closed-loop Bode graph, the error ratio graph, and the Nyquist diagram of the final design.

Press the "A" key to enter the analysis mode followed by the "C" key to display the closed-loop Bode diagram (shown in Figure 16.7). Press the "Z" key for a close-up view and the "U" key to return to the original view.

Press the "A" key followed by the "E" key and the "Z" key for a close-up view of the error-ratio graph (shown in Figure 16.8).

Press the "A" key followed by the "N" key and the "Z" key for a close-up view of the Nyquist diagram (shown in Figure 16.9).

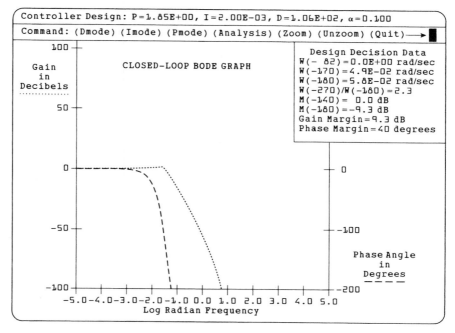

◆ **Figure 16.7** The closed-loop Bode graph for Example 16.3 as displayed by DESIGN after completion of design step 3.

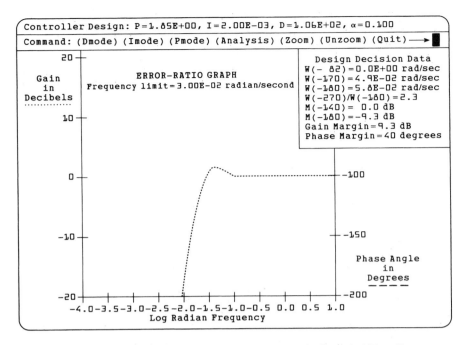

◆ **Figure 16.8** A close-up view of the error-ratio graph for Example 16.3 as displayed by DESIGN after completion of design step 3. The Zoom command was used to get the close-up view of the graph.

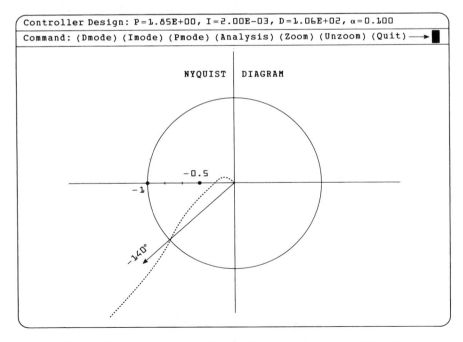

◆ **Figure 16.9** A close-up view of the Nyquist diagram for Example 16.3 as displayed by DESIGN after completion of design step 3. The Zoom command was used to get the close-up view of the graph.

(e) Design Summary

After entering the (Quit) command, the program presents the option to obtain a printed copy of the Bode data table and the design summary. The design summary is shown below:

```
                       DESIGN SUMMARY
Proportional gain: 1.85E+00
Integral action rate: 2.00E-03 1/second
Derivative action time constant: 1.06E+02 second
Derivative limiter: 0.100
M (-180): -9.3 decibel
ANGLE (ODB): -140.0 degree
Gain margin: 9.3 decibel
Phase margin: 40 degree
Frequency limit: 3.0E-02 radian/second
```

◆

Design Step 1: Derivative Mode

The derivative control mode is used to increase the maximum-frequency limit of the control system by moving the $-180°$ phase angle to a higher frequency. This means that the control system will be able to correct disturbances over a somewhat wider range of frequencies. The derivative mode also permits the use of a larger proportional gain setting, which increases the accuracy of the control system and reduces the residual offset.

Design step 1 begins with the determination of the usefulness of the derivative control mode. This determination is based on an estimate of how much the derivative mode will increase the $-180°$ frequency, which depends on the slope of the phase angle graph between the $-180°$ and $-270°$ frequencies. A steep slope results in less increase in the $-180°$ frequency than a shallow slope does. The ratio of the $-270°$ frequency divided by the $-180°$ frequency is used as a measure of the slope of the phase angle graph between those two frequencies. Tucker and Wills[*] suggest the following rule for determining whether the derivative control mode is useful.

$$\omega_{-180°} = \text{frequency for which the phase angle is } -180°$$
$$\omega_{-270°} = \text{frequency for which the phase angle is } -270°$$

1. The derivative mode is definitely useful if $\dfrac{\omega_{-270°}}{\omega_{-180°}} > 5$.

2. The derivative mode is probably useful if $2 < \dfrac{\omega_{-270°}}{\omega_{-180°}} < 5$.

3. The derivative mode is marginally useful if $\dfrac{\omega_{-270°}}{\omega_{-180°}} < 2$.

---

[*] G. K. Tucker and D. M. Wills, *A Simplified Technique of Control System Engineering* (Philadelphia: Minneapolis-Honeywell Regulator Company, 1962), p. 63.

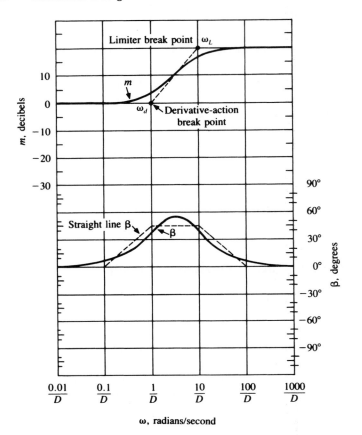

◆ **Figure 16.10** The Bode diagram of the derivative control mode displays the positive phase angle and the high frequency amplification that characterize this mode of control.

Figure 16.10 is the Bode diagram of a derivative control mode with a derivative limiter coefficient (α) equal to 0.1. The limiter break-point frequency (ω_L) is equal to $1/\alpha$ times the derivative action break-point frequency.

$$\omega_L = \frac{1}{\alpha D} = \frac{1}{\alpha}\omega_d = \frac{1}{0.1}\omega_d = 10\omega_d$$

Notice the positive phase angle of the derivative mode, especially between ω_d and ω_L. It is this positive phase angle that enables the derivative mode to increase the frequency at which the open-loop phase angle is $-180°$. Notice also the increase in the gain graph between ω_d and ω_L. This increase is called the *derivative amplitude*, and it is an unwanted side effect in the application of the derivative control mode. In Figure 16.10, the derivative amplitude is 20 dB, or a gain of 10. The phase-angle graph reaches a maximum value of 55° halfway between ω_d and ω_L. Other phase-angle values are given in Table 16.2.

The design of the derivative control mode consists of locating the derivative action break point such that the derivative action produces the maximum improvement in the control sys-

◆ **TABLE 16.2** Gain and Phase Values for the Derivative Control Mode $(Ds + 1)/(0.1Ds + 1)$

ω	m (dB)	β (deg)
0.1/D	0	5.1
0.2/D	0.2	10.2
0.5/D	1.0	23.7
1.0/D	3.0	39.3
2.0/D	6.8	52.1
3.16/D	10.0	55
5.0/D	13.2	52.1
10/D	17	39.3
20/D	19.0	23.7
50/D	19.8	10.2
100/D	20	5.1

tem. The positive phase angle of derivative action increases the phase margin. However, the gain of the derivative action reduces the gain margin, which nullifies some of the benefit of the increased phase margin. The designer must compromise between the benefit produced by the phase lead and the harm produced by the gain of the derivative mode.

In DESIGN, with $\alpha = 0.1$, the first $+50°$ point on the derivative phase graph is located at the $-180°$ frequency on the open-loop Bode diagram.[*] This results in a derivative break frequency equal to one-half the $-180°$ frequency.

$$\omega_d = 0.5\omega_{-180°} \qquad (16.1)$$

The derivative action time constant is given by

$$D = \frac{1}{\omega_d} = \frac{2}{\omega_{-180°}} \qquad (16.2)$$

This method tends to minimize the increase in gain without seriously reducing the increase in the $-180°$ frequency.

Design Step 2: Integral Mode

The integral control mode is used to increase the static accuracy of the control system by increasing the gain of the controller at low frequencies. The integral mode completely eliminates the residual offset that occurs in P and PD controllers. For stability, the integral mode gain is reduced to almost 0 dB as the frequency approaches the $-180°$ frequency and remains there for all higher frequencies. The integral mode also provides 90° of phase lag at low frequencies. This phase lag approaches 0° at about the same frequency that the gain approaches 0 dB. The high gain and 90° phase lag of the integral control mode have the potential to cause stability problems. The objective of the integral mode design is to select the integral rate, I, that provides the maximum accuracy benefit while minimizing the adverse effect on stability.

---

[*] Ibid., p. 70.

Design step 2 begins with the determination of the usefulness of the integral control mode. The integral mode is useful whenever the open-loop gain of the process does not increase as the frequency decreases toward zero. A process that displays this gain characteristic would not have an inherent integration term in its open-loop transfer function. If it did have an inherent integration term, the integral control mode would not be useful and should not be used. In fact, using the integral mode on a process that has an inherent integration term can cause serious stability problems.

The simplest way to determine the usefulness of the integral mode is to observe the process open-loop gain graph. The criteria for the usefulness of the integral mode can be stated as follows:

> If the open-loop gain of the process does not increase as the frequency approaches 0, the integral mode is useful and should be used to increase the static accuracy of the system.
>
> If the open-loop gain increases as the frequency approaches 0, the integral mode is not useful and should not be used because it could cause stability problems.

Another way to determine the usefulness of the integral mode is to observe the transfer function of each component in the process. If a component has an inherent integration term, then its transfer function will have the factor $1/s$. In other words, if $B(0) = 0$ in the transfer function of any component in the control system, the process has an integration term and the integral mode should not be used.

The criteria for the integral mode can also be stated as follows:

> If no control system component has a value of 0 for $B(0)$ in its transfer function, the integral mode is useful. If any component has a value of 0 for $B(0)$ in its transfer function, the integral mode is not useful and should not be used.[*]

Figure 16.11 is the Bode diagram of the integral control mode. Notice the shape of the gain graph. The gain increases steadily as the frequency decreases below the integral action break point. Above the break point, the gain is 1 (0 dB). Notice that the phase angle is $-90°$ at low frequencies and changes to $0°$ at high frequencies.

The design of the integral control mode involves a compromise between the benefit of the high gain at low frequencies and the negative effect of the phase lag, also at low frequencies. The designer applies the integral mode after the derivative mode, using the open-loop Bode diagram of the process plus the derivative control mode. The design objective is to place the integral action break point such that the negative effect of the phase lag is limited to about 5% at $\omega_{-180°}$. This is accomplished by setting the integral break frequency, ω_i, equal to about one-fifth of $\omega_{-170°}$ on the open-loop Bode diagram. The integral action rate, I, is equal to the integral action break-point frequency, ω_i.

---

[*]This criterion assumes that no component has a value of 0 for $A(0)$ in its transfer function. No component in this book has a value of 0 for $A(0)$, and it does not occur in normal processes, measuring transmitters, or final control elements.

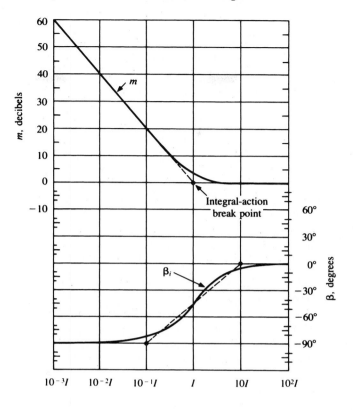

ω, radians/second or radians/minute

◆ **Figure 16.11**   The Bode diagram of the integral control mode displays the high gain and $-90°$ phase angle at low frequencies that characterize this mode of control.

$$\omega_i = 0.2\omega_{-170°} \qquad \textbf{(16.3)}$$

$$I = \omega_i = 0.2\omega_{-170°} \qquad \textbf{(16.4)}$$

However, Equation (16.4) does not always produce a satisfactory result. One exception occurs with first-order lag plus dead-time processes when the first-order time constant is greater than about 10 times the dead-time delay. If Equation (16.4) is used in this situation, the combined process plus derivative plus integral phase graph dips below $-140°$ at a very low frequency as illustrated in Figure 16.12. As the frequency increases, the phase graph rises and then drops rapidly below $-180°$ as the dead time takes over. The problem becomes evident in design step 3 when the phase margin criterion is used to determine the proportional gain, P. The phase margin requires the system gain to be 0 dB at the $-140°$ frequency. The result is an extremely overdamped, sluggish system.

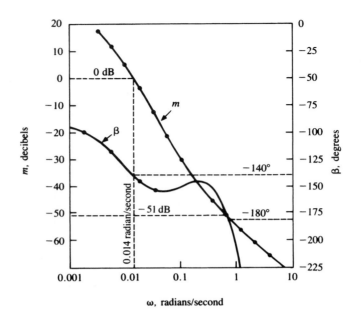

◆ **Figure 16.12** Open-loop Bode diagram of a first-order lag plus dead-time process and a PI controller (step 4). The integral mode was determined by $I = 0.2\omega_i$, which resulted in an unusually low proportional gain and a very overdamped control system.

Figure 16.12 illustrates the problem as it occurs in the design of a process with a first-order time constant of 100 s and a dead-time lag of 2 s. In design step 1, $\omega_{-270°}/\omega_{-180°} = 2$ and we elected not to use the derivative mode. Using Equation (16.4) in step 2 of the design procedure results in an integral action rate $I = 0.136 \text{ s}^{-1}$. In step 3, the phase margin criteria force a very low proportional gain, P. In step 4 the phase margin is 40° and the gain margin is 51.0 dB. The extremely large gain margin is a clear indication of the problem presented by the dip in the phase graph between 0.01 and 0.1 rad/s.

The preceding design can be improved by moving the integral break point to a lower frequency. The desired break point is obtained as follows:

1. Divide $-140°$ by 1.1 to provide a 10% safety factor:

$$\frac{-140°}{1.1} = -127°$$

2. Subtract the $-45°$ phase angle of the integral mode at its break point:

$$-127° - (-45°) = -82°$$

We therefore set the integral break-point frequency equal to the frequency at which the process plus derivative phase angle is $-82°$.

$$I = \omega_i = \omega_{-82°} \qquad \qquad \textbf{(16.5)}$$

Figure 16.13 shows the result of the redesign of the control system in Figure 16.12, using Equation (16.5) instead of Equation (16.4) to determine the integral action rate, T. Notice that the dip in the phase graph has been raised above $-140°$ and $\omega_{-140°}$ is about 0.39 rad/s (compared with 0.014 rad/s in Figure 16.12). The redesigned system has a gain margin of 6 dB, and a phase margin of 41°. The redesign has raised the gain by over 40 dB, and has an ideal gain and phase margin combination.

In some processes with high-order transfer functions, the $\omega_{-82°}$ value for I does not retard the integral mode break point enough to correct the excessive gain margin problem. Exercise 16.13 is an example of this type of process. In this situation, the designer may choose between a trial-and-examine procedure to determine the optimum integral mode setting and a compensation network to move the first-order break point to a higher frequency. The design of the compensation network is explained in Section 16.7. In either choice, DESIGN is a very useful tool to carry out the design procedure.

Equations (16.4) and (16.5) are combined in the following equation for integral action rate I:

$$I = \text{minimum of } \{\omega_{-82°} \text{ or } 0.2\omega_{-170°}\} \tag{16.6}$$

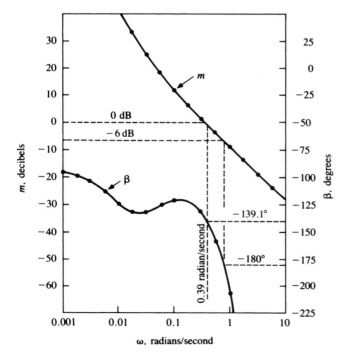

◆ **Figure 16.13** Open-loop Bode diagram of the first-order plus dead-time process from Figure 16.12 with a redesigned PI controller. The integral mode was determined by $I = \omega_{-82°}$, which resulted in a much higher proportional gain and a quarter amplitude underdamped control system.

Design Step 3: Proportional Mode

In step 3, the designer uses the open-loop Bode diagram of the process plus I and D to determine the value of the proportional gain, P, that will satisfy both the gain margin and the phase margin. The gain margin is satisfied if the sum of the controller decibel gain, P_{dB}, and the Bode diagram open-loop gain at $-180°$, $m_{-180°}$ is less than or equal to -6 dB.

$$\text{Gain margin: } P_{dB} \leq -m_{-180°} - 6 \tag{16.7}$$

The phase margin is satisfied if the sum of the controller decibel gain, P_{dB}, and the Bode diagram open-loop gain at $-140°$, $m_{-140°}$, is less than or equal to 0 dB.

$$\text{Phase margin: } P_{dB} \leq -m_{-140°} \tag{16.8}$$

Equations (16.7) and (16.8) can be combined into the following equation for the proportional gain, P.

$$P_{dB} = \min \{-m_{-140°} \text{ or } (-m_{-180°} - 6)\} \tag{16.9}$$

The controller gain, P, is

$$P = 10^{P_{dB}/20} \tag{16.10}$$

Design Step 4: Concluding Report

The final design step uses three diagrams of the process and PID controller. The open-loop Bode diagram is used to determine the gain margin and the phase margin. The error-ratio graph is used to determine the maximum frequency limit of the process (the frequency that separates zone 1 from zone 2). The closed-loop Bode diagram gives the frequency response of the control system.

PID CONTROLLER DESIGN SUMMARY

Comment: The following is a summary of the four-step procedure for the design of a PID controller using DESIGN.

Step 1: Derivative Mode

Effects:

1. Increase the maximum frequency limit.
2. Increase the phase margin.
3. Decrease the gain margin.

Usefulness:

1. Definitely useful if $\omega_{-270°}/\omega_{-180°} > 5$.
2. Probably useful if $2 < \omega_{-270°}/\omega_{-180°} < 5$.
3. Marginally useful if $\omega_{-270°}/\omega_{-180°} < 2$.

Design equation:

$$D = \frac{2}{\omega_{-180°}}$$

(16.2)

Step 2: Integral Mode

Effects:

1. Increase the low-frequency gain.
2. Reduce the maximum frequency limit.
3. Reduce the phase margin.

Usefulness:

1. Useful if the open-loop gain of the process does not increase as the frequency approaches 0.
2. Not useful if the open-loop gain of the process increases as the frequency approaches 0.

Design equation:

$$I = \text{minimum of } \{\omega_{-82°} \text{ or } 0.2\omega_{-170°}\}$$

(16.6)

Step 3: Proportional Mode

Effects:

1. Move the gain graph up or down.
2. Does not affect the phase graph.

Usefulness:
  Always useful.

Design equation:

$$P_{\text{dB}} = \min \{-m_{-140°} \text{ or } (-m_{-180°} -6)\}$$

(16.9)

$$P = 10^{P_{\text{dB}}/20}$$

(16.10)

Step 4: Concluding Report

  Open-loop Bode diagram
  Closed-loop Bode diagram
  Error-ratio graph
  Design summary

16.6 EXAMPLE DESIGN OF A THREE-LOOP CONTROL SYSTEM

In this section, DESIGN will be used to design three controllers for the blending and heating process shown in Figure 16.14. The process blends and heats a mixture of water and syrup, and has three control loops: concentration, temperature, and level. The concentration control loop manipulates the syrup input flow rate to maintain the desired concentration of syrup in the finished product. The concentration analyzer draws a sample of product from the outlet

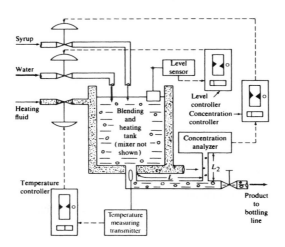

◆ **Figure 16.14** The blending and heating process has three loops to control concentration, temperature, and liquid level. Examples 16.4 to 16.6 use DESIGN to design PID controllers for these three loops.

line and measures the concentration of syrup in the sample. The level control loop manipulates the water input flow rate to maintain the level of liquid in the blending tank. The temperature control system manipulates the flow rate of the heating fluid to maintain the temperature of the liquid in the blending tank.

Blending and Heating Process Specifications

1. Product

Production rate:	2.5×10^{-4} m³/s (15 L/min)
Composition:	90° water and 10% syrup
Density, ρ:	1005 kg/m³
Viscosity, μ:	0.1 Pa · s
Specific heat, H_h:	4170 J/kg · K

2. Mixing Tank

Diameter:	0.75 m
Height:	1 m
Operating level:	0.75 m
Product temperature:	60°C
Heating fluid temperature:	100°C
Film coefficient $h_i = h_o$:	1349 W/m² · K
Outlet pipe diameter:	0.0351 m
Distance from outlet to temperature probe:	0.26 m
Wall thickness:	0.01 m
Wall material:	steel

3. Concentration Measuring Transmitter

Model:	First-order lag plus dead time
Time constant, ω_1:	200 s
Dead-time delay, t_{d1}:	15 s
Gain:	1 (% output/% input)

4. Level Measuring Transmitter
   Model: First-order lag plus dead time
   Time constant, ω_2:   2 s
   Dead-time delay, t_{d2}:   0.5 s
   Gain: 1 (% output/% input)
5. Temperature Measuring
   Transmitter
   Model: Overdamped second-order lag
   Time constants, τ_3:   50 s
                    τ_4:   240 s
   Gain: 1 (% output/% input)
6. Control Valves
   Model: Underdamped second-order lag
   Water valve, ω_{01}:   10.2 rad/s
                ζ_1:   0.75
        Gain: 5 (% output/% input)
   Syrup valve, ω_{02}:   2.4 rad/s
                ζ_2:   0.90
        Gain: 31.25 (% output/% input)
   Heating valve, ω_{03}:   21.6 rad/s
                ζ_3:   0.8
        Gain: 8 (% output/% input)
7. Concentration Process
   Model (τ_5):   First-order lag
   Gain (G):   1 (% output/% input)
8. Level Process
   Model: Integral (nonregulating)
   FS_{in}:   0.001 m³/s
   FS_{out}:   1 m
9. Thermal Process
   Model (τ_6, t_{d3}):   First-order lag plus dead time
   Gain (G):   1 (% output/% input)

EXAMPLE 16.4   Design a PID controller for the concentration control loop in the blending and heating process, Figure 16.14.

Solution   **(a)** Determine the concentration process time constant, τ_5.

$$\text{Liquid volume, } V = (\text{area})(\text{operating height})$$

$$= \frac{\pi D^2 h}{4}$$

$$= \frac{\pi (0.75)^2 (0.75)}{4}$$

$$= 0.3313 \text{ m}^3$$

Liquid flow rate, $Q = 2.5 \times 10^{-4}$ m^3 s

$$\text{Time constant, } \tau_5 = \frac{V}{Q} = \frac{0.3313}{2.5 \times 10^{-4}}$$

$$= 1325 \text{ s}$$

(b) Determine the syrup control valve transfer function coefficients.

$$B(1) = \frac{2\zeta_2}{\omega_{02}} = \frac{(2)(0.9)}{2.4} = 0.75$$

$$B(2) = \frac{1}{\omega_{02}^2} = \frac{1}{2.4^2} = 0.1736$$

(c) Determine the overall transfer function of the measuring transmitter, process, and control valve.

$$\text{TF} = \left(\frac{e^{-15s}}{1 + 200s} \right) \left(\frac{1}{1 + 1325s} \right) \left(\frac{31.25}{1 + 0.75s + 0.1736s^2} \right)$$

(d) *Note:* This is the same transfer function as the blending system in Example 16.3. The following summary was obtained from a run of DESIGN (see Example 16.3).

```
                        DESIGN SUMMARY

Proportional gain:                      1.85E+00
Integral action rate:                   2.00E-03 1/second
Derivative action time constant:        1.06E+02 second
Derivative limiter:                     0.100
M (-180):                               -9.3 decibel
ANGLE (ODB):                            -140.0 degree
Gain margin:                            9.3 decibel
Phase margin:                           40 degree
Frequency limit:                        3.0E-02 radian/second
```

◆

EXAMPLE 16.5

Design a PID controller for the temperature control loop in the blending and heating process, Figure 16.14.

Solution

(a) Determine the thermal process time constant, τ_6.

Thermal resistance (from the program THERMRES):

$$h_i = h_o = 1349 \text{ W/m}^2 \cdot \text{K}$$

$$A = \frac{\pi D^2}{4} + \pi Dh = \pi D \left(\frac{D}{4} + h \right)$$

$$= \pi(0.75) \left(\frac{0.75}{4 + 0.75} \right)$$

$$= 2.209 \text{ m}^2$$

$$x = 0.01 \text{ m}$$

$$K = 45 \text{ W/m} \cdot \text{K}$$

From THERMRES: $R_T = 7.718\text{E} - 4$ K/W

Thermal capacitance, $C = mH_h$:

$$m = \frac{\rho h \pi D^2}{4} = \frac{(1005)(0.75)\pi(0.75)^2}{4}$$

$$= 333 \text{ kg}$$

$$C = (333)(4170) = 1.389\text{E} + 6 \text{ J/K}$$

$$\tau_6 = RC = (7.71\text{E} - 4)(1.389\text{E} + 6)$$

$$= 1070 \text{ s}$$

(b) Determine the thermal dead-time lag, t_{d3}.

$$\text{Outlet flow rate} = U = \frac{Q}{A}$$

$$= \frac{2.5\text{E} - 4}{0.0351^2 \, \pi/4}$$

$$= 0.258 \text{ m/s}$$

$$\text{Dead-time lag} = t_{d3} = \frac{\text{distance}}{\text{velocity}}$$

$$= \frac{0.26}{0.258}$$

$$= 1 \text{ s}$$

(c) Determine the temperature measuring transmitter transfer function.

$$B(1) = 50 + 240 = 290$$

$$B(2) = (50)(240) = 12{,}000$$

(d) Determine the heating control valve transfer function coefficients.

$$B(1) = \frac{2\zeta_3}{\omega_{03}} = \frac{(2)(0.8)}{21.6} = 0.0741$$

$$B(2) = \frac{1}{\omega_{03}^2} = \frac{1}{21.6^2} = 0.00214$$

(e) Determine the overall transfer function of the measuring transmitter, process, and control valve.

$$\text{TF} = \left(\frac{1}{1 + 290s + 12{,}000s^2}\right)\left(\frac{e^{-s}}{1 + 1070s}\right)\left(\frac{8}{1 + 0.0741s + 0.00214s^2}\right)$$

(f) The following summary was obtained from a run of DESIGN.

```
                          DESIGN SUMMARY

Proportional gain:                      2.33E+00
Integral action rate:                   1.00E-03 1/second
Derivative action time constant:        1.97E+02 second
Derivative limiter:                     0.100
M(-180):                                -12.7 decibel
```

```
ANGLE (ODB):                    -140.0 degree
Gain margin:                    12.7 decibel
Phase margin:                   40 degree
Frequency limit:                1.6E-02 radian/second
```

◆

EXAMPLE 16.6

Design a PID controller for the level control loop in the blending and heating process, Figure 16.14.

Solution

(a) Determine the level process integral action time constant.

$$T_i = A \frac{FS_{out}}{FS_{in}}$$

$$A = \frac{\pi D^2}{4} = \frac{\pi(0.75)^2}{4} = 0.4418$$

$$T_i = 0.4418 \left(\frac{1}{0.001} \right) = 442 \text{ s}$$

(b) Determine the water control valve transfer function coefficients.

$$B(1) = \frac{2\zeta_1}{\omega_{01}} = \frac{(2)(0.75)}{10.2} = 0.147$$

$$B(2) = \frac{1}{\omega_{01}^2} = \frac{1}{10.2^2} = 0.00961$$

(c) Determine the overall transfer function of the measuring transmitter, process, and control valve.

$$TF = \left(\frac{e^{-0.5s}}{1 + 2s} \right) \left(\frac{1}{442s} \right) \left(\frac{5}{1 + 0.147s + 0.00961s^2} \right)$$

(d) The following summary was obtained from a run of DESIGN.

```
                        DESIGN SUMMARY

Proportional gain:                      7.44E+01
Integral action rate:                   0.00E+00 1/second
Derivative action time constant:        2.43E+00 second
Derivative limiter:                     0.100
M(-180):                                -6.0 decibel
ANGLE (ODB):                            -134.8 degree
Gain margin:                            6.0 decibel
Phase margin:                           45 degree
Frequency limit:                        1.3E-00 radian/second
```

◆

16.7 CONTROL SYSTEM COMPENSATION

The design methods discussed in this section are used in the design of servo control systems (i.e., position, velocity, or force control of mechanical loads). The component transfer functions in a servo loop are well defined and seldom involve dead-time delays. This is quite different from process control, where transfer functions are usually not well known and dead-time delays are common.

The design of a servomechanism begins with the establishment of the performance objectives (e.g., static accuracy, response time, overshoot, stability criteria, etc.) Next the servo designer selects the system components: a servo actuator, a power supply, a drive amplifier, and a feedback transducer. The component transfer functions are then used to construct the open-loop Bode diagram or root-locus plot of the system.

The designer begins the servo controller design with an analysis of a simple loop closure. If a loop gain adjustment (proportional control mode) satisfies the performance objectives, the designer simply provides the means of establishing the necessary loop gain. This may involve adjusting the gain of one or more components in the loop, or it may involve the addition of an amplifier in the loop.

Often the simple adjustment of loop gain does not satisfy the performance objectives. The designer must then modify the open-loop frequency response of the system. This can be done in one of two ways—either change or modify components in the loop, or insert additional components into the loop. In either case, the objective is to alter the open-loop frequency response so that the performance objectives can be met with a simple gain adjustment. The second option (adding components) is known as frequency compensation or simply *compensation. The objective of compensation is to alter the open-loop frequency response so that the performance objectives can be met by a subsequent gain adjustment.* The derivative and integral modes of a PID controller are examples of compensation.

Figure 16.15 shows the Bode diagrams of the following four simple compensation networks:

1. Integration-lead network compensation

$$\frac{V_{out}}{V_{in}} = \left(\frac{1 + T_i s}{T_i s} \right)$$

(16.11)

2. Lead-lag network compensation

$$\frac{V_{out}}{V_{in}} = \left(\frac{1 + T_{Lead} s}{1 + T_{Lag} s} \right)$$

(16.12)

$$T_{Lead} > T_{Lag}$$

3. Lag network compensation

$$\frac{V_{out}}{V_{in}} = \left(\frac{1}{1 + T_{Lag} s} \right)$$

(16.13)

4. Lag-lead network compensation

$$\frac{V_{out}}{V_{in}} = \left(\frac{1 + T_{Lead} s}{1 + T_{Lag} s} \right)$$

(16.14)

$$T_{Lead} < T_{Lag}$$

The *integration-lead compensation* network is actually the same as the integral control mode in a PID controller. To verify this fact, replace T_i in Equation 16.11 by $1/I$ and rearrange the result to get the equation of the integral control mode shown below:

$$\frac{V_{out}}{V_{in}} = \left(\frac{I + s}{s} \right)$$

(16.15)

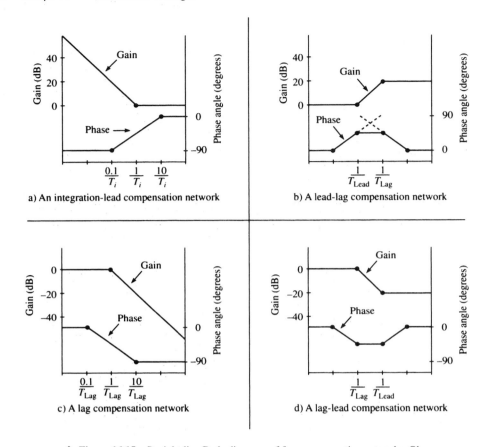

◆ **Figure 16.15** Straight-line Bode diagrams of four compensation networks. Observe the similarity between network (a) and the integral mode, between network (b) and the derivative mode, and between network (c) and a low-pass filter.

Integration-lead compensation is used for the same reason and in the same way as the integral control mode—to increase the static accuracy by increasing the gain at low frequencies while preserving stability by not changing the gain at the $-180°$ frequency or above. As with the integral control mode, integration-lead compensation must not be used when the loop has an inherent integration. Serious stability problems result when this rule is not observed. The integral-action time constant, T_i, is the reciprocal of the PID integral action rate, I (i.e., $T_i = 1/I$). The rules used to locate I for a PID controller also apply to the determination of T_i for the integration-lead compensation network (allowing for the reciprocal relationship between T_i and I, of course).

The *lead-lag compensation* network is very similar to the derivative control mode in a PID controller. To verify this fact, take Equation (16.12) and replace T_{Lead} by D and replace T_{Lag} by αD. The result is the following equation of the derivative control mode:

$$\frac{V_{\text{out}}}{V_{\text{in}}} = \left(\frac{1 + Ds}{1 + \alpha Ds}\right) \tag{16.16}$$

$$0 < \alpha < 1$$

In the lead-lag network, both the lead and the lag time constants are specified. In the derivative mode, the derivative action time constant, D, and the derivative limiter, α, are specified. In either case, the result is the specification of the two break-point frequencies: the lead network break-point, ω_{Lead}, and the lag break-point, ω_{Lag} (recall that $\omega_{Lead} = 1/T_{Lead}$ and $\omega_{Lag} = 1/T_{Lag}$). The lead-lag network is sometimes used for the same reason and in the same way as the derivative control mode. When this is done, the rules used to locate D for a PID controller also apply to the determination of T_{Lead} for the lead-lag compensation network. However, the servo designer frequently uses the lead-lag network to cancel a dominant first-order lag and replace it with a first-order lag that breaks at a higher frequency. This use of the lead-lag compensation network to cancel and replace a first-order lag is illustrated in Example 16.7.

The *lag compensation* network is actually a low-pass filter. It is used to remove undesirable high-frequency components of a signal. Often the high-frequency components are predominantly noise and are not part of the original signal.

The *lag-lead compensation* network is used to cancel dominant low-frequency lead components in much the same way that lead-lag compensation is used to cancel dominant low-frequency lag components. Note that the lead-lag and lag-lead networks have the same transfer function. The only difference is the relative size of the time constants. In the lead-lag network, the lead time constant is larger than the lag time constant ($T_{Lead} > T_{Lag}$). In the lag-lead network, the lag time constant is larger ($T_{Lag} > T_{Lead}$). The same design approach is used for both types of compensation.

In many systems, the open-loop frequency response is dominated by a few low-frequency lag elements. First-order lags are typical of temperature control systems and dc servos with short electrical time constants. Second-order lags are typical of dc servos with long electrical time constants. Faced with this type of process, a servo system designer will use lead-lag compensation to "cancel" a low-frequency lag and replace it with a higher-frequency lag in the compensation network. With reasonable care in the design, lead-lag compensation can increase the dominant lag break-point frequency by a factor of 10 or more (i.e., $T_{Lead} > 10T_{Lag}$). Example 16.7 illustrates the use of integration-lead and lead-lag compensation to improve the control of a system with three predominant first-order lags.

EXAMPLE 16.7

A control engineer analyzed a certain process and determined that it had three dominant first-order lags with time constants of 1000, 100, and 50 s. Use DESIGN to complete the following five controller designs for this system. Plot the error ratio for each design on a single graph, and determine the frequency limit for each design.

(a) Design a controller that uses only the proportional mode (simple gain adjustment).

(b) Design a controller that uses the proportional mode plus integration-lead compensation.

(c) Design a controller that uses the proportional mode plus integration-lead compensation plus lead-lag compensation that cancels the 50-s time constant and replaces it with a 5-s time constant.

(d) Design a controller that uses the proportional mode plus integration-lead compensation plus lead-lag compensation that cancels the 100-s time constant and replaces it with a 10-s time constant.

(e) Design a controller that uses the proportional mode plus integration-lead compensation plus two lead-lag compensation networks (one to replace the 50-s time constant with a 5-s time constant, and one to replace the 100-s time constant with a 10-s time constant).

Solution   We will use DESIGN to complete the design of the five servo controllers. The error-ratio graphs of the five designs are plotted in Figure 16.16. The five plots provide a very graphic way of comparing the accuracy of the five designs. We begin by writing the transfer function of the process:

$$\text{Process TF} = \left(\frac{1}{1 + 1000s} \right) \left(\frac{1}{1 + 100s} \right) \left(\frac{1}{1 + 50s} \right)$$

(a) In the first run of DESIGN, a Pmode value of 10.3 resulted in a gain margin of 10.8 dB, a phase margin of 40°, and a frequency limit of 0.0090 rad/s. The error-ratio graph of this design is labeled "*P*" in Figure 16.16.

Notice that the error ratio levels out at −40.4 dB for frequencies less than 0.0001 rad/s. This is the proportional offset error that characterizes the proportional control mode. There are two ways to reduce the proportional offset: increase the controller gain, *P*, or add integration-lead compensation. We could increase the gain by 4.8 dB and still have a marginally stable system (it would no longer meet the phase margin criteria). This would reduce the proportional offset, but proportional mode acting alone cannot eliminate the offset. The remaining four designs all include integration-lead compensation.

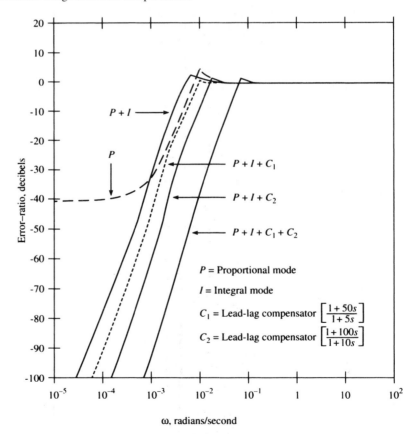

◆ **Figure 16.16** Error-ratio graphs for the five controller designs in Example 16.7. The error ratio is the closed-loop error divided by the open-loop error. A low error ratio means accurate control. The lower the error ratio is, the more accurate the control is. The frequency limit is the lowest frequency at which the error ratio crosses the 0-dB line.

(b) We will use the Imode setting in DESIGN to implement the integration-lead compensation network. This is valid, because the integration-lead compensator and the integral control mode have identical transfer functions. If we wish to report T_i rather than I, we will use the fact that T_i is the reciprocal of I.

In the second run of DESIGN, an Imode value of 0.00207 1/s ($T_i = 483.1$ s) and a Pmode value of 4.69 resulted in a gain margin of 14.8 dB, a phase margin of 40°, and a frequency limit of 0.0055. The error-ratio graph is labeled "$P + I$" in Figure 16.16. We can make two important observations about the error ratio graph. First, notice that the error ratio continues to decrease as the frequency decreases. As the frequency approaches zero, the error ratio also approaches zero. This is the reason we use integration-lead compensation—to give high-static and low-frequency accuracy. Second, notice that the "$P + I$" graph shows a larger error ratio than the "P" graph at frequencies above 0.001 rad/s. Also, the frequency limit is reduced from 0.009 to 0.0055 rad/s. This is the cost of integration-lead compensation—it reduces the frequency limit (bandwidth) of the system and increases the intermediate frequency error.

(c) In the third design, we will use the following lead-lag compensation network to cancel the 50-s time constant and replace it with a 5-s time constant:

$$\text{Compensation TF} = \left(\frac{1 + 50s}{1 + 5s}\right)$$

We can enter the compensation network in DESIGN as a fourth component, giving us the following process plus compensation transfer function:

$$\text{TF} = \left(\frac{1}{1 + 1000s}\right)\left(\frac{1}{1 + 100s}\right)\left(\frac{1}{1 + 50s}\right)\left(\frac{1 + 50s}{1 + 5s}\right)$$

A simpler approach is to complete the algebraic cancellation and enter the following equivalent transfer function:

$$\text{TF} = \left(\frac{1}{1 + 1000s}\right)\left(\frac{1}{1 + 100s}\right)\left(\frac{1}{1 + 5s}\right)$$

We used the second approach in the third run of DESIGN. An Imode value of 0.00244 1/s ($T_i = 409.8$ s) and a Pmode value of 9.05 resulted in a gain margin of 25.9 dB, a phase margin of 40°, and a frequency limit of 0.0096. The error-ratio graph is labeled "$P + I + C_1$" in Figure 16.16. Notice that the addition of the lead-lag compensation network regained the intermediate accuracy and bandwidth lost in design 2 and increased the low-frequency accuracy as well. The third design is more accurate than either of the first two designs.

(d) In the fourth design, we will use the following lead-lag compensation network to cancel the 100-s time constant and replace it with a 10-s time constant.

$$\text{Compensation TF} = \left(\frac{1 + 100s}{1 + 10s}\right)$$

The process plus compensation transfer function can be expressed in either of the following two forms:

$$\text{TF} = \left(\frac{1}{1 + 1000s}\right)\left(\frac{1}{1 + 100s}\right)\left(\frac{1}{1 + 50s}\right)\left(\frac{1 + 100s}{1 + 10s}\right)$$

or

$$\text{TF} = \left(\frac{1}{1 + 1000s}\right)\left(\frac{1}{1 + 10s}\right)\left(\frac{1}{1 + 50s}\right)$$

We used the second form of the transfer function in the fourth run of DESIGN. An Imode value of 0.00303 1/s ($T_i = 330.0$ s) and a Pmode value of 15.4 resulted in a gain margin of 16.8 dB, a phase margin of 40°, and a frequency limit of 0.017. The error-ratio graph is labeled "$P + I + C_2$" in Figure 16.16. Notice that the fourth design is more accurate than the first three designs.

(e) In the fifth design, we will use the following lead-lag compensation networks to cancel both the 50- and 100-s time constants, replacing them with 5- and 10-s time constants:

$$\text{Compensation TF} = \left(\frac{1 + 100s}{1 + 10s}\right)\left(\frac{1 + 50s}{1 + 5s}\right)$$

The process plus compensation transfer function can be expressed in either of the following two forms:

$$\text{TF} = \left(\frac{1}{1 + 1000s}\right)\left(\frac{1}{1 + 100s}\right)\left(\frac{1}{1 + 50s}\right)\left(\frac{1 + 100s}{1 + 10s}\right)\left(\frac{1 + 50s}{1 + 5s}\right)$$

or

$$\text{TF} = \left(\frac{1}{1 + 1000s}\right)\left(\frac{1}{1 + 10s}\right)\left(\frac{1}{1 + 5s}\right)$$

We used the second form of the transfer function in the fifth run of DESIGN. An Imode value of 0.00477 1/s ($T_i = 209.6$ s) and a Pmode value of 69.4 resulted in a gain margin of 12.5 dB, a phase margin of 40°, and a frequency limit of 0.082. The error-ratio graph is labeled "$P + I + C_1 + C_2$" in Figure 16.16. Notice that the fifth design is the most accurate design and has the highest bandwidth (frequency limit). ◆

In conclusion, we make the following observations from the results of Example 16.7:

1. If the process transfer function is dominated by two or three first-order lags, lead-lag compensation can effect significant improvements in accuracy and bandwidth.
2. The greatest improvement is achieved by lead-lag compensation networks that cancel and replace the two lags with the second- and third-largest time constants. These are the first order-lags that increase the system phase lag from 90 to 270°. We ignored the largest time constant and compensated the second- and third-largest time constants because we wanted to improve the frequency response in the region near the $-180°$ frequency.
3. If only one lead-lag compensation network is used, it should be applied to the second-largest time constant—the one that increases the phase lag from 90 to 180°.

◆ GLOSSARY

Bode design method: A design method that uses open-loop frequency response graphs of a control system to design control modes that alter the frequency response to obtain a stable, accurate system. (16.5)

Compensation: A design method that inserts networks into a control system to alter the open-loop frequency response of the system so that performance objectives can be met by a subsequent gain adjustment. (16.7)

Derivative amplitude: The high-frequency gain produced by the derivative control mode. (16.5)

Integration-lead compensation: Another name for the integral control mode. A network that increases static accuracy by increasing the gain at low frequencies while it preserves stability by not changing the gain at high frequencies. (16.7)

Lag compensation: Another name for a low-pass filter. Used to remove undesirable high-frequency components of a signal. (16.7)

Lag-lead compensation: A network that increases the phase lag over a range of frequencies and decreases the gain at high frequencies. Used to move the break points of first-order leads to higher frequencies. (16.7)

Lead-lag compensation: Similar to the derivative control mode. A network that decreases the phase lag over a range of frequencies and raises the gain at high frequencies. Used to increase the $-180°$ frequency. Also used to move the break points of first-order lags to higher frequencies. (16.7)

Model-based adaptive controller: A controller that uses an internal model of the process to determine the optimum control mode settings. Essentially an automated version of the process reaction method. (16.4)

Pattern-recognition adaptive controller: A controller that examines the response of the system to naturally occurring disturbances to determine the period of oscillation and the amplitude decay ratio. Formulas are then used to determine the optimum control mode settings. Essentially an automated version of the ultimate cycle method. (16.4)

Process reaction graph: The step response graph used to determine the first-order lag plus dead-time model of a process. (16.3)

Process reaction method: A method of tuning an installed controller in which the step response is used to determine a first-order lag plus dead-time model of the process. Formulas are then used to determine the control mode settings from the model of the process. (16.3)

Programmed adaptive controller: A controller that automatically adjusts the proportional gain, P, as a function of a number of process-related variables (e.g., controlled variable, setpoint, error, controller output, etc.). (16.4)

Self-tuning adaptive controllers: Controllers capable of changing their control mode settings to adapt to changes in the process. (16.4)

Tuning the controller: The process of determining the control mode settings of a controller during start-up. (16.1)

Ultimate cycle method: A method of tuning an installed controller in which the ultimate period and ultimate gain of the system are measured and used to determine the control mode settings. (16.2)

◆ EXERCISES

Section 16.2

16.1 Several process control systems are tested at start-up. The derivative modes are turned OFF, and the integral modes are set at the lowest setting. The gain of each controller is gradually increased until the control variable starts to oscillate. The gain setting and period of oscillation of each system are given below. Use the original ultimate cycle method to determine the PI and PID controller settings for each system.

System	Ultimate Gain	Ultimate Period (min)
1	0.42	20
2	6.3	6
3	0.8	2
4	1.2	18
5	2.0	0.5

16.2 Use the modified ultimate cycle method to determine the PI and PID controller settings for each system in Exercise 16.1. Compute the values of I and D, and explain how you would determine the value of P.

Section 16.3

16.3 During start-up, the manipulating element of a process control system is maintained constant until the controlled variable levels out at 20% of the full-scale value. A 10% change is produced in the manipulating element position at time 0, and the following data are obtained:

Time (min)	C_m (%)	Time (min)	C_m (%)
0	20.0	22	35.0
2	20.0	24	36.0
4	20.0	26	37.0
6	20.0	28	37.8
8	20.3	30	38.2
10	21.8	32	38.7
12	23.6	34	39.0
14	26.0	36	39.3
16	29.0	38	39.5
18	31.5	40	39.6
20	33.4		

Construct the process reaction graph and use the process reaction method to determine the settings of a PI and a PID controller.

Section 16.4

16.4 Select which of the following three types of adaptive controllers is described by each statement:

1. Programmed adaptive controller
2. Model-based adaptive controller
3. Pattern-recognition adaptive controller

(a) Introduces step changes and observes the reaction of the process to these changes.
(b) Constantly examines the response of the control system to naturally occurring disturbances.
(c) Uses an internal model to determine optimum PID values.
(d) Uses peaks in the error signal to compute the period of oscillation and amplitude decay ratio.
(e) Adjusts the P mode in response to changes in one or more process variables.
(f) Uses a graph of error vs. time to determine the optimum PID values.
(g) Uses a "self-learning" mode to determine the model of the process.
(h) Adjusts the P mode in response to changes in production rate.
(i) Uses the ultimate cycle method to determine the optimum PID values.
(j) Uses the process reaction method for computing the optimum PID values.

Sections 16.5 and 16.6

16.5 The maximum frequency limit, ω_{max}, of a control system is the frequency where the error ratio crosses the 0-dB line. If the error ratio crosses zero between 0.032 and 0.056 rad/s, we know the maximum frequency limit is between those two frequencies. Use the following linear interpolation formula and data to estimate the value for ω_{max}:

$$\omega_{max} = \omega_1 + (\omega_2 - \omega_1)\frac{0 - ER_1}{ER_2 - ER_1}$$

where ω_{max} = maximum frequency limit, rad/s
  ω_1 = 0.032 rad/s = frequency just before zero crossing
  ω_2 = 0.056 rad/s = frequency just after zero crossing

$ER_1 = -0.3$ = error ratio just before zero crossing
$ER_2 = 1.8$ error ratio just after zero crossing

16.6 A dc motor position control system has the following transfer function:

$$\frac{C_m}{V} = \frac{16.1}{s + 0.201s^2 + 3320s^3}$$

Use DESIGN to design a PID controller. Determine the maximum frequency limit from the error ratio graph.

16.7 An amplidyne position control system has the following open-loop transfer function:

$$\frac{C_m}{V} = \frac{8500}{(1 + 0.025s + 0.0001s^2)(s + 0.35s^2 + 0.015s^3)}$$

Use DESIGN to design a PID controller. Determine the maximum frequency limit from the error-ratio graph.

16.8 Design a PID controller for a thermal process with the following transfer function:

$$H(s) = \left(\frac{1}{1 + 700s}\right)e^{-20s}$$

16.9 Design a PID controller to control the level in a first-order lag plus dead-time process with the following transfer function:

$$H(s) = \left(\frac{1}{1 + 25s}\right)e^{-5s}$$

16.10 Design a PID controller for a dc motor positioning system with the following transfer function:

$$H(s) = \frac{55}{s + 0.744s^2 + 0.00965s^3}$$

16.11 Design a PID controller for a solid flow control system with the following transfer function:

$$H(s) = e^{-45s}$$

16.12 A first-order lag plus dead-time blending process has a time constant of 145 s and a dead-time delay of 35 s. Design a PID controller to control the process.

16.13 Design a PID controller to control the following process:

$$H(s) = \left(\frac{0.6}{1 + 0.08s + 0.01s^2}\right)e^{-0.05s}$$

16.14 Design a PID controller for a pressure system that has the following transfer functions:

$$\text{Measuring transmitter TF} = \frac{2}{1 + 0.25s}$$

$$\text{Control valve TF} = \frac{5}{1 + 0.1s + 0.0025s^2}$$

$$\text{Process TF} = \frac{15}{1 + 5s}$$

16.15 Design a PID controller for a blending system that has the following transfer functions:

$$\text{Measuring transmitter TF} = \frac{0.5}{1 + 85s}$$

$$\text{Control valve TF} = \frac{25}{1 + 0.4s + 0.016s^2}$$

$$\text{Process TF} = \frac{12}{1 + 90s}$$

16.16 The AJ Food Company uses a PID controller to control a jacketed kettle similar to Figure 14.6a. Design a PID controller to control the temperature of the product as it leaves the heat exchanger. The transfer functions of the system components are as follows:

$$\text{Measuring transmitter TF} = \frac{2}{1 + 60s + 500s^2}$$

$$\text{Process TF} = \left(\frac{4}{1 + 750s}\right)e^{-2.5s}$$

$$\text{Control valve} = \frac{12}{1 + 0.068s + 0.0044s^2}$$

16.17 Design PID controllers for the three control loops in the blending and heating process illustrated in Figure 16.16 for the following variation in the specifications given in Section 16.6.

Production rate:	3.5×10^{-4} m³/s
Mixing tank	
Diameter:	0.85 m
Operating level:	0.471 m
Film coefficient h_i:	90 W/m² · K
Film coefficient h_o:	1800 W/m² · K
Distance to temperature	
probe:	0.45 m
Wall thickness:	1.5 cm
Wall material:	aluminum

Concentration measuring	
transmitter	
Time constant τ_1:	100 s
Dead-time lag t_{d1}:	25 s
Gain:	2
Level measuring	
transmitter	
Time constant τ_2:	0.5 s
Dead-time lag t_{d2}:	0.8 s
Gain:	4
Temperature measuring	
transmitter	
Time constant τ_3:	80 s
Time constant τ_4:	220 s
Gain:	1
Water control valve	
Damping ratio ζ_1:	0.92
Resonant frequency ω_1:	18.6 rad/s
Gain:	20
Syrup control valve	
Damping ratio ζ_2:	0.84
Resonant frequency ω_2:	2.4 rad/s
Gain:	50
Heating control valve	
Damping ratio ζ_3:	0.90
Resonant frequency ω_3:	9.6 rad/s
Gain:	15
Concentration process	
Gain:	2
Level process	
FS_{in}:	0.002 m³/s
FS_{out}:	0.5 m
Thermal process	
Gain:	2

Section 16.7

16.18 Repeat Example 16.7 for a process that has three dominant first-order lags with time constants of 200, 50, and 10 s.

16.19 Figure 16.12 illustrates a problem that sometimes occurs when Equation 16.4 is used to determine the integral action rate, I, in the design of a PI controller for a first-order lag plus dead-time process with a time constant of 100 s and a dead-time delay of 2 s. Use DESIGN to complete the PI controller designs illustrated in Figures 16.12 and 16.13. Use Equation 16.4 to determine I in the first design (Figure 16.12). Use Equation 16.5 to determine I in the second design (Figure 16.13). Compare your results with Figures 16.12 and 16.13.

16.20 Repeat Exercise 16.19, but this time design a PID controller for the process. You will find that the large gain margin was not eliminated when Equation 16.5 was used to determine I. Use an iterative procedure to find a value of I that will result in a gain margin of about 6 dB. (Hint: Try the following values of I: 0.046, 0.069, 0.080, 0.078, and 0.079. Your final design should have a gain margin of 6.0 dB.)

16.21 Repeat the PI design in Exercise 16.19, but this time add lead-lag compensation to replace the 100-s time constant with a 10-s time constant. Compare the error-ratio graph and the frequency limit of your final design

with those of the final design in Exercise 16.19. What conclusion can you make from the comparison? Of the designs in Exercises 16.19, 16.20, and 16.21, which one has the highest frequency limit?

16.22 A dc motor position control system has the following transfer function:

$$H(s) = \left(\frac{55}{s}\right)\left(\frac{1}{1 + 0.731s}\right)\left(\frac{1}{1 + 0.0132s}\right)$$

Explain why a PID controller is not a good idea for this process. Use DESIGN to design a PD controller for this process. Repeat the design, but this time use a lead-lag compensation network to cancel the 0.731 time constant and replace it with a 0.0731 time constant. Then use DESIGN to design a PD controller for the compensated process. Compare the two designs.

◆ APPENDIX A

Properties of Materials

Properties of Solids

Solid	Density (kg/m³)	Thermal Conductivity (W/m · K)	Heat Capacity (J/kg · K)
Aluminum	2,700	204	910
Asbestos	2,400	0.16	815
Asphalt	1,041	0.17	1,675
Brass	8,470	100	370
Cast iron	7,400	47	460
Copper	8,940	380	400
Glass	2,600	1	490
Gold	19,300	294	130
Graphite	2,000	5	900
Ice	900	2.25	2,000
Insulation	—	0.036	—
Lead	11,340	35	130
Nickel	8,900	60	460
Paraffin	897	0.22	2,931
Rubber	1,500	0.2	2,000
Silver	10,500	400	234
Solder (50–50)	8,842	48	168
Steel	7,800	45	500
Wood (typical oak)	740	0.2	2,400
Wood (typical pine)	440	0.15	2,800

Melting Point and Latent Heat of Fusion

Solid/Liquid	Melting Point (°C)	Latent Heat of Fusion (kJ/kg)
Aluminum	660	393
Asphalt	121	93
Ice/water	0	333
Paraffin	56	147
Solder (50–50)	216	39.5

Properties of Liquids

Liquid	Density (kg/m^3)	Absolute Viscosity[a] (Pa · s)	Thermal Conductivity (W/m · K)	Heat Capacity (J/kg · K)
Ethyl alcohol	800	0.0013	0.18	2,300
Gasoline	740	0.0005	0.14	2,100
Glycerine	1,260	0.83	0.29	2,400
Kerosene	800	0.0024	0.15	2,070
Mercury	13,600	0.0015	8.0	140
Oil	880	0.160	0.16	2,180
Turpentine	870	0.0015	0.13	1,720
Water	1,000	0.001	0.6	4,190

[a] For a fluid temperature of 15°C.

Properties of Gases (at standard atmospheric conditions: 15°C and 76 cm of mercury)

Gas	Molecular Weight M[a]	Density (kg/m^3)	Absolute Viscosity (Pa · s)	Thermal Conductivity (W/m · K)	Heat Capacity[b] (J/kg · K)
Hydrogen (H$_2$)	2.016	0.0854	8.89×10^{-6}	0.163	14,200
Helium (He)	4.002	0.169	1.97×10^{-5}	0.140	
Carbon monoxide (CO)	28.0	1.19		0.023	1,015
Nitrogen (N$_2$)	28.016	1.19	1.77×10^{-5}	0.024	1,010
Air	28.8	1.22	1.81×10^{-5}	0.024	1,030
Oxygen (O$_2$)	32.0	1.36		0.024	910
Argon (A)	39.944	1.69	2.20×10^{-5}		515
Carbon dioxide (CO$_2$)	44.0	1.88	1.46×10^{-5}	0.014	906

[a] M is the gram molecular weight of the gas (i.e., the weight of 1 mol of the gas measured in grams). M is equal numerically to the molecular weight of the gas and has the units (g/mol).

[b] The heat capacity is for constant pressure.

STANDARD ATMOSPHERIC CONDITIONS

1. Temperature: 288 K
                  15°C
                  59°F
2. Pressure: 1.013×10^5 Pa
                  1.013×10^6 dynes/cm^2
                  14.7 lb/in^2
                  76 cm of mercury
                  29.92 in. of mercury
                  10.336 m of water
                  34 ft of water
3. Air density: 1.23 kg/m^3
                  1.23×10^{-3} g/cm^3
                  0.07651 lb/ft^3

REFERENCES

The entries in the tables of properties of solids, liquids, and gases were converted to SI units from data obtained from the following sources:

BINDER R. C. *Fluid Mechanics,* 2nd edition. Englewood Cliffs, N.J.: Prentice-Hall, Inc., 1950, pp. 6, 27, 62–65.

CARMICHAEL, COLIN. *Kent's Mechanical Engineers' Handbook: Design and Production,* 12th edition. New York: John Wiley & Sons, Inc., 1950, pp. 1–04, 1–29, 2–57, 2–58, 5–78, 19–06.

JENNINGS, BURGESS H., AND SAMUEL R. LEWIS. *Air Conditioning and Refrigeration,* 3rd edition. Scranton, Pa.: International Textbook Company, 1951, pp. 22, 99, 104–114.

LEMON, HARVEY B., AND MICHAEL FERENCE, JR. *Analytical Experimental Physics.* Chicago: The University of Chicago Press, 1946, pp. 37, 38, 134, 142, 177, 178, 181, 188, 219.

SALISBURY, J. KENNETH. *Kent's Mechanical Engineers' Handbook: Power,* 12th edition. New York: John Wiley & Sons, Inc., 1950, pp. 2–48, 2–59, 3–04, 3–05, 3–06, 3–14, 3–15, 3–16, 3–34, 3–37, 3–58, 5–03, 6–43, 6–44.

◆ APPENDIX B

Units and Conversion

SYSTEMS OF UNITS*

The different systems of units are best understood when applied to Newton's law of motion.

$$f = kma$$

where
f = force acting on a body
m = mass of the body
a = acceleration of the body
k = constant whose value depends on the system of units

1. The SI system, $k = 1$

$$f(\text{N}) = m(\text{kg}) \cdot a(\text{m/s}^2)$$

2. The cgs absolute system, $k = 1$

$$f(\text{dynes}) = m(\text{g}) \cdot a(\text{cm/s}^2)$$

3. The fps absolute system, $k = 1$

$$f(\text{poundals}) = m(\text{lb}) \cdot a(\text{ft/s}^2)$$

4. The engineering fps system, $k = 1/g_s = 1/32.174$

$$f(\text{lb force}) = m(\text{lb}) \cdot a(\text{ft/s}^2)/g_s$$

5. The engineering fss system, $k = 1$

$$f(\text{lb force}) = m(\text{slugs}) \cdot a(\text{ft/s}^2)$$

---

* From H. B. Lemon and M. Ference, Jr., *Analytical Experimental Physics* (Chicago: The University of Chicago Press, 1946), pp. 37–38.

◆◆◆◆

Conversion Factors (lbf = pound force, lbm = pound mass)

Quantity	To Obtain:	Multiply:	By:
Area, A	m^2	ft^2	0.09290
	ft^2	m^2	10.764
	m^2	$in.^2$	6.452E − 4
	$in.^2$	m^2	1550
Density, ρ	kg/m^3	lbm/ft^3	16.018
	lbm/ft^3	kg/m^3	0.06243
Energy (work)	J	ft lbf	1.356
	ft lbf	J	0.7376
	J	Btu	1055
	Btu	J	9.480E − 4
Flow rate, Q	m^3/s	gal/min	6.3088E − 5
	m^3/s	L/min	1.6667E − 5
	gal/min	m^3/s	15850.9
	gal/min	L/min	0.26418
	L/min	m^3/s	60.000
	L/min	gal/min	3.7853
Force, f	N	$kg \cdot m/s^2$	1.000
	N	poundal	1.383E − 1
	N	lbf	4.448
	N	ozf	2.780E − 1
	lbf	$lbm \cdot ft/s^2$	3.108E − 2
	ozf	$ozm \cdot in./s^2$	2.590E − 3
Latent heat	J/kg	Btu/lbm	2326
	Btu/lbm	J/kg	4.2992E − 4
Length, L	m	ft	0.3048
	ft	m	3.281
	m	in.	0.0254
	in.	m	39.37
Liquid resistance, R_L	$Pa \cdot s/m^3$	psi/gpm	1.093E + 8
	psi/gpm	$Pa \cdot s/m^3$	9.148E − 9
Mass, m	kg	$N \cdot s^2/m$	1.000
	kg	lbm	4.536E − 1
	kg	ozm	2.835E − 2
	lbm	$lbf \cdot s^2/ft$	32.174
	ozm	$ozf \cdot s^2/in.$	386.09
Moment of inertia, J	$N \cdot m \cdot s^2/rad$	$kg \cdot m^2/rad$	1.000
	$N \cdot m \cdot s^2/rad$	$ozf \cdot in. \cdot s^2/rad$	7.062E − 3
	$N \cdot m \cdot s^2/rad$	$lbf \cdot in. \cdot s^2/rad$	1.130E − 1
	$N \cdot m \cdot s^2/rad$	$lbf \cdot ft \cdot s^2/rad$	1.356
	$N \cdot m \cdot s^2/rad$	$ozm \cdot in.^2/rad$	1.829E − 5
	$N \cdot m \cdot s^2/rad$	$lbm \cdot in.^2/rad$	2.926E − 4
	$N \cdot m \cdot s^2/rad$	$lbm \cdot ft^2/rad$	4.214E − 2

Conversion Factors (*continued*)

Quantity	To Obtain:	Multiply:	By:
Power	W	hp	746
	hp	W	$1.341E - 3$
	W	Btu/hr	0.2931
	Btu/hr	W	3.4129
Pressure, P	Pa	lbf/ft^2	47.88
	Pa	$lbf/in.^2$	$6.895E + 3$
	lbf/ft^2	Pa	0.02088
	lbf/ft^2	$lbf/in.^2$	144
	$lbf/in.^2$	Pa	$1.45E - 4$
	$lbf/in.^2$	lbf/ft^2	$6.9444E - 3$
Thermal (heat) capacity, H_h	$J/kg \cdot K$	$Btu/lbm \cdot °F$	4187
	$Btu/lbm \cdot °F$	$J/kg \cdot K$	$2.388E - 4$
Thermal conductance, C	$W/m^2 \cdot K$	$Btu/hr \cdot ft^2 \cdot °F$	5.678
	$Btu/hr \cdot ft^2 \cdot °F$	$W/m^2 \cdot K$	0.17611
Thermal conductivity, K	$W/m \cdot K$	$Btu \cdot in./hr \cdot ft^2 \cdot °F$	0.1441
	$Btu \cdot in./hr \cdot ft^2 \cdot °F$	$W/m \cdot K$	6.938
Torque, T	$N \cdot m$	$ozf \cdot in.$	$7.062E - 3$
	$N \cdot m$	$lbf \cdot in.$	$1.130E - 1$
	$N \cdot m$	$lbf \cdot ft$	1.356
Torque constant, K_T	$N \cdot m/A$	$ozf \cdot in/A$	$7.062E - 3$
	$N \cdot m/A$	$lbf \cdot in/A$	$1.130E - 1$
	$N \cdot m/A$	$lbf \cdot ft/A$	1.356
Viscous friction, B (damping)	$N \cdot m \cdot s/rad$	$ozf \cdot in. \cdot s/rad$	$7.062E - 3$
	$N \cdot m \cdot s/rad$	$lbf \cdot in. \cdot s/rad$	$1.130E - 1$
	$N \cdot m \cdot s/rad$	$lbf \cdot ft \cdot s/rad$	1.356
	$N \cdot m \cdot s/rad$	$ozf \cdot in./krpm$	$6.744E - 5$
	$N \cdot m \cdot s/rad$	$lbf \cdot in./krpm$	$1.079E - 3$
	$N \cdot m \cdot s/rad$	$lbf \cdot ft/krpm$	$1.295E - 2$
Viscosity, absolute, μ	$Pa \cdot s$	$lbf \cdot s/ft^2$	47.88
	$Pa \cdot s$	$lbm/ft \cdot s$	1.488
	$Pa \cdot s$	poise	0.1
	$Pa \cdot s$	cp	0.001
	$lbf. s/ft^2$	$Pa \cdot s$	0.0209
	$lbm/ft \cdot s$	$Pa \cdot s$	0.672
Voltage constant, K_E	$V \cdot s/rad$	$V/krpm$	$9.549E - 3$
Volume, V	L	gal	3.7853
	gal	L	0.26418
	m^3	gal	$3.7853E - 3$
	gal	m^3	264.18
	gal	ft^3	7.480
	ft^3	gal	0.1337

◆ APPENDIX C

Binary Codes

◆ **TABLE C.1** Powers of 2

$2^0 = 1$	$2^9 = 512$
$2^1 = 2$	$2^{10} = 1,024$
$2^2 = 4$	$2^{11} = 2,048$
$2^3 = 8$	$2^{12} = 4,096$
$2^4 = 16$	$2^{13} = 8,192$
$2^5 = 32$	$2^{14} = 16,384$
$2^6 = 64$	$2^{15} = 32,768$
$2^7 = 128$	$2^{16} = 64,536$
$2^8 = 256$	

◆ **TABLE C.2** Octal and Binary Equivalents

Octal	Binary	Octal	Binary
0	000	4	100
1	001	5	101
2	010	6	110
3	011	7	111

◆ **TABLE C.3**   Decimal, Hexadecimal, and Binary Equivalents

Decimal	Hex	Binary	Decimal	Hex	Binary
0	0	0000	8	8	1000
1	1	0001	9	9	1001
2	2	0010	10	A	1010
3	3	0011	11	B	1011
4	4	0100	12	C	1100
5	5	0101	13	D	1101
6	6	0110	14	E	1110
7	7	0111	15	F	1111

◆ **TABLE C.4**   One's and Two's Complements

Decimal Number	Two's Complement		One's Complement	
	Binary	Hex	Binary	Hex
−1	1111 1111	FF	1111 1110	FE
−2	1111 1110	FE	1111 1101	FD
−3	1111 1101	FD	1111 1100	FC
−4	1111 1100	FC	1111 1011	FB
−5	1111 1011	FB	1111 1010	FA
−6	1111 1010	FA	1111 1001	F9
−7	1111 1001	F9	1111 1000	F8
−8	1111 1000	F8	1111 0111	F7
−9	1111 0111	F7	1111 0110	F6
−10	1111 0110	F6	1111 0101	F5
−11	1111 0101	F5	1111 0100	F4
−12	1111 0100	F4	1111 0011	F3

◆ **TABLE C.5**   The Gray Code

Decimal	Binary	Gray	Decimal	Binary	Gray
0	0000	0000	8	1000	1100
1	0001	0001	9	1001	1101
2	0010	0011	10	1010	1111
3	0011	0010	11	1011	1110
4	0100	0110	12	1100	1010
5	0101	0111	13	1101	1011
6	0110	0101	14	1110	1001
7	0111	0100	15	1111	1000

◆ **TABLE C.6** Binary Codes for Decimal Digits

Decimal	BCD	Excess-3	Decimal	BCD	Excess-3
0	0000	0011	5	0101	1000
1	0001	0100	6	0110	1001
2	0010	0101	7	0111	1010
3	0011	0110	8	1000	1011
4	0100	0111	9	1001	1100

◆ **TABLE C.7** Seven-Bit ASCII Code

ASCII Code (Hex)	Coded Symbol[a]	ASCII Code (Hex)	Coded Symbol	ASCII Code (Hex)	Coded Symbol	ASCII Code (Hex)	Coded Symbol
00	nul	20	space	40	@	60	`
01	soh	21	!	41	A	61	a
02	stx	22	"	42	B	62	b
03	etx	23	#	43	C	63	c
04	eot	24	$	44	D	64	d
05	enq	25	%	45	E	65	e
06	ack	26	&	46	F	66	f
07	bel	27	'	47	G	67	g
08	bs	28	(	48	H	68	h
09	ht	29	)	49	I	69	i
0A	lf	2A	*	4A	J	6A	j
0B	vt	2B	+	4B	K	6B	k
0C	ff	2C	,	4C	L	6C	l
0D	cr	2D	−	4D	M	6D	m
0E	so	2E	.	4E	N	6E	n
0F	si	2F	/	4F	O	6F	o
10	dle	30	0	50	P	70	p
11	dc1	31	1	51	Q	71	q
12	dc2	32	2	52	R	72	r
13	dc3	33	3	53	S	73	s
14	dc4	34	4	54	T	74	t
15	nak	35	5	55	U	75	u
16	syn	36	6	56	V	76	v
17	etb	37	7	57	W	77	w
18	can	38	8	58	X	78	x
19	em	39	9	59	Y	79	y
1A	sub	3A	:	5A	Z	7A	z
1B	esc	3B	;	5B	[	7B	{
1C	fs	3C	<	5C	\	7C	|
1D	gs	3D	=	5D	]	7D	}
1E	rs	3E	>	5E	∧	7E	~
1F	us	3F	?	5F	-	7F	rub

[a] ASCII codes from 00 to 1F are control characters.

◆ APPENDIX D

Instrumentation Symbols and Identification

The material in this section was abstracted from ANSI/ISA-S5.1-1984, "Instrumentation Symbols and Identification," copyright © Instrument Society of America, 1984, and is reprinted by permission.

1 PURPOSE

The purpose of this standard is to establish a uniform means of designating instruments and instrumentation systems used for measurement and control. To this end, a designation system that includes symbols and an identification code is presented.

2 SCOPE

2.1.2 Process equipment symbols are not part of this standard, but are included only to illustrate applications of instrumentation symbols.

2.2.1 The standard is suitable for use in the chemical, petroleum, power generation, air conditioning, metal refining, and numerous other process industries.

2.3.1 The standard is suitable for use whenever any reference to an instrument or a control system function is required for the purpose of symbolization and identification. Such references may be required for the following uses, as well as others:

Design sketches

Teaching examples

Technical papers, literature, and discussions

Instrumentation system diagrams, loop diagrams, and logic diagrams

Functional descriptions

◆◆◆◆

Flow diagrams: process, mechanical, engineering, systems, piping (process), and instrumentation

Construction drawing

Specifications, purchase orders, manifests, and other lists

Identification (tagging) of instruments and control functions

Installation, operating and maintenance instructions, drawings, and records

2.3.2 The standard is intended to provide sufficient information to enable anyone reviewing any document depicting process measurement and control (who has a reasonable amount of process knowledge) to understand the means of measurement and control of the process. The detailed knowledge of a specialist in instrumentation is not a prerequisite to this understanding.

3 DEFINITIONS

Balloon: synonym for *bubble.*

Bubble: the circular symbol used to denote and identify the purpose of an *instrument* or *function.* It may contain a tag number. Synonym for *balloon.*

Computing device: a device or *function* that performs one or more calculations or logic operations, or both, and transmits one or more resultant output signals. A computing device is sometimes called a *computing relay.*

Converter: a device that receives information in one form of an instrument signal and transmits an output signal in another form. An *instrument* that changes a sensor's output to a standard signal is properly designated as a *transmitter,* not a *converter.* Typically, a temperature element (TE) may connect to a transmitter (TT), not to a converter (TY). A converter is also referred to as a *transducer,* however, "transducer" is a completely general term, and its use specifically for signal conversion is not recommended.

4 OUTLINE OF THE IDENTIFICATION SYSTEM

Comment: Each instrument or function is identified by a *tag number* that is placed inside a *balloon.*

4.1.1 Each instrument or function to be identified is designated by an alphanumeric code or tag number, as shown in Figure D.1. The loop identification part of the tag number generally is common to all instruments or functions of the loop. A suffix or prefix may be added to complete the identification. Typical identification is shown in Figure D.1.

4.2.1 The *functional identification* of an instrument or its functional equivalent consists of letters from Table D.1 and includes one first letter (designating the measured or initiating variable) and one or more succeeding letters (identifying the functions performed).

4.2.2 The *functional identification* of an instrument is made according to the function and not according to the construction. Thus a differential-pressure recorder used for flow measurement is identified by FR; a pressure indicator and a pressure-actuated switch connected to the output of a pneumatic level transmitter are identified by LI and LS, respectively.

```
Typical Tag Number

TIC 103        Instrument Identification or Tag Number
T    103       Loop Identification
     103       Loop Number
TIC            Functional Identification
T              First Letter
  IC           Succeeding Letters

Expanded Tag Number

10-PAH-5A      Tag Number
10             Optional Prefix
         A     Optional Suffix

Note: Hyphens are optional as separators.
```

◆ **Figure D.1** Tag numbers.

4.2.3 In an instrument loop, the first letter of the *functional identification* is selected according to the measured or initiating variable, not according to the manipulated variable. Thus a control valve varying flow according to the dictates of a level controller is an LV, not an FV.

4.2.4 The succeeding letters of the *functional identification* designate one or more readout or passive functions and/or output functions. A modifying letter may be used, if required, in addition to one or more other succeeding letters. Modifying letters may modify either a first letter or succeeding letters, as applicable. The TDAL contains two modifiers. The letter D changes the measured variable T into a new variable, "differential temperature." The letter L restricts the readout function A, alarm, to represent a low alarm only.

4.2.5 The sequence of *identification letters* begins with a first letter selected according to Table D.1. Readout or passive functional letters follow in any order, and output functional letters follow these in any sequence, except that output letter C (control) precedes output letter V (valve) (e.g., PCV, a self-actuated control valve). However, modifying letters, if used, are interposed so that they are placed immediately following the letters they modify.

4.3.1 The *loop identification* consists of a first letter and a number. Each instrument within a loop has assigned to it the same loop number, and in the case of parallel numbering, the same first letter. Each instrument loop has a unique loop identification. An instrument common to two or more loops should carry the identification of the loop that is considered predominant.

4.3.2 Loop numbering may be parallel or serial. *Parallel numbering* involves starting a numerical sequence for each new first letter (e.g., TIC-100, FRC-100, LIC-100, AI-100, etc.). *Serial numbering* involves using a single sequence of numbers for a project or for large sections of a project, regardless of the first letter of the loop identification (e.g., TIC-100, FRC-101, LIC-102, AI-103, etc). A loop numbering sequence may begin with 1 or any other convenient number, such as 001, 301, or 1201. The number may incorporate coded information; however, simplicity is recommended.

4.4.1 The examples in this standard illustrate the symbols that are intended to depict instrumentation on diagrams and drawings. No inference should be drawn that the choice of any

◆ **TABLE D.1**   Identification Letters

First Letter		Succeeding Letters		
Measured or Initiating Variable	Modifier	Readout or Passive Function	Output Function	Modifier
A Analysis		Alarm		
B Burner, combustion		User's choice	User's choice	User's choice
C User's choice			Control	
D User's choice	Differential			
E Voltage		Sensor (primary element)		
F Flow rate	Ratio (fraction)			
G User's choice		Glass, viewing device		
H Hand				High
I Current (electrical)		Indicate		
J Power	Scan			
K Time, time schedule	Time rate of change		Control station	
L Level		Light		Low
M User's choice	Momentary			Middle, intermediate
N User's choice		User's choice	User's choice	User's choice
O User's choice		Orifice, restriction		
P Pressure, vacuum		Point (test) connection		
Q Quantity	Integrate, totalize			
R Radiation		Record		
S Speed, frequency	Safety		Switch	
T Temperature			Transmit	
U Multivariable		Multifunction	Multifunction	Multifunction
V Vibration, mechanical analysis			Valve, damper, louver	
W Weight, force		Well		
X Unclassified	X axis	Unclassified	Unclassified	Unclassified
Y Event, state or presence	Y axis		Relay, compute, convert	
Z Position, dimension	Z axis		Driver, actuator, unclassified final control element	

of the schemes for illustration constitutes a recommendation for the illustrated methods of measurement or control.

4.4.2 The bubble may be used to tag distinctive symbols, such as those for control valves, when such tagging is desired. In such instances, the line connecting the bubble to the instrument symbol is drawn close to, but not touching, the symbol. In other instances, the bubble serves to represent the instrument proper.

4.4.3 A distinctive symbol whose relationship to the remainder of the loop is easily apparent from a diagram need not be individually tagged on the diagram. For example, an orifice flange or a control valve that is part of a larger system need not be shown with a tag number on a diagram. Also, where there is a primary element connected to another instrument on a diagram, use of a symbol to represent the primary element on the diagram is optional.

4.4.4 A brief explanatory notation may be added adjacent to a symbol or line to clarify the function of an item.

Instrument Line Symbols

All lines are to be fine in relation to process piping lines.

1. Instrument supply* or connection to process
2. Undefined signal
3. Pneumatic signal†
4. Electric signal
5. Hydraulic signal
6. Capillary tube
7. Electromagnetic or sonic signal (guided)‡
8. Electromagnetic or sonic signal (not guided)
9. Internal system link (software or data link)
10. Mechanical link

The following are optional binary (ON/OFF) symbols.

11. Pneumatic binary signal
12. Electric binary signal

---

* "Or" means user's choice. Consistency is recommended.

The following abbreviations are suggested to denote the types of power supply. These designations may also be applied to purge fluid supplies.

AS	air supply			HS	Hydraulic supply
IA	Instrument air	} options		NS	Nitrogen supply
PA	Plant air			SS	Steam supply
				WS	Water supply

ES   Electric supply

GS   Gas supply

The supply level may be added to the instrument supply line (e.g., AS-100, a 100-psig air supply; ES-24DC, a 24-volt dc power supply).

† The pneumatic signal symbol applies to a signal using any gas as the signal medium. If a gas other than air is used, the gas may be identified by a note on the signal symbol or otherwise.

‡ Electromagnetic phenomena include heat, radio waves, nuclear radiation, and light.

Selected Symbols

Flow rate F

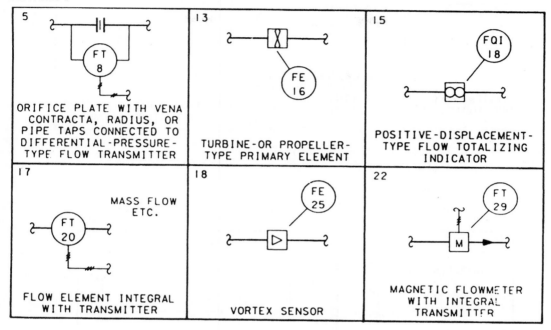

ORIFICE PLATE WITH VENA CONTRACTA, RADIUS, OR PIPE TAPS CONNECTED TO DIFFERENTIAL-PRESSURE-TYPE FLOW TRANSMITTER	TURBINE-OR PROPELLER-TYPE PRIMARY ELEMENT	POSITIVE-DISPLACEMENT-TYPE FLOW TOTALIZING INDICATOR
FLOW ELEMENT INTEGRAL WITH TRANSMITTER	VORTEX SENSOR	MAGNETIC FLOWMETER WITH INTEGRAL TRANSMITTER

Level L

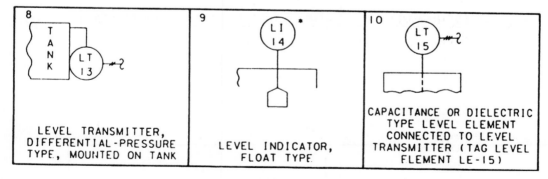

LEVEL TRANSMITTER, DIFFERENTIAL-PRESSURE TYPE, MOUNTED ON TANK	LEVEL INDICATOR, FLOAT TYPE	CAPACITANCE OR DIELECTRIC TYPE LEVEL ELEMENT CONNECTED TO LEVEL TRANSMITTER (TAG LEVEL ELEMENT LE-15)

Temperature T

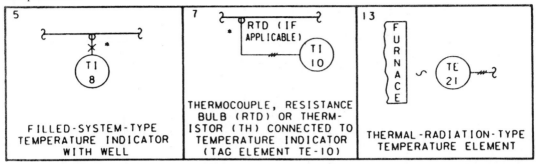

5 FILLED-SYSTEM-TYPE TEMPERATURE INDICATOR WITH WELL	**7** THERMOCOUPLE, RESISTANCE BULB (RTD) OR THERM-ISTOR (TH) CONNECTED TO TEMPERATURE INDICATOR (TAG ELEMENT TE-10)	**13** THERMAL-RADIATION-TYPE TEMPERATURE ELEMENT

Speed S   Control C

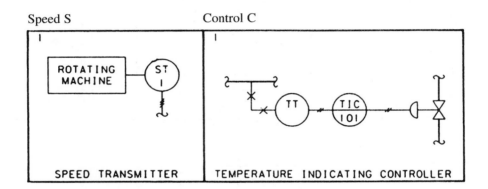

SPEED TRANSMITTER   TEMPERATURE INDICATING CONTROLLER

Control C

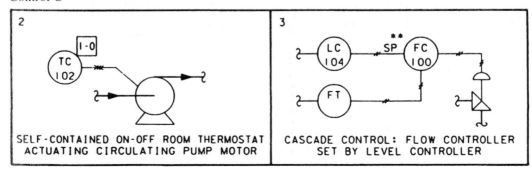

SELF-CONTAINED ON-OFF ROOM THERMOSTAT ACTUATING CIRCULATING PUMP MOTOR

CASCADE CONTROL: FLOW CONTROLLER SET BY LEVEL CONTROLLER

Complex Numbers

INTRODUCTION

Complex numbers were introduced to permit the extraction of the square root for negative numbers. For example, using the quadratic formula to solve the equation $x^2 - 8x + 25 = 0$ requires the extraction of the square root of the number -36. Because the square root of negative numbers is not allowed, the number -36 is written as $(-1)(36)$, the symbol j is defined as the square root of (-1), and the square root of $(-1)(36)$ is written as $j6$.* Thus the j operator permits the extraction of negative numbers, and the equation $x^2 - 8x + 25 = 0$ has the following solution:

$$x_1, x_2 = \frac{8 \pm \sqrt{(-8)^2 - 4(1)(25)}}{2(1)} = \frac{8 \pm \sqrt{-36}}{2} = 4 \pm j3$$

Thus $(4 + j3)$ and $(4 - j3)$ are the two values of x that satisfy the equation $x^2 - 8x + 25 = 0$. Values such as $(4 + j3)$ and $(4 - j3)$ belong to a set called complex numbers. Notice that complex numbers may have two components. The first component is a real number; it is called the real part of the complex number. The second component includes the j operator and a real number; the real number is called the imaginary part of the complex number (the j operator indicates the imaginary component, but it is not considered to be part of it). The complex number $(4 + j3)$ has the real part 4 and the imaginary part 3. A complex number may have both a real and an imaginary part, only a real part, or only an imaginary part. The following are examples of complex numbers:

$$(10 + j20) \qquad (10) \qquad (j20)$$

---

* In mathematics, the symbol i is used as the square root of (-1). In electrical engineering, i is used as the symbol for current, and j is used to denote $\sqrt{-1}$.

RECTANGULAR AND POLAR FORMS OF COMPLEX NUMBERS

There are two ways to designate complex numbers: the rectangular form and the polar form. In the introduction, the rectangular form was used to designate the complex numbers $(4 + j3)$ and $(4 - j3)$. The polar form of a complex number is based on Euler's identity, which is stated in the following two equations:

$$(e^{+j\theta} = \cos\theta + j\ \sin\theta)$$

and

$$(e^{-j\theta} = \cos\theta - j\ \sin\theta)$$

We will use Euler's identity and a rearrangement of the complex number $(N = 4 + j3)$ to obtain N in its polar form.

We begin by defining θ as the angle whose tangent is equal to the imaginary part of N divided by the real part of N.

$$\theta = \arctan(\tfrac{3}{4}) = 36.87° \tag{E.1}$$

Next we define c as the square root of the sum of the squares of the real and imaginary parts of N.

$$c = \sqrt{4^2 + (-3)^2} = \sqrt{25} = 5 \tag{E.2}$$

Then we multiply and divide N by $c = 5$.

$$N = 4 + j3 = 5(\tfrac{4}{5} + j\tfrac{3}{5}) = 5(0.8 + j0.6)$$

Now observe the following:

$$\tfrac{4}{5} = 0.8 = \cos(36.87°) = \cos\theta$$

$$\tfrac{3}{5} = 0.6 = \sin(36.87°) = \sin\theta$$

and

$$N = 5(\cos 36.87° + j\ \sin 36.87°) \tag{E.3}$$

Applying Euler's identity to Equation (E.3), we obtain the following polar form of N:

$$N = 5e^{j36.87°} \tag{E.4}$$

The symbol $\angle\theta$ is frequently used in place of $e^{j\theta}$, and the polar form of N is usually written as follows:

$$N = 5\underline{/36.87°} \tag{E.5}$$

Equation (E.5) is more convenient for writing and printing the polar form, but Equation (E.4) is more useful in determining how to multiply and divide complex numbers and how to raise a complex number to a power. These operations are all based on the rules of exponents and will be covered in the sections about operations on complex numbers.

CONVERSION OF COMPLEX NUMBERS

Euler's identity is also the basis for the conversion of complex numbers from rectangular to polar form or vice versa. The conversion formulas are summarized below:

Conversion of: $N = a + jb = c\angle\theta$
Rectangular to polar form:

$$c = \sqrt{a^2 + b^2} \qquad \text{(E.6)}$$

$$\theta = \arctan(b/a) \qquad \text{(E.7)}$$

Polar to rectangular form:

$$a = c(\cos\theta) \qquad \text{(E.8)}$$

$$b = c(\sin\theta) \qquad \text{(E.9)}$$

GRAPHICAL REPRESENTATION OF COMPLEX NUMBERS

Complex numbers are represented graphically by a point on a two-dimensional graph called the complex plane. Figure E.1 shows the graphical representation of the number $N = 4 + j3 = 5\angle36.87°$.

Either form of a complex number may be used to locate the point on the complex plane that represents the number, N. Using the rectangular form, the real part of N ($a = 4$) is plotted on the horizontal axis, which is called the real axis. The imaginary part of N ($b = 3$) is plotted on the vertical axis, which is called the imaginary axis.

Using the polar form, the angle of N ($\theta = 36.87°$) is measured in a counterclockwise direction from the positive real axis. The magnitude of N ($c = 5$) is measured from the origin to the point N, which lies on the line determined by the angle θ.

Complex numbers may be in any of the four quadrants of the complex plane as shown in Figure E.2. The complex number $N = a + jb$ will be in the first quadrant if both a and b are

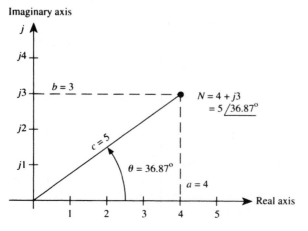

Figure E.1   Graphical representation of the complex number $N = 4 + j3$.

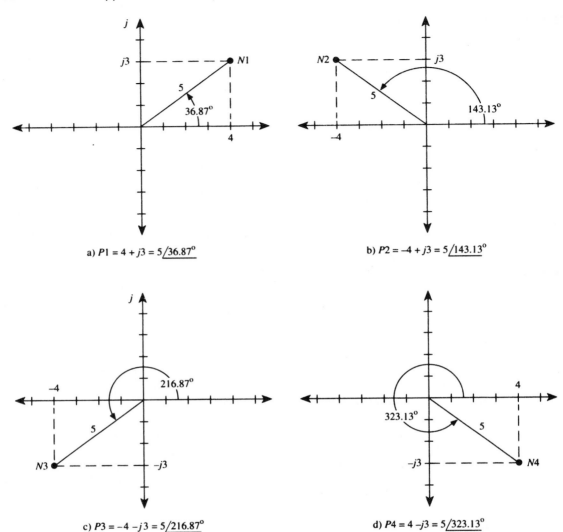

a) $P1 = 4 + j3 = 5\underline{/36.87^{\circ}}$

b) $P2 = -4 + j3 = 5\underline{/143.13^{\circ}}$

c) $P3 = -4 - j3 = 5\underline{/216.87^{\circ}}$

d) $P4 = 4 - j3 = 5\underline{/323.13^{\circ}}$

Figure E.2 The signs of the real and imaginary parts determine which quadrant the complex number will be in.

positive, in the second quadrant if a is negative and b is positive, in the third quadrant if a and b are both negative, and in the fourth quadrant if a is positive and b is negative.

The *conjugate* of a complex number is formed by reversing the sign of the imaginary part. In Figure E.2, $P1$ and $P4$ are conjugates of each other. Also, $P2$ and $P3$ are conjugates of each other. Graphically, the conjugate of a complex number is its mirror image about the real axis.

ADDITION AND SUBTRACTION OF COMPLEX NUMBERS

Complex numbers are most conveniently added when they are expressed in the rectangular form. The sum of two complex numbers is formed by adding the real parts to obtain the real part of the sum and then adding the imaginary parts to form the imaginary part of the sum. The following example illustrates the addition of two complex numbers:

$$N1 = 5 + j12$$

$$N2 = 4 - j3$$

$$N3 = N1 + N2 = (5 + 4) + j(12 - 3) = 9 + j9$$

Subtraction of $N2$ from $N1$ is accomplished by changing the sign of both parts of $N2$ and then adding the result to $N1$.

$$N4 = N1 - N2 = (5 - 4) + j(12 + 3) = 1 + j15$$

Numbers that are in polar form must be converted to rectangular form before they can be added or subtracted. A pocket calculator with the polar-to-rectangular and rectangular-to-polar functions is the most convenient way to convert complex numbers from one form to the other.

MULTIPLICATION AND DIVISION OF COMPLEX NUMBERS

Complex numbers are most conveniently multiplied or divided when they are expressed in polar form, but the operations can also be carried out in the rectangular form.

In polar form, the product of two complex numbers is obtained by multiplying the magnitudes of the two numbers and adding their angles. The quotient of two complex numbers is obtained by dividing their magnitudes and subtracting the angle of the divisor from the angle of the dividend. The following examples illustrate multiplication and division of complex numbers:

$$N5 = 5 + j12 = 13\underline{/67.38°}$$

$$N6 = 4 + j3 = 5\underline{/36.87°}$$

$$N5 \cdot N6 = (13 \cdot 5)\underline{/67.38° + 36.87°} = 65\underline{/104.25°}$$

$$\frac{N5}{N6} = (\tfrac{13}{5})\underline{/67.38° - 36.87°} = 2.6\underline{/30.51°}$$

The rules for multiplying and dividing complex numbers are based on the rules of exponents. This is more obvious if we use the exponential version of the polar form, as the following examples illustrate:

$$N5 \cdot N6 = (13e^{j67.38°}) \cdot (5e^{j36.87°}) = (13)(5)e^{j(67.38° + 36.87°)}$$

$$= 65e^{j104.25°}$$

$$N5/N6 = \frac{(13e^{j67.38°})}{(5e^{j36.87°})} = \left(\frac{13}{5}\right)e^{j(67.38° - 36.87°)}$$

$$= 2.6e^{+j30.51°}$$

The following example illustrates the multiplication of N5 times N6 in rectangular form:

$$N5 \cdot N6 = (5 + j12)(4 + j3)$$
$$= 5(4 + j3) + j12(4 + j3)$$
$$= 20 + j15 + j48 + j^2 36 \quad \{\text{Note: } j^2 = -1\}$$
$$= -16 + j63$$
$$= 65\underline{/104.25°}$$

The first step in dividing two complex numbers in rectangular form is to multiply both the numerator and the denominator by the conjugate of the denominator. This reduces the denominator to a real number. The final step is to divide this real number into the new numerator. The following example illustrates division of N5 by N6 in rectangular form:

$$\frac{N5}{N6} = \frac{5 + j12}{4 + j3} = \frac{(5 + j12)}{(4 + j3)} \cdot \frac{(4 - j3)}{(4 - j3)}$$

$$= \frac{56 + j33}{25} = 2.24 + j1.32 = 2.6\underline{/30.51°}$$

INTEGER POWER OF A COMPLEX NUMBER

Raising the complex number N to an integer power k can be viewed as the product of k factors, each equal to N. The result is a complex number whose magnitude is equal to the magnitude of N raised to the k power and whose angle is k times the angle of N.

$$N^k = (ce^{j\theta})^k = c^k e^{jk\theta}$$

For example,

$$(3\underline{/15°})^4 = 3^4\underline{/4 \cdot 15°} = 81\underline{/60°}$$

ROOTS OF A COMPLEX NUMBER

Finding the kth root of a complex number is equivalent to solving the equation

$$R^k = ce^{j\theta} \tag{E.10}$$

where R is the kth root of the complex number $ce^{j\theta}$. Since Equation (E.10) is an equation of degree k, there will be k roots, $R_1, R_2, \ldots, R_k$. To find the k roots, we note that complex numbers are circular numbers that repeat as the angle is increased by multiples of 2π radians (or $360°$). In other words,

$$ce^{j\theta} = ce^{j(\theta + 2\pi)} = ce^{j(\theta + 4\pi)} = ce^{j(\theta + 6\pi)} = \cdots \qquad \textbf{(E.11)}$$

The roots all have the same magnitude, $|R| = c^{1/k}$. The angles of the roots are obtained by dividing k into each of the angles of the complex numbers in Equation (E.11). Enough terms must be used to obtain the k roots. Additional terms will simply repeat the first k roots.

For example, let us find the fourth root of $16\angle 80°$. We know there will be four roots, so we write the following four equivalent complex numbers:

$$16\underline{/80°} = 16\underline{/(80 + 360)°} = 16\underline{/(80 + 720)°} = 16\underline{/(80 + 1080)°}$$

The magnitude of the roots will be 16 raised to the $\frac{1}{4}$ power.

$$|R| = 16^{1/4} = 2$$

The angles of the roots will be

$$\tfrac{80}{4} = 20°, \quad \frac{80 + 360}{4} = 110°$$

$$\frac{80 + 720}{4} = 200°, \text{ and } \frac{80 + 1080}{4} = 290°$$

The four roots are $2\underline{/20°}$, $2\underline{/110°}$, $2\underline{/200°}$, and $2\underline{/290°}$.

◆ APPENDIX F

Communications

COMMUNICATION INTERFACES

Communication is the orderly transmission of information from a sender to a receiver. A conversation between two individuals is a form of communication; so is reading a book, listening to a radio, or watching a program on television. Another form of communication takes place between electronic components such as computers, controllers, measuring instruments, display terminals, and printers. In this form of communication, information in the form of digital (or analog) signals is transmitted over a path called a *communication channel*.

If the communication channel consists of a single path, it is called a *serial channel*. If the channel consists of many paths, it is called a *parallel channel* or a bus. Parallel transfer is faster than serial transfer, but it is more expensive because more lines are required. Parallel data lines also have more crosstalk problems. Within a computer, data are transferred in parallel to maximize speed (the address bus and the data bus of a microcomputer system are examples of parallel communication channels). Beyond the computer, parallel transfer usually involves distances of a few feet and rarely more than 50 or 60 ft.

In either serial or parallel, the data may be transmitted synchronously or asynchronously. *Synchronous transmission* means that a timing signal is used to make the sender and receiver act together. *Asynchronous transmission* uses control signals between the sender and receiver to make sure the receiver accepts the data before the sender terminates the transmission. This procedure is sometimes referred to as "handshaking" or the "please-and-thank-you" routine.

A *communication channel* is a path for electrical transmission of information from a sending station to a receiving station. Communication channels are divided into three categories: simplex, half-duplex, or full-duplex. A *simplex channel* can communicate in only one direction. One station is always the sender; the other station is always the receiver. A *half-duplex channel* can communicate in both directions, but only one direction at a time. This is the normal mode of communication between two individuals. A *full-duplex channel* can communicate in both directions at once. Although this mode does not work well between individuals, it is very effective between electronic components.

◆◆◆◆

Information in digital form consists of a collection of binary digits that represent messages, numbers, or commands. A typical message might look like this: 0110 1001 1100 1011 1000 1000 1010 0011 0001. . . . The sending device must first convert the binary information into a group of signals; then it sends the information to the receiver, one signal at a time. The simplest conversion the sending device can use is a binary voltage signal for each binary digit (one voltage level for a binary 1 and another voltage level for a binary 0). However, other conversion methods may be used to increase the speed at which the information is transmitted. For example, a signal with four voltage levels can be used for each pair of binary digits as follows:

Voltage Level	Binary Digits
1	00
2	01
3	10
4	11

In a similar manner, a signal with eight voltage levels can be used for each three binary digits. In general, the number of voltage levels, (L), required to represent N binary digits is given by

$$L = 2^N \tag{F.1}$$

A voltage signal with 16 levels can be used to represent four binary digits. A serial path that uses a signal with 16 levels is similar to a 4-bit parallel bus. Both can transmit four binary digits at a time. The difference is that the parallel bus requires four paths to do the job, while the 16-level signal does it on one path.

Two important characteristics of a communication channel are the maximum number of bits it can transmit in 1 s (BPS_{max}) and the maximum number of signals it can transmit in 1 s (baud).

BPS$_{max}$ is the maximum number of binary digits a communication channel can transmit in 1 s.

Baud rate is the maximum number of signals a communication channel can transmit in 1 s.

If N is the number of binary digits represented by each signal, then

$$BPS_{max} = N(\text{baud rate}) \tag{F.2}$$

A communication channel has a frequency response that is similar to a band-pass filter. For example, a voice-grade telephone line will pass frequency components between 300 and 3300 Hz. It has a bandwidth of $3300 - 300 = 3000$ Hz. Frequency components below 300 Hz or above 3300 Hz will not pass through the channel. This band-pass characteristic of a communication channel has two significant consequences.

The first consequence has to do with the relationship between the bandwidth and the values of BPS_{max} and baud rate for the channel. In 1948, Claude Shannon showed that the maximum capacity of a channel in bits/second is proportional to the bandwidth of the channel. In

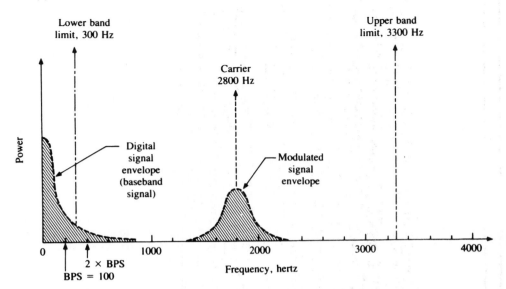

◆ **Figure F.1** The digital signal on the left has a bit transfer rate (BPS) of 100 BPS. Most of the power in the signal is in components with frequencies below 200 Hz (2 × BPS). The signal is translated by a modulator so that the power in the modulated signal is in components with frequencies between the band limits of the communication channel.

other words, the bandwidth of the channel sets an upper limit on the bit transfer rate, BPS_{max}, of a sending device.

The second consequence is that the digital signal must be frequency translated so that it falls between the band limits of the communication channel (see Figure F.1). It has been shown that a digital signal is composed of many frequency components. The components in the signal vary according to the pattern of 1's and 0's in the signal. For example, at 100 bits per second (BPS), the pattern 110011001100 . . . would produce a square wave with frequency components of 25, 75, 125, 175, 225, 275 Hz, and so on. The power in the components diminishes rapidly as the frequency increases. Figure F.1 shows the power envelope of a digital signal that has a bit transfer rate of 100 BPS. Most of the power is in components with a frequency of less than 200 Hz. A rule of thumb is that most of the power in a digital signal is in components with frequencies less than twice the bit transfer rate (2 × BPS).

Notice the positions of the digital signal envelope and the lower band limit of the channel in Figure F.1. The communication channel will not pass the digital signal because most of the power is below the lower band limit of the channel. A process called *modulation* can be used to translate the digital signal. Amplitude modulation is the simplest type of modulation. An amplitude modulator forms the product of two signals: the digital signal, which is called the *baseband signal,* and a sinusoidal signal, called the *carrier.* Amplitude modulation of a baseband signal produces two signals; one equal to the carrier plus the baseband signal, the other equal to the carrier minus the baseband signal. The two signals are called the *upper sideband* and the *lower sideband.* The envelopes of the two sideband signals are shown on either side of the carrier in Figure F.1. Notice that the envelopes are well within the band-pass limits of the signal. The communication channel will pass the modulated signal.

> Modulation of a digital signal translates a baseband signal to a frequency range that can be transmitted on a communication channel.

RS-232C Serial Interface

In 1969, the Electronic Industries Association published RS-232C, a specification that describes the signals in a 25-pin connector used for serial transmission of digital data. RS-232 has been almost universally accepted as the "standard" serial port for digital equipment. The full title of the specification is "Interface Between Data Terminal Equipment and Data Communication Equipment Employing Serial Binary Data Interchange." The term *data terminal equipment* refers to the sending or receiving equipment that is using the RS-232C serial port. This includes computers, terminals, printers, plotters, and so on. The term *data communication equipment* refers to a device called a *modem* that modulates the signal at the sending end of the communication channel and demodulates the signal at the receiving end. The term *modem* is a contraction of "modulator/demodulator." Figure F.2 illustrates an RS-232C communication circuit.

The RS-232C specification defines two communication channels, the primary channel and the secondary channel. Each channel has two carrier signals, one for each direction. The secondary channel is slower than the primary channel and may not be used in a particular application. The RS-232C specification describes the electrical characteristics of the signals in detail, but leaves many choices to the designer of the modem. Consequently, RS-232C does not guarantee compatibility between all terminals and all modems. RS-232C does not apply to communication speeds greater than 20,000 BPS, and the maximum cable length is 50 ft. The signals are restricted to two voltage regions: a positive region from 3 to 15 V dc, and a negative region from -3 to -15 V dc. Signals must pass through the transition region from -3 to 3 V in less than 1 ms. Table F.1 lists the signal names, pin assignments, and directions.

The following is a brief description of how the signals are used to transmit a message. When the sending terminal is ready to transmit a message, it turns ON *Request to Send;* the sending modem responds by turning ON its outgoing primary carrier. The receiving modem detects the carrier, turns ON its carrier, and turns ON *Carrier Detect* to alert the receiving ter-

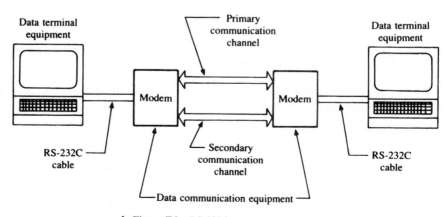

◆ **Figure F.2**   RS-232C communication circuit.

◆ **TABLE F.1** RS-232C Signals and Pin Assignments

Pin	Signal Name	Direction
1	Protective Ground (Frame)	
2	Transmitted Data	Terminal → modem
3	Received Data	Terminal ← modem
4	Request to Send	Terminal → modem
5	Clear to Send	Terminal ← modem
6	Data Set Ready	Terminal ← modem
7	Signal Ground (Common Return)	
8	Carrier Detect	Terminal ← modem
9		
10		
11		
12	Secondary Carrier Detect	Terminal ← modem
13	Secondary Clear to Send	Terminal ← modem
14	Secondary Transmitted Data	Terminal → modem
15	Transmit Clock (DCE Source)	Terminal ← modem
16	Secondary Received Data	Terminal ← modem
17	Receive Clock	Terminal ← modem
18		
19	Secondary Request to Send	Terminal → modem
20	Data Terminal Ready	Terminal → modem
21	Signal Quality Detector	Terminal ← modem
22	Ring Indicator	Terminal ← modem
23	Data Rate Selector	Terminal → modem
24	Transmit Clock (DTE Source)	Terminal → modem
25		

Source: EIA Standard RS-232C (Washington, D.C.: Electronic Industries Association, 1969).

minal that a message is forthcoming. The sending modem, 750 ms after sensing the receiver's carrier, turns ON *Clear to Send,* enabling transmission of data. When the sending terminal sense that Clear to Send is ON, it begins to send signals on its *Transmitted Data* line. The sending modem modulates the outgoing carrier with each signal. The receiving modem demodulates each signal and puts the demodulated signal on *Received Data.* The receiving terminal receives an exact replica of the transmitted message.

RS-449 Serial Interface

Between 1975 and 1977, the Electronic Industries Association published the following three specifications for an improved serial interface:

RS-442 defines balanced receivers and drivers.

RS-423 defines unbalanced receivers and drivers.

RS-449 defines the signals that make up the interface.

The RS-449 interface consists of two connectors: a nine-pin connector for the secondary channel signals, and a 37-pin connector for all other signals. The RS-449 specification includes all

◆ **TABLE F. 2**   RS-449 Maximum Cable Lengths and Data Rates

Maximum Cable Length (ft)	Data Rate (BPs)	
	Unbalanced Drivers and Receivers	Balanced Drivers and Receivers
4000	Below 900	Below 90,000
400	10,000	1,000,000
40	100,000	10,000,000

the RS-232C signals plus a few additional signals. The new signals provide features such as local loopback and remote loopback, which help pinpoint the location of a malfunction.

The addition of balanced drivers and receivers is the major improvement in the interface circuits. A balanced driver uses a separate return line for each signal. This is in contrast with the unbalanced circuits used in RS-232C, in which all signals share a common return line. Electrical improvements in the circuits allow higher operating speeds and longer cable lengths than RS-232C. Table F.2 lists the maximum cable lengths and data rates for balanced and unbalanced lines.

IEEE-488 Parallel Data Bus

The IEEE-488 data bus was originally developed in 1970 by Hewlett-Packard for use with its own minicomputers and instruments. The interface was well received and has been built into a wide range of instruments, peripherals, and calculators. In 1975, the Institute of Electrical and Electronic Engineers made it a standard with the designation IEEE-488. The International

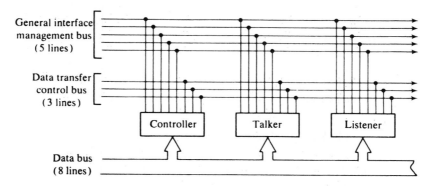

The bus controller uses the management and transfer control lines to handle bus traffic. The open-ended bus connects up to 15 devices, but total cable distance cannot exceed 66 ft. Device addresses are set by five-bit jumpers or switches on the back panel of each instrument.

◆ **Figure F.3**   IEEE-488 parallel data bus. (From J. Washburn, "Communications Interface Primer—Part II," *Instruments and Control Systems*, April 1978, pp. 62–63. Copyright Chilton Co., 1978.)

Electro-technical Commission also adopted the IEEE-488 standard with the designation IEC Standard 625-1. Industry often refers to the standard as the General Purpose Interface Bus (GPIB), and the names GPIB, HP-IB, IEEE-488, and IEC Standard 625-1 are used synonymously. Figure F.3 describes the IEEE-488 interface.

The IEEE-488 standard defines three device types: talkers, listeners, and controllers. *Talkers* are devices that send data, commands, and status information to listeners. *Listeners* are identified by an address so that a talker can send data to a specific listener. Talkers include sensing instruments of various types. Listeners include recording instruments such as printers and plotters. *Controllers* manage the communication between the talkers and the listeners.

IEEE-583 CAMAC Interface

The CAMAC (or IEEE-583) interface was originally designed for nuclear instrumentation laboratories, where instruments were constantly swapped among systems. IEEE-583 is a completely specified interface system. The standard defines the mechanical configuration, the electrical connectors, the data transmission paths, and the protocol. The mechanical configuration consists of a "crate" containing slots for 25 modules. The two slots on the right are occupied by a controller that manages all CAMAC communications. Five types of controllers are available: one for direct access to a computer, one for parallel transmission, two for different types of serial transmission, and one for standalone operation.

LOCAL AREA NETWORKS

A network is a communication channel that connects a large number of user stations to one or more central stations. The central stations provide access to large host computers, central databases, graphic printers, and other devices that are too expensive to be dedicated to a single station. If the stations are separated by considerable distance, such as branch offices in different cities, the network is called a *wide area network*. If all stations are located within a radius of about a mile, the network is called a *local area network* (LAN). LANs are an important means of communication in corporations, universities, government agencies, and many other organizations. Most of the effort in factory automation is directed at establishing a communication path between various units and devices in the plant. Indeed, the LAN is the central nervous system of the automated factory.

LANs usually use coaxial cable to provide the communication channel or channels. Some LANs provide a single communication channel, others provide many channels. A single-channel LAN may use an unmodulated *baseband* signal or a modulated *carrierband* signal. A multichannel LAN uses a *broadband* system that has many channels, each with its own carrier frequency. In a large LAN, a broadband network forms the *backbone* that interconnects many single channel subnets via interconnects called *bridges* or *gateways*.

IEEE-802.4 Single-Channel Systems

A single-channel network, whether baseband or carrierband, uses the entire bandwidth of the coaxial cable. Many single-channel networks are carrierband systems that use a version of frequency modulation called *frequency shift keying* (FSK). Two frequencies are selected, one for transmitting a binary 0, the other for transmitting a binary 1. If F0 is the frequency used for a 0, and F1 is the frequency used for a 1, the modulated signal appears to alternate between the two frequencies with a constant amplitude. A binary signal and the FSK modulated signal would appear as shown in Figure F.4.

Carrierband subnets are used to interconnect process controllers, programmable logic controllers, and other intelligent components in a factory network. The subnets are the branches of the factory network, each connecting to the broadband backbone network through an interconnecting bridge or gateway. All nodes in a subnet transmit and receive at the same frequency. The IEEE-802.4 networking standard specifies two carrierband signals: phase-continuous FSK and phase-coherent FSK.

An 802.4 *phase-coherent system* uses a data rate of either 5 or 10 MB/s. The binary signal is directly encoded using a frequency equal to the data rate for a binary 1, and a frequency equal to twice the data rate for a binary 0. The data rates and signal frequencies are listed in the following table.

Date Rate	Frequency (MHz)	
(MB/s)	Binary 1	Binary 0
5	5	10
10	10	20

The modulated signal will always consist of a full cycle of one of the two signals. A binary 1 is represented by one full cycle of the lower frequency; a binary 0 is represented by two full cycles of the higher frequency.

An 802.4 *phase-continuous system* has a data rate of 1 MB/s and uses a Manchester-encoded signal to modulate a 5-MHz carrier. The carrier is modulated to 6.25 MHz for a logic 1 and to 3.75 MHz for a logic 0. In the Manchester encoder, the signal bits are allocated equal time intervals, each equal to the reciprocal of the data transfer rate in bits/second. For the 802.4 phase-continuous system, the time interval for each bit is 1 μs. The value of each bit (0 or 1) is converted into a signal transition at the center of the interval. The signal is changed from 0 to 1 to represent a binary 1; and from 1 to 0 to represent a binary 0. At the end of each interval, the signal goes to the level required to encode the next bit. A Manchester-encoded

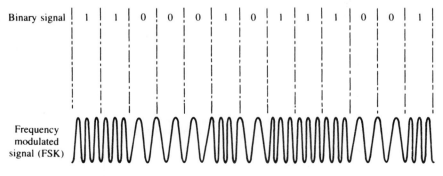

◆ **Figure F.4**   Example of a frequency-modulated binary signal using frequency shift keying (FSK).

message contains both the data and the synchronizing clock signal (the clock is defined by the signal transition).

IEEE-802.4 Broadband Networks

A broadband network provides many frequency channels that can accommodate several LANs plus voice and even video communications, all operating on the same cable. Each LAN uses two frequency channels. All LAN nodes transmit on one frequency and receive on another frequency. A device called a *headend repeater* translates the signal from the transmitting frequency to the receiving frequency. Figure F.5 depicts a broadband network serving two LANs. In the diagram, node A1 is sending a message to node A2 on LAN-A and node B3 is sending a message to node B2 on LAN-B. Node A1 modulates its signal on frequency F1. The LAN-A headend demodulates the F1 signal and remodulates the signal on frequency F2. The F2 signal goes to all nodes, but only node A2 accepts the message (the message is addressed to node A2). Node B3 modulates its signal on frequency F3. The LAN-B headend demodulates the F3 signal and remodulates the signal on frequency F4. All nodes receive the F4 signal, but only B2 accepts the message.

The IEEE-802.4 standard defines a multilevel duobinary AM/phase-shift keying broadband network. The frequency channels are predefined, and the signal is encoded to control modulation. The 802.4 token-bus protocol specifies the channel usage. The transmit and receive frequency channels can be changed or programmed. The standard designates three data rates: 1, 5, and 10 MB/s.

The Open Systems Interface

In 1979, the International Standards Organization (ISO) began developing a communication network model that became known as the *open systems interface* (OSI). A number of LANs

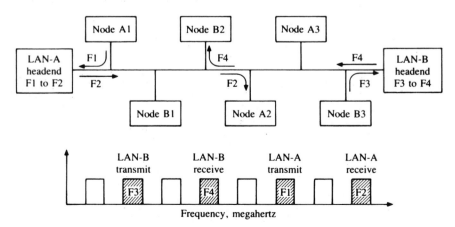

◆ **Figure F.5**   A broadband network can support more than one LAN. In the diagram, node A1 is sending a message to node A2 on LAN-A while node B3 is sending a message to node B2 on LAN-B. Each LAN is serviced by a headend that translates the transmitted signal (F1 or F3) into a receivable signal (F2 or F4).

are based on the OSI model, including Ethernet and *Manufacturing Automation Protocol* (MAP). The OSI model consists of the following seven functional levels:

Level	Name	Function
7	Application	Upper-level services
6	Presentation	Upper-level services
5	Session	Upper-level services
4	Transport	Upper-level services
3	Network	Routing between cables
2	Data link	Routing between nodes
1	Physical	Routing between nodes

Each layer is a module responsible for providing networking services to the level above it. The functions fall into one of three distinct divisions. The first two levels of the OSI model (physical and data link) deal with transmitting messages between nodes on the same cable. Level 3 (network) covers the transmission of messages from one cable to another. The upper four levels (transport, session, presentation, and application) handle upper-level services such as data formatting and access security.

The physical level of the OSI model is concerned with the data rate and the type of cable, connectors, and signaling methods used to transmit and receive the message. The data link level defines the formation of data frames, priorities, and procedures (protocol) to assure an orderly transfer of messages.

The network level specifies the routing of messages across a bridge or gateway from one cable to another cable. A *bridge* is used to connect two cables that have the same protocol. A *gateway* is used to connect two cables that have different protocols.

The transport level provides reliable end-to-end data transfer. The session level translates and synchronizes names and addresses. The presentation and application levels are concerned with factory management.

MAP Networks

The MAP is an emerging international standard for a multilevel communication network for factory automation. The purpose of a MAP network is to link together all the controllers, computers, workstations, production machines, and offices in an entire factory. The impetus for the development of MAP came in the late 1970s, when General Motors observed that half of their automation budget was used for custom interfaces for incompatible devices. The vast number of communication systems used by different vendors of automation equipment had created a modern-day Tower of Babel. In 1980, General Motors and Boeing established a task force to develop a public-domain communication standard for the factory environment. The specification was called the Manufacturing Automation Protocol, or simply MAP.

The MAP architecture is based on the seven-level OSI model. For the physical level, MAP specifies a 10-MB/s IEEE-802.4 broadband network for the backbone, and 5-MB/s IEEE-802.4 carrierband subnets. MAP specifies a token-passing protocol for the network. The backbone network connects various central resources such as mainframe computers, CAD/CAM workstations, databases, application computers, bridges to factory subnets, gate-

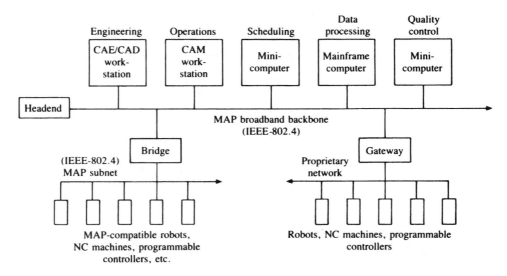

◆ **Figure F.6**   A MAP factory network uses bridges for MAP subnets and gateways for proprietary networks.

ways to proprietary networks, and wide area network (WAN) gateways to other sites. Most of the messages on the backbone network are high-volume database information movements. The carrier band subnets connect the robots, NC machines, programmable logic controllers, and process controllers on the plant floor. Some of the subnets are proprietary LANs that connect to the backbone via a gateway. Figure F.6 shows a typical MAP network.

Ethernet

In 1980, Digital Equipment Corporation, Intel, and Xerox jointly published the specification for the Ethernet LAN. Ethernet follows the seven-level OSI model. The physical layer specifies a 10-MB/s Manchester-encoded baseband signal. The data link layer operates under an access protocol called *carrier sense multiple access with collision detection* (CSMA/CD). It also handles error detection and addressing of the source and destination. Ethernet was integrated with DECnet, a proprietary network used with DEC computers. Figure F.7 shows an application of a DECnet/Ethernet network.

COMMUNICATION PROTOCOLS

A *protocol* is a set of rules that govern the operation of the devices that share a communication channel. The rules define the following functions:

1. *Framing*. Messages are transmitted in units called frames. A typical frame consists of a header section, a data section, and a trailer section. The header section defines the beginning of the frame and other information, such as the source address, the destination address, the length of the data section, and the purpose of the frame. The data

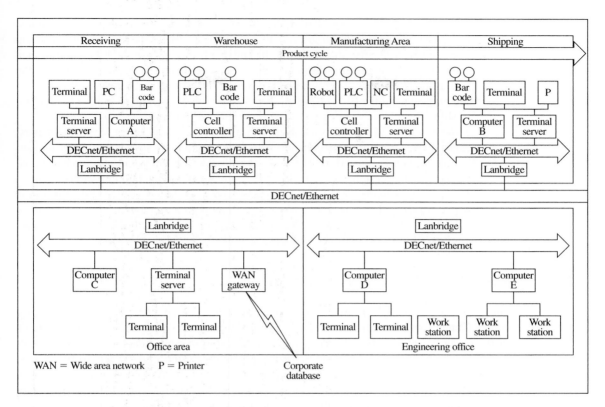

◆ **Figure F.7**   A DECnet/Ethernet baseband cable backbone extends a manufacturing network to all areas of a plant—even to a remote corporate database. (Reprinted from *Control Engineering* magazine with permission. Copyright Cahners Publishing Co., 1987.)

section contains the message to be transmitted. The trailer section contains an error-checking number and an end-of-frame mark.

2. *Error control.* The transmitter uses a predefined algorithm to compute a frame check sequence (a number) from the characters in the message. The cyclic redundancy check (CRC) is one such algorithm. In CRC, the entire message block is treated as a large binary number. This large number is divided by a predefined 16-bit number, and the remainder of this division (also 16 bits) is the frame check sequence that is included in the trailer. The receiver repeats the division process on the received message and checks its remainder with the frame check sequence. If the two numbers are not identical, an error has occurred, and a retransmission is requested. The probability that an error will be undetected by CRC is extremely low.

3. *Sequence control.* The messages are numbered in the order in which they occur. This sequence number is used to prevent the loss of messages, eliminate duplicate messages, and identify messages for retransmission. Sequencing is a type of error control.

4. *Access control.* Access control defines how each node on the network obtains access to the network to transmit data. There are two methods of access control: polling and con-

tention. In the *polling* method, the nodes on the network take turns being invited to transmit a message. In the *contention* system, each node competes for access to the network.

5. *Transparency*. The flag that indicates the end of a frame is a particular bit pattern, such as 01111110. It is possible that some messages will include the same bit pattern as the flag. Somehow the receiver must be able to tell the difference between a 01111110 in the message and the 01111110 flag that ends the message. The technique for accomplishing this objective is called *transparent text transmission*.

6. *Flow control*. Flow control involves starting and stopping the transmission of messages on a communication channel. It includes initiating transmissions after a channel has been idle, returning the channel to idle when there are no messages to transmit, and stopping transmission when the receiver cannot handle more information.

7. *Configuration control*. The configuration of a network consists of the addresses of the active nodes in the system. Occasionally, a node will be removed from a network or a new node will be added to the network. When this happens, the network must be reconfigured to incorporate the changes.

The CSMA/CD Access Protocol

The carrier sense multiple access with collision detection CSMA/CD) protocol allows a station to transmit whenever it is ready and it does not detect the presence of another station's carrier on the communication channel. If two stations start to transmit at the same time, a collision has occurred. When a collision occurs, a collision detection signal is sent to all stations. The two transmitting stations stop transmitting, wait for a random period of time, and then retry the transmission. The station delay is determined by an algorithm that uses a random number generator, so it is very unlikely that two stations will have the same delay time.

The CSMA/CD protocol works well as long as there are not a lot of collisions. As the volume of message traffic increases, so does the probability of collisions. As the channel approaches 50% of capacity, stations begin to experience long delays in obtaining access. Another disadvantage of a contention protocol is that it is not possible to calculate a worst-case time for a node to transmit its message over the network. For these reasons, a CSMA/CD system may not be the best choice for real-time environments such as process control.

The Token-Passing Access Protocol

The *token-passing* protocol is a polling-type access protocol. A special bit pattern called the token is passed from station to station in ascending order of the station address. The station with the highest address passes the token to the station with the lowest address, so the token continues to circulate through all stations in an unending loop.

The access rule is very simple. Only the station with the token can access the communication channel to transmit a message, and only one frame (or packet) can be transmitted before passing the token to the next station. If a station has several frames to transmit, it sends one frame each time it receives the token. If a station receives the token and has no message to transmit, it immediately passes the token to the next station. The size of a frame is defined, so it is possible to calculate the worst-case access time (i.e., every station is transmitting a full-sized frame). For this reason, token-passing systems are considered more practical for real-time control applications.

The HDLC Protocol

The High-Level Data Link Control (HDLC) is a standard communication link protocol established by the International Standards Organization (ISO) in 1977. HDLC very closely resembles the Synchronous Data Link Control (SDLC) introduced by IBM in 1974. Both HDLC and SDLC are bit-oriented, full-duplex communication protocols. HDLC has several different operation modes. The following is a description of the host computer/multiterminal operating mode. The host computer is called the primary station. The terminals are called the secondary stations.

A primary station controls the network and issues commands to the secondary stations according to a preassigned order of priority. The secondary stations respond by sending appropriate responses. The secondary stations can initiate a transmission only when given permission to do so by the primary station.

The HDLC uses three types of frames: an information frame used to transfer data, a supervisory frame used for control signals, and a nonsequenced frame used to initialize and control secondary stations. Figure F.8 shows the HDLC frame format.

The 8-bit flag (01111110) marks the beginning and the end of a frame. The flag pattern must never occur within the frame. If it does, the receiver will interpret the pattern as an end-of-frame flag and ignore the remainder of the message. A technique called *zero stuffing* is used to achieve transparent text transmission. When the transmitter encounters a sequence of five 1's in the frame data, it inserts a zero as the next bit in the message. Each time the receiver encounters five 1's followed by a zero, it removes the zero, knowing that it was stuffed by the transmitter. When the receiver encounters a zero followed by six 1's and a zero, it recognizes the end-of-frame flag, knowing that the transmitter would have stuffed a zero if it were not the flag.

The address byte identifies a secondary station. The control byte determines the type of frame (information, supervisory, or nonsequenced) and additional control information. The frame check sequence is the CRC error-checking number.

Opening flag	Secondary station address	Control	Information	Frame check sequence	Closing flag
01111110	8 bits	8 bits	Variable length	16 bits	01111110

◆ **Figure F.8**   HDLC information frame format.

◆ **GLOSSARY**

Asynchronous: A method of communication that does not use a timing signal to make the sending and receiving stations work together.

Backbone: A broadband network that connects various central resources, such as mainframe computers, CAD/CAM workstations, databases, application computers, bridges to factory subnets, gateways to proprietary networks, and wide area network (WAN) gateways to other sites.

Baseband: A network that uses the entire bandwidth of the cable for a single unmodulated signal.

Baud rate: The maximum number of signals a communication channel can transmit in 1 s. A signal may represent one or more binary bits in the message.

Bit transfer rate (BPS):   The number of bits a communication channel transmits in 1 s.

Bridge: A device used to carry messages between two distinct networks that use the same protocol.

Broadband: A network that provides many channels that can accommodate several local area networks plus other signals, such as voice and video.

Carrierband: A network that uses the entire bandwidth of the cable for a single carrier-modulated signal.

Communication channel: A path for electrical transmission of information from a sending station to a receiving station.

Contention: An access method in which the nodes transmit messages on a first-come, first-served basis.

CSMA/CD: An access protocol that allows a station to transmit whenever it is ready and it does not detect the presence of another carrier on the line. Collision detection is used to prevent two stations from transmitting at the same time.

Full-duplex: Simultaneous communication in both directions.

Gateway: A device used to carry messages between two distinct networks that use different protocols.

Half-duplex: Communication in both directions, but only one direction at a time.

Headend repeater: A device, located at the end of a network cable, that converts signals from one frequency to another frequency.

Manufacturing Automation Protocol (MAP):   A protocol for a factory communication network that uses a broadband backbone network serving a number of carrierband subnets.

Modulation: The alteration of a sinusoidal carrier by a data signal for the purpose of transmitting the data signal.

Open systems interface (OSI):   A seven-layer model for communications protocols developed by the International Standards Organization.

Parallel channel: A channel that provides multiple paths to transmit all bits of a binary message unit at the same time.

Polling: An access method in which the nodes take turns being invited to transmit a message.

Protocol: A set of rules that govern the operation of the devices that share a communication channel.

Serial channel: A channel that transmits binary messages one bit at a time.

Simplex: Communication in only one direction.

Synchronous: A method of communication that uses a timing signal to make the sending and receiving stations work together.

Token passing: An access protocol that uses polling with a circulating bit pattern called a token.

References

ARTHUR, K. *Transducer Measurements*. Beaverton, Ore.: Tektronix, Inc. 1970.

BATESON, R. N. *Motomatic Servo Control Course Manual*. Hopkins, Min.: Electro-Craft Corporation, 1971.

BINDER, R. C. *Fluid Mechanics*. Englewood Cliffs, N.J.: Prentice-Hall, Inc., 1949.

BOGART, T. F., JR. *Laplace Transforms and Control Systems Theory for Technology*. New York: John Wiley & Sons, Inc., 1982.

BOWER, J. L., AND P. M. SCHULTHEISS. *Introduction to the Design of Servomechanisms*. New York: John Wiley & Sons, Inc., 1964.

BRYAN, G. T. *Control Systems for Technicians*. Hart Publishing Company, Inc., 1967.

BUCKLEY, P. S. *Techniques of Process Control*. New York: John Wiley & Sons, Inc., 1964.

Chemical Engineering, Vol. 76, No. 12, June 2, 1969.

COX, E. *The Fuzzy Systems Handbook*. Chestnut Hill, Mass.: Academic Press, Inc., 1994.

DIEFENDERFER, A. J. *Principles of Electronic Instrumentation*. Philadelphia: W. B. Saunders Company, 1972.

ECKMAN, D. P. *Industrial Instrumentation*. New York: John Wiley & Sons, Inc., 1950.

ERICKSON, W. H., AND N. H. BRYANT. *Electrical Engineering Theory and Practice*. New York: John Wiley & Sons, Inc., 1952.

FLOYD, THOMAS L. *Digital Fundamentals*. New York: Macmillan Publishing Company, 1990.

Fundamentals of Industrial Instrumentation. Philadelphia: Honeywell, Inc., 1957.

HARRIOTT, P. *Process Control*. New York: McGraw-Hill Book Company, 1964.

HARRISON, H. L., AND J. G. BOLLINGER. *Introduction to Automatic Controls*. Scranton, Pa.: International Textbook Company, 1969.

HERMAN, S. L., AND W. N. ALERICH. *Industrial Motor Control*. Albany, N.Y.: Delmar Publishers, Inc., 1985.

HORDESKI, M. F. *Design of Microprocessor, Sensor and Control Systems*. Englewood Cliffs, N.J.: Reston Publishing Company, Inc., 1985.

JENNINGS, B. H., AND S. R. LEWIS. *Air Conditioning and Refrigeration*. Scranton, Pa.: International Textbook Company, 1949.

JOHNSON, C. D. *Process Control Instrumentation Technology.* Upper Saddle River, N.J.: Prentice Hall, Inc., 1997.

JONES, C. T., AND L. A. BRYAN. *Programmable Controllers: Concepts and Applications.* Atlanta: International Programmable Controls, Inc., 1983.

KAFRISSEN, E., AND M. STEPHANS. *Industrial Robots and Robotics.* Englewood Cliffs, N.J.: Reston Publishing Company, Inc., 1984.

KILLIAN, C. T. *Modern Control Technology Components and Systems.* St. Paul, Minn.: West Publishing Co., 1996.

LESA, A., AND R. ZAKS. *Microprocessor Interfacing Techniques.* Oakland, Calif.: SYBEX Inc., 1978.

LIPTAK, B. G., AND K. VENCZEL. *Instrument Engineers' Handbook Process Measurement.* Radnor, Pa.: Chilton Book Company, 1982.

MALONEY, T. J. *Industrial Solid-State Electronics Devices and Systems.* Englewood Cliffs, N.J.: Prentice-Hall, Inc., 1979.

McGLYNN, D. R. *Microprocessors: Technology, Architecture, and Applications.* New York: John Wiley & Sons, Inc., 1976.

McNAMARA, J. E. *Technical Aspects of Data Communication.* Maynard, Mass.: Digital Equipment Corporation, 1977.

MORRIS, N. M. *Control Engineering.* Maidenhead, Berkshire, England: McGraw-Hill Book Company (U.K.) Ltd., 1968.

NACHTIGAL, CHESTER L. *Instrumentation and Control.* New York: John Wiley & Sons, Inc., 1990.

NORTON, H. N. *Sensor and Analyzer Handbook.* Englewood Cliffs, N.J.: Prentice-Hall, Inc., 1982.

OLESTEIN, NILS O. *Numerical Control.* New York: John Wiley & Sons, Inc., 1970.

Principles of Automatic Process Control. Research Triangle Park, N.C.: Instrument Society of America, 1968 (text and film).

Principles of Frequency Response. Research Triangle Park, N.C.: Instrument Society of America, 1958 (text and film).

SEYER, M. D. *RS-232 Made Easy: Connecting Computers, Printers, Terminals, and Modems.* Englewood Cliffs, N.J.: Prentice-Hall, Inc., 1984.

SHINSKEY, F. G. *Process Control Systems.* New York: McGraw-Hill Book Company, 1967.

STEIN, D. H. *Introduction to Digital Data Communications.* Albany, N.Y.: Delmar Publishers, Inc., 1985.

TUCKER, G. K., AND D. M. WILLS. *A Simplified Technique of Control System Engineering.* Philadelphia: Minneapolis-Honeywell Regulator Company, 1962.

WEATHERS, TOM, JR., AND CLAUD C. HUNTER. *Automotive Computers and Control Systems.* Englewood Cliffs, N.J.: Prentice-Hall, Inc., 1984.

WEBB, JOHN W., AND R. A. REIS. *Programmable Logic Controllers,* 3rd edition. Englewood Cliffs, N.J.: Prentice Hall, Inc., 1995.

WILSON, H. S., AND L. M. ZOSS. *Control Theory Notebook.* Research Triangle Park, N.C.: Instrument Society of America. Reprint from *ISA Journal.*

ZEINES, B. *Automatic Control Systems.* Englewood Cliffs, N.J.: Prentice-Hall, Inc., 1972.

ZIEGLER, J. G., AND N. B. NICHOLS. "Optimum Settings for Automatic Controllers," *ASME Transactions,* Vol. 64, No. 8, 1942, pp. 759–768.

Answers
to Selected
Exercises

Chapter 1

1.4. Size and timing. The transfer function.

1.5. **(b)** Gain = 29.0

Phase difference = $-1°$

Transfer function = 29.0 $\underline{/-1°}$

(d) Gain = 10.0

Phase difference = $2°$

Transfer function = 10. $\underline{/-2°}$

1.7. Open-loop advantages: inexpensive, simple, trouble-free, no instability.

Disadvantages: cannot compensate for disturbances, must be calibrated.

1.11. **(b)** $C_m = 1.6C + 4$ mA

(c) $C_m = 13.6$ mA

(d) $C = 0$ L/min

1.18. Component one: nonlinearity

Component three: dead-band and saturation

1.19.

Component	Maximum Output Difference
1	24 at input = 4 to 20
2	40 at input = 12.5
3	Not applicable
4	63 at input = 12.5

1.21. **(e)** 4

(f) 2

(g) 3

(h) 1

1.26. Critical damping.

1.29. **(b)** $\dfrac{C}{R} = \dfrac{G + GH_1H_2}{1 + H_1H_2 + GH_1}$

Chapter 2

2.2.

Time	Temperature	Quantization Error
11:30	71.1°F	0.1°F
11:31	71.2°F	0.2°F
11:32	71.4°F	0.4°F
11:33	71.6°F	0.6°F
11:34	71.8°F	0.8°F
11:35	71.9°F	0.9°F
⋮	⋮	⋮
11:42	73.1°F	0.1°F

2.4. (Possible answers)

Regulator system: automobile cruise control

Follow-up system: automobile power steering

2.8. **(a)** $k_f = 7.07 \times 10^{-6}$

(b) $k = 2.38 \times 10^{-8}$

(c) $W = 4.46 \times 10^{-3}$ kg/s

2.12. Event-driven and time-driven

2.19. NC: Numerical control is a system that uses predetermined instructions to control a sequence of manufacturing operations.

CNC: Computerized numerical control uses a dedicated computer to accept the input of instructions and to perform the control functions required to produce a part.

DNC: Direct numerical control is a system in which a number of numerical control machines are connected to a central computer for real-time access to a common database of part programs and machine programs.

2.24. *Distributed control:* The controllers are close to their measuring transmitters and actuators. Advantage: Produces good control due to short loops. Disadvantage: Requires a communication network to keep the operator informed. *Centralized control:* The controllers are located in a central control room. Advantage: The operator can see all controllers at once and is able to monitor the process very closely. Disadvantage: Long transmission lines are required for each control loop, which is expensive and degrades control.

2.25. Process control: 2.14, 2.16, 2.17, 2.19, 2.20, 2.22
Servomechanism: 2.15, 2.18
Sequential control: 2.21

2.27. (b) Regulator
(d) Regulator
(f) Time-driven sequential
(h) Follow-up
(j) Event-driven sequential

Chapter 3

3.1. (a) $R = 4.33 \, \Omega$
(e) $R = 303 \, \Omega$
(i) $R = 7228 \, \Omega$

3.3. (a) $R_d = 4.33 \, \Omega$
(e) $R_d = 401 \, \Omega$
(i) $R_d = 9685 \, \Omega$

3.5. At $e = 10$ V, $R_d = 16.9 \, \Omega$

3.7. (a) $L = 42.0$ mH
(d) $L = 0.556$ H

3.9. (a) $R_e = 6480$. Flow is turbulent.
$R = 3.35\text{E} + 08 \, \text{Pa} \cdot \text{s}^2/\text{m}^3$
$\Delta p = 55.8$ KPa

3.11. (a) $L_L = 2.89\text{E} + 6 \, \text{Pa} \cdot \text{s}^2/\text{m}^3$

3.13. (a) $K_g = 1.715\text{E} + \text{Pa} \cdot \text{s}^2/\text{kg}^2$

3.15. (a) $p_2 = 104.2$ kPa

3.17. At $W = 0.002$ kg/s, $R_g = 1575 \, \text{kPa} \cdot \text{s/kg}$
check: $K_g = 4.0\text{E} + 8 \, \text{Pa} \cdot \text{s}^2/\text{kg}^2$
$R_g = 1600 \, \text{kPa} \cdot \text{s}$

3.19. (a) $C_g = 1.50\text{E} - 5$ kg/Pa

3.21. $T_d = 8°\text{C}$, Total resistance $= 0.286$ K/W
Total heat flow $= 52.9$ W

3.23. (a) $R_m = 84.4 \, \text{N} \cdot \text{s/m}$

3.25. (a) $C_m = 5.67\text{E} - 5$ m/N
$K = 17{,}650$ N/m

Chapter 4

4.1. (a) $579 \dfrac{dh}{dt} + h = 220q_{\text{in}}$

4.3. (a) $0.103 \dfrac{de_{\text{out}}}{dt} + e_{\text{out}} = e_{\text{in}}$

4.5. (a) $0.0182p_{\text{in}} = 0.561 \dfrac{d^2x}{dt^2} + 9.6 \dfrac{dx}{dt} + 60{,}206$

4.7. (a) $f(t) = 6.7t$
(c) $f(t) = 45e^{-77t}$
(e) $f(t) = 650te^{-8t}$
(g) $f(t) = 82(1 - e^{-t/5})$
(i) $f(t) = 28 \cos \omega t$

4.9. $\dfrac{F_o(s)}{F_i(s)} = e^{-245s}$

4.11. $\dfrac{X(s)}{I(s)} = \dfrac{0.3}{1 + 0.02s + 0.0001s^2}$

4.13. $\dfrac{H(s)}{Q(s)} = \dfrac{0.5}{s}$

4.15. $\dfrac{V(s)}{E(s)} = \dfrac{1.636 + 3.6s + 0.0288s^2}{s}$

4.19. (a) $v(0) = 200$
$v(\infty) = 20$
(b) $v(0) = 200$
$v(\infty) = 20$

4.23. Exercise 4.10: $\dfrac{I}{\Theta} = \dfrac{0.1}{1 + 8.6s}$

Exercise 4.12: $\dfrac{\Theta}{X} = \dfrac{125}{1 + 26s + 25s^2}$

Exercise 4.15: $\dfrac{V}{E} = \dfrac{1.636 + 3.6s + 0.0288s^2}{s}$

Chapter 5

5.1. (a) Mean $= 75.8°\text{C}$, $S_x = 0.84°\text{C}$

5.3. (a) Mean $= 14.70$ psi, $S_x = 0.06540$ psi
(b) Range $= 14.50$ psi to 14.90 psi

5.5. Span $= 20$ V
Avg. resolution $= 0.125\% = 0.025$ V/turn

5.7. Span $= 320°$, Output range $= 40$ V.
Sensitivity $= 0.125$ V/degree

5.9. (a) Sensitivity $(70°\text{F}) = 0.16$ mA/psi

5.11. Ideal pressure $= 135$ KPa
$134.5 \, \text{KPa} \leq \text{pressure} \leq 135.5 \, \text{KPa}$

5.13. (a) Bias $= 0.8°\text{C} = 0.8\%$
Repeatability $= 2°\text{C} = 2\%$

5.15. (a) Bias $= -0.30$ psi $= 2\%$
(b) Repeatability $= 0.19$ psi $= 1.3\%$

5.17. Max negative error $= 0$ psi $= 0\%$
Max positive error $= 0.61$ psi $= 4.1\%$

5.19. Max hysteresis $+$ deadband $= 14.0$

5.21. y (independent) $= 0.8925x + 3.55$
max error $= \pm 4.0$

y (terminal-based) $= 0.86x + 2.8$
max error $= -7.0, + 2.6$
y (zero-based) $= 0.9019x + 2.8$
max error $= -4.2, +4.2$

5.23. (a) Time constant $= 26$ s
95% response time $= 78$ s
10 to 90% rise time $= 57$ s

5.25. Time constant $= 89.3$ s

Chapter 6

6.1. (a) $+V_{sat} = 16$ V, $v_2 - v_1 = 1.6$ mV
$-V_{sat} = -16$ V, $v_2 - v_1 = -1.6$ mV

6.3. (a) CMRR $= 4E + 5$
CMR $= 112$ dB

6.6. (a) $R_{in} = 4$ kΩ

6.8. (a) $R_{in} = 6.67$ kΩ

6.9. $R_f = 180.8$ kΩ

6.10. (a) $V_{out} = -3$ V

6.12. $R_a = 2.92$ kΩ
$R_b = 10$ kΩ

6.14. (a) $V_{out} = -10$ V

6.15. (a) $R_e = 2.002$ Ω

6.17. (a) $R = 300$ Ω
(b) $R_{LOAD} = 250$ Ω
(c) $V_{CC} = 15$ V

6.19. $R_1 = 8$ kΩ, $R_2 = 40$ kΩ
$v_2 = v_1 = 8$ V

6.20. (a) $R_s = 855$ Ω

6.23. $R = 796$ Ω

6.27. Line 1: $C_m = 6.25p$, $0 \leq p \leq 3.2$
Line 3: $C_m = p + 245$, $16 \leq p \leq 36$
Line 5: $C_m = 0.5556p + 45.44$, $64 \leq p \leq 100$

6.31. f_s (min) $= 2400$ Hz

6.33. (a) Analog output of 0101 $= 6.25$ V

6.35. (a) Quantization error $= \pm 0.333$ V $= \pm 3.33\%$

Chapter 7

7.1. (c) $N = 500$ turns

7.3. (a) $\theta = 75°$, $e_{out} = 1.60 \sin 377t$ V
(c) $\theta = 150°$, $e_{out} = -5.37 \sin 377t$ V

7.5. (a) $\theta + \theta_d = 0°$, $e_{out} = 22.5 \sin 2513t$ V
(c) $\theta + \theta_d = 65°$, $e_{out} = 9.51 \sin 2513t$ V

7.8. (e) Number of positions $= 65,536$

7.9. (a) Displacement per pulse $= 1.571 \times 10^{-4}$ m
$N_T = 40,107$ pulses

7.11. (a) Selection: IP-3 (IP-4 or IP-5 could also be used, but IP-3 is smaller and less expensive)

7.14. (a) 1629 pulses
(b) 14 bits minimum

7.15. (a) $E = 64.8$ V

7.17. $K_E = 0.01728$ V/rpm
Calibration equation: $E = 0.01728S$

7.19. (a) $C = 600$

7.20. (a) $a_{max} = \pm 66.7$ m/s^2
(b) $f_0 = 26.0$ Hz
(c) $b = 2.35$ N · s/m
(d) $f_{max} = 10.4$ Hz

7.22. $A = 4.17$ in^2

Chapter 8

8.1. (a) vapor
(b) gas
(c) liquid
(d) mercury

8.3. $R = 129.6$ Ω
The value of R predicted by the equation is the same as the value given in Table 8.1 (within the accuracy given in Table 8.1).

8.5. Error in 15 feet of lead-wire

Wire Gage	Resistance (Ω/ft)	Error (°C)
12	0.00162	0.030
18	0.00651	0.120
28	0.0662	1.215

8.7. $i_s = 1$ mA,   Self-heating error $= 0.008°$C
$i_s = 10$ mA, Self-heating error $= 0.8°$C

8.9. (a) $\Delta R_{avg} (-70°C) = 78,350$ Ω/°C
Sensitivity $(-70°C) = 9.57\%$/°C

8.11. (a)

T (°C)	R_{Therm} (Ω)	V_{out} (1400 Ω)
0	8576	0.140
30	2116	0.398
60	653.3	0.682
90	240.5	0.853

8.13. $T(M) = 822.2$ °C

8.14. $E = 0.0515T + 1.20 \times 10^{-5}T^2$
$E(60) = 3.13$ mV

8.15. Terminal-based line: $e = 0.0539T$
Terminal-based linearity $= 0$ to 1.1%

8.19. (a) Terminal-based linearity $= 1.13\%$
(b) $i_{lin} = 1.046i_x - 0.183$ mA, 4 mA $\leq i_x \leq 7.06$ mA
$i_{lin} = 0.9726 + 0.5471$ mA, 16.71 mA $\leq i_x \leq 20$ mA

8.21. (a) FIC 101: $q = 0.0148 \sqrt{SP}$

(c) FIC 116: $q = 0.0275 \sqrt{SP}$
(e) FIC 134: $q = 0.0684 \sqrt{SP}$

8.22. (b) $N = 32{,}060$, $V = 1.09 \text{ m}^3$

8.24. Range = 0 to 200 kPa

8.26. $f_{min} = -132.4 \text{ N}$
$f_{max} = 63.8 \text{ N}$

8.29. (a) $p = 4.21 \text{ psi}$

Chapter 9

9.5. (1) Increase the anode voltage above the forward breakdown voltage.
(2) Apply a positive voltage pulse to the gate input of the SCR.
(3) Apply light to the gate/cathode junction.
(4) Rapidly increase the anode-to-cathode voltage, V_{AC}.

9.7. The major difference between the triac and the SCR is that the triac can conduct in both directions, whereas the SCR can conduct in only one direction. The triac is equivalent to two SCRs connected in parallel but oriented in opposite directions.

9.9. Voltage gain = -83.0
Current gain = -15.14

9.11. (a) $WP_2 = 1387 \text{ psi}$
(d) $WP_2 = 2246 \text{ psi}$
(g) $WP_2 = 1842 \text{ psi}$

9.13. (a) $Q_2 = 1.53 \text{ gpm}$
(d) $Q_2 = 0.89 \text{ gpm}$
(g) $Q_2 = 0.48 \text{ gpm}$

9.15. (a) 1.5 inch, $W_p = 141 \text{ psi}$
(b) 2.0 inch, $W_p = 79.6 \text{ psi}$
(c) 3.0 inch, $W_p = 35.4 \text{ psi}$

9.17. (a) $C_v = 0.25$
(d) $C_v = 32$
(g) $C_v = 380$

9.19. (a) $C_v = 2.5$
(d) $C_v = 39$
(g) $C_v = 319$

9.21. (a) $Q_1 = 1034 \text{ W}$
(d) $Q_1 = 250 \text{ W}$

9.23. (a) $Q_1 = 13{,}967 \text{ W}$
(d) $Q_1 = 4968 \text{ W}$

9.25. $Q_1 = 2278 \text{ W}$
$Q_2 = 2330 \text{ W}$
$Q_T = 4608 \text{ W}$

Chapter 10

10.1. Your paragraph should explain the following facts:

(1) A force is exerted on a current-carrying conductor in a transverse magnetic field.
(2) A voltage is induced in a conductor moving through a transverse magnetic field.

10.3. If the commutator of a dc generator is replaced by slip rings, an ac voltage will replace the dc voltage at the brushes. The dc generator becomes an ac alternator.

10.5. (a) $K_T = 7.65 \text{ ozf} \cdot \text{in.}/A$
$= 0.478 \text{ lbf} \cdot \text{in.}/A$
(d) $K_T = 78.3 \text{ ozf} \cdot \text{in.}/A$
$= 4.89 \text{ lbf} \cdot \text{in.}/A$

10.7. (a) $K_E = 0.06 \text{ V} \cdot \text{s/rad}$
(d) $K_E = 0.62 \text{ V} \cdot \text{s/rad}$

10.9. Velocity TF $= \dfrac{\Omega}{V_c} = \dfrac{4.252}{1 + 0.06378s} \text{ rad/V} \cdot \text{s}$

Position TF $= \dfrac{\Theta}{V_c} = \dfrac{4.252}{s + 0.06378s^2} \text{ rad/V}$

10.11. (a) $K_T = 0.06 \text{ N} \cdot \text{m}/A$
(d) $K_T = 0.62 \text{ N} \cdot \text{m}/A$

10.13. (a) $K_T = 8.50 \text{ ozf} \cdot \text{in.}/A$
$= 0.53 \text{ lbf} \cdot \text{in.}/A$
(d) $K_T = 87.8 \text{ ozf} \cdot \text{in.}/A$
$= 5.49 \text{ lbf} \cdot \text{in.}/A$

10.15. $e = 48.08 \text{ V}$
$\omega_2 = 308.83 \text{ rad/s}$
$S_2 = 2949 \text{ rpm}$

10.17.

Velocity TF $= \dfrac{\Omega_m(s)}{E_a(s)} = \dfrac{1.60}{1 + 0.023s + 6.1 \atop \times 10^{-5} s^2} \text{ rad/V} \cdot \text{s}$

Simplified TF $= \dfrac{\Omega_m(s)}{E_a(s)} = \dfrac{1.60}{1 + 0.023s} \text{ rad/V} \cdot \text{s}$

10.19. revised Equation (10.26):
$J_T = J_M + J_{N1} + (N_1/N_2)^2 (J_L + J_{N2})$

10.24. $R_1 = 8 \text{ k}\Omega$
$R_2 = 2 \text{ k}\Omega$
$R_3 = 0.08 \text{ }\Omega$

10.26. $\dfrac{\Omega}{SP} = \dfrac{KK_T}{(RB + KK_TK_m + K_EK_T)}$
$+ RB(\tau_m + \tau_e)s + RB\tau_m\tau_e s^2$

Chapter 11

11.7. $1M = CRM$
$= \text{NOT MASTER-STOP AND (START-MOTORS OR 1M) AND NOT OL}$

11.9. CR1 = NOT CR4 AND (START OR CR1) AND LS5
CR2 = CR1 AND NOT CR3 AND LS4 AND NOT LS2
CR3 = CR1 AND (LS2 OR CR3)
CR4 = NOT LS6 AND (CR3 AND LS1 OR CR4)
TD1 = CR4 AND NOT DELAY
Sol. A = CR2
Sol. B = CR3
Sol. C = CR1
Sol. D = TD1

11.15.

<table>
<tr><td colspan="8" align="center">Timing Diagram</td></tr>
<tr><td>Step:</td><td>1</td><td>2</td><td>3</td><td>4</td><td>5</td><td>6</td><td>7</td></tr>
<tr><td>CR1</td><td>xxxxx</td><td></td><td></td><td></td><td></td><td></td><td></td></tr>
<tr><td>CR2</td><td></td><td>xxxxx</td><td></td><td></td><td></td><td></td><td></td></tr>
<tr><td>CR3</td><td></td><td></td><td>xxxxx</td><td></td><td></td><td></td><td></td></tr>
<tr><td>CR4</td><td></td><td></td><td></td><td>xxxxx</td><td></td><td></td><td></td></tr>
<tr><td>CR5</td><td></td><td></td><td></td><td></td><td>xxxxx</td><td></td><td></td></tr>
<tr><td>CR6</td><td></td><td></td><td></td><td></td><td></td><td>xxxxx</td><td></td></tr>
<tr><td>CR7</td><td></td><td></td><td></td><td></td><td></td><td></td><td>xxxxx</td></tr>
</table>

Chapter 13

13.3. Error detector and control mode unit

13.5. Observe the controller output. If the output has changed, there has been a load change.

13.7. (a) Increase the capacitance of the process.
(b) Decrease the dead-time lag.
(c) Decrease the size of the load change.

13.9. Amplitude = 0.02 m
Period = 80 s
The amplitude was decreased from 0.03 m to 0.02 m.
The period was increased from 40 s to 60 s.
The range was reduced from 0.004 m^3/s to 0.002 m^3/s.

13.11. Floating control mode.

13.13. (a) Process capacitance is too small to allow use of two-position control.
(b) The process has fast-moving load changes.

13.14. (a) $e = 60\%$

13.16. (a) $v = 29.3\%$

13.17. The process has sudden load changes and the proportional mode is unable to keep the error within acceptable limits.

Chapter 12

12.5. (a) Logic 0
(d) Logic 0

12.6. (a) Logic 0 interpreted as FALSE
(d) Logic 1 interpreted as FALSE
(g) Logic 0 interpreted as FALSE

12.13. In rung 1, the NC T:4 Contact provides a path to turn OM:1-4 on when START is pressed and LL1 is open.

13.18. $v = 42.8\%$ at $t = 15$ s
13.20. (a) $v = 56.3\%$ at $t = 15$ s
13.23. PD mode
13.27. Two-position mode
13.31. $R_i = 185$ kΩ
$R_I = 28.9$ kΩ
13.33. $R_d = 320$ kΩ
$R_1 = 35.6$ kΩ
$R_i = 231$ kΩ
$C_i = 303$ μF
13.40. (a) 4 (d) 3 (e) 2 (h) 1
13.41. (a) 8 A
(c) 6 A
(f) 0 A
(i) −6 A
13.44. If POSITION is NS AND VELOCITY is NS, then CURRENT is PM.
If POSITION is NS AND VELOCITY is PS, then CURRENT is ZR.

Chapter 14

14.1. (e)
14.3. (d) and (e)

14.5. (a) $B_2 = 1.0\%$
(c) $B_2 = 25.0\%$

14.7.

Frequency (rad/s)	Gain	Output (%FS)	Output (m)
1.57E − 04	10	100	5
1.57E − 02	0.1	1	0.05
1.57E + 00	0.001	0.01	0.0005

14.9. (a) Output = 22.64%
(d) Output = 29.633%

14.11. (a) $g = 1$
(d) $g = 0.1$

14.13. (a) $C_L = 2.5 \times 10^{-4} \text{m}^3/\text{Pa}$
(b) $R_L = 6.68 \times 10^7 \text{ Pa} \cdot \text{s/m}^3$
(c) $\tau = 16{,}700$ s
(d) $G = 0.605$
(e) $16{,}700 \dfrac{dh^*}{dt} + h^* = 0.605 q_{in}^*$

(f) $\dfrac{H^*(s)}{Q_{in}^*(s)} = \dfrac{0.605}{1 + 16{,}700s}$

14.15. (a) Amplitude = 0.25 V, phase = 39.3°
(b) Amplitude = 0.354 V, phase = 0°
(c) Amplitude = 0.15 V, phase = −39.3°

14.18. (a) $C_g = 6.85 \times 10^{-6}$ kg/Pa
(b) $\tau = 1.23$ s
(c) $G = 4.50$
(d) $1.23 \dfrac{dp^*}{dt} + p^* = 4.5 w^*$

(e) $\dfrac{P^*(s)}{W^*(s)} = \dfrac{4.5}{1 + 1.23s}$

14.19. $\dfrac{C_o(s)}{C_i(s)} = \dfrac{1}{1 + 600s}$

14.23. (a) $m = 0$ dB, $g = 1$, $\beta = 0°$
(d) $m = -39$ dB, $g = 0.011$, $\beta = -176°$

14.26. (a) $0.0173 \dfrac{d^2h}{dt^2} + 0.029 \dfrac{dh}{dt} + h = 6.9 \times 10^{-4} f$

(b) $\dfrac{H(s)}{F(s)} = \dfrac{6.9 \times 10^{-4}}{1 + 0.029s + 0.0173s^2}$

(c) $\omega_o = 7.61$ rad/s
(d) $\zeta = 0.110$
(e) The process is underdamped.

14.27. (a) Amplitude = 0.01 m, phase = −1.2°
(b) Amplitude = 0.05 m, phase = −90°
(c) Amplitude = 0.0001 m, phase = −179°

14.29. $360{,}000 \dfrac{d^2h^*}{dt^2} + 1500 \dfrac{dh^*}{dt} + h^* = 0.561 q_{in}^*$

$\dfrac{H^*(s)}{Q_{in}^*(s)} = \dfrac{0.561}{1 + 1500s + 360{,}000s^2}$

14.31. $\tau_m = 1.6$ s, $\tau_e = 0.0625$ s

$0.00472 \dfrac{d^2\omega}{dt^2} + 0.0785 \dfrac{d\omega}{dt} + \omega = 4.33E$

$\dfrac{\Omega(s)}{E(s)} = \dfrac{4.33}{1 + 0.0785s + 0.00472s^2}$

$\omega_o = 14.6$ rad/s
$\zeta = 0.57$, underdamped

14.35. $f_o(t) = f_i(t - 18.6)$
$\dfrac{F_o(s)}{F_i(s)} = e^{-18.6s}$

14.38. $t_d = 6.91$ s, $\tau = 1220$ s
$\dfrac{H(s)}{M(s)} = \left(\dfrac{1}{1 + 1220s}\right) e^{-6.91s}$

Chapter 15

15.2. (a) $g = 21.1$, $\beta = -69°$
(c) $g = 3.0$, $\beta = -150°$

15.3. (a) Closed-loop: $m = 10$ dB, $\beta = -80°$
(d) Closed-loop: $m = 1$ dB, $\beta = -6°$
(i) Closed-loop: $m = -2.5$ dB, $\beta = -80°$

15.4. (a) g (closed-loop) = 2.88
m (closed-loop) = 9.2 dB
β (closed-loop) = −80°
(d) g (closed-loop) = 1.13
m (closed-loop) = 1.07 dB
β (closed-loop) = −6.6°
(g) g (closed-loop) = 0.32
m (closed-loop) = −9.9 dB
β (closed-loop) = −44°

15.6. TF $= \left(\dfrac{10}{1 + 0.5s}\right) e^{-0.05s}$

Gain margin = 4.2 dB, phase margin = 37°
The system does not satisfy the minimum criteria for gain margin or phase margin.
$\omega_{max} = 24$ rad/s
The system satisfies the Nyquist criterion for stability.

15.8. TF $= \left(\dfrac{0.628 + 9.1s}{s}\right)\left(\dfrac{1}{1 + 50s}\right) e^{-2s}$

Gain margin = 11.9 dB, phase margin = 53°
The system satisfies the minimum criteria for gain margin and phase margin.
$\omega_{max} = 0.4$ rad/s
The system satisfies the Nyquist criterion for stability.

15.10. $TF = \left(\dfrac{0.4 + 14.4s}{18s}\right)e^{-20s}$

Gain margin = 1.7 dB, phase margin = 98°
The system is marginally stable
$\omega_{max} = 0.16$ rad/s
The system satisfies the Nyquist criterion for stability.

15.12. $TF = \left(\dfrac{42.86 + 12s + 2.16s^2}{s + 0.018s^2}\right)$

$\times \left(\dfrac{0.1}{1 + 0.02s + 0.01s^2}\right)e^{-0.05s}$

No gain margin, no phase margin.
The system is unstable and does not satisfy the minimum criteria for gain margin or phase margin. The maximum frequency limit has no meaning when the system is unstable.
The system does not satisfy the Nyquist criterion for stability.

15.14. $TF =$

$\left(\dfrac{0.01 + 0.00067s}{s + 0.0067s}\right)\left(\dfrac{8500}{1 + 0.025s + 0.0001s^2}\right)$

$\times \left(\dfrac{1}{s + 0.35s^2 + 0.015s^3}\right)$

No gain margin, no phase margin.
The system is unstable and does not satisfy the minimum criteria for gain margin or phase margin. The maximum frequency limit has no meaning when the system is unstable.
The system does not satisfy the Nyquist criterion for stability.

15.16. (a) $M(-180) = -4.2$ dB, system is stable.
(g) $M(-180) = 2.3$ dB, system is unstable.
(j) $M(-180) = -10.9$ dB, system is stable.

15.17. (a) Gain margin = 4.2 dB
Phase margin = 37.2°
(g) Gain margin = -2.3 dB
Phase margin = -14.1°
(j) Gain margin = 10.9 dB
Phase margin = 37.2°

15.19. $s = -\left(\dfrac{12K + 4}{K + 1}\right)$

The loci of s are on the real axis from -4 to -12.

15.21. $s1 = -10 + \sqrt{84 - 10K}$, $s2 = -10 - \sqrt{84 - 10K}$

The loci of s are on the real axis from -0.835 to -19.165 and the vertical line from $s = -10 - j\infty$ to $s = -10 + j\infty$.

A value of $K = 14.03$ results in a damping ratio of $\zeta = 0.8$.

15.23. The root locus has two branches. One branch terminates at the zero at $s = -5$. The other branch goes to infinity along a single asymptote. (R1, R2, R3) The asymptote passes through the centroid, $c = -5$. (R4)
The asymptote angle, $\theta = 180°$. (R4)
The root locus lies on the portion of the real axis between $s = -5$ and $s = -\infty$, and at the point $s = 0$. (R5)
There is a breakaway point on the real axis at $s = 0$, and a breakin point at $s = -10$. (R6, R7)
Characteristic equation: $s^2 + Ks + 5K = 0$
Selected points on the root locus:

K	Characteristic Equation	s_1	s_2
0	$s^2 = 0$	0	0
5	$s^2 + 5s + 25 = 0$	$-2.5 + j4.33$	$-2.5 - j4.33$
10	$s^2 + 10s + 50 = 0$	$-5.0 + j5.0$	$-5.0 - j5.0$
15	$s^2 + 15s + 75 = 0$	$-7.5 + j4.33$	$-7.5 - j4.33$
20	$s^2 + 20s + 100 = 0$	-10	-10
30	$s^2 + 30s + 150 = 0$	-6.34	-23.66

15.26. The root locus has three branches that start at the three poles and go to infinity along the asymptotes. (R1, R2, R3)
The asymptote passes through the centroid, $c = -6.67$. (R4)
The asymptote angle, $\theta_1 = 60°$, $\theta_2 = -60°$, $\theta_3 = 180°$. (R4)
The root locus lies on the portion of the real axis between $s = 0$ and $s = -5$, and between $s = -15$ and $s = -\infty$. (R5)
There is a breakaway point on the real axis at $s = -2.26$, with $K = 78.89$. (R6, R7)
The branches cross the imaginary axis at $s = \pm j8.66$ with $K = 1500$. (R8)
Characteristic equation: $s^3 + 20s^2 + 75s + K = 0$
Selected points on the root locus:

K	s_1	s_2	s_3
78.89	-15.49	-2.26	-2.26
136.51	-15.80	$-2.1 + j2.06$	$-2.1 - j2.06$
138.62	-15.811	$-2.094 + j2.094$	$-2.094 - j2.094$

K	s_1	s_2		s_3	
176.0	-16.0	-2.0	$+ j2.65$	-2.0	$- j2.65$
408.0	-17.0	-1.5	$+ j4.66$	-1.5	$- j4.66$
702.0	-18.0	-1.0	$+ j6.16$	-1.0	$- j6.16$
1064	-19.0	-0.5	$+ j7.47$	-0.5	$- j7.47$
1500	-20.0		$+ j8.66$		$- j8.66$

Notes:

Operating point angle, $\theta = \arctan 0.707 = 45°$.

Operating point, $s_{\mathrm{op}} = -2.094 \pm j2.094$.

Gain at the operating point, $K_{\mathrm{op}} = 138.6$.

Natural frequency, $\omega_o = 2.96$ rad/s.

Chapter 16

16.1. System 2: $G_u = 6.3$, $P_u = 6$ min
    PI mode: $P = 2.8$, $I = 0.20$ min^{-1}
    PID mode: $P = 3.8$, $I = 0.33$ min^{-1}
       $D = 0.75$ min
    System 4: $G_u = 1.2$, $P_u = 18$ min
    PI mode: $P = 0.54$, $I = 0.067$ min^{-1}
    PID mode: $P = 0.72$, $I = 0.11$ min^{-1}
       $D = 2.25$ min

16.3. **(a)** PI mode: $P = 0.63$
       $I = 0.030$ min^{-1}
    (b) PID mode: $P = 0.84$
       $I = 0.051$ min^{-1}
       $D = 4.95$ min

16.4. **(a)** 2 **(d)** 3 **(h)** 1 **(j)** 2

16.5. $\omega_{\max} = 0.035$ rad/s

16.7. Design Summary
    Control modes used: PD
    Proportional gain = 0.00111
    Integral action rate = 0 (not used)
    Derivative action time constant = 0.31 s
    Gain margin = 8.5 dB
    Phase margin = 40.0°
    Frequency limit = 9.7 rad/s

16.9. Design Summary
    Control modes used: PI
    Proportional gain = 3.41
    Integral action rate = 0.0606
    Derivative action time constant = 0 (not used)
    Gain margin = 6.6 dB
    Phase margin = 40°
    Frequency limit = 0.17 rad/s

16.11. Design Summary
    Control modes used: PI
    Proportional gain = 0.490

Integral action rate = 0.0128 s^{-1}
Derivative action time constant = 0 (not used)
Gain margin = 6 dB
Phase margin = 101°
Frequency limit = 0.044 rad/s

16.13. Design Summary
    Control modes used: PID
    Proportional gain = 1.23
    Integral action rate = 4.63 s^{-1}
    Derivative action time constant = 0.128 s
    Gain margin = 6 dB
    Phase margin = 46°
    Frequency limit = 16 rad/s

16.15. Retard factor, R.F. = 0.25
    Design Summary
    Control modes used: PID
    Proportional gain = 4.13
    Integral action rate = 0.0025 s^{-1}
    Derivative action time constant = 8.35 s
    Gain margin = 13.0 dB
    Phase margin = 40.0°
    Frequency limit = 0.84 rad/s

16.17. Design Summary: Temperature Control Loop
    Control modes used: PID
    Proportional gain = 2.35
    Integral action rate = 0.000745 s^{-1}
    Derivative action time constant = 263 s
    Gain margin = 12.5 dB
    Phase margin = 40.0°
    Frequency limit = 0.012 rad/s

16.19. First Design Summary
    Proportional gain = 0.183
    Integral action rate = 0.136 s^{-1}
    Derivative action time constant = 0
    Gain margin = 51.0 dB
    Phase margin = 40°
    Frequency limit = 0.017 rad/s
    Second Design Summary
    Proportional gain = 36.5
    Integral action rate = 0.044 s^{-1}
    Derivative action time constant = 0
    Gain margin = 6.0 dB
    Phase margin = 41°
    Frequency limit = 0.46 rad/s

16.21. Design Summary
    Proportional gain = 3.50
    Integral action rate = 0.147 s^{-1}

Derivative action time constant = 0
Gain margin = 6.2 dB
Phase margin = 40°
Frequency limit = 0.46 rad/s
The gain margin, phase margin, frequency limit,

and error ratio graphs are almost identical to those obtained in Exercise 16.19, Design 2. The designs with the highest frequency limit are the two in Exercise 16.18 with $I = 0.046$ s^{-1} and $I = 0.069$ s^{-1}.

Index

A/D converter, 243
Absolute encoder, 264, 281
Absolute pressure, 309, 316
Acceleration measurement, 273
Accelerometer, 273–4, 281
  frequency requirement, 275
Accuracy, 157, 174
  measured, 160, 174
  measuring instrument, 153
AC motors, 364, 370–7, 411
  adjustable speed drives, 402–6
  induction, 370, 372–3, 411
  servo, 374–6, 411
  squirrel-cage, 370–2, 411
  synchronous, 374, 411
  wound rotor, 372, 411
Acquisition time, 227, 243
Actuator, 13, 30, 336, 357
  hydraulic, 336, 357
  pneumatic, 336, 357,
Adaptive controller, 510–1, 519
  model-based, 620, 649
  pattern-recognition, 620, 649
  programmed, 619, 649
  Self-tuning, 619, 649
Adaptive gain, 40
ADC (*see* Analog-to-digital converter)
Alias frequency, 229, 243
Aliasing, 229
Amplification, 182
Analog control, 39
Analog controller, 519
Analog signal, 38, 66
Analog-to-digital converter, 183, 230,
    234–41, 243
  selection, 242
Angle condition, root locus, 602, 608
Aperture time, 227, 243
Arithmetic functions, PLC, 465, 469
Armature, 364, 367, 411
ASCII code, 661
Asynchronous, 676, 690
Automated drilling machine, 425

Backbone, 690
Back emf, 369–70, 411
Backlash, 18
Balloon, 15, 663
Band heater, 354, 357
Band-pass filter, 243
Bandwidth, 214
Baseband, 678, 690
Batch process, 418, 420, 438, 457
Baud rate, 677, 690
Bellows, 310, 317
Benefits of automatic control, 20
Bias, 158, 174
Bimetallic thermostat, 287, 317
Binary codes for decimal digits, 661
Binary counter ADC, 243
Binary counter technique, 235
Binary-octal equivalents, 659
Binary-weighted DAC, 232–3, 243
Bit manipulation functions, 467–9
Bit transfer rate (BPS), 677, 690
Black body, 302, 317
Blending and heating process, 638
Block diagram, 3
  closed-loop, 8
  process control system, 9
  servo control system, 8
  simplification, 26
Bode design method, 620–1, 648
Bode diagram, 141–3, 147
  closed-loop, 571, 573
  dead-time process, 557
  derivative control mode, 630
  first-order lag process, 536
  integral control mode, 488, 633
  integral process, 530
  open-loop, 570, 573
  open-loop, final design, 626
  PI control mode, 491
  practical derivative mode, 496
  proportional control mode, 483
  second-order lag process, 173, 548
  thermometr, 171

Bonded strain gage, 277, 281
Boolean equation, 422, 438
Bourdon element, 309, 317
Braking:
  dynamic, 377, 411
  regenerative, 378, 411
Break point, 170
Break point frequency, 170, 536
Bridge, 690
Bridge circuit, 182, 205–13, 293
  current balance, 207
  three-wire, 293
  two-wire, 293
Broadband, 690
Brushes, 368
Bubble, 663
Bus network, 445

Calibration, 159, 174
Calibration curve, 160, 174
Calibration report, 159, 174
Capacitance, 69, 107
  defined, 72
  electrical, 75, 78, 107
  gas flow, 91–3, 107
  liquid flow, 85, 89, 107
  mechanical, 102, 104, 107
  thermal, 97, 99, 107
Carrierband, 690
Cartridge heater, 353–4, 357
Center-of-gravity method, 516
Central control room, 56
Central processing unit (CPU), 447, 469
Centroid, 602, 608
Centroid method, 516
Characteristic equation, closed-loop,
    596, 608
Closed-loop (*see* control system,
    closed-loop)
Cogging, 403, 411
Command line, 622
Common elements of systems, 69
Common-mode rejection (CMR), 187, 243

Common-mode rejection ratio (CMRR), 187, 243
Common-mode voltage, 187
Communication channel, 676, 690
Communications module, 449, 469
Commutation, 329, 368, 411
Commutator, 368, 411
Comparator, 188, 243
Compare functions, 465, 469
Compensation, 614, 642–8
  integration-lead, 643, 648
  lag, 643–5, 648
  lag-lead, 643–5, 649
  lead-lag, 643–4, 649
  networks, 643–5
  objective of, 643
Complex numbers, 669–75
Complex plane, 143
Composite wall, 94
Computer-aided design (CAD), 50
Computer-aided engineering (CAE), 53
Computer-integrated manufacturing (CIM), 51
Computerized numerical control (CNC), 51, 66
Contactor, 325, 358
Contact symbols, PLC, 452
Contention, 690
Control algorithms, 505
Control functions, PLC, 466, 469
Control modes, 13, 30, 473, 476–500, 520
  derivative (D), 13, 23, 30, 493, 520
    digital, 508
    when useful, 629
  floating, 479, 520
  integral (I), 13, 30, 486–7, 520
    digital, 507
    usefulness of, 632
  PD, 494–6, 520
  PI, 489–91, 520
  PID, 497–9, 520
  proportional (P), 13, 31, 480–6, 520
  two-position, 476–7, 520
Control system, 2
  advanced, 509–12
  cascade, 509, 519
  classifications, 37
  closed-loop, 5, 8, 15, 30, 473
  compensated mass flow, 42–4
  composition control, 65
  dc motor position, 46
  emergency generator, 59
  feedback, operations, 9
  hydraulic cylinder, 65
  hydraulic position, 45
  liquid flow rate, 62
  liquid level, 61
  manufacturing plant, 57
  objectives, defined, 24
  open-loop, 5–7
  pneumatic cylinder, 49
  pressure, 63
  sheet thickness, 64
  solid flow rate, 64
  speed, dc motor, 61
  speed, mechanical, 63
  temperature, 42

Control valve, 14, 120–1, 344–6, 358
  characteristics, 347
    equal-percentage, 348, 357
    inherent, 347, 357
    installed, 347, 349, 357
    linear, 348, 358
    quick-opening, 347, 358
  sizing, 350, 358
Control, analog and digital, 38
Controlled variable, 2, 10, 14, 30
Controller, 12, 30
  adaptive, 510
    model-based, 511
    pattern-recognition, 511
  analog, electronic, 500–4
  design, 614
  digital, 504–509
  fuzzy logic, 513
  output (V), 13, 15, 30
  self-tuning, 511
Convergent beam method, 269, 281
Conversion, 182
Conversion factors, 657–8
Converter:
  current-to-voltage, 204
  voltage-to-current, 205
Coulomb friction, 101, 107
Counter function, PLC, 458, 469
Counter status, PLC, 447
Criteria of good control, 24
Critical damping, 26, 30
Critically damped response, 23
CSMA/CD access protocol, 688, 690
Cylinder:
  hydraulic, 338, 358
  pneumatic, 338, 358
  selection of, 342

D/A converter, 243
DAC (*see* digital-to-analog converter)
Damping, 22, 30
  coefficient of, 167, 175, 542, 600
  output derivative, 22
  ratio, 167, 175, 542, 600
Data acquisition system, 224–5, 243
Data communication unit, PLC, 446
Data conversion, 226
Data distribution system, 224–5, 243
Data highway, 58
Data move functions, PLC, 466, 469
Data sampling, 183, 226, 243
DC motors, 364, 370, 377–92, 411
  adjustable speed drives, 407–11
  armature controlled, 379
  brushless, 380–1, 411
  compound, 379, 412
  moving coil, 379, 412
  pancake, 380, 412
  permanent magnet, 379, 412
  separately excited, 379, 412
  series, 378, 412
  shunt, 379, 412
Dead band, 18, 30, 156, 160, 175
Dead time, 69, 72–3, 107
  electrical, 107
  liquid flow, 107
  mechanical, 107

Dead-time delay:
  electrical, 77–8
  liquid flow, 87, 89
  mechanical, 103, 104
Dead-time process, 555–8, 560
Decay rate, 227, 243
Decimal-hexadecimal-binary equivalents, 660
Decision, 473
Defuzzification, 516, 520
Derivative, 147
Derivative action time constant, 474, 631
Derivative amplitude, 630, 648
Derivative break frequency, 631
Derivative limiter, 495
Derivative limiter coefficient, 474
Derivative mode (*see* control modes, derivative)
Derivative of controlled variable, 22
Design example, PID controller, computer aided, 620–9
Design example, three-loop control system, 637–42
Design step 1, derivative mode, 629–31
Design step 2, integral mode, 631–5
Design step 3, proportional mode, 636
Design step 4, concluding report, 636
Deviation ratio, 566, 577, 608
Diaphragm element, 311
Differential amplifier, 198, 243
Differential equation, 121–2, 147
Differential pressure, 309, 317
Differential pressure flow meter, 304, 317
Differentiator, 197, 243
Diffuse scan method, 268, 281
Digital conditioning, 242
Digital control, 39
Digital controller, 520
Digital signal, 38, 66
Digital-to-analog converter, 183, 230, 233–4, 244
Diode, 331, 358
Direct actuator, 345
Direct digital control (DDC), 56
Direct method, resistance measurement:
  four-wire, 291
  two-wire, 291
Direct numerical control (DNC), 51, 66
Direct scan method, 268, 281
Discrete-part manufacturing, 37
Discrete process, 417–8, 438
Displacement measurement, 252
Dissipation constant, 248
Distributed control, 58
Disturbance variables, 15, 30
Divide and conquer, 443
Drift, 157, 175
  sensitivity, 157, 175
  zero, 157, 175
Dual slope ADC, 244
Dual slope technique, 239
Dynamic characteristics, 154, 162
Dynamic error, 169, 175
Dynamic lag, 169, 175

Electric heating elements, 352–6
Electric motors, classification, 364
Emittance, 303, 317

Encoder, 262, 281
End effector, 53
Equations box:
  ac servomotor transfer function, 376
  closed-loop response, 572
  control valve Cv, 350
  control valve transfer function, 352
  dc motor steady state, 383–4
  dc motor transfer function, 390–2
  dead-time process, 557
  electrical elements, 78
  elements, summary, 105
  error and deviation ratio, 578
  first-order lag process, 537–8
  gas flow elements, 92
  ideal derivative mode, 494
  integral control mode, 487
  integral process, 528
  I/O summary, 122–3
  liquid flow elements, 88
  mechanical elements, 104
  open-loop response, 570
  PD control mode, 495
  PI control mode, 491
  PID controller design, 636–7
  PID control mode, 498
  proportional control mode, 484
  second-order lag process, 549–52
  thermal elements, 98
Error, 13–4, 30, 157, 175
Error detector, 10, 12, 30, 500
Error ratio, 566, 575, 608
Ethernet LAN, 685
Euler's identity, 670
Event-driven operation, 47, 66, 417
Event-driven process, 421, 438

Fanning equation, 81, 107
Farad, 75
Feedback, 3, 8, 30
Feedforward control, 510, 520
Field poles, 364, 367, 412
Filled thermal system, 287, 317
Film coefficient, 94, 98
Filter design, 216
Filters, 213–4
Final controlling element, 13
Final value theorem, 140, 147
First-order lag plus dead-time process, 558–9
First-order lag process, 531–42, 560
  blending, 534, 538
  electrical, 532, 537
  gas pressure, 534, 538
  liquid level, 532, 537
  thermal, 534, 538
Flash or parallel ADC, 244
Flash technique, 240
Flowchart, 419, 438
Flow rate measurement, 303–8
Flow rate sensors, 285
Flux vector control, 405
Follow-up system, 39, 66
Force measurement, 276–81
Frequency domain, 124, 147
Frequency limit, 566, 608
Frequency response, 6, 115, 141–7, 170, 566
  closed-loop, 566, 608
  determination of, 143

first order lag, 170–1
  graph, 526, 560
  open-loop, 566, 609
  overall, 569
  second order lag, 171
Frequency shift keying (FSK), 682
Frequency spectrum, 244
Friction factor, f, 81
Full-duplex, 676, 690
Full-step operation, 397, 412
Functional identification, 15, 663
Functional Laplace transform, 127–9, 147
Function generator, 201, 244
Fuzzification, 514, 520
Fuzzy logic, 513
Fuzzy logic controllers, 513–9
Fuzzy logic control module, 518
Fuzzy rule base, 515
Fuzzy set, 513, 520

Gage factor, 277, 281
Gage pressure, 309, 317
Gain, 5, 6, 30, 115, 141
  overall, 568–9, 609
Gain margin, 566, 588–90, 608
  when satisfied, 636
Gain margin point, 594, 608
Gateway, 690
Gray code, 660

Hagen-Poiseuille law, 81, 107
Half-duplex, 676, 690
Half-step operation, 399, 412
HDLC protocol, 689
Headend repeater, 690
Henry, 76
High-pass filter, 244
Hydraulic hoist, 424
Hydraulic motor, 340–1, 358
Hysteresis, 20, 30, 160, 175, 267

IC temperature sensor, 302
IEEE-488 data bus, 681
IEEE-583 interface, 682
Impedance transformation, 204
Incremental encoder, 263, 272, 281
Induced voltage, electric motor, 367
Inductance, 69, 76, 78, 107
  defined, 72
Inertance, 86, 89, 107
  defined, 72
Inertia, 69, 103–4, 107
  defined, 72
Initial condition, 122
Initial value theorem, 140, 147
Input image status, PLC, 447, 469
Input module:
  analog, 469
  discrete, 469
  PLC, 448
Input scan, PLC, 448, 453, 469
Instability, 30
Instrumentation amplifier, 199, 244
Instrumentation Symbols and Identification,
  15, 662–8
Instrument line symbols, 666
Integral, 147
Integral action rate, 474, 487, 489, 632, 635

Integral action time constant, 527, 560
Integral break point frequency, 634
Integral mode (see control modes, integral)
Integral process, 527–9, 560
Integrator, 195, 244
Integro-differential equation, 114, 122, 147
Interacting second-order process, 560
Interpolation formula, 83
Inverse Laplace transform, 132–6, 147
Inverse Laplace transformation, 114
Inverter, 412
  pulse-width modulated (PWM), 412
  variable-voltage (VVI), 403, 412
Inverting amplifier, 192, 244
ISA standard S51.1, 154
Isolation, 182, 204
Isolation amplifier, 204, 244

Ladder diagram, 322, 419, 422, 431, 438
  programming language (LD), 449, 469
Laminar flow, 79, 107
  when it occurs, 80
Laplace transform, 124–32, 147
Laplace transformation, 114
  diagram of, 125
Latent head of fusion, table of, 654
Least-squares fit, 162
Level indicator circuits, 190
Level sensor, 286
  capacitance probe, 315
  displacement float, 313
  static pressure, 314
Linear component, 17
Linear differential equation, 148
Linear operating region, 185, 244
Linear regression, 162
Linearity, 161, 175
  independent, 161, 175
  least-squares, 161,175
  terminal-based, 161, 175
  zero-based, 161, 175
Linearization 182, 218, 244
Linearizer, defined, 221
Linguistic variable, 513, 520
Liquid level measurement, 312
Liquid tank, nonregulating, 117–8
Liquid tank, self-regulating, 115–6
Load, 30
  change, 21, 474, 520
  on control system, defined, 21
  on process, 21
Load cell, 281
Loading error, 253–5, 281
Local area network (LAN), 58, 682
Logarithmic amplifier, 201
Logarithmic transformation, 114
Loop identification 15, 664
Low-pass filter, 244
LVDT, 256, 281

Machine control, 56
Machine program, 49, 66
Magnetic flow meter, 308, 317
Magnitude condition, root-locus, 602, 608
Maintained-action switch, 323, 358
Manchester encoder, 683
Manipulated variable, 2, 15, 30
Manipulating element, 13, 30

Manipulation, 473
Manipulator, 53
Manufacturing Automation Protocol (MAP), 58, 685, 690
Marginally stable, 609
Mass flow rate, 304
Maximum frequency limit, 575
Mean, 155
Measured variable, 14, 30
Measurement, 473
Measuring instrument model:
　first order, 165
　second order, 166
Measuring transmitter, 11, 30, 182, 251
Mechanical switches, 322
Melting point, table of, 654
Microcontrollers, 40–1
Microstep operation, 399, 412
Minimum integral of absolute error, 26, 30
Modes of control (*see* control modes)
Modulation, 678, 690
Module, 469
Modulus of elasticity (E), 277
Momentary-action switch, 322, 358
Motor starter, 325, 358
Multivariable control, 512, 520

Natural frequency, 167, 175
Nichols chart, 574
Noise prevention, 213
Noise reduction, 213
Noninteracting second-order process, 560
Noninverting amplifier, 193, 244
Nonlinearity, 17
Nonretentive timer, 454
Notch filter, 244
Numerical control (NC), 49–50, 66
Nyquist criterion
　data sampling, 227, 244
　stability, 114, 567, 592–3, 609

Objectives, of control, 23
Ohm's law, 73
One's and two's complements, 660
Open loop, 30
Open systems interface (OSI), 684, 690
Operate point, 281
Operating characteristics, 154–5
Operating conditions, 157, 175
Operation, 439
Operational amplifier, 182–3, 244
　ideal, 185
Operational Laplace transform, 129–30, 148
Operations, auxiliary, 418, 439
Operations, parallel, 418, 439
Operative limits, 157, 175
Optical encoder, 262–3
Order, of differential equation, 122, 148
Oscillation, conditions of, 584–6
Output functions, 431, 439
Output image status, PLC, 447, 469
Output module:
　analog, 469
　discrete, 469
　PLC, 446, 448
Output scan, PLC, 448, 453, 469
Overdamped response, 23

Overrange, 157, 175
Overrange limit, 157, 175
Overshoot, 175
Oversize factor, 342, 358

Parallel channel, 690
Parameters, table of, 71
Partial fraction expansion, 133, 148
Part program, 49, 66
Peak percent overshoot (PPO), 26
Perfect vacuum, 317
Phase angle, 30, 142
　overall, 568–9, 609
Phase difference, 6, 115
Phase margin, 566, 588–90, 609
　when satisfied, 636
Phase margin point, 594, 609
Photoelectric sensors, 268, 281
Pick-and-Place (PNP), 53
PID algorithm
　incremental, 505, 520
　positional, 505, 520
PID control function, 468–9
PID controller unit, 446
PID module, PLC, 449, 469
Piecewise-linear function, 221
PIP, 253, 281
PLC 443, 446
PLC program walkthrough, 453
Pneumatic motor, 340–1, 358
Point-to-point (PTP), 53
Polling, 690
Position measurement, 252
Potentiometer, 253, 281
Powers of two, 659
Pressure measurement, 308
Pressure sensors, 286
Primary element, 11, 182, 252
Process, 11, 30, 39
　control, 39, 55, 66
　load line, 474
　reaction graph, 617, 649
　reaction method, 616–7, 649
　timing diagram, 419, 428, 439
　variable, 11, 31
Processing, 37
Processor module, PLC, 446, 447, 469
Program:
　BODE, 145
　DESIGN, 115, 566
　　example of, 578–82
　　walkthrough of, 620–9
　LIQRESIS, 82, 84
　THERMRES, 95, 97
Programmable logic controller (PLC), 442–3, 469
Programmer, PLC, 449, 469
Programming unit, PLC, 446, 449
Program scan, PLC, 448, 453, 470
Properties, table of:
　gases, 654
　liquids, 654
　solids, 653
Proportional gain, 474, 481–2
Proportional mode (*see* control modes, proportional)
Proportional offset, 484, 520
Protocol, 685, 690

Proximity sensor, 281
　capacitive, 267
　inductive, 266
Pulse-width-modulation (PWM), 381, 404
Pushbutton switches, 323

Quantization error, 66, 231
Quarter amplitude decay, 26, 31, 616

R-2R ladder DAC, 234, 244
Radiation pyrometer, 302, 317
Ramp process, 527–9, 560
Ramp response, 169
Random access memory, 447
Range, 156, 175
Rangeability, 348
Reaction and blending process, 436–8
Read-only memory, 447
Recovery time, 157
References, 691
Regulator system, 39, 66
Relay, 324, 358
Relay, time-delay, 324–5, 358
Release point, 281
Reliability, 156, 175
Repeatability, 158, 175
Reproducibility, 158, 175
Residual error, 24, 31, 484, 520
Resistance, 69, 107
　defined, 71
　dynamic, 74
　electrical, 73, 78, 107
　gas flow, 89, 90, 92, 107
　liquid flow, 79, 88, 107
　mechanical, 100, 104, 107
　static, 74
　thermal, 93, 98, 107
　total, 93, 95
　unit, 93, 95
Resistance temperature detector (RTD), 203–4, 289
Resolution, 156, 176
　A/D converter, 230
　potentiometer, 253
Resolver, 260–1, 282
Resolver to digital converter, 262
Resonant frequency, 167, 176, 542
Response time, 164, 176
Retentive timer, 454
Retroreflective scan method, 268, 282
Reverse actuator, 345
Reynolds number, 79, 88, 92, 107
Rise time, 164, 176
Robot, 51–2, 66
Robotic work cell, 54
Robotics, 51–55
Root locus, 567, 595–609
Root locus rules, 602–4
Rotor, 364, 370, 412
RS-232C serial port, 679
RTD, 317
Rungs, 322

Sample-and-hold, 227, 244
Sampling, 505
　interval, 226, 244
　rate, 244
　minimum rate, 228

Saturation, 20, 31
Saturation region, 185, 244
Second-order parameters, 167
Second-order process, 542–55, 560
  dc motor, 546, 551
  electrical, 544, 550
  interacting, 546, 551
  mechanical, 543, 550
  noninteracting, 546, 550
Selection criteria, measuring
  instruments, 172
Self diagnositics, 40
Self-heating error, 295
Self-tuning, 40
Semi-star network, 445
Sensing range, 282
Sensitivity, 156, 176
Sensor, 11, 31, 252
Sequencer function, PLC, 459, 470
Sequencer, generic, 461–2
Sequential circuit design, 428–33
Sequential control, 47–9, 66
Sequential function chart, 419, 426, 439
Sequential process, 418, 439
Serial channel, 690
Servomechanism, 44, 66
Setpoint (SP), 10, 14, 31
Settling time, 24, 31, 165, 176
Signal conditioning, 182
  analog, 203
  digital, 183, 224
  RTD, 291–3
  thermocouple, 300
Signal lines, 3
Silicon-controlled rectifier (SCR), 326, 358
Simplex, 676, 690
Sine-coded PWM, 404
Singie-slope ADC, 244
Single-slope technique, 236
Size, input/output, 5
Slew rate, 186, 244
Slip, 371, 412
Sneak path, 451, 470
Solenoid valve, 338–9, 358
Solid state switching components, 326–7
Span, 156, 176
Specular scan method, 269, 282
Speed of response, 153
Stability, 566, 582–8
  conditions, 587–9, 609
Stability point, 593, 609
Standard atmospheric conditions, 655
Standard deviation, 155
Star network, 445
State chart, 419, 427, 439

Statement list, 419, 439
Static characteristics, 154, 157
Statistics, 154
Stator, 364, 370, 412
Status line, 621
Steady-state gain, 542
Step change, 126
Stepping motors, 396–401, 412
Step response, 24, 163, 168, 526, 560
  dead-time process, 556
  first-order lag process, 531
  first-order plus dead-time, 559
  five types, 23
  ideal derivative mode, 494
  integral process, 527
  PI control mode, 490
  second-order lag process, 543
Steps, 439
Strain gage, 282
Strain gage load cell, 277–8
Strouhal number, 307, 317
Successive approximation ADC, 236, 245
Summing amplifier, 194, 245
Switch, 322, 358
Synchro, 257, 282
Synchronous, 690
Synchronous speed, 371, 412
Synchronous transmission, 676
Systemic error, 158, 176
Systems of units, 656

Tachometer, 270, 282
  ac, 272
  dc, brushless, 271
  optical, 272
Tag number, 663
Teaching pendant, 54
Temperature measurement, 286
Temperature scales, 286
Temperature sensor, 285
Thermal sensitivity shift, 157, 176
Thermal zero shift, 157, 176
Thermistor, 296, 317
Thermocouple, 297–302, 317
  burnout proctection, 302
  type designations, 298
Thermometer, liquid-filled, 119
Three zones of control, 609
Threshold, 156
Time constant, 117, 148, 164, 176, 531
Time domain, 124, 148
  equation, 526, 560
Time-driven 47
  operations, 66, 417
  process, 419, 439

Time/event-driven process, 435–8
Timer function, PLC, 454, 470
Timer status, PLC, 447
Timing diagram, automatic washing
  machine, 48
Timing, input/output, 5
Token-passing, 688, 690
Torque:
  constant, 369
  dc motor, 369
  electric motor, 367
Transducer, 252
Transfer function, 5, 31, 114, 136–9, 148,
  526, 560
  closed-loop, 566, 608
  dc motor, 386–92
  defined, 6
  open-loop, 566, 609
  overall, 567–9
  process control system, 11
  servo control system, 10
Transistor, 332, 359
Transistor amplifiers, 333–5
Transition condition, 429, 431, 439
Transition function, 429
Triac, 329, 359
Tuning the controller, 614, 649
Turbine flow meter, 305, 317
Turbulent flow, 79, 107
  when it occurs, 80

Ultimate cycle method, 615, 649
Unbonded strain gage, 277, 282
Underdamped response, 22
Unijunction transistor (UJT), 330, 359
Unstable, 609
Unstable response, 23

Vacuum, 309, 317
Valve flow coefficient (Cv), 350, 359
Velocity, average, 88
Velocity measurement, 270
Velocity of propagation, 78
Viscous friction, 101, 107
Voltage constant, 370
Voltage follower, 191, 245
Vortex shedding flow meter, 307, 317

Way, 359
Wheatstone bridge, 245
  balanced, 206
  unbalanced, 210
Wide area network, 682
Work cell, training and industrial, 444
Working pressure, 342, 359